T0180065

Lecture Notes in Computer Science 12291

Founding Editors

Gerhard Goos
Karlsruhe Institute of Technology, Karlsruhe, Germany
Juris Hartmanis
Cornell University, Ithaca, NY, USA

Editorial Board Members

Elisa Bertino
Purdue University, West Lafayette, IN, USA
Wen Gao
Peking University, Beijing, China
Bernhard Steffen
TU Dortmund University, Dortmund, Germany
Gerhard Woeginger
RWTH Aachen, Aachen, Germany
Moti Yung
Columbia University, New York, NY, USA

More information about this series at http://www.springer.com/series/7407

François Boulier · Matthew England ·
Timur M. Sadykov · Evgenii V. Vorozhtsov (Eds.)

Computer Algebra in Scientific Computing

22nd International Workshop, CASC 2020
Linz, Austria, September 14–18, 2020
Proceedings

 Springer

Editors
François Boulier (iD)
University of Lille
Villeneuve d'Ascq, France

Matthew England (iD)
Coventry University
Coventry, UK

Timur M. Sadykov (iD)
Plekhanov Russian University of Economics
Moscow, Russia

Evgenii V. Vorozhtsov (iD)
Institute of Theoretical and Applied Mechanics
Novosibirsk, Russia

ISSN 0302-9743   ISSN 1611-3349 (electronic)
Lecture Notes in Computer Science
ISBN 978-3-030-60025-9   ISBN 978-3-030-60026-6 (eBook)
https://doi.org/10.1007/978-3-030-60026-6

LNCS Sublibrary: SL1 – Theoretical Computer Science and General Issues

© Springer Nature Switzerland AG 2020
This work is subject to copyright. All rights are reserved by the Publisher, whether the whole or part of the material is concerned, specifically the rights of translation, reprinting, reuse of illustrations, recitation, broadcasting, reproduction on microfilms or in any other physical way, and transmission or information storage and retrieval, electronic adaptation, computer software, or by similar or dissimilar methodology now known or hereafter developed.
The use of general descriptive names, registered names, trademarks, service marks, etc. in this publication does not imply, even in the absence of a specific statement, that such names are exempt from the relevant protective laws and regulations and therefore free for general use.
The publisher, the authors and the editors are safe to assume that the advice and information in this book are believed to be true and accurate at the date of publication. Neither the publisher nor the authors or the editors give a warranty, expressed or implied, with respect to the material contained herein or for any errors or omissions that may have been made. The publisher remains neutral with regard to jurisdictional claims in published maps and institutional affiliations.

This Springer imprint is published by the registered company Springer Nature Switzerland AG
The registered company address is: Gewerbestrasse 11, 6330 Cham, Switzerland

Sadly, Andreas Weber passed away on March 15, 2020. Andreas studied mathematics and computer science at the Universities of Tübingen, Germany, and Boulder, Colorado, USA. He then worked as a postdoc at the Institute for Computer Science at Cornell University, New York, USA; the University of Tübingen, Germany; and the Fraunhofer Institute for Computer Graphics Research, Germany. Since 2001, he was a professor at the University of Bonn, Germany.

Scientifically, he was considered an authority in the fields of physics-based modeling and simulation, as well as applications of computer algebra in the natural sciences and particularly biology. He was incredibly well-read, possessed an immense wealth of knowledge across different disciplines, and was highly committed to international and interdisciplinary networking. Due to his great hospitality and openness, his group in Bonn developed into a permanent lively meeting place of scientists from different fields from all over the world, where science could take place at its best.

Andreas made many contributions to CASC, attending almost all conferences in its history. He contributed many papers, hosted CASC in Bonn in 2007, was an invited speaker in 2010, a PC member during 2000–2013, and most recently served as publicity chair in 2014–2019. We will always remember his good-natured humor and his boundless compassion for science and for others. We will miss his presence very much.

Preface

The International Workshop on Computer Algebra in Scientific Computing (CASC), held annually since 1998, has established itself as a leading forum for presenting recent developments in the area of computer algebra and on pioneering applications of computer algebra methods in sciences such as physics, chemistry, celestial mechanics, life sciences, engineering, etc. The CASC 2020 International Workshop featured a balanced mix of high-level keynote speeches and concurrent oral sessions.

CASC 2020 Venue

It was initially decided, in the autumn of 2019, that the 22nd CASC International Workshop would be held at Johannes Kepler University (JKU), Linz, Austria, in September 2020. The choice of this university was particularly fitting since research in computer algebra has been conducted therein for many years.

The city of Linz, and maybe even more the little village of Hagenberg 25 km north of Linz, is well known in the computer algebra community, since the RISC (Research Institute for Symbolic Computation) is located in a medieval castle in Hagenberg. RISC was founded in 1987 by Bruno Buchberger as an institute of JKU and moved to Hagenberg in 1989. Since its foundation RISC has developed into one of the world's leading institutes in the area of symbolic computation. Buchberger's vision encloses the entire range from pioneering mathematical research to industry proven software engineering. The RISC Software Company was founded in 1992 as part of RISC and it embodies the duality of basic research and applications.

To mathematicians, Bruno Buchberger is known as the inventor of Gröbner bases theory and Buchberger's algorithm, which he developed in his PhD thesis in 1965. Nowadays, Gröbner bases are one of the fundamental pillars of symbolic computation and the applications range from algebraic geometry or applied mathematics to science and engineering. Every major computer algebra system has its implementation of Gröbner bases.

Presently, symbolic computation has a strong basis in Linz. RISC is currently directed by Peter Paule and consists of the following research groups: automated reasoning, computer algebra for combinatorics, computer algebra for differential equations, computer algebra for geometry, formal methods, rewriting-related techniques and applications, and symbolic methods in kinematics. The JKU Institute for Algebra, led by Manuel Kauers, also puts a strong emphasis on research in computer algebra. The main research areas covered are symbolic summation and integration, operator algebras, special functions identities and inequalities, and applications of computer algebra in combinatorics, experimental mathematics, and systems biology. Furthermore, there is a group on symbolic computation under the guidance of Josef Schicho at the Radon Institute for Computational and Applied Mathematics (RICAM), an institute of the Austrian Academy of Sciences (OeAW). Its focus is on computer algebra, algebraic geometry, differential algebra, holonomic functions, and kinematics.

The research activity of the mathematical departments of Linz actively promote the significant impact of computer algebra in scientific computing. For over 20 years they have been showing a strong commitment to interdisciplinary research. In 1998 the Special Research Program (SFB) "Numerical and Symbolic Scientific Computing" was launched with the participating institutes of Applied Geometry, Computational Mathematics, Industrial Mathematics, and RISC. This enterprise became a role model for interdisciplinary research and doctoral education. At the end of its runtime in 2008, it was succeeded by the doctoral program (DK) "Computational Mathematics: Numerical Analysis and Symbolic Computation." In addition to the four institutes of the SFB, currently the participating institutes are the institute for Algebra, Stochastics, and RICAM. One of the general goals of the doctoral education in the DK is to gain expertise in algorithmic mathematics. Two decades of interdisciplinary cooperation have also shaped the curriculum of the undergraduate education at JKU. There is a basic understanding that scientific computing and computer algebra go well together, which makes Linz a great place to meet for CASC, even if it is only virtually this year.

The Organizing Committee of CASC 2020 monitored the development of the COVID-19 pandemic. The safety and well-being of all conference participants was our priority. After studying and evaluating the announcements, guidance, and news released by relevant national departments, the decision was made to host CASC 2020 as an online event.

Overview of the Volume

This year, CASC 2020 had two categories of participation: (1) talks with accompanying papers to appear in the proceedings, and (2) talks with accompanying extended abstracts for distribution locally at the conference only. The latter was for work either already published, or not yet ready for publication, but in either case still new and of interest to the CASC audience. The former was strictly for new and original research results, ready for publication.

All papers submitted for the LNCS proceedings received a minimum of three reviews, and some received more, with the average number of reviews being 3.2 per paper. In addition, the whole Program Committee (PC) was invited to comment and debate on all papers. At the end of the review process, the PC chose to accept 28 papers. A further 6 papers were accepted later through a conditional path (the authors had to first provide a revised version to meet specific requirements set by the PC). Hence in total this volume contains 34 contributed papers, along with 2 papers to accompany our keynote talks.

The invited talk of Ovidiu Radulescu is devoted to the application of tropical geometry for the mathematical modeling of biological systems. Tropical geometry methods exploit a property of biological systems called **multiscaleness**, summarized by two properties: i) the orders of magnitude of variables and timescales are widely distributed, and ii) at a given timescale, only a small number of variables or components play a driving role, whereas large parts of the system have passive roles and can be reduced. Several models of biological systems and their reductions are presented. The change of variables is used to "tropicalize" biochemical networks. It is shown how to find the appropriate scalings for parameters with the aid of tropical geometry approaches. The conclusion is made that tropical geometry methods are possible ways to symbolic characterization of dynamics in high dimension, also to synthesize dynamical systems with desired features.

The other invited talk by Werner Seiler is accompanied by a joint paper with Matthias Seiß which gives an overview of their recent works on singularities of implicit ordinary or partial differential equations. This includes firstly the development of a general framework combining algebraic and geometric methods for dealing with general systems of ordinary or partial differential equations, and for defining the type of singularities considered here. An algorithm is also presented for detecting all singularities of an algebraic differential equation over the complex numbers. The adaptions required for the analysis over the real numbers are then discussed. The authors further outline, for a class of singular initial value problems, for a second-order ordinary differential equation, how geometric methods allow them to determine the local solution behavior in the neighborhood of a singularity, including the regularity of the solution. Finally, it is shown for some simple cases of algebraic singularities how such an analysis can be performed there.

Polynomial algebra, which is at the core of computer algebra, is represented by contributions devoted to establishing intrinsic complexity bounds for constructing zero-dimensional Gröbner bases, the implementation of power series arithmetic in the Basic Polynomial Algebra Subprograms (BPAS) Library, the investigation of the relations between the Galois group and the triviality of the exponent lattice of a univariate polynomial, multiplier verification with the aid of Nullstellensatz-proofs, the complexity analysis of sparse multivariate Hensel lifting algorithms for polynomial factorization, the new approximate GCD algorithm with the Bezout matrix, the computation of logarithmic vector fields along an isolated complete intersection singularity, the computation of parametric standard bases for semi-weighted homogeneous isolated hypersurface singularities with the aid of the CAS SINGULAR, acceleration of subdivision root-finders for real and complex univariate polynomials, the optimization of multiplying univariate dense polynomials with long integer unbalanced coefficients with the aid of Tom–Crook approach, the investigation of the Routh–Hurwitz stability of a polynomial matrix family under real perturbations, symbolic-numeric computation of the Bernstein coefficients of a polynomial from those of one of its partial derivatives, and the derivation with the aid of Gröbner bases of new optimal symplectic higher-order Runge–Kutta–Nyström methods for the numerical solution of molecular dynamics problems.

Several papers are devoted to linear algebra and its applications: finding good pivots for small sparse matrices, the presentation of a new linear algebra approach for detecting binomiality of steady state ideals of reversible chemical reaction networks, and parametric linear system solving by using the comprehensive triangular Smith normal form.

Two papers deal with applications of symbolic-numerical computations for: computing orthonormal bases of the Bohr–Mottelson collective model, implemented in the

CAS MATHEMATICA, and a symbolic-numeric study of geometric properties of adiabatic waveguide modes.

Two papers are devoted to the application of symbolic computations for investigating and solving ordinary differential equations (ODEs): contact linearizability of scalar ODEs of arbitrary order and the investigation of the invariance of Laurent solutions of linear ODEs under possible prolongations of the truncated series which represent the coefficients of the given equation.

Three papers deal with the investigation and solution of celestial mechanics problems: applications of the CAS MATHEMATICA, to the study of stationary motions of a system of two connected rigid bodies in a constant gravity field with the aid of Gröbner bases and to the analytic investigation of the translational-rotational motion of a non-stationary triaxial body in the central gravity field; the obtaining of periodic approximate solutions of the three-body problem with the aid of conservative difference schemes and the free open-source mathematics software system SAGE (www.sagemath.org).

The remaining topics include the new complexity estimates of computing integral bases of function fields, first-order tests for toric varieties, which arise, in particular, in chemical reaction networks, Hermite interpolation of a rational function with error correction, the improved balanced NUCOMP algorithm for the arithmetic in the divisor class group of a hyperelliptic curve, algebraic complexity estimates for an efficient method of removing all redundant inequalities in the input system, a multithreaded version of the robust tracking of one path of a polynomial homotopy, the improvement of the Lazard's method for constructing the cylindrical algebraic decomposition, new extensions implemented in the SCALA algebra system, a new MAPLE package that allows obtaining compatible routes in an overtaking railway station of any number of tracks, and the use of the LEGO digital designer for teaching algebraic curves in mathematical education via LEGO linkages.

August 2020

François Boulier
Matthew England
Timur M. Sadykov
Evgenii V. Vorozhtsov

Acknowledgments

The CASC 2020 workshop was monitored remotely by the Local Organizing Committee at JKU, Linz, which has provided excellent conference facilities.

The CASC 2020 workshop and proceedings were supported financially by the JKU doctoral program "DK Computational Mathematics" (Austria) and Maplesoft (Canada). Our particular thanks are due to the members of the CASC 2020 Local Organizing Committee at the JKU, i.e., Manuel Kauers, Veronika Pillwein (chair), Clemens Raab, and Georg Regensburger (chair), who ably handled all the local arrangements in Linz. In addition, they provided us with the information about the computer algebra activities at the JKU, RISC, JKU Institute for Algebra, and at the RICAM.

Furthermore, we want to thank all the members of the Program Committee for their thorough work. We also thank the external referees who provided reviews.

We are grateful to the members of the group headed by T. Sadykov for their technical help in the preparation of the camera-ready manuscript for this volume. We are grateful to Dr. Dominik Michels (King Abdullah University, Saudi Arabia) for the design of the conference poster. Finally, we are grateful to the CASC publicity chair Andreas Weber (University of Bonn, Germany) and his assistant Hassan Errami for the management of the conference web page http://www.casc-conference.org.

Organization

CASC 2020 was organized by the Johannes Kepler University (JKU), Linz, Austria.

Workshop General Chairs

François Boulier, Lille
Timur M. Sadykov, Moscow

Program Committee Chairs

Matthew England, Coventry
Evgenii V. Vorozhtsov, Novosibirsk

Program Committee

Changbo Chen, Chongqing
Jin-San Cheng, Beijing
Victor F. Edneral, Moscow
Jaime Gutierrez, Santander
Sergey A. Gutnik, Moscow
Thomas Hahn, Munich
Hui Huang, Waterloo, Canada
François Lemaire, Lille
Victor Levandovskyy, Aachen
Dominik L. Michels, Thuwal
Marc Moreno Maza, London, Canada
Johannes Middeke, Linz

Chenqi Mou, Beijing
Gleb Pogudin, Paris
Alexander Prokopenya, Warsaw
Eugenio Roanes-Lozano, Madrid
Valery Romanovski, Maribor
Timur M. Sadykov, Moscow
Doru Stefanescu, Bucharest
Thomas Sturm, Nancy
Akira Terui, Tsukuba
Elias Tsigaridas, Paris
Jan Verschelde, Chicago
Zafeirakis Zafeirakopoulos, Gebze

Local Organization (Linz)

Gabriela Danter   Johannes Kepler University, Linz, Austria
Manuel Kauers   Johannes Kepler University, Linz, Austria
Monika Peterseil   Johannes Kepler University, Linz, Austria
Veronika Pillwein   Johannes Kepler University, Linz, Austria
Clemens Raab   Johannes Kepler University, Linz, Austria
Georg Regensburger (Chair) Johannes Kepler University, Linz, Austria

Publicity Chairs

Dominik L. Michels, Thuwal
Andreas Weber†, Bonn

Advisory Board

Vladimir P. Gerdt, Dubna
Wolfram Koepf, Kassel
Ernst W. Mayr, Munich
Werner M. Seiler, Kassel
Andreas Weber†, Bonn

Website

http://casc-conference.org/2020/
(Webmaster: Dr. Hassan Errami)

Contents

Tropical Geometry of Biological Systems
(*Invited Talk*)

Ovidiu Radulescu[✉]

University of Montpellier, CNRS UMR5235 LPHI, Montpellier, France
`ovidiu.radulescu@umontpellier.fr`

Abstract. Tropical geometry ideas were developed by mathematicians that got inspired from very different topics in physics, discrete mathematics, optimization, algebraic geometry. In tropical geometry, tools like the logarithmic transformation coarse grain complex objects, drastically simplifying their analysis. I discuss here how similar concepts can be applied to dynamical systems used in biological modeling. In particular, tropical geometry is a natural framework for model reduction and for the study of metastability and itinerancy phenomena in complex biochemical networks.

Keywords: Tropical geometry · Chemical reaction networks · Model reduction · Singular perturbations · Metastability · Itinerancy

1 Introduction

Mathematical modelling of biological systems is a daunting challenge. In order to cope realistically with the biochemistry of cells, tissues and organisms, both in fundamental and applied biological research, systems biology models use hundreds and thousands of variables structured as **biochemical networks** [1,16,38]. Nonlinear, large scale network models are also used in neuroscience to model brain activity [7,29]. Ecological and epidemiological modelling cope with population dynamics of species organized in networks and interacting on multiple spatial and temporal scales [2,28]. Mathematical models of complex diseases such as cancer combine molecular networks with population dynamics [5].

Denis Noble, a pioneer of multi-cellular modelling of human physiology, advocated the use of **middle-out** approaches in biological modelling [23]. Middle-out is an alternative to bottom-up, that tries to explain everything from detailed first principles, and to top-down, that uses strongly simplified representations of reality. A middle-out model uses just enough details to render the essence of the overall system organization. Although this is potentially a very powerful principle, the general mathematical methods to put it into practice are still awaited.

Recently, we have used tropical geometry to extract the essence of biological systems and to simplify complex biological models [24,25,30,32,33]. Tropical

© Springer Nature Switzerland AG 2020
F. Boulier et al. (Eds.): CASC 2020, LNCS 12291, pp. 1–13, 2020.
https://doi.org/10.1007/978-3-030-60026-6_1

geometry methods exploit a feature of biological systems called **multiscaleness** [11,35], summarized by two properties: i) the orders of magnitude of variables and timescales are widely distributed; and ii) at a given timescale, only a small number of variables or components play a driving role, whereas large parts of the system have passive roles and can be reduced.

Modellers and engineers reduce models by introducing ad hoc small parameters in their equations. After scaling of variables and parameters, **singular perturbations** techniques such as asymptotic approximations, boundary layers, invariant manifolds, etc. can be used to cope with multiple time scales. These techniques, invented at the beginning of the 20th century for problems in aerodynamics and fluid mechanics [26] are also known in biochemistry under the name of quasi-equilibrium and quasi-steady state approximations [10,12,13,36]. The reduction of the model in a singular perturbation framework is traditionally based on a two time scales (slow and fast) decomposition: fast variables are slaved by the slow ones and can therefore be eliminated. Geometrically, this corresponds to fast relaxation of the system to a low dimensional invariant manifold. The mathematical bases of the slow/fast decomposition were set in [14,39,40] for the elimination of the fast variables, and in [8] for the existence of a low dimensional, slow invariant manifold. However, the slow/fast decomposition is neither unique, nor constant; it depends on model parameters and can also change with the phase space position on a trajectory. Despite of several attempts to automatically determine small parameters and slow/fast decompositions, the problem of finding a reduced model remains open. We can mention numerical approaches such as Computational Singular Perturbations [18], or Intrinsic Low Dimensional Manifold [21] that perform a reduction locally, in each point of the trajectory. Notwithstanding their many applications in reactive flow and combustion, these methods are simulation based and may not provide all the possible reductions. Furthermore, explicit reductions obtained by post-processing of the data generated by these numerical methods may be in conflict with more robust approaches [33]. A computer algebra approach to determine small parameters for the quasi-steady state reduction of biochemical models was proposed, based on Gröbner bases calculations, but this approach is limited to models of small dimension [9].

Perturbation approaches, both regular and singular, operate with orders of magnitude. In such approaches, some terms are much smaller than others and can, under some conditions, be neglected. Computations with orders of magnitudes follow maxplus (or, depending on the definition of orders, minplus) algebraic rules. The same rules apply to valuations, that are building blocks for tropical geometry [22]. We developed tropical geometry methods to identify subsystems that are dominant in certain regions of the phase- and/or parameter-space of dynamical systems [30,32]. Moreover, tropical geometry is a natural approach to find the scalings needed for slow/fast decompositions and perform model order reduction in the framework of geometric singular perturbation theory. Scaling calculations are based on finding solutions of the **tropical equilibration problem**, which is very similar to computing tropical prevarieties [25].

For such a problem we have effective algorithms that work well for medium size biochemical models (10–100 species) [20,33,37].

Tropical approaches provide a timescale for each biochemical species or relaxation process and one generally has not only two but **multiple timescales**. This situation is the rule rather than the exception in biology. For instance, cells or organisms use multiple mechanisms to adapt to changes of their environment. These mechanisms involve rapid metabolic or electrophysiological changes (seconds, minutes), slower changes of gene expression (hours), and even slower mutational genetic changes (days, months). Within each category, sub-mechanism timescales spread over several log scale decades. We are using tropical approaches also to cover such situations and obtain reductions for systems with more than two timescales [17].

Interestingly, the dominance relations unravelled by tropical geometry can highlight **approximate conservation laws** of biological systems, i.e. conservation laws satisfied by a dominant subsystem and not satisfied by the full system. These approximate conservations are exploited for the simplification of biological models [6]. Beyond their importance for model reduction, approximate conservations and tropical equilibrations can be used for computing **metastable states** of biochemical models, defined as regions of very slow dynamics in phase space [32,34,35]. Metastable states represent a generalization of the stable steady states commonly used in analyzing biological networks [29]. A dynamical system spends an infinite time in the neighbourhood of stable steady state, and a large but finite time in the neighborhood of metastable states. In biology, both steady and metastable states are important. The existence of metastable states leads to a property of biological systems called **itinerancy**, meaning that the system can pass from one metastable state to another one during its dynamics [15,29]. From a biological point of view itinerancy explains plasticity during adaptation, occurring in numerous situations: brain functioning, embryo development, cellular metabolic changes induced by changes of the environment. The study of the relation between the network structure and the metastable states is also a possible way to design dynamical systems with given properties. This is related to the direction suggested by O.Viro at the 3rd European Congress of Mathematics to use tropical geometry for constructing real algebraic varieties with prescribed properties in the sense of Hilbert's 16th problem [41,42].

2 Models of Biological Systems and Their Reductions

Chemical reaction networks (CRN) are bipartite graphs such as those represented in Fig. 1, where one type of node stands for chemical species and the other for reactions. Although mainly designed for modelling cell biochemistry, CRNs can also be used to describe interactions of the cell with its microenvironment in tissue models and also the population dynamics in compartment models in ecology and epidemiology. When the copy numbers of all molecular species are large, CRN dynamics is given by systems of ordinary differential equations, usually with polynomial or rational right hand side. For instance, in the Michaelis–Menten model, an archetype of enzymatic reactions, the differential equations for the concentrations of relevant chemical species (substrate and enzyme-substrate complex) have the form

$$\frac{dx_1}{dt} = -k_1 x_1 + k_2 x_1 x_2 + k_3 x_2$$
$$\frac{dx_2}{dt} = k_1 x_1 - k_2 x_1 x_2 - (k_3 + k_4) x_2, \tag{1}$$

where $k_1, \ldots, k_4$ are kinetic parameters.

The Michaelis–Menten model is already quite simple, however comprehensive models of cell biochemistry can be very large. By model reduction one transforms the system of differential equations into a system with less equations and variables that have approximately the same solutions. The variables of the full model missing in the reduced model should be also computable, for instance as functions of the reduced model variables. For applications it is handy when not only the full model but also the reduced model is a CRN, like in Fig. 1.

3   Tropical Geometry Approaches

In order to "tropicalize" biochemical networks, one replaces parameters and species with orders of magnitude. This is performed by the change of variables

$$x \mapsto a = \log(x)/\log \epsilon,$$

where ϵ is a small, positive parameter.

This logarithmic change of variables defines a map $V_\epsilon : \mathbb{R}_+ \to \mathbb{R} \cup \{-\infty\}$. It is easy to check that when $\epsilon \to 0$,

$$V_\epsilon(xy) = V_\epsilon(x) + V_\epsilon(y),$$
$$V_\epsilon(x + y) = \min(V_\epsilon(x), V_\epsilon(y)).$$

This mapping transforms the semifield $\mathbb{R}_+$ into the semifield $\mathbb{R}_{min}$ (or min-plus algebra) where multiplication, addition, $\mathbf{0}$ and $\mathbf{1}$ become addition, multiplication, $-\infty$ and 0, respectively. Furthermore, $V_\epsilon(x)$ represents the order of magnitude of x and can express dominance relations, because when $\epsilon \to 0$

$$V_\epsilon(x) < V_\epsilon(y) \implies x >> y.$$

Tropicalizing a biochemical model consists of keeping in the r.h.s. of the ODEs only the dominant terms and eliminating the other terms [24]. As can be seen in the Fig. 2 for the Michaelis–Menten model, generically there is only one dominant term, but there are special situations when more than two dominant terms exist. We called *tropical equilibration* the situation when at least two dominant terms, one positive and one negative exist [24]. Heuristically, the tropical equilibration corresponds to compensation of dominant terms and to slow dynamics, whereas the dynamics with uncompensated dominant terms is fast.

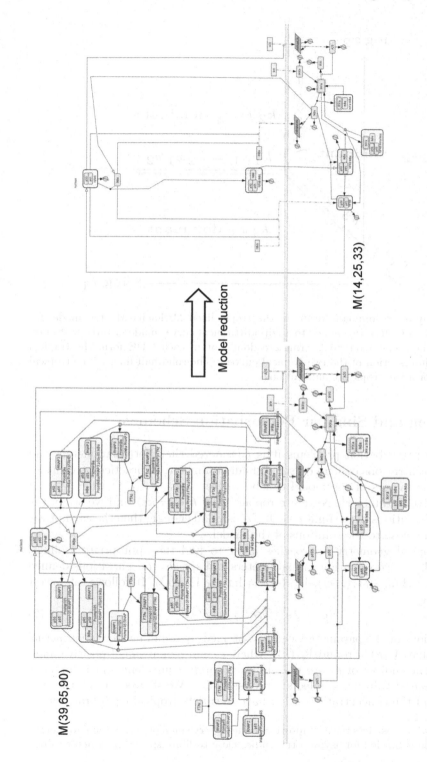

Fig. 1. Reduction of a biochemical network. A model $M(n, r, p)$ has n species, r reactions and p parameters. In the Systems Biology Graphical Notation, molecular species are represented by rectangles and reactions by dots [19]. Oriented edges (arrows) leave reactant species, enter reactions, and leave reactions, enter reaction products. After reduction, a network has less species, reactions and parameters. Modified from [30,31].

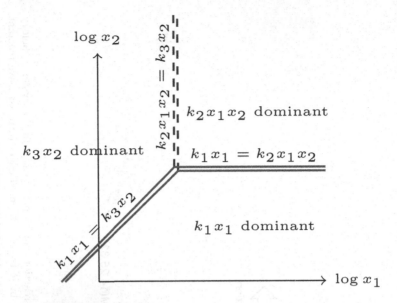

Fig. 2. Analysis of dominant terms in the tropicalized Michaelis–Menten model for $k_4 \ll k_3$. Each ODE corresponds to a tripod (tropical curve) made of three half-lines. The lines where two monomial terms are dominant in each ODE form the tropical prevariety (intersection of the two tropical curves). The solid half-lines of the tropical prevariety form the tropical equilibrations.

4   Scaling and Singular Perturbation Schemes

In singular perturbation problems it is considered that both parameters and species concentrations depend on some small parameter. In practice we consider that the parameters of the model can be written as $k = \bar{k}(\epsilon^*)^\gamma$ where ϵ^* is a small positive parameter[1]. Next, we replace k by $k(\epsilon) = \bar{k}\epsilon^\gamma$ and we study the asymptotic solutions in the limit $\epsilon \to 0$. If ϵ^* is small enough, then the asymptotic solutions are close to the solutions of the model.

The tropical geometry approaches allow to find the appropriate scalings. To this end, we use valuations of parameters and species concentrations defined as $V(x(\epsilon)) = \lim_{\epsilon \to 0} V_\epsilon(x(\epsilon))$, $V(k(\epsilon)) = \lim_{\epsilon \to 0} V_\epsilon(k(\epsilon))$. It follows, at lowest order that

$$x(\epsilon) = \bar{x}\epsilon^{V(x)}, k(\epsilon) = \bar{k}\epsilon^{V(k)}. \tag{2}$$

The valuations of the parameters can be obtained by rounding from their actual numeric values $V(k) = \text{round}(V_{\epsilon^*}(k))$, where ϵ^* can be any small positive number. Different choices of ϵ^* are only approximately equivalent and in practice one tries several values and selects a robust choice. We showed in [32] that the valuations of the concentrations $V(x)$ have to satisfy tropical equilibrations.

---

[1] this procedure restricts the asymptotic regime to very small or very large parameters; translation is needed for asymptotic studies close to finite special parameter values.

Let us illustrate how this can be used to define slow/fast decompositions and reduce the Michaelis–Menten model. Consider the case when $k_3 \ll k_4$ corresponding to the quasi-steady state approximation. In this case we have

$$\frac{\mathrm{d}\bar{x}_1}{\mathrm{d}t} = -\epsilon^{\gamma_1}\bar{k}_1\bar{x}_1 + \epsilon^{\gamma_2+a_2}\bar{k}_2\bar{x}_1\bar{x}_2 + \epsilon^{\gamma_3+a_2-a_1}\bar{k}_3\bar{x}_2,$$

$$\frac{\mathrm{d}\bar{x}_2}{\mathrm{d}t} = \epsilon^{\gamma_1+a_1-a_2}\bar{k}_1\bar{x}_1 - \epsilon^{\gamma_2+a_1}\bar{k}_2\bar{x}_1\bar{x}_2 - \epsilon^{\gamma_4}\bar{k}_4\bar{x}_2 - \epsilon^{\gamma_3}\bar{k}_3\bar{x}_2, \tag{3}$$

where $\gamma_i = V(k_i), 1 \le i \le 4, a_j = V(x_j), 1 \le j \le 2$.

Each tropical equilibration leads to a scaling and to a candidate reduced model. For instance, the tropical equilibration $\gamma_1 + a_1 - a_2 = \gamma_2 + a_1 = \gamma_4$ leads to

$$\bar{x}_1' = -\bar{k}_1\bar{x}_1 + \bar{k}_2\bar{x}_1\bar{x}_2 + \epsilon^{\gamma_3-\gamma_4}\bar{k}_3\bar{x}_2,$$

$$\epsilon^{\gamma_1-\gamma_4}\bar{x}_2' = \bar{k}_1\bar{x}_1 - \bar{k}_2\bar{x}_1\bar{x}_2 - \bar{k}_4\bar{x}_2 - \epsilon^{\gamma_3-\gamma_4}\bar{k}_3\bar{x}_2, \tag{4}$$

where the derivatives are with respect to the rescaled time $\tau = \epsilon^{\gamma_1}$ and $\gamma_3 - \gamma_4 > 0$ (because $k_3 \ll k_4$).

The case $\gamma_1 - \gamma_4 > 0$ is typically a singular perturbation case and the solution of (4) converges to the solution of

$$\bar{x}_1' = -\bar{k}_1\bar{x}_1 + \bar{k}_2\bar{x}_1\bar{x}_2,$$

$$0 = \bar{k}_1\bar{x}_1 - \bar{k}_2\bar{x}_1\bar{x}_2 - \bar{k}_4\bar{x}_2, \tag{5}$$

as $\epsilon \to 0$.

The justification of the convergence lies outside tropical geometry considerations and uses singular perturbations results; in this simple case it follows from [39]. Some general results of convergence can be found in [32] for the two time scale case and in [17] for the multiple timescale case.

The second equation of (5) is called quasi-steady state condition. Using this condition to eliminate $\bar{x}_2$, we obtain the reduced model

$$\bar{x}_1' = -\frac{V_{max}\bar{x}_1}{\bar{x}_1 + K_m}, \tag{6}$$

that is the Briggs-Haldane approximation to the Michaelis–Menten model, with $V_{max} = \bar{k}_1\bar{k}_4/\bar{k}_2$, $K_m = \bar{k}_4/\bar{k}_2$.

By this procedure a model of two differential equations and four parameters was reduced to a model of one differential equation and two parameters.

5   Approximate Conservation Laws

In the case when $k_3 \gg k_4$ using the same procedure as in the preceding section for the tropical equilibration $\gamma_1 + a_1 - a_2 = \gamma_2 + a_1 = \gamma_3$ leads to

$$0 = -\bar{k}_1\bar{x}_1 + \bar{k}_2\bar{x}_1\bar{x}_2 + \bar{k}_3\bar{x}_2,$$
$$0 = \bar{k}_1\bar{x}_1 - \bar{k}_2\bar{x}_1\bar{x}_2 - \bar{k}_3\bar{x}_2, \tag{7}$$

as $\epsilon \to 0$.

The quasi-steady state equations are indeterminate and one can not eliminate both fast variables x_1 and x_2 as usual in the quasi-steady state approximation. Furthermore, the slow variables whose dynamics have to be retained in the asymptotic limit are not explicit.

This degenerate case occurs quite often in practice. In order to obtain a reduction, we exploit approximate conservations. When $k_3 >> k_4$, the fast dynamics of the Michaelis–Menten model can be approximated by

$$\frac{dx_1}{dt} = -k_1x_1 + k_2x_1x_2 + k_3x_2,$$

$$\frac{dx_2}{dt} = k_1x_1 - k_2x_1x_2 - k_3x_2. \tag{8}$$

It can be easily checked that this system has a first integral $\frac{d(x_1+x_2)}{dt} = 0$. $x_1 + x_2$ is called an approximate conservation law because it is conserved by the fast approximated system and is not conserved by the full system.

By introducing the new variable $x_3 = x_1 + x_2 = \bar{x}_3\epsilon^{\min(a_1,a_2)}$ we get

$$\frac{d\bar{x}_1}{dt} = \epsilon^{\gamma_3+a_2-a_1}(-\bar{k}_1\bar{x}_1 + \bar{k}_2\bar{x}_1\bar{x}_2 + \bar{k}_3\bar{x}_2),$$

$$\frac{d\bar{x}_2}{dt} = \epsilon^{\gamma_3}(\bar{k}_1\bar{x}_1 - \bar{k}_2\bar{x}_1\bar{x}_2 - \bar{k}_3\bar{x}_2 - \epsilon^{\gamma_4-\gamma_3}\bar{k}_4\bar{x}_2),$$

$$\frac{d\bar{x}_3}{dt} = -\epsilon^{\gamma_4+a_2-\min(a_1,a_2)}\bar{k}_4\bar{x}_2. \tag{9}$$

As $\gamma_4 + a_2 - \min(a_1, a_2) > \gamma_3$ and $\gamma_4 + a_2 - \min(a_1, a_2) > \gamma_3$, it follows that the variable x_3 is slower than both x_1 and x_2, with no condition on the valuations of parameters and variables.

More generally, approximate conservations can be defined each time a scaling of the system by powers of ϵ is known. We proved, for polynomial ODEs, that any approximate linear or polynomial conservation law is a slow variable [6]. This result can be used for model reduction in the degenerate situation when the quasi-steady state equations are indeterminate.

6   Metastability

A typical trajectory of a multiscale system consists in a succession of qualitatively different slow segments separated by fast transitions (see Fig. 3). The slow segments, corresponding to metastable states or regimes, can be of several types such as attractive slow invariant manifolds, Milnor attractors, saddle connections, etc.

According to the famous conjecture of Jacob Palis, smooth dynamical systems on compact spaces should have a finite number of attractors whose basins cover the entire ambient space [27]. These conditions apply to biochemical reaction networks whose ambient space is compact because of conservation, or dissipativity. The conjecture could be extended to metastable states where smoothness of the vector fields and compactness of the ambient space should lead to a finite number of such states. In this case, symbolic descriptions of the trajectories as sequences of symbols, representing the metastable states that are visited, are possible.

Tropical equilibrations are natural candidates for slow attractive invariant manifolds and metastable states. Beyond the purely geometric conditions, hyperbolicity conditions are needed for the stability of such states [17, 32]. The set of tropical equilibrations is a polyhedral complex. The maximal dimension faces of such a complex are called branches (Fig. 4).

Because it is much easier to make calculations in polyhedral geometry than with high dimensional smooth dynamical systems, the computation of tropical branches represents a useful tool for understanding complex dynamical systems. Moreover, many dynamical properties such as timescales, are linear functions of parameter and species concentrations orders after tropicalization. Thus, polyhedral geometry can be used for expressing conditions for particular model behaviors. This opens fascinating directions for the synthesis of systems with desired properties.

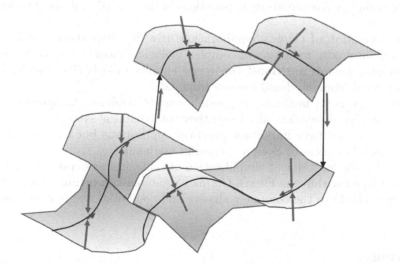

Fig. 3. Dynamics of multiscale systems can be represented as itinerant trajectory in a patchy phase space landscape made of slow attractive invariant manifolds. The term *crazy-quilt* was coined to describe such a patchy landscape [11]. In the terminology of the singular perturbations theory slow dynamics takes place on these slow manifolds, while fast transitions (layers) occur by following the flow of the fast vectorfield (long arrows) away from the slow manifolds. From [35].

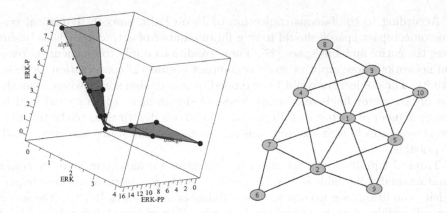

Fig. 4. Left: Polyhedral tropical branches for a model of MAPK cellular signaling in projection on directions of variation of three chemical species; a real trajectory spans several such branches in a well defined sequence, from alpha (initial condition) to omega (steady state). Right: The adjacency relations of the branches in the polyhedral complex are represented as a graph.

7   Conclusion

Tropical geometry has promising applications in the field of analysis of biological models.

The calculation of tropical equilibrations is a first important step in algorithms for automatic reduction of complex biological models. By model reduction, complex models are transformed into simpler models that can be more easily analyzed, simulated and learned from data.

Tropical geometry methods are possible ways to symbolic characterization of dynamics in high dimension, also to synthesis of dynamical systems with desired features. This may have important practical applications but can also provide at least partial answers to open questions in mathematics.

Several tools for tropical simplification of biological systems are currently being developed within the French-German Symbiont consortium [3,4] and will be made available to the computer algebra and computational biology communities.

References

1. Almquist, J., Cvijovic, M., Hatzimanikatis, V., Nielsen, J., Jirstrand, M.: Kinetic models in industrial biotechnology-improving cell factory performance. Metab. Eng. **24**, 38–60 (2014). https://doi.org/10.1016/j.ymben.2014.03.007
2. Barrat, A., Barthelemy, M., Vespignani, A.: Dynamic Processes on Complex Networks. Cambridge University Press, Cambridge (2008). https://doi.org/10.1017/CBO9780511791383

3. Boulier, F., et al.: The SYMBIONT project: symbolic methods for biological networks. F1000Research **7**, 1341 (2018). https://doi.org/10.7490/f1000research
4. Boulier, F., et al.: The SYMBIONT project: symbolic methods for biological networks. ACM Commun. Comput. Algebra **52**(3), 67–70 (2018). https://doi.org/10.1145/3313880.3313885
5. Deisboeck, T.S., Wang, Z., Macklin, P., Cristini, V.: Multiscale cancer modeling. Ann. Rev. Biomed. Eng. **13**, 127–155 (2011). https://doi.org/10.1146/annurev-bioeng-071910-124729
6. Desoeuvres, A., Iosif, A., Radulescu, O., Seiß, M.: Approximated conservation laws of chemical reaction networks with multiple time scales. preprint, April 2020
7. Eliasmith, C., et al.: A large-scale model of the functioning brain. Science **338**(6111), 1202–1205 (2012). https://doi.org/10.1126/science.1225266
8. Fenichel, N.: Geometric singular perturbation theory for ordinary differential equations. J. Differ. Equ. **31**(1), 53–98 (1979). https://doi.org/10.1016/0022-0396(79)90152-9
9. Goeke, A., Walcher, S., Zerz, E.: Determining "small parameters" for quasi-steady state. J. Differ. Equ. **259**(3), 1149–1180 (2015). https://doi.org/10.1016/j.jde.2015.02.038
10. Goeke, A., Walcher, S., Zerz, E.: Quasi-steady state – Intuition, perturbation theory and algorithmic algebra. In: Gerdt, V.P., Koepf, W., Seiler, W.M., Vorozhtsov, E.V. (eds.) CASC 2015. LNCS, vol. 9301, pp. 135–151. Springer, Cham (2015). https://doi.org/10.1007/978-3-319-24021-3_10
11. Gorban, A.N., Radulescu, O.: Dynamic and static limitation in multiscale reaction networks, revisited. Adv. Chem. Eng. **34**(3), 103–173 (2008). https://doi.org/10.1016/s0065-2377(08)00003-3
12. Gorban, A.N., Radulescu, O., Zinovyev, A.Y.: Asymptotology of chemical reaction networks. Chem. Eng. Sci. **65**(7), 2310–2324 (2010). https://doi.org/10.1016/j.ces.2009.09.005
13. Heineken, F.G., Tsuchiya, H.M., Aris, R.: On the mathematical status of the pseudo-steady state hypothesis of biochemical kinetics. Math. Biosci. **1**(1), 95–113 (1967). https://doi.org/10.1016/0025-5564(67)90029-6
14. Hoppensteadt, F.: Properties of solutions of ordinary differential equations with small parameters. Commun. Pure Appl. Math. **24**(6), 807–840 (1971). https://doi.org/10.1002/cpa.3160240607
15. Kaneko, K., Tsuda, I.: Chaotic itinerancy. Chaos Interdisc. J. Nonlinear Sci. **13**(3), 926–936 (2003). https://doi.org/10.1063/1.1607783
16. Kholodenko, B.N.: Cell-signalling dynamics in time and space. Nat. Rev. Mol. Cell Biol. **7**(3), 165 (2006). https://doi.org/10.1038/nrm1838
17. Kruff, N., Lüders, C., Radulescu, O., Sturm, T., Walcher, S.: Singular perturbation reduction of reaction networks with multiple time scales. In: Proceedings of CASC 2020, 14–18 September 2020
18. Lam, S., Goussis, D.: Understanding complex chemical kinetics with computational singular perturbation. In: Symposium (International) on Combustion. vol. 22, pp. 931–941. Elsevier (1989). https://doi.org/10.1016/S0082-0784(89)80102-X
19. Le Novere, N., et al.: The systems biology graphical notation. Nat. Biotechnol. **27**(8), 735–741 (2009). https://doi.org/10.1038/nbt.1558
20. Lüders, C.: Computing tropical prevarieties with satisfiability modulo theory (SMT) solvers, April 2020. arXiv preprint arXiv:2004.07058

21. Maas, U., Pope, S.B.: Implementation of simplified chemical kinetics based on intrinsic low-dimensional manifolds. In: Symposium (International) on Combustion, vol. 24, pp. 103–112. Elsevier (1992). https://doi.org/10.1016/S0082-0784(06)80017-2

22. Maclagan, D., Sturmfels, B.: Introduction to tropical geometry, Graduate Studies in Mathematics, vol. 161. American Mathematical Society, RI (2015). https://doi.org/10.1365/s13291-016-0133-6

23. Noble, D.: The future: putting humpty-dumpty together again. Biochem. Soc. Trans. **31**(1), 156–158 (2003). https://doi.org/10.1042/bst0310156

24. Noel, V., Grigoriev, D., Vakulenko, S., Radulescu, O.: Tropical geometries and dynamics of biochemical networks application to hybrid cell cycle models. In: Proceedings of SASB, ENTCS, vol. 284, pp. 75–91. Elsevier, June 2012. https://doi.org/10.1016/j.entcs.2012.05.016

25. Noel, V., Grigoriev, D., Vakulenko, S., Radulescu, O.: Tropicalization and tropical equilibration of chemical reactions. In: Litvinov, G., Sergeev, S. (eds.) Tropical and Idempotent Mathematics and Applications, Contemporary Mathematics, vol. 616, pp. 261–277. American Mathematical Society (2014). https://doi.org/10.1090/conm/616/12316

26. O'Malley, R.E.: Historical Developments in Singular Perturbations. Springer, New York (2014). https://doi.org/10.1007/978-3-319-11924-3

27. Palis, J.: A global view of dynamics and a conjecture on the denseness of finitude of attractors. Astérisque **261**(13–16), 335–347 (2000). http://www.numdam.org/item/?id=AST_2000__261__335_0

28. Pastor-Satorras, R., Castellano, C., Van Mieghem, P., Vespignani, A.: Epidemic processes in complex networks. Rev. Mod. Phys. **87**(3), 925 (2015). https://doi.org/10.1103/RevModPhys.87.925

29. Rabinovich, M.I., Varona, P., Selverston, A.I., Abarbanel, H.D.: Dynamical principles in neuroscience. Rev. Mod. Phys. **78**(4), 1213 (2006). https://doi.org/10.1103/RevModPhys.78.121

30. Radulescu, O., Gorban, A.N., Zinovyev, A., Noel, V.: Reduction of dynamical biochemical reactions networks in computational biology. Front. Genet. **3**, 131 (2012). https://doi.org/10.3389/fgene.2012.00131

31. Radulescu, O., Gorban, A.N., Zinovyev, A., Lilienbaum, A.: Robust simplifications of multiscale biochemical networks. BMC Syst. Biol. **2**(1), 86 (2008). https://doi.org/10.1186/1752-0509-2-86

32. Radulescu, O., Vakulenko, S., Grigoriev, D.: Model reduction of biochemical reactions networks by tropical analysis methods. Math. Mod. Nat. Phenom. **10**(3), 124–138 (2015). https://doi.org/10.1051/mmnp/201510310

33. Samal, S.S., Grigoriev, D., Fröhlich, H., Weber, A., Radulescu, O.: A geometric method for model reduction of biochemical networks with polynomial rate functions. Bull. Math. Biol. **77**(12), 2180–2211 (2015). https://doi.org/10.1007/s11538-015-0118-0

34. Samal, S.S., Krishnan, J., Esfahani, A.H., Lüders, C., Weber, A., Radulescu, O.: Metastable regimes and tipping points of biochemical networks with potential applications in precision medicine. In: Liò, P., Zuliani, P. (eds.) Automated Reasoning for Systems Biology and Medicine, pp. 269–295. Springer (2019). https://doi.org/10.1101/466714

35. Samal, S.S., Naldi, A., Grigoriev, D., Weber, A., Théret, N., Radulescu, O.: Geometric analysis of pathways dynamics: application to versatility of tgf-β receptors. Biosystem **149**, 3–14 (2016). https://doi.org/10.1016/j.biosystems.2016.07.004

36. Segel, L.A., Slemrod, M.: The quasi-steady-state assumption: a case study in per-
    turbation. SIAM Rev. **31**(3), 446–477 (1989). https://doi.org/10.1137/1031091
37. Soliman, S., Fages, F., Radulescu, O.: A constraint solving approach to model
    reduction by tropical equilibration. Algorithm Mol. Biol. (2014). https://doi.org/
    10.1186/s13015-014-0024-2
38. Stanford, N.J., Lubitz, T., Smallbone, K., Klipp, E., Mendes, P., Liebermeister, W.:
    Systematic construction of kinetic models from genome-scale metabolic networks.
    PloS One **8**(11), e79195 (2013). https://doi.org/10.1371/journal.pone.0079195
39. Tikhonov, A.N.: Systems of differential equations containing small parameters in
    the derivatives. Math. Sb. (N. S.) **73**(3), 575–586 (1952). https://www.mathnet.
    ru/links/d3a478d04c7ea72687899d99ec88ce5c/sm5548.pdf
40. Vasileva, A.B., Butuzov, V.: Singularly perturbed equations in critical cases.
    MIzMU (1978)
41. Viro, O.: Dequantization of real algebraic geometry on logarithmic paper. In: Euro-
    pean Congress of Mathematics, Progress in Mathematics, vol. 201, pp. 135–146.
    Springer (2001). https://doi.org/10.1007/978-3-0348-8268-2_8
42. Viro, O.: From the sixteenth hilbert problem to tropical geometry. Jpn. J. Math.
    3(2), 185–214 (2008). https://doi.org/10.1007/s11537-008-0832-6

Algebraic and Geometric Analysis of Singularities of Implicit Differential Equations (*Invited Talk*)

Werner M. Seiler[✉][iD] and Matthias Seiß

Institut Für Mathematik, Universität Kassel, 34109 Kassel, Germany
{seiler,mseiss}@mathematik.uni-kassel.de

Abstract. We review our recent works on singularities of implicit ordinary or partial differential equations. This includes firstly the development of a general framework combining algebraic and geometric methods for dealing with general systems of ordinary or partial differential equations and for defining the type of singularities considered here. We also present an algorithm for detecting all singularities of an algebraic differential equation over the complex numbers. We then discuss the adaptions required for the analysis over the real numbers. We further outline for a class of singular initial value problems for a second-order ordinary differential equation how geometric methods allow us to determine the local solution behaviour in the neighbourhood of a singularity including the regularity of the solution. Finally, we show for some simple cases of algebraic singularities how there such an analysis can be performed.

Keywords: Implicit differential equations · Algebraic differential equations · Singularities · Vessiot spaces · Regularity decomposition · Singular initial value problems

1 Introduction

Many different forms of "singular" behaviour appear in the context of differential equations and many different views have been developed for them. Most of them are related to singularities of individual solutions of a given differential equation like blow-ups or shocks, i. e., either a solution component or some derivative of it becomes infinite. Other interpretations are concerned with bifurcations, with multivalued solutions or with singular integrals. In dynamical systems theory, many authors call stationary points (or equilibria) singularities.

We will identify a (system of) differential equations with a geometric object and singularities are points on it which are "different" from the generic points. "Different" means e. g. that the dimensions of certain geometric structures jump. Therefore, from all the "singularities" mentioned above, stationary points are closest to our singularities. In fact, in the case of ordinary differential equations, we will analyze the local solution behaviour by constructing a dynamical system for which the singularity is a stationary point.

© Springer Nature Switzerland AG 2020
F. Boulier et al. (Eds.): CASC 2020, LNCS 12291, pp. 14–41, 2020.
https://doi.org/10.1007/978-3-030-60026-6_2

In this article, we will given an overview over some recent results of ours; for all details and in particular proofs, we must refer to the original works [22, 34,35]. In [22], a *general framework* for dealing with singularities of *arbitrary systems* of ordinary or partial differential equations was developed by combining methods from differential algebra, algebraic geometry and differential topology. Conceptually, it follows the classical geometric approach to define singularities that dates back at least to Clebsch and Poincaré (see [27] for a review and historical perspective) and extends it also to partial differential equations and to situations where one has no longer a manifold but a variety. Because of the use of algebraic methods, the theory could be made fully algorithmic and was implemented in MAPLE. We will present these results in the Sects. 2–4.

A central algebraic method used, the Thomas decomposition, assumes that the underlying field is algebraically closed. For our use of the differential Thomas decomposition the base field is largely irrelevant and we can continue to use it also for *real differential equations*. The actual identification of singularities is done via an algebraic Thomas decomposition and – as shown for some concrete examples in [35] – its application to real equations is problematic. As the key step for the detection of singularities is the analysis of a linear system of equations over an algebraic set, we can replace it in the real case by a parametric Gaussian elimination followed by a quantifier elimination. Simultaneously, this allows us to extend to semialgebraic equations, i. e. to systems comprising not only equations and inequations, but also more general inequalities like positivity constraints. These considerations from [35] are the topic of Sect. 5.

Once we have identified a singularity, we want to analyse the local solution behaviour. This cannot be done at the same level of generality as the detection. In Sect. 6, we study *geometric singularities* of real *ordinary* differential equations using methods from dynamical systems theory. Then we restrict further to quasilinear equations and show in Sect. 7 – following [34] – how, for a specific class of scalar second-order singular initial value problems, non-trivial existence, (non)uniqueness and regularity results can be obtained.

While the analysis of geometric singularities is a classical topic, *algebraic singularities* have been essentially ignored in the context of differential equations. One of the few exceptions is the work by Falkensteiner and Sendra [12] where the theory of plane algebraic curves is used to analyse first-order scalar autonomous ordinary differential equations. We will show in Sect. 8 how our geometric approach allows us to analyse certain simple situations with ad hoc methods.

2   Differential Systems and Algebraic Differential Equations

In this section we introduce most of the algebraic and geometric techniques used in this article. For lack of space, we cannot provide a completely self-contained introduction. For any unexplained terminology on the (differential) algebraic side we refer to [30] and on the geometric side to [32].

We begin with the algebraic point of view. Consider the polynomial ring $\mathcal{P} = \mathbb{C}[x_1, \ldots, x_n]$ with the ranking defined by $x_i < x_j$ for $i < j$. The largest variable appearing in a polynomial $p \in \mathcal{P}$ is called its *leader* $\mathrm{ld}\,p$. Considering p as a univariate polynomial in this variable, the *initial* $\mathrm{init}\,p$ is defined as leading coefficient and the *separant* $\mathrm{sep}\,p$ as the derivative $\partial p / \partial\,\mathrm{ld}\,p$. An *algebraic system* S is a finite set of polynomial equations and inequations

$$S = \{\ p_1 = 0,\ \ldots,\ p_s = 0,\ q_1 \neq 0,\ \ldots,\ q_t \neq 0\ \} \tag{A}$$

with polynomials $p_i, q_j \in \mathcal{P}$ and $s, t \in \mathbb{N}_0$. Its *solution set* $\mathrm{Sol}\,S = \{a \in \mathbb{C}^n \mid p_i(a) = 0,\ q_j(a) \neq 0\ \forall\,i,j\}$ is a locally Zariski closed set, namely the difference of the two varieties $\mathrm{Sol}\,(\{p_1 = 0, \ldots, p_s = 0\})$ and $\mathrm{Sol}\,(\{q_1 = 0, \ldots, q_t = 0\})$. The algebraic system (A) is *simple*, if (i) it is *triangular*, (ii) it has *non-vanishing initials*, i.e. for each $r \in \{p_1, \ldots, p_s, q_1, \ldots, q_t\}$, the equation $\mathrm{init}\,r = 0$ has no solution in $\mathrm{Sol}\,S$ and (iii) it is *square-free*, i.e. for each $r \in \{p_1, \ldots, p_s, q_1, \ldots, q_t\}$, the equation $\mathrm{sep}\,r = 0$ has no solution in $\mathrm{Sol}\,S$. Simple systems behave "better" in many respects than general systems. One can show that for a simple system S the saturated ideal

$$\mathcal{I}_{\mathrm{alg}}(S) := \langle p_1, \ldots, p_s \rangle : q^\infty \subset \mathcal{P} \qquad \text{where } q = \mathrm{init}\,p_1 \cdots \mathrm{init}\,p_s \tag{1}$$

is the vanishing ideal of the Zariski closure of $\mathrm{Sol}\,S$ [30, Prop. 2.2.7].

A *Thomas decomposition* of (A) consists of finitely many simple algebraic systems $S_1, \ldots, S_k$ such that $\mathrm{Sol}\,S$ is the *disjoint* union of $\mathrm{Sol}\,S_1, \ldots, \mathrm{Sol}\,S_k$. Any algebraic system admits a Thomas decomposition (which is not unique). This decomposition was introduced by Thomas [39,40] in the context of differential algebra. It follows the general philosophy of treating algebraic or differential systems via triangular sets (see [16,17] for a survey). A special feature of it is its disjointness. It had been largely forgotten, until it was revived by Gerdt [13]; a modern presentation can also be found in [30]. Concrete implementations have been provided in [2,3,14] and some more theoretical applications in [21,25].

In the differential case, we consider the ring of *differential polynomials* $K\{U\}$ where $K = \mathbb{C}(x_1, \ldots, x_n)$, $U = \{u_1, \ldots, u_m\}$ are finitely many differential indeterminates and where we take the partial derivatives $\delta_i = \partial/\partial x_i$ as derivations. Given some differential polynomials $p_1, \ldots, p_s \in K\{U\}$, we must distinguish between the algebraic ideal $\langle p_1, \ldots, p_s \rangle$ and the differential ideal $\langle p_1, \ldots, p_s \rangle_\Delta$ generated by them. The latter one contains in addition all differential consequences $\delta^\mu p$ of any element p of it. We also introduce the subring $\mathcal{D} \subset K\{U\}$ of those differential polynomials where also the coefficients are polynomials in the variables x^i. For any $\ell \in \mathbb{N}_0$, we define the finitely generated subalgebra

$$\mathcal{D}_\ell = \mathbb{C}\big[\,x^i,\ u_\mu^\alpha \mid 1 \leq i \leq n,\ 1 \leq \alpha \leq m,\ |\mu| \leq \ell\,\big]$$

which may be considered as the coordinate ring of a jet bundle (see below).

We choose on $K\{U\}$ an orderly Riquier ranking $<$. The notion of leader, initial and separant can be extended straightforwardly. A *differential system* S is a finite set of differential polynomial equations and inequations

$$S = \{\ p_1 = 0,\ \ldots,\ p_s = 0,\ q_1 \neq 0,\ \ldots,\ q_t \neq 0\ \} \tag{D}$$

with $p_i, q_j \in \mathcal{D}$ and $s, t \in \mathbb{N}_0$. As *solution set* $\mathrm{Sol}\, S$, we take for simplicity all formal power series solutions. The differential system (D) is *simple*, if (i) it is simple as an algebraic system in the finitely many jet variables u_μ^α which actually occur in S ordered according to $<$, (ii) its equation part forms a passive system in the sense of Janet–Riquier theory for the Janet division and (iii) no leader of an inequation q_j is an (iterated) derivative of the leader of an equation p_k.

A *Thomas decomposition* of the differential system (D) consists of finitely many simple differential systems $S_1, \ldots, S_k$ such that $\mathrm{Sol}\, S$ is the disjoint union of the solution sets $\mathrm{Sol}\, S_1, \ldots, \mathrm{Sol}\, S_k$. Any differential system admits such a decomposition which can be computed algorithmically by interweaving algebraic Thomas decompositions and the Janet–Riquier theory.

The key tool in the geometric theory of differential equations (see [32] and references therein) are *jet bundles*. For K being either $\mathbb{R}$ or $\mathbb{C}$, we set $\mathcal{X} = K^n$, $\mathcal{U} = K^m$ and consider maps $\phi : \mathcal{X} \to \mathcal{U}$ which in the real case are assumed to be smooth and in the complex case to be holomorphic and which need be defined only on some open subset of $\mathcal{X}$. The coordinates $\mathbf{x} = (x^1, \ldots, x^n)$ on $\mathcal{X}$ are the independent variables and the coordinates $\mathbf{u} = (u^1, \ldots, u^m)$ on $\mathcal{U}$ represent the dependent variables or unknown functions. The ℓth order jet bundle $J_\ell(\mathcal{X}, \mathcal{U})$ consist of all Taylor polynomials of degree ℓ of such maps ϕ. Coordinates on $J_\ell(\mathcal{X}, \mathcal{U})$ are therefore $(\mathbf{x}, \mathbf{u}^{(\ell)})$ where $\mathbf{x}$ gives the expansion point and the jet variables $\mathbf{u}^{(\ell)}$ represent the Taylor coefficients up to order ℓ which we may identify with the corresponding derivatives of ϕ at the point $\mathbf{x}$. For the components of $\mathbf{u}^{(\ell)}$ we use the usual multi-index notation $u_\mu^\alpha = \partial^{|\mu|} \phi^\alpha / \partial \mathbf{x}^\mu$ with $1 \leq \alpha \leq m$ and $\mu \in \mathbb{N}_0^n$ satisfying $0 \leq |\mu| \leq \ell$. Hence we find that $J_\ell(\mathcal{X}, \mathcal{U}) \cong K^{d_\ell}$ with $d_\ell = n + m\binom{m+\ell}{\ell}$. For our purposes, two topologies on $J_\ell(\mathcal{X}, \mathcal{U})$ are relevant: one is induced by the Euclidean metric on K^{d_ℓ} and the other one is the Zariski topology on K^{d_ℓ} with varieties as closed sets. Finally, we introduce the canonical projection maps $\pi_k^\ell : J_\ell(\mathcal{X}, \mathcal{U}) \to J_k(\mathcal{X}, \mathcal{U})$ with $\pi_k^\ell(\mathbf{x}, \mathbf{u}^{(\ell)}) = (\mathbf{x}, \mathbf{u}^{(k)})$ for $\ell > k \geq 0$ and $\pi^\ell : J_\ell(\mathcal{X}, \mathcal{U}) \to \mathcal{X}$ with $\pi^\ell(\mathbf{x}, \mathbf{u}^{(\ell)}) = \mathbf{x}$.

Definition 1. *An* algebraic jet set *of order ℓ is a locally Zariski closed subset $\mathcal{J}_\ell \subseteq J_\ell(\mathcal{X}, \mathcal{U})$ (i. e. $\mathcal{J}_\ell$ is the difference of two varieties in $J_\ell(\mathcal{X}, \mathcal{U})$). Such a set $\mathcal{J}_\ell$ is an* algebraic differential equation *of order ℓ, if in addition the Euclidean closure of $\pi^\ell(\mathcal{J}_\ell)$ equals $\mathcal{X}$. An algebraic jet set or an algebraic differential equation is called* irreducible, *if it is an irreducible locally Zariski closed subset.*

We define here differential equations as a geometric object and do not distinguish between scalar equations and systems. An algebraic jet set is obtained by considering the solution set of an algebraic system on $J_\ell(\mathcal{X}, \mathcal{U})$. The additional condition for an algebraic differential equation ensures that the independent variables are indeed independent. It excludes equations like $x^1 + x^2 = 0$ which obviously is not a differential equation. Admitted is an equation like $xu' = 1$ where $x = 0$ is not contained in the projection but in its closure. We use here the Euclidean closure, as we would like to be able to express each point outside of the set $\pi^\ell(\mathcal{J}_\ell)$ as the limit of a sequence of points inside.

Any map $\phi : \mathcal{X} \to \mathcal{U}$ induces a map $j_0\phi : \mathcal{X} \to J_0(\mathcal{X}, \mathcal{U}) = \mathcal{X} \times \mathcal{U}$ defined by $j_0\phi(\mathbf{x}) = (\mathbf{x}, \phi(\mathbf{x}))$. The graph Γ_ϕ of ϕ is the image of $j_0\phi$. For any order $\ell > 0$, we may consider the *prolongations* $j_\ell\phi : \mathcal{X} \to J_\ell(\mathcal{X},\mathcal{U})$ given by $j_\ell\phi(\mathbf{x}) = \big(\mathbf{x}, \phi(\mathbf{x}), \partial_\mathbf{x}\phi(\mathbf{x}), \dots, \partial_\mathbf{x}^\ell\phi(\mathbf{x})\big)$ where $\partial_\mathbf{x}^k\phi(\mathbf{x})$ represents all derivatives of ϕ of order k. The next definition reformulates geometrically the classical one.

Definition 2. *A* (classical) solution *of the algebraic differential equation* $\mathcal{J}_\ell \subseteq J_\ell(\mathcal{X},\mathcal{U})$ *is a map* $\phi : \mathcal{X} \to \mathcal{U}$ *such that its prolongation satisfies* $\operatorname{im} j_\ell\phi \subseteq \mathcal{J}_\ell$.

Example 3. Let us consider the ordinary differential equation $u' = xu^2$ from this geometric view point. It represents a typical differential equation with singular solutions, as an elementary integration yields the general solution $\phi_c(x) = 2/(2c - x^2)$ parametrised by an arbitrary constant $c \in \mathbb{R}$.

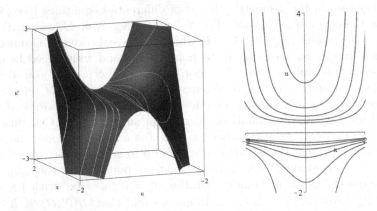

Fig. 1. Ordinary differential equation with singular solutions. Left: prolonged solutions in $J_1(\mathbb{R},\mathbb{R})$. Right: classical solution graphs in x-u plane (Color figure online).

The left half of Fig. 1 shows in dark blue the corresponding algebraic differential equation $\mathcal{J}_1 \subset J_1(\mathbb{R},\mathbb{R})$ and in light blue for some values of c the prolonged solutions $j_1\phi_c(x)$. In the right half, traditional graphs of these solutions show clearly the poles for positive values of c. We will see later that this differential equation has *no* singularities in the sense relevant for this article.

If one tries to combine the algebraic and the geometric point of view, one has to note some fundamental differences between the two. A differential ideal automatically contains all differential consequences of its generators. By choosing a jet bundle of a certain order ℓ, we immediately restrict to equations of order at most ℓ. On the other hand, geometric notions like the singularities we will consider in the next section cannot even be formulated in differential algebra. Thus moving from one point of view to the other one requires some care, as otherwise information is lost.

In applications, one usually starts with a differential system S like (D). The following approach appears to be very natural to associate with it for any given order $\ell \in \mathbb{N}_0$ an algebraic jet set in $J_\ell(\mathcal{X},\mathcal{U})$. We take the *differential* ideal

$$\hat{\mathcal{I}}_{\mathrm{diff}}(S) = \langle p_1, \dots, p_s \rangle_\Delta \subseteq \mathcal{D}$$

generated by the equations in S. It induces the *algebraic* ideal

$$\hat{\mathcal{I}}_\ell(S) = \hat{\mathcal{I}}_{\text{diff}}(S) \cap \mathcal{D}_\ell \subseteq \mathcal{D}_\ell$$

as the corresponding finite-dimensional truncation. It automatically contains all hidden integrability conditions up to order ℓ. The inequations in S are also used to define an *algebraic* ideal: $\mathcal{K}_\ell(S) = \langle \hat{Q}_\ell \rangle_{\mathcal{D}_\ell}$ with $\hat{Q}_\ell = \prod_{\text{ord}\,(q_j) \le \ell} q_j$. We then define the algebraic jet set

$$\hat{\mathcal{J}}_\ell(S) = \text{Sol}\,(\hat{\mathcal{I}}_\ell(S)) \setminus \text{Sol}\,(\mathcal{K}_\ell(S)) \subseteq J_\ell(\mathcal{X},\mathcal{U})$$

consisting of all points of $J_\ell(\mathcal{X},\mathcal{U})$ satisfying both the equations and the inequations in S interpreted as algebraic (in)equations in $J_\ell(\mathcal{X},\mathcal{U})$.

However, this procedure leads to many problems. The ideals $\hat{\mathcal{I}}_\ell(S)$ are often too small (not radical) and the algebraic jet sets $\hat{\mathcal{J}}_\ell(S)$ are not necessarily algebraic differential equations. Furthermore, the effective determination of $\hat{\mathcal{I}}_\ell(S)$ is difficult. Finally, the sets $\hat{\mathcal{J}}_\ell(S)$ are possibly too small, as an algebraic interpretation of inequations is much stronger than a differential one. Differentially, the inequation $u' \ne 0$ simply excludes the zero function; algebraically, it excludes all points with a vanishing u'-coordinate and thus e. g. all critical points of solutions. A more extensive discussion of these problems can be found in [22].

The situation improves, if one assumes that S is a *simple* differential system. Taking – following the reasoning behind (1) – the saturated differential ideal

$$\mathcal{I}_{\text{diff}} = \hat{\mathcal{I}}_{\text{diff}}(S) : \left(\prod_{j=1}^{s} \text{init}\,(p_j)\,\text{sep}\,(p_j) \right) \tag{2}$$

instead of $\hat{\mathcal{I}}_{\text{diff}}(S)$ and then using the same procedure as above to define algebraic ideals $\mathcal{I}_\ell(S)$ and algebraic jet sets $\mathcal{J}_\ell(S)$, one can show that these ideals are automatically radical and that explicit generators of the algebraic ideals $\mathcal{I}_\ell(S)$ are easily computable (see [22] for details).

The saturation in (2) leads to a Zariski closure. The inequations of a simple differential system exclude all points where an initial or separant vanishes and thus most of the singularities studied later. The saturation restores some of them and we only exclude irreducible components completely consisting of such points.

Example 4. Consider the system $\hat{S}$ consisting of the two partial differential equations $p_1 = uu_x - yu - y^2$ and $p_2 = yu_y u$ which is *not* simple. A differential Thomas decomposition yields only one simple system S obtained by augmenting $\hat{S}$ by the inequation $q = \text{sep}\,p_1 = u$. The algebraic ideal $\hat{\mathcal{I}}_1(S)$ obtained by truncating the differential ideal $\langle p_1, p_2 \rangle_\Delta$ has the prime decomposition $\hat{\mathcal{I}}_1(S) = \langle p_2, p_3 \rangle \cap \langle u, y \rangle$ where $p_3 = u_x u_y - u - y$ implying that also the differential ideal is not prime. Saturating with respect to $Q = yu$ removes the prime component $\langle u, y \rangle$ and we find that $\mathcal{I}_{\text{diff}}(S) = \langle p_2, p_3 \rangle_\Delta$ and $\mathcal{I}_1(S) = \langle p_2, p_3 \rangle \subset \mathcal{D}_1$.

Definition 5. *An algebraic differential equation $\mathcal{J}_\ell \subset J_\ell(\mathcal{X},\mathcal{U})$ is locally integrable, if there exists a Zariski open and dense subset $\mathcal{R}_\ell \subseteq \mathcal{J}_\ell$ such that $\mathcal{J}_\ell$ possesses for each point $\rho \in \mathcal{R}_\ell$ at least one solution ϕ with $\rho \in \text{im}\,j_\ell\phi$.*

Local integrability is for many purposes an important concept. If a point $\rho \in \mathcal{J}_\ell$ is "far away" from any (prolonged) solution, then one can argue how relevant such a point is. In fact, the existence of such points is a clear indication that $\mathcal{J}_\ell$ has not been well chosen. A typical problem are overlooked hidden integrability conditions. As any simple differential system is passive, we obtain via the existence theorem of Riquier the following result for our construction.

Proposition 6 ([22, Prop. 3.6, Lemma 3.7]). *Let S be a simple differential system. The the Zariski closure $\overline{\mathcal{J}_\ell(S)} = \mathrm{Sol}\left(\mathcal{I}_\ell(S)\right)$ is a locally integrable algebraic differential equation.*

3   Singularities of Algebraic Differential Equations

In the affine space K^{d_ℓ} all coordinates are equal. In the jet bundle $J_\ell(\mathcal{X}, \mathcal{U})$ we distinguish different types like independent and dependent variables or derivatives. The *contact distribution* $\mathcal{C}_\ell \subset TJ_\ell(\mathcal{X}, \mathcal{U})$ encodes these different roles and is generated by the vector fields

$$C_i^{(\ell)} = \partial_{x^i} + \sum_{\alpha=1}^{m} \sum_{0 \le |\mu| < \ell} u_{\mu+1_i}^\alpha \partial_{u_\mu^\alpha}, \qquad\qquad 1 \le i \le n, \qquad (3)$$

$$C_\alpha^\mu = \partial_{u_\mu^\alpha}, \qquad\qquad 1 \le \alpha \le m, \; |\mu| = \ell \qquad (4)$$

where $\mu + 1_i$ is obtained by increasing the ith entry of μ by one.

Definition 7. *Let $\mathcal{J}_\ell \subseteq J_\ell(\mathcal{X}, \mathcal{U})$ be an algebraic jet set. The* Vessiot cone $\mathcal{V}_\rho[\mathcal{J}_\ell]$ *at $\rho \in \mathcal{J}_\ell$ is the intersection of the tangent cone $C_\rho \mathcal{J}_\ell$ with the contact space $\mathcal{C}_\ell|_\rho$.*

At a smooth point, the tangent cone $C_\rho \mathcal{J}_\ell$ and the tangent space $T_\rho \mathcal{J}_\ell$ coincide and thus the Vessiot cone becomes a *Vessiot space*, i. e. a K-linear space, which can be computed by linear algebra. Since the Vessiot spaces are contained in the contact distribution, we make for any vector $\mathbf{v} \in \mathcal{V}_\rho[\mathcal{J}_\ell]$ the ansatz

$$\mathbf{v} = \sum_i a^i C_i^{(\ell)}|_\rho + \sum_{|\mu|=\ell} \sum_\alpha b_\mu^\alpha C_\alpha^\mu|_\rho \qquad (5)$$

with yet to be determined coefficients $a^i, b_\mu^\alpha \in K$. Let the jet set $\mathcal{J}_\ell$ be given as the solution set of an algebraic system on $J_\ell(\mathcal{X}, \mathcal{U})$ with equations $p_\tau = 0$. At a smooth point ρ, $\mathbf{v}$ is tangential to $\mathcal{J}_\ell$, if and only if $dp_\tau|_\rho(\mathbf{v}) = 0$ for all τ leading to a homogeneous linear system for the coefficient vectors $\mathbf{a}, \mathbf{b}$,

$$D(\rho)\mathbf{a} + M_\ell(\rho)\mathbf{b} = 0, \qquad (6)$$

where the entries of the matrices D, M_ℓ are given by $D_{i\tau}(\rho) = C_i^{(\ell)}(p_\tau)(\rho)$ and $(M_\ell)_{\alpha\tau}^\mu(\rho) = C_\alpha^\mu(p_\tau)(\rho)$. The rank of (6) and thus the dimension of $\mathcal{V}_\rho[\mathcal{J}_\ell]$ may vary over $\mathcal{J}_\ell$. Considered as functions of ρ, the solutions of (6) are smooth outside of a Zariski closed set and – by potentially enlarging this set – we may even assume that the dimension remains constant, since dimension is an upper semi-continuous function. Thus on a Zariski open and dense set we obtain a smooth regular distribution.

The projection $\pi^\ell_{\ell-1} : J_\ell(\mathcal{X},\mathcal{U}) \to J_{\ell-1}(\mathcal{X},\mathcal{U})$ induces at any point $\rho \in J_\ell(\mathcal{X},\mathcal{U})$ the vertical space $V_\rho\pi^\ell_{\ell-1} = \ker T_\rho\pi^\ell_{\ell-1}$ spanned by the vectors $C^\mu_\alpha|_\rho$. The vertical part of the Vessiot cone at a point $\rho \in \mathcal{J}_\ell$ is the *symbol cone* $\mathcal{N}_\rho[\mathcal{J}_\ell] = \mathcal{V}_\rho[\mathcal{J}_\ell] \cap V_\rho\pi^\ell_{\ell-1}$. At smooth points, we will speak of the *symbol space*. Again, on a Zariski open subset of $\mathcal{J}_\ell$ the symbol spaces $\mathcal{N}_\rho[\mathcal{J}_\ell]$ define a smooth regular distribution $\mathcal{N}[\mathcal{J}_\ell]$.

At a smooth point $\rho \in \mathcal{J}_\ell$, the symbol space $\mathcal{N}_\rho[\mathcal{J}_\ell]$ consists of those solutions of (6) where all coefficients **a** vanish: it is the kernel of the *symbol matrix* $M_\ell(\rho)$. Hence, we can write the Vessiot space as a direct sum $\mathcal{V}_\rho[\mathcal{J}_\ell] = \mathcal{N}_\rho[\mathcal{J}_\ell] \oplus \mathcal{H}_\rho$ with some π^ℓ-transversal complement $\mathcal{H}_\rho$ which is not uniquely determined. $\mathcal{J}_\ell$ is a *differential equation of finite type*, if on a Zariski open and dense subset the symbol cones vanish. For such equations, we expect that generically to every point $\rho \in \mathcal{J}_\ell$ there exists a unique solution ϕ with $\rho \in \operatorname{im} j_q\phi$, i.e. we may consider ρ as initial data for an initial value problem.

Remark 8. Computing the Vessiot space via (6) can be seen as a "projective" version of prolongation. Indeed, the formal derivative with respect to x^i of a differential equation $p_\tau = 0$ of order ℓ is

$$D_i p_\tau = C^{(\ell)}_i(p_\tau) + \sum_{\alpha=1}^{m} \sum_{|\mu|<\ell} C^\mu_\alpha(p_\tau) u^\alpha_{\mu+1_i}. \tag{7}$$

For an *ordinary* differential equation of finite type **a** $= a$ is scalar and we have one coefficient b^α for each unknown function u^α. If $(a,\mathbf{b})$ is a solution of (6), then the unique solution ϕ with $\rho \in \operatorname{im} j_q\phi$ satisfies $\phi^{(\ell+1)}(x_0) = \mathbf{b}/a$ where $x_0 = \pi^\ell(\rho)$, i.e. the Vessiot space contains information about the derivatives in the next order. Obviously, if $a = 0$, then $\phi^{(\ell+1)}$ blows up, as x approaches x_0.

We will denote the family of Vessiot cones by $\mathcal{V}[\mathcal{J}_q]$ and call it briefly the *Vessiot distribution* of $\mathcal{J}_\ell$, although strictly speaking we obtain a distribution only on a subset of $\mathcal{J}_q$. But the considerations above justify this slight abuse of language. The Vessiot distribution can be interpreted as a kind of "infinitesimal solution space" of $\mathcal{J}_\ell$: if ϕ is any solution of $\mathcal{J}_\ell$ and ρ lies on $\operatorname{im} j_\ell\phi$, then the tangent space $T_\rho \operatorname{im} j_\ell\phi$ lies in the Vessiot cone $\mathcal{V}_\rho[\mathcal{J}_\ell]$.

Definition 9. *A* generalized solution *of the algebraic differential equation* $\mathcal{J}_\ell \subseteq J_\ell(\mathcal{X},\mathcal{U})$ *with* $\dim \mathcal{X} = n$ *is an n-dimensional submanifold* $\mathcal{N} \subseteq \mathcal{J}_\ell$ *such that* $T_\rho\mathcal{N} \subseteq \mathcal{V}_\rho[\mathcal{J}_\ell]$ *at every point* $\rho \in \mathcal{N}$. *The projection* $\pi^\ell_0(\mathcal{N}) \subset J_0(\mathcal{X},\mathcal{U})$ *of a generalized solution is called a* geometric solution.

If ϕ is a classical solution of $\mathcal{J}_\ell$, then $\operatorname{im} j_\ell\phi$ is a generalized solution and the graph $\Gamma_\phi = \operatorname{im} j_0\phi$ of ϕ the corresponding geometric solution. Furthermore, at any point $\rho \in \operatorname{im} j_\ell\phi$ we find that $\mathcal{V}_\rho[\mathcal{J}_\ell] = \mathcal{N}_\ell[\mathcal{J}_\rho]\oplus T_\rho \operatorname{im} j_\ell\phi$. As we will see later, an algebraic differential equation may possess further generalized solutions.

Definition 10 ([22, Def. 4.1]). *Let $\mathcal{J}_\ell \subseteq J_\ell(\mathcal{X}, \mathcal{U})$ be a locally integrable algebraic differential equation and* $\dim \mathcal{X} = n$. *A point $\rho \in \mathcal{J}_\ell$ is an algebraic singularity of $\mathcal{J}_\ell$, if ρ is a non-smooth point of $\mathcal{J}_\ell$ in the sense of algebraic geometry. A smooth point $\rho \in \mathcal{J}_\ell$ is called*

(i) *regular, if ρ possesses a Euclidean open neighbourhood $\mathcal{U} \subseteq \mathcal{J}_\ell$ such that the Vessiot cones form on $\mathcal{U}$ a regular distribution which is decomposable as $\mathcal{V}[\mathcal{J}_\ell]|_{\mathcal{U}} = \mathcal{N}[\mathcal{J}_\ell]|_{\mathcal{U}} \oplus \mathcal{H}$ with an n-dimensional, transversal, involutive, smooth distribution $\mathcal{H} \subseteq T\mathcal{U}$;*

(ii) *regular singular, if ρ possesses a Euclidean open neighbourhood $\mathcal{U} \subseteq \mathcal{J}_\ell$ such that the Vessiot cones form on $\mathcal{U}$ a regular distribution but where $\dim \mathcal{V}_\rho[\mathcal{J}_\ell] - \dim \mathcal{N}_\rho[\mathcal{J}_\ell] < n$;*

(iii) *irregular singular, if there does not exist a Euclidean open neighbourhood $\mathcal{U} \subseteq \mathcal{J}_\ell$ such $\mathcal{V}[\mathcal{J}_\ell]|_{\mathcal{U}}$ is a regular distribution, i. e. any such neighbourhood contains a point $\bar{\rho}$ such that $\dim \mathcal{V}_\rho[\mathcal{J}_\ell] > \dim \mathcal{V}_{\bar{\rho}}[\mathcal{J}_\ell]$.*

An irregular singularity ρ is purely irregular, *if $\dim \mathcal{V}_\rho[\mathcal{J}_\ell] - \dim \mathcal{N}_\rho[\mathcal{J}_\ell] = n$. Regular and irregular singular points are also called* geometric singularities.

Algebraic singularities are not considered in the differential topological theory and one finds there much simpler definitions (see e. g. [1] or [27]), as only *ordinary* differential equations are considered where it is not necessary to consider neighbourhoods. One knows in advance the "right" dimension of the Vessiot spaces and can thus compare pointwise with this value. For *partial* differential equations, this is generally no longer the case and Definition 10 represents to our knowledge the first definition of geometric singularities for general systems of partial differential equations (the much simpler intermediate case of partial differential equations of finite type was already considered in [18]).

Remark 11. The three cases distinguished in Definition 10 for smooth points correspond essentially to an analysis of the linear system (6). At an irregular singular point, its rank does not take the maximal possible value attained in the other two cases. At a regular point, the symbol matrix alone is already of this rank. Thus geometric singularities are characterized by a rank drop of the symbol matrix. At non-singular points of an *ordinary* differential equations, the complement $\mathcal{H}$ is always one-dimensional and thus trivially involutive. In this case (or more generally for any locally integrable differential equation of finite type), the taxonomy of Definition 10 is complete. For *partial* differential equations, it is still an open question whether points can exist on $\mathcal{J}_\ell$ which satisfy all conditions for a regular point except the involutivity of $\mathcal{H}$ (see [22] for a more extensive discussion of this topic).

Example 12. We consider the algebraic differential equation $\mathcal{J}_2 \subset J_2(\mathbb{C}^2, \mathbb{C})$ for one unknown function u in two independent variables x, y defined by:

$$x^2 u_{xx} + x u_x + (x-1)^2 u = 0, \qquad (1-y^2) u_{yy} + 2y u_y + 2u = 0.$$

Seven cases arise in the analysis of the linear system (6) for the Vessiot spaces:

1. *Regular points* on $\mathcal{J}_2$ are characterized by the conditions $x \neq 0$ and $y^2 - 1 \neq 0$. They have a three-dimensional Vessiot space.
2. Points where $x = 0$, $y^2 - 1 \neq 0$ and either $u_x \neq 0$ or $u_y \neq 0$ are *regular singular*. They also possess a three-dimensional Vessiot space. As the coefficients a_1 and a_2 in (5) must satisfy the equation $2u_x a_1 + u_y a_2 = 0$, only a one-dimensional transversal complement exists.
3. Basically the same holds for points where $y^2 - 1 = 0$, $x \neq 0$ and either $yu_x + u_{xy} \neq 0$ or $u \neq 0$: they are *regular singular* and have a three-dimensional Vessiot space with a one-dimensional transversal complement defined by the equation $(yu_x + u_{xy})a_1 - 2ua_2 = 0$.
4. Points where $x = 0$, $y^2 - 1 = 0$ and either $u_x \neq 0$ or $yu_{xy} + u_x \neq 0$ are *irregular singularities* which are not purely irregular: the Vessiot space is four-dimensional with a one-dimensional transversal complement defined by the condition $a_1 = 0$.
5. Points where $x = 0$, $u_x = 0$, $u_y = 0$ and $y^2 - 1 \neq 0$ are *purely irregular singular* and possess a four-dimensional Vessiot space defined by the equation $(y^2 - 1)b_{02} - 2yu_{xy}a_1 = 0$.
6. The same behaviour is shown by points with $y^2 - 1 = 0$, $u = 0$, $u_y = 0$, $x \neq 0$, but with the Vessiot space defined by the equation $x^2 b_{20} + (x^2 - xy - 2x - 1)u_x a_1 = 0$.
7. Finally, the points where $x = 0$, $y^2 - 1 = 0$, $u_x y = 0$ and $u = 0$ are also *purely irregular singular* but now with a five-dimensional Vessiot space.

Note that the cases 2, 3 and 4 do not correspond to an algebraic jet set but the union of two such sets, because of the disjunctions in their defining conditions. Hence, if one applies the algorithm we will present in the next section to this example, then one obtains actually $10 = 7 + 3$ cases.

Any definition of a "singularity" is only meaningful, if generic points are regular. For equations of finite type, this is obvious and not even discussed in the literature. However, for general partial differential equations, such a statement becomes highly non-trivial and its proof requires major results from the geometric theory of differential equations. The key issue is to prove the involutivity of the complement $\mathcal{H}$ over a Zariski open and dense subset.

Theorem 13 ([22, Thm. 4.7]). *Let S be a simple differential system which contains no equation of an order greater than $\ell \in \mathbb{N}$ and $\mathcal{J}_\ell(S)$ the associated algebraic differential equation. Then the regular points in the Zariski closure $\overline{\mathcal{J}_\ell(S)}$ contain a Zariski open and dense subset.*

4 Regularity Decompositions

(Geometric) singularities are points where the dimensions of some geometric structures like symbol or Vessiot spaces jump. An algebraic jet set $\mathcal{J}_\ell \subseteq J_\ell(\mathcal{X}, \mathcal{U})$ is *regular*, if it consists only of smooth points and both its Vessiot distribution

$\mathcal{V}[\mathcal{J}_\ell]$ and its symbol $\mathcal{N}[\mathcal{J}_\ell]$ define smooth vector bundles over $\mathcal{J}_\ell$. The solution space of the linear system (6) behaves uniformly over a regular algebraic jet set and thus all points on such a set are classified identically by Definition 10.

Definition 14. *Let $S \subset \mathcal{D}$ be a simple differential system and $\overline{\mathcal{J}_\ell(S)} \subset J_\ell(\mathcal{X},\mathcal{U})$ the associated algebraic jet set in a sufficiently high order ℓ. Let furthermore $\overline{\mathcal{J}_\ell(S)} = \mathcal{J}_{\ell,1} \cup \cdots \cup \mathcal{J}_{\ell,t}$ be its decomposition into irreducible varieties. A regularity decomposition of the variety $\mathcal{J}_{\ell,k}$ represents it as a disjoint union of finitely many regular algebraic jet sets $\mathcal{J}_{\ell,k}^{(1)}, \ldots, \mathcal{J}_{\ell,k}^{(r)}$, the regularity components of $\mathcal{J}_{\ell,k}$, and of the set $\mathrm{ASing}\big(\overline{\mathcal{J}_\ell(S)}\big)$ of algebraic singularities.*

A constructive proof of the existence of regularity decompositions for any simple differential system is provided by Algorithm 1 below. Regularity decompositions are not unique and thus this algorithm simply returns one possible decomposition. The first two lines represent an algebraic preprocessing. In Line 2, the algebraic ideal $\mathcal{I}_\ell(S)$ is constructed explicitly via Janet–Riquier and Gröbner theory – for details see [22, Rem. 3.8]. The determination of a prime decomposition in Line 3 is a standard task in commutative algebra. Then the algorithm loops over each prime component. In Line 5, two simultaneous linear systems are set up over each prime component, i. e. we consider the combined system

$$
\begin{cases}
\mathbf{J}(p_{k,j}) = 0\,, \\
\mathbf{v}(p_{k,j}) = 0\,, \\
\quad p_{k,j} = 0\,,
\end{cases}
\qquad j = 1, \ldots, s_k\,.
\tag{8}
$$

Here the polynomials $p_{k,j}$ form a basis of the kth prime ideal $\mathcal{I}_{\ell,k}(S)$. The two linear systems are obtained by applying two vector fields to these generators. The first one, $\mathbf{J} = \sum_\mu \sum_\alpha c_\mu^\alpha \partial_{u_\mu^\alpha} + \sum_i d^i \partial_{x^i}$ represents a general tangent vector in the jet bundle with yet undetermined coefficients $\mathbf{c}$ and $\mathbf{d}$. The first linear system in (8) encodes the condition that $\mathbf{J}$ is tangential to the kth prime component. By the Jacobian criterion, a jump in its rank characterizes algebraic singularities. The second linear system is constructed with a general contact vector (5) and thus represents (6) for determining the Vessiot spaces. Changes in its behaviour indicate geometric singularities.

The undetermined coefficients $\mathbf{a}, \mathbf{b}, \mathbf{c}, \mathbf{d}$ represent the unknowns of the linear systems and we consider the left hand sides of the equations as elements of $\mathcal{D}_\ell^{\mathrm{ex}} = \mathcal{D}_\ell[\mathbf{a}, \mathbf{b}, \mathbf{c}, \mathbf{d}]$. As changes in the behaviour of the linear systems indicate singularities, these can be detected by an algebraic Thomas decomposition of (8) for a suitably chosen ordering. More precisely, we must have: (i) $\mathbf{d} > \mathbf{c} > \mathbf{b} > \mathbf{a} > \mathbf{u} > \mathbf{x}$, (ii) restricted to the jet variables $\mathbf{u}$ it must correspond to an orderly ranking and (iii) the variables c_μ^α and b_μ^α are ordered among themselves in the same way as the derivatives u_μ^α.

Given a simple algebraic system $S_{k,\ell}^{\mathrm{ex}}$ in the obtained decomposition, the subsystem $S_{k,\ell}$ obtained by eliminating all equations and inequations containing some of the auxiliary variables $\mathbf{a}, \mathbf{b}, \mathbf{c}, \mathbf{d}$ describes a regular jet set. In practice, one is nevertheless strongly interested in getting the extended systems $S_{k,\ell}^{\mathrm{ex}}$, as the appearing leaders allow us to deduce the dimensions of the Vessiot and

---

Algorithm 1: Regularity Decomposition for a Simple Differential System

Input: a simple differential system S over the ring $K\{U\}$ of differential
    polynomials and a sufficiently high order $\ell \in \mathbb{N}$

Output: a regularity decomposition for each prime component $\mathcal{I}_{\ell,k}(S)$ of the
    algebraic ideal $\mathcal{I}_\ell(S) \subset \mathcal{D}_\ell$

1 **begin**
2   compute a generating set $\{p_1, \ldots, p_s\}$ of the radical ideal $\mathcal{I}_\ell(S)$
3   compute a prime decomposition $\mathcal{I}_\ell(S) = \mathcal{I}_{\ell,1}(S) \cap \ldots \cap \mathcal{I}_{\ell,t}(S)$ of $\mathcal{I}_\ell(S)$ and
         a generating set $\{p_{k,1}, \ldots, p_{k,s_k}\}$ for each prime component $\mathcal{I}_{\ell,k}(S)$
4   **for** $k = 1$ **to** t **do**
5   compute an algebraic Thomas decomposition $S_{k,1}^{\text{ex}}, \ldots, S_{k,r_k}^{\text{ex}}$ of the
         algebraic system defined over $\mathcal{D}_\ell^{\text{ex}}$ for an ordering as described above
6   **return** *the systems $S_{k,i}$ consisting of those equations $p = 0$ and inequations*
         $q \neq 0$ in $S_{k,i}^{\text{ex}}$ with $p \in \mathcal{D}_\ell$ and $q \in \mathcal{D}_\ell$

---

symbol spaces and thus to classify automatically the points on the jet set [22, Prop. 5.10]. The proof of the correctness of Algorithm 1 requires a number of rather technical issues and cannot be discussed here – see [22, Thm. 5.13].

Example 15. The *hyperbolic gather* is one of the elementary catastrophes. Interpreted as a first-order ordinary differential equation, it is given by $\mathcal{J}_1 = \{(u')^3 + uu' - x = 0\}$. A real picture of it is contained in Fig. 2 presented in Example 19 below. Despite its simplicity, it well illustrates some of the problems appearing in the practical use of Algorithm 1. Using the implementation of the Thomas decomposition described in [2], our algorithm returns a regularity decomposition with seven components all consisting of smooth points. One component contains the two irregular singularities, namely the points $(2, -3, -1)$ and $(-2, -3, 1)$ shown in Fig. 2 in red. The regular singularities fill three components. Two of them correspond to the fold line shown in Fig. 2 in white which arises as the common zero set of our equation and its separant. The "tip" of the fold line is put in a separate component. The third component contains only complex points and is thus not visible in Figure 2. Finally, there are three components with regular points. A closer analysis of the Vessiot spaces at these points (presented in [22, Ex. 7.2]) reveals that they can be combined into a single regularity component; the splitting into three separate components is solely an artifact of the Thomas decomposition due to its internal use of projections along each coordinate axis.

It represents a general problem of Algorithm 1 that it performs implicitly a Thomas decomposition of the considered irreducible varieties. Some singularities indeed arise from the geometry of these varieties: it was no coincidence that in Example 15 all singularities lie on the fold line. But the Thomas decomposition automatically also puts all points on the differential equation lying under or over the fold line in separate components, although this is generally unnecessary for the singularity analysis. In systems with several unknown functions $\mathbf{u}$ (and corresponding derivatives), this effect can be much more pronounced and its size depends generally on the ordering of the entries of $\mathbf{u}$ (and the induced effect on the ordering of their derivatives $\mathbf{u}^{(\ell)}$), although this ordering is irrelevant for the analysis of the differential equation.

5   Semialgebraic Differential Equations

So far, we have exclusively considered the case of complex differential equations, as the Thomas decomposition assumes that the underlying field is algebraically closed. As in applications real equations dominate, we discuss now following [35] an extension of the ideas of the last sections to differential equations over the real numbers. We will not provide a complete solution to this problem. The approach presented above consists essentially of three phases. In the first phase, we perform a differential Thomas decomposition. Here the underlying field plays only a minor role, as a crucial point is the completion to a passive system.

In the second phase, we construct from an obtained simple differential system a suitable algebraic differential equation. This step involves some problematic operations. Firstly, we perform a saturation which provides us with a radical ideal. Over the real numbers, we should actually strive for the real radical according to the real nullstellensatz (see e.g. [6, Sect. 4.1] for a discussion). An algorithm for determining the real radical was proposed by Becker and Neuhaus [5, 24]; an implementation over the rational numbers exists in SINGULAR [37]. Secondly, we need irreducible varieties and hence a prime decomposition. Since computing such a decomposition is related to factorization, it strongly depends on the underlying field. Again, effective methods exist only over the rational numbers. Thus we conclude that for arbitrary simple differential systems the second phase cannot be done completely algorithmically.

It is unclear whether all steps of the second phase are really necessary. For example, for determining the tangent space at a smooth point, one does not need the real radical. Many problems in practise automatically lead to prime ideals. We will in the sequel assume that we are able to perform all required computations by saying that we are dealing with a *well-prepared system* and concentrate on a real variant of the third phase.

In the third phase, we solve two linear systems over a locally Zariski closed set. In the previous section, we simply threw all (in)equations together and computed an algebraic Thomas decomposition. Now we will present an alternative approach making stronger use of the "staggered" structure of the problem and the partial linearity. Furthermore, we also extend the class of differential equations considered. An algebraic differential equation was essentially defined as a locally Zariski closed sets, i.e. it was described by equations $p_i = 0$ and inequations $q_j \neq 0$. Over the real numbers, it is desirable to include also inequalities $q_j \diamond 0$ where $\diamond$ stands for some relation in $\{<, >, \leq, \geq, \neq\}$. Thus we replace the condition "locally Zariski closed" by "semialgebraic" (see e.g. [6, Chap. 2]).

Definition 16. *A semialgebraic jet set of order ℓ is a semialgebraic subset $\mathcal{J}_\ell \subseteq J_\ell(\mathcal{X}, \mathcal{U})$. Such a set $\mathcal{J}_\ell$ is a semialgebraic differential equation of order ℓ, if in addition the Euclidean closure of $\pi^\ell(\mathcal{J}_\ell)$ equals $\mathcal{X}$.*

We call a semialgebraic jet set $\mathcal{J}_\ell \subseteq J_\ell \pi$ *basic*, if it can be described by a finite set of equations $p_i = 0$ and a finite set of inequalities $q_j > 0$. We call such a pair of sets a *basic semialgebraic system* on $J_\ell \pi$. It follows from an elementary result in real algebraic geometry [6, Prop. 2.1.8] that any semialgebraic jet set can be expressed as a union of finitely many basic semialgebraic jet sets. We will always assume that our sets are given in this form and study each basic semialgebraic system separately, as for some steps in our analysis it is crucial that at least the equation part of the system is a pure conjunction.

The basic idea underlying the approach of [35] is to treat the system (8) in stages. So far, we computed an algebraic Thomas decomposition of the full system for a suitably chosen ordering. It was not really relevant that parts of the system were linear (although it makes the determination of the Thomas decomposition faster). Now we study first only the linear parts of (8) as a parametric linear system in the unknowns **a**, **b**, **c** and **d** with the jet variables **x** and $\mathbf{u}^{(q)}$ considered as parameters (appearing in polynomial form).

Parametric Gaussian elimination has been studied for more than 30 years, see e. g. [4,15,36]. A parametric Gaussian elimination returns a finite set of pairs $(\gamma, \mathbf{H})$ where the *guard* γ describes the conditions for this particular case and $\mathbf{H}$ represents the corresponding solution of the linear system. The guard γ is basically a conjunction of equations and inequations describing the choices made for the various pivots arising during the solution process. A key point is the application of advanced logic and decision procedures for an efficient heuristic handling of the potentially exponentially large number of arising cases. We used a reimplementation of the REDLOG [10] package PGAUSS – see [35, Sect. 3]. It applies strong heuristic simplification techniques [11] and quantifier elimination-based decision procedures [19,31,42,43].

The two linear subsystems of (8) are independent of each other. The analysis of the Jacobian criterion is straightforward: changes in the rank of the matrix are automatically delivered by a parametric Gauss algorithm. The analysis of the system for the Vessiot spaces is a bit more involved. For simplicity, we restrict to the case of ordinary differential equations where the vector **a** contains only a single entry a. Here we can give pointwise criteria: a point is regular, if the Vessiot space is one- and the symbol space zero-dimensional; it is regular singular, if both spaces are one-dimensional and irregular singular if the Vessiot space has a dimension higher than one. Thus here it does not suffice to look only at the rank of the matrix; one must also analyse the relative position of the Vessiot space to the vertical space of the jet bundle or more prosaically whether there are non-trivial solutions for which $a = 0$. In [35, Sect. 3], we developed a variant of parametric Gaussian elimination which takes as additional input a sublist of variables for which such considerations are taken into account.

Once all the different cases appearing in the solution of the linear systems have been determined, we must check which of them actually occur on our

---

Algorithm 2: RealSingularities

Input: $\Sigma_\ell = \big((p_a = 0)_{a=1,...,A}, (q_b > 0)_{b=1,...,B}\big)$ well-prepared, basic
semialgebraic system with $p_a, q_b \in \mathcal{D}_\ell \cap \mathbb{Z}[t, \mathbf{u}, \dots, \mathbf{u}^{(\ell)}]$
Output: finite system $(\Gamma_i, \mathrm{H}_i)_{i=1,...,I}$ with
  (i) each Γ_i is a disjunctive normal form of polynomial equations, inequations,
     and inequalities over $\mathcal{D}_\ell$ describing a semialgebraic subset $\mathcal{J}_{\ell,i} \subseteq \mathcal{J}_\ell$
  (ii) each H_i describes the Vessiot spaces of all points on $\mathcal{J}_{\ell,i}$
 (iii) all sets $\mathcal{J}_{\ell,i}$ are disjoint and their union is $\mathcal{J}_\ell$

1 **begin**
2   set up the matrix A of the second linear part of (8) using the equations
     $(p_a = 0)_{a=1,...,A}$
3     $\Pi = \big(\gamma_\tau, \mathrm{H}_\tau\big)_{\tau=1,...,t} \leftarrow \mathtt{ParametricGauss}\big(A, (\mathbf{b}, a), (a), \mathbb{R}\big)$
4   **for** $\tau = 1$ **to** t **do**
5   let Γ_τ be a disjunctive normal form of $\gamma_\tau \wedge \bigwedge \Sigma_\ell$
6   check satisfiability of Γ_τ using real quantifier elimination on
        $\exists t\, \exists \mathbf{u} \dots \exists \mathbf{u}^{(\ell)}\, \Gamma_\tau$
7   **if** Γ_τ is unsatisfiable **then**
8 | delete $(\gamma_\tau, \mathrm{H}_\tau)$ from Π
9   **else**
10   replace $(\gamma_\tau, \mathrm{H}_\tau)$ by $(\Gamma_\tau, \mathrm{H}_\tau)$ in Π

11   **return** Π

---

semialgebraic jet set. Thus for each case (γ, H) obtained we must verify whether or not the conjunction of its guard γ with the semialgebraic description of the jet set possesses a solution. Such a check represents a classical task for *real quantifier elimination* [7]. If the answer is yes, then the conjunction gives a semialgebraic description of one component in our regularity decomposition. For various reasons like a better readability of the results, we always return a disjunctive normal form of the semialgebraic description (for more details see [35]). In a more formal language we arrive thus at Algorithm 2.

Example 17. We study again the *hyperbolic gather*. In Example 15, the Thomas decomposition yielded unnecessarily many components in the regularity decomposition and some of them contained only complex points making them irrelevant for a real analysis. Using the above outlined approach, one obtains a real regularity decomposition consisting of exactly three components corresponding to the regular, the regular singular and the irregular singular points [35, Ex. 15]. This is an effect of the reversal of the analysis: we first study the different cases arising in the linear systems, then we check where on the differential equation the cases occur. This strategy should generally lead to a lower number of cases.

In [35, Ex. 15] the *elliptic gather* given by $(u')^3 - uu' - x = 0$ (i.e., it differs only by the sign of the middle term) is considered. In a complex analysis, one obtains for both gathers essentially the same result. Over the real numbers, the elliptic gather has no irregular singularities (they are now complex). Consequently, our real approach yields a regularity decomposition with only two components.

6   Analysis of Geometric Singularities

After the detection of singularities, we will now discuss the local solution behaviour around them – but only for ordinary differential equations of finite type, as not much is known for partial differential equations. We also consider only the real case using methods from dynamical systems theory. Let $\rho = (\bar{x}, \bar{\mathbf{u}}^{(\ell)})$ be a smooth point on an algebraic differential equation $\mathcal{J}_\ell$. We consider it as initial data for an initial value problem: we search for solutions ϕ of $\mathcal{J}_\ell$ such that $\phi(\bar{x}) = \bar{\mathbf{u}}$, $\phi'(\bar{x}) = \bar{\mathbf{u}}', \ldots, \phi^{(\ell)}(\bar{x}) = \bar{\mathbf{u}}_\ell$. We distinguish between *two-sided solutions* which exist in an interval $(\bar{x} - \epsilon, \bar{x} + \epsilon)$, i. e., for which $\operatorname{im} j_\ell\phi$ goes through the point ρ, and *one-sided solutions* which either begin in ρ, i. e., exist on an interval $[\bar{x}, \bar{x}+\epsilon)$, or end in ρ, i. e., exist on an interval $(\bar{x}-\epsilon, \bar{x}]$. We are interested in the existence, (non)uniqueness and regularity of such solutions. Away from irregular singularities, the theory is rather simple, as the following generalization of standard results for explicit ordinary differential equations shows.

Theorem 18 ([18, Thm. 4.1]). *Let $\mathcal{J}_\ell$ be a smooth algebraic ordinary differential equation of order ℓ such that at every point $\rho \in \mathcal{J}_\ell$ the Vessiot space $\mathcal{V}_\rho[\mathcal{J}_\ell]$ is one-dimensional. If ρ is a regular point, then there exists a unique smooth classical two-sided solution ϕ with $\rho \in \operatorname{im} j_\ell\phi$. More precisely, it can be extended in both directions until $\operatorname{im} j_\ell\phi$ reaches either the boundary of $\mathcal{J}_\ell$ or a regular singular point. If ρ is a regular singular point, then either two smooth classical one-sided solutions ϕ_1, ϕ_2 exist with $\rho \in \overline{\operatorname{im} j_\ell\phi_i}$ which either both start or both end in ρ or only one classical two-sided solution exists whose $(\ell+1)$th derivative blows up at $x = \pi^\ell(\rho)$.*

Proof. By the made assumptions, $\mathcal{V}[\mathcal{J}_\ell]$ can be generated in an open neighbourhood of ρ by a smooth vector field X. The standard existence and uniqueness theorems guarantee for each point $\rho \in \mathcal{J}_\ell$ the existence of a unique integral curve of X defining a unique generalized solution $\mathcal{N}_\rho$ with $\rho \in \mathcal{N}_\rho$. This generalized solution is a smooth curve which can be extended until it reaches the boundary of $\mathcal{J}_\ell$ and around each regular point $\bar{\rho} \in \mathcal{N}_\rho$ it projects onto the graph of a strong solution ϕ, since $\mathcal{V}_{\bar{\rho}}[\mathcal{J}_\ell]$ is transversal to π^ℓ by definition of a regular point.

If ρ is a regular singular point, then X_ρ is vertical for π^ℓ, i. e. its ∂_x-component vanishes. The behaviour of the corresponding geometric solution $\tilde{\mathcal{N}}_\rho = \pi_0^\ell(\mathcal{N}_\rho)$ depends on whether or not the ∂_x-component changes its sign at ρ. If the sign changes, then $\tilde{\mathcal{N}}_\rho$ has two branches corresponding to two classical solutions which either both end or both begin at $\hat{\rho} = \pi_0^\ell(\rho)$. Otherwise $\tilde{\mathcal{N}}_\rho$ is around $\hat{\rho}$ the graph of a classical solution, but Remark 8 implies that the $(\ell+1)$th derivative of this solution at $x = \pi^\ell(\rho)$ must be infinite.

Example 19. We continue our study of the *hyperbolic gather* by looking at the local solution behaviour. Figure 2 shows on the left hand side a number of generalized solutions in cyan and on the right hand side the corresponding geometric solutions in blue. One can see that whenever a generalized solution crosses transversely the white fold line outside of an irregular singularity, then the geometric solution reverses its direction (more precisely, the curve defining it has a cusp there). At these "reversal points", the geometric solution cannot be interpreted as the graph of a function. Hence one obtains in the classical picture two one-sided solutions. The curve in magenta shows the generalized solution that goes through the tip of the fold line. The corresponding geometric solution is still a classical one, but only $\mathcal{C}^1$: one can see that it is not smooth at the origin, as by Theorem 18 its second derivative blows up.

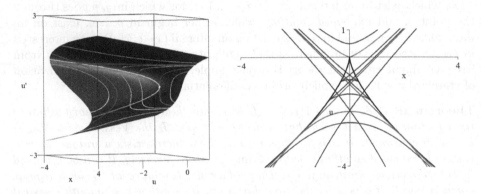

Fig. 2. Generalized solutions of the hyperbolic gather. Left: situation in $J_1(\mathbb{R}, \mathbb{R})$. Right: projection to x-u plane (Color figure online).

At irregular singularities, the solution behaviour can be more complicated. It follows from their definition that they form an algebraic jet set of codimension at least 2. Hence, if $\rho \in \mathcal{J}_\ell$ is an irregular singularity, then we can find an open, simply connected submanifold $\mathcal{U} \subset \mathcal{J}_\ell$ such that $\rho \in \overline{\mathcal{U}}$ and everywhere in $\mathcal{U}$ the Vessiot spaces are one-dimensional. On $\mathcal{U}$ the Vessiot distribution $\mathcal{V}[\mathcal{J}_\ell]$ can be generated by a single smooth vector field X. In principle, it is straightforward to construct such a vector field by solving (6), but one must exclude certain degeneracies appearing e. g. in the presence of singular integrals.

If $\mathcal{J}_\ell$ is locally integrable, then we may assume without loss of generality by (the proof of) [32, Prop. 9.5.10] that $\mathcal{J}_\ell$ is described by a square system $p_\tau = 0$ with as many equations as unknowns. Thus the symbol matrix M_ℓ is square and for ordinary differential equations the matrix D becomes a vector $\mathbf{d}$. Let $M^\dagger = \mathrm{adj}\,(M_\ell)$ be the adjugate of M. On $\mathcal{U}$, the Vessiot distribution is generated by the vector field $X = \det\,(M) C_1^{(\ell)} - \sum_{\alpha=1}^m (M^\dagger \mathbf{d})^\alpha C_\alpha^\ell$ which can be smoothly extended to a neighbourhood of ρ, as all its coefficients are polynomials.

Proposition 20. *Let $\mathcal{J}_\ell$ be a locally integrable differential equation and assume that on $\mathcal{U}$ the vector $\mathbf{d}$ does not vanish and that $\det(M)$ and the components of $M^\dagger \mathbf{d}$ do not possess a non-trivial joint common divisor. Then any smooth extension of the vector field X vanishes at ρ.*

Proof. The made assumptions ensure that X is a "minimal" generator of the Vessiot distribution on $\mathcal{U}$. At the irregular singularity ρ, the rank of $M(\rho)$ drops and thus $\det(M) = 0$. If it drops by more than one, then $M^\dagger(\rho) = 0$ and the claim is trivial. If the rank drops only by one, then the vector $\mathbf{d}(\rho)$ must lie in the column space of $M(\rho)$ for an irregular singularity. It follows now by Cramer's rule that $M^\dagger(\rho)\mathbf{d}(\rho) = 0$ and hence $X_\rho = 0$.

Thus at least generically we can analyse the local solution behaviour by using dynamical systems theory: we are given a smooth vector field X on $\mathcal{J}_\ell$ for which ρ is a stationary point. If ρ is a hyperbolic stationary point, then the eigenvalues of $\mathrm{Jac}(X)$ completely determine the local phase portrait. Otherwise, one must resort to more advanced techniques like blow-ups.

Example 21. In the case of the *hyperbolic gather*, the Vessiot distribution is generated by the vector field $X = (3(u')^2 + u)(\partial_x + u'\partial_u) + (1 - (u')^2)\partial_{u'}$. The Jacobian at the irregular singularity $\rho = (2, -3, 1)$ (the analysis of the other irregular singularity proceeds analogously) is $J = \left(\begin{smallmatrix} 0 & 1 & -6 \\ 0 & -1 & 6 \\ 0 & 0 & 2 \end{smallmatrix}\right)$ with the three eigenvalues 2, -1 and 0. Although $\mathcal{J}_1$ is a two-dimensional submanifold, we are computing here with all three jet coordinates in $J_1(\mathbb{R}, \mathbb{R})$. Thus we must decide which eigenvalue is irrelevant. This is straightforward: we only have to check which eigenvector is not tangential to $\mathcal{J}_1$. It turns out that in our case 0 is irrelevant. Hence ρ is a saddle point of X, as one can also clearly see on the left hand side of Fig. 2. The red curves there are two invariant manifolds tangent to the eigenspaces which for us represent two generalized solutions which intersect at the irregular singularity.

If an irregular singularity ρ is a node of the vector field X, then infinitely many (two-sided) generalized solutions intersect there. At a focus, all generalized solutions are one-sided, as they do not possess a well-defined tangent when spiralling into ρ and hence cannot be combined to a smooth curve through ρ. For higher-dimensional equations, the analysis in particular of non-hyperbolic stationary points can be arbitrarily complicated. For scalar first-order equations a complete classification of generic irregular singularities was given in [8,9]. The typical behaviour at an irregular singularity is thus that the usual uniqueness statements break down and several general solutions intersect there. There are, however, also degenerate situations where one still obtains a unique solution (see e. g. [34, Ex. 3.5]); in this case one speaks of an *apparent singularity*.

7   Quasilinear Equations

Quasilinear ordinary differential equations have their own theory, which somewhat surprisingly seems to have been overlooked in the differential topological literature. By contrast, in the context of differential algebraic equations, authors have studied almost exclusively the quasilinear case – see e. g. [26,28,38,41] – using analytic methods. For simplicity, we study here following [34] only the case of a scalar ordinary differential equation ([33] treats first-order systems)

$$g(x, u^{(\ell-1)})u_\ell = f(x, u^{(\ell-1)}) \tag{9}$$

where u_ℓ denotes the ℓth derivative of u and $u^{(\ell)}$ all derivatives up to order ℓ. We further assume that f, g are polynomials of their arguments. Let $\mathcal{J}_\ell$ be the corresponding algebraic jet set.

Whether or not a point $\rho = (\bar{x}, \bar{u}^{(\ell)}) \in \mathcal{J}_\ell$ is a singularity does not depend on the value of $\bar{u}_\ell$ in this special case, as it does not appear in (6). The key property of quasilinear equations is that they can be studied at one order less. More precisely, outside of the irregular singularities the Vessiot distribution $\mathcal{V}[\mathcal{J}_\ell]$ can be generated by a vector field X. Denoting by $C_t^{(\ell)} = \partial_x + \sum_{i=0}^{\ell-1} u_{i+1}\partial_{u_i}$ the transversal contact field on $J_\ell(\mathbb{R}, \mathbb{R})$ and by $C_v^{(\ell)} = \partial_{u_\ell}$ the vertical one, we may choose $X = gC_t^{(\ell)} + \big(C_t^{(\ell)}(g)u_\ell - C_t^{(\ell)}(f)\big)C_v^{(\ell)}$. Expanding X, one sees that it is projectable to the field $Y = gC_t^{(\ell-1)} + fC_v^{(\ell-1)}$ on $J_{\ell-1}(\mathbb{R}, \mathbb{R})$. Strictly speaking, Y is only defined outside the projections of the irregular singularities of $\mathcal{J}_\ell$. But as we assume that f, g are polynomials, Y can obviously be extended smoothly to the whole jet bundle $J_{\ell-1}(\mathbb{R}, \mathbb{R})$.

Definition 22. *A point $\tilde{\rho} \in J_{\ell-1}(\mathbb{R}, \mathbb{R})$ is an* impasse point *for $\mathcal{J}_\ell$, if Y is not transversal at $\tilde{\rho}$ (i. e. if $g(\tilde{\rho}) = 0$). Otherwise, it is a* regular point. *An impasse point is* proper, *if Y vanishes there, and* improper *otherwise. A* weak generalized solution *of $\mathcal{J}_\ell$ is a one-dimensional manifold $\tilde{\mathcal{N}} \subset J_{\ell-1}(\mathbb{R}, \mathbb{R})$ such that $Y_{\tilde{\rho}} \in T_{\tilde{\rho}}\tilde{\mathcal{N}}$ for all points $\tilde{\rho} \in \tilde{\mathcal{N}}$. A* weak geometric solution *is the projection $\pi_0^{\ell-1}(\tilde{\mathcal{N}})$ of a weak generalized solution $\tilde{\mathcal{N}}$.*

We use here the terminology "impasse points" to distinguish them from the singularities of $\mathcal{J}_\ell$ which are always points on $\mathcal{J}_\ell$. Singularities always project on impasse points, but there may be impasse points without a point on $\mathcal{J}_\ell$ above them (see [34, Prop. 5.4] for a more precise analysis). This is the deeper reason why quasilinear equations require their own theory. Like a singularity, an impasse point can be only apparent – see [34, Ex. 6.3]. We speak about "weak" generalized solutions, as even in the case that they are the prolongations of a function it is not guaranteed that this function is ℓ times differentiable. Hence it can be considered as a solution only in a weak sense. One can provide an existence and (non)uniqueness theorem analogous to Theorem 18 for equations without proper impasse points. For lack of space, we omit the details and refer to [34, Thm. 6.5].

To indicate the wide variety of phenomena that may appear around impasse points of quasilinear equations, we now specialise to the following class of singular second-order initial value problems

$$g(x)u'' = f(x, u, u'), \qquad u(y) = c_0, \; u'(y) = c_1 \tag{10}$$

where we assume that y is *simple* zero of g. Liang [23] studied it for the special case $g(x) = x$ and $y = 0$ with analytical techniques. We showed in [34] that all his results can be recovered with geometric means in our slightly more general situation in a much more transparent way. Here we can only sketch some basic ideas of our approach; for all details we refer to [34, Sect. 8].

Key questions are the (non)uniqueness and the regularity of the solutions of (10). For the latter point, it does not suffice to study only the differential equation $\mathcal{J}_2 \subset J_2(\mathbb{R}, \mathbb{R})$ corresponding to (10), but one must also analyse its *prolongations* $\mathcal{J}_\ell \subset J_\ell(\mathbb{R}, \mathbb{R})$ for all $\ell > 2$ which are obtained by differentiating the given equation (10). We set $F_2(x, u^{(2)}) = g(x)u'' - f(x, u, u')$ and write for any order $\ell > 2$

$$F_\ell(x, u^{(\ell)}) = g(x)u_\ell + \Big[(\ell - 2)g'(x) - f_{u'}(x, u^{(1)})\Big]u_{\ell-1} - h_\ell(x, u^{(\ell-2)})$$

where the contributions of the lower-order terms can be recursively computed as $h_3(x, u^{(1)}) = C_t^{(1)} f(x, u^{(1)})$ and for $\ell > 3$ as

$$h_\ell(x, u^{(\ell-2)}) = C_t^{\ell-2}\Big(h_{\ell-1}(x, u^{(\ell-3)}) - [(\ell - 3)g'(x) - f_{u'}(x, u^{(1)})]u_{\ell-2}\Big).$$

Then the equation $\mathcal{J}_\ell$ is the zero set of $F_2, \dots, F_\ell$. If we apply the above idea of projecting the Vessiot distribution to one order less, then $\mathcal{J}_2$ yields the vector field $Y^{(1)} = g(x)\partial_x + g(x)u'\partial_u + f(x, u^{(1)})\partial_{u'}$ on $J_1(\mathbb{R}, \mathbb{R})$ and for any $\ell \geq 2$ we get from $\mathcal{J}_{\ell+1}$ the vector field

$$Y^{(\ell)} = g(x)C_t^{(\ell)} + \Big(h_{\ell-1}(x, u^{(\ell-3)}) - [(\ell - 1)g'(x) - f_{u'}(x, u^{(1)})]u_\ell\Big)C_v^{(\ell)} \quad (11)$$

defined on the three-dimensional submanifold $\mathcal{J}_\ell \subset J_\ell(\mathbb{R}, \mathbb{R})$.

Our initial data define a point $\rho_1 = (y, c_0, c_1) \in J_1(\mathbb{R}, \mathbb{R})$. It turns out that for the analysis of our initial value problem at each relevant prolongation order ℓ there exists a unique irregular singularity $\rho_\ell \in \mathcal{J}_\ell$ above ρ_1, i.e. with $\pi_1^\ell(\rho_\ell) = \rho_1$. We will find that the existence, (non)uniqueness and regularity of solutions depend solely on two values: $\delta = g'(y)$ and $\gamma = f_{u'}(\rho_1)$. Because of our assumption that y is a simple zero, δ cannot vanish. Our initial value problem has a *resonance* at order $q \in \mathbb{N}$, if $q\delta = \gamma$. If this is the case, we consider the *resonance parameter* $A_q = h_{q+2}(\rho_q)$ and speak of a *smooth* resonance for $A_q = 0$ and of a *critical* resonance otherwise.

Our approach consists of analyzing the phase portraits of the vector fields $Y^{(\ell)}$ around their stationary points ρ_ℓ. This requires in particular to determine the eigenvalues of the Jacobians $\mathrm{Jac}\,(Y^{(\ell)})(\rho_\ell)$. This is fairly simple except that we face again the problem that (11) is written in all jet variables up to order ℓ, although $Y^{(\ell)}$ lives only on a three-dimensional manifold. Fortunately, it can be overcome with a little trick and one obtains as eigenvalues δ, 0 and $\gamma - (\ell - 1)\delta$. If δ and γ have different signs, then we find at any prolongation order one negative, one zero and one positive eigenvalue and thus qualitatively the same phase portrait. If δ and γ have the same sign, then at a certain prolongation order the phase portrait changes qualitatively, as one eigenvalue changes its sign. In the case of a resonance, one finds at a certain prolongation order a double eigenvalue. It depends on the vanishing of the resonance parameter whether or not the Jacobian is diagonalisable. A deeper study of the invariant manifolds of the stationary point leads to the following result.

Theorem 23. *If there is no resonance, then three cases arise:*

$\delta\gamma < 0$: *The initial value problem* (10) *possesses a unique smooth two-sided solution and no additional one-sided solutions.*

$\delta\gamma > 0$: *The initial value problem* (10) *possesses a one-parameter family of two-sided solutions and no additional one-sided solutions. One member of the family is smooth, all others are in $C^k \setminus C^{k+1}$ with $k = \lceil \gamma/\delta \rceil$.*

$\gamma = 0$: *The initial value problem* (10) *possesses a unique smooth two-sided solution and possibly further additional one-sided solutions.*

If there is a resonance at order $k > 0$, then the initial value problem (10) *possesses a one-parameter family of two-sided solutions and no additional one-sided solutions. If the resonance is smooth, all solutions are smooth. For a critical resonance, all solutions are in $C^k \setminus C^{k+1}$.*

As demonstrated by an explicit example in [34], there are many possibilities in the case $\gamma = 0$: there could be no one-sided solutions at all or there could be infinitely many which either come from both sides or only from one side. The exact behaviour depends on further values besides δ and γ and no complete classification is known. The situation becomes much more complicated, if one drops the assumption that y is a simple zero. In this case $\delta = 0$ and if in addition $\gamma = 0$, then the Jacobian has a triple eigenvalue 0. The analysis of such a stationary point is rather difficult, as it requires a blow-up in three dimensions. For the subsequent blow-down, one must understand the *global* dynamics of a two-dimensional dynamical system which can be very complicated.

8 Analysis of Algebraic Singularities

Singularities of varieties have been extensively studied in algebraic geometry, but not much is known about their effect on differential equations. As algebraic differential equations are locally Zariski closed sets, we cannot avoid dealing with them. In the complex case, their detection is straightforward using the Jacobian criterion (over the real numbers the situation is somewhat different, as at a singularity the variety may still be locally a manifold). Thus the main point is to analyse the local solution behaviour in their neighbourhood. Ritt provides several examples of algebraic differential equations with singular integrals where the singular integrals consist entirely of algebraic singularities (see e. g. [29, II.§19]). However, he does not comment on this fact.

We will not develop a general theory for handling algebraic singularities, but we will indicate with two concrete examples some phenomena that can show up. We will use a rather ad hoc approach which probably can be extended to more general situations, but we refrain here from any formalities. In the first example, we study the local solution behaviour near an isolated algebraic singularity using the Vessiot spaces of neighbouring points.

Example 24. Let $\mathcal{J}_1$ be the two-dimensional cone in the three-dimensional jet bundle $J_1(\mathbb{R}, \mathbb{R})$ given by $(u')^2 - u^2 - x^2 = 0$. The vertex is an isolated algebraic singularity representing one component of a regularity decomposition while all other points are regular and form the second component. We are interested in how many solutions go through the vertex and their regularity.

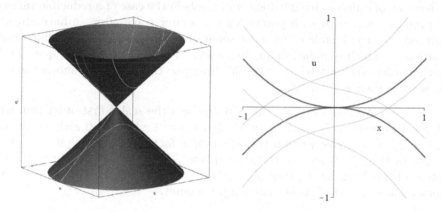

Fig. 3. Generalized solutions going through an algebraic singularity of a real first-order differential equation. Left: situation in $J_1(\mathbb{R}, \mathbb{R})$. Right: projection to x-u plane (Color figure online).

Consider the Vessiot spaces of the regular points. They are generated by the vector field $X = u'\partial_x + (x^2 + u^2)\partial_u + (x - uu')\partial_{u'}$. By restricting to either the lower or the upper half cone, we can express u' by x and u and project to the x-u plane obtaining the vector fields $Y_\pm = \pm\sqrt{x^2 + u^2}\partial_x + (x^2 + u^2)\partial_u$ which can trivially be continued to the origin where they vanish. As they are not differentiable there, the origin cannot be studied using the Jacobian.

By transforming to polar coordinates, i. e. by performing a blow-up of the stationary point in the origin, one can show that the dynamical system defined by Y possesses a unique invariant curve going through the origin and within a sufficiently small neighbourhood of the origin all nearby trajectories look similar to this manifold. Furthermore, on one side of the origin the invariant curve corresponds to a trajectory going into the origin, while on the side we have an outgoing trajectory (hence the stationary point is not really visible).

Recall that such an invariant curve corresponds to a generalized solution and we obtain one such curve for each half cone, i. e. from each of the fields $Y_\pm$ (see the red curves in Fig. 3). As the graphs of both solutions possess a horizontal tangent at the origin, it is possible to "switch" at the singularity from one to the other. Hence, we find that our equation possesses exactly four $\mathcal{C}^1$ solutions for the initial condition $u(0) = 0$ and $u'(0) = 0$. By analyzing the prolongations of our equation, it is not difficult to verify that the solutions that stay inside of one half cone are even smooth, whereas the "switching" solutions are only $\mathcal{C}^1$, as their second derivative jumps from 1 to -1 or vice versa at $x = 0$. Figure 3 also shows in white the Vessiot cone at the algebraic singularity which consists of two intersecting lines. One sees that they are indeed the tangents to the prolonged solutions through the singularity.

In Algorithm 1, we perform a prime decomposition so that we always work with (subsets of) irreducible varieties. One reason for this in practise rather expensive step is to avoid the algebraic singularities automatically given by the intersection of different irreducible components in the case of a reducible variety. We prefer to deal with such points only in a later stage after we have already analyzed each irreducible component separately. Given an algebraic differential equation $\mathcal{J}_q$ which is reducible in this sense, an obvious interesting question is whether solutions exist which "switch" from one component to another and if yes, what is their regularity?

Example 25. We consider over the real numbers the scalar first-order ordinary differential equation $\mathcal{J}_1$ given by $(u' - c)\big((u')^2 + u^2 + x^2 - 1\big) = 0$ with a constant $c \in [-1, 1]$. As we have written the equation in factored form, one immediately recognizes that $\mathcal{J}_1$ is simply the unit sphere $\mathcal{J}_{1,1}$ in the jet bundle $J_1(\mathbb{R}, \mathbb{R})$ united with a horizontal plane $\mathcal{J}_{1,2}$ at height c (see Fig. 4). For $|c| \neq 1$, the intersection is a circle $\mathcal{C}$, otherwise simply a point.

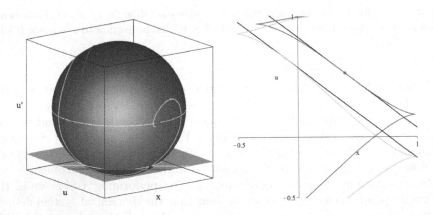

Fig. 4. First-order differential equation with two irreducible components. Left: generalized solutions in $J_1(\mathbb{R}, \mathbb{R})$. Right: solution graphs in x-u plane (Color figure online).

The differential equation $\mathcal{J}_{1,2}$: $u' = c$ is of course trivial to analyse: as it is explicit, all points on the corresponding plane are regular and all generalized solutions are straight lines. The differential equation $\mathcal{J}_{1,1}$: $(u')^2 + u^2 + x^2 = 1$ has already been studied at many places (see e. g. [32, Ex. 9.1.12] or [35, Ex. 10]) and its singularities form the equator. A point $\rho = (\bar{x}, \bar{u}, \bar{p})$ lies on the intersection and thus is an algebraic singularity of $\mathcal{J}_1$, if $\bar{p} = c$ and $\bar{u}^2 + \bar{x}^2 = 1 - c^2$.

Consider first the case $c \neq 0$. Then ρ is for each component $\mathcal{J}_{1,i}$ a regular point and we have on each component a unique generalized solution curve γ_i through ρ. Without loss of generality, assume $\gamma_i(0) = \rho$. We may form a further generalized solution like $\gamma(t) = \gamma_1(t)$ for $t \leq 0$ and $\gamma(t) = \gamma_2(t)$ for $t \geq 0$ (and yet another by swapping the indices 1 and 2). While this curve γ is trivially continuous at 0, it is in general not differentiable there, as the tangent vectors $\gamma_1'(0)$ and $\gamma_2'(0)$ disagree. However, the corresponding geometric solution, i. e. the projection of $\mathrm{im}\,\gamma$ to the x-u plane, is the graph of an everywhere differentiable function $u = f(x)$ with $f'(x) = \bar{p}$. This can be seen on the right hand side of Fig. 4. The green and the red curve represent geometric solutions of $\mathcal{J}_{1,1}$; the black lines the corresponding ones of $\mathcal{J}_{1,2}$. One can see that the latter ones are exactly the tangents of the former one and hence connecting "half" a curve with "half" a line yields still the graph of a function which is at least $\mathcal{C}^1$.

The tangent vectors $\gamma_1'(0)$ and $\gamma_2'(0)$ generate the Vessiot spaces $\mathcal{V}_\rho[\mathcal{J}_{1,1}]$ and $\mathcal{V}_\rho[\mathcal{J}_{1,2}]$. As their slopes correspond to the second derivatives of the solutions leading to the curves γ_1 and γ_2, our "composed" solutions can be $\mathcal{C}^2$, if and only if these Vessiot spaces coincide. A simple computation shows that $\mathcal{V}_\rho[\mathcal{J}_{1,1}]$ is generated by the vector $\bar{p}(\partial_x + \bar{p}\partial_u) - (\bar{x} + \bar{u}\bar{p})\partial_p$ while $\mathcal{V}_\rho[\mathcal{J}_{1,2}]$ is spanned by $\partial_x + c\partial_u$. It is straightforward to show that the Vessiot spaces coincide only at two special points $\rho_\pm$ on the intersection $\mathcal{C}$, namely at

$$\rho_\pm = \left(\mp c \sqrt{\frac{1-c^2}{1+c^2}}, \pm \sqrt{\frac{1-c^2}{1+c^2}}, c \right).$$

In Fig. 4 this corresponds to the red curve, as one can see on the right hand side that the intersection of the red and the black graph happens at an inclination point of the red graph. By analyzing the next prolongation, one can show that even for these two special points the "composed" solutions are only $\mathcal{C}^2$. We conclude that our differential equation $\mathcal{J}_1$ possesses four solutions through any point $\rho \in \mathcal{C}$. Two of them are smooth (the ones corresponding to γ_1 and γ_2), the two "composed" ones are only $\mathcal{C}^1$ respectively $\mathcal{C}^2$, if ρ is one of the points $\rho_\pm$.

In the case $c = 0$, the intersection $\mathcal{C}$ is the equator and thus the singular locus of $\mathcal{J}_{1,1}$. Here we have an example where the classification of a point lying on several irreducible components differs for the different components. $\mathcal{J}_{1,1}$ has two irregular singularities, namely the points $(0, \pm 1, 0)$, and both are a folded focus. This means that no generalized solution of $\mathcal{J}_{1,1}$ approaches them with a well-defined tangent and this would be necessary for going through them. Thus through each of these two points there exists only one generalized solution of $\mathcal{J}_1$, namely the one of $\mathcal{J}_{1,2}$. Any other point $\rho \in \mathcal{C}$ is a regular singularity of $\mathcal{J}_{1,1}$. Thus on $\mathcal{J}_{1,1}$ there are only two one-sided solutions starting or ending at ρ. However, we can combine each of them with "half" a solution of $\mathcal{J}_{1,2}$ as discussed for $c \neq 0$ to generate two additional $\mathcal{C}^1$ solutions so that $\mathcal{J}_1$ has three solutions through ρ one of which is smooth. As at all regular singularities the Vessiot space is vertical, we do not find any special points where the "composed" solutions possess a higher regularity than $\mathcal{C}^1$.

9   Conclusions

We presented a mixture of geometric and algebraic techniques for studying singularities of differential equations. For the basic concepts, we followed essentially the differential topological approach to geometric singularities and extended it also to differential equations which are not of finite type. We augmented this approach by algebraic ideas to extend its range of applicability, as many differential equations in applications do not lead to manifolds, but only to varieties. This implies that we must furthermore deal with algebraic singularities.

In the first half of this article, we concentrated on the algorithmic detection of singularities. We used the differential Thomas decomposition from differential algebra to obtain simple differential systems from which we can extract in a well-defined manner an algebraic differential equation to which the geometric theory can be applied. The actual detection of the singularities is then performed with an algebraic Thomas decomposition. Although there is still a gap in the theory for differential equations which are not of finite type, as it is not clear whether there might appear further types of singularities, it is remarkable that the classification can be performed completely algorithmically.

This is possible only, because we searched for singularities at a prescribed order. For lack of space, we did not discuss here the question how singularities at different prolongation orders are related. It is easy to see that if we prolong, then every point on the prolonged equation lying over a singularity of the original equation must be again a singularity. Furthermore, there can never be a point over a regular singularity. For the existence of (formal) power series solutions, it is therefore necessary that we can construct an infinite tower of irregular singularities lying above each other. An example due to Lange-Hegermann [20, Ex. 2.93] (see also the discussion in [35, Ex. 16]) shows that generally it is not possible to decide the existence of such an infinite tower at any finite order.

It should have become apparent that for the study of singularities it makes a great difference whether we work over the real or the complex numbers. The detection of singularities over the complex numbers is simpler, as they form an algebraically closed field which is algorithmically a great advantage. Any analysis of singularities was performed in this work over the real numbers, as it was based on techniques from dynamical systems theory. One should note that also the questions studied differ considerably in dependence of the base field. The regularity of solutions or the difference between one- and two-sided solutions is an issue only over the real numbers. Over the complex numbers, there exists already an extensive theory of singularities of *linear* differential equations going back at least to Fuchs and Frobenius which is nowadays often considered as a part of differential Galois theory. Here the determination of monodromy or the Stokes phenomenon are of great importance and have no real counterpart.

Acknowledgments. This work has been supported by the bilateral project ANR-17-CE40-0036 and DFG-391322026 SYMBIONT. We thank our co-authors on this subject, Markus Lange-Hegermann, Daniel Robertz and Thomas Sturm, for a pleasant collaboration. The first author furthermore thanks the organizers of CASC 2020 for the honor to present an invited talk on this subject.

References

1. Arnold, V.: Geometrical Methods in the Theory of Ordinary Differential Equations, Grundlehren der mathematischen Wissenschaften, 2nd edn. Springer-Verlag, New York (1988)
2. Bächler, T., Gerdt, V., Lange-Hegermann, M., Robertz, D.: Algorithmic Thomas decomposition of algebraic and differential systems. J. Symb. Comput. **47**(10), 1233–1266 (2012)
3. Bächler, T., Lange-Hegermann, M.: AlgebraicThomas and DifferentialThomas: Thomas Decomposition for algebraic and differential systems (2008–2012). http://www.mathb.rwth-aachen.de/go/id/rnab/lidx/1
4. Ballarin, C., Kauers, M.: Solving parametric linear systems: an experiment with constraint algebraic programming. ACM SIGSAM Bull. **38**, 33–46 (2004)
5. Becker, E., Neuhaus, R.: Computation of real radicals of polynomial ideals. In: Eyssette, F., Galligo, A. (eds.) Computational Algebraic Geometry, pp. 1–20. Progress in Mathematics 109, Birkhäuser, Basel (1993)
6. Bochnak, J., Conte, M., Roy, M.: Real Algebraic Geometry. Ergebnisse der Mathematik und ihrer Grenzgebiete 36, Springer-Verlag, Berlin (1998)
7. Caviness, B., Johnson, J. (eds.): Quantifier Elimination and Cylindrical Algebraic Decomposition. Texts and Monographs in Symbolic Computation. Springer-Verlag, Vienna (1998)
8. Dara, L.: Singularités génériques des équations différentielles multiformes. Bol. Soc. Bras. Mat. **6**, 95–128 (1975)
9. Davydov, A., Ishikawa, G., Izumiya, S., Sun, W.Z.: Generic singularities of implicit systems of first order differential equations on the plane. Jpn. J. Math. **3**, 93–119 (2008)
10. Dolzmann, A., Sturm, T.: REDLOG: Computer algebra meets computer logic. ACM SIGSAM Bull. **31**, 2–9 (1997)
11. Dolzmann, A., Sturm, T.: Simplification of quantifier-free formulae over ordered fields. J. Symb. Comput. **24**, 209–231 (1997)
12. Falkensteiner, S., Sendra, J.: Solving first order autonomous algebraic ordinary differential equations by places. Math. Comput. Sci. **14**, 327–337 (2020)
13. Gerdt, V.: On decomposition of algebraic PDE systems into simple subsystems. Acta Appl. Math. **101**, 39–51 (2008)
14. Gerdt, V., Lange-Hegermann, M., Robertz, D.: The Maple package TDDS for computing Thomas decompositions of systems of nonlinear PDEs. Comput. Phys. Commun. **234**, 202–215 (2019)
15. Grigoriev, D.: Complexity of deciding Tarski algebra. J. Symb. Comput. **5**, 65–108 (1988)
16. Hubert, E.: Notes on Triangular Sets and Triangulation-Decomposition Algorithms II: Differential Systems. In: Winkler, F., Langer, U. (eds.) SNSC 2001. LNCS, vol. 2630, pp. 40–87. Springer, Heidelberg (2003). https://doi.org/10.1007/3-540-45084-X_2

17. Hubert, E.: Notes on Triangular Sets and Triangulation-Decomposition Algorithms II: Differential Systems. In: Winkler, F., Langer, U. (eds.) SNSC 2001. LNCS, vol. 2630, pp. 40–87. Springer, Heidelberg (2003). https://doi.org/10.1007/3-540-45084-X_2

18. Kant, U., Seiler, W.: Singularities in the geometric theory of differential equations. In: Feng, W., Feng, Z., Grasselli, M., Lu, X., Siegmund, S., Voigt, J. (eds.) Dynamical Systems, Differential Equations and Applications Proceedings 8th AIMS Conference, Dresden 2010, vol. 2, pp. 784–793. AIMS (2012)

19. Košta, M.: New Concepts for Real Quantifier Elimination by Virtual Substitution. Doctoral dissertation, Saarland University, Germany (2016)

20. Lange-Hegermann, M.: Counting Solutions of Differential Equations. Ph.D. thesis, RWTH Aachen, Germany (2014). http://publications.rwth-aachen.de/record/229056

21. Lange-Hegermann, M.: The differential counting polynomial. Found. Comput. Math. **18**, 291–308 (2018)

22. Lange-Hegermann, M., Robertz, D., Seiler, W., Seiß, M.: Singularities of algebraic differential equations. Preprint Kassel University (2020). (arXiv:2002.11597)

23. Liang, J.: A singular initial value problem and self-similar solutions of a nonlinear dissipative wave equation. J. Differ. Equ. **246**, 819–844 (2009)

24. Neuhaus, R.: Computation of real radicals of polynomial ideals II. J. Pure Appl. Algorithm **124**, 261–280 (1998)

25. Plesken, W.: Counting solutions of polynomial systems via iterated fibrations. Arch. Math. (Basel) **92**(1), 44–56 (2009)

26. Rabier, P., Rheinboldt, W.: Theoretical and numerical analysis of differential-algebraic equations. In: Ciarlet, P., Lions, J. (eds.) Handbook of Numerical Analysis, vol. VIII, pp. 183–540. North-Holland, Amsterdam (2002)

27. Remizov, A.: Multidimensional Poincaré construction and singularities of lifted fields for implicit differential equations. J. Math. Sci. **151**, 3561–3602 (2008)

28. Riaza, R.: Differential-Algebraic Systems. World Scientific, Hackensack (2008)

29. Ritt, J.: Differential Algebra. Dover, New York (1966), (Original: AMS Colloquium Publications, vol. XXXIII, 1950)

30. Robertz, D.: Introduction. Formal Algorithmic Elimination for PDEs. LNM, vol. 2121, pp. 1–4. Springer, Cham (2014). https://doi.org/10.1007/978-3-319-11445-3_1

31. Seidl, A.: Cylindrical Decomposition Under Application-Oriented Paradigms. Doctoral dissertation, Universität Passau, Germany (2006)

32. Seiler, W.: Involution - The Formal Theory of Differential Equations and its Applications in Computer Algebra. Algorithms and Computation in Mathematics. Springer-Verlag, Berlin (2010)

33. Seiler, W.M.: Singularities of Implicit Differential Equations and Static Bifurcations. In: Gerdt, V.P., Koepf, W., Mayr, E.W., Vorozhtsov, E.V. (eds.) CASC 2013. LNCS, vol. 8136, pp. 355–368. Springer, Cham (2013). https://doi.org/10.1007/978-3-319-02297-0_29

34. Seiler, W., Seiß, M.: Singular initial value problems for scalar quasi-linear ordinary differential equations. Preprint Kassel University (2018). (arXiv:2002.06572)

35. Seiler, W., Seiß, M., Sturm, T.: A logic based approach to finding real singularities of implicit ordinary differential equations. Math. Comp. Sci. (2020). (arXiv: 2003.00740)

36. Sit, W.: An algorithm for solving parametric linear systems. J. Symb. Comput. **13**, 353–394 (1992)

37. Spang, S.: On the Computation of the Real Radical. Diploma thesis, Technical University Kaiserslautern, Department of Mathematics (2007)
38. Thomas, G.: The problem of defining the singular points of quasi-linear differential-algebraic systems. Theor. Comput. Sci. **187**, 49–79 (1997)
39. Thomas, J.: Differential Systems. Colloquium Publications XXI, American Mathematical Society, New York (1937)
40. Thomas, J.: Systems and Roots. W. Byrd Press, Richmond (1962)
41. Tuomela, J.: On singular points of quasilinear differential and differential-algebraic equations. BIT **37**, 968–977 (1997)
42. Weispfenning, V.: The complexity of linear problems in fields. J. Symb. Comput. **5**, 3–27 (1988)
43. Weispfenning, V.: Quantifier elimination for real algebra-the quadratic case and beyond. Appl. Algebra Eng. Commun. **8**, 85–101 (1997)
44. Winkler, F., Langer, U. (eds.): SNSC 2001. LNCS, vol. 2630. Springer, Heidelberg (2003). https://doi.org/10.1007/3-540-45084-X

On the Complexity of Computing Integral Bases of Function Fields

Simon Abelard[✉]

Laboratoire d'informatique de l'École polytechnique (LIX, UMR 7161) CNRS,
Institut Polytechnique de Paris, Palaiseau, France
abelard@lix.polytechnique.fr

Abstract. Let $\mathcal{C}$ be a plane curve given by an equation $f(x, y) = 0$ with $f \in K[x][y]$ a monic irreducible polynomial. We study the problem of computing an integral basis of the algebraic function field $K(\mathcal{C})$ and give new complexity bounds for three known algorithms dealing with this problem. For each algorithm, we study its subroutines and, when it is possible, we modify or replace them so as to take advantage of faster primitives. Then, we combine complexity results to derive an overall complexity estimate for each algorithm. In particular, we modify an algorithm due to Böhm et al. and achieve a quasi-optimal runtime.

Keywords: Puiseux series · Linear algebra · Polynomial matrices

1 Introduction

When handling algebraic function fields, it is often helpful –if not necessary– to know an integral basis. Computing such bases has a wide range of applications from symbolic integration to algorithmic number theory and applied algebraic geometry. It is the function field analogue of a well-known and difficult problem: computing rings of integers in number fields. As often, the function field version is easier: the algorithm of Zassenhaus [25] described for number fields in the late 60's can indeed be turned into a polynomial-time algorithm for function fields which was later precisely described by Trager [23].

However, there are very few complexity results going further than just stating a polynomial runtime. Consequently, most of the existing algorithms in the literature are compared based on their runtimes on a few examples and this yields no consensus on which algorithm to use given an instance of the problem. In this paper, we provide complexity bounds for three of the best-known algorithms to compute integral bases and provide complexity bounds based on state-of-the art results for the underlying primitives.

In this paper, we focus on the case of plane curves given by equations of the form $f(x, y) = 0$ with $f \in K[x, y]$ irreducible. Without loss of generality, we also assume that f is monic in y. We set the notation $n = \deg_y f$ and $d_x = \deg_x f$. The associated function field is $K(\mathcal{C}) = \mathrm{Frac}\,(K(x)[y]/f(x, y))$. It is an algebraic extension of degree n of $K(x)$. An element $h(x, y)$ of $K(\mathcal{C})$ is

© Springer Nature Switzerland AG 2020
F. Boulier et al. (Eds.): CASC 2020, LNCS 12291, pp. 42–62, 2020.
https://doi.org/10.1007/978-3-030-60026-6_3

integral (over $K[x]$) if there exists a monic bivariate polynomial $P(x, y)$ such that $P(x, h(x, y))$ equals 0 in $K(\mathcal{C})$. The set of such elements forms a free $K[x]$-module of rank n and a basis of this module is called an integral basis of $K(\mathcal{C})$.

The irreducibility of f is required to make sure that the function field $K(\mathcal{C})$ is indeed a field and not a product of fields. If this hypothesis fails it will be detected during the factorization process in Algorithm 3 and integral bases for each factor will be computed, while Algorithm 1 will return a basis for the integral closure of $K[x]$ in $K(x)[y]/\langle f \rangle$ (see the beginning of [12, Section 6]). However, these algorithms will both fail if f is not squarefree because it means that $\mathrm{Disc}(f) = \mathrm{Res}_y \left(f, \frac{\partial f}{\partial y} \right) = 0$.

Computing integral bases of algebraic function fields has applications in symbolic integration [23] but more generally an integral basis can be useful to handle function fields. For instance, the algorithm of van Hoeij and Novocin [13] uses such a basis to "reduce" the equation of function fields and thus makes them easier to handle. The algorithm of Hess [11] to compute Riemann-Roch spaces is based on the assumption that integral closures have been precomputed. This assumption is sufficient to establish a polynomial runtime, but a more precise complexity estimate for Hess' approach requires to assess the cost of computing integral closures as well.

Our Contribution. We provide complexity estimates for three algorithms dedicated to computing integral bases of algebraic function fields in characteristic 0 or greater than n. This assumption serves two purposes: it ensures the existence of Puiseux expansions used in Algorithms 1 and 3 and is also used in technical hypothesis in Algorithm 2 to compute the radical of an ideal. To the best of our knowledge, no previous bounds were given for these algorithms. Another approach which has received a lot of attention is the use of Montes' algorithm. We do not tackle this approach in the present paper, a complexity estimate has been given by Bauch in [2, Lemma 3.10] in the case of number fields. Using the Montes algorithm, a local integral basis of a Dedeking domain A at a prime ideal $\mathfrak{p}$ is computed in $O \left(n^{1+\varepsilon} \delta \log q + n^{1+\varepsilon} \delta^{2+\varepsilon} + n^{2+\varepsilon} \delta^{1+\varepsilon} \right)$ $\mathfrak{p}$-small operations, with δ the $\mathfrak{p}$-valuation of $\mathrm{Disc}(f)$, q the cardinal of $A/\mathfrak{p}$ and ε any positive real number.

Our contribution is actually not limited to a complexity analysis: the algorithms that we present have been modified so that we could establish better complexity results. We also discuss possible improvements to van Hoeij's algorithm in a particular case which is not uncommon in the literature. Our main complexity results are Theorems 1, 2 and 3. Note that we count field operations and do not take into account the coefficient growth in case of infinite fields nor the field extensions incurred by the use of Puiseux series. We also made the choice not to delve into probabilistic aspects: all the algorithms presented here are "at worst" Las Vegas due to the use of Poteaux and Weimann's algorithm, see for instance [21, Remark 3].

We decided to give worst-case bounds and to only involve n and $\mathrm{Disc}(f)$ in our theorems so as to give ready-to-use results. Our proofs, however, are meant to allow the interested reader to derive sharper bounds involving more precise parameters such as the regularity and ramification indices of Puiseux series.

We summarize these complexity estimates in Table 1 in a simpler context: we ignore the cost of factorizations and bound both n and $d_x = \deg_x f$ by D. In this case, the input size is in $O(D^2)$ and output size in $O(D^4)$. The constant $2 \le \omega \le 3$ refers to a feasible exponent for matrix multiplication, see [18] for the smallest value currently known. Translating the above bound, the complexity of the Montes approach is at best in $\widetilde{O}(D^5)$ but only for computing a local integral basis at one singularity, while the algorithm detailed in Sect. 4 computes a global integral basis for a quasi-optimal arithmetic complexity (i.e. in $\widetilde{O}(D^4)$).

Although we hope that this will change in a near future, our contribution remains of purely theoretical nature because we crucially rely on primitives (fast computation of Popov and Hermite forms, Puiseux series and factorization over $K[[x]][y]$) which have not been implemented yet. This is the reason why we do not provide timings in the present paper and redirect to [4, Section 8] for runtime comparisons of state-of-the-art implementations.

Organization of the Paper. We sequentially analyze the three algorithms: Sect. 2 is dedicated to van Hoeij's algorithm [12], Sect. 3 to Trager's algorithm [23] and Sect. 4 to an algorithm by Böhm et al. introduced in [4]. In each section, we first give an overview of the corresponding algorithm and insist on the parts where we perform some modifications. The algorithms we describe are variations of the original algorithms so we give no detailed proof of exactness and refer to the original papers in which they were introduced. Then, we establish complexity bounds for each algorithm by putting together results from various fields of computer algebra. We were especially careful about how to handle linear algebra, Puiseux series and factorization over $K[[x]][y]$.

Table 1. Simplified complexity estimates for computing integral bases.

Algorithm	Worst-case complexity
Trager's algorithm [23]	$\widetilde{O}(D^7)$
Van Hoeij's algorithm [12]	$\widetilde{O}(D^{\omega+4})$
Böhm et al.'s algorithm [4]	$\widetilde{O}(D^4)$

2 Van Hoeij's Algorithm

2.1 Puiseux Series

We recall some basic concepts about Puiseux series and refer to [24] for more details. Assuming that the characteristic of K is either 0 or $> n$, the Puiseux theorem states that $f \in K[x][y]$ has n roots in the field of Puiseux series $\bigcup_{e \ge 1} \overline{K}((x^{1/e}))$.

Following Duval [10], we group these roots into irreducible factors of f. First, one can write $f = \prod_{i=1}^{r} f_i$ with each f_i irreducible in $K[[x]][y]$. Then, for

$1 \leq i \leq r$ we write $f_i = \prod_{j=1}^{\varphi_i} f_{ij}$, where each f_{ij} is irreducible in $\overline{K}[[x]][y]$. Finally, for any $(i,j) \in \{1,\ldots,r\} \times \{1,\ldots,\varphi_i\}$ we write

$$f_{ij} = \prod_{k=0}^{e_i-1} \left(y - S_{ij}(x^{1/e_i}\zeta_{e_i}^k)\right),$$

where $S_{ij} \in \overline{K}((x))$ and ζ_{e_i} is a primitive e_i-th root of unity.

Definition 1. The n fractional Laurent series $S_{ijk}(x) = S_{ij}(x^{1/e_i}\zeta_{e_i}^k)$ are called the classical Puiseux series of f above 0. The integer e_i is called the ramification index of S_{ijk}.

Proposition 1. For a fixed i, the f_{ij}'s all have coefficients in K_i, a degree φ_i extension of K and they are conjugated by the associated Galois action. We have $\sum_{i=1}^{r} e_i\varphi_i = n$.

Definition 2. [21, Definition 2] A system of rational Puiseux expansions over K (K-RPE) of f above 0 is a set $\{R_i\}_{1 \leq i \leq r}$ such that

- $R_i(T) = (X_i(T), Y_i(T)) \in K_i((T))^2$,
- $R_i(T) = (\gamma_i T^{e_i}, \sum_{j=n_i}^{\infty} \beta_{ij} T^j)$, where $n_i \in \mathbb{Z}$, $\gamma_i \neq 0$ and $\beta_{in_i} \neq 0$,
- $f_i(X_i(T), Y_i(T)) = 0$,
- the integer e_i is minimal.

In the above setting, we say that R_i is centered at $(X_i(0), Y_i(0))$. We may have $Y_i(0) = \infty$ if $n_i < 0$ but this cannot happen if f is monic.

Definition 3. [21, Definition 3] The regularity index of a Puiseux series S of f with ramification index e is the smallest $N \geq \min(0, ev_x(S))$ such that no other Puiseux series S' have the same truncation up to exponent N/e. The truncation of S up to its regularity index is called the singular part of S.

It can be shown that two Puiseux series associated to the same RPE share the same regularity index so we can extend this notion (and the notion of singular part) to RPE's.

2.2 Description of van Hoeij's Algorithm

We will be looking for an integral basis of the form $p_i(x,y)/d_i(x)$, where the p_i are degree i monic polynomials in y. It is known that the irreducible factors of the denominators d_i are among the irreducible factors of the discriminant with multiplicity at least 2. We can treat these factors one by one by first looking for local integral bases at each of these factors, i.e. bases whose denominators can only be powers of such an irreducible factor. A global integral basis is then recovered from these local bases by CRT.

To compute a local integral basis at a fixed factor ϕ, van Hoeij [12] follows the following strategy. Start from $(1, y, \ldots, y^{n-1})$ and update it so that it generates a larger module, until this module is integrally closed. This basis is modified by multiplying it by an appropriate triangular matrix in the following way. Let us fix a d, then b_d must be a linear combination of the b_0, ..., b_{d-1} such that $(yb_{d-1} + \sum_{i=0}^{d-1} a_i b_i)/\phi^j$ is integral with j as large as possible.

To this end, the coefficients of the linear combination are first set to be variables and we write equations enforcing the fact that the linear combination divided by ϕ has to be integral. If a solution of this system is found, the value of b_d is updated and we repeat the process so as to divide by the largest possible power of ϕ. Note that a solution is necessarily unique, otherwise the difference of two solutions would be an integral element with numerator of degree $d - 1$, which means that the j computed in the previous step was not maximal. When there is no solution, we have reached the maximal exponent and move on to computing b_{d+1}.

For the sake of completeness, we give a description of van Hoeij's algorithm but we refer to van Hoeij's original paper [12] for a proof that this algorithm is correct. This algorithm is originally described for fields of characteristic 0 but also works in the case of positive characteristic provided that we avoid wild ramification (see [12, Section 6.2.]). To deal with this issue, we make the assumption that we are either considering characteristic zero or greater than n.

2.3 Complexity Analysis

In this section, we prove the following theorem.

Theorem 1. *Let $f(x, y)$ be a degree n monic irreducible polynomial in y. Algorithm 1 returns an integral basis for the corresponding function field and costs the factorization of $\mathrm{Disc}(f)$ and $\widetilde{O}(n^{\omega+2} \deg \mathrm{Disc}(f))$ field operations, where $2 \leq \omega \leq 3$ is a feasible exponent for linear algebra.*

Proof. First, we need to compute the discriminant and recover its square factors, which costs a factorization of a univariate polynomial of degree $\leq nd_x$.

Then, we need to compute the Puiseux expansions η_i of f at one root of each factor in S_{fac}, up to precision $N = \max_i \sum_{i \neq j} v(\eta_i - \eta_j)$. Using the algorithm of Poteaux and Weimann [21], the Puiseux expansions are computed up to precision N in $\widetilde{O}(n(\delta + N))$ field operations, where δ stands for the valuation of $\mathrm{Disc}(f)$. Indeed, these expansions are computed throughout their factorization algorithm, which runs in $\widetilde{O}(n(\delta + N))$ field operations as stated in [21, Theorem 3]. Therefore, in theory, we will see that computing the Puiseux expansions has a negligible cost compared to other parts of the algorithm since $N \leq n^2$.

Another problem coming from the use of Puiseux expansions is that we have to evaluate bivariate polynomials (the b_i's) at the Puiseux expansions of f. However this matter can be dealt with by keeping them in memory and updating them along the computations. This way, for a fixed d we first initialize $b_d = yb_{d-1}$ so we just have to perform a product of Puiseux expansions at precision $O(n^2)$ and then each time b_d is updated it will amount to performing a linear combination of Puiseux expansions. Since we fix precision at $N \leq n^2$, taking into account the denominator in the exponents of the Puiseux series this amounts to

Input   : A monic irreducible polynomial $f(y)$ over $K[x]$
Output : An integral basis for $K[x, y]/\langle f \rangle$
$n \leftarrow \deg_y f$;
$S_{fac} \leftarrow$ set of factors P such that $P^2|\mathrm{Disc}(f)$;
for ϕ **in** S_{fac} **do**

    Compute α a root of ϕ (possibly in an extension);
    Compute η_i the singular parts of the n Puiseux expansions of f at α ;
    $N \leftarrow \max_i \sum_{i \neq j} v(\eta_i - \eta_j)$;
    Extend the precision of the η_i's up to N;
    $b_0 \leftarrow 1$;
    for $d \leftarrow 1$ *to* $n - 1$ **do**

        $b_d \leftarrow y b_{d-1}$;
        solutionfound $\leftarrow$ true;
        Let $a_0, \ldots a_{d-1}$ be variables;
        $a \leftarrow (b_d + \sum_{i=0}^{d-1} a_i b_i)/(x - \alpha)$;
        while *solutionfound* **do**

            Write the equations, i.e. the coefficients of $a(x, \eta_i(x))$ with negative
            power of $(x - \alpha)$ for any i;
            Solve this linear system in the a_i's;
            if *no solution* **then**

                solutionfound $\leftarrow$ false;
            end
            else

                There is a unique solution (a_i) in $K(\alpha)^d$;
                Substitute α by x in each a_i;
                $b_d \leftarrow (b_d + \sum_{i=0}^{d-1} a_i b_i)/\phi$;
            end
        end
    end
end
From all the local bases perform CRT to deduce B an integral basis;
return B;

$$\textbf{Algorithm 1: Van Hoeij's algorithm [12]}$$

handling polynomials of degrees $\leq n^3$. Thus, in our case, arithmetic operations on Puiseux series can be performed in $\widetilde{O}(n^3)$ field operations.

The main task in this algorithm is to solve a linear system of c equations in d variables over the extension $K(\alpha)$, where c is the total number of terms of degrees < 1 in the n Puiseux expansions. In the worst case, each Puiseux series has n terms of degree < 1 and so c can be bounded above by n^2. More precisely, we can bound it by ne, where e is the maximum of the ramification indices of the classical Puiseux expansions of f.

In most cases, this system will be rectangular of size $c \times d$ so we solve it in time $\widetilde{O}(cd^{\omega-1})$ using [6, Theorem 8.6]. This step is actually the bottleneck for each iteration and using the bounds on d and c it runs in $\widetilde{O}(n^{\omega+1} \deg \phi)$ field operations, since the extension $K(\alpha)$ of K has degree $\leq \deg \phi$.

This process is iterated over the irreducible factors of the discriminant appearing with multiplicity at least 2, and for ϕ such a factor we have to solve at most $n + M(\phi)/2$ systems, where $M(\phi)$ is the multiplicity of ϕ in $\mathrm{Disc}(f)$. Indeed, each time a solution to a system is found the discriminant is divided by ϕ^2 so that cannot happen more than $M(\phi)/2$ times, but since we need to make sure that we have no solution before incrementing d we will have to handle n additional systems. Thus, for a fixed factor ϕ the cost of solving the systems is bounded by $O(n \cdot n^{\omega+1} \deg \phi + n^{\omega+1} \deg \phi M(\phi))$. Thus, the complexity is in $\widetilde{O}\left(\sum_{\phi \in S_{fac}} n^{\omega+1} M(\phi) \deg \phi + n^{\omega+2} \sum_{\phi \in S_{fac}} \deg \phi\right)$.

Remark 1. If the base field is a finite field $\mathbb{F}_q$, factoring the discriminant is done in $\widetilde{O}((nd_x)^{1.5} \log q + nd_x (\log q)^2)$ bit operations [16].

Remark 2. The above formula shows how the size of the input is unsufficient to give an accurate estimate of the runtime of van Hoeij's algorithm. Indeed, in the best possible case $\#S_{fac}$, $\deg \phi$ and $M(\phi)$ might be constant, and all the $c_{\phi,i}$'s might be equal to d, leading to an overall complexity in $O(n^{\omega+2})$. In the worst possible case however, the sum $\sum_{\phi \in S_{fac}} \deg \phi$ is equal to the degree of the discriminant, leading to an overall complexity in $\widetilde{O}(n^{\omega+2} \deg \mathrm{Disc}(f))$.

2.4   An Improvement in the Case of Low-Degree Singularities

Instead of incrementally computing the b_i's, it is possible to compute one b_k by solving the exact same systems, except that this time the previous b_i's may not have been computed (and are thus set to their initial values y^i). The apparent drawback of this strategy is that it computes b_k without exploiting previous knowledge of smaller b_i's and therefore leads to solving more systems. More precisely, if we already know b_{k-1} then we have to solve $e_k - e_{k-1} + 1$ systems otherwise we may have to solve up to $e_k + 1$ systems. Using the complexity analysis above, we can bound the complexity of finding a given b_k without knowing any other b_i by $\widetilde{O}(n^2 k^{\omega-1} (e_k + 1) \deg \phi)$.

However, we know that for a fixed ϕ, the b_i's can be taken of the form $p_i(x, y)/\phi^{e_i}$ where the exponents e_i's are non-decreasing and bounded by $M(\phi)$. Therefore, when $M(\phi)$ is small enough compared to n, it makes sense to pick a number k and compute b_k. If $b_k = y^k$ then we know that $b_i = y^i$ for any i smaller than k. If $b_k = p_k(x, y)/\phi^{M(\phi)}$ then we know that we can take $b_i = y^{i-k} b_k$ for i greater than k. In most cases neither of this will happen but then we can repeat the process recursively and pick one number between 1 and $k - 1$ and another one between $k + 1$ and n and repeat.

In the extreme case where we treat $M(\phi)$ as a constant (but $\deg \phi$ is still allowed to be as large as $\deg \mathrm{Disc}(f)/2$) this approach saves a factor $\widetilde{O}(n)$ compared to the iterative approach computing the b_i's one after another. This is summarized by the following proposition.

Proposition 2. *Let $f(x, y)$ be a degree n monic irreducible polynomial in y such that irreducible factors of $\mathrm{Disc}(f)$ only appear with exponent bounded by an* *absolute constant. The above modification of van Hoeij's algorithm returns an integral basis for the corresponding function field and costs a univariate factorization of degree $\leq nd_x$ and $\widetilde{O}(n^{\omega+1} \deg \text{Disc}(f))$ field operations, where ω is a feasible exponent for linear algebra.*

Proof. Let us first assume that $M(\phi) = 1$, then the problem is just to find the smallest k such that $e_k = 1$. Since the e_i's are non-decreasing, we can use binary search and find this k after computing $O(\log n)$ basis elements b_i's, for a total cost in $\widetilde{O}(n^{\omega+1} \deg \phi)$ and we indeed gain a quasi-linear factor compared to the previous approach. As long as $M(\phi)$ is constant, a naive way to get the same result is to repeat binary searches to find the smallest k such that $e_k = 1$, then the smallest k such that $e_k = 2$ and so on.

Remark 3. Such extreme cases are not uncommon among the examples presented in the literature and we believe that beyond this extreme, there will be a trade-off between this strategy and the classical one for non-constant but small multiplicities. We do not investigate this trade-off further because finding proper turning points should be addressed in practice as it depends both on theory and implementation.

3   Trager's Algorithm

3.1   A Description of Trager's Algorithm

We first need to introduce the notion of discriminant of a module as this will give us a measure of the "size" of $K[x]$-modules as well as a stopping criterion in the following algorithm.

Consider n elements $v_1, \ldots, v_n$ in $K(x)[y]$, since this is a degree n separable extension of $K(x)$, we can define n distinct embeddings σ_i into an algebraic closure. The matrix $(\sigma_i(v_j)_{i,j})$ is called the conjugate matrix of $(v_1, \ldots, v_n)$.

Definition 4. *The discriminant of $(v_1, \ldots, v_n)$ is the square of the determinant of the conjugate matrix of $(v_1, \ldots, v_n)$.*

Definition 5. *Let V be a $K[x]$-module of rank n and $(v_1, \ldots, v_n)$ a $K[x]$-basis for V. The discriminant of V, denoted by $\text{Disc}(V)$, is the ideal generated by the discriminant of $(v_1, \ldots, v_n)$ defined above.*

Computing an integral basis amounts to computing the integral closure of the $K[x]$-module generated by the powers of y. Trager's algorithm [23] computes such an integral closure iteratively using the following integrality criterion to decide when to stop. Note that there exists many similar algorithms like Round 2 and Round 4 using various criteria for integrality. A more precise account on these algorithms and their history is given in the final paragraphs of [9, Section 2.7].

Proposition 3. *[23, Theorem 1] Let R be a principal domain ($K[X]$ in our case) and V a domain that is a finite integral extension of R. Then V is integrally closed if and only if the idealizer of every prime ideal containing the discriminant equals V.*

Proof. See [23].

More precisely, Trager's algorithm uses the following corollary to the above proposition:

Proposition 4. *[23, Corollary 2] The module V is integrally closed if and only if the idealizer of the radical of the discriminant equals V.*

Starting from any basis of integral elements generating a module V the idea is to compute $\hat{V}$ the idealizer of the radical of the product of all such ideals in V. Either $\hat{V}$ is equal to V and we have found an integral basis, or $\hat{V}$ is strictly larger and we can repeat the operation. We therefore build a chain of modules whose length has to be finite. Indeed, the discriminant of each V_i has to be a strict divisor of that of V_{i-1}.

Input : A degree n monic irreducible polynomial $f(y)$ over $K[x]$
Output : An integral basis for $K[x,y]/\langle f \rangle$
$D \leftarrow \mathrm{Disc}(f)$;
$B \leftarrow (1, y, \ldots, y^{n-1})$;
while *true* **do**
$\quad$ Set V the $K[x]$-module generated by B;
$\quad$ $Q \leftarrow \prod P_i$, where $P_i^2 | D$;
$\quad$ If Q is a unit then return B;
$\quad$ Compute $J_Q(V)$ the Q-trace radical of V;
$\quad$ Compute $\hat{V}$ the idealizer of $J_Q(V)$;
$\quad$ Compute M the change of basis matrix from $\hat{V}$ to V;
$\quad$ Compute $\det M$, if it is a unit then return V;
$\quad$ Update B by applying the change of basis;
$\quad$ $D \leftarrow D/(\det M)^2$;
$\quad$ $V \leftarrow \hat{V}$;
end

Algorithm 2: A bird's eye view of Trager's algorithm [23]

Computing the Radical. Following Trager, we avoid computing the radical of the ideal generated by $\mathrm{Disc}(f)$ directly. First, we note that this radical is the intersection of the radical of the prime ideals generated by the irreducible factors of $\mathrm{Disc}(f)$. Let P be such a factor, we then use the fact that in characteristic zero or greater than n, the radical of $\langle P \rangle$ is exactly the so-called P-trace radical of V (see [23]) i.e. the set $J_P(V) = \{u \in V | \forall w \in V, P | \mathrm{tr}(uw)\}$, where the trace $\mathrm{tr}(w)$ is the sum of the conjugates of a $w \in K(x)[y]$ viewed as a degree n algebraic extension of $K(x)$.

The reason we consider this set is that it is much easier to compute than the radical. Note that Ford and Zassenhaus' Round 2 algorithm is designed to handle the case where this assumption fails but we do not consider this possibility because if it should happen it would be more suitable to use another algorithm designed by van Hoeij for the case of small characteristic [14]. This latter algorithm is different from the one we detailed in Sect. 2 but follows the same principle, replacing Puiseux series by a criterion for integrality based on the Frobenius endomorphism.

Finally, for $Q = \prod P_i$ we define $J_Q(V)$, the Q-trace radical of V, to be the intersection of all the $J_{P_i}(V)$. Here, we further restricted the P_i's to be the irreducible factors of $\mathrm{Disc}(f)$ whose square still divide $\mathrm{Disc}(f)$. In what follows, we summarize how $J_Q(V)$ is computed in Trager's algorithm. Once again, we refer to [23] for further details and proofs.

Let M be the trace matrix of the module V, i.e. the matrix whose entries are the $(\mathrm{tr}(w_i w_j))_{i,j}$, where the w_i's form a basis of V. An element u is in the Q-trace radical if and only if Mu is in $Q \cdot K[x]^n$. In Trager's original algorithm, the Q-trace radical is computed via a $2n \times n$ row reduction and one $n \times n$ polynomial matrix inversion.

We replace this step and compute a $K[x]$-module basis of the Q-trace radical by using an algorithm of Neiger [20] instead. Indeed, given a basis w_i of the $K[x]$-module v, the Q-trace radical can be identified to the set

$$\left\{ f_1, \ldots, f_n \in K[x]^n \,\middle|\, \forall 1 \le j \le n, \ \sum_{i=1}^{n} f_i \mathrm{tr}(w_i w_j) = 0 \bmod Q(x) \right\}.$$

Using [20, Theorem 1.4] with $n = m$ and the shift $s = 0$, there is a deterministic algorithm which returns a basis of the Q-trace radical in Popov form for a cost of $\tilde{O}(n^\omega \deg Q)$ field operations.

Computing the Idealizer. The idealizer of an ideal $\mathfrak{m}$ of V is the set of $u \in \mathrm{Frac}(V)$ such that $u\mathfrak{m} \subset \mathfrak{m}$. Let M_i represent the multiplication matrix by m_i with input basis $(v_1, \ldots, v_n)$ and output basis $(m_1, \ldots, m_n)$. We define M to be the concatenation of such matrices, namely $M = (M_1^t, \ldots, M_n^t)^t$. Then to find the elements $\sum_{i=1}^{n} u_i v_i$ in the idealizer we have to find all $u = (u_1, \ldots, u_n)^t \in K(x)^n$ such that $Mu \in K[x]^{n^2}$. Note that building these multiplication matrices has negligible cost (in $O(n^2)$ field operations) using the technique of [22].

Following Trager, we row-reduce the matrix M and consider $\hat{M}$ the top left $n \times n$ submatrix and the elements of the idealizer are now exactly the u such that $\hat{M}u \in K[x]^n$. Thus, the columns of $\hat{M}^{-1}$ form a basis of the idealizer. Furthermore, the transpose of $\hat{M}^{-1}$ is the change of basis matrix from V_i to V_{i+1}.

3.2 Complexity Analysis

The purpose of this section is to prove the following theorem.

Theorem 2. *Consider f a degrees n monic irreducible polynomial in $K[x][y]$, then Algorithm 2 returns an integral basis for the cost of factoring $\mathrm{Disc}(f)$ and $\widetilde{O}(n^5 \deg \mathrm{Disc}(f))$ operations in K.*

Proof. The dominant parts in this algorithm are the computations of radicals and idealizers, which have been reduced to linear algebra operations on polynomial matrices. First, we have already seen how to compute the Q-trace radical $J_Q(V)$ in $\widetilde{O}(n^\omega \deg Q)$ field operations using the algorithm presented in [20].

To compute the idealizer of $J_Q(V)$, we row-reduce a $n^2 \times n$ matrix with entries in $K[x]$ using naive Gaussian elimination. This costs a total of $O(n^4)$ operations in $K(x)$.

Then we extract the top $n \times n$ square submatrix $\hat{M}$ from this row-reduced $n^2 \times n$ matrix and invert it for $\widetilde{O}(n^\omega)$ operations in $K(x)$. The inverse $\hat{M}^{-1}$ is a basis of a module $\hat{V}$ such that $V \subset \hat{V} \subset \overline{V}$.

To translate operations in $K(x)$ into operations in K, one can bound the degrees of all the rational fractions encountered, however it is quite fastidious to track degree growth while performing the operations described above. In fact, we exploit the nature of the problem we are dealing with.

Our first task is to row-reduce a matrix M built such that a $u = \sum_{i=1}^{n} \rho_i v_i$ is in $\hat{V}$ if and only if $M(\rho_1, \ldots, \rho_n)^t \in K[x]^n$. The ρ_i's are rational fractions but their denominators divide Q. Therefore, we fall back to finding solutions of $M(\tilde{u}_1, \ldots, \tilde{u}_n)^t \in (Q(x) \cdot K[x])^n$, where the $\tilde{u}_i$'s are polynomials. In this case, it does no harm to reduce the entries of the matrix M modulo Q, however performing Gaussian elimination will induce a degree growth that may cause us to handle polynomials of degree up to $n \deg Q$ instead of $\deg Q$. With this bound, the naive Gaussian elimination costs a total of $O(n^5 \deg Q)$ operations in K.

After elimination, we retrieve a $n \times n$ matrix $\hat{M}$ whose entries have degrees bounded by $n \deg Q$. Inverting it will cause another degree increase by a factor at most n. Thus, the inversion step has cost in $\widetilde{O}(n^{\omega+2} \deg Q)$. Since $\omega \leq 3$, each iteration of Trager's algorithm has cost bounded by $O(n^5 \deg Q)$.

Now, let us assess how many iterations are necessary. Let us assume that we are exiting step i and have just computed V_{i+1} from V_i. Let us consider P a square factor of $\mathrm{Disc}(V_i)$. Let $\mathfrak{m}$ be a prime ideal of V_i containing P. Let us consider $u \in V_{i+1}$, then by definition $uP \in \mathfrak{m}$ because $P \in \mathfrak{m}$ and therefore $u \in \frac{1}{P}\mathfrak{m} \subset \frac{1}{P}V_i$. Thus, $V_{i+1} \subset \frac{1}{P}V_i$. This means that at each step i we have $\mathrm{Disc}(V_{i+1}) = \mathrm{Disc}(V_i)/Q_i^2$, where Q_i is the product of square factors of $\mathrm{Disc}(V_i)$. Thus, the total number of iterations is at most half the multiplicity of the largest factor of $\mathrm{Disc}(f)$.

More precisely, if we assume that the irreducible factors of $\mathrm{Disc}(f)$ are r polynomials of respective degrees d_i and multiplicity ν_i, then the overall complexity of Trager's algorithm is in

$$\widetilde{O}\left(\sum_{i=1}^{\nu} n^5 \sum_{j \leq r,\, \nu_j \geq 2i} d_j\right),$$

where $\nu = \lfloor \max \nu_i / 2 \rfloor$.

Since $\sum_{i=1}^{r} \nu_i d_i \leq \deg \mathrm{Disc}(f)$, the above bound is in $\tilde{O}(n^5 \deg \mathrm{Disc}(f))$, which ranges between $\tilde{O}(n^6 d_x)$ and $\tilde{O}(n^5)$ depending on the input f.

Remark 4. In the above proof, our consideration of degree growth seems quite pessimistic given that the change of basis matrix has prescribed determinant. It would be appealing to perform all the computations modulo Q but it is unclear to us whether the algorithm remains valid. Another possibility of improvement would be to apply a more sophisticated technique than Gaussian elimination. However, Trager's algorithm manipulates $n^2 \times n$ polynomial matrices of degrees up to $\deg \mathrm{Disc}(f)$ so it runs in time $\Omega(n^3 \deg \mathrm{Disc}(f))$, which is no better than the bound we give in next section.

4 Integral Bases Through Weierstrass Factorization and Truncations of Puiseux Series

Like van Hoeij's algorithm, this algorithm due to Böhm et al. [4] relies on computing local integral bases at each "problematic" singularity and then recovering a global integral basis. But this algorithm splits the problem further into computing a local contributions to an integral basis at each branch of each singularity.

More precisely, given a reduced Noetherian ring A we denote by $\overline{A}$ its normalization i.e. the integral closure of A in its fraction field $\mathrm{Frac}(A)$. In order to compute the normalization of $A = K[x,y]/\langle f(x,y)\rangle$ we use the following result to perform the task locally at each singularity.

Proposition 5. *[4, Proposition 3.1] Let A be a reduced Noetherian ring with a finite singular locus $\{P_1, \ldots, P_s\}$. For $1 \leq i \leq s$, let an intermediate ring $A \subset A^{(i)} \subset \overline{A}$ be given such that $A_{P_i}^{(i)} = \overline{A_{P_i}}$. Then $\sum_{i=1}^{s} A^{(i)} = \overline{A}$.*

Proof. See the proof of [5, Proposition 3.2].

Each of these intermediate rings is respectively called a *local contribution* to $\overline{A}$ at P_i. In the case where $A_{P_j}^{(i)} = A_{P_j}$ for any $j \neq i$, we say that $A^{(i)}$ is a *minimal local contribution* to $\overline{A}$ at P_i. Here, we consider the case $A = K[x,y]/\langle f(x,y)\rangle$ and will compute minimal local contributions at each singularity of f. This is summarized in Algorithm 3.

In this section, we revisit the algorithm presented by Böhm et al. in [4] and replace some of its subroutines in order to derive a complexity bound stated in Theorem 3. Note that these modifications are performed solely for the sake of complexity and rely on algorithms for which implementations may not be available. Our new description makes this algorithm both simpler and more efficient because Hensel lifting remains "hidden" within Poteaux and Weimann's factorization algorithm over $K[[x]][y]$. Some useful quantities are also shown to come as byproducts of the factorization so we avoid recomputing them.

Theorem 3. *Let $f(x,y)$ be a degree n irreducible monic polynomial in y. Then Algorithm 3 returns an integral basis of $K[x,y]/\langle f\rangle$ and costs a factorization of $\mathrm{Disc}(f)$ over K, at most n factorizations of degree n polynomials over an extension of K of degree $\leq \deg \mathrm{Disc}(f)$ and $\tilde{O}(n^2 \deg \mathrm{Disc}(f))$ operations in K.*

Input : A monic irreducible polynomial $f(y)$ over $K[x]$
Output : An integral basis for $K[x, y]/\langle f \rangle$
$n \leftarrow \deg_y f$;
$S_{fac} \leftarrow$ set of factors ϕ such that $\phi^2 | \mathrm{Disc}(f)$;
for ϕ **in** S_{fac} **do**
> Compute α a root of ϕ (possibly in an extension);
> Apply a linear transform to fall back to the case of a singularity at $x = 0$;
> Compute the maximal integrality exponent $E(f)$;
> Using Proposition 9, factor f over $K[[x]][y]$;
> Compute the Bézout relations of Proposition 7;
> Compute integral bases for each factor as in Section 4.1;
> As in Section 4.2, recover the local contribution corresponding to ϕ;
> (For this, use Proposition 7 and Proposition 11)

end
From all the local contributions, use CRT to deduce an integral basis B;
return B;

Algorithm 3: Adaptation of the algorithm by Böhm et al. [4]

Remark 5. The factorizations incurred by the use of Poteaux and Weimann's algorithm are only necessary to ensure that quotient rings are actually fields, this cost can be avoided by using the D5 principle [8] at the price of a potential complexity overhead. However, using directed evaluation [15] yields the same result without hurting our complexity bounds.

4.1   Computing Normalization at One Branch

Let us first address the particular case when $f(x, y)$ is an irreducible Weierstrass polynomial. This way, we will be able to compute integral bases for each branches at a given singularity. The next section will then show how to glue this information first into a local integral basis and then a global integral basis can be computed using CRT as in van Hoeij's algorithm. The main result of this section is the following proposition.

Proposition 6. *Let g be an irreducible Weierstrass polynomial of degree m whose Puiseux expansions have already been computed up to sufficiently large precision ρ. An integral basis for the normalization of $K[[x]][y]/\langle g \rangle$ can be computed in $\widetilde{O}(\rho m^2)$ operations in K.*

As in van Hoeij's algorithm, the idea is to compute for any $1 \leq d < m$ a polynomial $p_d \in K[x][y]$ and an integer e_d such that $p_d(x, y)/x^{e_d}$ is integral and e_d is maximal. We clarify this notion of maximality in the following definition.

Definition 6. *Let $P \in K[x][y]$ be a degree d monic polynomial (in y). We say that P is d-maximal if there exists an exponent e_d such that $P(x, y)/x^{e_d}$ is integral and there is no degree d monic polynomial Q such that $Q(x, y)/x^{e_d + 1}$ is integral.*

Remark 6. To the best of our knowledge this notion has not received a standard name in the literature and was often referred to using only the word maximal.

Let us consider the m Puiseux expansions γ_i of g. Since g is irreducible, these expansions are conjugated but let us first make a stronger assumption: there exists a $t \in \mathbb{Q}$ such that all the terms of degree lower than t of the expansions γ_i are equal and the terms of degree t are conjugate. We truncate all these series by ignoring all terms of degree greater or equal to t. This way, all the expansions share the same truncation $\overline{\gamma}$.

Lemma 1. *[4, Lemma 7.5] Using the notation and hypotheses of previous paragraph, for any $1 \le d < m$ the polynomial $p_d = (y - \overline{\gamma})^d$ is d-maximal.*

Proof. See [4].

In a more general setting, more truncations are iteratively performed so as to fall back in the previous case. We recall below the strategy followed in [4] for the sake of completeness.

Initially we have $g_0 = g = \prod_{i=1}^{m}(y - \gamma_i)$. We compute the smallest exponent t such that the expansions γ_i are pairwise different. We truncate the expansions to retain only the exponents smaller than t and denote these truncations $\gamma_j^{(1)}$. Among these m expansions, we extract a set of r mutually distinct expansions which we denote by η_i. Note that by local irreducibility, each of these expansions correspond to exactly m/r identical $\gamma_j^{(1)}$'s. We further denote $\overline{g_0} = \prod_{i=1}^{m}(y - \gamma_i^{(1)})$ and $g_1 = \prod_{i=1}^{r}(y - \eta_i)$ and $u_1 = m/r$. We actually have $\overline{g_0} = g_1^{u_1}$.

We recursively repeat the operation: starting from a polynomial $g_{j-1} = \prod_{i=1}^{r_{j-1}}(y - \eta_i)$, we look for the first exponent such that all the truncations of the η_i are pairwise different. Truncating these expansions up to exponent strictly smaller, we compute $\overline{g_{j-1}} = \prod_{i=1}^{m_j}(y - \gamma_i^{(j)})$. Once again we retain only one expansion per set of identical truncations and we define a $g_j = \prod_{i=1}^{r_j}(y - \eta_i)$ and $u_j = m_j/r_j$.

The numerators of the integral basis that the algorithm shall return are products of these g_i's. Loosely speaking, the g_i have decreasing degrees in y and decreasing valuations so for a fixed d the denominator p_d is chosen of the form $\prod g_i^{\nu_i}$ where the ν_i's are incrementally built as follows: ν_1 is the largest integer such that $\deg_y(g_1^{\nu_1}) \le d$ and $\nu_1 \le u_1$, then ν_2 is the largest integer such that $\deg_y(g_1^{\nu_1} g_2^{\nu_2}) \le d$ and $\nu_2 \le u_2$, and so on. This is Algorithm 6 of [4], we refer to the proof of [4, Lemma 7.8] for exactness.

Since we assumed that we are treating a singularity at 0, the denominators are powers of x. The proper exponents are deduced in the following way: for each g_i we keep in memory the set of expansions that appear, we denote this set by N_{g_i}. Then for any γ in the set Γ of all Puiseux expansions of g we compute $\sigma_i = \sum_{\eta \in N_{g_i}} v(\gamma - \eta)$ which does not depend on the choice of $\gamma \in \Gamma$. For any j, if $p_j = \prod_k g_k^{\nu_k}$ then the exponent e_j of the denominator is given by $\lfloor \sum_k \nu_k \sigma_k \rfloor$. Detailed justifications of this are given in [4].

Complexity Analysis. Let us now give a proof of Proposition 6. To do so, remark that the g_k's are polynomials whose Puiseux series are precisely the truncation η_i's of the above $\gamma_j^{(i)}$. Equivalently, one can say that the g_k's are the norms of the Puiseux expansions η_i's.

To compute them, we can appeal to Algorithm NormRPE of Poteaux and Weimann [21, Section 4.1]. Suppose we know all the expansions involved up to precision ρ sufficiently large. These expansions are not centered at $(0, \infty)$ because g is monic. Therefore, the hypotheses of [21, Lemma 8] are satisfied and Algorithm NormRPE compute each of the g_i's above in time $\widetilde{O}(\rho \deg_y(g_i)^2)$.

Then we remark that the total number of such g_i's is in $O(\log m)$. Indeed, at each step the number of expansions to consider is at least halved (Puiseux expansions are grouped according to their truncations being the same, at least two series having the same truncation). Since the degree of each g_i is no greater than $m - 1$, all these polynomials can be computed in $\widetilde{O}(m^2 \rho)$ operations in K.

Once the g_i's are known we can deduce the numerators p_i's as explained above. Building them incrementally starting from p_1 each p_i is either equal to a g_j or can be expressed as one product of quantities that were already computed (either a g_j or a p_k for $k < i$). Therefore, computing all the numerators amounts to computing at most m products of polynomials whose degrees are bounded by m over $K[x]/\langle x^\rho \rangle$. Using Schönhage-Strassen's algorithm for these products the total cost is in $\widetilde{O}(\rho m^2)$ operations in K. The cost of computing denominators is negligible so this concludes the proof.

4.2   Branch-Wise Splitting for Integral Bases

Once again, let us assume that we are treating the local contribution at the singularity $x = 0$. In the setting of van Hoeij's algorithm, this corresponds to dealing with a single irreducible factor of the discriminant. We further divide the problem by considering the factorization $f = f_0 \prod_{i=1}^r f_i$, where f_0 is a unit in $K[[x]][y]$ and the other f_i's are irreducible Weierstrass polynomials in $K[[x]][y]$.

We can apply the results from the previous section to each f_i for $i > 0$ in order to compute an integral basis of $K[[x]][y]/\langle f_i \rangle$. In this section, we deal with two problems: we explain how to compute the factorization of f and how to efficiently perform an analogue of the Chinese Remainder Theorem to compute an integral basis of $K[[x]][y]/\langle f_1 \cdots f_r \rangle$ from the integral bases at each branch. For the sake of completeness, we recall in Sect. 4.3, how Böhm et al. take f_0 into account and deduce a minimal local contribution at any given singularity.

Proposition 7. *[4, Proposition 5.9] Let f_1, ..., f_r be the irreducible Weierstrass polynomials in $K[[x]][y]$ appearing in the factorization of f into branches. Let us set $h_i = \prod_{j=1, j \neq i} f_j$. Then f_i and h_i are coprime in $K((x))[y]$ so that there exist polynomials a_i, b_i in $K[[x]][y]$ and positive integers c_i such that $a_i f_i + b_i h_i = x^{c_i}$ for any $1 \leq i \leq r$.*

Furthermore, the normalization of $K[[x]][y]/(f_1 \cdots f_r)$ splits as

$$\overline{K[[x]][y]/\langle f_1 \cdots f_r \rangle} \cong \bigoplus_{i=1}^{r} \overline{K[[x]][y]/\langle f_i \rangle}$$

and the splitting is given explicitly by

$$(t_1 \bmod f_1, \ldots, t_r \bmod f_r) \mapsto \sum_{i=1}^{r} \frac{b_i h_i t_i}{x^{c_i}} \bmod f_1 \cdots f_r.$$

Proof. See [7, Theorem 1.5.20].

The following corollary is used to recover an integral basis for $K[[x]][y]/\langle f_1 \cdots f_r \rangle$.

Proposition 8. *[4, Corollary 5.10] With the same notation, let*

$$\left(1, \frac{p_1^{(i)}(x, y)}{x^{e_1^{(i)}}}, \ldots, \frac{p_{m_i-1}^{(i)}(x, y)}{x^{e_{m_i-1}^{(i)}}} \right)$$

represent an integral basis for f_i, where each $p_j^{(i)} \in K[x][y]$ is a monic degree j polynomial in y. For $1 \leq i \leq r$, set

$$\mathcal{B}^{(i)} = \left(\frac{b_i h_i}{x^{c_i}}, \frac{b_i h_i p_1^{(i)}}{x^{c_i+e_1^{(i)}}}, \ldots, \frac{b_i h_i p_{m_i-1}^{(i)}}{x^{c_i+e_{m_i-1}^{(i)}}} \right).$$

Then $\mathcal{B}^{(1)} \cup \cdots \cup \mathcal{B}^{(r)}$ is an integral basis for $f_1 \cdots f_r$.

In [4], these results are not used straightforwardly because the authors remarked that it was time-consuming in practice. Instead, the c_i's are computed from the singular parts of the Puiseux expansions of f and the polynomials β_i replace the b_i's, playing a similar role but being easier to compute.

Indeed, these β_i's are computed in [4, Algorithm 8] and they are actually products of the polynomials g_i's already computed by [4, Algorithm 7], which is the algorithm that we detailed above to describe the computation of an integral basis for each branch. The only new data computed in order to deduce the β_i's are the suitable exponents of the g_i's. This is achieved through solving linear congruence equations. This step can be fast on examples considered in practice and we also note that the β_i's seem more convenient to handle because they are in $K[x][y]$ and they contain less monomials than the b_i's. However the complexity of this problem (often denoted LCON in the literature) has been widely studied, see for example [1,3] but, to the best of our knowledge, none of the results obtained provide bounds that we could use here.

For the sake of complexity, we therefore suggest another way which is based on computing the b_i's of Proposition 7. We also compute the factorization of f into branches in a different way: instead of following the algorithms of [4, Section 7.3 & 7.4] we make direct use of the factorization algorithm of Poteaux and Weimann [21] so we also invoke their complexity result [21, Theorem 3] which is recalled below. Another advantage to this is that we will see that the b_i's can actually be computed using a subroutine involved in the factorization algorithm, which simplifies even further the complexity analysis.

Proposition 9. *[21, Theorem 3] There exists an algorithm that computes the irreducible factors of f in $K[[x]][y]$ with precision N in $\widetilde{O}(\deg_y(f)(\delta + N))$ expected operations in K plus the cost of one univariate factorization of degree at most $\deg_y(f)$, where δ stands for the valuation of $\mathrm{Disc}(f)$.*

Proof. See [21, Section 7].

Let us now get back to the first steps of Algorithm 3: we have to compute $E(f)$ to assess up to what precision we should compute the Puiseux series and then compute the factorization of f, the integers c_i and the polynomials b_i.

In each section, we tried to keep the notation of the original papers as much as we could which is why we introduced $E(f)$ but the definition given in [4, Section 4.8] is exactly the same as the N in van Hoeij's paper [12]. This bound can be directly computed from the singular part of the Puiseux expansions of f. We recall its definition: $E(f) = \max_i \sum_{i \neq j} v(\gamma_i - \gamma_j)$, where the γ_i's are the Puiseux expansions of f.

Following [4], we need to compute the factorization of f into branches up to precision $E(f) + c_i$. Using Poteaux and Weimann's factorization algorithm from Proposition 9, we can compute the factors f_i up to the desired precision.

Furthermore, using a subroutine contained within this algorithm, we can compute the Bézout relation $a_i f_i + b_i h_i = x^{c_i}$ up to precision $E(f) + c_i$. This is detailed in [21, Section 4.2], where our c_i is the lifting order κ and our f_i and h_i are respectively the H and G of Poteaux and Weimann. The algorithm used to compute the Bézout relations is due to Moroz and Schost [19].

Complexity Analysis. We analyze the cost of the computations performed in this section and summarize them by the following proposition.

Proposition 10. *Let $f(x, y)$ be a degree n monic irreducible polynomial in y and let δ be the x-valuation of $\mathrm{Disc}(f)$. Then the integers c_i's and $E(f)$, a factorization in branches $f = f_0 \prod_{i=1}^{r} f_i$ as well as the polynomials a_i's and b_i's of Proposition 7 can be computed up to precision $E(f) + c_i$ for a univariate factorization degree n over K and a total of $\widetilde{O}(n^2 \delta)$ field operations.*

Proof. First, the singular parts of the Puiseux series of f above 0 are computed for $\widetilde{O}(n\delta)$ field operations by [21, Theorem 1]. This allows us to compute $E(f)$.

Then we compute the factorization in branches up to a sufficient precision to compute the c_i's. We then extend the precision further so as to compute the factorization and the Bézout relations $a_i f_i + b_i h_i = x^{c_i}$ up to precision $E(f) + c_i$.

Invoking [19, Corollary 1], computing a single Bézout relation up to precision $E(f) + c_i$ costs $\widetilde{O}(n(E(f) + c_i))$ field operations. Computing the factorization of f in branches up to the same precision with Proposition 9 accounts for $\widetilde{O}(n(\delta + c_i + E(f)))$ operations in K and one univariate factorisation of degree n over K.

Using [4, Definition 4.14], we note that $E(f)$ can also be seen as e_{n-1}, which is bounded by the valuation δ of the discriminant because we assumed that we were handling a singularity at $x = 0$. Thanks to [21, Proposition 8] we can bound c_i by the valuation of $\frac{\partial f}{\partial y}$ which is itself bounded by δ.

Putting these bounds together, the overall cost is one univariate factorization of degree n over K and $\widetilde{O}(n\delta)$ operations in K for the factorization step while the n Bézout relations requires $\widetilde{O}(n^2\delta)$ operations in K. This concludes the proof.

4.3 Contribution of the Invertible Factor f_0

To deal with this problem, we reuse the following result without modification.

Proposition 11. *[4, Proposition 6.1] Let $f = f_0 g$ be a factorization of f with f_0 and g in $K[[x]][y]$, f_0 a unit and g a Weierstrass polynomial of y-degree m. Let $\left(p_0 = 1, \frac{p_1}{x^{e_1}}, \ldots, \frac{p_{m-1}}{x^{e_{m-1}}}\right)$ be an integral basis for $K[[x]][y]/\langle g \rangle$ such that the p_i's are degree i monic polynomials in $K[x][y]$ and let $\overline{f_0}$ be a monic polynomial in $K[x][y]$ such that $\overline{f_0} = f_0 \bmod x^{e_{m-1}}$. Let us denote $d_0 = \deg_y(\overline{f_0})$.
Then*

$$\left(1, y, \ldots, y^{d_0-1}, \overline{f_0}p_0, \frac{\overline{f_0}p_1}{x^{e_1}}, \ldots, \frac{\overline{f_0}p_{m-1}}{x^{e_{m-1}}}\right)$$

is an integral basis for the normalization of $K[[x]][y]/\langle f \rangle$.

Proof. See [4]. □

Since we handle a single singularity at 0, the previous basis is also a $K[x]$-module basis of the minimal local contribution at this singularity by [4, Corollary 6.4].

Complexity Analysis. This step involves a truncation of f_0 modulo $x^{e_{m-1}}$ and m products of polynomials in $K[[x]][y]/\langle x^{e_{m-1}} \rangle$ whose y-degrees are bounded by $n = \deg_y(f)$. This incurs $\widetilde{O}(mne_{m-1})$ field operations. Since we are treating a singularity at $x = 0$, we have $e_{m-1} = O(\delta)$ with δ the valuation of $\mathrm{Disc}(f)$ so that we can simplify the above bound as $\widetilde{O}(n^2\delta)$ field operations.

4.4 Proof of Theorem 3

In this section, we put all the previous bounds together and prove Theorem 3.

Proof. As in van Hoeij's algorithm, we first compute $\mathrm{Disc}(f)$ and factor it in order to recover its irreducible square factors. For each irreducible factor ϕ such that $\phi^2|\mathrm{Disc}(f)$, we compute the corresponding minimal local contribution. For each of them, we first perform a translation so as to handle a singularity at $x = 0$. If there are several conjugated singularities we can handle them like in van Hoeij's algorithm, at the price of a degree $\deg \phi$ extension of K which we denote by K' in this proof. Also note that through this transform the multiplicity $M(\phi)$ corresponds to the valuation δ of the discriminant.

First, we split f into branches using Proposition 10 for a cost in $\widetilde{O}(n^2 M(\phi))$ operations in K' and one univariate factorization of degree $\leq n$ over K'.

Then, at each branch f_i, we apply Proposition 6 with precision $\rho = E(f) + c_i$. Therefore, the complexity of computing an integral basis at each branch f_i is in $\widetilde{O}(M(\phi) \deg_y(f_i)^2)$ operations in K'. Since $\sum_i \deg f_i \leq n$, computing the integral bases at all the branches costs $\widetilde{O}(n^2 M(\phi))$ operations in K'.

At the end of this step, we have integral bases $\mathcal{B}_i$ of the form

$$\left(1, \frac{p_1(x,y)}{x^{e_1}}, \ldots, \frac{p_{m_i-1}(x,y)}{x^{e_{m_i-1}}} \right)$$

with $m_i = \deg_y f_i$ but the p_i's are in $K'[[x]][y]$.

At first glance, this is a problem because Proposition 8 requires the p_i's to be in $K'[x][y]$. However, the power of x in the denominators is bounded a priori by $E := E(f) + \max_{1 \leq i \leq r} c_i$ so we can truncate all series beyond this exponent. Indeed, forgetting the higher order terms amounts to subtracting each element of the basis by a polynomial in $K'[x]$. Such polynomials are obviously integral elements so they change nothing concerning integrality.

We can thus apply Proposition 8 to get an integral basis for $f_1 \cdots f_r$. This costs $O(n)$ operations in $K'[[x]][y]/\langle x^E, f(x,y) \rangle$. Each such operation amounts to nE operations in K'. We have previously seen that E is in $O(M(\phi))$ so the overall cost of applying Proposition 8 is in $O(n^2 M(\phi))$ operations in K'.

After this process, the basis that we obtained must be put in "triangular form" (i.e. each numerator p_i should have degree i in y in order for us to apply Proposition 11). To do this, we first reduce every power of y greater or equal to n using the equation $f(x,y) = 0$. For a fixed i, by the Bézout relations, h_i has y-degree $\leq n - m_i$ and b_i has y-degree $< m_i$, so we have to reduce a total of $O(n)$ bivariate polynomials whose degrees in y are in $O(n)$. Using a fast Euclidean algorithm, this amounts to $\widetilde{O}(n^2)$ operations in $K'[x]/\langle x^E \rangle$, hence a cost in $O(n^2 M(\phi))$ operations in K'.

Once done, every element in the basis can be represented by a vector of polynomials in $K'[x]$ whose degrees are bounded by E. To put the above integral basis in triangular form, it suffices to compute a Hermite Normal Form of a full rank $n \times n$ polynomial matrix. Using [17, Theorem 1.2] an algorithm by Labahn, Neiger and Zhou performs this task in $\widetilde{O}(n^{\omega-1} M(\phi))$ operations in K'.

We can finally apply Proposition 11 and deduce the minimal local contribution for the factor ϕ in $\widetilde{O}(n^2 M(\phi))$ operations in K'.

Overall, given a factor ϕ, computing the corresponding minimal local contribution to the normalization of $K[\mathcal{C}]$ costs the factorization of $\mathrm{Disc}(f)$, one univariate factorization of degree $\leq n$ over K and $\widetilde{O}(n^2 M(\phi))$ operations in K'. Computing all the local contributions can therefore be done for the factorization of $\mathrm{Disc}(f)$, $\#S_{fac}$ univariate factorization of degree $\leq n$ over extensions of K of degree $\leq \max_{\phi \in S_{fac}} \deg \phi$ and $\widetilde{O}(n^2 \deg \mathrm{Disc}(f))$ operations in K.

In the case of conjugate singularities, we follow the idea of van Hoeij rather than [4, Remark 7.17] and simply replace α by x in the numerators and $(x - \alpha)$ by ϕ in the denominators because it does not harm our complexity bound.

In this process, some coefficients of the numerators are multiplied by polynomials in x, which clearly preserves integrality. Since the numerators are monic in y, no simplification can occur and the basis property is also preserved.

Finally, a global integral basis for $K[x, y]/\langle f \rangle$ is deduced by a Chinese remainder theorem. This can be achieved in quasi-linear time in the size of the local bases. Each of them being in $O(n^2 \deg \mathrm{Disc}(f))$, this last CRT does not increase our complexity bound. This concludes the proof.

5 Conclusion

In the setting of Table 1, the best bound given in this paper is in $\tilde{O}(D^4)$ which is quasi-quadratic in the input size, but quasi-linear in the output size. It is surprising that we are able to reach optimality without even treating the local factors f_i through a divide-and-conquer approach like in [21]. This would allow us to work at precision δ/n instead of δ most of the time, but this does not affect the worst-case complexity of the whole algorithm. From an implementation point of view, however, this approach will probably make a significant difference.

Note that we are still relatively far from having implementations of algorithms actually reaching these complexity bounds because we lack implementations for primitives involved in computing Popov/Hermite forms, Puiseux series and factorizations over $K[[x]][y]$. In some experiments we performed, Puiseux series were actually the most time-consuming part, which is why Trager's algorithm may remain a competitive choice despite our complexity results. We refer to [4, Section 8] for more detailed timings and experiments.

Acknowledgments. Part of this work was completed while the author was at the Symbolic Computation Group of the University of Waterloo. This paper is part of a project that has received funding from the French Agence de l'Innovation de Défense. The author is grateful to Grégoire Lecerf, Adrien Poteaux and Éric Schost for helpful discussions and to Grégoire Lecerf for feedback on a preliminary version of this paper. The author also wishes to thank the anonymous reviewers for their comments.

References

1. Arvind, V., Vijayaraghavan, T.C.: The complexity of solving linear equations over a finite ring. In: Diekert, V., Durand, B. (eds.) STACS 2005. LNCS, vol. 3404, pp. 472–484. Springer, Heidelberg (2005). https://doi.org/10.1007/978-3-540-31856-9_39
2. Bauch, J.D.: Computation of integral bases. J. Number Theory **165**, 382–407 (2016)
3. de Beaudrap, N.: On the complexity of solving linear congruences and computing nullspaces modulo a constant. arXiv preprint arXiv:1202.3949 (2012)
4. Böhm, J., Decker, W., Laplagne, S., Pfister, G.: Computing integral bases via localization and Hensel lifting. arXiv preprint arXiv:1505.05054 (2015)
5. Böhm, J., Decker, W., Laplagne, S., Pfister, G., Steenpaß, A., Steidel, S.: Parallel algorithms for normalization. J. Symb. Comput. **51**, 99–114 (2013)

6. Bostan, A., et al.: Algorithmes efficaces en calcul formel (2017)
7. De Jong, T., Pfister, G.: Local Analytic Geometry: Basic Theory and Applications. Springer, Wiesbaden (2013). https://doi.org/10.1007/978-3-322-90159-0
8. Della Dora, J., Dicrescenzo, C., Duval, D.: About a new method for computing in algebraic number fields. In: Caviness, B.F. (ed.) EUROCAL 1985. LNCS, vol. 204, pp. 289–290. Springer, Heidelberg (1985). https://doi.org/10.1007/3-540-15984-3_279
9. Diem, C.: On arithmetic and the discrete logarithm problem in class groups of curves. Habilitation, Universität Leipzig (2009)
10. Duval, D.: Rational Puiseux expansions. Compositio Mathematica **70**(2), 119–154 (1989)
11. Hess, F.: Computing Riemann-Roch spaces in algebraic function fields and related topics. J. Symb. Comput. **33**(4), 425–445 (2002)
12. van Hoeij, M.: An algorithm for computing an integral basis in an algebraic function field. J. Symb. Comput. **18**(4), 353–363 (1994)
13. van Hoeij, M., Novocin, A.: A reduction algorithm for algebraic function fields (2008)
14. van Hoeij, M., Stillman, M.: Computing an integral basis for an algebraic function field (2015). https://www.math.fsu.edu/~hoeij/papers/2015/slides.pdf
15. van der Hoeven, J., Lecerf, G.: Directed evaluation, December 2018. Working paper or preprint. https://hal.archives-ouvertes.fr/hal-01966428
16. Kedlaya, K.S., Umans, C.: Fast polynomial factorization and modular composition. SIAM J. Comput. **40**(6), 1767–1802 (2011)
17. Labahn, G., Neiger, V., Zhou, W.: Fast, deterministic computation of the Hermite normal form and determinant of a polynomial matrix. J. Complexity **42**, 44–71 (2017)
18. Le Gall, F.: Powers of tensors and fast matrix multiplication. In: Proceedings of the 39th International Symposium on Symbolic and Algebraic Computation, pp. 296–303 (2014)
19. Moroz, G., Schost, É.: A fast algorithm for computing the truncated resultant. In: Proceedings of the ACM on International Symposium on Symbolic and Algebraic Computation, pp. 341–348 (2016)
20. Neiger, V.: Fast computation of shifted Popov forms of polynomial matrices via systems of modular polynomial equations. In: Proceedings of the ACM on International Symposium on Symbolic and Algebraic Computation, pp. 365–372 (2016)
21. Poteaux, A., Weimann, M.: Computing Puiseux series: a fast divide and conquer algorithm. arXiv preprint arXiv:1708.09067 (2017)
22. Trager, B.M.: Algorithms for manipulating algebraic functions. SM thesis MIT (1976)
23. Trager, B.M.: Integration of algebraic functions. Ph.D. thesis, Massachusetts Institute of Technology (1984)
24. Walker, R.J.: Algebraic curves (1950)
25. Zassenhaus, H.: Ein algorithmus zur berechnung einer minimalbasis über gegebener ordnung. In: Collatz, L., Meinardus, G., Unger, H. (eds.) Funktionalanalysis Approximationstheorie Numerische Mathematik, pp. 90–103. Springer, Basel (1967). https://doi.org/10.1007/978-3-0348-5821-2_10

Truncated and Infinite Power Series in the Role of Coefficients of Linear Ordinary Differential Equations

Sergei A. Abramov$^{(\boxtimes)}$ [iD], Denis E. Khmelnov [iD], and Anna A. Ryabenko [iD]

Dorodnicyn Computing Centre, Federal Research Center "Computer Science and Control" of the Russian Academy of Sciences, Vavilova, 40, Moscow 119333, Russia
sergeyabramov@mail.ru, dennis_khmelnov@mail.ru, anna.ryabenko@gmail.com

Abstract. We consider linear ordinary differential equations, each of the coefficients of which is either an algorithmically represented power series, or a truncated power series. We discuss the question of what can be learned from equations given in this way about its Laurent solutions, i.e., solutions belonging to the field of formal Laurent series. We are interested in the information about these solutions, that is invariant with respect to possible prolongations of the truncated series which are the coefficients of the given equation.

Keywords: Differential equations · Truncated power series · Algorithmically represented infinite power series · Laurent series · Computer algebra systems

1 Introduction

We will consider operators and differential equations written using the operation $\theta = x\frac{d}{dx}$. In the original operator

$$L = \sum_{i=0}^{r} a_i(x)\theta^i, \tag{1}$$

as well as in the equation $L(y) = 0$, for each $a_i(x)$, $i = 0, 1, \ldots, r$, one of two possibilities is allowed: $a_i(x)$ can be

- an infinite series represented algorithmically: the series $\sum a_n x^n$ is defined by an algorithm computing a_n by n,
  or
- a truncated series

$$a_i(x) = \sum_{j=0}^{t_i} a_{ij} x^j + O(x^{t_i+1}), \tag{2}$$

Supported in part by RFBR grant, project No. 19-01-00032.

© Springer Nature Switzerland AG 2020
F. Boulier et al. (Eds.): CASC 2020, LNCS 12291, pp. 63–76, 2020.
https://doi.org/10.1007/978-3-030-60026-6_4

where t_i is an integer such that $t_i \geq -1$ (if $t_i = -1$ then the sum in (2) is 0). We call t_i *the truncation degree* of the coefficient $a_i(x)$ represented in the form (2). Note that a coefficient in (1) can be of the form $O(x^m)$, $m \geq 0$.

We assume that at least one of the constant terms of $a_0(x), \ldots, a_r(x)$ is nonzero.

The coefficients of series belong to a field K of characteristics 0. The following notations are standard:

- $K[x]$ and $K[x, x^{-1}]$ are the rings of polynomials and, resp., Laurent polynomials with coefficients from K.
- $K[[x]]$ is the ring of formal power series with coefficients from K.
- $K((x))$ is the quotient field of the ring $K[[x]]$; the elements of this field are formal Laurent series with coefficients from K.

Definition 1. *The degree* $\deg f(x)$ *of a polynomial* $f(x)$ *from* $K[x]$ *or* $K[x, x^{-1}]$, *is defined as the largest degree of* x *belonging to* $f(x)$ *(*$\deg 0 = -\infty$ *by convention). Note that the degree of a Laurent polynomial is in some cases nonpositive, even when this polynomial is not a constant:* $\deg(2x^{-2} + x^{-1}) = -1$, $\deg(3x^{-1} + 1) = 0$, *etc.*

The solutions we are interested in belong to the field of formal Laurent series with the coefficients from K. We will call such solutions as *Laurent*. A more exact specification for the problem of finding such solutions will be given later in this introductory section. We will not discuss the questions of convergence of series.

A discussion of the algorithmic aspect of problems of this kind involves considering the question of representing infinite series, in particular, the series which play the role of the coefficients of the equation.

In [1–3], an algorithmic representation was considered. It was detected that some problems associated with the solutions of equations given in this way turn out to be algorithmically unsolvable, though, at the same time, the other part is successfully solvable. For example, the problem of finding Laurent solutions is solvable: these solutions can be represented algorithmically in the same sense as the representation of the coefficients of the equation. (In the mentioned papers, not only scalar equations were discussed, but also systems of equations.)

In [3,8], the authors considered the problems of constructing solutions under the assumption that all series playing the role of the coefficients of a given equation or system are represented in the truncated form. In [8], it was found out which truncation of the system coefficients will be sufficient to calculate a given number of initial terms of the series, included in the exponentially-logarithmic solutions of the system. In [5,6], we considered this problem as the task of constructing truncated solutions; it was shown how to construct the maximum possible number of initial terms of the series included in the Laurent and regular solutions of the equation.

In this paper, we admit the presence in the original equation of such coefficients that are of two different kinds indicated below formula (1). We are interested in information on Laurent solutions that is invariant with respect to possible prolongations of the truncated series representing the coefficients of the equation. But everything is not so simple already with the definition of the concept of "solution" for an equation which, in the presence of truncated coefficient, is, in fact, not completely specified. We introduce the concept "truncated Laurent solution".

Definition 2. *Let an operator L have the form (1) and*

$$L(y) = 0 \tag{3}$$

be the equation corresponding to this operator. An expression

$$f(x) + O(x^{k+1}), \tag{4}$$

in which $f(x) \in K[x^{-1}, x] \setminus \{0\}$ and k is an integer greater than or equal to $\deg f(x)$, is called a truncated Laurent solution *of (3), if for any specification of all $O(x^{t_i+1})$ (that is, for any replacement of the symbols $O(...)$ by concrete series having corresponding valuation) included in the coefficients $a_i(x)$ of equation (3), such a specification of the series $O(x^{k+1})$ in (4) is possible that the specified expression (4) becomes a Laurent solution to the specified equation. This k in (4) is* the truncation degree *of the solution.*

We propose an algorithm that allows for an equation $L(y) = 0$ represented in the explained form and an integer k to construct all such truncated Laurent solutions of this equation that have a truncation degree not exceeding k. If the equation does not have such truncated Laurent solutions then the result of the algorithm will indicate this.

The algorithm is described in Sect. 5. But first it is shown in Sect. 2 that both checking the finiteness of the set of those k for which the formulated problem has a solution, and finding the maximum possible value of k if it exists, are algorithmically undecidable problems. Our algorithm works with a specific given k.

Section 6 describes the implementation of the algorithm in Maple [9].

2   The Equation Threshold

Definition 3. *Let L be of the form (1). Consider the set N of all integers n such that the equation $L(y) = 0$ has a truncated Laurent solution whose truncation degree is n. Let N be nonempty and have the maximal element. We will call this element the* threshold *of the equation $L(y) = 0$. If the set N contains arbitrarily large integers, then we say that the threshold of the equation is ∞. If this set is empty, then the threshold is conventionally $-\infty$.*

Remark 1. In Sect. 3 it will be, in particular, shown that if the set N considered in the previous definition is nonempty, then the subset of its negative elements is finite.

For demonstrating an example, we will need the notions of the series valuation and the prolongation of an equation.

Definition 4. *For a nonzero formal Laurent series* $a(x) = \sum a_i x^i \in K((x))$ *its valuation is defined as* $\operatorname{val} a(x) = \min\{i \mid a_i \neq 0\}$, *with* $\operatorname{val} 0 = \infty$. *A prolongation of the operator L of the form (1) is defined as an operator*

$$\tilde{L} = \sum_{i=0}^{r} b_i(x)\theta^i \in K[[x]][\theta]$$

such that $b_i(x) = a_i(x)$ *if all terms of $a_i(x)$ are known and*

$$b_i(x) - a_i(x) = O(x^{t_i+1}) \tag{5}$$

(i.e., $\operatorname{val}(b_i(x) - a_i(x)) > t_i$*) for truncated* $a_i(x)$, $i = 0, 1, \ldots, r$.

Example 1. Consider the equation

$$(1 + O(x))\theta y + a_0(x)y = 0, \tag{6}$$

where

$$a_0(x) = \sum_{j=k}^{\infty} a_{0j} x^j,$$

Set $k = \operatorname{val} a_0(x)$. For $k = 0$, i.e., for $a_{00} \neq 0$, truncated Laurent solutions exist only when a_{00} is an integer. We will consider the case

$$k = \operatorname{val} a_0(x) \geq 1. \tag{7}$$

Here, any prolongation of equation (6) has Laurent solutions. If the leading coefficient of this equation is $1 + \sum_{j=1}^{\infty} a_{1j} x^j$ then each Laurent solution is of the form

$$C\left(1 - \frac{a_{0k}}{k}x^k + \frac{a_{0k}a_{11} - a_{0,k+1} + a_{01}^2}{k+1}x^{k+1} + O(x^{k+2})\right),$$

where C is an arbitrary constant.

Thus, for equations of the form (6) with $\operatorname{val} a_0(x) > 0$, the coefficient of x^{k+1} in all nonzero Laurent solutions depends on coefficients of prolongation of the original equation (6). Consequently, the number $k = \operatorname{val} a_0(x)$ is the threshold.

If $\operatorname{val} a_0(x) = \infty$, in other words, if $a_0(x) = 0$, then all extensions of the equation (6) will have Laurent solutions $y(x) = C$, and the threshold of the equation (6) is $k = \operatorname{val} a_0(x) = \infty$.

Proposition 1. *There exists no algorithm that, for an arbitrary equation $L(y) = 0$ with an operator L of the form (1), finds out whether its threshold is finite or infinite.*

Proof. It is known that there is no algorithm that allows for an arbitrary series represented algorithmically to check whether this series is zero — this fact follows from the fundamental results of A. Turing [10]. If there was an algorithm that allows us to solve the problem formulated in the condition of the present proposition, then applying it to equation (6), in which the series $a_0(x)$ is represented algorithmically, would allow us to determine whether the series $a_0(x)$ is zero (first it is necessary to verify that the constant term of this series is zero: see the assumption made by us when considering Example 1 (7); if this constant term is not equal to zero, then, of course, the series is non-zero). The impossibility of this algorithm follows.

Corollary 1. *There is no algorithm that allows for an arbitrary equation $L(y) = 0$ with the operator L of form (1), to calculate the value (an integer or one of the symbols ∞, $-\infty$) of its threshold.*

Proposition 2. *Let $L(y) = 0$ be an equation with an operator L of form (1) and $k \in \mathbb{Z}$. It can be tested algorithmically whether k exceeds the threshold of the equation or not; if the answer is positive then the threshold h of this equation can be found. In addition, all such truncated Laurent solutions whose truncation degree does not exceed h can be constructed.*

Proof. We check the existence of invariant initial segments of solutions up to x^k by actually constructing these segments. The construction is considered as successful, if the resulting coefficients for powers of x do not include unknown coefficients of the prolongation of the equation (this approach was used in [5] when considering equations with all coefficients represented by truncated series). If it was not possible to reach x^k then, first, it was established that the value of k exceeds the threshold of the equation, and, second, it is possible to find the threshold value and find all the invariant initial segments of the Laurent solutions.

Remark 2. Thus, if the considered k is such that for a given equation $L(y) = 0$ with the operator L having the form (1) there does not exist truncated Laurent solutions of truncation degree k, then this circumstance opens, in particular, the opportunity of finding the threshold of the original equation — a quantity which is by Corollary 1 of the Proposition 1 non-computable algorithmically, in case if one is based only on the original equation.

3   Induced Recurrence Equations

Let σ denote the shift operator such that $\sigma c_n = c_{n+1}$ for any sequence (c_n). The transformation

$$x \to \sigma^{-1}, \quad \theta \to n$$

assigns to a differential equation

$$\sum_{i=0}^{r} a_i(x)\theta^i y(x) = 0, \tag{8}$$

where $a_i(x) \in K[[x]]$, the induced recurrent equation

$$u_0(n)c_n + u_{-1}(n)c_{n-1} + \cdots = 0. \tag{9}$$

Equation (8) has a Laurent solution $y(x) = c_v x^v + c_{v+1}x^{v+1} + \ldots$ if and only if the two-sided sequence $\ldots, 0, 0, c_v, c_{v+1}, \ldots$ satisfies equation (9) (see [4]). In our assumption, for the given operator (1), some of whose coefficients are truncated series, at least one of the constant terms of $a_0(x), \ldots, a_r(x)$ is not equal to zero. Thus,

$$u_0(n) = \sum_{i=0}^{r} a_{i,0} \, n^i \tag{10}$$

is a non-zero polynomial which is independent of any prolongations of the given operator L. It can be considered as a version of the *indicial* polynomial of the given equation. The finite set of integer roots of this polynomial contains all possible valuations v of Laurent solutions of all prolongations of the equation $L(y) = 0$.

If the polynomial $u_0(n)$ has no integer roots, then no prolongation of $L(y) = 0$ has nonzero Laurent solutions. In this case, set the threshold of the equation $L(y) = 0$ to be $-\infty$.

Let $\alpha_1 < \ldots < \alpha_s$ be all integer roots of the polynomial $u_0(n)$. Then, the set N from Definition 3 has no element which is less than α_1. All prolongations of the equation $L(y) = 0$ have Laurent solutions with valuation α_s (see, e.g., [5]). Thus, $\alpha_s \in N$, and as a consequence, the threshold is greater than or equal to α_s. The threshold is $-\infty$ if and only if the polynomial $u_0(n)$ has no integer root.

4   Computing Coefficients of Truncated Laurent Solutions

Computing elements of the sequence (c_n) of coefficients of Laurent solutions can be performed by successively increasing n by 1, starting with $n = \alpha_1$ which is the minimum integer root of the polynomial $u_0(n)$. Set $c_n = 0$ for $n < \alpha_1$. If $u_0(n) \neq 0$ for some integer n then (9) allows us to find c_n by $c_{n-1}, c_{n-2}, \ldots$ Since $c_n = 0$ when $n < \alpha_1$, relation (9) has a finite number of non-zero terms. If $u_0(n) = 0$, we declare c_n an *unknown constant*. The previously calculated $c_{n-1}, c_{n-2}, \ldots, c_{\alpha_1}$ satisfy the relation

$$u_{-1}(n)c_{n-1} + u_{-2}(n)c_{n-2} + \cdots + u_{-n+\alpha_1}(n)c_{\alpha_1} = 0. \tag{11}$$

These relations allow us to calculate the values of some previously introduced unknown constants. After the value of n exceeds the greatest integer root of $u_0(n)$, new unknown constants and relations of form (11) will not occur any longer.

If L has truncated coefficients, then it is possible that for some $n \geq \alpha_1$, the left-hand side of (9) depends on those unspecified coefficients that are hidden in (1) in the symbols O (some of coefficients of L may be of the form (2)). These unspecified coefficients will be called *literals*. For $u_0(n) \neq 0$, the calculated value of c_n depends on literals. For $u_0(n) = 0$, if relation (11) depends on literals then computing the previously introduced unknown constants is postponed until n reaches α_s. When $n = \alpha_s$, we obtain

(a) the values of the coefficients c_{α_1}, $c_{\alpha_1+1}, \ldots$, c_{α_s} (all of which depend on unknown constants, some of which may depend on literals as well);

(b) the set of unknown constants;

(c) the set of relations for unknown constants containing literals.

By the set (c) we can find values of unknown constants which are invariant to all prolongations of the given truncated equation (see [6] for details). We declare the unknown constants that did not get values involved into the Laurent solution of the differential equation the *arbitrary constants*.

5   Algorithm

Input data:

- a differential operator L of the form (1), whose each coefficient is either an algorithmically represented power series or a truncated power series,
- an integer number k.

Output result:

- The answer to the question of the existence of truncated Laurent solutions for the equation $L(y) = 0$. If there are no such solutions, then the output is the empty list [ ].
- If the answer to the question is positive then the algorithm computes all the truncated Laurent solutions, whose truncation degrees do not exceed k; it is possible that some solutions are computed with bigger truncation degree (such solutions are found by the algorithm due to the general computation strategy). If the algorithm finds out that k exceeds the threshold of the equation $L(y) = 0$ then the algorithm computes the value h of the threshold (see Remark 2) and constructs all the truncated Laurent solutions, whose truncation degrees do not exceed h.

The steps:

1. By (10), compute $u_0(n)$. Find the set

$$\alpha_1 < \cdots < \alpha_s$$

   of all integer roots of $u_0(n)$. If the set is empty then there are no truncated Laurent solutions; stop the work with the result [ ].

2. $d := \alpha_s - \alpha_1$; compute the coefficients

$$u_{-j}(n) := \sum_{i=0}^{r} a_{i,j}\,(n-j)^i, \quad j = 1, \ldots, \max\{d,\, k - \alpha_1\},$$

   of the induced recurrent equation (9).

3. Compute the coefficients c_n, $n = \alpha_1, \alpha_1 + 1, \ldots, \alpha_s$, of the truncated Laurent solution using (9) as it is described in Sect. 4.
4. If $k > \alpha_s$ then continue computing c_n using (9) with n subsequently increased by 1 while the both following conditions hold
   (a) $n \leq k$,
   (b) a non trivial set of the values of the arbitrary constants exists such that c_n is independent of the literals (it is detailed in [6, Sect. 4.1]).
   If (a) is true, but (b) is false for the current n then the threshold of the equation is computed as $h = n - 1$. In the latter case, substitute k by smaller value: $k := h$. Report the substitution with the value h.
5. Construct the list of all truncated Laurent solutions

$$c_v x^v + c_{v+1} x^{v+1} + \cdots + c_m x^m + O(x^{m+1}), \quad v \in \{\alpha_1, \ldots, \alpha_s\}, \quad m \leq k, \quad (12)$$

containing no literals as described in [6, Sect. 4.1]. (Some elements of the set $\{\alpha_1, \ldots, \alpha_s\}$ might be not used in the truncated Laurent solutions (12)).

6   Implementation; Examples of Use

We have implemented the algorithm in Maple [9] as an extension of `LaurentSolution` procedure from the package `TruncatedSeries` [7]. The first argument of the procedure is a differential equation $L(y) = 0$ where L is an operator of form (1). Previously, the procedure worked for the case where all the series, which are the coefficients of the equation, are represented as truncated series. Now it is also possible to represent them (or part of them) algorithmically. The application of θ^k to the unknown function $y(x)$ is written as `theta(y(x),x,k)`. The truncated coefficients of the equation, i.e., the coefficients of the form (2) are written as `a_i(x)+O(x^(t_i+1))`, where `a_i(x)` is a polynomial of the degree not higher than `t_i` over the field of algebraic numbers. Algorithmically represented series might be specified either as a polynomial or as a finite or infinite power series in integer powers of x, or as a sum of a polynomial and such power series. The power series is written in a usual Maple form as `Sum(f(i)*x^i,i=a..b)`, where `f(i)` is an expression or a function that implements an algorithm for computing the number coefficient of the series with the index `i`, the specified `a` and `b` are the lower and the upper bounds of summation, the upper bound might be infinite which is designated as `infinity`. The coefficients of both polynomials and series, as in the case of truncated series, are the elements of the field of algebraic numbers. Irrational algebraic numbers are represented in Maple as the expression `RootOf(p(_Z), index = k)`, where `p(_Z)` is an irreducible polynomial, whose k-th root is the given algebraic number. For example, `RootOf(_Z^2-2, index=2)` represents $-\sqrt{2}$. An unknown function of the equation is specified as the second argument of the procedure.

   Concerning the implementation of the algorithm from Sect. 5, the procedure has got two new optional parameters:

– `'top'=k` — where k is an integer number, for which it is needed to determine whether it exceeds the threshold of the given equation (by default, k equals the maximum integer root of the indicial polynomial if at least one coefficient of the equation is non-truncated and k equals the threshold otherwise);

– `'threshold'='h'` — where h specifies the name of the variable, which will be assigned to the value of the threshold in the case if it is computed, or to the value FAIL, if the threshold is not determined, i.e., if it exceeds the given value k.

The result of the procedure is a list of truncated Laurent solutions with different valuations. Each element of the list is represented as

$$c_{v_i}x^{v_i} + c_{v_i+1}x^{v_i+1} + \cdots + c_{m_i-1}x^{m_i-1} + O(x^{m_i}), \tag{13}$$

where v_i is the valuation for which the existence of a truncated Laurent solution is determined; m_i has the previous meaning, c_i are the calculated coefficients of the truncated Laurent solution, which can be linear combinations of arbitrary constants of the form $\_c_j$.

The implementation and a session of Maple with examples of using the procedure LaurentSolution are available at the address

http://www.ccas.ru/ca/truncatedseries

in the section "The next version of the procedure LaurentSolution".

Below we present six examples, which we combine into one, containing paragraphs 1–6.

Example 2

1. In the equation two coefficients are given as a truncated series and one coefficient is represented algorithmically as the sum of the polynomial and the power series:

```
> eq1 := (-1+x+x^2+O(x^3))*theta(y(x), x, 2)+
>        (-2+O(x^3))*theta(y(x), x, 1)+
>        (1+x+Sum(x^i/i!, i = 2 .. infinity))*y(x);
```

$$eq1 := \left(-1 + x + x^2 + O(x^3)\right)\theta(y(x), x, 2) + \left(-2 + O(x^3)\right)\theta(y(x), x, 1)$$

$$+ \left(1 + x + \sum_{i=2}^{\infty} \frac{x^i}{i!}\right)y(x)$$

```
> LaurentSolution(eq1, y(x), 'top' = 2, 'threshold' = h1_2);
```

$$[\,]$$

The output means that the equation has no truncated Laurent solutions. The threshold:

```
> h1_2;
```

$$-\infty$$

The value of the threshold is $-\infty$, and it confirms that there are no truncated Laurent solutions.

2. The equation has both truncated and algorithmically represented coefficients. The separate function f is used to specify power series:

```
> f := (i -> i^2+2*i+1-(i+1)^2):
> eq2 := (-1+x+x^2+O(x^3))*theta(y(x), x, 2)+
>         (-2+O(x^3))*theta(y(x), x, 1)+
>         (Sum(f(i)*x^i, i = 0 .. infinity))*y(x);
```

$$eq2 := \left(-1 + x + x^2 + O(x^3)\right)\theta(y(x), x, 2) + \left(-2 + O(x^3)\right)\theta(y(x), x, 1)$$

$$+ \left(\sum_{i=0}^{\infty}(i^2 + 2i + 1 - (i+1)^2)x^i\right)y(x)$$

```
> LaurentSolution(eq2, y(x), 'top' = 2, 'threshold' = h2_2);
```

$$\left[\frac{\_c_1}{x^2} - \frac{4\_c_1}{x} + \_c_2 + O(x), \_c_2 + O(x^3)\right]$$

The truncated Laurent solutions with valuations -2 and 0 and with different truncation degrees are found. The threshold:

```
> h2_2;
```

$$\text{FAIL}$$

It means that the given value $k = 2$ does not exceed the threshold.

Apply the procedure to the given equation again with $k = 5$:

```
> LaurentSolution(eq2, y(x), 'top' = 5, 'threshold' = h2_5);
```

$$\left[\frac{\_c_1}{x^2} - \frac{4\_c_1}{x} + \_c_2 + O(x), \_c_2 + O(x^6)\right]$$

It is seen that the truncated solution with the valuation -2 is not changed, and the truncation degree of the one with the valuation 0 is increased. The threshold:

```
> h2_5;
```

$$\text{FAIL}$$

It means that the given value $k = 5$ does not exceed the threshold. The function f, which is used to specify the series coefficient of $y(x)$, computes 0 coefficient for any index value i. Therefore, the coefficient of $y(x)$ equals 0. The threshold of the equation is ∞, and any value k will not exceed the threshold. Note that the zero series might be specified just as the polynomial 0, or the term with $y(x)$ might be absent in the equation.

3. The equation has also both truncated and algorithmically represented coefficients. The algorithmically represented coefficient of $y(x)$ is written as the polynomial:

```
> eq3 := (-1+x+x^2+O(x^3))*theta(y(x), x, 2)+
```

```
>        (-2+O(x^3))*theta(y(x), x, 1)+(x+6*x^2)*y(x);
```

$$eq3 := \left(-1 + x + x^2 + O(x^3)\right)\theta(y(x), x, 2) + \left(-2 + O(x^3)\right)\theta(y(x), x, 1)$$
$$+ \left(x + 6x^2\right)y(x)$$

```
> LaurentSolution(eq3, y(x), 'top' = 2, 'threshold' = h3_2);
```

$$\left[\frac{\_c_1}{x^2} - \frac{5\_c_1}{x} + \_c_2 + O(x), \_c_2 + \frac{1}{3}x\_c_2 + \frac{5}{6}x^2\_c_2 + O(x^3)\right]$$

The truncated Laurent solutions with valuations -2 and 0 and with different truncation degrees are found again. The threshold:

```
> h3_2;
```

$$\text{FAIL}$$

It means that the given value $k = 2$ does not exceed the threshold.

Apply the procedure to the given equation with $k = 5$:

```
> LaurentSolution(eq3, y(x), 'top' = 5, 'threshold' = h3_5);
```

$$\left[\frac{\_c_1}{x^2} - \frac{5\_c_1}{x} + \_c_2 + O(x), \_c_2 + \frac{1}{3}x\_c_2 + \frac{5}{6}x^2\_c_2 + \frac{13}{30}x^3\_c_2 + O(x^4)\right]$$

The threshold:

```
> h3_5;
```

$$3$$

It is seen that the threshold is achieved in the computed truncated solutions with the valuation 0.

4. The equation is a prolongation of the equation eq3:

```
> eq4 := (-1+x+x^2+9*x^3+O(x^4))*theta(y(x), x, 2)
         +(-2+(x^3)/2+O(x^4))*theta(y(x), x, 1)+(x+6*x^2)*y(x);
```

$$eq4 := \left(-1 + x + x^2 + 9x^3 + O(x^4)\right)\theta(y(x), x, 2)$$
$$+ \left(-2 + \frac{1}{2}x^3 + O(x^4)\right)\theta(y(x), x, 1) + \left(x + 6x^2\right)y(x)$$

```
> LaurentSolution(eq4, y(x), 'top' = 5, 'threshold' = h4_5);
```

$$\left[\frac{\_c_1}{x^2} - \frac{5\_c_1}{x} + \_c_2 + \frac{1}{3}x\_c_2 + O(x^2), \_c_2 + \frac{1}{3}x\_c_2 + \frac{5}{6}x^2\_c_2 + \frac{13}{30}x^3\_c_2 \right.$$
$$\left. + \frac{95}{144}x^4\_c_2 + O(x^5)\right]$$

The truncated Laurent solutions with valuations -2 and 0 and with different truncation degrees are found again. These truncated solutions are the prolongations of the computed truncated solutions of the equation eq3. The threshold:

```
> h4_5;
```

$$4$$

It is seen that the threshold is achieved in the computed truncated solutions with the valuation 0.

5. The equation is another prolongation of the equation eq3:

```
> eq5 := (-1+x+x^2+RootOf(z^2-2, z, index = 2)*x^3+O(x^4))*
> theta(y(x), x, 2)+(-2+2*RootOf(z^2-2, z, index = 2)*x^3+
> O(x^4))*theta(y(x), x, 1)+(x+6*x^2)*y(x);
```

$$eq5 := \left(-1 + x + x^2 + \mathrm{RootOf}\left(\_Z^2 - 2, index = 2\right)x^3 + O\left(x^4\right)\right)\theta(y(x), x, 2)$$

$$+\left(-2 + 2\mathrm{RootOf}\left(\_Z^2 - 2, index = 2\right)x^3 + O\left(x^4\right)\right)\theta(y(x), x, 1)$$

$$+\left(x + 6x^2\right)y(x)$$

```
> LaurentSolution(eq5, y(x), 'top' = 5, 'threshold' = h5_5);
```

$$\left[\frac{\_c_1}{x^2} - \frac{5\_c_1}{x} + \_c_2 + x\left(\frac{1}{3}\_c_2 - \frac{35}{3}\_c_1\right) + O\left(x^2\right), \_c_2 + \frac{1}{3}x\_c_2 + \frac{5}{6}x^2\_c_2\right.$$

$$\left. + \frac{13}{30}x^3\_c_2 + x^4\left(\frac{19}{36}\_c_2 + \frac{1}{24}\mathrm{RootOf}\left(\_Z^2 - 2, index = 2\right)\_c_2\right) + O\left(x^5\right)\right]$$

The truncated Laurent solutions with valuations -2 and 0 and with different truncation degrees are found again. These truncated solutions are the prolongations of the computed truncated solutions of the equation eq3, but are different from the computed truncated solutions of the equation eq4. The threshold:

```
> h5_5;
```

$$4$$

It is seen that the threshold is achieved again in the computed truncated solutions with the valuation 0.

The results of the application of the procedure to the equations eq4 and eq5 show that the earlier computed truncated Laurent solutions of the equation eq3 contain the maximum possible number of initial terms, since two different prolongations of the equation eq3 have different truncated Laurent solutions, which are the prolongations of the found truncated solutions of the equation eq3.

6. The equation is a prolongation of the equation eq3 as well, and all its coefficients are represented algorithmically:

```
> eq6 := (-1+x+x^2+Sum((-1)^i*x^i/i!, i = 3 .. infinity))*
> theta(y(x), x, 2)+(-2+2*(Sum((-1)^i*x^i/i!, i = 3 ..
> infinity)))*theta(y(x), x, 1)+(x+6*x^2)*y(x);
```

$$eq6 := \left(-1 + x + x^2 + \sum_{i=3}^{\infty} \frac{(-1)^i x^i}{i!}\right)\theta(y(x), x, 2)$$

$$+\left(-2+2\left(\sum_{i=3}^{\infty}\frac{(-1)^i x^i}{i!}\right)\right)\theta(y(x),x,1)+\left(x+6x^2\right)y(x)$$

```
> LaurentSolution(eq6, y(x), 'top' = 5, 'threshold' = h6_5);
```

$$\left[\frac{\_c_1}{x^2}-\frac{5\_c_1}{x}+\_c_2+x\left(\frac{1}{3}\_c_2-\frac{35}{3}\_c_1\right)\right.$$

$$+x^2\left(\frac{5}{6}\_c_2-\frac{145}{48}\_c_1\right)+x^3\left(\frac{13}{30}\_c_2-\frac{103}{16}\_c_1\right)+x^4\left(\frac{25}{48}\_c_2-\frac{2131}{576}\_c_1\right)$$

$$+x^5\left(\frac{2057}{5040}\_c_2-\frac{4303}{960}\_c_1\right)+O\left(x^6\right),\_c_2+\frac{1}{3}x\_c_2+\frac{5}{6}x^2\_c_2$$

$$\left.+\frac{13}{30}x^3\_c_2+\frac{25}{48}x^4\_c_2+\frac{2057}{5040}x^5\_c_2+O\left(x^6\right)\right]$$

The two truncated Laurent solutions with the same truncation degree are found, which are the prolongations of the computed truncated solutions of the equation eq3. The threshold:

```
> h6_5;
```

FAIL

Thus, $k=5$ does not exceed the threshold. The case when all the coefficients are represented algorithmically is the case when the threshold of the equation is ∞, and any value k does not exceed the value of the threshold.

7 Concluding Remarks

This study is a continuation of the studies started in [2,7], in which it was assumed that either all the coefficients of a differential equation are represented algorithmically, and in this sense, are given completely, or are represented in the truncated form. In the current paper, the presence of both types of coefficients is allowed.

The presence of infinite series in the input data of a problem is a source of difficulties (the algorithmic impossibility of answering certain natural questions). This is, e.g., related to the fact that if sequences of coefficients of series can be specified by arbitrary algorithms, then it is impossible to test algorithmically the equality of such series to zero (this is a consequence of the classical results of A. Turing on the undecidability of the problem of terminating of an algorithm [10]).

There is nowhere to go from this in the problem considered above, — see Proposition 1. However, along with this, we must admit that in the situation we are faced, in a certain sense, with a lighter version of the algorithmic undecidability. This undecidability is, so to speak, not too burdensome.

Indeed, we cannot indicate the greatest degree of the truncated Laurent solution existing for a given equation (the threshold of the equation). However, if we are interested in all solutions of a truncation degree not exceeding a given integer k then the algorithm proposed in Sect. 5 allows us to construct all of them.

It would be interesting to try to obtain similar results for the solutions of a more general form — the so-called regular and exponentially logarithmic solutions, and generalize this to the systems of differential equations. We will continue to investigate this line of enquiry.

Acknowledgments. The authors are grateful to anonymous referees for their helpful comments.

References

1. Abramov, S., Barkatou, M.: Computable infinite power series in the role of coefficients of linear differential systems. In: Gerdt, V.P., Koepf, W., Seiler, W.M., Vorozhtsov, E.V. (eds.) CASC 2014. LNCS, vol. 8660, pp. 1–12. Springer, Cham (2014). https://doi.org/10.1007/978-3-319-10515-4_1
2. Abramov, S., Barkatou, M., Khmelnov, D.: On full rank differential systems with power series coefficients. J. Symb. Comput. **68**, 120–137 (2015)
3. Abramov, S., Barkatou, M., Pflügel, E.: Higher-order linear differential systems with truncated coefficients. In: Gerdt, V.P., Koepf, W., Mayr, E.W., Vorozhtsov, E.V. (eds.) CASC 2011. LNCS, vol. 6885, pp. 10–24. Springer, Berlin, Heidelberg (2011) https://doi.org/10.1007/978-3-642-23568-9_2
4. Abramov, S., Bronstein, M., Petkovšek, M.: On polynomial solutions of linear operator equations. In: ISSAC '95. Proc. 1995 International Symposium on Symbolic and Algebraic Computation, pp. 290–296. ACM Press, New York (1995)
5. Abramov, S., Khmelnov, D., Ryabenko, A.: Linear ordinary differential equations and truncated series. Comput. Math. Math. Phys. **49**(10), 1649–1659 (2019)
6. Abramov, S., Khmelnov, D., Ryabenko, A.: Regular solutions of linear ordinary differential equations and truncated series. Comput. Math. Math. Phys. **60**(1), 2–15 (2020)
7. Abramov, S., Khmelnov, D., Ryabenko, A.: Procedures for searching Laurent and regular solutions of linear differential equations with the coefficients in the form of truncated power series. Progr. and Comp. Soft. **46**(2), 67–75 (2020)
8. Lutz, D.A., Schäfke, R.: On the identification and stability of formal invariants for singular differential equations. Linear Algebra and Its Applications **72**, 1–46 (1985)
9. Maple online help, http://www.maplesoft.com/support/help/
10. Turing, A.: On computable numbers, with an application to the Entscheidungsproblem. London Math. Soc. **42**(2), 230–265 (1937)

On Periodic Approximate Solutions of the Three-Body Problem Found by Conservative Difference Schemes

Edic A. Ayryan[1]($\boxtimes$)(iD), Mikhail D. Malykh[2](iD), Leonid A. Sevastianov[1,2](iD), and Yu Ying[2,3](iD)

[1] Joint Institute for Nuclear Research (Dubna), Joliot-Curie, 6,
Dubna, Moscow Region 141980, Russia
ayrjan@jinr.ru

[2] Department of Applied Probability and Informatics,
Peoples' Friendship University of Russia (RUDN University),
6 Miklukho-Maklaya St, Moscow 117198, Russia
{malykh_md,sevastianov_la}@rudn.university, yingy6165@gmail.com

[3] Department of Algebra and Geometry, Kaili University, 3 Kaiyuan Road,
Kaili 556011, Russia

Abstract. The possibility of using implicit difference schemes to study the qualitative properties of solutions of dynamical systems, primarily the periodicity of the solution, is discussed.

An implicit difference scheme for the many-body problem that preserves all algebraic integrals of motion is presented based on the midpoint scheme. In this concern, we consider the finite-difference analogue of the Lagrange problem: using the midpoint scheme to find all approximate solutions of the three-body problem on a plane in which the distances between the bodies do not change. It is shown that this problem can be solved by purely algebraic methods. Two theorems are proved that reduce this problem to the study of the midpoint scheme properties for a system of coupled oscillators. It is shown that in the case when the bodies form a regular triangle, the approximate solution inherits the periodicity property of the exact Lagrange solution.

Keywords: Finite difference method · Algebraic integrals of motion · Dynamical system

Introduction

Algebraic methods for studying dynamic systems are traditionally used in problems of symbolic integration of such systems. Among the most interesting results

The computations presented in the paper were performed in Sage computer algebra system (www.sagemath.org). The contribution of E. A. Ayryan (Investigation), M. D. Malykh (Investigation, proofs of theorems), and L. A. Sevastianov (Conceptualization, writing) is supported by the Russian Science Foundation (grant no. 20-11-20257).

© Springer Nature Switzerland AG 2020
F. Boulier et al. (Eds.): CASC 2020, LNCS 12291, pp. 77–90, 2020.
https://doi.org/10.1007/978-3-030-60026-6_5

of the application of algebraic methods to dynamical systems are the Singer theorems on elementary integrals [5, 15]. However, it is well known that the vast majority of dynamical systems cannot be integrated in elementary functions and numerical methods are used to study them alongside with a qualitative analysis of the solution behavior.

The standard numerical method for studying dynamical systems suggests replacing the derivatives in differential equations with finite differences, and thus reduces the numerical analysis of a dynamical system to solving systems of algebraic equations. Traditionally, the focus of numerical analysis is on the accuracy of approximation, the proximity of an approximate and exact solution, and its stability with respect to rounding errors [12]. Despite the importance of these questions, one cannot fail to notice that the finite difference method produces algebraization of the problem, that is, it brings the problem to the form most convenient for using purely algebraic tools.

The ultimate goal of our research is to clarify how to use the finite difference method for a qualitative analysis of dynamical systems. We believe that the approach that provides a simple and effective tool for the approximate calculation of solution parameters can be no less effective in a qualitative analysis.

For inheritance by an approximate solution of the properties of an exact solution, it is extremely important that the approximate solution is found using conservative schemes, i.e., schemes that preserve all algebraic integrals of motion. We made sure that in the general case, such schemes should be implicit [1]. We use numerical calculations using implicit schemes only to illustrate theoretical results, being fully aware of the difficulties encountered in developing effective numerical methods based on implicit schemes [17]. It is likely that explicit schemes are good for developing purely numerical methods, and implicit ones for combining numerical and symbolic methods.

First of all, we considered the simplest example, namely, a linear oscillator and showed that the midpoint scheme not only preserves all the algebraic integrals of this system, but also gives periodic approximate solutions (Example 3). This result can be transferred to a system of coupled oscillators, since the matrix of this system can always be reduced to a diagonal form. One class of solutions traditionally associated with the name of Lagrange, can be obtained by assuming that all the distances a_{ij} are equal to one constant a [4]. The next natural step is to consider the inheritance of periodicity in nonlinear problems that have a lot of algebraic integrals but are not reducible to quadratures. Therefore, as an object of study in the present paper, we chose periodic solutions of the planar three-body problem.

In Sect. 1 we introduce the necessary notation. Inheritance of the exact solution properties is discussed in Sect. 2, where we investigate a linear oscillator. Section 3 describes our finite difference schemes preserving the integrals of the three-body problem. In Sect. 4 we formulate the analogue of Lagrange problem for approximate solutions. One class of its solutions (triangular solutions) is described in Sect. 5.

1   Basic Definitions

Consider a dynamical system

$$\dot{x} = f(x),\qquad(1)$$

where $x = (x_1, \ldots x_n)$, and f is a rational function of its argument. Any finite-difference scheme for this system is a system (S) of algebraic equations that determine the relation between the value of the variable x, taken at a certain moment of time t, and the value $\hat{x}$ of this variable taken at the moment of time $t + dt$. In this case, $x, \hat{x}$ can be considered as two lists of symbolic variables $[x_1, \ldots, x_n]$ and $[\hat{x}_1, \ldots, \hat{x}_n]$. Even the step dt can be treated as a symbolic variable. It is quite acceptable to restrict ourselves to a case when the coefficients of this scheme are rational numbers. In this case the investigation of the scheme (S) can be performed using the toolkit developed for operating with the ideals of polynomial rings over the field $\mathbb{Q}$.

Example 1. An explicit Euler scheme for the system

$$\dot{x} = -y, \quad \dot{y} = x,\qquad(2)$$

describing a harmonic oscillator, is written as a system of two algebraic equations:

$$\hat{x} - x = -ydt, \quad \hat{y} - y = xdt$$

Similar to Ref. [1], we do not require that at $dt \to 0$ the system should turn into a differential equation. Instead, we assume a purely algebraic condition: the system (S) allows a solution for $\hat{x}$ in the form of Puiseux series in powers of dt, and

$$\hat{x} = x + f(x)dt + \ldots .$$

The conservation of the integral $I(x)$ of the system (1) means that from the system (S), the equation

$$I(\hat{x}) = I(x)$$

follows. We will call the system (S) conservative, if it preserves all algebraic integrals of motion of the considered dynamical system.

Example 2. The system (2) has an algebraic integral

$$x^2 + y^2 = C.\qquad(3)$$

The standard Euler scheme does not preserve it, while the midpoint scheme

$$\hat{x} - x = -(\hat{y} + y)\frac{dt}{2}, \quad \hat{y} - y = (\hat{x} + x)\frac{dt}{2}\qquad(4)$$

preserves it according to Cooper theorem [2, 11].

Now let us take algebraic numbers for the coordinate of the point x_0 and the step dt. We will understand the approximate solution of the initial-value problem

$$\frac{dx}{dt} = f(x), \quad x(0) = x_0, \tag{5}$$

found according to the finite-difference scheme (S) as a finite or infinite sequence $\{x_0, x_1, x_2, \dots\}$, whose elements are determined recursively. To find x_{n+1}, one has to do the following:

- substitute $x = x_n$ into the system (S),
- find the solution $\hat{x}$ of this system that tends to x_n when $dt \to 0$, and accept this solution for x_{n+1}.

In this treatment, we will interpret x_n as approximate solutions of the initial-value Problem (5) at point $t = ndt$. It is quite possible that at the chosen value of dt, the desired root x_n does not exist. Then we truncate the sequence at the nth step and say that for $t = ndt$ there is a singular point of the approximate solution.

Quantitatively, the relation between the approximate solution of the Problem (5) and the exact one has always been in the focus of researchers' attention [12]. Let $x = x(t)$ be an exact solution considered on a segment $0 \le t \le T$ and $dt = T/N$, $N \in \mathbb{N}$. Then, under the above assumptions about the scheme, there are such constants M and λ that

$$\|x_n - x(ndt)\| \le M e^{\lambda T} dt.$$

We now proceed to the subject of inheritance of the exact properties by an approximate solution.

2   Inheritance of the Exact Solution Properties

One of the most noticeable qualitative properties of an exact solution is its periodicity. Can an approximate solution inherit it? To understand how to determine the concept of periodicity of an approximate solution, consider the simplest example.

Example 3. The exact solution of the system (2), first, has a period of 2π and, second, describes a circle (3) on the phase plane xy. The approximate solution found using the Euler scheme, does not inherit these properties. The sequence $x_n^2 + y_n^2$ grows monotonically, and the points (x_n, y_n) lie on an expanding spiral, see Fig. 1. On the contrary, the solution found using the midpoint scheme (4) inherits the above properties. From the conservation of the quadratic integral $x^2 + y^2 = C$ it immediately follows that the points (x_n, y_n) lie on a circle. If we take for α the minimal positive root of the equation

$$(1 + i\alpha)^{2N} = (1 + \alpha^2)^N, \tag{6}$$

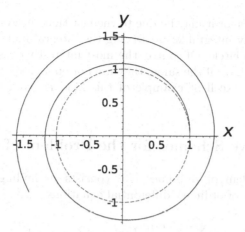

Fig. 1. Solution of the intial-value problem for Eq. (2) using the Euler method (solid) and the midpoint method (dashed), $\Delta t = 0.1$, 100 steps are made

which in terms of trigonometric functions can be expressed as

$$\alpha = \tan \frac{\pi}{N},$$

then the calculation according to the midpoint finite-difference scheme (4) with the step

$$\Delta t = 2\alpha = 2 \tan \frac{\pi}{N},$$

in N steps leads to the initial values of x, y. The proof of this theorem is presented in [4, th. 2].

This example prompts how to define the notion of period in a general case. The approximate solution $\{x_0, x_1, \ldots\}$, found using the scheme (S) with a certain step dt, will be called periodic if this sequence is periodic, i.e., if there exists such a natural number N that $x_N = x_0$. The number Ndt will be referred to as the period of this solution. We emphasize that when the step dt is changed the sequence $\{x_0, x_1, \ldots\}$ is no more periodic.

Example 4. The midpoint scheme (4) yields periodic solutions at a number of step values that form a descending sequence converging to zero. The corresponding sequence of periods

$$N\Delta t = 2N \tan \frac{\pi}{N} = 2\pi + \frac{2\pi^3}{3} \frac{1}{N^2} + \ldots$$

converges to the exact solution period 2π.

As can be seen, in the case of the simplest dynamical system, the approximate solution found by the conservative scheme inherits the basic qualitative properties of the exact solution. This circumstance may be a general property of conservative schemes, or it may be due to the fact that the system under consideration is completely integrable: preserving the integral makes all points of the phase plane lie on a circle. Of course, the most interesting cases are when there are many integrals, but all the same, they are not enough to reduce the problem to quadratures. An excellent example of this kind is given by the problem of many bodies.

3   Conservative Schemes for the Problem of Many Bodies

The classical problem of n bodies [13] consists in finding solutions of the autonomous system of ordinary differential equations

$$m_i \ddot{\boldsymbol{r}}_i = \sum_{j=1, j \neq i}^{n} \gamma \frac{m_i m_j}{r_{ij}^3} (\boldsymbol{r}_j - \boldsymbol{r}_i), \quad i = 1, \ldots, n \tag{7}$$

Here $\boldsymbol{r}_i$ is the position vector of the i-th body, and r_{ij} is the distance between the i-th and j-th body. For brevity we denote the velocity components of the i-th body as $\dot{x}_i = u_i$, $\dot{y}_i = v_i$ and $\dot{z}_i = w_i$ and the appropriate velocity vector as $\boldsymbol{v}_i$. The first finite-difference scheme for the many-body problem, preserving all classical integrals of motion, was proposed in 1992 by Greenspan [7–10] and independently in somewhat different form by J.C. Simo and O. González [6,14].

Our approach is close to the invariant energy quadratization method (IEQ method) which was used by Zhang et al. [17] to conserve the energy at the discretization of Hamiltonian systems including Kepler two-body problem. Like in IEQ method we "transform the energy into a quadratic form of a new variable via a change of variables" [17]. We introduce the following additional variables

$$r_{ij} = \sqrt{(x_i - x_j)^2 + (y_i - y_j)^2 + (z_i - z_j)^2}, \quad i, j = 1, 2, \ldots, N, i < j$$

and

$$\rho_{ij} = \frac{1}{r_{ij}}, \quad i, j = 1, 2, \ldots, N, i < j.$$

We arrive at a system of ordinary differential equations that incorporates the system for coordinates

$$\dot{\boldsymbol{r}}_i = \boldsymbol{v}_i, \quad i = 1, \ldots, n, \tag{8}$$

the system for velocities

$$m_i \dot{\boldsymbol{v}}_i = \sum_{j=1}^{n} \gamma \frac{m_i m_j \rho_{ij}}{r_{ij}^2} (\boldsymbol{r}_j - \boldsymbol{r}_i), \quad i = 1, \ldots, n, \tag{9}$$

the system for distances

$$\dot{r}_{ij} = \frac{1}{r_{ij}} (\boldsymbol{r}_i - \boldsymbol{r}_j) \cdot (\boldsymbol{v}_i - \boldsymbol{v}_j), \quad i, j = 1, \ldots, n; i \neq j, \tag{10}$$

and the system for inverse distances

$$\dot{\rho}_{ij} = -\frac{\rho_{ij}}{r_{ij}^2}(\boldsymbol{r}_i - \boldsymbol{r}_j) \cdot (\boldsymbol{v}_i - \boldsymbol{v}_j), \quad i,j = 1,\ldots,n; \; i \neq j. \tag{11}$$

This system possesses all algebraic integrals of a many-body problem and additional integrals

$$r_{ij}^2 - (x_i - x_j)^2 - (y_i - y_j)^2 - (z_i - z_j)^2 = \text{const}, \quad i \neq j \tag{12}$$

and

$$r_{ij}\rho_{ij} = \text{const}, \quad i \neq j. \tag{13}$$

The solutions of this extended system, on which

$$r_{ij}\rho_{ij} = 1, \quad i \neq j, \tag{14}$$

and

$$r_{ij}^2 - (x_i - x_j)^2 - (y_i - y_j)^2 - (z_i - z_j)^2 = 0, \quad i \neq j \tag{15}$$

correspond to the solutions of the many-body problem.

According to the Cooper theorem, the midpoint scheme

$$\frac{\hat{x} - x}{dt} = f\left(\frac{\hat{x} + x}{2}\right), \tag{16}$$

written for this extended system, preserves all quadratic integrals of motion exactly, therefore, it is a conservative scheme for the many-body problem. To investigate the inheritance by the approximate solution of the properties of the exact solution, we consider the particular periodic solutions. For the many-body problem many such solutions have been found, and we will consider the simplest family of such solutions, discovered as early as by Euler and Lagrange [13].

4   Lagrange Problem

Lagrange found all the exact solutions to the three-body problem at which the distances between the bodies do not change. These solutions fall into two families. In the first case, the bodies form a regular triangle, which rotates around its center of gravity with a constant angular velocity. In the second case, the bodies lie on one straight line that rotates around the center of gravity. Conventionally, the first case is associated with the name of Lagrange, and the second with the name of Euler [13]. We will find out if there are approximate solutions that inherit these properties.

Problem 1. Using the midpoint scheme, find all approximate solutions of the planar three-body problem (8)–(11), in which the distances between the bodies are unchanged

$$\hat{r}_{ij} = r_{ij}, \quad i \neq j, \tag{17}$$

and the constraint integrals have their natural values

$$r_{ij}^2 - (x_i - x_j)^2 - (y_i - y_j)^2 = 0, \quad r_{ij}\rho_{ij} = 1. \tag{18}$$

In theory, this problem is algorithmically solvable; moreover, an algorithm for eliminating the unknowns based on the calculation of Gröbner bases and implemented in any computer algebra system is quite enough to solve it. Indeed, we add to the system of algebraic equations (S) defining a midpoint scheme for differential equations (8)–(11), 9 more equations (17) and (18). Our system comprises $3 \cdot 4 + 3 \cdot 2 = 18$ variables, to which we must also add the 19-th one, dt. In the 19-dimensional affine space these equations define a certain algebraic set M. Let us denote its projection on the 10-dimensional space of initial values and dt by V and the projection on the 10-dimensional space of the finite values and dt by $\hat{V}$. To make the approximate solution satisfy the conditions of the Problem 1, it is necessary that at each step its points belong to both V and $\hat{V}$. This condition is close to the sufficient one, if we specially exclude the case of singular points.

Unfortunately, we could not apply the standard computer algebra tools implemented in Sage to find the manifolds V and $\hat{V}$. Probably, the number of variables is still too large for calculations on a common computer (see the Appendix 5). Therefore, the problem will have to be solved partially by hand.

Assume from the beginning, that the solution $x_0, x_1, \ldots$, satisfying the conditions of the Problem 1, exists. Then r_{ij} and $\hat{r}_{ij}$ are equal to one and the same number; we denote it by a_{ij}, i.e., we assume

$$r_{ij} = a_{ij}, \quad \rho_{ij} = \frac{1}{a_{ij}}. \tag{19}$$

In this case the coordinates and velocities can be found using a simpler scheme. The projections of the coordinates and velocities on the Ox-axis are described by the scheme

$$\begin{cases} \hat{x}_i - x_i = \dfrac{dt}{2}(\hat{u}_i + u_i), & i = 1, 2, 3 \\[2mm] m_i(\hat{u}_i - u_i) = \dfrac{\gamma dt}{2} \displaystyle\sum_{j \neq i} \dfrac{m_i m_j}{a_{ij}^3}(\hat{x}_j + x_j - \hat{x}_i - x_i), & i = 1, 2, 3 \end{cases} \tag{20}$$

An analogous scheme is obtained for the projection on the Oy-axis

$$\begin{cases} \hat{y}_i - y_i = \dfrac{dt}{2}(\hat{v}_i + v_i), & i = 1, 2, 3 \\[2mm] m_i(\hat{v}_i - v_i) = \dfrac{\gamma dt}{2} \displaystyle\sum_{j \neq i} \dfrac{m_i m_j}{a_{ij}^3}(\hat{y}_j + y_j - \hat{y}_i - y_i), & i = 1, 2, 3 \end{cases} \tag{21}$$

These equations define the midpoint scheme for the dynamical system

$$\begin{cases} \dot{z}_i = w_i, & i = 1, 2, 3 \\[2mm] m_i \dot{w}_i = -\dfrac{\partial U}{\partial z_i}, & i = 1, 2, 3, \end{cases} \tag{22}$$

where the potential is given by the expression

$$U = \frac{\gamma m_1 m_2}{2 a_{12}^3}(z_1 - z_2)^2 + \frac{\gamma m_1 m_3}{2 a_{13}^3}(z_1 - z_3)^2 + \frac{\gamma m_2 m_3}{2 a_{23}^3}(z_2 - z_3)^2.$$

This observation allows replacing the finite-difference schemes (20) and (21) with much more common system of differential equations (22), describing oscillations in a system of coupled oscillators.

Theorem 1. *If the Problem 1 allows a solution, then the projections of the coordinates and velocities onto the axes Ox and Oy can be found as solutions of the Problem (22), found using the midpoint scheme.*

Due to (17), the equations for the distances yield three additional equations

$$(\hat{x}_j + x_j - \hat{x}_i - x_i)(\hat{u}_j + u_j - \hat{u}_i - u_i) + (\hat{y}_j + y_j - \hat{y}_i - y_i)(\hat{v}_j + v_j - \hat{v}_i - v_i) = 0 \quad (23)$$

The equations for reciprocal distances with constraint (18) taken into account yield the same three equations. To clarify the role of these relations in Sage is already an easy problem. We have expressed from the linear systems (20) and (21) the velocities via the coordinates $x_1, \ldots, \hat{x}_1, \ldots$. Then we have substituted these equations into (23), and as a result, we have obtained again equalities (18). From this a theorem inverse to the Theorem 1 immediately follows.

Theorem 2. *Let a pair of solutions $\{z_i = x_i, w_i = u_i\}$ and $\{z_i = y_i, w_i = v_i\}$ to the Problem (22) is found using the midpoint scheme and is constrained by three equations*

$$(x_i - x_j)^2 + (y_i - y_j)^2 = a_{ij}^2, \quad i \neq j. \tag{24}$$

Then it may be raised to an approximate solution of the Lagrange problem, accepting (19).

We emphasize that constraints (24) are imposed on the approximate solution rather than on the exact one. The complete solution of Problem 1 is somewhat cumbersome because of a large number of complex solutions ignored in mechanics [13], therefore, here we restrict ourselves to an expressive particular case.

5   Triangular Solution

One class of solutions traditionally associated with the name of Lagrange can be obtained by assuming that all the distances a_{ij} are equal to one constant a. Then system (22) can be rewritten as a system of three equations

$$\begin{cases} \dfrac{d^2(z_2 - z_1)}{dt^2} + \omega^2(z_2 - z_1) = 0, \\[2mm] \dfrac{d^2(z_3 - z_1)}{dt^2} + \omega^2(z_3 - z_1) = 0, \\[2mm] \dfrac{d^2(m_1 z_1 + m_2 z_2 + m_3 z_3)}{dt^2} = 0. \end{cases} \tag{25}$$

Here an auxiliary variable has been introduced

$$\gamma \frac{m_1 + m_2 + m_3}{a^3} = \omega^2, \tag{26}$$

that has the meaning of the frequency squared. Therefore, this system possesses three quadratic integrals

$$(w_i - w_j)^2 + \omega^2(z_i - z_j)^2 = C_{ij},$$

exactly preserved by the midpoint scheme. Of course, only two of them are independent.

For any solution $\{z_i = x_i, w_i = u_i\}$

$$(u_i - u_j)^2 + \omega^2(x_i - x_j)^2 = C_{ij}.$$

To obtain from them the relations (24), we define the triple y_1, y_2, y_3 as a solution to a system of linear equations

$$\begin{cases} u_2 - u_1 = \omega(y_2 - y_1), \\ u_3 - u_1 = \omega(y_3 - y_1), \\ m_1 u_1 + m_2 u_2 + m_3 u_3 = \omega(m_1 y_1 + m_2 y_2 + m_3 y_3). \end{cases}$$

Velocities v_1, v_2, v_3 are defined to ensure the validity of the following relations

$$\begin{cases} -\omega(x_2 - x_1) = v_2 - v_1, \\ -\omega(x_3 - x_1) = v_3 - v_1, \\ m_1 v_1 + m_2 v_2 + m_3 v_3 = 0, \end{cases}$$

which in the case of differential equations are obtained as derivatives of the preceding equalities. As a result, we arrive at the set $\{z_i = y_i, w_i = v_i\}$ that satisfies the system (22) and is coupled with the first set by the relation (24). This means that in the present case, all conditions of Theorem 2 are satisfied.

Corollary 1. *There exists a family of approximate solutions to the three-body problem, on which the bodies form a regular triangle with the constant side a. This solution can be raised from two solutions of a linear dynamical system (22) found using the midpoint scheme.*

Example 5. We implemented the calculation of the solution of dynamical systems according to the midpoint scheme in Sage. To solve the system of nonlinear equations we use the method of simple iterations with the control of the conservation of all classical integrals. As applied to the Lagrange case with $m_1 = m_2 = m_3 = 1$, we have obtained a circular motion even when choosing an intentionally coarse step (Fig. 2).

As we saw above (Example 3), the step dt in the midpoint scheme for the system (22) can be chosen in such a way that the approximate solution becomes periodic, and its frequency tends to the frequency of the exact solution, i.e., to the value (26). This solution inherited the periodicity of the exact solution, in which three bodies, while maintaining the shape and dimensions of a regular triangle, rotate around the center of gravity with an angular frequency (26) [13].

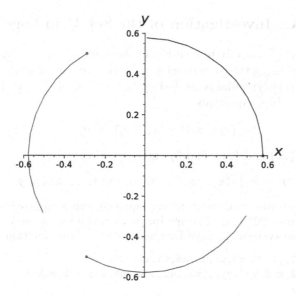

Fig. 2. Solution to the three-body problem in the Lagrange case, $\Delta t = 0.1$

Conclusion

In the present paper, we have formulated the Lagrange Problem 1 of finding all *approximate* solutions of the three-body problem at which the distances between the bodies remain constant. The Theorems 1 and 2 reduce the problem to the study of the properties of linear oscillators. In particular, it turns out that the periodicity of the exact solution in the Lagrange case, when the bodies form a regular triangle, is inherited by the approximate solution (Corollary 1).

We have so far avoided a complete solution of the Problem 1, since it obviously has more complex solutions discarded in mechanics. It does not seem to us that these solutions are not interesting, so we will return to this issue later. It is quite natural to try further to consider finite difference analogues of other classical problems of the theory of three bodies, including triangular solutions with the points moving along ellipses.

At the moment, our experiments with the three-body problem are limited not so much by present-day computer capacities, which are still not enough to work with dozens of symbol variables, but rather by very incomplete kit of tools implemented to exclude unknowns. The fact is that the midpoint scheme for the many-body problem is rich in discrete symmetries. The exclusion method implemented in Sage is based on the Gröbner bases with lex ordering and does not take the symmetry of the system into account, but sometimes the usage of such symmetries speeds up computing [3, 16]. We believe that the "factorization" of the system by these symmetries can reduce the number of unknowns by n times, where n is the number of bodies in the system.

Appendix A   Investigation of the Set V in Sage

We have tried unsuccessfully to investigate the algebraic set V introduced above in Sect. 4 of Sage, using the tools developed by W. Stein for working with ideals in multidimensional polynomial rings. Using the notation of (19), we have rewritten system (S) as 1) three equations

$$(x_i - x_i)^2 + (y_i - y_j)^2 = a_{ij}^2,$$

2) two systems (20) and (21), and 3) three equations

$$(\boldsymbol{r}_i - \boldsymbol{r}_j) \cdot (\boldsymbol{v}_i - \boldsymbol{v}_j) = 0, \quad i, j = 1, \ldots, n; \ i \neq j,$$

which in our case are equivalent to six equations that approximate Eqs. (10) and (11). Systems (20) and (21) are linear in variables marked with hats. We have solved these systems in Sage by means of standard function `solve`:

```
x=var('x1,x2,x3,y1,y2,y3,u1,u2,u3,v1,v2,v3')
xx=var('xx1,xx2,xx3,yy1,yy2,yy3,uu1,uu2,uu3,vv1,vv2,vv3')
var('dt,a,b,c')
K=QQ[a,b,c,dt,x1,x2,x3,y1,y2,y3,u1,u2,u3,v1,v2,v3]

def lagrange_hat():
    eqs=[
xx1-x1==(uu1+u1)*dt/2, xx2-x2==(uu2+u2)*dt/2, xx3-x3==(uu3+u3)*dt/2,
yy1-y1==(vv1+v1)*dt/2, yy2-y2==(vv2+v2)*dt/2, yy3-y3==(vv3+v3)*dt/2,
uu1-u1==dt/2*(1/a^3*(xx2+x2-xx1-x1) + 1/b^3*(xx3+x3-xx1-x1)),
uu2-u2==dt/2*(1/a^3*(xx1+x1-xx2-x2) + 1/c^3*(xx3+x3-xx2-x2)),
uu3-u3==dt/2*(1/c^3*(xx2+x2-xx3-x3) + 1/b^3*(xx1+x1-xx3-x3)),
vv1-v1==dt/2*(1/a^3*(yy2+y2-yy1-y1) + 1/b^3*(yy3+y3-yy1-y1)),
vv2-v2==dt/2*(1/a^3*(yy1+y1-yy2-y2) + 1/c^3*(yy3+y3-yy2-y2)),
vv3-v3==dt/2*(1/c^3*(yy2+y2-yy3-y3) + 1/b^3*(yy1+y1-yy3-y3)), ]
    return solve(eqs,xx)
```

Then we have constructed an ideal of the algebraic set V (in our code called Lagrange ideal):

```
def lagrange_ideal():
    S=lagrange_hat()
    Q1=(xx1+x1-xx2-x2)*(uu1+u1-uu2-u2)+(yy1+y1-yy2-y2)*(vv1+v1-vv2-v2)
    Q2=(xx3+x3-xx2-x2)*(uu3+u3-uu2-u2)+(yy3+y3-yy2-y2)*(vv3+v3-vv2-v2)
    Q3=(xx1+x1-xx3-x3)*(uu1+u1-uu3-u3)+(yy1+y1-yy3-y3)*(vv1+v1-vv3-v3)
    Q1=Q1.subs(S).numerator()
    Q2=Q2.subs(S).numerator()
    Q3=Q3.subs(S).numerator()
    global K
    J=K*[Q1,Q2,Q3, (x1-x2)^2+(y1-y2)^2-a^2,
(x1-x3)^2+(y1-y3)^2-b^2,
(x3-x2)^2+(y3-y2)^2-c^2,
u1+u2+u3, v1+v2+v3,
x1+x2+x3, y1+y2+y3]
    return J
```

For simplicity, we have added to the equations that determine the set V a condition of coincidence of the gravity center and the origin of coordinates. Thus we obtained an ideal J in the ring

$$\mathbb{Q}[a, b, c, dt, x_1, x_2, x_3, y_1, y_2, y_3, u_1, u_2, u_3, v_1, v_2, v_3].$$

Sage is unable to answer even such trivial questions about this ideal as belonging of element $y_1 + y_2 + y_3$ to ideal J:

```
sage: J=lagrange_ideal()
sage: K(y1 + y2 + y3) in J
```

References

1. Ayryan, E.A., Malykh, M.D., Sevastianov, L.A., Ying, Yu.: On Explicit Difference Schemes for Autonomous Systems of Differential Equations on Manifolds. In: England, M., Koepf, W., Sadykov, T.M., Seiler, W.M., Vorozhtsov, E.V. (eds.) CASC 2019. LNCS, vol. 11661, pp. 343–361. Springer, Cham (2019). https://doi.org/10.1007/978-3-030-26831-2_23
2. Cooper, G.J.: Stability of Runge-Kutta methods for trajectory problems. IMA J. Numer. Anal. **7**, 1–13 (1987)
3. Faugère, J.C., Svartz, J.: Gröbner bases of ideals invariant under a commutative group: the non-modular case. In: The 38th International Symposium on Symbolic and Algebraic Computation, ISSAC' 2013, Boston, United States, pp. 347–354, June 2013. https://doi.org/10.1145/2465506.2465944
4. Gerdt, V.P., Malykh, M.D., Sevastianov, L.A., Ying, Y.: On the properties of numerical solutions of dynamical systems obtained using the midpoint method. Discrete Continuous Models Appl. Comput. Sci. **27**(3), 242–262 (2019). https://doi.org/10.22363/2658-4670-2019-27-3-242-262
5. Goriely, A.: Integrability and Nonintegrability of Dynamical Systems. World Scientific, Singapore, River Edge, NJ (2001)
6. Graham, E., Jelenić, G., Crisfield, M.A.: A note on the equivalence of two recent time-integration schemes for n-body problems. Commun. Numer. Methods Eng. **18**, 615–620 (2002). https://doi.org/10.1002/cnm.520
7. Greenspan, D.: Completely conservative and covariant numerical methodology for n-body problems with distance-dependent potentials, Technical Report no. 285 (1992). http://hdl.handle.net/10106/2267
8. Greenspan, D.: Completely conservative, covariant numerical methodology. Comput. Math. Appl. **29**(4), 37–43 (1995). https://doi.org/10.1016/0898-1221(94)00236-E
9. Greenspan, D.: Completely conservative, covariant numerical solution of systems of ordinary differential equations with applications. Rendiconti del Seminario Matematico e Fisico di Milano **65**, 63–87 (1995). https://doi.org/10.1007/BF02925253
10. Greenspan, D.: N-Body Problems and Models. World Scientific, River Edge (2004)
11. Hairer, E., Wanner, G., Lubich, C.: Geometric Numerical Integration: Structure-Preserving Algorithms for Ordinary Differential Equations. Springer, Berlin, Heidelberg, New York (2000)
12. Hairer, E., Wanner, G., Nørsett, S.P.: Solving Ordinary Differential Equations, 3rd edn. Springer, New York (2008)

13. Pars, L.A.: A Treatise on Analytical Dynamics. Ox Bow Press, Newark (1979)
14. Simo, J.C., González, M.A.: Assessment of energy-momentum and symplectic schemes for stiff dynamical systems. American Society of Mechanical Engineers (1993)
15. Singer, M.F.: Liouvillian first integral of differential equations. Trans. Amer. Math. Soc. **333**(2), 673–688 (1992)
16. Steidel, S.: Gröbner bases of symmetric ideals. J. Symb. Comput. **54**, 72–86 (2013). https://doi.org/10.1016/j.jsc.2013.01.005
17. Zhang, H., Qian, X., Song, S.: Novel high-order energy-preserving diagonally implicit runge-kutta schemes for nonlinear hamiltonian odes. Appl. Math. Lett. **102**, 106091 (2020). https://doi.org/10.1016/j.aml.2019.106091

Univariate Polynomials with Long Unbalanced Coefficients as Bivariate Balanced Ones: A Toom–Cook Multiplication Approach

Marco Bodrato[1] and Alberto Zanoni[2][(✉)]

[1] Ministero dell'Istruzione, Torino, Italy
bodrato@mail.dm.unipi.it
[2] I.S.S. "Ferrari–Hertz", Via Grottaferrata 76, Roma, Italy
alberto.zanoni@posta.istruzione.it

Abstract. Multiplication of univariate dense polynomials with long integer unbalanced (having different lengths) coefficients is considered. By reducing the problem to the product of bivariate polynomials with balanced coefficients, Toom–Cook approach is shown, pointing out some optimizations in order to reduce the computational cost. As a byproduct, univariate sparse Toom–Cook is also sketched. Lastly, some experimental results concerning performance comparisons are presented.

Keywords: Long integer multiplication · Toom-Cook · Polynomial multiplication · Interpolation

1 Introduction

This paper deals with the product of univariate polynomials with unbalanced long integer coefficients: each of them having many digits (hundreds, thousands or even more bits) and differing by small ratios (e.g. one coefficient is two, three, or $4/3$ times longer than another one).

Let $a(x), b(x), c(x) = a(x) \cdot b(x) \in \mathbb{Z}[x]$, with

$$a(x) = \sum_{i=0}^{d_a} a_i x^i \quad ; \quad b(x) = \sum_{i=0}^{d_b} b_i x^i \quad ; \quad c(x) = \sum_{i=0}^{d=d_a+d_b} c_i x^i$$

We analyze the problem by using the point of view of Toom–Cook paradigm (in the following simply TC, for short). To make things clear, for $z \in \mathbb{Z}$ let $\delta(z)$ be its number of digits (in whatever but fixed "long" base $1 \ll \mathcal{B} \in \mathbb{N}$: for example, in computer science $\mathcal{B}$ is usually $2^{32\mu}$ or $2^{64\mu}$ for a certain $0 \neq \mu \in \mathbb{N}$).

The subquadratic multiplication algorithms due to Karatsuba [14], Toom and Cook [6,19] were proposed more than 50 years ago and are widely used today for long integer products when the length of the factors lies in a certain

© Springer Nature Switzerland AG 2020
F. Boulier et al. (Eds.): CASC 2020, LNCS 12291, pp. 91–107, 2020.
https://doi.org/10.1007/978-3-030-60026-6_6

range [17], while Schönhage–Strassen algorithm [16], based on FFT, is used for bigger ones. Actually they are formulated in terms of univariate polynomials, with the underlying hypothesis that the coefficients appearing in the core of the methods are more or less balanced, that is to say with the same number of "digits". Other multiplication methods exist, see [7,9] and the more recent results [11,12], but their interest is purely theoretical, as they became efficient only for really huge numbers.

TC–k approach is based on the Evaluate–(recursive) Multiply–Interpolate (EMI) paradigm, with two more steps (splitting and recomposition) when used recursively or to compute the product of two natural numbers. The idea is to obtain the $2k - 1$ coefficients of $c(x)$ with as many non-linear operations at the price of some linear ones more than schoolbook method (SB). For example, for long integer multiplication, let $u, v \in \mathbb{N}$, with $h = \delta(u)$ and $k = \delta(v)$: we indicate with $M(h, k)$ the computational cost of their product, setting $M_k = M(k, k)$ and $M = M_1$. To compute the product $u \cdot v = w \in \mathbb{N}$, follow the five steps indicated below.

1) **Splitting:** Fix an appropriate base $\mathcal{B} \in \mathbb{N}$ and represent the two operands by two homogeneous polynomials $\bar{a}, \bar{b} \in \mathbb{N}[x, h]$ with degree d_1, d_2 respectively and coefficients $0 \leqslant \bar{a}_i, \bar{b}_i < \mathcal{B}$, with $\bar{a}_{d_1}, \bar{b}_{b_2} \neq 0$ (base $\mathcal{B}$ expansion).

$$\bar{a}(x, h) = \sum_{i=0}^{d_1} \bar{a}_i x^i h^{d_1 - i} \quad ; \quad \bar{b}(x, h) = \sum_{i=0}^{d_2} \bar{b}_i x^i h^{d_2 - i}$$

This way $u = \bar{a}(\mathcal{B}, 1)$ and $v = \bar{b}(\mathcal{B}, 1)$. Let $\bar{c}(x, h) = \bar{a}(x, h)\bar{b}(x, h)$. For the classical TC–$k$ method one has $d_1 = d_2 = k - 1$, while in general, for possibly unbalanced operands, $d_1 + d_2 = 2(k - 1)$, k being an arbitrary multiple of $1/2$.

2) **Evaluation:** Choose $2k - 1$ values $\nu_i = (\nu_i', \nu_i'') \in \mathbb{Z}^2$ (the corresponding homogeneous value is ν_i'/ν_i'', with $(1, 0)$ usually represented by ∞) with ν_i' and ν_i'' coprime and $\nu_i \neq \pm \nu_j$ for $i \neq j$: evaluate both operands at all of them, obtaining the values $\bar{a}(\nu_i), \bar{b}(\nu_i)$.

3) **Recursion:** Compute $w_i = \bar{a}(\nu_i) \cdot \bar{b}(\nu_i)$ recursively. Let $\mathbf{w} = (w_i)$ be the resulting vector of values.

4) **Interpolation:** Solve the interpolation problem $\bar{c}(\nu_i) = w_i$ inverting the pseudo–Vandermonde matrix A_k generated by the ν_i values, computing $\bar{\mathbf{c}} = A_k^{-1}\mathbf{w}$, where $\bar{\mathbf{c}} = (\bar{c}_i)$ is the vector of $\bar{c}(x, h)$ coefficients.

5) **Recomposition:** Once all the coefficients have been computed, it is sufficient to evaluate back $w = \bar{c}(\mathcal{B}, 1)$.

For example, interpolation matrices for TC–2.5 and TC–3 are

$$A_{2.5} = \begin{pmatrix} 1 & 0 & 0 & 0 \\ -1 & 1 & -1 & 1 \\ 1 & 1 & 1 & 1 \\ 0 & 0 & 0 & 1 \end{pmatrix} \quad ; \quad A_3 = \begin{pmatrix} 1 & 0 & 0 & 0 & 0 \\ 16 & 8 & 4 & 2 & 1 \\ 1 & -1 & 1 & -1 & 1 \\ 1 & 1 & 1 & 1 & 1 \\ 0 & 0 & 0 & 0 & 1 \end{pmatrix}$$

TC–k algorithm core complexity is based on the relation $M_k = (2k - 1)M$. Standard analysis shows that the computational asymptotic complexity of the method is $O(n^{\log_k(2k-1)})$. The multiplicative constant hidden by the $O(\cdot)$ notation absorbs the complexity of the first two and last two phases. In order to minimize it, a careful choice of ν_i values and of the operations sequence for Evaluation, Interpolation and Recomposition phases helps in reducing the extra overhead.

We underline explicitly that the TC–k algorithm optimality is restricted to a limited range of factor lengths, that in practice depends on the computer architecture, available amount of memory, used data structures, and other technical details. Therefore the following considerations are in general not applicable in an asymptotic sense, when the sizes of the coefficients have no growth limit.

2   Multiplication of Polynomials with Unbalanced Long Coefficients

Considering univariate dense polynomial with long coefficients, TC algorithms can be doubly involved: in the polynomials themselves and in their coefficients. In regard to the Kronecker trick [15], reducing polynomial multiplication to long integer multiplication (see e.g. [8]), we point out a small note, but we will literally go in the opposite direction: adding instead one variable more.

The starting point of our analysis is the fact that when coefficients are unbalanced, straightforward use of TC–k for polynomials can sometimes be counterproductive.

Example (1) : consider TC–3 case, with $\delta(a_2) = \delta(b_2) = 4$ and $\delta(a_i) = \delta(b_j) = 1$ otherwise (represented with the notation $[1, 1, 4]^2$). Classical TC–3 recursive multiplication phase

$$
\begin{aligned}
c_4 = w_\infty &= a_2 b_2 & \Rightarrow M_4 = 7M \\
w_2 &= (4a_2 + 2a_1 + a_0)(4b_2 + 2b_1 + b_0) & \Rightarrow M_4 = 7M \\
w_1 &= (a_2 + a_1 + a_0)(b_2 + b_1 + b_0) & \Rightarrow M_4 = 7M \\
w_{-1} &= (a_2 - a_1 + a_0)(b_2 - b_1 + b_0) & \Rightarrow M_4 = 7M \\
c_0 = w_0 &= a_0 b_0 & \Rightarrow M
\end{aligned}
$$

would require $\mathcal{P} = 4 \cdot 7 + 1 = 29$ "basic" products, not to mention interpolation cost. By applying straightforwardly the schoolbook method the following is obtained instead:

$$
\begin{aligned}
c_4 &= a_2 b_2 & \Rightarrow M_4 & = 7M \\
c_3 &= a_2 b_1 + a_1 b_2 & \Rightarrow M(4,1) + M(1,4) & = 8M \\
c_2 &= a_2 b_0 + a_1 b_1 + a_0 b_2 & \Rightarrow M(4,1) + M + M(1,4) & = 9M \\
c_1 &= a_1 b_0 + a_0 b_1 & \Rightarrow M + M & = 2M \\
c_0 &= a_0 b_0 & \Rightarrow M
\end{aligned}
$$

so that just $\mathcal{P}' = 7 + 8 + 9 + 2 + 1 = 27 < \mathcal{P}$ products would be needed. This suggests that unbalance can sometimes make some difference. At a first glance, a mix of TC and schoolbook methods could be considered, computing c_1 instead of e.g. w_2, saving thus 5 products and simplifying the interpolation phase: this way $\mathcal{P}$ would be lowered to 24. In the following we will show how only 17 products are sufficient.

The point is that the classical TC–k preprocessing idea is to split factors into as balanced as possible $(k-1)$-degree "polynomials", precisely because of this: TC–k methods exploit the, so to say, mixing of coefficient products. If $i_1, j_1, i_2, j_2, h \in \mathbb{N}$ are such that $i_1 + i_2 = j_1 + j_2 = h$ then we obtain as (partial) result $(a_{i_1} b_{i_2} + a_{j_1} b_{j_2}) x^h$, and the more pairs "converge" to x^h, the more efficient the TC method is. Consider on the other side the product

$$(a_2 x^2 + a_0)(b_1 x + b_0) = (a_2 b_1) x^3 + (a_2 b_0) x^2 + (a_0 b_1) x + (a_0 b_0)$$

There is no coefficient mixing at all here: it is an ordinary polynomial product, so no TC–optimization can be done. In this paper we study multiplications for which there is at least one mix, so that TC methods can effectively be applied.

2.1   A Note About the Kronecker Trick

Consider the following case: let $a_2 = b_2 = \mathcal{B}^4 - 1$ and $a_i = b_j = \mathcal{B} - 1$ otherwise in example (1) above (the maximum possible values). Applying blindly the upper bound of [8] for $c(x)$ coefficients $|c_i| \leqslant L = (1 + \min(d_a, d_b)) N(a) N(b)$, where $N(a) = \max_i \{|a_i|\}$, the value $L = 3(\mathcal{B}^4 - 1)^2$ is obtained. Working with binary representations (as it is usually the case with computers), if $\mathcal{B} = 2^n$, then $L < 2^{8n+2}$, and the splitting parameter K_p for Kronecker trick results therefore to be $K_p = 8n + 2$.

Better upper bounds could be computed by considering how c_i depends on a_j, b_h coefficients and their exact size, but we take the opportunity to point out a completely general idea: we simply don't take into account the head coefficient c_d in the analysis, because it has never overlapping issues with other coefficients to take care of.

Here the new bound is $L' = 2(\mathcal{B}^4 - 1)(\mathcal{B} - 1) = 2(2^{4n} - 1)(2^n - 1) < 2^{5n+1}$, with a smaller splitting parameter $K_p' = 5n + 1$.

3   Sparse Univariate Toom–Cook

Classical balanced and more recent unbalanced TC algorithms [1–3,20] are now quite well established fast multiplication methods for dense polynomials (with balanced coefficients). To the best of our knowledge, their sparse versions have still not been deeply analyzed, but they may appear in the bivariate approach (BZ), that can benefit from their use – see next section. We present here some basic versions, with some details to give an idea, considering $\deg(a) \geqslant \deg(b)$ and $a_0, b_0 \neq 0$. In general, sparsity can reduce the overhead of evaluation and interpolation phases or lower the actual value of k.

To begin with, Karatsuba algorithm core is a clever way of computing $c(x) = (a_1 x + a_0)(b_1 x + b_0) = (a_1 b_1) x^2 + (a_1 b_0 + a_0 b_1) x + a_0 b_0$ with just three – instead of four – coefficient products, as

$$c(x) = (a_1 b_1) x^2 + [(a_1 + a_0)(b_1 + b_0) - a_1 b_1 - a_0 b_0] x + a_0 b_0$$

Note that it can be easily generalized for the case $c(x) = (a_2 x^r + a_0)(b_2 x^r + b_0)$ (with $r > 1$), which strictly speaking is sparse, but in practice it can be treated as a dense one with a simple change of variable[1]. Three cases of sparsity follow, in which some basic products and one "interlaced" Karatsuba emerge and suffice.

$\boxed{1}$ $(a_3 x^3 + a_1 x + a_0)(b_1 x + b_0)$
$= (a_3 b_1)x^4 + (a_3 b_0)x^3 + (a_1 b_1)x^2 + (a_1 b_0 + a_0 b_1)x + (a_0 b_0)$
$= (a_3 b_1)x^4 + (a_3 b_0)x^3 + (a_1 b_1)x^2 + [(a_1 + a_0)(b_1 + b_0) - a_1 b_1 - a_0 b_0]x + (a_0 b_0)$

$\boxed{2}$ $(a_2 x^2 + a_1 x + a_0)(b_2 x^2 + b_0)$
$= (a_2 b_2)x^4 + (a_1 b_2)x^3 + (a_2 b_0 + a_0 b_2)x^2 + (a_1 b_0)x + (a_0 b_0)$
$= (a_2 b_2)x^4 + (a_1 b_2)x^3 + [(a_2 + a_0)(b_2 + b_0) - a_2 b_2 - a_0 b_0]x^2 + (a_1 b_0)x + (a_0 b_0)$

$\boxed{3}$ $(a_3 x^3 + a_2 x^2 + a_0)(b_1 x + b_0)$
$= (a_3 b_1)x^4 + (a_3 b_0 + a_2 b_1)x^3 + (a_2 b_0)x^2 + (a_0 b_1)x + (a_0 b_0)$
$= (a_3 b_1)x^4 + [(a_3 + a_2)(b_1 + b_0) - a_3 b_1 - a_2 b_0]x^3 + (a_2 b_0)x^2 + (a_0 b_1)x + (a_0 b_0)$

Two distinct simultaneous applications of Karatsuba method are also possible:

$(a_3 x^3 + a_2 x^2 + a_1 x + a_0)(b_2 x^2 + b_0) =$
$(a_3 b_2)x^5 + (a_2 b_2)x^4 + (a_3 b_0 + a_1 b_2)x^3 + (a_2 b_0 + a_0 b_2)x^2 + (a_1 b_0)x + (a_0 b_0) =$
$(a_3 b_2)x^5 + (a_2 b_2)x^4 + [(a_3 + a_1)(b_2 + b_0) - a_3 b_2 - a_1 b_0]x^3 +$
$$[(a_2 + a_0)(b_2 + b_0) - a_2 b_2 - a_0 b_0]x^2 + (a_1 b_0)x + (a_0 b_0)$$

These basic cases show that the generic EMI approach (involved matrices and number of operations) becomes simpler for sparse factors. In the last following example a Karatsuba and a TC 2.5 are at the same time interlaced.

$(a_4 x^4 + a_3 x^3 + a_2 x^2 + a_1 x + a_0)(b_2 x^2 + b_0) =$
$a_4 b_2 x^6 + a_3 b_2 x^5 + (a_4 b_0 + a_2 b_2)x^4 + (a_3 b_0 + b_2 a_1)x^3 + (a_2 b_0 + b_2 a_0)x^2 +$
$$a_1 b_0 x + a_0 b_0 =$$
$a_4 b_2 x^6 + a_3 b_2 x^5 + \left[\frac{(a_4 + a_2 + a_0)(b_0 + b_2) + (a_4 - a_2 + a_0)(b_0 - b_2)}{2} - a_0 b_0\right]x^4 +$
$$[(a_3 + a_1)(b_2 + b_0) - a_3 b_2 - a_1 b_0]x^3 +$$
$$\left[\frac{(a_4 + a_2 + a_0)(b_0 + b_2) - (a_4 - a_2 + a_0)(b_0 - b_2)}{2} - a_4 b_2\right]x^2 + a_1 b_0 x + a_0 b_0$$

A bit more generally, if e.g. $a(x) = a_e(x^2)$ (one factor is even) and $b(x) = b_e(x^2) + x b_o(x^2)$ (the other one is not even nor odd) the product can be split into two independent subproducts – $c(x) = (a_e b_e)(x^2) + x(a_e b_o)(x^2)$ – and the resulting interpolation matrix (phase 4 of EMI) into two submatrices, making thus the inversion sequence of linear basic operations shorter.

---

[1] The same idea can, of course, be applied to the general case $a(x^r)b(x^r)$.

4 Using Toom–Cook Bivariate Multiplication

Multivariate TC methods have not been so deeply studied as their univariate versions (refer to [13] and, for a detailed study on Karatsuba, TC–2.5 and TC–3, to [4]). We let them enter the game by considering univariate polynomials $p(x)$ as bivariate ones $p(x,t)$, representing their (long) coefficients as polynomials themselves in t (similarly as in phase 1 of EMI, but without the homogenizing h variable – with the underlying meaning $t \equiv \mathcal{B}$), see [5]. This is actually the definition of a general injective mapping $\tau : \mathbb{Z}[x] \ni p(x) \overset{\tau}{\hookrightarrow} p(x,t) \in \mathbb{Z}[x,t]$. In most software implementations the split of the generic a_i (or b_i) coefficient is almost for free, as it can be simply identified by a pointer to its first "digit" (word, in computer science terms) and its length, as is the case with GMP library. Example (1) would imply e.g. for $a(x)$ the following representation

$$a(x,t) = (a_{2,3}t^3 + a_{2,2}t^2 + a_{2,1}t + a_{2,0})x^2 + a_{1,0}x + a_{0,0}$$

with, in this case, $a_{1,0} = a_1$ and $a_{0,0} = a_0$: similarly for $b(x,t)$. A graphical representation of the supports follows, where each number in the support of $c(x,t)$ indicates the number of products $a_{i,j}b_{h,k}$ appearing as addend in the corresponding coefficient (we define as *r-point* a product support monomial whose coefficient expression has r addends and as $r_>$-*point* one with more than r).

$$
t^3 \begin{bmatrix} 1 \\ 1 \\ 1 \\ 1 & 1 & 1 \end{bmatrix}
\times
\begin{bmatrix} 1 \\ 1 \\ 1 \\ 1 & 1 & 1 \end{bmatrix}
=
\begin{bmatrix} & & & & 1 \\ & & & & 2 \\ & & & & 3 \\ & & 2 & 2 & 4 \\ & & 2 & ② & 3 \\ & & 2 & 2 & 2 \\ 1 & 2 & 3 & 2 & 1 \end{bmatrix}
\begin{matrix} t^6 \\ t^5 \\ t^4 \\ t^3 \\ t^2 \\ t \\ 1 \end{matrix}
$$

t^2, t, 1 … $1 \ x \ x^2$ (left), $1 \ x \ x^2$ (middle), $1 \ x \ x^2 \ x^3 \ x^4$ (right)

As you can see, $c(x,t)$ support cardinality is 17. For example, $c_{3,2} = a_{2,2}b_{1,0} + a_{1,0}b_{2,2}$ (circled in the representation) is a 2-point. Generally speaking, the bigger the numbers, the more effective bivariate TC is.

The results of simple calculations à la TC of the number of needed products – not considering linear operations – for some classes of examples are shown in the table alongside.

In general, for what concerns the

Structure	SB	TC	BZ
$[1,1,k]^2$	$6k+3$	$8k-3$	$4k+1$
$[1,k,1]^2$	$6k+3$	$6k-1$	$4k+1$
$[k,1,k]^2$	$12k-3$	$10k-5$	$8k-3$

computational complexity, the products $[u_0, ..., u_{d_a}] \times [v_0, ..., v_{d_b}]$ and its symmetric $[u_{d_a}, ..., u_0] \times [v_{d_b}, ..., v_0]$ are obviously equivalent.

It can be useful to consider terms $x^i t^j$ as points $(i,j) \in \mathbb{N}^2$, in order to visualize operations and have thus a double point of view, helping to better visualize and understand ideas and procedures. What follows is essentially a high level description of our bivariate BZ method proposal, with details on some of its possible optimizations made explicit.

4.1 Evaluation Phase

In order to have an as-easy-as-possible interpolation matrix (structured and with many zeroes) to be inverted in phase 4 of EMI, we distinguish between sides, base, border and inner points of the product support.

Let $c(x,t) = \tau(c(x)) = \sum_{i,j} c_{ij} x^i t^j \in \mathbb{Z}[x,t]$, let d be its degree in the x variable and $J_i = \max\{j \mid c_{ij} \neq 0\}$ for $i = 0, \ldots, d$. The two *sides* of the "tower" are the sets $S_1 = \{t^j \mid j = 0, \ldots, J_0\}, S_2 = \{x^d t^j \mid j = 0, \ldots, J_d\}$: the *base* is the set $B = \{x^i \mid i = 0, \ldots, d\}$, with $\sigma_1 = |S_1|, \sigma_2 = |S_2|, \beta = |B|$.

We call *border points* the terms $x^i t^{J_i}$ and *border* the set of all border points. Graphically speaking, they correspond to the topmost points in each column of the tower. Border points of $c(x,t)$ derive from composition of border points of $a(x,t)$ and $b(x,t)$: they can be used as subsets of evaluations for the bivariate case just like univariate TC methods. We will also consider the convex hull of the tower. Its interesting part is actually the one relative to the border: we will focus on it and call it simply *hull*.

Let $\varphi(n) : \mathbb{N} \ni n \rightarrow (n+1)/2 \in \mathbb{Q}$. To individuate the needed linear combinations giving a less dense as possible interpolation matrix we consider the sides, the base, the border (hull) and finally the inner points:

Algorithm :

1. Apply (possibly unbalanced) univariate TC–$\varphi(\sigma_1)$, TC–$\varphi(\sigma_2)$, TC–$\varphi(\beta)$ to the sides and the base of the tower, respectively.
2. Divide the hull into "segments" s_i: sequences of hull points (sub-polynomials, so to say) belonging to the same line (considered in the $\mathbb{N}^2$ space), with slope m_i. Segments may or may not contain all points between extremes (for example, in the presented case the segment $[1, x^2 t^3, x^4 t^6]$ has only the point corresponding to $x^2 t^3$ between its extremes). Let l_i be the actual number of points of s_i.
3. Apply the appropriate univariate (possibly unbalanced or sparse) TC–$\varphi(l_i)$ to each s_i – see details below.
4. Consider the remaining points on the border and check if some other univariate TC can be applied – see some possible cases in the examples in (1), at the end of the section.
5. Eventually apply "true" bivariate TC to recover the set of remaining inner points (values) of the tower. Theoretically it is possible to recycle the evaluation values of $a(x,t), b(x,t)$ obtained in the precedent points for border evaluations as partial evaluation results for this phase.

Univariate TC evaluations for the three cases in step 1) above – base and borders – require in total just $\sigma_1 + \sigma_2 + \beta - 2$ products: the values $a_{0,0} b_{0,0}$ and $a_{d_a,0} b_{d_b,0}$ can in fact be shared between the evaluation of the base and the left (right) side, respectively, and can be computed just once.

Consider the segments of the hull of $c(x,t)$. The ordered sequence of its slopes can in general be easily deduced by the hulls of $a(x,t)$ and $b(x,t)$ by first merging their set of slopes $m^{(a)}, m^{(b)}$ and ordering the resulting set in decreasing order.

Let $^\#s_i'$, $^\#s_i''$ be the discrete lengths of the segments of $a(x,t)$ and $b(x,t)$ (1 if there is no such segment) having slope m_i, respectively. The corresponding segment s_i of $c(x,t)$ will have length $l_i \geqslant {}^\#s_i' + {}^\#s_i'' - 1$. Hull segments can be treated as univariate polynomials, but – geometrically speaking – they are embedded in a bivariate set along oblique lines.

In fact, nontrivial product hull segments derive from (product of) segments with the same slope $m = m_1/m_2$ (with $m_2 > 0$) correspondingly belonging to the hulls of the factors. They can be detected with a weighted homogenization (with homogenizing variable H). Focusing e.g. on $a(x,t)$, let its total weighted degree be $\delta_a = \max\limits_{i=0,\ldots,d_a} \{m_2 J_i - m_1 i\}$.

Theoretically speaking, the "partial" polynomial $a_s(x,t)$ whose terms contribute to the hull points on the segment s can be obtained first by computing

$$\bar{a}_s(x,t,H) = a\left(xH^{m_1}, \frac{t}{H^{m_2}}\right) H^{\delta_a}$$

and then setting $H = 0$ – and similarly for $b(x,t)$. By knowing this, it is in practice simply sufficient to detect the coefficients of $a_s(x,t)$ and $b_s(x,t)$ to work with and compute the needed evaluations for the corresponding segment by using them directly.

4.2   Interpolation Phase

By ordering opportunely the terms of the product support, the initial lines of the interpolation phase matrix – for the part related to sides and base – show three interlaced block submatrices corresponding to univariate TCs, while remaining lines (relative to inner points) are in general full. In order to have a well defined matrix structure, one can order terms e.g. in the following way:

$$\begin{aligned} &[\, x^d t^{J_d},\ldots,x^d t,x^d,x^{d-1},\ldots,x,1,t,\ldots,t^{J_0}, \begin{cases} \text{sides} \\ \text{base} \end{cases} \\ &x^i t^{J_i}, \quad \text{hull vertices (1-points)} \\ &x^{i'} t^{J_{i'}}, \quad \text{other border 1-points (left-right)} \\ &x^{i''} t^{J_{i''}}, \text{ border } 1_>\text{-points (left-right)} \\ &x^{i'''} t^j \,] \ \text{ inner points (left-right, bottom-up)} \end{aligned}$$

In the interpolating phase one can take advantage of the sparsity of the interpolation matrix, if evaluation was done as explained. Some examples follow, in which submatrices corresponding to sides, base and border appear. The last full lines correspond to the inner points.

$$[1,1,3] \times [1,1,3]$$

$$
\begin{array}{ccc}
 & & 1 \\
 & & 2 \\
1 \quad\ 1 & \Rightarrow & 223 \\
\ \ 1\times\ 1 & & 222 \\
111\ \ 111 & & 12321
\end{array}
$$
(3 inner points)

$$
\left(
\begin{array}{ccccccccccccc}
1 \\
16 & 8 & 4 & 2 & 1 \\
1 & 1 & 1 & 1 & 1 \\
1 & -1 & 1 & -1 & 1 \\
 & & & & 1 \\
 & & & & 16 & 8 & 4 & 2 & 1 \\
 & & & & 1 & 1 & 1 & 1 & 1 \\
 & & & & 1 & -1 & 1 & -1 & 1 \\
 & & & & & & & & 1 \\
1 & & & & & & & & & 1 & 1 \\
1 & 1 & 1 & 1 & 1 & 1 & 1 & 1 & 1 & 1 & 1 & 1 & 1 \\
1 & -1 & 1 & -1 & 1 & -1 & 1 & -1 & 1 & 1 & -1 & 1 & -1 \\
1 & -1 & 1 & -1 & 1 & 1 & 1 & 1 & 1 & 1 & -1 & -1 & 1
\end{array}
\right)
$$

$$[1,1,4] \times [1,1,4]$$

$$
\begin{array}{ccc}
 & & 1 \\
 & & 2 \\
 & & 3 \\
1 \quad\ 1 & \Rightarrow & 224 \\
\ \ 1\times\ 1 & & 223 \\
1 \quad\ 1 & & 222 \\
1111\ \ 111 & & 12321
\end{array}
$$
(5 inner points)

$$
\left(
\begin{array}{ccccccccccccccccc}
1 \\
64 & 32 & 16 & 8 & 4 & 2 & 1 \\
1 & 1 & 1 & 1 & 1 & 1 & 1 \\
1 & -1 & 1 & -1 & 1 \\
1 & 2 & 4 & 8 & 16 & 32 & 64 \\
1 & -2 & 4 & -8 & 16 & -32 & 64 \\
 & & & & & & 1 \\
 & & & & & & 1 & 1 & 1 & 1 & 1 \\
 & & & & & & 1 & -1 & 1 & -1 & 1 \\
 & & & & & & 1 & 2 & 4 & 8 & 16 \\
 & & & & & & & & & & 1 \\
1 & & & & & & & & & & 1 & 1 \\
64 & 32 & 16 & 8 & 4 & 2 & 1 & 1 & 1 & 1 & 1 & 8 & 8 & 2 & 4 & 2 & 4 \\
1 & -1 & 1 & -1 & 1 & -1 & 1 & 1 & 1 & 1 & 1 & -1 & -1 & -1 & 1 & -1 & 1 \\
1 & 1 & 1 & 1 & 1 & 1 & 1 & -1 & 1 & -1 & 1 & 1 & -1 & -1 & -1 & 1 & 1 \\
1 & 1 & 1 & 1 & 1 & 1 & 1 & 1 & 1 & 1 & 1 & 1 & 1 & 1 & 1 & 1 & 1 \\
1 & -1 & 1 & -1 & 1 & -1 & 1 & -1 & 1 & -1 & 1 & 1 & -1 & -1 & -1 & 1
\end{array}
\right)
$$

As the number of inner points grows in general quadratically with the x,t-degrees of the factors, the benefit of matrices sparsity given by border analysis is someway limited. It is therefore convenient to use this kind of approach with polynomials with low x-degrees and split the coefficients in a reasonable number of "pieces", with a clever choice of $\mathcal{B}$, letting possibly recursion do its job on subproducts.

Note that, beyond having trivial products on the sides or border, sometimes inner points (coefficients) can be as well directly obtained with a single multiplication. This someway permits to bypass and shorten the general interpolation scheme. For example, they correspond to the three underlined "1"s in the below shown case.

$$\boxed{[3,1,1] \times [1,1,2]} \qquad \begin{pmatrix} 1 & & & & & & & & & & & \\ 1 & & & & & & & & & & & \\ 16 & 8 & 4 & 2 & 1 & & & & & & & \\ 1 & 1 & 1 & 1 & 1 & & & & & & & \\ 1 & -1 & 1 & -1 & 1 & & & & & & & \\ & & & & 1 & & & & & & & \\ & & & & & 1 & & & & & & \\ & & & & & & 1 & & & & & \\ & & & & & & & 1 & & & & \\ & & & & & & & & \underline{1} & & & \\ & & & & & & & & & \underline{1} & & \\ -1 & 1 & 1 & 1 & 1 & -1 & 1 & -1 & 1 & -1 & 1 & -1 & -1 \\ 1 & 1 & 1 & 1 & 1 & 1 & 1 & 1 & 1 & 1 & 1 & 1 & 1 \\ & & & & & & & & & & & & \underline{1} \end{pmatrix}$$

$$
\begin{array}{ccc}
 & 1 & \\
1 & 1\underline{1}2 & \\
1 \quad \times \quad 1 & \Rightarrow 1\underline{1}2\underline{1}1 \\
111 \quad 111 & 12321
\end{array} \quad ;
$$

We point out for the sake of completeness that in some cases, depending on the shape of factor supports, "univariate" TCs can be present on the border but not on the hull (underlined entries in the following cases).

$$
\text{Inner Karatsuba} \qquad\qquad \begin{array}{cc} & \underline{1} \\ 1 & 1\ \underline{2}2 \\ 1 & 11 & \Rightarrow 12\underline{1}32 \\ 1 \times 1 & 11 & 12232 \\ 11 & 1111 & 12221 \end{array}
$$

$$\boxed{[1,2] \times [3,1,3,4]}$$

Inner TC–3, TC–2.5

$$\boxed{[2,2,2,1,1,4,5,6] \times [2,2,2,1,1,4,5]}$$

$$
\begin{array}{l}
\qquad\qquad\qquad\qquad\qquad \underline{1} \\
\qquad\qquad\qquad\qquad\qquad \underline{2}2 \\
\qquad\qquad\qquad\qquad\qquad \underline{2}43 \\
\qquad\qquad\qquad\quad 1\ 11\underline{1}464 \\
\qquad 1 \qquad\qquad\qquad 2\ 4\ 423785 \\
\quad 11 \qquad\qquad 1 \Rightarrow\quad 2\ 6\ 8\ 646995 \\
\ \ 111 \qquad\qquad 11 \qquad 4\ 8\ 10867984 \\
\ \ 111 \times \qquad 11 \qquad 1\underline{2}3\underline{2}14\ 8\ 10866763 \\
111 \quad 111 \quad 111 \quad 11 \quad 24666810\,10865542 \\
11111111 \quad 1111111 \quad 123456\ 7\ 7654321
\end{array}
\tag{1}
$$

5   Experimental Results

Some results comparing four multiplication methods, implemented as scripts for PARI/GP [18], follow. We present in Figs. 1, 2, 3 graphics for the three cases $[1,1,n]^2$ with $n = 2, 3, 4$, reporting what happens with greater and greater coefficient unbalance. Timing comparisons are made with respect to schoolbook polynomial multiplication method (SB) and percentage results are shown.

In the keys, TC stands for Toom-Cook–3, KR for the Kronecker trick (keeping in mind the idea in note 2.1) and BZ for the bivariate approach. Logarithmic scale is used for the axis of the abscissas (number of bits of the smallest coefficient).

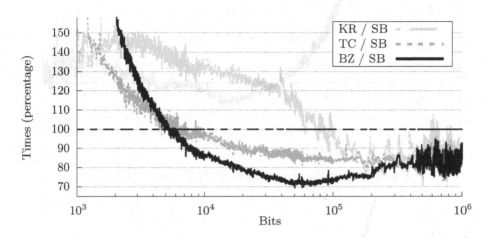

Fig. 1. Time ratios in percent for $[1, 1, 2]^2$ case (PARI–GP)

SB (the 100% line) reveals to be quite effective compared to both TC and KR under a certain threshold, and the different ranges of integer coefficient lengths for which BZ is competitive can be noticed. Note also how the graphic shapes depend on coefficient unbalancedness, but the general behavior is anyway sufficiently well defined.

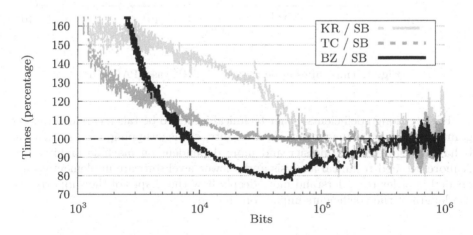

Fig. 2. Time ratios in percent for $[1, 1, 3]^2$ case (PARI–GP)

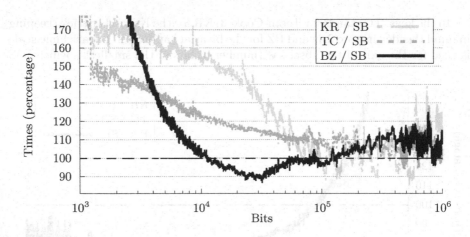

Fig. 3. Time ratios in percent for $[1, 1, 4]^2$ case (PARI–GP)

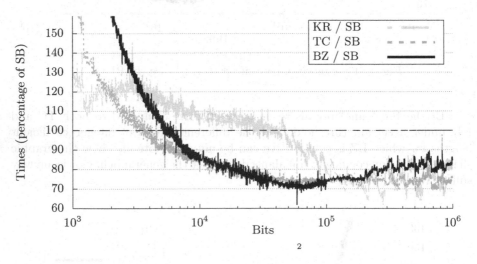

Fig. 4. Time ratios in percent for $[1, 2, 1]^2$ case (PARI–GP)

For very long coefficients, other technical issues to be taken care of enter the game, as memory management, data locality, cache swapping, etc., so that the behavior begins to get messier and more dependent on machine architecture. A more detailed technical analysis and a lower level software implementation is needed in order to understand more deeply how the shapes of the supports and the length of the coefficients impact on timings.

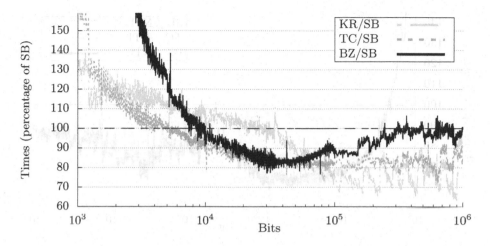

Fig. 5. Time ratios in percent for $[1,3,1]^2$ case (PARI–GP)

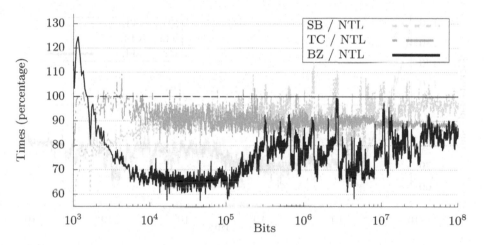

Fig. 6. Time ratios in percent for $[3,1,1]^2$ case (NTL)

Figures 4 and 5 refer to the more symmetric cases $[1,2,1]^2$ and $[1,3,1]^2$. In this case the benefits are inferior and the best performance seems not to overcome TC's.

Comparisons with respect to NTL C++ library [10] (version 11.4.3, using GMP-6.2.0) are shown in Figs. 6 ($[3,1,1]^2$ case), 7 ($[4,1,1]^2$), 8 ($[1,2,1]^2$) and 9 ($[1,3,1]^2$). Currently there is no TC implementation in NTL: subquadratic methods "jump" from Karatsuba directly to FFT. We therefore implemented in C++ (BZ and) an ad-hoc TC–3 method for NTL, by using the mpz-layer of GMP as well. Time ratios (in percent) of SB (called *plain* in NTL), TC and BZ are computed with respect to default NTL polynomial multiplication method.

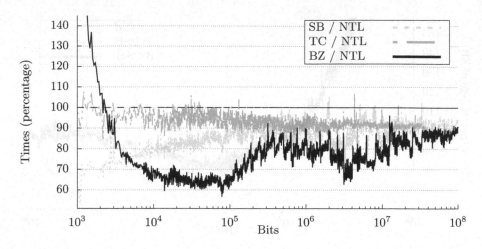

Fig. 7. Time ratios in percent for $[4, 1, 1]^2$ case (NTL)

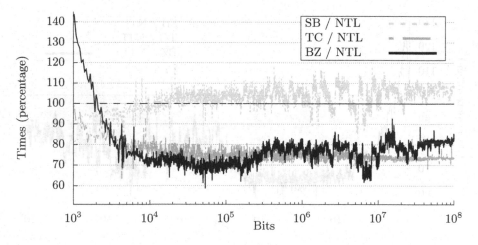

Fig. 8. Time ratios in percent for $[1, 2, 1]^2$ case (NTL)

The gains that can be observed seem to indicate that the implementation of special code handling unbalanced multiplication could give significant improvements. Moreover, with NTL the effect we highlighted with our example (**1**) emerges to evidence: a clever algorithm (Karatsuba in NTL) can be even slower than SB on unbalanced coefficients.

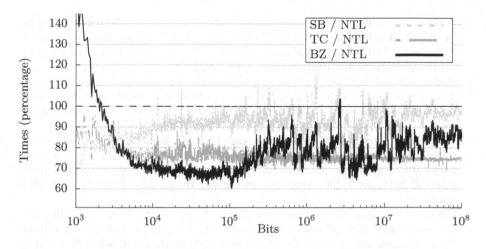

Fig. 9. Time ratios in percent for $[1, 3, 1]^2$ case (NTL)

6   Conclusions

An analysis of univariate dense polynomial multiplication with unbalanced long coefficients was presented, highlighting computational particularities of different approaches, comparing them both from the theoretical and (in some case) experimental point of view. Switching from the unbalanced univariate to the balanced bivariate paradigm seems to be quite promising for coefficients having a certain length/ratio and for some polynomial support shape.

The proposed multiplication algorithm was detailed for some degrees of unbalance (3:1, 4:1) managed by small-degree TCs. The used EMI evaluation/inversion sequences, nevertheless, have a wider use: by padding the smaller coefficients, they can be applied also when the ratio of the sizes of numbers is even larger (5:1, 6:1, ...).

For small degrees of x, t, complete case studies seem to reveal good chances of still not completely explored optimization possibilities for implementation.

Acknowledgments. The authors would like to deeply thank all the referees for their precious comments that helped to improve the paper.

References

1. Bodrato, M.: Towards optimal toom-cook multiplication for univariate and multivariate polynomials in characteristic 2 and 0. In: Carlet, C., Sunar, B. (eds.) WAIFI 2007. LNCS, vol. 4547, pp. 116–133. Springer, Heidelberg (2007). https://doi.org/10.1007/978-3-540-73074-3_10

2. Bodrato, M.: High degree Toom'n'half for balanced and unbalanced multiplication. In: Antelo, E., Hough, D., Ienne, P. (eds.) 20th IEEE Symposium on Computer Arithmetic, ARITH 2011, Tübingen, Germany, 25–27 July 2011, pp. 15–22. IEEE Computer Society, Washington D.C., USA (2011)

3. Bodrato, M., Zanoni, A.: Integer and polynomial multiplication: towards optimal Toom-Cook matrices. In: Wang, D. (ed.) Symbolic and Algebraic Computation, International Symposium, ISSAC 2007, Waterloo, Ontario, Canada, July 28 - 1 August 2007, Proceedings, pp. 17–24. ACM, New York, USA (2007)

4. Bodrato, M., Zanoni, A.: Karatsuba and Toom-Cook methods for multivariate polynomials. In: Breaz, D., Breaz, N., Ularu, N. (eds.) Proceedings of the International Conference on Theory and Applications of Mathematics and Informatics ICTAMI 2011, pp. 11–60. Aeternitas Publishing House, Alba Iulia, Romania (July 2011)

5. Chen, C., Covanov, S., Mansouri, F., Maza, M.M., Xie, N., Xie, Y.: Parallel integer polynomial multiplication. In: Davenport, J.H., et al. (eds.) 18th International Symposium on Symbolic and Numeric Algorithms for Scientific Computing, SYNASC 2016, Timisoara, Romania, 24–27 September 2016, pp. 72–80. IEEE Computer Society, Washington D.C., USA (2016)

6. Cook, S.A.: On the minimum computation time of functions. Ph.D. thesis, Harvard University (1966)

7. De, A., Kurur, P.P., Saha, C., Saptharishi, R.: Fast integer multiplication using modular arithmetic. In: Proceedings of the Fortieth Annual ACM Symposium on Theory of Computing, pp. 499–506. STOC 2008. Association for Computing Machinery, New York, USA (2008)

8. Fateman, R.: Can you save time in multiplying polynomials by encoding them as integers? / revised 2010 (2010). https://people.eecs.berkeley.edu/~fateman/papers/polysbyGMP.pdf

9. Fürer, M.: Faster integer multiplication. SIAM J. Comput. **39**(3), 979–1005 (2009)

10. Hart, W.B.: Fast library for number theory: an introduction. In: Proceedings of the Third International Congress Conference on Mathematical Software, pp. 88–91. ICMS 2010. Springer, Berlin, Heidelberg (2010)

11. Harvey, D., van der Hoeven, J.: Integer multiplication in time $O(n \log n)$ March 2019. https://hal.archives-ouvertes.fr/hal-02070778/document

12. Harvey, D., van der Hoeven, J., Lecerf, G.: Even faster integer multiplication. J. Complexity **36**, 1–30 (2016)

13. van der Hoeven, J., Lecerf, G.: On the complexity of multivariate blockwise polynomial multiplication. In: Proceedings of the 37th International Symposium on Symbolic and Algebraic Computation, pp. 211–218. ISSAC 2012, ACM, New York, USA (2012)

14. Karatsuba, A.A., Ofman, Y.: Multiplication of multidigit numbers on automata. Soviet Phys. Doklady **7**, 595–596 (1963)

15. Kronecker, L.: Grundzüge einer arithmetischen theorie der algebraischen Grössen. J. für die reine und angewandte Mathematik **92**, 1–122 (1882)

16. Schönhage, A., Strassen, V.: Schnelle multiplikation großer zahlen. Computing **7**(3–4), 281–292 (1971)

17. The GMP team: Bodrato, M., Glisse, M., Granlund, T., Möller, N., et al.: GNU MP: The GNU Multiple Precision Arithmetic Library, 6.2.0 edn. (2020). https://gmplib.org

18. The PARI Group, Univ. Bordeaux: PARI/GP version 2.11.2 (2019). http://pari.math.u-bordeaux.fr/

19. Toom, A.L.: The complexity of a scheme of functional elements realizing the multiplication of integers. Soviet Math. Doklady **3**, 714–716 (1963)
20. Zanoni, A.: Iterative Toom-Cook methods for very unbalanced long integer multiplication. In: Koepf, W. (ed.) Symbolic and Algebraic Computation, International Symposium, ISSAC 2010, Munich, Germany, July 25–28, 2010, Proceedings, pp. 319–323. ACM, New York, USA (2010)

Power Series Arithmetic with the BPAS Library

Alexander Brandt$^{(\boxtimes)}$ (iD), Mahsa Kazemi, and Marc Moreno-Maza

Department of Computer Science, The University of Western Ontario,
London, Canada
{abrandt5,mkazemin}@uwo.ca, moreno@csd.uwo.ca

Abstract. We discuss the design and implementation of multivariate power series, univariate polynomials over power series, and their associated arithmetic operations within the Basic Polynomial Algebra Subprograms (BPAS) Library. This implementation employs lazy variations of Weierstrass preparation and the factorization of univariate polynomials over power series following Hensel's lemma. Our implementation is lazy in that power series terms are only computed when explicitly requested. The precision of a power series is dynamically extended upon request, without requiring any re-computation of existing terms. This design extends into an "ancestry" of power series whereby power series created from the result of arithmetic or Weierstrass preparation automatically hold on to enough information to dynamically update themselves to higher precision using information from their "parents".

Keywords: Lazy power series · Weierstrass preparation · Hensel's lemma

1 Introduction

Power series are polynomial-like objects with, potentially, an infinite number of terms. They play a fundamental role in theoretical computer science, functional analysis, computer algebra, and algebraic geometry. Of course, the fact that power series may have an infinite number of terms presents interesting challenges to computer scientists. How to represent them on a computer? How to perform arithmetic operations effectively and efficiently with them?

One standard approach is to implement power series as *truncated power series*, that is, by setting up in advance a sufficiently large accuracy, or precision, and discarding any power series term with a degree equal or higher to that accuracy. Unfortunately, for some important applications, not only is such accuracy problem-specific, but sometimes cannot be determined before calculations start, or later may be found to not go far enough. This scenario occurs, for instance, with modular methods [16] for polynomial system solving [7] based on Hensel lifting and its variants [11]. It is necessary then to implement power

© Springer Nature Switzerland AG 2020
F. Boulier et al. (Eds.): CASC 2020, LNCS 12291, pp. 108–128, 2020.
https://doi.org/10.1007/978-3-030-60026-6_7

series with data structures and techniques that allow for reactivity and dynamic updates.

Since a power series has potentially infinitely many terms, it is natural to represent it as a function, that we shall call a *generator*, which computes the terms of that power series for a given accuracy. This point of view leads to natural algorithms for performing arithmetic operations (addition, multiplication, division) on power series based on *lazy evaluation*.

Another advantage of this functional approach is the fact that it supports concurrency in a natural manner. Consider a procedure which takes some number of power series as input and returns a number of power series. Assume the generators of the outputs can be determined in essentially constant time, which is often the case. Subsequent computations involving those output power series can then start almost immediately. In other words, the first procedure call is essentially non-blocking, and the output power series can (*i*) be used immediately as input to other procedure calls, and (*ii*) have their terms computed only as needed. This approach allows for power series terms to be computed or "produced" while concurrently being "consumed" in subsequent computations. These procedure calls can be seen as the stages of a *pipelined* computation [17, Ch. 9].

In this work, we present our implementation of *multivariate power series* (Sect. 3) and *univariate polynomials over multivariate power series* "UPoPS" (Sect. 4) based on the ideas of lazy evaluation. Factoring such polynomials, by means of Hensel's lemma and its extensions and variants, like the extended Hensel construction (EHC) [1, 20] and the Jung-Abhyankar Theorem [19], is our driving application. We discuss a lazy implementation of factoring via Hensel's lemma (Sect. 6) by means of lazy Weierstrass preparation (Sect. 5).

Our implementation is part of the Basic Polynomial Algebra Subprograms (BPAS) library [3], a free and open-source computer algebra library for polynomial algebra. The library's core, of which our power series and UPoPS are a part, is carefully implemented in C for performance. The library also has a C++ interface for better usability. Such an interface for power series is forthcoming. Our current implementation is both sequential and over the field of rational numbers. However, the BPAS library has the necessary infrastructure, in particular asynchronous generators, see [4], to take advantage of the concurrency opportunities (essentially pipelining) created by our design based on lazy evaluation.

Existing implementations of multivariate power series are also available in MAPLE's PowerSeries[1] library [2, 15] and SAGEMATH [22]. The former is similarly based on lazy evaluation, while the latter uses the truncated power series approach mentioned above. Our experimental results show that our implementation in BPAS outperforms its counterparts by several orders of magnitude.

Lazy evaluation in computer algebra has some history, see the work of Karczmarczuk [14] (discussing different mathematical objects with an "infinite" length) and the work of Monagan and Vrbik [18] (discussing sparse polynomial arithmetic). Lazy univariate power series, in particular, have been implemented

---

[1] This library is accessible, yet undocumented, in MAPLE 2020 as RegularChains:-PowerSeries. See www.regularchains.org/documentation.html.

by Burge and Watt [6] and by van der Hoeven [13]. However, up to our knowledge, our implementation is the first for *multivariate* power series in a compiled code.

2   Background

This section gathers basic concepts about multivariate formal power series. We suggest the book of G. Fischer [9] for an introduction to the subject. We start with formal power series arithmetic. Let $\mathbb{K}$ be an algebraic number field and $\overline{\mathbb{K}}$ its algebraic closure. We denote by $\mathbb{K}[[X_1, \ldots, X_n]]$ the ring of formal power series with coefficients in $\mathbb{K}$ and with variables $X_1, \ldots, X_n$.

Let $f = \sum_{e \in \mathbb{N}^n} a_e X^e$ be a formal power series and $d \in \mathbb{N}$. The *homogeneous part* and *polynomial part* of f in degree d are denoted by $f_{(d)}$ and $f^{(d)}$, and defined by $f_{(d)} = \sum_{|e|=d} a_e X^e$ and $f^{(d)} = \sum_{k \le d} f_{(k)}$. Note that $e = (e_1, \ldots, e_n)$ is a multi-index, X^e stands for $X_1^{e_1} \cdots X_n^{e_n}$, $|e| = e_1 + \cdots + e_n$, and $a_e \in \mathbb{K}$ holds.

Let $f, g \in \mathbb{K}[[X_1, \ldots, X_n]]$. Then the *sum*, *difference*, and *product* of f and g are given by $f \pm g = \sum_{d \in \mathbb{N}} (f_{(d)} \pm g_{(d)})$ and $fg = \sum_{d \in \mathbb{N}} \left(\Sigma_{k+\ell=d} (f_{(k)} g_{(\ell)}) \right)$. The *order* of a formal power series $f \in \mathbb{K}[[X_1, \ldots, X_n]]$, denoted by $\mathrm{ord}(f)$, is defined as $\min\{d \mid f_{(d)} \ne 0\}$, if $f \ne 0$, and as ∞ otherwise. We recall several properties. First, $\mathbb{K}[[X_1, \ldots, X_n]]$ is an integral domain. Second, the set $\mathcal{M} = \{f \in \mathbb{K}[[X_1, \ldots, X_n]] \mid \mathrm{ord}(f) \ge 1\}$ is the only maximal ideal of $\mathbb{K}[[X_1, \ldots, X_n]]$. Third, for all $k \in \mathbb{N}$, we have $\mathcal{M}^k = \{f \in \mathbb{K}[[X_1, \ldots, X_n]] \mid \mathrm{ord}(f) \ge k\}$.

Krull Topology. Let $(f_n)_{n \in \mathbb{N}}$ be a sequence of elements of $\mathbb{K}[[X_1, \ldots, X_n]]$ and $f \in \mathbb{K}[[X_1, \ldots, X_n]]$. We say that $(f_n)_{n \in \mathbb{N}}$ *converges* to f if for all $k \in \mathbb{N}$ there exists $N \in \mathbb{N}$ s.t. for all $n \in \mathbb{N}$ we have $n \ge N \Rightarrow f - f_n \in \mathcal{M}^k$. We say that $(f_n)_{n \in \mathbb{N}}$ is a *Cauchy sequence* if for all $k \in \mathbb{N}$ there exists $N \in \mathbb{N}$ s.t. for all $n, m \in \mathbb{N}$ we have $n, m \ge N \Rightarrow f_m - f_n \in \mathcal{M}^k$. The following results hold: we have $\bigcap_{k \in \mathbb{N}} \mathcal{M}^k = \langle 0 \rangle$. Moreover, if every Cauchy sequence in $\mathbb{K}$ converges, then every Cauchy sequence of $\mathbb{K}[[X_1, \ldots, X_n]]$ converges too.

Inverse of a Power Series. Let $f \in \mathbb{K}[[X_1, \ldots, X_n]]$. Then, the following properties are equivalent: (*i*) f is a unit, (*ii*) $\mathrm{ord}(f) = 0$, (*iii*) $f \notin \mathcal{M}$. Moreover, if f is a unit, then the sequence $(u_n)_{n \in \mathbb{N}}$, where $u_n = 1 + g + g^2 + \cdots + g^n$ and $g = 1 - f/f_{(0)}$, converges to the inverse of $f/f_{(0)}$.

Assume $n \ge 1$. Denote by $\mathbb{A}$ the ring $\mathbb{K}[[X_1, \ldots, X_{n-1}]]$ and by $\mathcal{M}$ be the maximal ideal of $\mathbb{A}$. Note that $n = 1$ implies $\mathcal{M} = \langle 0 \rangle$.

Lemma 1. *Let $f, g, h \in \mathbb{A}$ such that $f = gh$ holds. Assume $n \ge 2$. We write $f = \sum_{i=0}^{\infty} f_i$, $g = \sum_{i=0}^{\infty} g_i$ and $h = \sum_{i=0}^{\infty} h_i$, where $f_i, g_i, h_i \in \mathcal{M}^i \setminus \mathcal{M}^{i+1}$ holds for all $i > 0$, with $f_0, g_0, h_0 \in \mathbb{K}$. We note that these decompositions are uniquely defined. Let $r \in \mathbb{N}$. We assume that $f_0 = 0$ and $h_0 \ne 0$ both hold. Then the term g_r is uniquely determined by $f_1, \ldots, f_r, h_0, \ldots, h_{r-1}$.*

Lemma 1 is essential to our implementation of Weierstrass Preparation Theorem (WPT). Hence, we give a proof by induction on r. Since $g_0 h_0 = f_0 = 0$

and $h_0 \neq 0$ both hold, the claim is true for $r = 0$. Now, let $r > 0$ and we can assume that $g_0, \ldots, g_{r-1}$ are uniquely determined by $f_1, \ldots, f_{r-1}, h_0, \ldots, h_{r-2}$. Observe that to determine g_r, it suffices to expand $f = gh$ modulo $\mathcal{M}^{r+1}$:

$$f_1 + f_2 + \cdots + f_r = g_1 h_0 + (g_2 h_0 + g_1 h_1) + \cdots + (g_r h_0 + g_{r-1} h_1 + \cdots + g_1 h_{r-1}).$$

g_r is then found by polynomial multiplication and addition and a division by h_0.

Now, let $f \in \mathbb{A}[[X_n]]$, written as $f = \sum_{i=0}^{\infty} a_i X_n^i$ with $a_i \in \mathbb{A}$ for all $i \in \mathbb{N}$. We assume $f \not\equiv 0 \mod \mathcal{M}[[X_n]]$. Let $d \geq 0$ be the smallest integer such that $a_d \notin \mathcal{M}$. Then, WPT states the following.

Theorem 1. *There exists a unique pair (α, p) satisfying the following:*

(i) α is an invertible power series of $\mathbb{A}[[X_n]]$,
(ii) $p \in \mathbb{A}[X_n]$ is a monic polynomial of degree d,
(iii) writing $p = X_n^d + b_{d-1} X_n^{d-1} + \cdots + b_1 X_n + b_0$, we have: $b_{d-1}, \ldots, b_1, b_0 \in \mathcal{M}$,
(iv) $f = \alpha p$ holds.

Moreover, if f is a polynomial of $\mathbb{A}[X_n]$ of degree $d + m$, for some m, then α is a polynomial of $\mathbb{A}[X_n]$ of degree m.

PROOF. If $n = 1$, then writing $f = \alpha X_n^d$ with $\alpha = \sum_{i=0}^{\infty} a_{i+d} X_n^i$ proves the existence of the claimed decomposition. Now assume $n \geq 2$. Let us write $\alpha = \sum_{i=0}^{\infty} c_i X_n^i$ with $c_i \in \mathbb{A}$ for all $i \in \mathbb{N}$. Since we require α to be a unit, we have $c_0 \notin \mathcal{M}$. We must then solve for $b_{d-1}, \ldots, b_1, b_0, c_0, c_1, \ldots, c_d, \ldots$ such that for all $m \geq 0$ we have:

$$a_0 = b_0 c_0$$
$$a_1 = b_0 c_1 + b_1 c_0$$
$$a_2 = b_0 c_2 + b_1 c_1 + b_2 c_0$$
$$\vdots$$
$$a_{d-1} = b_0 c_{d-1} + b_1 c_{d-2} + \cdots + \cdots + b_{d-2} c_1 + b_{d-1} c_0$$
$$a_d = b_0 c_d + b_1 c_{d-1} + \cdots + \cdots + b_{d-1} c_1 + c_0$$
$$a_{d+1} = b_0 c_{d+1} + b_1 c_d + \cdots + \cdots + b_{d-1} c_2 + c_1$$
$$\vdots$$
$$a_{d+m} = b_0 c_{d+m} + b_1 c_{d+m-1} + \cdots + \cdots + b_{d-1} c_{m+1} + c_m$$
$$\vdots$$

We will compute each of $b_{d-1}, \ldots, b_1, b_0, c_0, c_1, \ldots, c_d, \ldots$ modulo each of the successive powers of $\mathcal{M}$, that is, $\mathcal{M}, \mathcal{M}^2, \ldots, \mathcal{M}^r, \ldots$. We start modulo $\mathcal{M}$. By definition of d, the left hand sides of the first d equations above are all 0 mod $\mathcal{M}$. Since c_0 is a unit, each of $b_0, b_1, \ldots, b_{d-1}$ is $0 \mod \mathcal{M}$. Plugging this into the remaining equations we obtain $c_i \equiv a_{d+i} \mod \mathcal{M}$, for all $i \geq 0$. Therefore, we have solved for each of $b_{d-1}, \ldots, b_1, b_0, c_0, c_1, \ldots, c_d, \ldots$ modulo $\mathcal{M}$. Let $r > 0$ be an integer. We assume that we have inductively determined each of $b_{d-1}, \ldots, b_1, b_0, c_0, c_1, \ldots, c_d, \ldots$ modulo each of $\mathcal{M}, \ldots, \mathcal{M}^r$. We wish to determine them modulo $\mathcal{M}^{r+1}$. Consider the first equation, namely $a_0 = b_0 c_0$, with $a_0, b_0, c_0 \in \mathbb{A}$. It follows from the hypothesis and Lemma 1 that we can compute

b_0 modulo $\mathcal{M}^{r+1}$. Consider the second equation, that we re-write $a_1 - b_0 c_1 = b_1 c_0$. A similar reasoning applies and we can compute b_1 modulo $\mathcal{M}^{r+1}$. Continuing in this manner, we can compute each of $b_2, \ldots, b_{d-1}$ modulo $\mathcal{M}^{r+1}$. Finally, using the remaining equations, determine $c_i \mod \mathcal{M}^{r+1}$, for all $i \geq 0$.   □

This theorem allows for three remarks. First, the assumption of the theorem, namely $f \not\equiv 0 \mod \mathcal{M}[[X_n]]$, can always be met, for any $f \neq 0$, by a suitable linear change of coordinates. Second, WPT can be used to prove that $\mathbb{K}[[X_1, \ldots, X_n]]$ is both a unique factorization domain (UFD) and a Noetherian ring. Third, in the context of the theory of analytic functions, WPT implies that any analytic function (namely f in our context) resembles a polynomial (namely p in our context) in the vicinity of the origin.

Now, let $f = a_k Y^k + \cdots + a_1 Y + a_0$ with $a_k, \ldots, a_0 \in \mathbb{K}[[X_1, \ldots, X_n]]$. We define $\overline{f} = f(0, \ldots, 0, Y) \in \mathbb{K}[Y]$. We assume that f is monic in Y ($a_k = 1$). We further assume $\mathbb{K}$ is algebraically closed. Thus, there exist positive integers $k_1, \ldots, k_r$ and pairwise distinct elements $c_1, \ldots, c_r \in \mathbb{K}$ such that we have $\overline{f} = (Y - c_1)^{k_1}(Y - c_2)^{k_2} \cdots (Y - c_r)^{k_r}$.

Theorem 2 (Hensel's Lemma).   *There exists* $f_1, \ldots, f_r \in \mathbb{K}[[X_1, \ldots, X_n]]$ $[Y]$, *all monic in* Y, *such that we have:*

1. $f = f_1 \cdots f_r$,
2. $\deg(f_j, Y) = k_j$, *for all* $j = 1, \ldots, r$,
3. $\overline{f_j} = (Y - c_j)^{k_j}$, *for all* $j = 1, \ldots, r$.

PROOF. The proof is by induction on r. Assume first $r = 1$. Observe that $k = k_1$ necessarily holds. Now define $f_1 := f$. Clearly f_1 has all the required properties. Assume next $r > 1$. We apply a change of coordinates sending c_r to 0. That is: $g(X_1, \ldots, X_n, Y) := f(X_1, \ldots, X_n, Y + c_r) = (Y + c_r)^k + a_1(Y + c_r)^{k-1} + \cdots + a_k$. WPT applies to g. Hence there exist $\alpha, p \in \mathbb{K}[[X_1, \ldots, X_n]][Y]$ such that α is a unit, p is a monic polynomial of degree k_r, with $\overline{p} = Y^{k_r}$, and we have $g = \alpha p$. Then, we set $f_r(Y) = p(Y - c_r)$ and $f^* = \alpha(Y - c_r)$. Thus f_r is monic in Y and we have $f = f^* f_r$. Moreover, we have The induction hypothesis applied to f^* implies the existence of $f_1, \ldots, f_{r-1}$.   □

3   The Design and Implementation of Lazy Power Series

Our power series implementation is both lazy and high-performing. To achieve this, our design and implementation has two goals:

(i) compute only terms of the series which are truly needed; and
(ii) have the ability to "resume" a computation, in order to obtain a higher precision power series without restarting from the beginning.

Of course, the lazy nature of our implementation refers directly to (i), while the high-performance nature is due in part to (ii) and in part to other particular implementation details to be discussed.

Facilitating both of these aspects requires the use of some sort of generator function—a function which returns new terms for a power series to increase its precision. Such a *generator*, is the key to high-performance in our implementation, yet also the most difficult part of the design.

Our goal is to define a structure encoding power series so that they may be dynamically updated on request. Each power series could then be represented as a polynomial alongside some generator function. A key element of this design is to "hide" the updating of the underlying polynomial. In our C implementation this is done through a functional interface comprising of two main functions: (*i*) getting the homogeneous part of a power series, and (*ii*) getting the polynomial part of a power series, each for a requested degree. These functions call some underlying generator to produce terms until the requested degree is satisfied.

```
geometric_series_ps := proc(vars::list)
   local homog_parts := proc(vars::list)
      return d -> sum(vars[i], i=1..nops(vars))^d;
   end proc;
   ps := table();
   ps[DEG] := 0;
   ps[GEN] := homog_parts(vars); #capture vars in closure, return a function
   ps[POLY] := ps[GEN](0);
   return ps;
end proc;
```

Listing 1. The geometric series as a lazy power series

As a first example, consider, the construction of the geometric series as a lazy power series, in MAPLE-style pseudo-code, in Listing 1. A power series is a data structure holding a polynomial, a generator function, and an integer to indicate up to which degree the power series is currently known. In this simple example, we see the need to treat functions as first-class objects. The manipulation of such functions is easy in functional or scripting languages, where dynamic typing and first-class function objects support such manipulation. This manipulation becomes further interesting where the generator of a power series must invoke other generators, as in the case of arithmetic (see Sect. 3.2).

In support of high-performance we choose to implement our power series in the strongly-typed and compiled C programming language rather than a scripting language. On one hand, this allows direct access to our underlying high-performance polynomial implementation [5], but on the other hand creates an impressive design challenge to effectively handle the need for dynamic function manipulation. In this section we detail our resulting solution, which makes use of a so-called *ancestry* in order for the generator function of a newly created power series to "remember" from where it came. We begin by discussing the power series data structure, and our solution to generator functions in C. Then, Sect. 3.2 examines power series multiplication and division using this structure, and evaluates our arithmetic performance against SAGEMATH and MAPLE.

3.1   The Power Series Data Structure, Generators, and Ancestors

The organization of our power series data structure is focused on supporting incremental generation of new terms through continual updates. To support this, the first fundamental design element is the storage of terms of the power series. The current polynomial part, i.e. the terms computed so far, of a power series are stored in a *graded representation*. A dense array of (pointers to) polynomials is maintained whereby the index of a polynomial in this array is equal to its (total) degree. Thus, this is an array of homogeneous polynomials representing the homogeneous parts of the power series, called the *homogeneous part array*. The power series data structure is a simple C `struct` holding this array, as well as integer numbers indicating the degree up to which homogeneous parts are currently known, and the allocation size of the homogeneous part array.

Using our graded representation, the generator function is simply a function returning the homogeneous part of a power series for a requested degree. Unfortunately, in the C language, functions are not readily handled as objects. Hence, we look to essentially create a *closure* for the generator function (see, e.g., [21, Ch. 3]), by storing a function pointer along with the values necessary for the function. For simplicity of implementation, these captured values are passed to the function as arguments. We first describe this function pointer.

In an attempt to keep the generators as simple as possible, we enforce some symmetry between all generators and thus the stored function pointers. Namely: (*i*) the first parameter of each generator must be an integer, indicating the degree of the homogeneous polynomial to be generated; and (*ii*) they must return that homogeneous polynomial. For some generator functions, e.g. the geometric series, this single integer argument is enough to obtain a particular homogeneous part. However, this is insufficient for most cases, particularly for generating a homogeneous part of a power series which resulted from an arithmetic operation.

Therefore, to introduce some flexibility in the generators, we extend the previous definition of a generator function to include a finite number of `void` pointer parameters following the first integer parameter. The use of `void` pointer parameters is a result of the fact that function pointers must be declared to point to a function with a particular number and type of parameters. Since we want to store this function pointer in the power series `struct`, we would otherwise need to capture all possible function declarations, which is a very rigid solution. Instead, `void` pointer parameters simultaneously allow for flexibility in the types of the generator parameters, as well as limit the number of function pointer types which must be captured by the power series `struct`. Flexibility arises where these `void` pointers can be cast to any other pointer type, or even cast to any machine-word-sized plain data type (e.g. `long` or `double`). In our implementation these so-called *void generators* are simple wrappers, casting each `void` pointer to the correct data type for the particular generator, and then calling the *true generator*. Sect. 3.2 provides an example in Listing 4.

Our implementation, which supports power series arithmetic, Weierstrass preparation, and factorization via Hensel's lemma, currently requires only 4 unique types of function pointers for these generators. All of these function pointers return a polynomial and take an integer as the first parameter. They differ in taking 0–3 void pointer parameters as the remaining parameters. We call the number of these void pointer parameter the generator's *order*. We then create a union type for these 4 possible function pointers and store only the union in the power series struct. The generator's order is also stored as an integer to be able to choose the correct generator from the union type at runtime.

Finally, these void pointers are also stored in the struct to eventually be passed to the generator. When the generator's order is less than maximum, these extra void pointers are simply set to NULL. The structure of these generators, the generator union type, and the power series struct itself is shown in Listing 2. In our implementation, these generators are used generically, via the aforementioned functional interface. In the code listings which follow, these functions are named homogPart_PS and polynomialPart_PS, to compute the homogeneous part and polynomial part of a power series, respectively.

```
typedef Poly_ptr (*homog_part_gen)(int);
typedef Poly_ptr (*homog_part_gen_unary)(int, void*);
typedef Poly_ptr (*homog_part_gen_binary)(int, void*, void*);
typedef Poly_ptr (*homog_part_gen_tertiary)(int, void*, void*, void*);

typedef union HomogPartGenerator {
    homog_part_gen nullaryGen;
    homog_part_gen_unary unaryGen;
    homog_part_gen_binary binaryGen;
    homog_part_gen_tertiary tertiaryGen;
} HomogPartGenerator_u;

typedef struct PowerSeries {
    int deg;
    int alloc;
    Poly_ptr* homog_polys;
    HomogPartGenerator_u gen;
    int genOrder;
    void *genParam1, *genParam2, *genParam3;
} PowerSeries_t;
```

Listing 2. A first implementation of the power series struct in C and function pointer declarations for the possible generator functions. Poly_ptr is a pointer to a polynomial.

In general, these void pointer generator parameters are actually pointers to existing power series struct. For example, the operands of an arithmetic operation would become arguments to the generator of the result. This relation then yields a so-called *ancestry* of power series. In this indirect way, a power series "remembers" from where it came, in order to update itself upon request via its generator. This may trigger a cascade of updates where updating a power series requires updating its "parent" power series, and so on up the *ancestry tree*. Section 3.2 explores this detail in the context of power series arithmetic, meanwhile it is also discussed as a crucial part of a lazy implementation of Weierstrass preparation (Sect. 5) and factorization via Hensel's lemma (Sect. 6).

The implementation of this ancestry requires yet one more additional feature. Since our implementation is in the C language, we must manually manage memory. In particular, references to parent power series (via the void pointers) must remain valid despite actions from the user. Indeed, the underlying updating mechanism should be transparent to the end-user. Thus, it should be perfectly valid for an end-user to obtain, for example, a power series product, and then free the memory associated with the operands of the multiplication.

In support of this we have established a reference counting scheme. Whenever a power series is made the parent of another power series its reference count is incremented. Therefore, users may "free" or "destroy" a power series when it is no longer needed, but the memory persists as long as some other power series has reference to it. Destruction is then only a decrement of a reference counter. However, once the counter falls to 0, the data is actually freed, and moreover, a child power series will decrement the reference count of its parents. In a final complication, we must consider the case when a void pointer parameter is not pointing to a power series. We resolve this by storing, in the power series struct, a value to identify the actual type of a void parameter. A simple if condition can then check this type and conditionally free the generator parameter, if it is not plain data. For example, a power series or a UPoPS, see Listing 3.

```
typedef enum GenParamType {
    PLAIN_DATA = 0,
    POWER_SERIES = 1,
    UPOPS = 2,
    MPQ_LIST = 3
} GenParamType_e;

void destroyPowerSeries_PS(PowerSeries_t* ps) {
    --(ps->refCount);
    if (ps->refCount <= 0) {
        for (int i = 0; i <= ps->deg; ++i) {
        freePolynomial(homog_polys[i]);
        }
        if (ps->genParam1 != NULL && ps->paramType1 == POWER_SERIES) {
            destroyPowerSeries_PS((PowerSeries_t*) ps->genParam1);
        }
        // repeat for other parameters.
    }
}
```

Listing 3. Extending the power series struct to include reference counting and proper management of reference counts to parent power series via **destroyPowerSeries_PS**

3.2   Implementing Power Series Arithmetic

With the power series structure fully defined, we are now able to see examples putting its generators to use. Given the design established in the previous section, implementing a power series operation is as simple as defining the unique generator associated with that operation. In this section we present power series multiplication and division using this design. Let us begin with the former.

As we have seen in Sect. 2, the power series product of $f, g \in \mathbb{K}[[X_1, \ldots, X_n]]$ is defined simply as $h = fg = \sum_{d \in \mathbb{N}} \left(\sum_{k+\ell=d} (f_{(k)} g_{(\ell)}) \right)$. In our graded representation, continually computing new terms of h requires simply computing homogeneous parts of increasing degree. Indeed, for a particular degree d we have $(fg)_{(d)} = \sum_{k+\ell=d} f_{(k)} g_{(\ell)}$. Through our use of an ancestry and generators, the power series h can be constructed lazily, by simply defining its generator and generator parameters, and instantly returning the resulting struct. The generator in this case is exactly a function to compute $(fg)_{(d)}$ from f and g.

In reality, the generator stored in the struct encoding h is the *void generator* homogPartVoid_prod_PS which, after casting parameters, simply calls the *true generator*, homogeneousPart_prod_PS. This is shown in Listing 4. There, multiplyPowerSeries_PS is the actual power series operator, returning a lazily constructed power series product. There, the parents f and g are reserved (reference count incremented) and assigned to be generator parameters, and the generator function pointer set. Finally, a single term of the product is computed.

```
Poly_ptr homogPart_prod_PS(int d, PowerSeries_t* f, PowerSeries_t* g) {
    Poly_ptr sum = zeroPolynomial();
    for (int i = 0; i <= d; i++) {
        Poly_ptr prod = multiplyPolynomials(
                            homogPart_PS(d-i, f), homogPart_PS(i, g));
        sum = addPolynomials(sum, prod);
    }
    return sum;
}


Poly_ptr homogPartVoid_prod_PS(int d, void* param1, void* param2) {
    return homogPart_prod_PS(d, (PowerSeries_t*) param1,
                                (PowerSeries_t*) param2);
}


PowerSeries_t* multiplyPowerSeries_PS(PowerSeries_t* f, PowerSeries_t* g) {
    if (isZeroPowerSeries_PS(f) || isZeroPowerSeries_PS(g)) {
        return zeroPowerSeries_PS();
    }
    reserve_PS(f); reserve_PS(g);
    PowerSeries_t* prod = allocPowerSeries(1);
    prod->gen.binaryGen= &(homogPartVoid_prod_PS)
    prod->genParam1 = (void*) f;
    prod->genParam2 = (void*) g;
    prod->paramType1 = POWER_SERIES;
    prod->paramType2 = POWER_SERIES;
    prod->deg = 0;
    prod->homogPolys[0] = homogPart_prod_PS(0, f, g);
    return prod;
}
```

Listing 4. Computing the multiplication of two power series, where homogPart_prod_PS is the generator of the product

Now consider finding the quotient $h = \sum_e c_e X^e$ which satisfies $f = gh$ for a given power series $f = \sum_e a_e X^e$ and an invertible power series $g = \sum_e b_e X^e$. One could proceed by equating coefficients in $f = gh$, with b_0 being the constant term of g, to obtain $c_e = 1/b_0 \left(a_e - \sum_{k+\ell=e} b_k c_\ell \right)$. This formula can easily be rearranged in order to find the homogeneous part of h for a given degree d: $h_{(d)} = 1/g_{(0)} \left(f_{(d)} - \sum_{k=1}^{d} g_{(k)} h_{(d-k)} \right)$. This formula is possible since to compute $h_{(d)}$ we need only $h_{(i)}$ for $i = 1, \ldots, d-1$. Moreover, the base case is simply $h_{(0)} = f_{(0)}/g_{(0)}$, a valid division in $\mathbb{K}$ since $g_{(0)} \neq 0$. The rest follows by induction.

In our graded representation, this formula yields a generator for a power series quotient. The realization of this generator in code is simple, as shown in Listing 5. Not shown is the void generator wrapper and a top-level function to return the lazy quotient, which is simply symmetric to the previous multiplication example.

The only trick to this generator for the quotient is that it requires a reference to the quotient itself. This creates an issue of a circular reference in the power series ancestry. To avoid this, we abuse our parameter typing and label the quotient's reference to itself as plain data.

```
Poly_ptr homogPart_quo_PS(int d, PowerSeries_t* f, PowerSeries_t* g,
    PowerSeries_t* h) {
    if (d == 0) {
        return dividePolynomials(homogPart_PS(0, f), homogPart_PS(0, g));
    }
    Poly_ptr s = homogPart_PS(d, f);
    for (int i = 1; i <= deg; ++i) {
        Poly_ptr p = multiplyPolynomials(homogPart_PS(i, g),
                                         homogPart_PS(d-i, h));
        s = subPolynomials(s, p);
    }
    return divideByRational(s, homogPart(0, g))
}
```

Listing 5. Computing the division of two power series, where `homogPart_quo_PS` is the genrator of the quotient

We now look to compare our implementation against SAGEMATH [22], and MAPLE 2020. In MAPLE, the `PowerSeries` library [2,15] provides lazy multivariate power series, meanwhile the built-in `mtaylor` command provides *truncated* multivariate taylor series. Similarly, SAGEMATH includes only truncated power series. In these latter two, an explicit precision must be used and truncations cannot be extended once computed. Consequently, our experimentation only measures computing a particular precision, thus not using our implementation's ability to resume computation. We compare against all three; see Figs. 1, 2 and 3.

In SAGEMATH, the multivariate power series ring $R[[X_1, \ldots, X_n]]$ is implemented using the univariate power series ring $S[[T]]$ with $S = R[X_1, \ldots, X_n]$. In $S[[T]]$, the subring formed by all power series f such that the coefficient of T^i in f is a homogeneous polynomial of degree i (for all $i \geq 0$) is isomorphic to $R[[X_1, \ldots, X_n]]$. By default, Singular [8] underlies the multivariate polynomial ring S while Flint [12] underlies the univariate polynomials used in univariate power series. Python 3.7 interfaces and joins these underlying implementations. To see exactly how SAGEMATH works consider $f \in \mathbb{K}[[X_1, X_2]]$ with the goal is to compute $\frac{1}{f}$ and $f \cdot \frac{1}{f}$ to precision d. One begins by constructing the power series ring in X_1, X_2 over $\mathbb{Q}$ with the default precision set to d as `R.<x,y> = PowerSeriesRing(QQ, default_prec=d)`. Then `g = f^-1` returns the inverse, and `h = f * g` the desired product, to precision d.

Throughout this paper our benchmarks were collected with a time limit of 1800 seconds on a machine running Ubuntu 18.04.4 with an Intel Xeon X5650 processor running at 2.67 GHz, with 12x4GB DDR3 memory at 1.33 GHz.

The first set of benchmarks are presented in Fig. 1 where the power series $f = 1 + X_1 + X_2$ is both inverted and multiplied by its inverse. Figures 2 and 3 present the same but for $f = 1 + X_1 + X_2 + X_3$ and $f = 2 + \frac{1}{3}(X_1 + X_2)$, respectively. In all cases, $f \cdot \frac{1}{f}$ includes the time to compute the inverse. It is clear that our implementation is orders of magnitude faster than existing implementations. This is due in part to the efficiency of our underlying polynomial arithmetic implementation [5], but also to our execution environment. Our implementation is written in the C language and fully compiled, meanwhile, both SAGEMATH and MAPLE have a level of interpreted code, which surely impacts performance. We note that, through truncated power series as polynomials, the dense multiplication of a power series by its inverse is trivial for SAGEMATH and mtaylor.

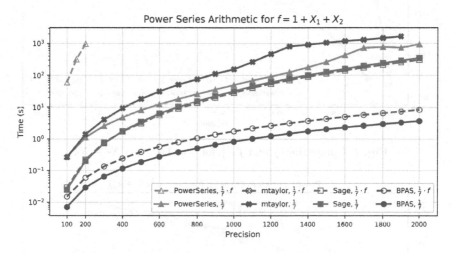

Fig. 1. Computing $\frac{1}{f}$ and $f \cdot \frac{1}{f}$ for $f = 1 + X_1 + X_2$

4   Univariate Polynomials over Lazy Power Series

A univariate polynomial with multivariate power series coefficients, i.e. a univariate polynomial over power series (UPoPS), is implemented as a simple extension of our existing power series. Following a simple dense univariate polynomial design, our UPoPS are represented as an array of coefficients, each being a pointer to a power series, where the index of the coefficient in the array implies the degree of the coefficient's associated monomial. Integers are also stored for the degree of the polynomial and the allocation size of the coefficient array. In support of the underlying lazy power series, we also add reference counting to UPoPS.

The arithmetic of UPoPS is inherited directly from its coefficient ring (our lazy power series) and follows a naive implementation of univariate polynomials (see, e.g. [10, Ch. 2]). Through the use of our lazy power series, our implementation of UPoPS is automatically lazy through each individual coefficient's ancestry. Lazy UPoPS addition, subtraction, and multiplication follow easily.

One important operation on UPoPS which is not inherited directly from our power series implementation is Taylor shift. This operation takes a UPoPS $f \in \mathbb{K}[[X_1, \ldots, X_n]][Y]$ and returns $f(Y + c)$ for some $c \in \mathbb{K}$. Normally, the shift operator would be defined for any element of the ground ring $\mathbb{K}[[X_1, \ldots, X_n]]$, however our use of Taylor shift in applying Hensel's lemma (see Sect. 6), requires

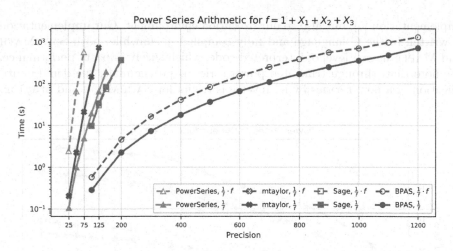

Fig. 2. Computing $\frac{1}{f}$ and $f \cdot \frac{1}{f}$ for $f = 1 + X_1 + X_2 + X_3$

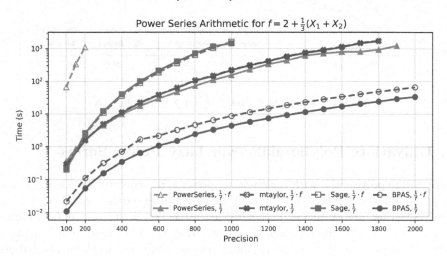

Fig. 3. Computing $\frac{1}{f}$ and $f \cdot \frac{1}{f}$ for $f = 2 + \frac{1}{3}(X_1 + X_2)$

only shifting by elements of $\mathbb{K}$, and we thus specialize to that case. Since the coefficients of f are lazy power series, our goal is to compute $f(Y + c)$ lazily as well. Since our UPoPS are represented in a dense fashion, we compute the coefficients of $f(Y+c)$ as a polynomial in Y. Let $\mathbf{S} = (s_{i,j})$ be the lower triangular matrix such that $s_{i,j}$ is the coefficient of Y^j in the binomial expansion $(Y + c)^i$, for $i = 0, \ldots, k$, and $j = 0, \ldots, i$, where $k = \deg(f)$. Let $\mathbf{A} = (a_i)$ be the vector of the coefficients of f and $\mathbf{B} = (b_i)$ that of the coefficients of $f(Y + c)$, so that we have $f(Y) = \sum_{0 \le i \le k} a_i Y^i$ and $f(Y + c) = \sum_{0 \le i \le k} b_i Y^i$. Then we can verify that b_i is the inner product of the i-th sub-diagonal of $\mathbf{S}$ with the lower $k + 1 - i$ elements of $\mathbf{A}$, for $i = 0, \ldots, k$. In particular for $i = 0$, the coefficient b_0 is the inner product of the diagonal of $\mathbf{S}$ and the vector $\mathbf{A}$.

Recalling that $c \in \mathbb{K}$, the construction of b_i can be performed in a graded fashion from the linear combinations of homogeneous parts of a_j for $j \leq i$. The homogeneous part $b_{i_{(d)}}$ of degree d, can be computed from only $a_{j_{(d)}}$, for $j \leq i$. Therefore, a generator for b_i is easily constructed from the homogeneous parts of a_j, for $j \leq i$, using multiplication by elements of $\mathbb{K}$ and polynomial addition. Therefore, we can construct the entire UPoPS $f(Y+c)$ in a lazy manner through initializing each coefficient b_i with a so-called linear combination generator. Since the main application of Taylor shift is factorization via Hensel's lemma, we leave its evaluation to Sect. 6 where benchmarks for factorization are presented.

5 Lazy Weierstrass Preparation

In this section we consider the application of Weierstrass Preparation Theorem (WPT) to univariate polynomials over power series. Let f, p, α be elements of $\mathbb{K}[[X_1, \ldots, X_n]][Y]$, where $f = \sum_{i=0}^{d+m} a_i Y^i$, $p = Y^d + \sum_{i=0}^{d-1} b_i Y^i$, and $\alpha = \sum_{i=0}^{m} c_i Y^i$. From the proof of WPT (Theorem 1), we have that $f = \alpha p$ implies the following equalities:

$$
\begin{aligned}
a_0 &= b_0 c_0 \\
a_1 &= b_0 c_1 + b_1 c_0 \\
&\vdots \\
a_{d-1} &= b_0 c_{d-1} + b_1 c_{d-2} + \cdots + b_{d-2} c_1 + b_{d-1} c_0 \\
a_d &= b_0 c_d + b_1 c_{d-1} + \cdots + b_{d-1} c_1 + c_0 \\
&\vdots \\
a_{d+m-1} &= b_{d-1} c_m + c_{m-1} \\
a_{d+m} &= c_m
\end{aligned}
\tag{1}
$$

Following the proof, we wish to solve these equations modulo successive powers of $\mathcal{M}$, the maximal ideal of $\mathbb{K}[[X_1, \ldots, X_n]]$. This implies that we will be iteratively updating each power series $b_0, \ldots, b_{d-1}, c_0, \ldots, c_m$ by adding homogeneous polynomials of increasing degree, precisely as we have done for all lazy power series operations thus far. To solve these equations modulo $\mathcal{M}^{r+1}$, both the proof of WPT and the algorithm operate in two phases. First, the coefficients $b_0, \ldots, b_{d-1}$ of p are updated using the equations from a_0 to a_{d-1}, one after the other. Second, the coefficients $c_0, \ldots, c_m$ of α are updated.

Let us begin with the first phase. Rearranging the equations that express a_0 to a_{d-1} shows their successive dependency where b_{i-1} is needed for b_i:

$$
\begin{aligned}
a_0 &= b_0 c_0 \\
a_1 - b_0 c_1 &= b_1 c_0 \\
a_2 - b_0 c_2 - b_1 c_1 &= b_2 c_0 \\
&\vdots \\
a_{d-1} - b_0 c_{d-1} - b_1 c_{d-2} + \cdots - b_{d-2} c_1 &= b_{d-1} c_0
\end{aligned}
\tag{2}
$$

Consider that $b_0, \ldots, b_{d-1}, c_0, \ldots, c_m$ are known modulo $\mathcal{M}^r$ and $a_0, \ldots, a_{d-1}$ are known modulo $\mathcal{M}^{r+1}$. Using Lemma 1 the first equation $a_0 = b_0 c_0$ can then be solved for b_0 modulo $\mathcal{M}^{r+1}$. From there, the expression $a_1 - b_0 c_1$ then becomes known modulo $\mathcal{M}^{r+1}$. Notice that the constant term of b_0 is 0 by definition, thus the product $b_0 c_1$ is known modulo $\mathcal{M}^{r+1}$ as long as b_0 is known modulo $\mathcal{M}^{r+1}$. Therefore, the entire expression $a_1 - b_0 c_1$ is known modulo $\mathcal{M}^{r+1}$ and Lemma 1 can be applied to solve for b_1 in the equation $a_1 - b_0 c_1 = b_1 c_0$. This argument follows for all equations, therefore solving for all $b_0, \ldots, b_{d-1}$ modulo $\mathcal{M}^{r+1}$.

In the second phase, we look to determine $c_0, \ldots, c_m$ modulo $\mathcal{M}^{r+1}$. Here, we have already computed $b_0, \ldots, b_{d-1}$ modulo $\mathcal{M}^{r+1}$. A rearrangement of the remaining equations of (1) shows that each c_i may be computed modulo $\mathcal{M}^{r+1}$:

$$
\begin{aligned}
c_m &= a_{d+m} \\
c_{m-1} &= a_{d+m-1} - b_{d-1} c_m \\
c_{m-2} &= a_{d+m-2} - b_{d-2} c_m - b_{d-1} c_{m-1} \\
&\vdots \\
c_0 &= a_d - b_0 c_d - b_1 c_{d-1} - \cdots - b_{d-1} c_1
\end{aligned}
\tag{3}
$$

Consider the second equation. Observe that a_{d+m-1} and b_{d-1} are known modulo $\mathcal{M}^{r+1}$ and that $b_{d-1} \in \mathcal{M}$ holds. Then, the product $b_{d-1} c_m$ is known modulo $\mathcal{M}^{r+1}$ and we deduce c_{m-1} modulo $\mathcal{M}^{r+1}$. The same follows for $c_{m-2}, \ldots, c_0$.

With these two sets of re-arranged equations, we have seen how the coefficients of p and α can be updated modulo successive powers of $\mathcal{M}$. That is to say, how they can be updated by adding homogeneous parts of successive degrees. This design lends itself to be implemented as generator functions.

The first challenge to this design is that each power series coefficient of p is not independent, and must be updated in a particular order. Moreover, to generate homogeneous parts of degree d for the coefficients of p, the coefficients of α must also be updated to degree $d - 1$. Therefore, it is a required side effect of each generator of $b_0, \ldots, b_{d-1}, c_0, \ldots, c_m$ that all other power series are updated. To implement this, the generators of the power series of p are a mere wrapper of the same underlying updating function which updates all coefficients simultaneously. This so-called *Weierstrass update* follows two phases as just explained.

In the first phase one must use Lemma 1 to solve for the homogeneous part of degree r for each $b_0, \ldots, b_{d-1}$. To achieve this effectively, our implementation follows two key points. The first is an efficient implementation of Lemma 1 itself. Consider again the equations of Lemma 1 for $f = gh$ modulo $\mathcal{M}^{r+1}$:

$$
\begin{aligned}
f_{(1)} + f_{(2)} + \cdots + f_{(r)} &= (g_{(1)} + g_{(2)} + \cdots + g_{(r)})(h_{(0)} + h_{(1)} + \cdots + h_{(r)}) \\
&= \big(g_{(1)} h_{(0)}\big) + \big(g_{(2)} h_{(0)} + g_{(1)} h_{(1)}\big) + \cdots + \\
&\quad \big(g_{(r)} h_{(0)} + g_{(r-1)} h_{(1)} + \cdots + g_{(1)} h_{(r-1)}\big).
\end{aligned}
\tag{4}
$$

The goal is to obtain $g_{(r)}$. What one should realize is that computing $g_{(r)}$ requires only a fraction of this formula. In particular, we have

---

Algorithm 1. WeierstrassUpdate(f, p, α, $\mathcal{F}$)

Input: $f = \sum_{i=0}^{d+m} a_i Y^i$, $p = \sum_{i=0}^{d} b_i Y^i$, $\alpha = \sum_{i=0}^{m} c_i Y^i$, $a_i, b_i, c_i \in \mathbb{K}[[X_1, \ldots, X_n]]$
  satisfying Theorem 1, $\mathcal{F} = \{F_i \mid F_i = a_i - \sum_{k=0}^{i} b_k c_{i-k}, i = 0, \ldots, d-1\}$, with
  $b_0, \ldots, b_{d-1}, c_0, \ldots, c_m$ known modulo $\mathcal{M}^r$, the maximal ideal of $\mathbb{K}[[X_1, \ldots, X_n]]$.
Output: $b_0, \ldots, b_{d-1}, c_0, \ldots, c_m$ known modulo $\mathcal{M}^{r+1}$, updated in-place.

   ▷ phase one
1: **for** $i = 0$ to $d - 1$ **do**
2:    | $s := 0$
3:    | **for** $k = 1$ to $r - 1$ **do**
4:    | | $s := s +$ homogPart_PS$(r - k, b_i) \times$ homogPart_PS(k, c_0)
5:    | homogPart_PS$(r, b_i) := ($ homogPart_PS$(r, F_i) - s\ )/$ homogPart_PS$(0, c_0)$

   ▷ phase two
6: **for** $i = 0$ to m **do**
7:    | homogPart_PS(r, c_i) ▷ force an update of c_i for next update.

$$f_{(r)} = g_{(r)}h_{(0)} + g_{(r-1)}h_{(1)} + \cdots + g_{(1)}h_{(r-1)}, \tag{5}$$

and $g_{(r)}$ can be computed with simply polynomial addition and multiplication, followed by the division of a single element of $\mathbb{K}$, since $h_{(0)}$ has degree 0.

The second key point is that, in order to compute $g_{(r)}$, i.e. the homogeneous parts of degree r of $b_0, \ldots, b_{d-1}$, we must first find $f_{(r)}$, i.e. the homogeneous parts of degree r of $a_0, a_1 - b_0 c_1, a_2 - b_0 c_2 - b_1 c_1$, etc. from (2). A nice result of our existing power series design is that we can define some lazy power series, say F_i, such that $F_i = a_i - \sum_{k=0}^{i} b_k c_{i-k}$. These F_i can then be automatically updated via its generators when the b_k are updated. The implementation of phase one of Weierstrass update is then simply a loop over solving Eq. (5), where $f_{(r)}$ is automatically obtained through the use of generators on the power series F_i.

Phase two of Weierstrass update follows the same design as in the definition of those F_i power series. In particular, from (3) we can see that each $c_m, \ldots, c_0$ is merely the result of some power series arithmetic. Hence, we simply rely on the underlying power series arithmetic generators to be the generators of $c_m, \ldots, c_0$.

With the above discussion, we have fully defined a lazy implementation of Weierstrass preparation. It begins with an initialization, which simply uses lazy power series arithmetic to create $F_0, \ldots, F_{d-1}, c_m, \ldots, c_0$, and initializes each $b_0, \ldots, b_{d-1}$ to 0. Then, the generators for $b_0, \ldots, b_{d-1}$ all call the same underlying Weierstrass update function. This function is shown in Algorithm 1, which is split into two phases as our discussion has suggested.

In our implementation, we store a pointer to the array of $F_0, \ldots, F_{d-1}$ in the UPoPS **struct** of p for ease of calling Weierstrass update. This, along with the circular references between coefficients of p and α, creates a delicate situation for reference counting. Readers may refer to our code in BPAS [3] for our solution.

Notice also that, although phase one requires updating each b_i in order from $i = 0$ to $d - 1$, the same is not true for $c_0, \ldots, c_m$. This second phase is embarrassingly parallel. Structuring Weierstrass preparation as a lazy operation also naturally exposes further concurrency opportunities, such a parallel pipeline structure in the case of factorization via Hensel's lemma, see Sect. 6.

Finally, we report on experimental results for Weierstrass preparation against the `PowerSeries` library. We note that the latter is not a lazy implementation, returning only a truncated UPoPS. We have studied two families of examples:

(i) $\frac{1}{1+X_1+X_2}Y^k + Y^{k-1} + \cdots + Y^2 + X_2Y + X_1$ and

(ii) $\frac{1}{1+X_1+X_2}Y^k + Y^{k-1} + \cdots + Y^{\lceil k/2 \rceil} + X_2Y^{\lceil k/2 \rceil -1} + \cdots + X_2Y + X_1$.

The first results in p of degree 2, while the second results in p of degree $\lceil k/2 \rceil$, thus emphasizing the performance of phase two and phase one of the algorithm, respectively. The results of this experiment are summarized in Figs. 4 and 5.

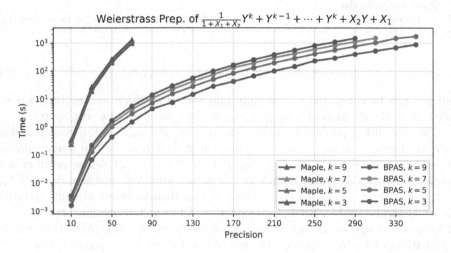

Fig. 4. Applying Weierstrass preparation on family (i) for increasing precisions

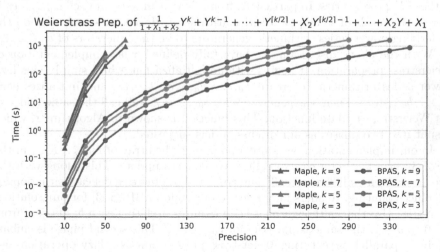

Fig. 5. Applying Weierstrass preparation on family (ii) for increasing precisions

Not only is our implementation orders of magnitude faster than MAPLE, but the difference in computation time further increases with increasing precision (total degree in X_1, X_2). This can be attributed to our efficient underlying power series arithmetic, as well as our smart implementation of Lemma 1.

6 Lazy Factorization via Hensel's Lemma

In Sect. 2 we have seen the description of Hensel's lemma for univariate polynomial over power series. Specifically, that the proof by construction provides a mechanism to factor UPoPS. We now look to make that construction lazy.

Recall that the proof of Theorem 2 provides a mechanism to factor a UPoPS $f \in \mathbb{K}[[X_1, \ldots, X_n]][Y]$ into factors $f_1, \ldots, f_r$ based on Taylor shift and repeated applications of Weierstrass preparation. The construction begins by first factorizing the polynomial $\bar{f} = f(0, \ldots, 0, Y) \in \mathbb{K}[Y]$, obtained by evaluating all variables in power series coefficients to 0, into linear factors. This can be performed with a suitable (algebraic) factorization algorithm for $\mathbb{K}$. For simplicity of presentation, let us assume that $\bar{f}$ factorizes into linear factors over $\mathbb{K}$, thus returning a list of roots $c_1, \ldots, c_r \in \mathbb{K}$ with respective multiplicities $k_1, \ldots, k_r$. The construction then proceeds recursively, obtaining one factor at a time.

Let us describe one step of the recursion, where f^* describes the current polynomial to factorize, initially being set to f. For a root c_i of $\bar{f}$, we perform a Taylor shift to obtain $g = f^*(Y + c_i)$ such that g has order k_i (as a polynomial in Y). The Weierstrass preparation theorem can then be applied to obtain p and $\alpha \in K[[X_1, \ldots, X_n]][Y]$ where p is monic and of degree k_i. A Taylor shift is then applied in reverse to obtain $f_i = p(Y - c_r)$, a factor of f, and $f^* = \alpha(Y - c_r)$, the UPoPS to factorize in the next step. The full procedure for obtaining all factors of f is shown as an iterative process, instead of recursive, in Algorithm 2.

The beauty of this algorithm is that it is immediately a lazy algorithm with no additional effort. Using the underlying lazy operations of Taylor shift (Sect. 4) and Weierstrass preparation (Sect. 5), the entire factorization is performed lazily, returning a factorization nearly instantly. The power series coefficients of these factors can automatically be updated later using their generators, which are simply Taylor shift operations on top of a Weierstrass update.

Notice too the opportunities for concurrency exposed from a lazy Taylor shift and lazy Weierstrass. The factors $f_1, \ldots, f_r$ are created from successive applications of Weierstrass preparation. They in essence form a *pipeline* of processes [17, Ch. 9]. Updating one factor simultaneously causes its associated α from Weierstrass preparation to be updated. This in turn allows the next factor to be updated since this α is the input into the next Weierstrass preparation. This concurrency is on top of that available within a single Weierstrass preparation.

We now compare our implementation of factorization via Hensel's lemma in BPAS against that of MAPLE's PowerSeries library. In the latter, two functions are available for this operation: ExtendedHenselConstruction (EHC) and FactorizationViaHenselLemma (FVHL). FVHL has the same specifications as Algorithm 2 while EHC factorizes UPoPS over the field of Puiseux series in

---

Algorithm 2. HenselFactorization(f)

Input: $f = \sum_{i=0}^{k} a_i Y^i, a_i \in \mathbb{K}[[X_1, \ldots, X_n]]$.
Output: $f_1, \ldots, f_r$ satisfying Theorem 2.
1: $\bar{f} = f(0, \ldots, 0, Y)$
2: $c_1, \ldots, c_r := $ obtain roots of $\bar{f}$ ▷ by some appropriate factorization algorithm
3: $f^* = f$
4: **for** $i = 1$ to r **do**
5:     $g := f^*(Y + c_i)$
6:     $p, \alpha := $ WeierstrassPreparation(g)
7:     $f_i := p(Y - c_i)$
8:     $f^* := \alpha(Y - c_i)$
9: **return** $f_1, \ldots, f_r$

$X_1, \ldots, X_n$, see [1]. Our tests use two UPoPS f, one of degree 3 and one of degree 4, such that $\bar{f}$ splits into linear factors over $\mathbb{Q}$; in this way the output is the same for our BPAS code, EHC, and FVHL.

The results of this experiment are summarized in Fig. 6 for the two UPoPS. Our implementation is orders of magnitude faster. We observe that the gap between our implementation and EHC increases both as UPoPS degree increases and as power series precision increases. A theoretical comparison, in terms of complexity analysis, between the EHC and Algorithm 2 is work in progress.

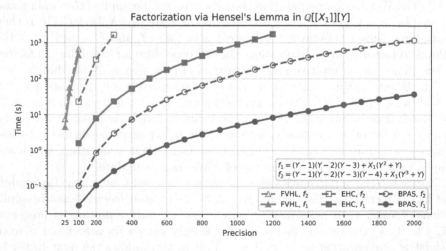

Fig. 6. Applying factorization via Hensel's lemma to the UPoPS $f_1 = (Y-1)(Y-2)(Y-3) + X_1(Y^2 + Y)$ and $f_2 = (Y-1)(Y-2)(Y-3)(Y-4) + X_1(Y^3 + Y)$

7   Conclusions and Future Work

Throughout this work we have explored the design and implementation of lazy multivariate power series, employing them in Weierstrass preparation and the factorization of univariate polynomials over power series via Hensel's lemma. Our implementation in the C language is orders of magnitude faster than existing implementations in SAGEMATH and MAPLE's PowerSeries library. In part, this is due to overcoming the challenge of working with dynamic generator functions in a compiled language, rather than using a more simplistic scripting language.

Yet, still more work can be done to further improve the performance of our implementation. The implementation of our arithmetic follows naive quadratic algorithms; instead, relaxed algorithms [13] should be integrated into our implementation to improve its algebraic complexity. Further, as mentioned in the case of Weierstrass preparation and in factorization via Hensel's lemma, there are opportunities for concurrency in their implementation as lazy operations. This concurrency can be exploited with parallel programming techniques, including a parallel map and parallel pipeline, to yield further improved performance.

Acknowledgments. The authors would like to thank NSERC of Canada (award CGSD3-535362-2019), Robert H. C. Moir, and the reviewers for their helpful comments.

References

1. Alvandi, P., Ataei, M., Kazemi, M., Moreno Maza, M.: On the extended Hensel construction and its application to the computation of real limit points. J. Symb. Comput. **98**, 120–162 (2020)
2. Alvandi, P., Kazemi, M., Moreno Maza, M.: Computing limits with the regular-chains and powerseries libraries: from rational functions to zariski closure. ACM Commun. Comput. Algebra **50**(3), 93–96 (2016)
3. Asadi, M., et al.: Basic Polynomial Algebra Subprograms (BPAS) (2020). http://www.bpaslib.org/
4. Asadi, M., Brandt, A., Moir, R.H.C., Moreno Maza, M., Xie, Y.: On the parallelization of triangular decomposition of polynomial systems. In: Proceedings of International Symposium on Symbolic and Algebraic Computation, ISSAC 2020, pp. 22–29. ACM (2020)
5. Asadi, M., Brandt, A., Moir, R.H.C., Moreno Maza, M.: Algorithms and data structures for sparse polynomial arithmetic. Mathematics **7**(5), 441 (2019)
6. Burge, W.H., Watt, S.M.: Infinite structures in scratchpad II. In: Davenport, J.H. (ed.) EUROCAL 1987. LNCS, vol. 378, pp. 138–148. Springer, Heidelberg (1989). https://doi.org/10.1007/3-540-51517-8_103
7. Dahan, X., Moreno Maza, M., Schost, É., Wu, W., Xie, Y.: Lifting techniques for triangular decompositions. In: Proceedings of ISSAC 2005, Beijing, China, 2005, pp. 108–115 (2005)
8. Decker, W., Greuel, G.M., Pfister, G., Schönemann, H.: Singular 4-1-1 – a computer algebra system for polynomial computations (2018). http://www.singular.uni-kl.de
9. Fischer, G.: Plane Algebraic Curves. AMS (2001)

10. von zur Gathen, J., Gerhard, J.: Modern Computer Algebra, 2nd edn. Cambridge University Press, Cambridge (2003)
11. von zur Gathen, J.: Hensel and Newton methods in valuation rings. Math. Comput. **42**(166), 637–661 (1984)
12. Hart, W., Johansson, F., Pancratz, S.: FLINT: Fast Library for Number Theory (2015), version 2.5.2. http://flintlib.org
13. van der Hoeven, J.: Relax, but don't be too lazy. J. Symb. Comput. **34**(6), 479–542 (2002)
14. Karczmarczuk, J.: Generating power of lazy semantics. Theor. Comput. Sci. **187**(1–2), 203–219 (1997)
15. Kazemi, M., Moreno Maza, M.: Detecting singularities using the PowerSeries library. In: Gerhard, J., Kotsireas, I. (eds.) MC 2019. CCIS, vol. 1125, pp. 145–155. Springer, Cham (2020). https://doi.org/10.1007/978-3-030-41258-6_11
16. Lauer, M.: Computing by homomorphic images. In: Buchberger, B., Collins, G.E., Loos, R., Albrecht, R. (eds.) Computer Algebra, pp. 139–168. Springer, Vienna (1983). https://doi.org/10.1007/978-3-7091-7551-4_10
17. McCool, M., Reinders, J., Robison, A.: Structured Parallel Programming: Patterns for Efficient Computation. Elsevier, Amsterdam (2012)
18. Monagan, M., Vrbik, P.: Lazy and forgetful polynomial arithmetic and applications. In: Gerdt, V.P., Mayr, E.W., Vorozhtsov, E.V. (eds.) CASC 2009. LNCS, vol. 5743, pp. 226–239. Springer, Heidelberg (2009). https://doi.org/10.1007/978-3-642-04103-7_20
19. Parusiński, A., Rond, G.: The Abhyankar-Jung theorem. J. Algebra **365**, 29–41 (2012)
20. Sasaki, T., Kako, F.: Solving multivariate algebraic equation by Hensel construction. Japan J. Indust. Appl. Math. **16**(2), 257–285 (1999)
21. Scott, M.L.: Programming Language Pragmatics, 3rd edn. Academic Press, Cambridge (2009)
22. The Sage Developers: SageMath, the Sage Mathematics Software System (Version 9.1) (2020). https://www.sagemath.org

Enhancements to Lazard's Method for Cylindrical Algebraic Decomposition

Christopher W. Brown[1($\boxtimes$)] and Scott McCallum[2]

[1] United States Naval Academy, Annapolis, U.S.A.
wcbrown@usna.edu
[2] Macquarie University, Sydney, Australia
scott.mccallum@mq.edu.au

Abstract. In 1994 Daniel Lazard proposed an improved method for constructing a cylindrical algebraic decomposition (CAD) from a set of polynomials, which recent work has, finally, fully validated. Lazard's method works for any set of input polynomials, but is less efficient than the method of Brown (2001) which, however, fails for input sets that are not "well-oriented". The present work improves Lazard's method so that it is as efficient for well-oriented input as Brown's method, while retaining its infallibility. Justifying these improvements requires novel and non-trivial mathematics.

1 Introduction

In 1994 Lazard [13] proposed an improved method for construction of cylindrical algebraic decomposition (CAD). The method comprised a simplified projection operation together with a generalized cell lifting process. However a flaw in some of the essential underlying theory was subsequently noticed [3,7]. Two recent papers [17,18] addressed the validity of the method. This paper describes a number of enhancements to the method. For some of these improvements, nontrivial mathematics is required.

Developed by George Collins in the early 1970s [6], CAD is a data structure that provides an explicit geometric representation of a *semi-algebraic* set in $\mathbb{R}^n$— a set that is defined by boolean combinations of n-variate integral polynomial equalities and inequalities over the real numbers (these expressions are called *Tarski formulas*). This data structure supports a number of important operations on semi-algebraic sets/Tarski formulas, including satisfiability of formulas (equivalently determining whether a semi-algebraic set is non-empty), determining the dimension of a semi-algebraic set, quantifier elimination for quantified Tarski formulas, and more. Constructing a CAD from a formula is computationally hard, in fact its worst-case is doubly exponential in n [4]. None-the-less, variations of CAD have been implemented in a number of systems, and improvements to CAD construction are still the subject of on-going research.

Collins' original algorithm followed this basic scheme: start with a (typically quantified) formula F, extract the set A of n-variate integral polynomials

© Springer Nature Switzerland AG 2020
F. Boulier et al. (Eds.): CASC 2020, LNCS 12291, pp. 129–149, 2020.
https://doi.org/10.1007/978-3-030-60026-6_8

appearing in F, compute P, a superset of the elements of A referred to as the *projection* of A, and construct the CAD data structure based on P (referred to as *lifting*). Generally in this scheme, the smaller one is able to make the projection P, the smaller the data-structure and the quicker it can be constructed. So several improved *projection operators* have been developed in the intervening years, such as Hong's projection [10], the McCallum projection [15] and the Brown-McCallum projection [3].

The last 30 years have seen many improvements to CAD construction and its application to various problems that break with the basic scheme above by connecting CAD construction more tightly with the initial formula F and/or intertwining projection and the construction of the CAD data structure. This includes "partial" CAD [8], in which the CAD is built in such a way that the formula F is leveraged for early termination; "open" CAD [19], which modifies the projection for formulas with only strict inequalities; "equational constraints" [16], which optimizes projection for formulas that include constraints that are equations; divide-and-conquer [20] and incremental CAD construction [9,12]; truth-table invariant CAD [2]; "conflict-driven" approaches to CAD [11]; and many more.

The present work is concerned with the Lazard projection operator, defined in Subsect. 2.1. We present a number of improvements to this operator in the form of certain commonly occurring situations in which the projection set can be made smaller than what was originally proposed by Lazard. Our main contribution is an analogue of the improvement to the McCallum projection introduced in [3]. Let $A \subset \mathbb{Z}[x_1, \ldots, x_n]$ be an irreducible basis. We show that, for every $f \in A$ for which f is nullified by only finitely many points in $\mathbb{R}^{n-1}$, the trailing coefficient of f can be omitted from the Lazard projection set provided that certain CAD method modifications are made. This and other improvements are discussed in the context of following the basic scheme in which projection is followed by constructing the CAD data structure, as outlined above, and the present work limits itself to that scope. However, it is important to note that these improvements will carry forward into Lazard projection versions of all of the ideas referenced in the previous paragraph. So, for example, a Lazard version of the equational constraints projection will be able to leverage the improvements we describe here, just as a Lazard version of divide-and-conquer or incremental CAD would.

We believe our contribution here will be useful for symbolic computation researchers and practitioners, as well as related communities. Here is our reasoning. CAD is widely applicable, and the recently validated Lazard method uses the smallest (hence most efficient) infallible projection operator for CAD amongst all the different projection schemes that have been proposed. Therefore we expect the Lazard method to be adopted by those wishing to use CAD for specific applications of interest. It follows that improvements to the original Lazard method which result in computational savings are important. Our contribution herein offers an improvement to this method for so-called well-oriented input sets A—the same conditions under which the Brown-McCallum projection applies. (This means roughly that polynomials in A should have at most finitely many nullifying points, and a similar condition should hold recursively for the projection of A.) It has been pointed out [15] that such input is expected to occur frequently, and a survey of the QE/CAD literature reveals that indeed many, if not most, problems considered therein have well-orientation. Moreover, our improvement is applicable under much weaker conditions as well. It can be applied to projecting any polynomial p that has at most finitely many nullifying points, even if A as a whole fails to be well-oriented, and even if elements of the projection of p are not well-oriented. Hence we expect our enhancement to remain useful in principle and practice in the years to come.

The remainder of this paper is organized as follows. Section 2 gives an overview of Lazard's projection and its supporting theory. The presentation includes some novel aspects. Section 3 presents the main contribution of the paper, as stated above. The mathematics required to make this improvement to the Lazard projection is rather different from the earlier result of [3], and more details are involved. Section 4 describes several additional improvements that apply in lower-dimensional cases. Section 5 contains some brief concluding remarks.

2   Synopsis of Lazard's Method and Supporting Theory

2.1   Lazard's Projection and Valuation

Background material on CAD, and in particular its projection operation, can be found in [1, 6–8, 14, 15]. We present a precise definition of the projection operator P_L for CAD introduced by Lazard [13]. Put $R_0 = \mathbb{Z}$ and, for $n \geq 1$, put $R_n = R_{n-1}[x_n] = \mathbb{Z}[x_1, \ldots, x_n]$. Elements of the ring R_n will usually be considered to be polynomials in x_n over R_{n-1}. We shall call a subset A of R_n whose elements are irreducible polynomials of positive degree and pairwise relatively prime an *irreducible basis*. (This concept is analogous to that of *squarefree basis* which is used in the CAD literature, for example [14].)

Definition 1 (Lazard projection). *Let A be a finite irreducible basis in R_n, with $n \geq 2$. The Lazard projection $P_L(A)$ of A is the subset of R_{n-1} comprising the following polynomials:*

1. *all leading coefficients of the elements of A,*
2. *all trailing coefficients (i.e. coefficients independent of x_n) of the elements of A,*
3. *all discriminants of the elements of A, and*
4. *all resultants of pairs of distinct elements of A.*

Remark 1. Let A be an irreducible basis. Lazard's projection $P_L(A)$ is the smallest (hence optimal) "general purpose" projection set that has been proposed for CAD. Now $P_L(A)$ contains and is usually strictly larger than the Brown-McCallum projection $P_{BM}(A)$ [3] since the latter set omits the trailing coefficients of the elements of A. However, the Brown-McCallum projection is not "general purpose" in the sense that it may fail in case A is not well-oriented.

Lazard [13] outlined a claimed CAD algorithm for $A \subset R_n$ and $\mathbb{R}^n$ which uses the projection set $P_L(A)$. The specification of his algorithm requires the following concept of his valuation:

Definition 2 (Lazard valuation). *Let K be a field. Let $n \geq 1$, $f \in K[x_1, \ldots, x_n]$ nonzero, and $\alpha = (\alpha_1, \ldots, \alpha_n) \in K^n$. The Lazard valuation (valuation, for short) $v_\alpha(f)$ of f at α is the element $\mathbf{v} = (v_1, \ldots, v_n)$ of $\mathbb{N}^n$ least (with respect to $\leq_{lex}$) such that f expanded about α has a term*

$$c(x_1 - \alpha_1)^{v_1} \cdots (x_n - \alpha_n)^{v_n}$$

with $c \neq 0$. (Note that $\leq_{lex}$ denotes the lexicographic order on $\mathbb{N}^n$.)

With K, n and f as in the above definition, and $S \subset K^n$, we say f is *valuation-invariant* in S if the valuation of f is the same at every point of S. Lazard's proposed CAD algorithm also uses a technique for "evaluating" a polynomial $f \in R_n$ at a sample point in $\mathbb{R}^{n-1}$. This technique is described in slightly more general and structured terms in the next subsection.

2.2 Lazard Evaluation

Let K be a field which supports explicit arithmetic computation. Let $n \geq 2$, $1 \leq i < n$, take a nonzero element f_i in $K[x_i, \ldots, x_n]$, and let $\alpha_i \in K$. The *Lazard evaluation* of f_i at α_i is a certain element f_{i+1} of $K[x_{i+1}, \ldots, x_n]$ which (together with a nonnegative integer v_i) is the result of the following process (subalgorithm):

Algorithm 1 (LazardEvalStep)
Input: $f_i \in K[x_i, \ldots, x_n]$ such that $f_i \neq 0$, $\alpha_i \in K$.
Output: $f_{i+1} \in K[x_{i+1}, \ldots, x_n]$, $v_i \geq 0$ such that $f_i = (x_i - \alpha_i)^{v_i} g_i$, with $g_i \in K[x_i, \ldots, x_n]$ and $g_i(\alpha_i, x_{i+1}, \ldots, x_n) \neq 0$ and $f_{i+1} = g_i(\alpha_i, x_{i+1}, \ldots, x_n)$.

Set $v_i \leftarrow$ the greatest integer v s.t. $(x_i - \alpha_i)^v | f_i$.
Set $g_i \leftarrow f_i/(x_i - \alpha_i)^{v_i}$.
Set $f_{i+1} \leftarrow g_i(\alpha_i, x_{i+1}, \ldots, x_n)$.

There are two immediate applications of the subalgorithm LazardEvalStep. Here is the first. Suppose we are given $f \in K[x_1, \ldots, x_n]$ with $f \neq 0$, and $\alpha = (\alpha_1, \ldots, \alpha_n) \in K^n$, and that we wish to compute $v_\alpha(f)$. We could repeatedly call subalgorithm LazardEvalStep as follows:

Algorithm 2 (LazardVal)
Input: $f \in K[x_1, \ldots, x_n]$, $f \neq 0$, $\alpha \in K^n$.
Output: $v = (v_1, \ldots, v_n)$, the Lazard valuation of f at α.

Set $f_1 \leftarrow f$.
For $i \leftarrow 1$ to n do
    Call LazardEvalStep with inputs f_i and α_i, obtaining outputs f_{i+1} and v_i.

Set $v \leftarrow (v_1, \ldots, v_n)$.

With $\alpha^{(i)} = (\alpha_i, \ldots, \alpha_n)$, the assertion "$v_\alpha(f)$ = the result of concatenating $(v_1, \ldots, v_{i-1})$ and $v_{\alpha^{(i)}}(f_i)$" is an invariant of Algorithm 2, from which the correctness of this algorithm follows. Here is the second application of Algorithm 1. Suppose we are given f as above and $\alpha = (\alpha_1, \ldots, \alpha_{n-1}) \in K^{n-1}$, and that we wish to compute a certain element $f_\alpha(x_n)$ of $K[x_n]$ called the *Lazard evaluation* of f at α [18]. The following method could be used to compute $f_\alpha(x_n)$ together with an $(n-1)$-tuple v called the *Lazard valuation* of f on α.

Algorithm 3 (LazardEval)
Input: $f \in K[x_1, \ldots, x_n]$, $f \neq 0$, $\alpha \in K^{n-1}$.
Output: $f_\alpha(x_n)$, *the Lazard evaluation of* f *at* α, *and* $v = (v_1, \ldots, v_{n-1})$, *the Lazard valuation of* f *on* α.

Set $f_1 \leftarrow f$.
For $i \leftarrow 1$ *to* $n - 1$ *do*
    Call LazardEvalStep with inputs f_i *and* α_i, *obtaining outputs* f_{i+1} *and* v_i.
Set $f_\alpha(x_n) \leftarrow f_n$. *Set* $v \leftarrow (v_1, \ldots, v_{n-1})$.

Remark 2. With K, n and f as above, $v = (v_1, \ldots, v_{n-1})$, and $\alpha \in K^{n-1}$, the second output of Algorithm 3, $f(\alpha, x_n) = 0$ (identically) if and only if $v_i > 0$, for some i in the range $1 \leq i \leq n - 1$. With $\alpha_n \in K$ arbitrary, the integers v_i, with $1 \leq i \leq n - 1$, are the first $n - 1$ coordinates of $v_{(\alpha, \alpha_n)}(f)$.

2.3 Lazard's CAD Algorithm

One more definition is needed before we can state Lazard's main claim and his algorithm based on it. This definition is not explicit in [13] – it was introduced in [17] to help clarify and highlight Lazard's main claim:

Definition 3. *[Lazard delineability] With $K = \mathbb{R}$ and x denoting $(x_1, \ldots, x_{n-1})$, let f be a nonzero element of $\mathbb{R}[x, x_n]$ and S a subset of $\mathbb{R}^{n-1}$. We say that f is* Lazard delineable *on S if*

1. *the Lazard valuation of f on α is the same for each point $\alpha \in S$;*
2. *there exist finitely many continuous functions $\theta_1 < \cdots < \theta_k$ from S to $\mathbb{R}$, with $k \geq 0$, such that, for all $\alpha \in S$, the set of real roots of $f_\alpha(x_n)$ is $\{\theta_1(\alpha), \ldots, \theta_k(\alpha)\}$ (where in the case $k = 0$, this means that, for all $\alpha \in S$, the set of real roots of $f_\alpha(x_n)$ is empty);*
3. *in the case $k > 0$, there exist positive integers $m_1, \ldots, m_k$ such that, for all $\alpha \in S$ and all i, m_i is the multiplicity of $\theta_i(\alpha)$ as a root of $f_\alpha(x_n)$.*

When f is Lazard delineable on S we refer to the graphs of the θ_i as the Lazard sections *of f over S, and to the m_i as the* associated multiplicities *of these sections. The regions between successive Lazard sections, together with the region below the lowest Lazard section and that above the highest Lazard section, are called* Lazard sectors.

It is shown in [18] that if f is Lazard delineable on S then f is valuation-invariant in every Lazard section and sector of f over S. We remark that if f vanishes identically at no point of S, then f is Lazard delineable on S if and only if f is delineable on S in the usual sense [14,17]. Lazard delineability addresses the partitioning of the cylinder over S in relation to f in case f vanishes identically on S, in which case the usual delineability does not apply.

We express Lazard's main claim, essentially the content of his Proposition 5 and subsequent remarks, as follows (as in [17]):

Let A be a finite irreducible basis in R_n, where $n \geq 2$. Let S be a connected subset of $\mathbb{R}^{n-1}$. Suppose that each element of $P_L(A)$ is valuation-invariant in S. Then each element of A is Lazard delineable on S, and the Lazard sections over S of the elements of A are pairwise disjoint.

This claim concerns valuation-invariant lifting in relation to $P_L(A)$: it asserts that the condition, "each element of $P_L(A)$ is valuation-invariant in S", is sufficient for an A-valuation-invariant stack in $\mathbb{R}^n$ to exist over S. We can now describe Lazard's proposed CAD algorithm (as in [17]):

Algorithm 4 (Valuation-invariant CAD, Lazard projection (VCADL))
Input: *A is a list of integral polynomials in* $x_1, \ldots, x_n$.
Output: *$\mathcal{I}$ and $\mathcal{S}$ are lists of indices and sample points, respectively, of the cells comprising an A-valuation-invariant CAD of* $\mathbb{R}^n$.

---

1. *If $n > 1$ then go to (2). Isolate the real roots of the irreducible factors of the nonzero elements of A. Construct cell indices $\mathcal{I}$ and sample points $\mathcal{S}$ from the real roots and return.*
2. *Set $B \leftarrow$ the finest squarefree basis for* prim(A). *That is, B is assigned the set of ample irreducible factors of elements of the set* prim(A) *of primitive parts of elements of A of positive degree. Set $P \leftarrow$* cont(A) $\cup P_L(B)$ *(where* cont(A) *denotes the set of contents of elements of A). Call VCADL with input P, obtaining outputs $\mathcal{I}'$ and $\mathcal{S}'$. Set $\mathcal{I} \leftarrow$ the empty list. Set $\mathcal{S} \leftarrow$ the empty list.*
   For each $\alpha = (\alpha_1, \ldots, \alpha_{n-1})$ in $\mathcal{S}'$ do
      Let i be the index of the cell containing α. Set $f^*(x_n) \leftarrow \prod_{f \in B} f_\alpha(x_n)$. (For each $f \in B$, if $f(\alpha, x_n) \neq 0$ then set $f_\alpha(x_n) \leftarrow f(\alpha, x_n)$, else construct $f_\alpha(x_n)$ using Algorithm 3, with $K = \mathbb{Q}(\alpha)$.) Isolate the real roots of $f^*(x_n)$. Construct cell indices and sample points for Lazard sections and sectors of elements of B from i, α and the real roots of $f^*(x_n)$. Add the cell indices to $\mathcal{I}$ and the sample points to $\mathcal{S}$.

The correctness of the above algorithm – namely, the claim that, given $A \subset R_n$, it produces a CAD of $\mathbb{R}^n$ such that each cell of the CAD is valuation-invariant with respect to each element of A – follows from Lazard's main claim by induction on n. (However, recall that Lazard's original proof sketch of his main claim contained a serious flaw.)

We note that it is remarked already in [18], page 64, that when the leading coefficient of polynomial f in the variable x_n is nowhere vanishing (e.g. constant), the trailing coefficient does not need to be included in the projection. For the remainder of the paper we will assume this optimization as part of the "original" Lazard projection.

2.4 Example of Lazard's Projection

Consider computing a CAD in which the quadratic polynomial

$$q := x^2 - ((a+c)d - (a - bc))$$

is sign-invariant, where the variable order is $a \prec b \prec c \prec d \prec x$ (meaning project with respect to x first, then with respect to d, etc.). Because the irreducible polynomial $f := (a+c)d - (a - bc)$ is the discriminant of q with respect to x, and by the remark recalled above, the projection $P_L(\{q\})$ is $\{f\}$, which indeed is the case regardless of what projection operator you use. The interesting point is the next projection, i.e. $P_L(\{f\})$, because f is nullified on the positive-dimensional region $a = -c \wedge b = -1$. Lifting into x-space is justified by Theorem 1 below (with $f = q$), which requires that the discriminant f of q is Lazard-valuation invariant in any region we lift over. Were we to rely on the McCallum projection, we would be requiring the *order-invariance* of f, and this nullification on a positive dimensional region would be a problem. Thus, it is precisely in this case that it is most interesting to see how the Lazard projection and the associated lifting method work. $P_L(\{f\})$ eliminating d produces $\{a+c, bc - a\}$, and the remaining projections produce $\{b, b+1\}$ and $\{a\}$. After some lifting steps, we will have a positive dimensional cell $C := 0 < a \wedge b = -1 \wedge c = -a$ with sample point, for example, $(a, b, c) = (2, -1, -2)$. Following VCADL, we set $f^*(d)$ to the Lazard evaluation of f at the sample point, which we compute as:

i	f_i	α_i	f_{i+1}	v_i
1	$cd + ad + bc - a$	$a = 2$	$cd + 2d + bc - 2$	0
2	$cd + 2d + bc - 2$	$b = -1$	$cd + 2d - c - 2$	0
3	$(c+2)^1(d-1)$	$c = -2$	$d - 1$	1
Lazard evaluation yields:	$d - 1$		$(0, 0, 1)$	

So lifting into d-space over the cell C partitions $C \times \mathbb{R}$ into the region in which $d - 1 = 0$ (where the Lazard valuation is $(0, 0, 1, 1)$), and the regions above and below (where the Lazard valuation is $(0, 0, 1, 0)$). Thus we have Lazard-valuation invariance, as required.

2.5 Summary of Supporting Theory

A special form of Lazard's main claim, in which S is assumed to be a connected submanifold of $\mathbb{R}^{n-1}$, and each element of A, an irreducible basis, is concluded to be Lazard analytic delineable on S etc., is proved in [18]. This special form of the claim is sufficient to validate Lazard's method. The key result used in proving this claim is the following:

Theorem 1. *Let $f(x, x_n) \in \mathbb{R}[x, x_n]$ have positive degree d in x_n, where $x = (x_1, \ldots, x_{n-1})$. Let $D(x)$, $l(x)$ and $t(x)$ denote the discriminant, leading coefficient and trailing coefficient (that is, the coefficient independent of x_n) of f, respectively, and suppose that each of these polynomials is nonzero (as an element of $\mathbb{R}[x]$). Let S be a connected analytic submanifold of $\mathbb{R}^{n-1}$ in which D, l and t are all valuation-invariant. Then f is Lazard analytic delineable on S, hence valuation-invariant in every Lazard section and sector over S. Moreover, the same conclusion holds for the polynomial $f^*(x, x_n) = x_n f(x, x_n)$.*

This is Theorem 5.1 of [18]. The main ingredients of the proof are certain results on transforming the valuation of f at a point into the order of f at the origin along a certain curve, a "wing" argument which involves application of the valuation transformation results, and a corollary of the classical Puiseux with parameter theorem well known to singularity theorists.

We state the Puiseux with parameter theorem here as it will be used in the next section of the present paper. We use the following notation (as in [18]): with $\varepsilon = (\varepsilon_1, \ldots, \varepsilon_k)$, $U_{\varepsilon,r} = U_\varepsilon \times U_r$, where $U_\varepsilon = \{x = (x_1, \ldots, x_k) \in \mathbb{C}^k : |x_i| < \varepsilon_i, \forall i\}$, $U_r = \{y \in \mathbb{C} : |y| < r\}$. In the statement of the theorem "analytic" means "complex analytic".

Theorem 2. (Puiseux with parameter)
Let

$$f(x, y, z) = z^d + \sum_{i=0}^{d-1} a_i(x, y) z^i, \tag{1}$$

be a monic polynomial in z with coefficients $a_i(x, y)$ analytic in $U_{\varepsilon,r}$. Suppose that the discriminant of f is of the form $D_f(x, y) = y^m u(x, y)$ with analytic function u non vanishing on $U_{\varepsilon,r}$. Then, there are a positive integer N (we may take $N = d!$) and analytic functions $\xi_i(x, t) : U_{\varepsilon, r^{1/N}} \to \mathbb{C}$ such that for all $(x, t) \in U_{\varepsilon, r^{1/N}}$,

$$f(x, t^N, z) = \prod_{i=1}^{d} (z - \xi_i(x, t)).$$

Note that "parameter" in the name "Puiseux with parameter" refers to the k-tuple x, which "parametrizes" the Puiseux roots $\xi_i(x, t)$. Note also that the appendix of [18] contains a concise proof of the Puiseux with parameter theorem.

3   Refinement to Lazard's Projection and Method

In this section we first present a corollary to Puiseux with parameter which is slightly different from Corollary 4.2 of [18]. Our corollary, unlike the one in [18], omits any hypothesis about the trailing coefficient of f, and its conclusion is slightly weaker. Yet it is sufficient for our main purpose here. (It is in essence a complex analytic analogue of Corollary 3.15 of [17].) Next we state and prove the key theorem which underlies our refinement to Lazard's projection. Our key theorem is at once an analogue of the main result (that is, Theorem 3.1) of [3] (which is stated in slightly expanded form as Theorem 3.9 of [17]), and an analogue of the main theorem (Theorem 5.1) of [18]. Its proof uses our variant corollary of Puiseux with parameter. Finally we describe circumstances in which we may use our refined Lazard projection in CAD construction, and the modifications to CAD such use entails.

3.1   New Corollary to Puiseux with Parameter

We use the same notation as used in the presentation of Puiseux with parameter above. Puiseux with parameter assumes that $f(x, y, z)$ is a monic polynomial with analytic coefficients. We shall prove the following corollary of Puiseux with parameter which can be applied to a nonmonic polynomial provided that its leading coefficient is of a suitable form. Here again, "analytic" means "complex analytic".

Corollary 1. (Corollary to Puiseux with parameter)
Let

$$f(x, y, z) = a_d(x, y)z^d + \sum_{i=0}^{d-1} a_i(x, y)z^i, \tag{2}$$

be a polynomial in z with coefficients $a_i(x, y)$ analytic in $U_{\varepsilon, r}$. Suppose that the discriminant $D(x, y)$ and leading coefficient $a_d(x, y)$ of f satisfy $D(x, y) = y^m u(x, y)$ and $a_d(x, y) = y^h v(x, y)$, for some integers $m, h \geq 0$ and some analytic functions u and v nonvanishing in $U_{\varepsilon, r}$. Then there exists integers $N > 0$ and $\kappa \geq 0$, analytic functions $\xi'_1, \ldots, \xi'_\kappa : U_{\varepsilon, r^{1/N}} \to \mathbb{C}$ and $\xi_{\kappa+1}, \ldots, \xi_d : U_{\varepsilon, r^{1/N}} \to \mathbb{C}$, and integers $\delta \geq 0$ and $\mu_1, \ldots, \mu_\kappa > 0$ such that for all $(x, t) \in U_{\varepsilon, r^{1/N}}$

$$f(x, t^N, z) = t^\delta v(x, t^N) \prod_{i=1}^{\kappa} (t^{\mu_i} z - \xi'_i(x, t)) \prod_{j=\kappa+1}^{d} (z - \xi_j(x, t)).$$

Proof. Define the polynomial $\tilde{f}(x, y, \tilde{z})$ by

$$\tilde{f}(x, y, \tilde{z}) = \tilde{z}^d + \sum_{i=0}^{d-1} a_d^{d-1-i} a_i \tilde{z}^i. \tag{3}$$

We may apply Theorem 2 to this polynomial because its discriminant equals $a_d^{(d-1)(d-2)} D(x, y)$, and hence satisfies the hypothesis of Theorem 2. Therefore, by this theorem, there is a positive integer N and analytic functions $\eta_i(x, t)$: $U_{\varepsilon, r^{1/N}} \to \mathbb{C}$ such that

$$\tilde{f}(x, t^N, \tilde{z}) = \prod_{i=1}^{d} (\tilde{z} - \eta_i(x, t))$$

for all $(x,t) \in U_{\varepsilon, r^{1/N}}$. Then we have

$$a_d^{d-1} f(x, t^N, z) = \tilde{f}(x, t^N, a_d z) = \prod_{i=1}^{d} (a_d z - \eta_i(x,t)) \tag{4}$$

for all $(x,t) \in U_{\varepsilon, r^{1/N}}$ where, for simplicity, a_d denotes $a_d(x, t^N)$ in Eq. 4 above. Let $m_0 = hN$ and let κ denote the number of i such that m_0 strictly exceeds the order m_i of $\eta_i(x,t)$ in t. (That is, m_i is the least power of t which occurs amongst all the nonzero terms of the power series expansion of η_i at the origin, but if η_i vanishes identically then $m_i = \infty$. So κ is the number of indices i for which $m_0 > m_i$.) Renumber the roots $\eta_i(x,t)$ so that $m_0 > m_i$ for $1 \leq i \leq \kappa$, and hence $m_0 \leq m_j$ for $\kappa + 1 \leq j \leq d$. For $\kappa + 1 \leq j \leq d$, put $\xi_j(x,t) = a_d(x, t^N)^{-1} \eta_j(x,t)$. Then the ξ_j are analytic in $U_{\varepsilon, r^{1/N}}$.

From Eq. 4, after multiplying both sides by $a_d(x, t^N)$, dividing both sides by $a_d(x, t^N)^{d-\kappa}$, and applying the definition of $\xi_j(x,t)$, we obtain:

$$a_d^{\kappa} f(x, t^N, z) = a_d \prod_{i=1}^{\kappa} (a_d z - \eta_i(x,t)) \prod_{j=\kappa+1}^{d} (z - \xi_j(x,t)) \tag{5}$$

for all $(x,t) \in U_{\varepsilon, r^{1/N}}$, where (again) a_d denotes $a_d(x, t^N)$ in Eq. 5 above. For $1 \leq i \leq \kappa$, let $\mu_i = m_0 - m_i$. Then each $\mu_i > 0$, as we have noted. Now it is not too difficult to show that $m_0 - (\mu_1 + \mu_2 + \cdots + \mu_\kappa) \geq 0$. (The basic idea of the proof of this inequality is to equate the term $a_d(x, t^N)^\kappa a_{d-\kappa}(x, t^N) z^{d-\kappa}$ on the left hand side of Eq. 5 with the corresponding term on the right hand side of this equation: in particular, the order in t of this term must be considered.) Putting $\delta = m_0 - (\mu_1 + \cdots + \mu_\kappa)$, it follows that $\delta \geq 0$.

For $1 \leq i \leq \kappa$, put $\xi_i'(x,t) = t^{\mu_i} a_d(x, t^N)^{-1} \eta_i(x,t)$. Then the ξ_i' are analytic in $U_{\varepsilon, r^{1/N}}$. Finally from Eq. 5, after dividing both sides by $a_d(x, t^N)^\kappa$, redistributing certain powers of t, and applying the definition of ξ_i', we obtain the desired expression for $f(x, t^N, z)$.

3.2   Key Result for Projection Refinement

Our main result is stated as follows.

Theorem 3. *Let $f(x, x_n) \in \mathbb{R}[x, x_n]$ have positive degree d in x_n, where $x = (x_1, \ldots, x_{n-1})$. Let $D(x)$ and $l(x)$ denote the discriminant and leading coefficient of f with respect to x_n, respectively, and suppose that $D(x)$ is a nonzero polynomial. (Of course $l(x)$ is a nonzero polynomial, by definition.) Let S be a connected analytic submanifold of $\mathbb{R}^{n-1}$ in which both D and l are valuation-invariant, and at no point of which f vanishes identically. Then f is analytic delineable on S, hence valuation-invariant in every section and sector over S.*

Proof. Recall from [18] that an $(n-1)$-tuple of positive integers $\mathbf{c} = (c_1, \ldots c_{n-1})$ is termed an *evaluator* for a set $V \subset \mathbb{N}^{n-1}$ if for every $i = n-2, \ldots, 2, 1$ we have

$$c_i \geq 1 + \max_{\mathbf{v} \in V} \sum_{j > i} c_j v_j.$$

Let $\mathbf{c}$ be an evaluator for $V_D \cup V_l$, where $V_D = \{v_p(D) \mid p \in S\}$ and $V_l = \{v_p(l) \mid p \in S\}$. (Recall that $\mathbf{c}$ exists because both V_D and V_l are finite.) By hypotheses, D and l are valuation-invariant in S. Consider $\psi : S \times \mathbb{R} \to \mathbb{R}^{n-1}$ defined by

$$\psi(p, y) = p + (y^{c_1}, \ldots, y^{c_{n-1}}).$$

By Proposition 5.4 of [18], for $p \in S$ fixed, $D(\psi(p,y))$ and $l(\psi(p,y))$ (as polynomials in y) have finite orders $m := \langle \mathbf{c}, v_p(D) \rangle$ and $h := \langle \mathbf{c}, v_p(l) \rangle$ at $y = 0$, respectively. Since D and l are valuation-invariant in S, these orders are thus independent of $p \in S$. With $f(x, x_n) = a_d(x)x_n^d + a_{d-1}(x)x_n^{d-1} + \cdots + a_0(x)$, where $a_d(x) = l(x)$, let

$$f_\psi(p, y, z) = f(\psi(p,y), z) = a_d(\psi(p,y))z^d + \cdots + a_0(\psi(p,y)),$$

for all $p \in S$ and $y \in \mathbb{R}$. As observed, for every fixed $p \in S$, $l(\psi(p,y))$, as a polynomial in y, has finite order h at $y = 0$, and hence is a nonzero polynomial. Therefore, for every fixed $p \in S$, the degree in z of $f_\psi(p, y, z)$ equals d, and $l(\psi(p,y))$ and $D(\psi(p,y))$ are equal to the leading coefficient $l_\psi(p,y)$ and discriminant $D_\psi(p,y)$ of $f_\psi(p, y, z)$ with respect to z, respectively. Let p_0 be fixed. We shall show that, with respect to local coordinates on S near p_0, and after complexification, f_ψ (and D_ψ, l_ψ) satisfy the hypotheses of Corollary 1. We shall use the conclusion of this Corollary to show that f is analytic delineable on a neighbourhood of $p_0 \in S$. We now present this argument in detail.

Denote the dimension of S by k, and choose local coordinates $\hat{x} = (\hat{x}_1, \ldots, \hat{x}_k)$ on S near p_0. That is, choose an analytic coordinate system $\Phi : U \to V$, with $\Phi = (\phi_1, \ldots, \phi_{n-1})$, where $U \subset \mathbb{R}^{n-1}$ is a neighbourhood of p_0, $V \subset \mathbb{R}^{n-1}$ is a neighbourhood of 0, and $\Phi(p_0) = 0$, such that $S \cap U$ is defined by $\phi_{k+1}(x) = 0, \ldots, \phi_{n-1}(x) = 0$ within U [14]. Denote by $\phi : S \cap U \to W$ the homeomorphism (chart) defined by $\phi(x) = (\phi_1(x), \ldots, \phi_k(x))$, which maps $S \cap U$ onto the neighbourhood $W \subset \mathbb{R}^k$ of 0, and identify $(\phi_1, \ldots, \phi_k)$ with $(\hat{x}_1, \ldots, \hat{x}_k)$, denoted by $\hat{x}$. Note that ϕ^{-1} is an analytic mapping.

Denote by $\hat{f}_\psi(\hat{x}, y, z)$ the function f_ψ expressed in these coordinates. That is,

$$\hat{f}_\psi(\hat{x}, y, z) = a_d(\psi(\phi^{-1}(\hat{x}), y))z^d + \cdots + a_0(\psi(\phi^{-1}(\hat{x}), y)).$$

Then each coefficient of $\hat{f}_\psi$ with respect to z is analytic in $W \times \mathbb{R}$ and the same is true of its discriminant $\hat{D}_\psi$. Indeed, for every fixed $\hat{x} \in W$, $\hat{D}_\psi(\hat{x}, y) = D(\psi(\phi^{-1}(\hat{x}), y))$, as polynomials in y. Hence, for every fixed $\hat{x} \in W$, the order in y of $\hat{D}_\psi(\hat{x}, y)$ equals m, and this order is independent of $\hat{x} \in W$. Hence, by Lemma 4.4 of [18],

$$\hat{D}_\psi(\hat{x}, y) = y^m \hat{u}(\hat{x}, y), \tag{6}$$

for some analytic $\hat{u}$ defined near the origin, with $\hat{u}(0,0) \neq 0$. Similarly, the leading coefficient $\hat{l}_\psi$ of $\hat{f}_\psi$ satisfies

$$\hat{l}_\psi(\hat{x}, y) = y^h \hat{v}(\hat{x}, y), \tag{7}$$

for some analytic $\hat{v}$ defined near the origin, with $\hat{v}(0,0) \neq 0$. Denoting the coefficient of z^i in $\hat{f}_\psi$ by $\hat{a}_i(\hat{x}, y)$, notice that $\hat{a}_i$ has a complexification: that is, a complex analytic extension, essentially unique, to a polydisk $U_{\varepsilon,r}$ in $\mathbb{C}^{k+1}$, which we shall also denote by $\hat{a}_i(\hat{x}, y)$. Thus and similarly, $\hat{f}_\psi$, $\hat{D}_\psi$, etc., are all considered to be defined in $U_{\varepsilon,r}$, and Eqs. 6, 7 are valid therein. By refining $U_{\varepsilon,r}$ as necessary, we may assume that $\hat{u}$ and $\hat{v}$ are both nonvanishing in $U_{\varepsilon,r}$. In summary, $\hat{f}_\psi$ satisfies the hypotheses of Corollary 1 in $U_{\varepsilon,r}$. Therefore, by this Corollary, there are integers $N > 0$ and $\kappa \geq 0$, analytic functions $\hat{\xi}_i'$, $\hat{\xi}_j$ defined in $U_{\varepsilon,r^{1/N}}$, and integers $\delta \geq 0$ and $\mu_i > 0$, such that for all $(x, t) \in U_{\varepsilon,r^{1/N}}$

$$\hat{f}_\psi(\hat{x}, t^N, z) = t^\delta \hat{v}(\hat{x}, t^N) \prod_{i=1}^{\kappa} (t^{\mu_i} z - \hat{\xi}_i'(\hat{x}, t)) \prod_{j=\kappa+1}^{d} (z - \hat{\xi}_j(\hat{x}, t)). \tag{8}$$

We claim that, for all i and j, with $\kappa + 1 \leq i \neq j \leq d$, $\hat{R}_{i,j}(\hat{x}, t) := \hat{\xi}_i(\hat{x}, t) - \hat{\xi}_j(\hat{x}, t)$ is valuation-invariant in $U_\varepsilon \times \{0\}$. This claim follows by Proposition 3.4 of [18], since $\hat{R}_{i,j}(\hat{x}, t)$ is a factor of $\hat{D}_\psi(\hat{x}, t^N)$ and the latter function is valuation-invariant in this set. An important consequence of this is that either $\hat{R}_{i,j}(\hat{x}, 0)$ vanishes throughout U_ε or $\hat{R}_{i,j}(\hat{x}, 0)$ vanishes at no point of U_ε. In other words, either the roots $\hat{\xi}_i(\hat{x}, 0)$ and $\hat{\xi}_j(\hat{x}, 0)$ coincide for all $\hat{x} \in U_\varepsilon$ or these roots remain different for all $\hat{x} \in U_\varepsilon$.

Now restricting Eq. 8 to real space, and transforming the restricted Equation into our original coordinates, we have

$$f_\psi(p, t^N, z) = t^\delta v(p, t^N) \prod_{i=1}^{\kappa} (t^{\mu_i} z - \xi_i'(p, t)) \prod_{j=\kappa+1}^{d} (z - \xi_j(p, t)) \tag{9}$$

in $(S \cap U) \times I_{r^{1/N}}$, where $I_{r^{1/N}} = (-r^{1/N}, r^{1/N})$. The complex valued functions ξ_i' and ξ_j are defined in $(S \cap U) \times I_{r^{1/N}}$ by the equations $\xi_i'(p, t) = \hat{\xi}_i'(\phi(p), t)$ and $\xi_j(p, t) = \hat{\xi}_j(\phi(p), t)$, and are hence continuous.

By hypothesis, $f_\psi(p, 0, z) = f(p, z) \neq 0$, as a polynomial in z, for all $p \in S \cap U$. Hence $\delta = 0$ and $\xi_i'(p, 0) \neq 0$ for all $p \in S \cap U$. By Eq. 9 we may therefore write

$$f(p, z) = (-1)^\kappa v(p, 0) \prod_{i=1}^{\kappa} (\xi_i'(p, 0)) \prod_{j=\kappa+1}^{d} (z - \xi_j(p, 0)) \tag{10}$$

in $S \cap U$. Hence, in particular, $f(p, z)$ has degree $d - \kappa$ for each fixed $p \in S \cap U$.

Let $\alpha_1 < \alpha_2 < \cdots < \alpha_s$ denote the real roots of $f(p_0, z)$ and let $\alpha_{s+1}, \ldots, \alpha_t$ denote the distinct nonreal roots of $f(p_0, z)$. (This is a fresh use of variable t, having now dispensed with its previous use.) For each i, $1 \leq i \leq t$, let m_i

denote the multiplicity of α_i. Then $\sum_{i=1}^{t} m_i = d - \kappa$. Let σ denote the minimum separation between the roots α_i if $t > 1$, and let $\sigma = 1$ if $t = 1$. For $1 \leq i \leq t$, let C_i denote the circle of radius $\sigma/2$ centred at α_i.

For each i, with $1 \leq i \leq t$, let $\mathcal{S}_i = \{j \mid \kappa + 1 \leq j \leq d$ and $\xi_j(p_0, 0) = \alpha_i\}$. Then $\mathcal{S}_i$ has exactly m_i elements. By continuity of the functions $\xi_j(p, 0)$, with $p \in S \cap U$, and since the degree of $f(p, z)$ equals $d - \kappa$ for all $p \in S \cap U$, as previously noted, there exists a refinement U_0 of the neighbourhood U of p_0 such that for all $p \in S \cap U_0$ and for all i with $1 \leq i \leq t$, all $\xi_j(p, 0)$ with $j \in \mathcal{S}_i$ lie in the interior of circle C_i, and this interior contains no other roots of $f(p, z)$. Since the nonreal roots of $f(p_0, z)$ occur in conjugate pairs, each C_i's interior, with $s + 1 \leq i \leq t$, contains no real points. By our observation above concerning the functions $\hat{R}_{i,j}$, either the roots $\xi_i(p, 0)$ and $\xi_j(p, 0)$ coincide for all $p \in S \cap U_0$ or they remain different for all such p. Therefore, for each i, with $1 \leq i \leq s$, the roots $\xi_j(p, 0)$ with $j \in \mathcal{S}_i$ remain coincident for all $p \in S \cap U_0$, and we denote the common value of such roots by $\theta_i(p)$. Then $\theta_i(p)$ is necessarily real for all $p \in S \cap U_0$, since any nonreal roots of $f(p, z)$ occur in conjugate pairs. Also, each function $\theta_i : S \cap U_0 \rightarrow \mathbb{R}$ is analytic since the composite $\theta_i \phi^{-1}$ is in $\phi(S \cap U_0)$. Therefore f is analytic delineable on $S \cap U_0$. That f is analytic delineable on the whole of S follows by connectedness of f.

3.3   Method Modifications

Theorem 3 justifies leaving the trailing coefficient out of the projection of a polynomial f whenever we can ensure that f is not nullified on the regions we lift over. Also, noting that any polynomial is clearly Lazard-valuation invariant on a single-point cell, we may leave the trailing coefficient of a polynomial f out of its projection if f is not nullified over any of the regions we lift over, except possibly single-point regions. This suggests a modification of VCADL, which we describe in this section.

At the outset we assume we have a useful and efficient finite zero test, with some additional features, for the coefficients of a given element $f \in R_n$. More precisely, we will use a procedure, which we shall denote simply by T, that takes an element f of R_n as input and outputs a finite set $S \subset \mathbb{R}^{n-1}$ or FAIL, with the requirement that if $S \subset \mathbb{R}^{n-1}$ is returned, then S contains the set of all points in $\mathbb{R}^{n-1}$ on which the input f is nullified. Such a procedure could be obtained as follows. Consider the set of $(n-1)$-variate coefficients of f. First use a "quick test" that tries to prove that the system of equations defined by the coefficients is unsatisfiable over the reals. There are a number of possibilities for this, including simple substitution of linear equations. If such a "quick test" fails to prove unsatisfiability over $\mathbb{R}$, then one could apply any suitable finite zero test over $\mathbb{C}$ – an example of such a test is described in Sect. 6 of [15]. If this test does not succeed in showing that the coefficients of f have only finitely many common zeros, then T returns FAIL. If one of the above tests succeeded, then T calls a selected algorithm which computes a desired set $S \subset \mathbb{R}^{n-1}$.

Definition 4 (Modified Lazard projection). *Let A be a finite irreducible basis in R_n, with $n \geq 2$. Let Γ be a finite set of points in $\mathbb{R}^n$ with algebraic number components. The modified Lazard projection $P_L^{mod}(A, \Gamma)$ of A, Γ is the finite subset of R_{n-1} comprising the following polynomials:*

1. *all leading coefficients of the elements of A,*
2. *all discriminants of the elements of A, and*
3. *all resultants of pairs of distinct elements of A,*
4. *all trailing coefficients (i.e. coefficients independent of x_n) of elements of A for which T returns* FAIL,

paired with the set of points

$$\{(\gamma_1, \ldots, \gamma_{n-1}) \mid (\gamma_1, \ldots, \gamma_{n-1}, \gamma_n) \in \Gamma\} \cup \bigcup_{f \in \overline{A}} T(f),$$

where $\overline{A} = \{f \mid f \in A \wedge T(f) \neq$ FAIL$\}$.

Use of the modified Lazard projection in CAD construction requires the description of the following subalgorithm which takes as inputs both $A \subset R_n$ and Γ (as defined above), and returns the sample data for an associated CAD of $\mathbb{R}^n$. (Γ is used to pass nullification point data from one recursive call to the next, and is intended to be empty for the initial call.)

Algorithm 5 (VCADL modified subalgorithm (VCADLmodsub))
Input: *A is a list of integral polynomials in $x_1, \ldots, x_n$, and Γ a set of algebraic points in $\mathbb{R}^n$.*
Output: *$\mathcal{I}$ and $\mathcal{S}$ are lists of indices and sample points, respectively, of the cells comprising an A-valuation-invariant CAD of $\mathbb{R}^n$.*

---

1. *If $n > 1$ then go to (2). Isolate the real roots of the irreducible factors of the nonzero elements of A. Construct cell indices $\mathcal{I}$ and sample points $\mathcal{S}$ from the real roots and the points in Γ and return.*
2. *Set $B \leftarrow$ the finest squarefree basis for $\mathrm{prim}(A)$. Set $(P, \Gamma') \leftarrow P_L^{mod}(B, \Gamma)$. Call VCADLmodsub with inputs $(\mathrm{cont}(A) \cup P, \Gamma')$, obtaining outputs $(\mathcal{I}', \mathcal{S}')$. Set $\mathcal{I} \leftarrow$ the empty list. Set $\mathcal{S} \leftarrow$ the empty list. For each $\alpha = (\alpha_1, \ldots, \alpha_{n-1})$ in $\mathcal{S}'$ do*
   Let i be the index of the cell containing α. Set $f^*(x_n) \leftarrow \prod_{f \in B} f_\alpha(x_n)$.
   Let D be a list of the isolated real roots of $f^*(x_n)$.
   Set $E \leftarrow \{\gamma_n \mid (\alpha_1, \ldots, \alpha_{n-1}, \gamma_n) \in \Gamma\}$.
   Construct cell indices and sample points for Lazard sections and sectors of elements of B from i, α and $D \cup E$.
   Add the cell indices to $\mathcal{I}$ and the sample points to $\mathcal{S}$.

A main algorithm VCADLmod, with input only $A \subset R_n$, could be obtained by simply invoking VCADLmodsub with inputs A and ϕ (empty set of points).

3.4   An Example Illustrating the Modified Lazard Projection

Here we consider one concrete example showing how the modified Lazard projection results in reduced projection factor sets. Consider the polynomial set

$$A := \{f := (x-1)^2 + (ac-1)dx + (b^2 + ac), g := ad + ac + 1\}$$

and variable order $a \prec b \prec c \prec d \prec x$.

Projecting x: since $\deg_x(g) = 0$, we only project f, which has leading coefficient $\mathrm{ldcf}_x(f) = 1$ and trailing coefficient $\mathrm{trcf}_x(f) = b^2 + ac + 1$. Because the leading coefficient is constant, the projection consists only of the discriminant $d_f := \mathrm{discr}_x(f) = (ac-1)^2 d^2 - 4(ac-1)d - 4(ac+b^2)$. (Note that the original and modified Lazard projections produce the same thing, according to our remark at the end of Sect. 2.3.)

Projecting d: the set to be projected is $\{g, d_f\}$. Since $a = ac + 1 = 0$ is not satisfiable, g is nowhere nullified, and since the discriminant of g is a constant, the modified Lazard projection of g consists solely of $\mathrm{ldcf}_d(g) = a$. It is also clear that d_f is nowhere nullified, since its leading and trailing coefficients cannot be zero simultaneously[1]. Thus its projection consists of the $\mathrm{ldcf}_d(d_f) = (ac-1)^2$ and $\mathrm{discr}_d(d_f) = 16(ac-1)^2(ac+b^2+1)$. Finally the projection also includes $\mathrm{res}_d(df, g) = a^4 c^4 + 2a^2(2a-1)c^2 - 4a^3 c - 4a^2 b^2 + 4a + 1$. The irreducible factors of the projection are:

$$\{a, ac - 1, ac + b^2 + 1, a^4 c^4 + 2a^2(2a-1)c^2 - 4a^3 c - 4a^2 b^2 + 4a + 1\}$$

which differs from the original Lazard projection in that the two trailing coefficients $ac + 1$ and $ac + b^2$ are missing.

If we were to follow through with projecting c and then b, we would arrive at a projection set in which there are eight irreducible factors at level one (i.e. in variable a only). In comparison, using the original Lazard projection would have produced a projection set in which there are 21 irreducible factors at level one. The table below shows how the degrees of these factors are distributed.

degree	1	2	3	4
num. 1-level polys, original Lazard	7	6	2	6
num. 1-level polys, modified Lazard	3	1	2	2

Note that since the given set A is well-oriented, the Brown-McCallum projection could also be applied here. However, a key point of the discussion is to compare the modified and original Lazard projections.

3.5   Complexity Analysis

Let $A \subset \mathbb{Z}[x_1, \ldots, x_n]$. We denote by $P_M(A)$ the McCallum projection of A, as defined in Sect. 6 of [15] and Sect. 2.1 of [2]. We define the *enhanced Lazard*

---

[1] The system $ac - 1 = 0 \wedge ac + b^2 = 0$ is easily shown to be unsatisfiable by a number of means, including simply substituting for linearly occurring variables.

projection of A, denoted by $P_{LE}(A)$, to be the set $\mathrm{cont}(A) \cup P$ constructed in algorithm VCADLmodsub. Following the presentation in Sect. 2.3 of [2] we undertake a preliminary complexity analysis for a variation of VCADLmod applied to A under a certain assumption. In particular, we will derive a bound for the number of cells in the CAD of $\mathbb{R}^n$ produced by VCADLmod (the *cell count*) under the assumption that for each element f of the irreducible basis B for $\mathrm{prim}(A)$, $T(f) \neq \mathrm{FAIL}$ and $T(f) = \phi$, and with the variation that the original Lazard projection P_L is to be used instead for projections subsequent to the first, if any.

As in [2], the key parameters for the analysis are, apart from n, variations on the number m of polynomials in A and their maximum degree d in any variable.

Definition 5. *A has* partition bounds (m, d) *if it can be partitioned into no more than m subsets such that the product of the elements of each subset has degree in any variable at most d.*

Lemma 1. *Suppose that A has partition bounds (m, d) and that for each $f \in B$ $T(f) \neq \mathrm{FAIL}$, then $P_{LE}(A)$ has partition bounds $(M, 2d^2)$, with $M = m(m+1)/2$.*

Proof. This result is an analogue of Lemma 11 of [2] which applies to $P_M(A)$. The proof of Lemma 11 can be adjusted to yield the justification needed here. Steps 1 and 2 (the first and second claims) remain valid here. Since $P_{LE}(A)$ includes no non-leading coefficients of B by hypothesis, step 3 is not required here. Hence $P_{LE}(A)$ can be partitioned into $m + (m(m-1)/2) = m(m+1)/2$ subsets each with partition bounds $(1, 2d^2)$. ∎

We can now use Table 1 of [2], in conjunction with the lemma above and Corollary 12 of [2], to estimate the growth in the number and degrees of projection polynomials when using operator P_{LE} for the first projection and operator P_L for subsequent projections, if any. Under the assumption stated above the cell count of the CAD produced depends only on the number of real roots of projection polynomials subject to standard or Lazard evaluation. Hence we have:

Theorem 4. *Let $A \subset \mathbb{Z}[x_1, \ldots, x_n]$, with irreducible basis B for $\mathrm{prim}(A)$, have partition bounds (m, d). Suppose that for each $f \in B$, $T(f) \neq \mathrm{FAIL}$ and $T(f) = \phi$. Then, letting $M = m(m+1)/2$, the cell count of the CAD of $\mathbb{R}^n$ produced by the variant of VCADLmod described above is bounded by*

$$(2Md + 1) \prod_{r=1}^{n-1} (2^{2^r} d^{2^r} M^{2^{r-1}} + 1).$$

This bound has the same form as that stated for CAD construction using operator P_M on page 11 of [2]. However, in that context, the parameter M has the slightly larger value $\lfloor (m+1)^2/2 \rfloor$. Both bounds are likely to be pessimistic.

3.6   An Example Contrasting Various Projection Operators

Consider the formula $y \geq 0 \wedge y + (a-b)x^2 + (2a - 3b + c)ax + c - b \leq 0$ and the variable order $a \prec b \prec c \prec x \prec y$. We attempted to construct CADs for this formula using QEPCAD B's implementation of the Hong, Brown-McCallum and McCallum projections. Note that QEPCAD B implements a number of optimizations to these operators to reduce the projection set sizes and, for the McCallum/Brown-McCallum operators, to deduce that correctness can be guaranteed for many kinds of non well-oriented input. We then used interactive features of QEPCAD B to simulate the original and improved Lazard methods to determine the number of cells in the CADs they would produce. All the results are summarized in the following table.

Method	Number of 5-level cells
Improved Lazard	1389
McCallum/Brown-McCallum	FAIL
Original lazard	5125
Hong	4933

We note first that Brown-McCallum & McCallum fail on this because the resultant of the two input polynomials is nullified on the positive dimensional region $a = b = c$, and none of the checks that QEPCAD B has can verify that we have order-invariance none-the-less. We next note that our improved Lazard clearly outperforms regular Lazard because we discover that the projection factor $a^2c^2 + (4a^3 - 6a^2b - 4a + 4b)c + 4a^4 - 12a^3b + 9a^2b^2 + 4ab - 4b^2$ as a polynomial in c is only nullified at $a = b = 0$. So we add that as a projection point, and do not include the trailing coefficient, which has higher degree and more terms than the other coefficients. Interestingly, when using Hong's method, QEPCAD B deduces that the leading and degree-one coefficients of this polynomial are sufficient, and so avoids adding the large trailing coefficient. This is why Hong's method actually outperforms the original Lazard projection on this example. So we see that, without the optimizations from this paper, Lazard's method may be outperformed by the optimized version of Hong's method.

4   Further Improvements

There are special cases in which the Lazard projection may be reduced beyond what is suggested by the results of the previous section. In this section we describe several such situations with regards to polynomials in two or three variables. This is more useful than it might seem at first, because these improvements also hold for polynomials in two or three variables "embedded" in higher dimensional problems. This is made precise in Theorem 9 below. It will be useful in the following to define the "level" of a polynomial.

Definition 6. *Let V be a set of variables, and let $x_1 \prec x_2 \prec \cdots \prec x_n$ be an ordering of the elements of V. The* level *of polynomial f over V with respect to the ordering is the largest i such that $deg_{x_i}(f) > 0$.*

4.1   Reduced Projections for Two or Three Variables

The following theorems justify leaving out the trailing coefficients for 2-level and, in some cases, 3-level polynomials. In the theorem statements, polynomial primitivity, degree, etc. are understood to be with respect to the variable of highest order (that is, the main variable), as usual.

Theorem 5. *With $x \prec y$, let $f \in \mathbb{R}[x, y]$ be primitive, of positive degree and squarefree. Let $D(x)$ and $l(x)$ denote the discriminant and leading coefficient, respectively, of f. On any connected subset S of $\mathbb{R}$ in which $D(x)$ and $l(x)$ are sign-invariant, f is Lazard delineable.*

This is a consequence of an observation in [6].

Theorem 6. *Let f be a primitive and squarefree polynomial in the variables, x, y and z, ordered $x \prec y \prec z$, such that $deg_z(f) > 1$. Let $D(x, y)$ and $l(x, y)$ denote the discriminant and leading coefficient, respectively, of f. If D and l are relatively prime, then on any connected region $S \subseteq \mathbb{R}^2$ in which D and l are Lazard-valuation invariant, f is Lazard delineable.*

Proof. Since D and l are relatively prime, their set of common zeros is a finite set of isolated points. Thus if S has positive dimension, the Lazard invariance of D and l implies that at least one of them is non-zero throughout S. If l is non-zero, clearly f is not nullified in S. The same conclusion holds if D is non-zero, since at any point that nullifies f, the discriminant of f must be zero. So, if S has positive dimension, f is nullified nowhere on S, and thus Theorem 3 implies that f is Lazard delineable. This leaves only the case in which S is a single-point cell, but of course any polynomial is vacuously Lazard delineable over a single point cell.   □

For our final special case, we consider three-level polynomials that are linear in the main variable—since the preceding theorem does not apply to that case.

Theorem 7. *Let f be a primitive and squarefree polynomial in the variables, x, y and z, ordered $x \prec y \prec z$, of the form $a_1^{d_1} a_2^{d_2} \cdots a_m^{d_m} z + b_1^{e_1} b_2^{e_2} \cdots b_n^{e_n}$, where the a_i's and b_j's are irreducible polynomials in x and y. Assume that all of the a_i's have positive degree in y. Let $P = \{r_{i,j} \mid deg_y(b_j) > 0 \wedge r_{i,j} = res_y(a_i, b_j) \vee deg_y(b_j) = 0 \wedge r_{i,j} = b_j\}$. On any connected region $S \subseteq \mathbb{R}^2$ in which the elements of P are Lazard-valuation invariant, f is Lazard delineable.*

Proof. If polynomial f is nullified at a point (α_1, α_2), there must be i and j such that $a_i(\alpha_1, \alpha_2) = b_j(\alpha_1, \alpha_2) = 0$. This means $r_{i,j}(\alpha_1, \alpha_2) = 0$. Since f is primitive, a_i and $r_{i,j}$ are relatively prime, so their common zeros are a finite set of isolated points. Thus, the Lazard-valuation invariance of both in S implies that S is a single-point, which means f is vacuously Lazard-valuation invariant on S. If S has positive dimension, f is no nullified on S, and therefore Theorem 3 implies that f is Lazard delineable.   □

4.2 Carrying over Improvements

In this section we prove that improvements to projection that we produce for two and three level special cases can be carried over into problems that include polynomials in two and three variables embedded in more dimensions. This further increases the applicability of problems from this section.

Theorem 8. *Fix variable ordering $x_1 \prec \cdots \prec x_n$, which we call ord. Consider suborder $x_{i_1} \prec x_{i_2} \prec \ldots \prec x_{i_k}$, which we call ord'. Let $p \in R[x_{i_1}, \ldots x_{i_k}]$, be a primitive squarefree polynomial. Let $m : \mathbb{R}^n \to \mathbb{R}^k$ be given by $m(a_1, \ldots, a_n) = (a_{i_1}, a_{i_2}, \ldots, a_{i_k})$. Denote by $v_{ord}(p, a)$ the Lazard valuation of p at point a with respect to ord. Let $(r_1, \ldots, r_n) = v_{ord}(p, a)$.*

1. *If x_j is not in $\{x_{i_1}, \ldots, x_{i_k}\}$ then $r_j = 0$, and*
2. *$v_{ord'}(p, m(a)) = m(v_{ord}(p, a))$ for any $a \in \mathbb{R}^n$.*

Proof. This follows directly from the Lazard evaluation process, since a variable in ord that does not appear in polynomial p just results in a zero entry in the Lazard valuation, and otherwise has no effect.   □

Theorem 9. *Continuing with the assumptions from Theorem 8. Let S be a region in $\mathbb{R}^n$, and let $S' = \{m(\alpha) \mid \alpha \in S\}$. Polynomial p is Lazard valuation invariant in S with respect to ord if and only if p is Lazard valuation invariant in S' with respect to ord'.*

Proof. Assume that p is Lazard valuation invariant in S with respect to ord, and let a' and b' be points in S'. Then by the definition of S', there exist points a and b in S such that $m(a) = a'$ and $m(b) = b'$. By assumption, $v_{ord}(p, a) = v_{ord}(p, b)$, so by Theorem 8 we have $v_{ord'}(p, m(a)) = v_{ord'}(p, m(b))$. Thus, $v_{ord'}(p, a') = v_{ord'}(p, b')$, and we get that p is Lazard valuation invariant in S' with respect to ord'.

Assume p is not Lazard valuation invariant in S with respect to ord. Then there exist points a and b in S such that $v_{ord}(p, a) \neq v_{ord}(p, b)$. Thus we have $v_{ord'}(p, m(a)) \neq v_{ord'}(p, m(b))$. Since $m(a)$ and $m(b)$ are both in S', we get that p is not Lazard valuation invariant in S' with respect to ord'.   □

5 Conclusion

To summarize, we have proved that we can omit from the Lazard projection of A the trailing coefficient of every element f of A for which f can be determined to have only finitely many nullifying points in $\mathbb{R}^{n-1}$, provided certain modifications to the CAD method are made. This makes Lazard's method competitive in principle with the Brown-McCallum method in the well-oriented case. Lazard's method is already known to be infallible in the non-well-oriented case (unlike Brown-McCallum). It will be interesting to compare the performance of the two methods experimentally for a range of well-oriented sample problems. We have provided also a selection of further enhancements to Lazard's method.

It will be great to see progress on the adaptation of Lazard's method to the other approaches to CAD mentioned in the Introduction. As remarked there, we expect that the results of the present paper will be beneficial for such efforts.

References

1. Arnon, D.S., Collins, G.E., McCallum, S.: Cylindrical algebraic decomposition I: the basic algorithm. SIAM J. Comput. **13**, 865–877 (1984)
2. Bradford, R., Davenport, J.H., England, M., McCallum, S., Wilson, D.: Truth table invariant cylindrical algebraic decomposition. J. Symbolic Comput. **76**, 1–35 (2016)
3. Brown, C.W.: Improved projection for cylindrical algebraic decomposition. J. Symbolic Comput. **32**, 447–465 (2001)
4. Brown, C. W., Davenport, J.H.: The complexity of quantifier elimination and cylindrical algebraic decomposition. In: ISSAC 2007: Proceedings of the 2007 International Symposium on Symbolic and Algebraic Computation, pp. 54–60. ACM, New York (2007)
5. Caviness, B., Johnson, J.R. (Eds.) Quantifier Elimination and Cylindrical Algebraic Decomposition. Texts and Monographs in Symbolic Computation. Springer, Wien (1998). https://doi.org/10.1007/978-3-7091-9459-1
6. Collins, G.E.: Quantifier elimination for real closed fields by cylindrical algebraic decompostion. In: Brakhage, H. (ed.) GI-Fachtagung 1975. LNCS, vol. 33, pp. 134–183. Springer, Heidelberg (1975). https://doi.org/10.1007/3-540-07407-4_17
7. Collins, G.E.: Quantifier elimination by cylindrical algebraic decomposition - twenty years of progress. In [5]
8. Collins, G.E., Hong, H.: Partial cylindrical algebraic decomposition for quantifier elimination. J. Symbolic Comput. **12**, 299–328 (1991)
9. Cowen-Rivers, A.I., England, M.: Towards incremental cylindrical algebraic decomposition. In: Bigatti, A., Brain, M. (Eds.) Proceedings of 3rd Workshop on Satisfiability Checking and Symbolic Computation 2018. No. 2189 in CEUR Workshop Proceedings, pp. 3–18. http://ceur-ws.org/Vol-2189/
10. Hong, H.: An improvement of the projection operator in cylindrical algebraic decomposition. In: ISSAC 1990: Proceedings of the 1990 International Symposium on Symbolic and Algebraic Computation, pp. 261–264. ACM Press, New York (1990). Reprinted in [5]
11. Jovanović, D., de Moura, L.: Solving non-linear arithmetic. In: Gramlich, B., Miller, D., Sattler, U. (eds.) IJCAR 2012. LNCS (LNAI), vol. 7364, pp. 339–354. Springer, Heidelberg (2012). https://doi.org/10.1007/978-3-642-31365-3_27
12. Kremer, G., Abraham, E.: Fully incremental cylindrical algebraic decomposition. J. Symbolic Comput. **100**, 11–37 (2020)
13. Lazard, D.: An improved projection for cylindrical algebraic decomposition. Algebraic Geometry and its Applications. Springer, New York (1994). https://doi.org/10.1007/978-1-4612-2628-4_29
14. McCallum, S.: An improved projection operation for cylindrical algebraic decomposition of three-dimensional space. J. Symbolic Comput. **5**, 141–161 (1988)

15. McCallum, S.: An improved projection operation for cylindrical algebraic decomposition. In [5]
16. McCallum, S.: On projection in CAD-based quantifier elimination with equational constraint. In: Dooley, S. (ed.) Proceedings of International Symposium on Symbolic and Algebraic Computation ISSAC 1999, pp. 145–149. ACM Press, New York (1999)
17. McCallum, S., Hong, H.: On using Lazard's projection in CAD construction. J. Symbolic Comput. **72**, 65–81 (2016)
18. McCallum, S., Parusiński, A., Paunescu, L.: Validity proof of Lazard's method for CAD construction. J. Symbolic Comput. **92**, 52–69 (2019)
19. Strzebonski, A.: Solving systems of strict polynomial inequalities. J. Symbolic Comput. **29**, 471–480 (2000)
20. Strzebonski, A.: Divide-and-conquer computation of cylindrical algebraic decomposition. arXiv:1402.0622 (2014)

The Complexity and Parallel Implementation of Two Sparse Multivariate Hensel Lifting Algorithms for Polynomial Factorization

Tian Chen and Michael Monagan[⊠]

Department of Mathematics, Simon Fraser University, Burnaby,
British Columbia V5A 1S6, Canada
tca71@sfu.ca, mmonagan@cecm.sfu.ca

Abstract. Sparse multivariate Hensel lifting (SHL) algorithms are used in multivariate polynomial factorization as efficient randomized algorithms. They improve on Wang's classical multivariate Hensel lifting which can be exponential in the number of variables for sparse factors.

In this work, we present worst case complexity analyses and failure probability bounds for two recently developed SHL algorithms. One of the algorithms solves the multivariate Diophantine equations using sparse interpolation, and the other interpolates the factors directly from bivariate images obtained using bivariate Hensel lifting.

We have observed that a linear expression swell occurs in both approaches. We have modified the second approach to eliminate the expression swell. Our improvement also injects more parallelism into the sparse interpolation step.

We have made a high-performance parallel implementation of our new SHL algorithm in Cilk C. We present timing benchmarks comparing our Cilk C implementation with the factorization algorithms in Maple and Magma. We obtain good parallel speedup and our algorithm is much faster than Maple and Magma on our benchmarks.

Keywords: Sparse multivariate Hensel lifting · Sparse interpolation · Multivariate Diophantine equations · Polynomial factorization · Bivariate Hensel lifting · Cilk C

1 Introduction

Polynomial factorization has been a central topic in computer algebra, and it continues to play a critical role in other fields such as algebraic coding theory, cryptography, number theory and algebraic geometry [5]. In this work, we focus on the main tool used to factor multivariate polynomials, namely, multivariate Hensel lifting (MHL). MHL was initially developed by Yun [19] and Wang [18] to factor polynomials with integer coefficients, but it can be applied to polynomials with coefficients in other domains, for example, finite fields [1,4] and algebraic number fields [15,16,20].

© Springer Nature Switzerland AG 2020
F. Boulier et al. (Eds.): CASC 2020, LNCS 12291, pp. 150–169, 2020.
https://doi.org/10.1007/978-3-030-60026-6_9

To factor a multivariate polynomial $a \in \mathbb{Z}[x_1, x_2, \cdots, x_n]$, Wang's multivariate Hensel lifting [18] first chooses integers $\alpha_2, \ldots, \alpha_n$ and factors the univariate image $a(x_1, \alpha_2, \ldots, \alpha_n)$ in $\mathbb{Z}[x_1]$. Then it recovers the multivariate factors from their images one variable at a time. A key step in Wang's MHL is the solution of a sequence of multivariate polynomial Diophantine equations (MDPs). Wang's MHL has been implemented in many computer algebra systems including Maple, Magma, Macsyma, Mathematica and Singular. For a detailed description of Wang's MHL we refer the reader to Chap. 6 of [5].

It is known that when factors are sparse and the evaluation points $\alpha_2, \ldots, \alpha_n$ are mostly non-zero, Wang's method for solving MDPs can be exponential in the number of variables [9,12]. To resolve this, Zippel [22] introduced the first polynomial-time probabilistic algorithm in 1981 that takes advantage of sparsity. Other sparse Hensel lifting (SHL) algorithms were developed by Kaltofen in 1985 [6] and Kaltofen and Trager in 1990 [7].

In 2016, Monagan and Tuncer [9] proposed a new sparse Hensel lifting algorithm called MTSHL. The authors made a key observation which they call the strong SHL assumption (see Lemma 1 of [9]) which is applied to solve the MDPs that appear in Wang's MHL in random polynomial time. A detailed complexity analysis for MTSHL was completed for the average-case in [12]. MTSHL was integrated into Maple 2019 [13].

In 2018, Monagan and Tuncer [11] introduced another approach that does not solve MDPs. Instead, at each Hensel lifting step, it interpolates the factors from many bivariate images which are obtained using bivariate Hensel lifting. Classical bivariate Hensel lifting (BHL) costs $O(d^4)$ where $d = \deg(a)$ is the total degree of the input polynomial. The cost of BHL is improved to $O(d^3)$ by Monagan in [14]. This approach is appropriate for multivariate Hensel lifting because the degree of the factors is rarely 100, and often 10 or lower.

Our work is motivated by the following observation. In the main Hensel lifting step (see Algorithm 1) which is used in MTSHL ([12]), in [11] and also in Wang's multivariate Hensel lifting [18], when the evaluation point α_j is non-zero, an expression swell occurs in each factor as it is recovered (in line 13). This increases the cost of the error computation (in line 14).

Our first contribution is a new algorithm CMSHL which reorganizes the sparse Hensel lifting algorithm in [11] to eliminate the expression swell. Our second contribution is a worst case complexity analysis for CMSHL and for MTSHL and bounds for the failure probability of both algorithms. Our third contribution is a high-performance parallel implementation of CMSHL using Cilk C [3] for multi-core computers.

Our paper is organized as follows. In Sect. 2, we present the two sparse multivariate Hensel lifting algorithms from [9,12] and [11]. We study the expression swell and give examples of it for the worst case and then we present our new algorithm CMSHL which eliminates the expression swell. In Sect. 3 we give worst case complexity analyses for MTSHL and CMSHL along with their failure probabilities. In Sect. 4 we present timings comparing our Cilk C implementation of CMSHL with Maple and Magma's factorization commands for a variety of input problems. The timings (see Tables 3, 4 and 5) demonstrate good parallel speedup. In Sect. 5 we give some details of our Cilk C implementation.

2   Two Algorithms of Sparse Multivariate Hensel Lifting

Suppose we seek the factors of a multivariate polynomial $a \in \mathbb{Z}[x_1, \cdots, x_n]$. Similar to Wang's multivariate Hensel lifting (MHL), a few preliminary steps are done before sparse multivariate Hensel lifting (SHL) [10]:

The first step is to compute and remove the content of a in a chosen main variable, say x_1. For $a = \sum_{i=0}^{d} a_i(x_2, \cdots, x_n)x_1^i$, the content of a is $\gcd(a_0, \cdots, a_d)$, a polynomial with one fewer variable which can be factored recursively. Let us assume this has already been done.

The second step is to identify any repeated factor in a by doing a *square-free factorization* (see ch.8 in [5]). After this, we obtain the factorization $a = b_1 b_2^2 \cdots b_k^k$ such that each factor b_i is square-free and $\gcd(b_i, b_j) = 1$ for $i \neq j$. Without loss of generality, suppose this has also been done and let $a = f_1 f_2 \cdots f_r$ be the irreducible factorization of a over $\mathbb{Z}$.

Next, an evaluation point $\boldsymbol{\alpha} = (\alpha_2, \cdots, \alpha_n) \in \mathbb{Z}^{n-1}$ is chosen and then $a(x_1, \boldsymbol{\alpha})$ is factored over $\mathbb{Z}$. The evaluation point $\boldsymbol{\alpha}$ must satisfy the following conditions: (i) $L(\boldsymbol{\alpha}) \neq 0$ where L is the leading coefficient of a in x_1, (ii) $a(x_1, \boldsymbol{\alpha})$ must have no repeated factor in x_1, and (iii) $f_i(x_1, \boldsymbol{\alpha})$ must be irreducible. Conditions (i) and (ii) can be enforced in advance whereas (iii) can be ensured with high probability by choosing the integers α_i from a sufficiently large set.

For simplicity, throughout this paper we only consider two irreducible factors f and g both monic in x_1. For multi-factor cases, we refer the reader to [10]. Let $a = fg$ where f and g are monic irreducible polynomials in $\mathbb{Z}[x_1, \cdots, x_n]$. We define $h_j := h(x_1, \cdots, x_j, \alpha_{j+1}, \cdots, \alpha_n)$ for a polynomial $h \in \mathbb{Z}[x_1, \cdots, x_n]$. To factor a, the image a_1 is first factored over $\mathbb{Z}$. From Hilbert's irreducibility theorem (see e.g. [8]), f_1 and g_1 are irreducible with high probability.

Now we start the process of sparse multivariate Hensel lifting to recover f and g from a, f_1, g_1. The inputs are $a, f_1, g_1, \boldsymbol{\alpha}$ and a prime p such that $\gcd(f_1, g_1) = 1$ in $\mathbb{Z}_p[x_1]$. The algorithm lifts (f_1, g_1) to (f_2, g_2), then lifts (f_2, g_2) to (f_3, g_3) etc. until (f_n, g_n) is obtained. At each step, $a_j - f_j g_j \bmod p = 0$ so that at the final step, $a_n - f_n g_n \bmod p = 0$. To recover the integer coefficients in the factors one may either use a sufficiently large p or perform the Hensel lifting using a machine prime (we use 63 bit primes) and, if necessary, do a subsequent p-adic lift [10].

2.1   MTSHL

The jth Hensel lifting step for both approaches of sparse multivariate Hensel lifting in [9,12] and [11] is presented. Our presentation includes worst case complexity bounds for the main steps as an aid for the reader and for later reference. We use the notation $\#f$ to be the number of non-zero terms of a polynomial f.

The first approach (MTSHL [9,12]) is presented in Algorithms 1 and 2. Algorithm 2 is called from Algorithm 1 in a loop to solve the MDPs via sparse interpolation. Note that in Algorithm 2, if $\max(t_i)$ is much larger (or much smaller) than $\max(s_i)$ then it will be faster to interpolate the smaller of σ and τ only and obtain the larger of σ and τ using $\sigma u + \tau w = c$. The second approach in [11] is shown in Algorithm 3.

Algorithm 1. MTSHL: Hensel lift x_j with MDPs via sparse interpolation.

1: **Input:** A prime p, $\alpha_j \in \mathbb{Z}_p$, $a_j \in \mathbb{Z}_p[x_1, \cdots, x_j]$ monic in x_1,
   $f_{j-1}, g_{j-1} \in \mathbb{Z}_p[x_1, \cdots, x_{j-1}]$ s.t. $a_j(x_1, ..., x_{j-1}, \alpha_j) = f_{j-1}g_{j-1}$ with $j > 2$.
2: **Output:** $f_j, g_j \in \mathbb{Z}_p[x_1, \cdots, x_j]$ s.t. $a_j = f_jg_j$ where $f_j(x_j = \alpha_j) = f_{j-1}$ and
   $g_j(x_j = \alpha_j) = g_{j-1}$; Otherwise, FAIL.
3: $(\sigma_0, \tau_0) \leftarrow (f_{j-1}, g_{j-1})$; $(f_j, g_j) \leftarrow (f_{j-1}, g_{j-1})$.
4: error $\leftarrow a_j - f_jg_j$; monomial $\leftarrow 1$.
5: **for** $i = 1, 2, \cdots$ **while** error $\neq 0$ and $\deg(f_j, x_j) + \deg(g_j, x_j) < \deg(a_j, x_j)$ **do**
6:     monomial $\leftarrow$ monomial $\cdot (x_j - \alpha_j)$.
7:     $c_i \leftarrow$ coeff(error, $(x_j - \alpha_j)^i$).
8:     **if** $c_i \neq 0$ **then**
9:         // Solve the MDP $\sigma_i g_{j-1} + \tau_i f_{j-1} = c_i$ for $\sigma_i, \tau_i \in \mathbb{Z}_p[x_1, \cdots, x_{j-1}]$.
10:        $\sigma_f \leftarrow \sigma_{i-1}$; $\tau_f \leftarrow \tau_{i-1}$.
11:        $(\sigma_i, \tau_i) \leftarrow$ SparseInterp($g_{j-1}, f_{j-1}, c_i, \sigma_f, \tau_f$) // Algorithm 2
12:        **if** $(\sigma_i, \tau_i) =$ FAIL **then return** FAIL **end if**
13:        $(f_j, g_j) \leftarrow (f_j + \sigma_i \cdot$ monomial, $g_j + \tau_i \cdot$ monomial$)$.
14:        error $\leftarrow a_j - f_jg_j$.
15:    **end if**
16: **end for**
17: **if** error $= 0$ **then return** (f_j, g_j) **else return** FAIL **end if**

Algorithm 2. SparseInterp: solve an MDP using sparse interpolation.

1: **Input:** $u, w, c, \sigma_f, \tau_f \in \mathbb{Z}_p[x_1, \cdots, x_{j-1}]$ where u, w are monic in x_1.
2: **Output:** The solution (σ, τ) to the MDP $\sigma u + \tau w = c \in \mathbb{Z}_p[x_1, \cdots, x_{j-1}]$ or FAIL.
3: Let $d\sigma = \deg(\sigma_f, x_1)$ and $\sigma = \sum_{i=0}^{d\sigma} \zeta_i(x_2, \cdots, x_{j-1})x_1^i$ with $\zeta_i = \sum_{l=1}^{s_i} a_{il}M_{il}$
   and $d\tau = \deg(\tau_f, x_1)$ and $\tau = \sum_{i=0}^{d\tau} \eta_i(x_2, \cdots, x_{j-1})x_1^i$ with $\eta_i = \sum_{l=1}^{t_i} b_{il}N_{il}$, where
   a_{il}, b_{il} are to be determined, $x_1^i M_{il}, x_1^i N_{il}$ are monomials in σ_f, τ_f.
4: Let s be the maximum of s_i and t_i.
5: Pick $\boldsymbol{\beta} = (\beta_2, \cdots, \beta_{j-1}) \in (\mathbb{Z}_p \setminus \{0\})^{j-2}$ at random.
6: Evaluate monomials at $\boldsymbol{\beta}$: $\dots\dots\dots\dots\dots\dots\dots\dots$ $\mathcal{O}((j-2)(\#f + \#g + d_{max}))$
   $\mathcal{S} = \{S_i = \{m_{il} = M_{il}(\boldsymbol{\beta}) : 1 \leq l \leq s_i\}, 0 \leq i \leq d\sigma\}$ and
   $\mathcal{T} = \{T_i = \{n_{il} = N_{il}(\boldsymbol{\beta}) : 1 \leq l \leq t_i\}, 0 \leq i \leq d\tau\}$.
7: **if** any $|S_i| \neq s_i$ or $|T_i| \neq t_i$ **then return** FAIL **end if**
8: **for** k from 1 to s in parallel **do**
9:     Let $Y_k = (x_2 = \beta_2^k, \cdots, x_{j-1} = \beta_{j-1}^k)$.
10:    Evaluate u, w, c at Y_k: $u(x_1, Y_k), w(x_1, Y_k), c(x_1, Y_k)$. $\dots\dots$ $\mathcal{O}(s(\#f + \#g + \#a))$
11:    **if** $\gcd(u(x_1, Y_k), w(x_1, Y_k)) \neq 1$ **then return** FAIL **end if**
12:    Solve $\sigma_k(x_1)u(x_1, Y_k) + \tau_k(x_1)w(x_1, Y_k) = c(x_1, Y_k) \in \mathbb{Z}_p[x_1]$. $\dots\dots\dots$ $\mathcal{O}(s\,d_1^2)$
13: **end for**
14: **for** i from 0 to $d\sigma$ in parallel **do**
15:    Construct and solve the $s_i \times s_i$ linear system for a_{il}: $\dots\dots\dots\dots\dots\dots$ $\mathcal{O}(s\#f)$
$$\left\{ \sum_{l=1}^{s_i} a_{il}m_{il}^k = \text{coeff}(\sigma_k(x_1), x_1^i) \text{ for } 1 \leq k \leq s_i \right\}.$$
16: **end for**
17: Substitute the solution a_{il} into σ.
18: Similarly, construct τ. $\dots\dots\dots\dots\dots\dots\dots\dots\dots\dots\dots\dots\dots\dots$ $\mathcal{O}(s\#g)$
19: **if** $\sigma u + \tau w = c$ **then return** (σ, τ) **else return** FAIL // wrong σ_f or τ_f

---

Algorithm 3. Hensel lift x_j via bivariate Hensel lifting [11].

1: **Input:** A prime p, $\alpha_j \in \mathbb{Z}_p$, $a_j \in \mathbb{Z}_p[x_1, \cdots, x_j]$ monic in x_1,
   $f_{j-1}, g_{j-1} \in \mathbb{Z}_p[x_1, \cdots, x_{j-1}]$ s.t. $a_j(x_1, ..., x_{j-1}, \alpha_j) = f_{j-1}g_{j-1}$ with $j > 2$.

2: **Output:** $f_j, g_j \in \mathbb{Z}_p[x_1, \cdots, x_j]$ s.t. $a_j = f_jg_j$ where $f_j(x_j = \alpha_j) = f_{j-1}$ and
   $g_j(x_j = \alpha_j) = g_{j-1}$; Otherwise, FAIL.

3: Let $\sigma_0 = f_{j-1}$, $f_j = \sum_{h=0}^{df_j} \sigma_h(x_1, ..., x_{j-1})(x_j - \alpha_j)^h$ with $\sigma_h = \sum_{i=0}^{df} \left(\sum_{l=1}^{s_i} c_{hil} M_{il}\right) x_1^i$.
   Let $\tau_0 = g_{j-1}$, $g_j = \sum_{h=0}^{dg_j} \tau_h(x_1, ..., x_{j-1})(x_j - \alpha_j)^h$ with $\tau_h = \sum_{i=0}^{dg} \left(\sum_{l=1}^{t_i} d_{hil} N_{il}\right) x_1^i$.
   $M_{il}x_1^i$, $N_{il}x_1^i$ are monomials in σ_0, τ_0. $df = \deg(f_{j-1}, x_1)$, $dg = \deg(g_{j-1}, x_1)$.
   We are to determine: c_{hil}, d_{hil}, $df_j = \deg(f_j, x_j)$, $dg_j = \deg(g_j, x_j)$.

4: Pick $\boldsymbol{\beta} = (\beta_2, \cdots, \beta_{j-1}) \in \mathbb{Z}_p^{j-2}$ at random.

5: Evaluate monomials at $\boldsymbol{\beta}$ .... $\mathcal{O}((j-2)(\#f + \#g + d_{max}))$
   $\mathcal{S} = \{S_i = \{m_{il} = M_{il}(\boldsymbol{\beta}), 1 \le l \le s_i\}, 0 \le i \le df - 1\}$ and
   $\mathcal{T} = \{T_i = \{n_{il} = N_{il}(\boldsymbol{\beta}), 1 \le l \le t_i\}, 0 \le i \le dg - 1\}$.

6: **if** any $|S_i| \ne s_i$ or any $|T_i| \ne t_i$ **then return** FAIL **end if**

7: Let s be the maximum of s_i and t_i.

8: **for** k from 1 to s **in parallel do**

9:      Let $Y_k = (x_2 = \beta_2^k, \cdots, x_{j-1} = \beta_{j-1}^k)$.

10:     $A_k, F_k, G_k \leftarrow a_j(x_1, Y_k, x_j), f_{j-1}(x_1, Y_k), g_{j-1}(x_1, Y_k)$. .... $\mathcal{O}(s(\#f + \#g + \#a))$

11:     **if** $\gcd(F_k, G_k) \ne 1$ **then return** FAIL **end if** // unlucky evaluation

12:     Call *BivariateHenselLift*$(A_k, F_k, G_k, \alpha_j, p)$ to compute $\sigma_{hk}(x_1)$ and $\tau_{hk}(x_1)$ s.t.
       $A_k = f_kg_k$ where $f_k = \sum_{h=0}^{df_j} \sigma_{hk}(x_j - \alpha_j)^h$ and $g_k = \sum_{h=0}^{dg_j} \tau_{hk}(x_j - \alpha_j)^h$.

13: **end for**

14: **for** h from 1 to df_j **do**

15:     **for** i from 0 to df **do**

16:        Construct and solve the $s_i \times s_i$ linear system for c_{hil} .... $\mathcal{O}(sd_j\#f)$

$$\left\{ \sum_{l=1}^{s_i} c_{hil}m_{il}^k = \text{coeff}(\sigma_{hk}(x_1), x_1^i) \text{ for } 1 \le k \le s_i \right\}$$

17:     **end for**

18: **end for**

19: Substitute the solution c_{hil} into σ_h and expand to get f_j. .... $\mathcal{O}(d_j^2\#f)$

20: Similarly to construct g_j. .... $\mathcal{O}(sd_j\#g)$

21: **if** $a_j = f_jg_j$ **then return** (f_j, g_j) **else return** FAIL **end if**

2.2 Intermediate Expression Swell

In Algorithm 1, an expression swell may occur in line 13 and 14. In Algorithm 3 an expression swell may occur at the final expansion step (line 19). To illustrate the expression swell, we consider the partial sums of f_j. Let

$$f_j^{(i)} = \sum_{k=0}^{i} \sigma_k(x_1, \cdots, x_{j-1})(x_j - \alpha_j)^k \text{ for } 0 \le i \le d_j, \text{ and}$$

$$f_{jH}^{(i)} = (((\sigma_{d_j}(x_j - \alpha_j) + \sigma_{d_j-1})(x_j - \alpha_j) + \cdots)(x_j - \alpha_j)) + \sigma_{d_j-i},$$

Table 1. Number of terms in $f_j^{(i)}$ and $f_{jH}^{(i)}$ with a randomly generated polynomial.

i	0	1	2	3	4	5	6	7	8	9	10	11	12	13	14
$\#\sigma_i$	925	737	584	459	352	268	196	134	94	64	48	24	13	7	3
$\#f_j^{(i)}$	925	1512	1851	1999	**2021**	1934	1768	1628	1486	1411	1226	1130	1071	1028	989
$\#f_{jH}^{(i)}$	3	10	23	47	95	159	253	387	583	851	1203	1662	2246	**2983**	989

Table 2. Number of terms in $f_j^{(i)}$ and $f_{jH}^{(i)}$ in a worst case.

i	0	1	2	3	4
$\#\sigma_i$	7	7	7	7	7
$\#f_j^{(i)}$	7	14	21	28	7
$\#f_{jH}^{(i)}$	7	14	21	28	7

where $f_{jH}^{(i)}$ is the expansion in Horner's form and $d_j = \deg(f_j, x_j)$. $f_j^{(i)}$ and $f_{jH}^{(i)}$ correspond to the intermediate steps in line 13 of Algorithm 1 and in line 19 of Algorithm 3 respectively. We are interested in $\#f_j^{(i)}$ and $\#f_{jH}^{(i)}$ for $0 \le i \le d_j$.

An example of a randomly generated polynomial with $p = 2^{31} - 1$, $j = 5$, $d_j = 14$, $d = 20$ and $\#f_j = 989$ is shown in Table 1. The density ratio $\#f_j / \binom{d+j}{j} \approx 0.0186$. The ratios $\max(\#f_j^{(i)})/\#f_j$ and $\max(\#f_{jH}^{(i)})/\#f_j$ are 2.043 and 3.016 respectively. This example shows a typical trend in an average case where $\max(\#f_j^{(i)})/\#f_j \lesssim 1 + d/j$ [12]. We observe that $\#\sigma_i$ decreases as i increases from 0 to d_j. The number of terms $\#f_j^{(i)}$ increases to a peak in the first few expansions and gradually shrinks back to $\#f_j$, whereas $\#f_{jH}^{(i)}$ increases to a higher peak than $\max(\#f_j^{(i)})$ and drops down to $\#f_j$ at the last iteration.

The following example illustrates the worst case where $\#f_j^{(i)}$ increases linearly to its maximum, $d_j \# f_j$.

$$f_j = (31x_3 + 100x_3^2 + (49 + 36x_2^2 + (x_1^4 + 44x_1^2 + 28)x_2^3)x_3^3)x_4^4,$$

with $p = 101$, $j = 4$ and $\#f_j = 7$. Table 2 shows the number of terms in $f_j^{(i)}$. $\max(\#f_j^{(i)})$ equals to $d_j \# f_j$. This is because

$$\sigma_i = \frac{1}{i!}\frac{\partial^{(i)} f_j}{\partial x_j^i}(x_j = \alpha_j) \text{ for } 0 \le i \le d_j$$

and in this example f_j only contains the terms with $x_j^{d_j}$. In this case, $\#\sigma_i$ is never reduced as i increases from 0 to d_j and we have $\max(\#f_j^{(i)})/\#f_j = d_j$.

2.3   Our New Algorithm: CMSHL

We present a new approach which eliminates the expression swell in Algorithm 3. The idea is depicted in Fig. 1. Consider one of the factors f_j at the jth Hensel lifting step:

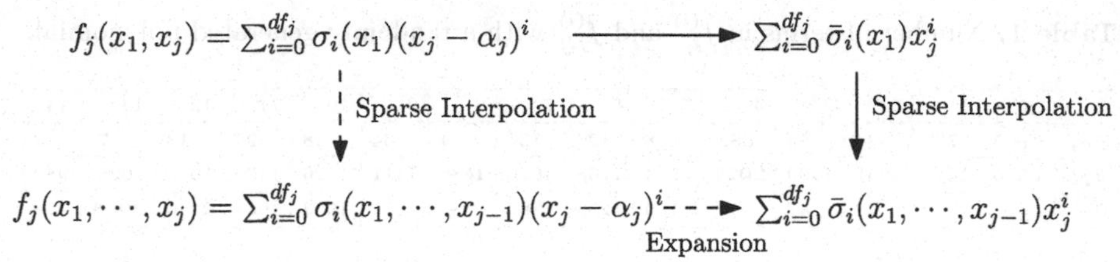

$$f_j(x_1, x_j) = \sum_{i=0}^{df_j} \sigma_i(x_1)(x_j - \alpha_j)^i \qquad \longrightarrow \qquad \sum_{i=0}^{df_j} \bar{\sigma}_i(x_1)x_j^i$$

```
        ┊ Sparse Interpolation              │ Sparse Interpolation
        ▼                                    ▼
```

$$f_j(x_1, \cdots, x_j) = \sum_{i=0}^{df_j} \sigma_i(x_1, \cdots, x_{j-1})(x_j - \alpha_j)^i \dashrightarrow \sum_{i=0}^{df_j} \bar{\sigma}_i(x_1, \cdots, x_{j-1})x_j^i$$

Expansion

Fig. 1. Dashed arrows: Algorithm 3 [11], expression swell occurs at the expansion step. Lined arrows: CMSHL (Algorithm 4).

$$f_j(x_1, \cdots, x_j) = \sum_{i=0}^{df_j} \sigma_i(x_1, \cdots, x_{j-1})(x_j - \alpha_j)^i = \sum_{i=0}^{df_j} \bar{\sigma}_i(x_1, \cdots, x_{j-1})x_j^i, \quad (1)$$

where $df_j = \deg(f_j, x_j)$. There are two routes to recover $\bar{\sigma}_i(x_1, \cdots, x_{j-1})$ in (1) from its bivariate image $f_j(x_1, x_j)$. One route is to first recover $\sigma_i(x_1, \cdots, x_{j-1})$ from $\sigma_i(x_1)$ using sparse interpolation and then expand to get $\bar{\sigma}_i(x_1, \cdots, x_{j-1})$ in (1) (through the dashed arrows in Fig. 1). This has been done previously in [11]. In our new algorithm (CMSHL), bivariate images are expanded first and then the coefficients $\bar{\sigma}_i(x_1, \cdots, x_{j-1})$ are recovered directly from $\bar{\sigma}_i(x_1)$ to get the final expanded form. This is through the lined arrows in Fig. 1. Multivariate polynomial expansions are avoided where expression swells can occur.

Our solution is presented in Algorithm 4 (CMSHL). The correctness of this algorithm is based on the following. Since all loop ranges are finite, algorithm CMSHL terminates. When CMSHL terminates it outputs either two factors f_j, g_j or FAIL. Since CMSHL tests if $a_j = f_j g_j$ in line 21 the output (f_j, g_j) is the correct factorization. The failure probability is presented in Sect. 3.

Algorithm CMSHL also has a significant advantage for parallelization, however, it only uses the weak SHL assumption during a sparse interpolation. It cannot use the strong SHL assumption as in MTSHL to reduce the number of terms in a loop for a typical average case. The strong and the weak SHL assumptions are both defined in Sect. 3.1.

3   Complexity Analyses

For both MTSHL and CMSHL, the number of arithmetic operations in $\mathbb{Z}_p$ are bounded for the worst-case, along with the failure probabilities. We first need the Schwartz-Zippel Lemma [17, 21]:

Lemma 1. *Let F be a field and $f \neq 0$ be a polynomial in $F[x_1, x_2, \cdots, x_n]$ with total degree d and let $S \subseteq F$. Then the number of roots of f in S^n is at most $d|S|^{n-1}$. Hence if β is chosen at random from S^n then $\Pr[f(\beta) = 0] \leq \frac{d}{|S|}$.*

Algorithm 4. CMSHL: Hensel lifting x_j via bivariate Hensel lifting.

1: **Input:** A prime p, $\alpha_j \in \mathbb{Z}_p$, $a_j \in \mathbb{Z}_p[x_1, \cdots, x_j]$ monic in x_1,
$f_{j-1}, g_{j-1} \in \mathbb{Z}_p[x_1, \cdots, x_{j-1}]$ s.t. $a_j(x_1, ..., x_{j-1}, \alpha_j) = f_{j-1}g_{j-1}$ with $j > 2$.

2: **Output:** $f_j, g_j \in \mathbb{Z}_p[x_1, \cdots, x_j]$ s.t. $a_j = f_j g_j$ where $f_j(x_j = \alpha_j) = f_{j-1}$ and $g_j(x_j = \alpha_j) = g_{j-1}$; Otherwise, FAIL.

3: Let $f_{j-1} = x_1^{df} + \sum_{i=0}^{df-1} \sigma_i(x_2, ..., x_{j-1})x_1^i$ with $\sigma_i = \sum_{k=1}^{s_i} c_{ik} M_{ik}$
and $g_{j-1} = x_1^{dg} + \sum_{i=0}^{dg-1} \tau_i(x_2, ..., x_{j-1})x_1^i$ with $\tau_i = \sum_{k=1}^{t_i} d_{ik} N_{ik}$,
where M_{ik}, N_{ik} are the monomials in σ_i, τ_i respectively.

4: Pick $\boldsymbol{\beta} = (\beta_2, \cdots, \beta_{j-1}) \in \mathbb{Z}_p^{j-2}$ at random.

5: Evaluate monomials at $\boldsymbol{\beta}$: .... $\mathcal{O}((j-2)(\#f + \#g + d_{max}))$
$\mathcal{S} = \{S_i = \{m_{ik} = M_{ik}(\boldsymbol{\beta}), 1 \le k \le s_i\}, 0 \le i \le df - 1\}$ and
$\mathcal{T} = \{T_i = \{n_{ik} = N_{ik}(\boldsymbol{\beta}), 1 \le k \le t_i\}, 0 \le i \le dg - 1\}$.

6: **if** any $|S_i| \ne s_i$ **or** any $|T_i| \ne t_i$ **then return** FAIL **end if**

7: Let s be the maximum of s_i and t_i.

8: **for** k from 1 to s **in parallel do**

9:     Let $Y_k = (x_2 = \beta_2^k, \cdots, x_{j-1} = \beta_{j-1}^k)$.

10:     $A_k, F_k, G_k \leftarrow a_j(x_1, Y_k, x_j), f_{j-1}(x_1, Y_k), g_{j-1}(x_1, Y_k)$. .... $\mathcal{O}(s(\#f + \#g + \#a))$

11:     **if** $\gcd(F_k, G_k) \ne 1$ **then return** FAIL **end if** // unlucky evaluation

12:     $f_k, g_k \leftarrow BivariateHenselLift(A_k, F_k, G_k, \alpha_j, p)$. .... $\mathcal{O}(s(d_1^2 d_j + d_1 d_j^2))$

13: **end for**

14: Let $f_k = x_1^{df} + \sum_{l=1}^{\mu} \alpha_{kl} \tilde{M}_l(x_1, x_j)$ for $1 \le k \le s$, where $\mu \le d_1 d_j$.

15: **for** l from 1 to μ **in parallel do**

16:     $i \leftarrow \deg(\tilde{M}_l, x_1)$.

17:     Solve the $s_i \times s_i$ linear system for c_{lk} .... $\mathcal{O}(sd_j\#f)$

$$\left\{ \sum_{k=1}^{s_i} m_{ik}^n c_{lk} = \alpha_{nl} \text{ for } 1 \le n \le s_i \right\}$$

18: **end for**

19: Construct $f_j \leftarrow x_1^{df} + \sum_{l=1}^{\mu} (\sum_{k=1}^{s_i} c_{lk} M_{ik}(x_2, ..., x_{j-1})) \tilde{M}_l(x_1, x_j)$.

20: Similarly, construct g_j. .... $\mathcal{O}(sd_j\#g)$

21: **if** $a_j = f_j g_j$ **then return** (f_j, g_j) **else return** FAIL **end if**

3.1 MTSHL

MTSHL uses the strong SHL assumption to solve the multivariate Diophantine equations (MDPs) in a loop. The following lemma was proved in [9]:

Lemma 2. *Let* $f \in \mathbb{Z}_p[x_1, \cdots, x_n]$ *and let* α *be a randomly chosen element in* $\mathbb{Z}_p$. *Let* $f = \sum_{i=0}^{d_n} \sigma_i(x_1, \cdots, x_{n-1})(x_n - \alpha)^i$ *where* $d_n = \deg(f, x_n)$. *Then,*

$$\Pr[\text{Supp}(\sigma_{i+1}) \nsubseteq \text{Supp}(\sigma_i)] \le |\text{Supp}(\sigma_{i+1})| \frac{d_n - i}{p - d_n + i + 1} \text{ for } 0 \le i < d_n.$$

The assumption that $\text{Supp}(\sigma_i) \subseteq \text{Supp}(\sigma_{i-1})$ for $1 \le i \le d_n$ is called the **strong SHL assumption** in [9,12]. In Sect. 3.2 our new algorithm will assume $\text{Supp}(\sigma_i) \subseteq \text{Supp}(\sigma_0)$ for $1 \le i \le d_n$. This assumption is called the **weak SHL assumption** in [9,12].

Step 11 of Algorithm 1 applies the strong SHL assumption by employing $\mathrm{Supp}(\sigma_f) = \mathrm{Supp}(\sigma_{i-1})$ and $\mathrm{Supp}(\tau_f) = \mathrm{Supp}(\tau_{i-1})$ as the supports for σ_i and τ_i. Therefore we only solve systems of linear equations for the coefficients. This is the key feature to solve the MDPs via Algorithm 2, which we shall analyze in the following.

3.1.1   The Failure Probability of the MDPs

There are two places where Algorithm 2 can return FAIL intermediately: line 7 and line 11. The failure probabilities are bounded as follows. Proofs follow [12].

Proposition 1. *Let p be a large prime, $d = \deg(a)$ and s be the number defined in line 4 of Algorithm 2. When Algorithm 1 calls Algorithm 2 with inputs $(u, w, c, \sigma_f, \tau_f) = (g_{j-1}, f_{j-1}, c_i, \sigma_{i-1}, \tau_{i-1})$, if $\mathrm{Supp}(\sigma_i) \subseteq \mathrm{Supp}(\sigma_{i-1})$ and $\mathrm{Supp}(\tau_i) \subseteq \mathrm{Supp}(\tau_{i-1})$, for $i = 1, 2, 3, \cdots$, then Algorithm 2 fails to compute (σ_i, τ_i) for the MDP $\sigma_i g_{j-1} + \tau_i f_{j-1} = c_i$ with a probability less than*

$$\underbrace{\frac{d\,s(\#f_{j-1} + \#g_{j-1})}{2(p-1)}}_{\text{line 7}} + \underbrace{\frac{d^2 s^2}{p-1}}_{\text{line 11}}. \tag{2}$$

Proof. For line 7, let $\Delta_i = \prod_{1 \le l < k \le s_i} (M_{il} - M_{ik})$, where M_{il}, M_{ik} are monomials in S defined in line 6. Let $\Delta = \prod_{i=0}^{d_\sigma} \Delta_i$. Then $\Delta(\beta) = 0$ implies $\Delta_i(\beta) = 0$ for some i so that not all monomial evaluations are distinct. Also, $\deg(M_{il}) < d$ for each monomial in S. Thus,

$$\deg(\Delta) < \sum_{i=0}^{d_\sigma} d\binom{s_i}{2} \le \frac{d\,s}{2} \sum_{i=0}^{d_\sigma} (s_i - 1) < \frac{d\,s\#f_{j-1}}{2}.$$

By Lemma 1,

$$\Pr[\Delta(\beta) = 0] \le \frac{\deg(\Delta)}{p-1} < \frac{d\,s\#f_{j-1}}{2(p-1)}.$$

Similarly, the monomial evaluations for τ are considered.
To solve the Diophantine equation in line 12, we need

$$\gcd(u(x_1, Y_k), w(x_1, Y_k)) = \gcd(g_{j-1}(x_1, Y_k), f_{j-1}(x_1, Y_k)) = 1.$$

Let $R = \mathrm{res}(g_{j-1}, f_{j-1}, x_1) \in \mathbb{Z}_p[x_2, \cdots, x_{j-1}]$. Since f_{j-1} and g_{j-1} are monic in x_1, the univariate Diophantine solver returns FAIL if

$$\gcd(g_{j-1}(x_1, Y_k), f_{j-1}(x_1, Y_k)) \ne 1 \iff R(Y_k) = 0.$$

Let $S = \prod_{k=1}^{s} R(x_2^k, x_3^k, \cdots, x_{j-1}^k)$. Since $\deg(f_{j-1}) < d$ and $\deg(g_{j-1}) < d$, $\deg(R) < d^2$ and

$$\deg(S) = \sum_{k=1}^{s} k \deg(R) < \sum_{k=1}^{s} k d^2 = \frac{d^2 s(s+1)}{2}.$$

By Lemma 1,

$$\Pr[R(Y_k) = 0 \text{ for some } k] = \Pr[S(\beta) = 0] \leq \frac{\deg(S)}{p-1} < \frac{d^2 s^2}{p-1}.$$

Adding the failure probabilities at line 7 and 11, we obtain the result. □

At the end of Algorithm 2, $\sigma u + \tau w = c$ can be checked probabilistically with a single evaluation point. If Algorithm 2 returns FAIL at line 19, the support in either σ or τ was wrong (strong SHL assumption fails). By Lemma 2, Algorithm 1 fails at the jth Hensel lifting step due to a wrong support in σ_i with a probability no more than

$$\sum_{i=0}^{d_j-1} |\text{supp}(\sigma_{i+1})| \frac{d_j - i}{p - d_j + i + 1} \leq \#f_{j-1} \sum_{i=0}^{d_j-1} \frac{d_j - i}{p - d_j + i + 1} < \frac{d_j(d_j + 1)\#f_{j-1}}{2(p - d_j + 1)},$$

where $d_j = \deg(a_j, x_j)$.

Note that the number s in Proposition 1 varies since MDPs are called in a loop from Algorithm 1. We denote $s_{j,i}$ as the maximum number of monomials in the coefficients of σ_{i-1} and τ_{i-1} in x_1 for the ith call of the MDP in the jth Hensel lifting step. Let $s_j = \max_i(s_{j,i})$ and $T_{fg_{j-1}} = \max(\#f_{j-1}, \#g_{j-1})$. We have $d_j \leq d$. Adding up the failure probabilities at line 7, 11 and 19, we obtain the failure probability at the jth Hensel lifting step:

Proposition 2. *Let p be a large prime, $d = \deg(a)$, $s_j = \max_i(s_{j,i})$ and $T_{fg_{j-1}} = \max(\#f_{j-1}, \#g_{j-1})$. Algorithm 1 (MTSHL) fails to compute f_j, g_j from f_{j-1}, g_{j-1} at the jth Hensel lifting step $(j > 2)$ via Algorithm 2 with a probability less than*

$$\frac{d^2 s_j (T_{fg_{j-1}} + d\, s_j) + d^2 T_{fg_{j-1}} + d T_{fg_{j-1}}}{p - d + 1}. \tag{3}$$

For the whole MTSHL process (for $2 \leq j \leq n$), we have $\#f_{j-1} \leq \#f$, $\#g_{j-1} \leq \#g$.

Proposition 3. *Let p be a large prime, n be the number of variables in a, $d = \deg(a)$, $s_{\max} = \max(s_j)$ and $T_{fg} = \max(\#f, \#g)$. MTSHL (the jth Hensel lifting step as in Algorithm 1) fails to solve the MDP via sparse interpolation (Algorithm 2) with a probability less than*

$$\frac{(n-2)\left(d^2 s_{\max}(T_{fg} + d\, s_{\max}) + d^2 T_{fg} + d T_{fg}\right)}{p - d + 1}. \tag{4}$$

We illustrate the probability in Proposition 3 for a typical large factorization problem. Let $n = 10$, $d = 10^2$, $T_{fg} = 10^4$ and $s_{\max} = 10^2$. If p is a 64-bit prime $\approx 1.8 \times 10^{19}$, then MTSHL fails with probability less than 8.72×10^{-9}. Thus for p sufficiently large, the failure probability is low.

3.1.2   The Complexity of the MDP

After discussing the failure probabilities, it remains to bound the number of arithmetic operations in $\mathbb{Z}_p$. We have the complexity of the MDP as follows:

Theorem 1. *Let p be a large prime, s be the number defined in line 4 of Algorithm 2, $d_1 = \deg(a, x_1)$ and a_j $(j > 2)$ be monic in x_1. When Algorithm 1 calls Algorithm 2, if the strong SHL assumption holds, then with a failure probability less than in (2), the number of arithmetic operations in $\mathbb{Z}_p$ for solving the MDP $\sigma_i g_{j-1} + \tau_i f_{j-1} = c_i$ for $i = 1, 2, \cdots$ in the worst case is*

$$\mathcal{O}(s(\#a_j + d_1^2)). \tag{5}$$

Proof. Appendix A.

3.1.3   The Complexity of MTSHL

Now we return to the analysis of Algorithm 1 – Hensel lifting x_j with multivariate Diophantine equations. One bottleneck of Algorithm 1 is the error computation step at line 14. There is an expression swell of f_j and g_j at line 13 of up to a factor of $d_j = \deg(a, x_j)$. We have the complexity at the jth Hensel lifting step:

Theorem 2. *Let p be a large prime, $d_1 = \deg(a, x_1)$, $d_j = \deg(a, x_j)$ and $s_j = \max_i(s_{j,i})$. With a failure probability less than in (3), the number of arithmetic operations in $\mathbb{Z}_p$ for the jth Hensel lifting step (via Algorithm 1) in the worst case is*

$$\mathcal{O}(\underbrace{d_j^2 \#a_j}_{\text{line7,13}} + \underbrace{d_j s_j(\#a_j + d_1^2)}_{\text{MDP}} + \underbrace{d_j^3 \#f_{j-1} \#g_{j-1}}_{\text{error comp.}}). \tag{6}$$

Proof. To compute coeff(error, $(x_j - \alpha_j)^i$) in step 7, using repeated differentiation and evaluation costs $O(i\#\text{error})$. The total cost is $\mathcal{O}(d_j^2 \#a_j)$.

The total cost of sparse interpolation in step 11 is $\mathcal{O}(d_j s_j(\#a_j + d_1^2))$, from Theorem 1.

The total cost of adding the factors in step 13 is $\mathcal{O}\left(\sum_{i=1}^{d_j} i(\#f_{j-1} + \#g_{j-1})\right)$, which is $\mathcal{O}(d_j^2(\#f_{j-1} + \#g_{j-1}))$.

The total cost of error computation in step 14 is $\mathcal{O}\left(\sum_{k=1}^{d_j} \#f_j^{(k)} \#g_j^{(k)}\right) = \mathcal{O}\left(\sum_{i=1}^{d_j}(i\#f_{j-1})(i\#g_{j-1})\right) \subseteq \mathcal{O}(d_j^3 \#f_{j-1} \#g_{j-1})$.

Assuming $\#f_{j-1} \leq \#a_j$ and $\#g_{j-1} \leq \#a_j$, the total cost for Algorithm 1 is

$$\mathcal{O}(\underbrace{d_j^2 \#a_j}_{\text{line 7,13}} + \underbrace{d_j s_j(\#a_j + d_1^2)}_{\text{MDP}} + \underbrace{d_j^3 \#f_{j-1} \#g_{j-1}}_{\text{error comp.}}). \quad \square$$

In Theorem 2, the expression swell appears as the factor of d_j^2. On average the expression swell is much less. For sparse factors, we know that $\#f_{j-1} \lesssim \#f_j$ for $n/2 \leq j \leq n$ [12]. The complexity of the whole MTSHL process is given in the following:

Theorem 3. *Let p be a large prime, $a \in \mathbb{Z}_p[x_1, \cdots, x_n]$ monic in x_1, $\boldsymbol{\alpha} = (\alpha_2, \cdots, \alpha_n) \in \mathbb{Z}_p^{n-1}$ be a random evaluation point, f, g be the monic irreducible factors of a, $f_1 = f(x_1, \boldsymbol{\alpha})$, $g_1 = g(x_1, \boldsymbol{\alpha})$ be the image polynomials with $\gcd(f_1, g_1) = 1$. Then with a failure probability less than in (4), the total number of arithmetic operations in $\mathbb{Z}_p$ for lifting f_1, g_1 to f_n, g_n in $n - 1$ steps using MTSHL (Algorithm 1) in the worst case is*

$$\mathcal{O}(\underbrace{d_1^2 d_2 + d_1 d_2^2}_{\text{first BHL}} + (n-2)\underbrace{(d_{\max}^2 \#a + s_{\max} d_{\max}(\#a + d_1^2) + d_{\max}^3 \#f \#g)}_{\text{MTSHL}}), \quad (7)$$

where $d_i = \deg(a, x_i)$ for $1 \leq i \leq n$, $d_{\max} = \max_{i=3}^n(d_i)$ and $s_{\max} = \max(s_j)$.

3.2   CMSHL

In Algorithm CMSHL, the weak SHL assumption is used instead of the strong SHL assumption (see Sect. 3.1 for the definition). Similar to Lemma 2, we have the following (proof follows [9]):

Lemma 3. *Let $f \in \mathbb{Z}_p[x_1, \cdots, x_n]$ and let α be a randomly chosen element in $\mathbb{Z}_p$. Let $f = \sum_{i=0}^{d_n} \sigma_i(x_1, \cdots, x_{n-1})(x_n - \alpha)^i$ where $d_n = \deg(f, x_n)$. Then,*

$$\Pr[\mathrm{Supp}(\sigma_i) \nsubseteq \mathrm{Supp}(\sigma_0)] \leq |\mathrm{Supp}(\sigma_i)| \frac{d_n}{p - d_n + i} \text{ for } 1 \leq i \leq d_n.$$

3.2.1   The Failure Probability of CMSHL

For the jth Hensel lifting step, by Lemma 3, the failure probability due to a wrong support in either f_j or g_j (Algorithm 4 fails at line 21) is bounded by

$$(\#f_{j-1} + \#g_{j-1}) \sum_{i=1}^{d_j} \frac{d_j}{p - d_j + i} \leq \frac{d_j^2(\#f_{j-1} + \#g_{j-1})}{p - d_j + 1}.$$

The number s defined in line 7 of Algorithm 4 is equivalent to $s_j = \max(s_{j,i})$ in MTSHL. We denote s_j as the number s in line 7 of Algorithm 4 at the jth Hensel lifting step. Identical to MTSHL (Proposition 1), the failure probabilities at line 6 and 11 are

$$\underbrace{\frac{d\, s_j(\#f_{j-1} + \#g_{j-1})}{2(p-1)}}_{\text{line 6}} + \underbrace{\frac{d^2 s_j^2}{p-1}}_{\text{line 11}}.$$

Adding the failure probabilities at line 6, 11 and 21, we have the failure probability at the jth Hensel lifting step for Algorithm 4 (CMSHL):

Proposition 4. *Let p be a large prime, $d = \deg(a)$, $T_{fg_{j-1}} = \max(\#f_{j-1}, \#g_{j-1})$ and s_j be the number s defined in line 7 of Algorithm 4 at the jth Hensel lifting step. Then Algorithm 4 fails to compute f_j, g_j from f_{j-1}, g_{j-1} at the jth Hensel lifting step $(j > 2)$ with a probability less than*

$$\frac{d\, s_j(T_{fg_{j-1}} + d\, s_j) + 2d^2 T_{fg_{j-1}}}{p - d + 1}. \quad (8)$$

3.2.2 The Complexity of CMSHL

Theorem 4. *Let p be a large prime, $d_1 = \deg(a, x_1)$, $d_j = \deg(a, x_j)$ and s_j be the number s defined in line 7 of Algorithm 4 for the jth Hensel lifting step. With a failure probability less than in (8), the number of arithmetic operations in $\mathbb{Z}_p$ for the jth Hensel lifting step (via Algorithm 4) in the worst case is*

$$\mathcal{O}(d_j s_j(\#f_{j-1} + \#g_{j-1} + d_1^2 + d_1 d_j) + s_j \#a_j). \tag{9}$$

Proof. Appendix B.

For the whole process of CMSHL (for $2 \le j \le n$), we have the following:

Theorem 5. *Let p be a large prime, $a \in \mathbb{Z}_p[x_1, \cdots, x_n]$ monic in x_1, $\boldsymbol{\alpha} = (\alpha_2, \cdots, \alpha_n)$ be a randomly chosen evaluation point from $\mathbb{Z}_p^{n-1}$, f, g be the monic irreducible factors of a, $f_1 = f(x_1, \boldsymbol{\alpha})$, $g_1 = g(x_1, \boldsymbol{\alpha})$ be the image polynomials with $\gcd(f_1, g_1) = 1$. With a failure probability less than*

$$\frac{(n-2)(d\,s_{\max}(T_{fg} + d\,s_{\max}) + 2d^2 T_{fg})}{p - d + 1}, \tag{10}$$

the number of arithmetic operations in $\mathbb{Z}_p$ for lifting f_1, g_1 to f_n, g_n in $n-1$ steps using CMSHL (Algorithm 4) in the worst case is

$$\mathcal{O}(\underbrace{d_1^2 d_2 + d_1 d_2^2}_{\text{first BHL}} + (n-2)\underbrace{(s_{\max} d_{\max}(\#f + \#g + d_1^2 + d_1 d_{\max}) + s_{\max} \#a))}_{\text{CMSHL}}, \tag{11}$$

where $d = \deg(a)$, $d_i = \deg(a, x_i)$ for $1 \le i \le n$, $d_{\max} = \max_{i=3}^{n}(d_i)$, $s_{\max} = \max(s_j)$ and $T_{fg} = \max(\#f, \#g)$.

4   Experimental Results

We have implemented our Hensel lifting algorithm in the C programming language and parallelized parts of it for multi-core computers using Cilk C [3]. Cilk uses the fork-join idiom for parallel programming. Our C code and Cilk C code is freely available on the web at http://www.cecm.sfu.ca/CAG/code/CASC2020

Following the recommendation in [12] we interpolate $f(x_1, \ldots, x_j)$ (using sparse interpolation) from trivariate images $f(x_1, x_2, \beta^i, x_j)$ instead of from bivariate images $f(x_1, \beta^i, x_j)$. To obtain a trivariate image we interpolate x_2 using dense interpolation from bivariate images $f(x_1, \gamma_k, \beta^i, x_j)$ which we obtain using bivariate Hensel lifting. Although this increases the cost of computing images by a factor of $\deg(f, x_2)$, using trivariate images typically reduces s_j in equation (9) which speeds up all other parts of our algorithm. We refer the reader to [12] for an analysis of the expected reduction in s_j.

We give three sets of timings for our factorization code. The first set (see Table 3) is for factors with a low degree of 7 and an increasing number of terms t ($\#f = \#g = t$). For this case evaluating $a(x_1, x_2, \beta^i, x_j)$ is the bottleneck of our algorithm. The second set (see Table 4) is for factors with a fixed

Table 3. Real timings in CPU seconds for low degree d and increasing terms t. NA: not attempted; > 32GB: out of memory.

n	d	t	s	$f \times g$	New times (1 core)			New times (16 cores)			Maple	Maple	Magma
					Total	Hensel	Eval	Total	Hensel	Eval	2019	2017	V2.25-5
6	7	500	17	0.025	0.084	0.012	0.029	0.081	0.016	0.007	1.897	33.77	43.21
6	7	1000	30	0.107	0.340	0.021	0.170	0.169	0.028	0.027	4.540	95.48	50.38
6	7	2000	47	0.451	1.199	0.033	0.768	0.321	0.044	0.114	97.21	186.7	195.6
6	7	4000	81	1.932	3.583	0.055	2.632	0.543	0.065	0.281	139.4	325.4	777.0
6	7	8000	144	8.249	8.778	0.101	7.107	1.248	0.125	0.830	201.1	470.5	1958.0
9	7	500	14	0.025	0.119	0.013	0.031	0.094	0.013	0.005	4.699	2794.2	849.2
9	7	1000	28	0.108	0.493	0.021	0.204	0.232	0.066	0.031	15.57	16094	915.8
9	7	2000	50	0.449	2.169	0.034	1.313	0.433	0.042	0.160	3597.7	>32GB	9082.8
9	7	4000	99	1.963	13.94	0.067	10.47	1.570	0.076	0.816	>32GB	NA	15444
9	7	8000	178	8.244	88.39	0.121	74.14	8.313	0.138	5.575	NA	NA	>32GB

number of terms and an increasing degree d. For these problems Hensel lifting becomes the bottleneck. To address this we use Monagan's $O(d^3)$ method [14] for Hensel lifting in $\mathbb{Z}_p[x, y]$. The third set (see Table 5) is for polynomials where the factor f has a lot more terms than g. For these problems evaluation and solving Vandermonde systems are the bottlenecks. To solve the Vandermonde systems we use Zippel's linear space quadratic time method in [23].

All experiments were performed on a server with two Intel E5-2660 8 core CPUs running at 2.2GHz (base) and 3.0GHz (turbo) hence the maximum theoretical parallel speedup is a factor of $16 \times 2.2/3.0 = 11.7$.

In Tables 3, 4 and 5 the factors f and g are of the form $x_1^d + \sum_{i=2}^{t-1} a_i \prod_{j=1}^{n} x_j^{e_{ji}}$ with coefficients a_i chosen randomly from $[1, 999]$ and exponents e_{ji} chosen randomly from $[0, d-1]$. The time in column $f \times g$ is the time our C code takes to multiply $a = f \times g$ using an algorithm with arithmetic complexity $O(\#f \#g)$.

Because the factors are monic and have many terms, almost all of the factorization time is in multivariate Hensel lifting. The timings for our algorithm are for Hensel lifting x_n the last variable only, which is most of the time. The quantity s in column 4 is the number of images needed to interpolate $x_3, \ldots, x_n$.

For Maple we report timings for Maple 2017 and Maple 2019. Maple 2017 and Magma 2.25-5 are both using Wang's organization of MHL as described in Chap. 6 of [5]. Maple 2019 is using Monagan and Tuncer's sparse Hensel lifting from [9,12]. These algorithms do many computations with multivariate polynomials in $\mathbb{Z}_p[x_1, \ldots, x_j]$ including many multiplications and divisions. In contrast, our algorithm does no arithmetic with multivariate polynomials.

In Tables 3, 4 and 5 we report the total time of our new algorithm in column total, the time evaluating $a(x_1, x_2, \beta^i, x_n)$ for $1 \le i \le s$ in column eval, the time in bivariate Hensel lifting in column hensel, and for Table 5, the time solving the Vandermonde systems in column solve. Timings are given for our Cilk C code for 1 core and 16 cores.

Table 4. Real timings in CPU seconds for increasing degree d and fixed t. NA: not attempted; > 32GB: out of memory.

n	d	t	s	$f \times g$	New times (1 core)			New times (16 cores)			Maple	Maple	Magma
					Total	Hensel	Eval	Total	Hensel	Eval	2019	2017	v2.25-5
6	10	500	10	0.026	0.079	0.022	0.017	0.069	0.020	0.004	3.068	466.8	134.7
6	15	500	6	0.025	0.101	0.051	0.011	0.094	0.036	0.004	8.206	11002	610.1
6	20	500	5	0.025	0.168	0.117	0.012	0.101	0.036	0.004	18.77	51325	27317
6	40	500	3	0.025	0.669	0.617	0.011	0.272	0.205	0.008	148.7	NA	29.04
6	60	500	3	0.025	2.083	2.025	0.014	0.583	0.519	0.010	545.2	NA	371.4
6	80	500	3	0.025	5.644	5.586	0.014	0.950	0.892	0.010	1210.9	NA	1242.1
6	100	500	2	0.025	7.740	7.687	0.008	1.375	1.303	0.008	NA	NA	NA
6	10	2000	30	0.455	1.434	0.070	0.737	0.258	0.043	0.056	675.11	1889.3	908.0
6	15	2000	18	0.455	1.327	0.168	0.488	0.341	0.100	0.060	3905.7	63082	9317.1
6	20	2000	12	0.455	1.336	0.329	0.332	0.335	0.136	0.042	4677.2	$> 10^5$	17339
6	40	2000	6	0.455	2.853	1.999	0.183	0.686	0.472	0.038	>32GB	NA	$> 10^5$
6	60	2000	6	0.455	8.940	8.071	0.203	1.313	1.106	0.052	NA	NA	NA
6	80	2000	4	0.455	15.17	14.34	0.158	2.565	2.279	0.084	NA	NA	NA
6	100	2000	3	0.455	21.77	20.92	0.173	2.644	2.357	0.086	NA	NA	NA

Table 5. Real timings in CPU seconds for increasing $\#f = t$ and $\#g = 20$.

n	d	t	$\#g$	s	$f \times g$	Total	Hensel	Eval	Solve	Total	Hensel	Eval	Solve
9	7	10000	20	212	0.043	0.871	0.156	0.340	0.287	0.350	0.155	0.060	0.039
9	7	20000	20	409	0.076	2.641	0.254	1.107	1.108	0.663	0.256	0.122	0.096
9	7	40000	20	789	0.135	9.465	0.475	4.243	4.175	1.917	0.477	0.480	0.361
9	7	80000	20	1503	0.258	34.16	0.920	15.68	16.33	4.782	0.913	1.362	1.373
9	7	160000	20	2984	0.499	132.3	1.791	62.13	64.37	13.67	1.844	5.586	5.244

Tables 3 and 5 show good parallel speedup for the evaluations $a(x_1, x_2, \beta^i, x_n)$. Table 4 shows that for higher degree polynomials the Hensel lifting dominates. To obtain the parallel speedups for the Hensel lifting in Table 4 we parallelize the evaluations of $a(x_1, x_2, \beta^i, x_n)$ at $x_2 = \gamma_k$ for different k as well as the bivariate Hensel Lifts in $\mathbb{Z}_p[x_1, x_n]$.

Table 5 shows that when one factor is much larger than the other, the time solving Vandermonde systems becomes significant. The solving time is not reported in Tables 3 and 4 because it is insignificant.

The timings in Tables 3, 4, and 5 agree with our analysis for CMSHL in Theorem 4. In Table 3, for example, when $n = 9$ and t increases from 2000 to 4000, $\#a_j$ is quadrupled and s_j is doubled, we see the evaluation time for 1 core increases by a factor of $10.47/1.313 = 7.97 \approx 8$. This agrees with the term $s_j \#a_j$ in (9). In Table 4, when $t = 2000$ and d increases from 60 to 100, we expect the time for Hensel lifting at $d = 100$ to be $\frac{1}{2}(100/60)^3 \cdot 8.071 = 18.68$ which is close to the result 20.92. In Table 5, when t and s are doubled, both timings for evaluations and solving Vandermonde systems are quadrupled as expected.

5   Implementation Notes

To store the multivariate polynomial $a = \sum_{i=1}^{t} a_i M_i(x_1, \ldots, x_n)$ we encode the monomials M_i in 64 bit integers m_i. We store a as the triple (A, X, t) where

$$A = \boxed{a_1 \; a_2 \; \ldots \; a_t} \quad \text{and} \quad X = \boxed{m_1 \; m_2 \; \ldots \; m_t}$$

are stored as arrays. For n variables we use $q = 64/n$ bits per variable which limits the maximum degree in each variable to $2^q - 1$. Although monomial packing limits the degree and number of variables that our software can handle it significantly improves the speed. For polynomials with more variables and/or higher degrees, we are experimenting with the 128 bit integer type `__int128` supported by the gcc compiler.

One of the advantages of our algorithm is that there are no multivariate polynomial multiplications and divisions. The most time consuming operation is evaluation which is linear in the number of terms. We compute $a(x_1, x_2, \beta^i, x_n)$ for $\beta \in \mathbb{Z}_p^{n-3}$ for $1 \leq i \leq s$. We have parallelized these evaluations. We parallelize each evaluation $a(x_1, x_2, \beta^i, x_n)$ in blocks and do two evaluations at a time.

We also execute the bivariate Hensel lifts in parallel and we solve the Vandermonde linear systems in parallel. To avoid memory bottlenecks, we use in-place algorithms for all parallel tasks. A routine is in-place if it, and all the subroutines it calls, allocate no memory. They work in the memory of the input and output. This means that our Cilk tasks are not simultaneously trying to allocate and de-allocate memory. We give an example of an in-place algorithm.

The following conversion occurs at the end of bivariate Hensel lifting. We have polynomials $a_0(x), a_1(x), \ldots, a_d(x)$ in $\mathbb{Z}_p[x]$, a non-zero element $\alpha \in \mathbb{Z}_p$, and we want to expand the bivariate polynomial

$$f(x, y) = \sum_{i=0}^{d} a_i(x)(y - \alpha)^i,$$

that is, we want to compute new polynomials $\bar{a}_i \in \mathbb{Z}_p[x]$ such that $f(x, y) = \sum_{i=0}^{d} \bar{a}_i(x) y^i$ in $\mathbb{Z}_p[x, y]$. One way to expand $f(x, y)$ is to use Horner's rule

$$f(x, y) = a_0(x) + (y - \alpha)\left[a_1(x) + (y - \alpha)\left[a_2(x) + \cdots + (y - \alpha)a_d(x)\ldots\right]\right].$$

Coding this in Maple or Magma will cause $2d$ pieces of memory to be allocated for the intermediate products and sums. To code this in C we have to handle the memory explicitly. How we do this depends the data structure we use for storing polynomials in $\mathbb{Z}_p[x, y]$.

Let $f \in \mathbb{Z}_p[x, y]$, $dx = \deg(f, x)$, $dy = \deg(f, y)$ and $d_i = \deg(f, x_i)$. We store f as a pair (D, A) where D is an array of integers storing the degree information $[d_y, d_0, d_1, \ldots, d_{dy}]$ and A is an array of arrays storing $[a_0(x), a_1(x), \ldots, a_d(x)]$. To do the conversion we use this version of Horner's rule

```
for i = d − 1, d − 2, . . . , 0 do
    for j = i, i + 1, . . . , d − 1 do
        a_j(x) := a_j(x) − α a_{j+1}(x).
```

To implement this in-place we use the inplace routine polsubmul$(a, da, b, db, \alpha, p)$ from our $\mathbb{Z}_p[x]$ library which computes $a(x) := a(x) - \alpha\, b(x)$ in the memory of a and returns the degree of the result. The routine polsubmul assumes the size of the array a is big enough to hold b. Assuming each array A_i has space for $1 + dx$ coefficients in $\mathbb{Z}_p$, we can do the conversion in the memory of (D, A) with

```
for( i=D[0]-1; i>=0; i-- )
    for( j=i; j<d; j++ )
        D[j+1] = polsubmul(A[j],D[j],A[j+1],D[j+1],alpha,p);
```

We have coded every subroutine in our bivariate Hensel lift to run in-place so that our bivariate Hensel lift can also be made in-place. In this way, when we run bivariate Hensel lifts in parallel, they all run in their own pre-allocated memory.

Our C implementation of Algorithm CMSHL for Hensel lifting is implemented for machine primes $p < 2^{63}$ for efficiency. It obtains the factors modulo p. To recover factors with larger integer coefficients, one may, starting with the factors modulo p, do a p-adic lift (see Monagan and Tuncer [10]) to obtain the factors modulo p^k for sufficiently large k.

6   Conclusion

Algorithm 1 is the basis for several polynomial factorization algorithms, including Wang's method from [18] which is used in most computer algebra systems today, and Monagan and Tuncer's method [9,12] which is now used in Maple. In this work we observed an expression swell in Algorithm 1 that is linear in the worst case. We presented a new sparse Hensel lifting algorithm CMSHL that avoids the expression swell. CMSHL, which is based on the method in [11], is suited for parallelization because it reduces multivariate polynomial factorization to many polynomial evaluations, many bivariate Hensel lifts, and solving many Vandermonde systems.

Our Cilk C implementation of CMSHL shows good parallel speedup for these three steps. The code is also much faster than the Maple and Magma factorization algorithms for the large factorization problems we tested, mainly because it does not do any multivariate polynomial arithmetic. We have also given a worst case complexity analysis for CMSHL and have determined its failure probability. For factors with many terms and not too high degree, as in Table 3, our experiments show that evaluation is the bottleneck. This agrees with the term $s_j \# a_j$ in equation (9) in our complexity analysis. Thus further improvement will need to consider these evaluations.

For future work, we would like to use CMSHL to factor polynomials represented by black boxes in the spirit of Kaltofen and Trager [7] and Diaz and Kaltofen [2].

Appendix A   Proof of Theorem 1

We bound the total number of arithmetic operations in $\mathbb{Z}_p$ for the worst case in Algorithm 2. Let s be the number defined in line 4, $d_{\max} = \max_{i=2}^{j-1} \deg(a, x_i)$ and $df_{\max} = \max_{i=2}^{j-1} \deg(f, x_i)$.

For step 6, one way to evaluate the monomials is to create a table of powers for each variable $x_2, \cdots, x_{j-1}$, as shown in Fig. 2. It takes $\sum_{i=2}^{j-1}(d\sigma_{f_i} - 1) \leq (j-2)(df_{\max}-1)$ multiplications to compute the table, where $d\sigma_{f_i} = \deg(\sigma_f, x_i)$. After creating the table, it takes $\mathcal{O}\left((j-3)\sum_{i=0}^{d\sigma} s_i\right) = \mathcal{O}((j-3)\#\sigma_f)$ multiplications to evaluate monomials in $\mathcal{S}$. Similarly for the evaluations in $\mathcal{T}$. Thus, the total cost is $\mathcal{O}((j-2)(\#\sigma_f + \#\tau_f + d_{\max}))$.

$$
\boxed{\beta_2} \boxed{\beta_2^2} \boxed{\beta_2^3} \quad \cdots \qquad \boxed{\beta_2^{d\sigma_{f_2}}}
$$

$$
\boxed{\beta_3} \boxed{\beta_3^2} \boxed{\beta_3^3} \quad \cdots \qquad \boxed{\beta_3^{d\sigma_{f_3}}}
$$

$$
\vdots
$$

$$
\boxed{\beta_{j-1}} \boxed{\beta_{j-1}^2} \boxed{\beta_{j-1}^3} \quad \cdots \qquad \boxed{\beta_{j-1}^{d\sigma_{f_{j-1}}}}
$$

Fig. 2. Evaluation table for variables $x_2, \cdots, x_{j-1}$.

In step 7, it costs $\mathcal{O}\left(\sum_{i=0}^{d\sigma} s_i \log(s_i) + \sum_{i=0}^{d\tau} t_i \log(t_i)\right)$ number of comparisons to sort the monomial evaluations and search for identical values along the sorted arrays. This is $\mathcal{O}(\log(s)(\#\sigma_f + \#\tau_f))$.

For step 10, monomial evaluations and its coefficients are stored in two arrays, say M and C. At the first iteration, each entry in M is squared and then multiplied by the corresponding coefficient in C to compute the sum. Each iteration costs $3(\#u + \#w + \#c)$ arithmetic operations. The total cost is $\mathcal{O}(s(\#f_{j-1} + \#g_{j-1} + \#a_j))$.

In step 12, each univariate Diophantine solver costs $\mathcal{O}(d_1^2)$.

In step 14 to 16, the Vandermonde solver costs $\sum_{i=0}^{d\sigma} \mathcal{O}(s_i^2) \subseteq \mathcal{O}(s\#\sigma_f)$.

We have $\#\sigma_f \leq \#f_{j-1}$ and $\#\tau_f \leq \#g_{j-1}$. Assuming $j-2 \lesssim s$, $\#f_{j-1} \leq \#a_j$ and $\#g_{j-1} \leq \#a_j$, the total cost of Algorithm 2 is

$$
\mathcal{O}(\underbrace{s(\#f_{j-1} + \#g_{j-1} + \#a_j)}_{\text{Eval in line 10}} + \underbrace{s\,d_1^2}_{\text{line 12}} + \underbrace{s(\#\sigma_f + \#\tau_f)}_{\text{Solve in line 14}-16}) \subseteq \mathcal{O}(s(\#a_j + d_1^2)). \quad \square
$$

Appendix B   Proof of Theorem 4

Similar to the analysis of Algorithm 2, we bound the total number of arithmetic operations in $\mathbb{Z}_p$ for the worst case in Algorithm 4. Let $d_{\max} = \max_{i=2}^{j-1} \deg(a, x_i)$.

The total cost of evaluations in step 5 is $\mathcal{O}((j-2)(\#f_{j-1} + \#g_{j-1} + d_{\max}))$.
The **if** statement in step 6 costs $\mathcal{O}\left(\sum_{i=0}^{df-1} s_i \log(s_i) + \sum_{i=0}^{dg-1} t_i \log(t_i)\right) \subseteq$
$\mathcal{O}(\log(s)(\#f_{j-1} + \#g_{j-1}))$ comparisons to sort and search for identical values.
The total cost of step 10 is $\mathcal{O}(s(\#f_{j-1} + \#g_{j-1} + \#a_j))$.
Each bivariate Hensel lift in line 12 costs $\Theta(d_1 d_j^2 + d_j d_1^2)$ [14].
Using Zippel [23] the total cost of the Vandermonde solver in step 17 is
$\sum_{i=0}^{df-1} d_j \mathcal{O}(s_i^2) \subseteq \mathcal{O}(d_j s \# f_{j-1})$ for f_j. Similarly, for g_j, we have $\mathcal{O}(d_j s \# g_{j-1})$.
Assuming $j - 2 \lesssim s$, the total cost of Algorithm 4 is

$$\mathcal{O}(\underbrace{s(\#f_{j-1} + \#g_{j-1} + \#a_j)}_{\text{Eval in line 10}} + \underbrace{s(d_1^2 d_j + d_1 d_j^2)}_{\text{BHL in line 12}} + \underbrace{s\, d_j(\#f_{j-1} + \#g_{j-1})}_{\text{Solve in line 17}})$$

$$\subseteq \mathcal{O}(d_j s(\#f_{j-1} + \#g_{j-1} + d_1^2 + d_1 d_j) + s\#a_j). \quad \square$$

References

1. Bernardin, L., Monagan, M.B.: Efficient multivariate factorization over finite fields. In: Mora, T., Mattson, H. (eds.) AAECC 1997. LNCS, vol. 1255, pp. 15–28. Springer, Heidelberg (1997). https://doi.org/10.1007/3-540-63163-1_2
2. Diaz A., Kaltofen E.: FOXBOX: a system for manipulating symbolic objects in black box representation. In: Proceedings of ISSAC 1998, pp. 30–37. ACM (1998)
3. Frigo, M., Leiserson, C.E., Randall K.H.: The implementation of the Cilk-5 multi-threaded language. In: Proceedings of PLDI 1998, pp. 212–223. ACM (1998)
4. von zur Gathen, J., Kaltofen, E.: Factorization of multivariate polynomials over finite fields. Math. Comp. **45**(7), 251–261 (1985)
5. Geddes, K.O., Czapor, S.R., Labahn, G.: Algorithms for Computer Algebra. Publ, Kluwer Acad, Dordrecht (1992)
6. Kaltofen, E.: Sparse Hensel lifting. In: Caviness, B.F. (ed.) EUROCAL 1985. LNCS, vol. 204, pp. 4–17. Springer, Heidelberg (1985). https://doi.org/10.1007/3-540-15984-3_230
7. Kaltofen, E., Trager, B.M.: Computing with polynomials given by black boxes for their evaluations: greatest common divisors, factorization, separation of numerators and denominators. J. Symb. Comput. **9**(3), 301–320 (1990)
8. Lang, S.: Fundamentals of Diophantine Geometry. Springer, New York (1983). https://doi.org/10.1007/978-1-4757-1810-2
9. Monagan, M., Tuncer, B.: Using Sparse interpolation in Hensel lifting. In: Gerdt, V.P., Koepf, W., Seiler, W.M., Vorozhtsov, E.V. (eds.) CASC 2016. LNCS, vol. 9890, pp. 381–400. Springer, Cham (2016). https://doi.org/10.1007/978-3-319-45641-6_25
10. Monagan, M., Tuncer, B.: Factoring multivariate polynomials with many factors and huge coefficients. In: Gerdt, V.P., Koepf, W., Seiler, W.M., Vorozhtsov, E.V. (eds.) CASC 2018. LNCS, vol. 11077, pp. 319–334. Springer, Cham (2018). https://doi.org/10.1007/978-3-319-99639-4_22
11. Monagan, M., Tuncer, B.: Sparse multivariate Hensel lifting: a high-performance design and implementation. In: Davenport, J.H., Kauers, M., Labahn, G., Urban, J. (eds.) ICMS 2018. LNCS, vol. 10931, pp. 359–368. Springer, Cham (2018). https://doi.org/10.1007/978-3-319-96418-8_43

12. Monagan, M., Tuncer, B.: The complexity of sparse Hensel lifting and sparse polynomial factorization. J. Symb. Comput. **99**, 189–230 (2020)
13. Monagan, M., Tuncer, B.: Polynomial factorization in maple 2019. In: Gerhard, J., Kotsireas, I. (eds.) MC 2019. CCIS, vol. 1125. Springer, Cham (2020). https://doi.org/10.1007/978-3-030-41258-6
14. Monagan, M.: Linear Hensel lifting for $\mathbb{F}[x, y]$ and $\mathbb{Z}[x]$ with cubic cost. In: Proceedings of ISSAC 2019, pp. 299–306. ACM (2019)
15. Javadi, S.M.M., Monagan, M.: On factorization of multivariate polynomials over algebraic number and function fields. In: Proceedings of ISSAC 2009, pp. 199–206. ACM (2009)
16. Lee, M.M.: Factorization of multivariate polynomials. Ph.D. Thesis (2013)
17. Schwartz, J.T.: Fast probabilistic algorithms for verification of polynomial identities. J. ACM **27**, 701–717 (1980)
18. Wang, P.S., Rothschild, L.P.: Factoring multivariate polynomials over the integers. Math. Comp. **29**, 935–950 (1975)
19. Yun, D.Y.Y.: The Hensel Lemma in algebraic manipulation. Ph.D. Thesis (1974)
20. Zhi, L.: Optimal algorithm for algebraic factoring. J. Comput. Sci. Technol. **12**(1), 1–9 (1997)
21. Zippel, R.: Probabilistic algorithms for sparse polynomials. In: Ng, E.W. (ed.) Symbolic and Algebraic Computation. LNCS, vol. 72, pp. 216–226. Springer, Heidelberg (1979). https://doi.org/10.1007/3-540-09519-5_73
22. Zippel, R.E.: Newton's iteration and the sparse Hensel algorithm. In: Proceedings of the ACM Symposium on Symbolic Algebraic Computation, pp. 68–72 (1981)
23. Zippel, R.E.: Interpolating polynomials from their values. J. Symb. Comput. **9**(3), 375–403 (1990)

The GPGCD Algorithm with the Bézout Matrix

Boming Chi[1] and Akira Terui[2(✉)] [iD]

[1] Graduate School of Pure and Applied Sciences, University of Tsukuba,
Tsukuba 305-8571, Japan
`hakumei-t@math.tsukuba.ac.jp`
[2] Faculty of Pure and Applied Sciences, University of Tsukuba,
Tsukuba 305-8571, Japan
`terui@math.tsukuba.ac.jp`
`https://researchmap.jp/aterui`

Abstract. For a given pair of univariate polynomials with real coefficients and a given degree, we propose a modification of the GPGCD algorithm, presented in our previous research, for calculating approximate greatest common divisor (GCD). In the proposed algorithm, the Bézout matrix is used in transferring the approximate GCD problem to a constrained minimization problem, whereas, in the original GPGCD algorithm, the Sylvester subresultant matrix is used. Experiments show that, in the case that the degree of the approximate GCD is large, the proposed algorithm computes more accurate approximate GCDs than those computed by the original algorithm. They also show that the computing time of the proposed algorithm is smaller than that of the SNTLS algorithm, which also uses the Bézout matrix, with a smaller amount of perturbations of the given polynomials and a higher stability.

Keywords: Approximate GCD · GPGCD algorithm · Bézout matrix · Modified newton method

1 Introduction

With the progress of algebraic computation with polynomials and matrices, we are paying more attention to approximate algebraic algorithms. Algorithms for calculating approximate GCD, which are approximate algebraic algorithms, consider a pair of given polynomials f and g that are relatively prime in general, and find $\tilde{f}$ and $\tilde{g}$ which are close to f and g, respectively, in the sense of polynomial norm, and have the greatest common divisor of a certain degree. These algorithms can be classified into two categories: 1) for a given tolerance (magnitude) of $\|f - \tilde{f}\|$ and $\|g - \tilde{g}\|$, make the degree of approximate GCD as large as possible, and 2) for a given degree d, minimize the magnitude of $\|f - \tilde{f}\|$ and $\|g - \tilde{g}\|$.

In both categories, algorithms based on various methods have been proposed including the Euclidean algorithm ([1,17,18]), low-rank approximation of the

© Springer Nature Switzerland AG 2020
F. Boulier et al. (Eds.): CASC 2020, LNCS 12291, pp. 170–187, 2020.
https://doi.org/10.1007/978-3-030-60026-6_10

Sylvester matrix or subresultant matrices ([5,6,9–11,19,24,26]), Padé approximation ([15]), and optimizations ([3,12,20,25]). Among them, the second author of the present paper has proposed the GPGCD algorithm based on low-rank approximation of subresultant matrices by optimization ([22,23]), which belongs to the second category above.

In this paper, we propose another formulation of the GPGCD algorithm by using the Bézout matrix, which is also used in zerofinding [14], while subresultant matrices have been used in the original algorithm. Using the Bézout matrix, we get higher accuracy for computed approximate GCD in the case of given polynomials of high degrees. In comparison with the SNTLS algorithm which also uses the Bézout matrix, the proposed algorithm calculates an approximate GCD with a smaller amount of computing time, smaller perturbations of the given polynomials, and higher stability.

The rest of the paper is organized as follows. In Sect. 2, we give a formulation of the transformation of the approximate GCD problem to the optimization problem using the Bézout matrix. In Sect. 3, we review the modified Newton method used for optimization. In Sect. 4, we illustrate the proposed algorithm by examples. In Sect. 5, the results of experiments are shown.

2 Transformation of the Approximate GCD Problem Using the Bézout Matrix

Let $F(x)$ and $G(x)$ be univariate polynomials with real coefficients:

$$
\begin{aligned}
F(x) &= f_m x^m + \cdots + f_0 x^0, \\
G(x) &= g_n x^n + \cdots + g_0 x^0,
\end{aligned}
\tag{1}
$$

with $m \geq n > 0$. Throughout this paper, the norm $\|F(x)\|$ denotes the 2-norm $\|F(x)\|_2 := (f_m^2 + f_{m-1}^2 + \cdots + f_0^2)^{\frac{1}{2}}$. For a given integer d with $n \geq d > 0$, let us find polynomials $\tilde{F}(x)$ and $\tilde{G}(x)$ whose degrees respectively do not exceed those of $F(x)$ and $G(x)$, such that

$$
\begin{aligned}
\tilde{F}(x) &= \tilde{f}_m x^m + \cdots + \tilde{f}_0 x^0 = \bar{F}(x) \times \tilde{H}(x), \\
\tilde{G}(x) &= \tilde{g}_n x_n + \cdots + \tilde{g}_0 x^0 = \bar{G}(x) \times \tilde{H}(x),
\end{aligned}
\tag{2}
$$

where $\tilde{H}(x)$ is a polynomial of degree d and $\bar{F}(x)$ and $\bar{G}(x)$ are relatively prime polynomials. In this case, we call $\tilde{H}(x)$ an *approximate GCD* of polynomials $F(x)$ and $G(x)$. For a pair of polynomials $F(x)$ and $G(x)$ and a degree d, we consider the problem which finds the approximate GCD with the degree d while minimizing the norm of the perturbations

$$
\Delta := \sqrt{\|F(x) - \tilde{F}(x)\|^2 + \|G(x) - \tilde{G}(x)\|^2}.
\tag{3}
$$

Without loss of generality, for a polynomial G of degree n, we also represent G as $G(x) = g_m x^m + \cdots + g_0 x^0$, where $g_m = \cdots = g_{n+1} = 0$. Here, we choose the Bézout matrix to formulate the problem.

Definition 1 (Bézout Matrix [8]). *Let $F(x)$ and $G(x)$ be two real polynomials with the degree at most m. Then, the matrix $\mathrm{Bez}(F,G) = (b_{ij})_{i,j=1...m}$, where*

$$\frac{F(x)G(y) - F(y)G(x)}{x - y} = \sum_{i,j=1}^{m} b_{ij} x^{i-1} y^{j-1},$$

is called the Bézout matrix associated to $F(x)$ and $G(x)$.

Note that the $\mathrm{Bez}(F,G)$ is $m \times m$ symmetric matrix, and its element (b_{ij}) is represented in terms of the coefficients of $F(x)$ and $G(x)$ as

$$b_{ij} = \sum_{k=1}^{m_{ij}} f_{j+k-1} g_{i-k} - f_{i-k} g_{j+k-1}, \tag{4}$$

where $m_{ij} = \min\{i, m+1-j\}$ [8]. Furthermore, we have the following relationship between the degree of the GCD, the rank, and the elements of the Bézout matrix.

Theorem 1 (Barnett's theorem [8]). *Let $F(x)$ and $G(x)$ be two real polynomials with the degree at most m. Let $d = \deg(\gcd(F,G))$ and $(\boldsymbol{b}_1, \ldots, \boldsymbol{b}_m) = \mathrm{Bez}(F,G)$. Then, the vectors $\boldsymbol{b}_{d+1}, \ldots, \boldsymbol{b}_m$ are linearly independent, and there exists coefficients $c_{i,j}$ such that*

$$\boldsymbol{b}_i = \sum_{j=1}^{m-d} c_{i,j} \boldsymbol{b}_{d+j}, \quad 1 \le i \le d, \tag{5}$$

Furthermore, the monic form of the GCD of $F(x)$ and $G(x)$ is represented as

$$\gcd(F,G) = x^d + c_{d,1} x^{d-1} + \cdots + c_{1,1} x^0. \tag{6}$$

From the Barnett's theorem, obviously we have

$$\mathrm{rank}(\mathrm{Bez}(F,G)) + \deg(\gcd(F,G)) = m. \tag{7}$$

Let $\tilde{B} = \mathrm{Bez}(\tilde{F}, \tilde{G})$. In the case that $\tilde{F}(x)$ and $\tilde{G}(x)$ have a GCD of degree d, we have $\mathrm{rank}(\tilde{B}) = m - d$. Let $\tilde{B} = \tilde{U}\tilde{\Sigma}\tilde{V}^T$ be the singular value decomposition (SVD) of $\tilde{B}$, where $\tilde{V} = (\tilde{\boldsymbol{v}}_1, \ldots, \tilde{\boldsymbol{v}}_m)$. Then, we have

$$\tilde{B}\tilde{\boldsymbol{v}} = \boldsymbol{0}, \tag{8}$$

where $\tilde{\boldsymbol{v}} = \tilde{\boldsymbol{v}}_{m-d+1} = (\tilde{v}_1, \ldots, \tilde{v}_m)$. Namely, we have the constraints

$$G_i = \tilde{b}_{i1}\tilde{v}_1 + \cdots + \tilde{b}_{im}\tilde{v}_m = 0, \quad i \le m, \tag{9}$$

where

$$(\tilde{b}_{ij})_{i,j} = \tilde{B}, \tag{10}$$

which can be represented by Eq. (4) with the coefficients $\tilde{f}_i$ and $\tilde{g}_i$.

Let $B = \text{Bez}(F, G)$ and let $B = U\Sigma V^T$ be the SVD of B, with $V = (v_1, \ldots, v_m)$. For solving the problem more effectively, we also make the perturbation of

$$v = v_{m-d+1} = (v_1, \ldots, v_m) \tag{11}$$

small. Thus, the objective function is represented as

$$F = \sum_{i=0}^{m} (\tilde{f}_i - f_i)^2 + \sum_{i=0}^{n} (\tilde{g}_i - g_i)^2 + \sum_{i=1}^{m} (\tilde{v}_i - v_i)^2. \tag{12}$$

Here, we let the variables of the objective function be

$$\begin{aligned} x &= (x_1, x_2, \ldots, x_{2m+n+2})^T \\ &= (\tilde{f}_0, \ldots, \tilde{f}_m, \tilde{g}_0, \ldots, \tilde{g}_n, \tilde{v}_1, \ldots, \tilde{v}_m)^T. \end{aligned} \tag{13}$$

Therefore, the problem of finding an approximate GCD is formulated as a constrained minimization problem: find a minimizer of the objective function F in Eq. (12), subject to $g(x) = (G_1(x), \ldots, G_m(x))^T = 0$ in Eq. (9).

3   The Modified Newton Method

We consider a constrained minimization problem of minimizing an objective function $f(x) : \mathbb{R}^s \to \mathbb{R}$ which is twice continuously differentiable, subject to the constraints $g(x) = (G_1(x), \ldots, G_t(x))^T$, where $G_i(x)$ is a function of $\mathbb{R}^s \to \mathbb{R}$ and is also twice continuously differentiable. We use the modified Newton method by Tanabe ([21]), which is a generalization of the Gradient Projection method ([16]), used in the original GPGCD algorithm ([23]). For x_k which satisfies $g(x_k) = 0$, we calculate the search direction d_k and the Lagrange multipliers λ_k by solving the following linear system

$$\begin{pmatrix} I & -(J_g(x_k))^T \\ J_g(x_k) & O \end{pmatrix} \begin{pmatrix} d_k \\ \lambda_{k+1} \end{pmatrix} = - \begin{pmatrix} \nabla f(x_k) \\ g(x_k) \end{pmatrix}, \tag{14}$$

where $J_g(x)$ is the Jacobian matrix represented as

$$J_g(x) = \frac{\partial g_i}{\partial f_j}. \tag{15}$$

4   The Algorithm for Calculating Approximate GCD

In this section, we give an algorithm for calculating approximate GCD using the Bézout matrix. In the modified Newton method, the Jacobian matrix and the initial values are represented as follows.

4.1 Representation of the Jacobian Matrix

From the constraints (9) and the objective function (12), the elements of the Jacobian matrix are represented as follows:

$$\frac{\partial G_i}{\partial \tilde{x}_j} = \sum_{l=1}^{m} \frac{\partial \tilde{b}_{il}}{\partial \tilde{x}_j} \tilde{v}_l$$

$$= \begin{cases} -\sum_{k=0}^{i-1} \tilde{g}_k \tilde{v}_{j-i+k} & 1 \leq i < j \leq m+1 \\ \sum_{k=0}^{m-i} \tilde{g}_{i+k} \tilde{v}_{j+k} & 1 \leq j \leq i \leq m \\ \sum_{k=0}^{i-1} \tilde{f}_k \tilde{v}_{j-m-i+k-1} & 1 \leq i < j - m - 1 \\ & \qquad \leq n+1 \\ -\sum_{k=0}^{m-i} \tilde{f}_{i+k} \tilde{v}_{j-m+k-1} & 1 \leq j - m - 1 \\ & \qquad \leq \min\{i, n+1\} \leq m \\ \tilde{b}_{i,j-m-n-2} & m+n+2 < j \\ & \qquad \leq 2m+n+2 \end{cases}, \qquad (16)$$

where $\tilde{b}_{ij}$ is represented as in Eq. (10). The size of the Jacobian matrix is $m \times (2m + n + 2)$.

4.2 Setting the Initial Values

For the given polynomials (1), and the singular vectors (11), we give the initial value x_0 with the coefficients and elements of the singular vector as

$$x_0 = (f_0, \ldots f_m, g_0, \ldots, g_n, v_1, \ldots, v_m). \qquad (17)$$

4.3 Calculating the Approximate GCD

Let x^* be the minimizer calculated by the modified Newton method, corresponding to the coefficients of $\tilde{F}(x)$ and $\tilde{G}(x)$. Then, we calculate the GCD of $\tilde{F}(x)$ and $\tilde{G}(x)$ using the Bézout matrix $\tilde{B} = \mathrm{Bez}(\tilde{F}, \tilde{G})$ with Theorem 1.

4.4 The Algorithm

Summarizing the above, we give the algorithm for calculating approximate GCD as follows:

Algorithm 1 (The GPGCD algorithm with the Bézout matrix).
Inputs:

- $F(x)$, $G(x) \in \mathbb{R}(x)$: the given polynomials with $\deg F(x) \geq \deg G(x) > 0$,
- $d \in \mathbb{N}$: the given degree of approximate GCD with $d \leq \deg(G)$,
- $\epsilon > 0$: the stop criterion with the modified Newton method,
- $0 < \alpha \leq 1$: the step width with the modified Newton method.

Outputs:

- $\tilde{H}(x)$: *the approximate GCD, with* $\deg(\tilde{H}) = d$,
- $\tilde{F}(x)$, $\tilde{G}(x)$: *the polynomials which are close to* F *and* G, *respectively, with the GCD* $\tilde{H}$.

Step 1 *Calculate the Bézout matrix* $B = \text{Bez}(F, G)$, *the Jacobian matrix (16), and the singular vector* $\boldsymbol{v}$ *as in (11).*
Step 2 *Set the initial values* $\boldsymbol{x}_0$ *as in (17).*
Step 3 *Solve the linear system (14) to find the search direction* $\boldsymbol{d}_k$.
Step 4 *If* $\|\boldsymbol{d}_k\| < \epsilon$, *obtain the* $\boldsymbol{x}^*$ *as* $\boldsymbol{x}_k$, *calculate polynomials* $\tilde{F}(x)$ *and* $\tilde{G}(x)$, *then go to Step 1. Otherwise, let* $\boldsymbol{x}_{k+1} = \boldsymbol{x}_k + \alpha \boldsymbol{d}_k$ *and calculate the Bézout matrix and the Jacobian matrix with* $\boldsymbol{x}_{k+1}$, *then go to Step 3.*
Step 5 *Calculate the approximate GCD* $\tilde{H}(x)$ *with Theorem 1. Return* $\tilde{F}(x)$, $\tilde{G}(x)$ *and* $\tilde{H}(x)$.

4.5 Running Time Analysis

We give an analysis of the arithmetic running time of Algorithm 1.

In Step 1, we set the initial values by the construction of the Bézout matrix, the construction of the Jacobian matrix, and the SVD of the Bézout matrix. Since the dimension of the Bézout matrix and the Jacobian matrix is m and $m(2m+n+2)$, respectively, we can estimate the running time of the construction of the Bézout matrix, the construction of the Jacobian matrix and the SVD of the Bézout matrix is $O(m^2)$ ([4]), $O(m(2m + n + 2))$ (the Jacobian matrix is computed by (16)) and $O(m^3)$ ([7]), respectively.

In Step 3, since the dimension of the Jacobian matrix is $m(2m + n + 2)$, the running time for solving the linear system is $O((3m + n + 2)^3)$ ([7]).

In Step 4, the running time of the construction of the Bézout matrix and the Jacobian matrix is $O(m^2)$ and $O(m(2m + n + 2))$, respectively.

In Step 5, the running time for calculating the approximate GCD $\tilde{H}(x)$ is $O(m^2)$.

As a consequence, the running time of Algorithm 1 is the number of iteration times $O((3m + n + 2)^3)$.

5 Experiments

We have implemented our GPGCD algorithm on Maple 2016 ([2]). The test polynomials $F(x)$ and $G(x)$ are generated as follows:

$$
\begin{aligned}
F(x) &= F_0(x)H(x) + \frac{e_F}{\|F_N(x)\|}F_N(x), \\
G(x) &= G_0(x)H(x) + \frac{e_G}{\|G_N(x)\|}G_N(x).
\end{aligned}
\tag{18}
$$

Here $F_0(x)$, $G_0(x)$, and $H(x)$ are polynomials of degrees $m - d$, $n - d$, and d, respectively, where $F_0(x)$ and $G_0(x)$ are relatively prime polynomials. They are generated as polynomials that their coefficients are floating-point numbers and their absolute values are not greater than 10.[1] The noise polynomials $F_N(x)$ and $G_N(x)$ are polynomials of degrees $m - 1$ and $n - 1$, respectively, which are randomly generated with coefficients given as the same as for $F_0(x)$, $G_0(x)$ and $H(x)$.

We have generated 15 groups of test polynomials, each group comprising 100 tests. The degrees of polynomials in each test are shown in Table 1. In the first 10 groups, we set the degree of input polynomials to be twice the degree of the approximate GCD. In the last 5 groups, we have changed the degree of approximate GCD for the same degree of input polynomials. In our tests, the norm of the noise e_F and e_G are set as $e_F = e_G = 0.01$. The stop criterion ϵ and the stop width α in Algorithm 1 are set as 10^{-8} and 1, respectively. The stop tolerance of the SNTLS algorithm is 10^{-5}.

We have carried out the tests on CPU Intel(R) Core(TM) i5-6600 at 3.30 GHz with RAM 8.00 GB, under Windows 10.

5.1 Setting the Criteria for Classifying Successful Computation of an Approximate GCD

By generating the test polynomials in Eq. (18), the perturbation is set as

$$\sqrt{e_F^2 + e_G^2} = \sqrt{0.01^2 + 0.01^2} \approx 1.414 \times 10^{-2}, \tag{19}$$

which is called "the given perturbation". For setting the criteria for classifying successful computation of an approximate GCD, we first give a criterion

$$\sqrt{\|F(x) - \tilde{F}(x)\|^2 + \|G(x) - \tilde{G}(x)\|^2} \leq 1.414 \times 10^{-2},$$

as the perturbations Δ in Eq. (3) is not supposed to exceed the given perturbation (19). However, we have found out that this criterion is not enough with examples that were found in our experiments with the proposed algorithm, shown in Eqs. (21) and (22). In Examples (21) and (22), F and G are input polynomials, while $\tilde{F}$, $\tilde{G}$, and $\tilde{H}$ are the outputs of Algorithm 1.

In both examples, the perturbations Δ in Eq. (3) do not exceed the given perturbation (19). In the examples, we have also calculated $\hat{F}$ and $\hat{G}$ which are the remainders[2] of $\tilde{F}$ and $\tilde{G}$ divided by $\tilde{H}$, respectively ($\tilde{F} = A\tilde{H} + \hat{F}, \tilde{G} = B\tilde{H} + \hat{G}$). If the approximate GCD is successfully calculated, the norm of the remainders

$$R := \sqrt{\|\hat{F}\|^2 + \|\hat{G}\|^2} \tag{20}$$

is supposed to be close to 0. In Example (21), the norm of the remainders (20) is sufficiently small, while, in Example (22), the norm of the remainders (20) is not sufficiently small.

---

[1] The coefficients are generated with the Mersenne Twister algorithm [13] by built-in function `Generate` with `RandomTools:-MersenneTwister` in Maple, which approximates a uniform distribution on $[-10, 10]$.

[2] Remainders are calculated with built-in function `SNAP:-Remainder`.

The examples show that the calculated approximate GCD may not be accurate even though the perturbation is small. Therefore, to classify whether an approximate GCD is successfully calculated or not, the criteria are set as follows:

1. $\sqrt{\|F(x) - \tilde{F}(x)\|^2 + \|G(x) - \tilde{G}(x)\|^2} \leq 1.414 \times 10^{-2}$,

2. $R = \sqrt{\|\hat{F}\|^2 + \|\hat{G}\|^2} \leq 10^{-5}$.

In Criterion 1, the perturbation is supposed not to exceed the given perturbation (19). In Criterion 2, the norm of the remainders $R = \sqrt{\|\hat{F}\|^2 + \|\hat{G}\|^2}$ is supposed to be sufficiently small.

5.2   The Experimental Results

We have carried out the tests with the GPGCD algorithm with the Bézout matrix. For comparison, we have also carried out the tests with the original GPGCD algorithm from Group 1 to Group 15, and the SNTLS algorithm [20], which also uses the Bézout matrix, from Group 1 to Group 5 and from Group 11 to Group 15.[3]

Based on the criteria, the number of successful tests for each group is shown in Table 2. Columns 'Bézout', 'Sylvester', 'SNTLS', 'All' represent the number of successful tests for the GPGCD algorithm with the Bézout matrix, the original GPGCD algorithm, the SNTLS algorithm and for all of the 3 algorithms, respectively.

Remark 1. To keep fairness, we have compared the data only for successful tests for all of the algorithms.

The average computing time, the norm of perturbations, the norm of remainders, and the number of iterations for successful tests for all of the algorithms are shown in Tables 3, 4, 5, and 6, respectively.

In Tables 3, 4, 5, and 6, columns 'Bézout', 'Sylvester', and 'SNTLS' represent the data for the GPGCD algorithm with the Bézout matrix, the original GPGCD algorithm, and the SNTLS algorithm, respectively.

For the average of the norm of the remainders, the arithmetic mean value may be affected by some of the extreme values. In Table 5, we also show the geometric mean of the norm of the remainders, which shows the average of the order of magnitude.

5.3   Comparison with the Original GPGCD Algorithm

In Table 2, we see that in the 15 groups of tests, the number of successfully calculated tests for the GPGCD algorithm with the Bézout matrix is smaller

---

[3] We have excluded test polynomials from Group 6 to Group 10, because, for those polynomials, the computing time for the SNTLS algorithm is too long (over 100 min with tests in Group 6 whose degree of input polynomials is 60).

than that for the original GPGCD algorithm in the case of input polynomials and approximate GCD of low degrees, and larger than that for the original GPGCD algorithm in the case of input polynomials and approximate GCD of high degrees.

By comparing the data between the first 10 groups of tests, we see that for the successful tests for both algorithms, though the computing time of the proposed algorithm is not as small as that of the original GPGCD algorithm (shown in Table 3), the norm of remainders of the proposed algorithm is smaller than that of the original GPGCD algorithm in the case of polynomials of high degrees (shown in Table 5).

By comparing the data between the last 5 groups of tests, we see that for the successful tests for both algorithms, the norm of remainders of the proposed algorithm is smaller than that of the original GPGCD algorithm in the case of approximate GCD of high degrees (shown in Table 5).

The running time of the original algorithm is $O((3(m+n-d))^3)$ ([23]). The running time of the proposed algorithm, $O((3m+n+2)^3)$, is smaller than that of the original GPGCD algorithm in the case that $d < \frac{2n+4}{3}$. The computing time of the proposed algorithm is supposed to be smaller than that of the original GPGCD algorithm with the conditions in the experiments, but the results of the experiments show the opposite.

5.4   Comparison with the SNTLS Algorithm

Table 2 shows that, in the first 5 groups and the last 5 groups of tests, the number of successfully calculated tests for the proposed algorithm is smaller than that for the SNTLS algorithm.

Through the experiments, the SNTLS algorithm tends to have higher accuracy in the sense of the norm of the remainders than the proposed algorithm and the original GPGCD algorithm (shown in Table 5), while perturbation of the SNTLS algorithm is not as small as that of the two GPGCD algorithms (shown in Table 4). A typical example that was found in our experiments is shown in Eq. (23). We see that in Example (23), the norm of the remainders (R) is sufficiently small, while the perturbation (Δ) is larger than the given perturbation (19).

We have estimated that the running time for each iteration in the SNTLS algorithm is $O((3m+n-d+2)^3)$, which is smaller than that of the proposed algorithm, while the computing time of the proposed algorithm is smaller than that of the SNTLS algorithm (shown in Table 3).

Through the experiments, we have found out that the computation for some of the tests with the SNTLS algorithm have not finished within one hour, while for the same tests with both of the GPGCD algorithm have finished successfully. For these cases with the SNTLS algorithm, we have interrupted the computation after 50 iterations and computed the perturbed polynomials and the approximate GCD with the data from the iterations which have been interrupted. We see that in all of these cases, the perturbation (Δ) is larger than the given perturbation (19), which is similar to Example (23). In this sense, it shows that the proposed algorithm is more stable than the SNTLS algorithm.

6   Conclusions

We have proposed an algorithm that uses the Bézout matrix based on the GPGCD algorithm. The proposed algorithm has a smaller running time in the case that the degree of approximate GCD is not large, and through the experiments, the proposed algorithm shows higher accuracy than the original GPGCD algorithm in the case when the degree of the approximate GCD is large. In comparison with the SNTLS algorithm which also uses the Bézout matrix, the proposed algorithm calculates an approximate GCD with a smaller amount of computing time, smaller perturbations of the input polynomials, and higher stability through experiments.

Though the estimated running time of the proposed algorithm is smaller than that of the original GPGCD algorithm in the case when the degree of approximate GCD is not large, the actual computing time of the proposed algorithm is larger than that of the original GPGCD algorithm through the experiments. We have found out that the actual computing time needed for the construction of the Bézout matrix and the Jacobian matrix is larger than that for solving the linear system (14), despite the estimate of running time which shows that the running time for solving the linear system (14) is $O(m^3)$ which is larger than that for the construction of the Bézout matrix and the Jacobian matrix which is $O(m^2)$. Making the proposed algorithm more efficient will be one of directions of our future research. Furthermore, an extension of our algorithm using weighted norms ([14]) will also be one of our future research topics.

Despite the fact that the estimated running time of the proposed algorithm is larger than that of the SNTLS algorithm, the actual computing time of the proposed algorithm is smaller than that of the SNTLS algorithm. The result suggests that these implementations deserve a more detailed analysis.

Acknowledgments. We thank Professor Lihong Zhi for providing the source code of the SNTLS algorithm. We also thank Kosaku Nagasaka, Masaru Sanuki, Takuya Kitamoto and Professor Hiroshi Sekigawa for their helpful comments with this research.

This research was supported in part by JSPS KAKENHI Grant Number 16K05035.

Table 1. Degrees of test polynomials (18)

Group	$m = \deg(F)$	$n = \deg(G)$	$d = \deg(H)$
1	10	10	5
2	20	20	10
3	30	30	15
4	40	40	20
5	50	50	25
6	60	60	30
7	70	70	35
8	80	80	40
9	90	90	45
10	100	100	50
11	50	50	5
12	50	50	10
13	50	50	15
14	50	50	20
15	50	50	25

Table 2. The number of tests that meets the criteria

Group	Number of tests			
	Bézout	Sylvester	SNTLS	All
1	97	99	98	97
2	82	91	85	81
3	60	82	77	60
4	46	79	66	46
5	47	39	66	36
6	30	22	—	20
7	27	18	—	18
8	18	14	—	12
9	14	9	—	6
10	14	9	—	9
11	17	57	47	17
12	22	61	52	22
13	30	20	50	18
14	34	27	49	25
15	45	41	62	38

Table 3. Comparison of computing time (See Remark 1 for detail)

Group	Average time (sec.)		
	Bézout	Sylvester	SNTLS
1	7.504×10^{-2}	5.961×10^{-2}	1.535×10^{-1}
2	7.958×10^{-1}	4.367×10^{-1}	1.181
3	1.653	8.089×10^{-1}	3.242
4	2.985	1.347	11.730
5	4.426	1.958	22.546
6	6.651	2.954	—
7	7.981	3.646	—
8	10.569	5.185	—
9	12.440	4.393	—
10	17.592	5.984	—
11	4.729	1.106	30.673
12	4.645	1.011	24.877
13	4.792	9.696×10^{-1}	24.739
14	4.564	8.630×10^{-1}	23.627
15	5.118	7.550×10^{-1}	22.771

Table 4. Comparison of the norm of perturbations (3) (See Remark 1 for detail)

Group	Average		
	Bézout	Sylvester	SNTLS
1	6.346931×10^{-3}	6.346940×10^{-3}	6.346997×10^{-3}
2	7.035965×10^{-3}	7.036040×10^{-3}	7.111237×10^{-3}
3	6.795088×10^{-3}	6.795097×10^{-3}	8.742038×10^{-3}
4	6.706903×10^{-3}	6.706909×10^{-3}	6.706925×10^{-3}
5	6.824742×10^{-3}	6.824741×10^{-3}	6.824750×10^{-3}
6	6.715120×10^{-3}	6.715120×10^{-3}	—
7	7.118744×10^{-3}	7.118732×10^{-3}	—
8	6.840067×10^{-3}	6.840078×10^{-3}	—
9	7.256039×10^{-3}	7.256032×10^{-3}	—
10	7.124559×10^{-3}	7.124564×10^{-3}	—
11	2.634340×10^{-3}	2.634942×10^{-3}	2.634340×10^{-3}
12	4.617887×10^{-3}	4.617956×10^{-3}	4.617887×10^{-3}
13	5.332392×10^{-3}	5.332505×10^{-3}	5.332392×10^{-3}
14	6.452837×10^{-3}	6.452873×10^{-3}	6.452837×10^{-3}
15	6.798100×10^{-3}	6.798127×10^{-3}	6.798108×10^{-3}

Table 5. Comparison of the norm of remainders (20) (See Remark 1 for detail)

Gr.	Arithmetic mean			Geometric mean		
	Bézout	Sylvester	SNTLS	Bézout	Sylvester	SNTLS
1	2.585×10^{-7}	5.591×10^{-11}	6.910×10^{-11}	1.563×10^{-11}	7.148×10^{-13}	6.939×10^{-13}
2	4.514×10^{-6}	3.002×10^{-10}	1.719×10^{-10}	4.026×10^{-10}	3.905×10^{-12}	3.522×10^{-12}
3	3.106×10^{-6}	2.594×10^{-10}	2.939×10^{-10}	6.765×10^{-9}	1.617×10^{-11}	1.506×10^{-11}
4	4.138×10^{-6}	1.816×10^{-10}	1.506×10^{-10}	2.962×10^{-8}	2.989×10^{-11}	2.587×10^{-11}
5	1.059×10^{-6}	1.693×10^{-5}	1.572×10^{-10}	2.054×10^{-7}	7.708×10^{-6}	7.428×10^{-11}
6	1.647×10^{-6}	2.923×10^{-5}	—	4.546×10^{-7}	1.460×10^{-5}	—
7	3.753×10^{-6}	2.906×10^{-5}	—	5.408×10^{-7}	1.762×10^{-5}	—
8	9.924×10^{-7}	3.267×10^{-5}	—	6.056×10^{-7}	2.383×10^{-5}	—
9	2.582×10^{-6}	3.970×10^{-5}	—	1.065×10^{-6}	3.184×10^{-5}	—
10	5.170×10^{-6}	6.185×10^{-5}	—	1.328×10^{-6}	5.090×10^{-5}	—
11	4.217×10^{-6}	1.624×10^{-9}	6.863×10^{-10}	1.760×10^{-8}	3.654×10^{-11}	3.410×10^{-11}
12	4.749×10^{-6}	5.795×10^{-10}	5.110×10^{-10}	3.255×10^{-7}	1.197×10^{-10}	1.241×10^{-10}
13	3.340×10^{-6}	3.330×10^{-5}	7.452×10^{-10}	3.908×10^{-7}	1.766×10^{-5}	2.980×10^{-10}
14	4.246×10^{-7}	1.528×10^{-5}	2.529×10^{-10}	1.717×10^{-7}	9.999×10^{-6}	1.205×10^{-10}
15	6.482×10^{-7}	1.610×10^{-5}	2.366×10^{-10}	2.313×10^{-7}	8.445×10^{-6}	1.016×10^{-10}

Table 6. Comparison of the number of iterations (See Remark 1 for detail)

Group	Average number of iterations		
	Bézout	Sylvester	SNTLS
1	2.237	3.093	3.031
2	2.370	3.062	3.716
3	2.35	3.05	5.867
4	2.434	3.022	4.761
5	2.583	3.028	2.778
6	2.6	3.05	—
7	2.833	3	—
8	2.833	3	—
9	2.5	3	—
10	2.556	3	—
11	2.765	1	3.706
12	2.546	1	3.864
13	2.556	1	4.111
14	2.4	1	3.72
15	2.579	1	2.974

$$F(x) := 24.35467659x^{10} - 16.98261439x^9 - 74.84052127x^8$$
$$+ 45.72321036x^7 + 25.24635127x^6 - 6.618882298x^5$$
$$+ 28.00488463x^4 - 23.01650876x^3 + 39.85773244x^2$$
$$- 17.87827873x - 28.59581303,$$

$$G(x) := 18.07487567x^{10} - 0.5207310267x^9 - 45.95049998x^8$$
$$- 24.03693723x^7 + 30.52699312x^6 \quad + 96.43514331x^5$$
$$- 12.98476016x^4 - 63.09966135x^3 - 42.21071223x^2$$
$$+ 2.244520517x + 43.86799924,$$

$$\tilde{F}(x) = 24.3518562750250x^{10} - 16.9810481741899x^9 - 74.8412673761213x^8$$
$$+ 45.7239877362181x^7 \quad + 25.2453891983873x^6 - 6.61806019040930x^5$$
$$+ 28.0049393854550x^4 - 23.0170430598979x^3 + 39.8579457401336x^2$$
$$- 17.8774686643460x - 28.5967141490441,$$

$$\tilde{G}(x) = 18.0763796579614x^{10} - 0.522186160911192x^9 - 45.9486461908583x^8$$
$$- 24.0379730412682x^7 + 30.52681977720611x^6 + 96.4361328610033x^5$$
$$- 12.9845682284478x^4 - 63.1010698450796x^3 - 42.2093213135731x^2$$
$$+ 2.24532874696889x + 43.8660092385339,$$

$$\tilde{H}(x) = x^5 - 0.0198739343264150x^4 - 1.67341575817232x^3$$
$$- 0.0826245996156027x^2 - 0.531150763766694x + 1.40628969338512,$$

$$\Delta = \sqrt{\|F(x) - \tilde{F}(x)\|^2 + \|G(x) - \tilde{G}(x)\|^2} = 0.00578261308305222,$$

$$\mathring{F}(x) = 2.30926389122033 \times 10^{-13}x^4 - 3.05533376376843 \times 10^{-13}x^3$$
$$+ 1.42108547152020 \times 10^{-13}x^2 - 2.20268248085631 \times 10^{-13}x$$
$$+ 3.05533376376843 \times 10^{-13},$$

$$\hat{G}(x) = 5.50670620214078 \times 10^{-14}x^4 - 3.55271367880050 \times 10^{-13}x^3$$
$$+ 1.56319401867222 \times 10^{-13}x^2 - 1.66089364483923 \times 10^{-13}x$$
$$+ 2.27373675443232 \times 10^{-13},$$

$$R = \sqrt{\|\hat{F}\|^2 + \|\hat{G}\|^2} = 7.22300963376246 \times 10^{-13}.$$

$$(21)$$

$$F(x) := 0.1766713555x^{10} + 61.40696577x^9 + 16.97723836x^8$$
$$- 98.53372181x^7 - 33.29739487x^6 - 53.01237461x^5$$
$$- 107.2631922x^4 - 44.39926514x^3 - 65.76778029x^2$$
$$+ 49.07634648x + 27.73066658,$$

$$G(x) := -0.0799481570x^{10} - 26.37832559x^9 - 16.98213985x^8$$
$$+ 33.28081373x^7 + 90.2104557x^6 - 27.34128713x^5$$
$$- 20.66348810x^4 + 12.71888631x^3 + 81.21689912x^2$$
$$+ 91.05587991x + 23.63244802,$$

$$\tilde{F}(x) = 0.177716772952069x^{10} + 61.4048417594346x^9 + 16.9760552139165x^8$$
$$- 98.5341789215241x^7 - 33.2981170063200x^6 - 53.0129121533162x^5$$
$$- 107.262870697403x^4 - 44.3995689180854x^3 - 65.7681756996879x^2$$
$$+ 49.0761750948950x + 27.7326910319257,$$

$$\tilde{G}(x) = -0.0762724095297875x^{10} - 26.3817204683292x^9 - 16.9848862314274x^8$$
$$+ 33.2795003835884x^7 + 90.2100574768607x^6 - 27.3423956063330x^5$$
$$- 20.6642442968902x^4 + 12.7198843248631x^3 + 81.2156355194280x^2$$
$$+ 91.0567877271623x + 23.6284289478540,$$

$$\tilde{H}(x) = x^5 + 344.630262139286x^4 - 210.982672870932x^3$$
$$- 181.719887733887x^2 - 433.650477152187x - 158.328743852362,$$

$$\Delta = \sqrt{\|F(x) - \tilde{F}(x)\|^2 + \|G(x) - \tilde{G}(x)\|^2} = 0.00826656291248941,$$

$$\hat{F}(x) = 3.47523730496033 \times 10^8 x^4 - 2.11852203486360 \times 10^8 x^3$$
$$- 1.81658961482431 \times 10^8 x^2 - 4.36057563872763 \times 10^8 x$$
$$- 1.59376135103876 \times 10^8,$$

$$\hat{G}(x) = -1.48991716316745 \times 10^8 x^4 + 9.08260951212297 \times 10^7 x^3$$
$$+ 7.78815317646166 \times 10^7 x^2 + 1.86948283392243 \times 10^8 x$$
$$+ 6.83283523549969 \times 10^7,$$

$$R = \sqrt{\|\hat{F}\|^2 + \|\hat{G}\|^2} = 7.00237743622062 \times 10^8.$$

$$(22)$$

$$F(x) := 0.2444489655x^{10} - 18.92615417x^9 + 63.91966715x^8$$
$$- 3.013484233x^7 - 70.45198569x^6 + 88.67816884x^5$$
$$- 115.7540570x^4 - 44.02256569x^3 + 68.62533154x^2$$
$$- 57.13421559x + 45.63944175,$$

$$G(x) := -0.2968329607x^{10} + 22.04593872x^9 - 55.89650876x^8$$
$$- 49.40533781x^7 + 61.74422687x^6 - 56.67607764x^5$$
$$+ 20.76671672x^4 + 23.11312753x^3 - 85.16586593x^2$$
$$- 49.75412158x - 86.87574838,$$

$$\tilde{F}(x) = -0.0921867419910542x^{10} + 25.1711846426639x^9 - 48.3791835554963x^8$$
$$- 42.3226815168449x^7 + 0.587145137610493x^6 + 87.9881298102165x^5$$
$$- 115.806492504772x^4 - 43.9219670362163x^3 + 68.5088816347304x^2$$
$$- 57.0592112917500x + 45.6388192929715,$$

$$\tilde{G}(x) = 0.971366003301017x^{10} - 26.3817204683292x^9 - 16.9848862314274x^8$$
$$+ 33.2795003835884x^7 + 78.9365949745973x^6 - 33.0045642782866x^5$$
$$+ 44.4446856175190x^4 + 37.1210544363078x^3 + 0.770946756608030x^2$$
$$+ 22.0716689225433x - 21.3709987313362,$$

$$\tilde{H}(x) = x^5 - 1.29900475944232x^4 - 0.494680213464673x^3$$
$$- 181.719887733887x^2 - 0.643083948588068x + 0.511524452719042,$$

$$\Delta = \sqrt{\|F(x) - \tilde{F}(x)\|^2 + \|G(x) - \tilde{G}(x)\|^2} = 111.265101285836,$$

$$\hat{F}(x) = -1.13686837721616 \times 10^{-13}x^4 - 2.84217094304040 \times 10^{-14}x^3$$
$$+ 5.68434188608080 \times 10^{-14}x^2 - 3.55271367880050 \times 10^{-14}x$$
$$+ 4.26325641456060 \times 10^{-14},$$

$$\hat{G}(x) = 1.27897692436818 \times 10^{-13}x^4 + 7.10542735760100 \times 10^{-15}x^3$$
$$- 3.55271367880050 \times 10^{-14}x^2 + 1.77635683940025 \times 10^{-14}x$$
$$- 4.26325641456060 \times 10^{-14},$$

$$R = \sqrt{\|\hat{F}\|^2 + \|\hat{G}\|^2} = 1.41574637134818 \times 10^{-13}.$$

$$(23)$$

References

1. Beckermann, B., Labahn, G.: A fast and numerically stable euclidean-like algorithm for detecting relatively prime numerical polynomials. J. Symb. Comput. **26**(6), 691–714 (1998). https://doi.org/10.1006/jsco.1998.0235
2. Chi, B., Terui, A.: ct1counter/bezout-gpgcd: Initial release (2020). https://doi.org/10.5281/zenodo.3965389

3. Chin, P., Corless, R.M., Corliss, G.F.: Optimization strategies for the approximate GCD problem. In: Proceedings of the 1998 International Symposium on Symbolic and Algebraic Computation. pp. 228–235. ACM (1998). https://doi.org/10.1145/281508.281622

4. Chionh, E.W., Zhang, M., Goldman, R.N.: Fast computation of the Bezout and dixon resultant matrices. J. Symb. Comput. **33**(1), 13–29 (2002). https://doi.org/10.1006/jsco.2001.0462

5. Corless, R.M., Gianni, P.M., Trager, B.M., Watt, S.M.: The singular value decomposition for polynomial systems. In: Proceedings of the 1995 International Symposium on Symbolic and Algebraic Computation. pp. 195–207. ACM (1995). https://doi.org/10.1145/220346.220371

6. Corless, R.M., Watt, S.M., Zhi, L.: QR factoring to compute the GCD of univariate approximate polynomials. IEEE Trans. Signal Process. **52**(12), 3394–3402 (2004). https://doi.org/10.1109/TSP.2004.837413

7. Demmel, J.W.: Applied numerical linear algebra. Society for Industrial and Applied Mathematics (1997). https://doi.org/10.1137/1.9781611971446

8. Diaz-Toca, G.M., Gonzalez-Vega, L.: Barnett's theorems about the greatest common divisor of several univariate polynomials through bezout-like matrices. J. Symb. Comput. **34**(1), 59–81 (2002). https://doi.org/10.1006/jsco.2002.0542

9. Emiris, I.Z., Galligo, A., Lombardi, H.: Certified approximate univariate GCDs. J. Pure Appl. Algebra **117**(118), 229–251 (1997). https://doi.org/10.1016/S0022-4049(97)00013-3

10. Kaltofen, E., Yang, Z., Zhi, L.: Approximate greatest common divisors of several polynomials with linearly constrained coefficients and singular polynomials. In: Proceedings of the 2006 International Symposium on Symbolic and Algebraic Computation. pp. 169–176. ACM, New York, NY, USA (2006). https://doi.org/10.1145/1145768.1145799

11. Kaltofen, E., Yang, Z., Zhi, L.: Structured low rank approximation of a Sylvester matrix. In: Wang, D., Zhi, L. (eds.) Symbolic-Numeric Computation, pp. 69–83. Trends in Mathematics, Birkhäuser (2007), https://doi.org/10.1007/978-3-7643-7984-1_5

12. Karmarkar, N.K., Lakshman, Y.N.: On approximate GCDs of univariate polynomials. J. Symb. Comput. **26**(6), 653–666 (1998). https://doi.org/10.1006/jsco.1998.0232

13. Matsumoto, M., Nishimura, T.: Mersenne twister: a 623-dimensionally equidistributed uniform pseudo-random number generator. ACM Trans. Model. Comput. Simul. **8**(1), 3–30 (1998). https://doi.org/10.1145/272991.272995

14. Nakatsukasa, Y., Noferini, V., Townsend, A.: Computing the common zeros of two bivariate functions via Bézout resultants. Numer. Math. **129**(1), 181–209 (2014). https://doi.org/10.1007/s00211-014-0635-z

15. Pan, V.Y.: Computation of approximate polynomial GCDs and an extension. Inf. and Comput. **167**(2), 71–85 (2001). https://doi.org/10.1006/inco.2001.3032

16. Rosen, J.B.: The gradient projection method for nonlinear programming. II. Nonlinear constraints. J. Soc. Indust. Appl. Math. **9**, 514–532 (1961). https://doi.org/10.1137/0109044

17. Sasaki, T., Noda, M.T.: Approximate square-free decomposition and root-finding of ill-conditioned algebraic equations. J. Inf. Process. **12**(2), 159–168 (1989)

18. Schönhage, A.: Quasi-gcd computations. J. Complexity **1**(1), 118–137 (1985). https://doi.org/10.1016/0885-064X(85)90024-X

19. Schost, É., Spaenlehauer, P.-J.: A quadratically convergent algorithm for structured low-rank approximation. Found. Comput. Math. **16**(2), 457–492 (2015). https://doi.org/10.1007/s10208-015-9256-x
20. Sun, D., Zhi, L.: Structured low rank approximation of a bezout matrix. Math. Comput. Sci. **1**(2), 427–437 (2007). https://doi.org/10.1007/s11786-007-0014-6
21. Tanabe, K.: A geometric method in nonlinear programming. J. Optimiz. Theory Appl. **30**(2), 181–210 (1980). https://doi.org/10.1007/BF00934495
22. Terui, A.: An iterative method for calculating approximate GCD of univariate polynomials. In: Proceedings of the 2009 International Symposium on Symbolic and Algebraic Computation - ISSAC 2009. pp. 351–358. ACM Press, New York, New York, USA (2009). https://doi.org/10.1145/1576702.1576750
23. Terui, A.: GPGCD: an iterative method for calculating approximate gcd of univariate polynomials. Theor. Comput. Sci. **479**, 127–149 (2013). https://doi.org/10.1016/j.tcs.2012.10.023
24. Zarowski, C.J., Ma, X., Fairman, F.W.: QR-factorization method for computing the greatest common divisor of polynomials with inexact coefficients. IEEE Trans. Signal Process. **48**(11), 3042–3051 (2000). https://doi.org/10.1109/78.875462
25. Zeng, Z.: The numerical greatest common divisor of univariate polynomials. In: Gurvits, L., Pébay, P., Rojas, J.M., Thompson, D. (eds.) Randomization, Relaxation, and Complexity in Polynomial Equation Solving, Contemporary Mathematics, vol. 556, pp. 187–217. AMS (2011). https://doi.org/10.1090/conm/556
26. Zhi, L.: Displacement structure in computing approximate GCD of univariate polynomials. In: Computer mathematics: Proceedings Six Asian Symposium on Computer Mathematics (ASCM 2003), Lecture Notes Series on Computing, vol. 10, pp. 288–298. World Scientific Publishing, River Edge, NJ (2003), https://doi.org/10.1142/9789812704436_0024

On Parametric Linear System Solving

Robert M. Corless[1], Mark Giesbrecht[2], Leili Rafiee Sevyeri[1(✉)],
and B. David Saunders[3]

[1] Ontario Research Centre for Computer Algebra, School of Mathematical and
Statistical Sciences, University of Western Ontario, London, ON, Canada
{rcorless,lrafiees}@uwo.ca
[2] David R. Cheriton School of Computer Science, University of Waterloo,
Waterloo, ON, Canada
mwg@uwaterloo.ca
[3] University of Delaware, Newark, Delaware, USA
saunders@udel.edu

Abstract. Parametric linear systems are linear systems of equations in which some symbolic parameters, that is, symbols that are not considered to be candidates for elimination or solution in the course of analyzing the problem, appear in the coefficients of the system. In this work we assume that the symbolic parameters appear polynomially in the coefficients and that the only variables to be solved for are those of the linear system. The consistency of the system and expression of the solutions may vary depending on the values of the parameters. It is well-known that it is possible to specify a covering set of regimes, each of which is a semi-algebraic condition on the parameters together with a solution description valid under that condition.

We provide a method of solution that requires time polynomial in the matrix dimension and the degrees of the polynomials when there are up to three parameters. In previous methods the number of regimes needed is exponential in the system dimension and polynomial degree of the parameters. Our approach exploits the Hermite and Smith normal forms that may be computed when the system coefficient domain is mapped to the univariate polynomial domain over suitably constructed fields. Our approach effectively identifies *intrinsic singularities* and *ramification points* where the algebraic and geometric structure of the matrix changes. Parametric eigenvalue problems can be addressed as well: simply treat λ as a parameter in addition to those in $\mathbf{A}$ and solve the parametric system $(\lambda \mathbf{I} - \mathbf{A})\mathbf{u} = 0$. The algebraic conditions on λ required for a nontrivial nullspace define the eigenvalues. We do not directly address the problem of computing the Jordan form, but our approach allows the construction of the algebraic and geometric eigenvalue multiplicities revealed by the Frobenius form, which is a key step in the construction of the Jordan form of a matrix.

Keywords: Hermite form · Smith form · Frobenius form · Parametric linear systems.

© Springer Nature Switzerland AG 2020
F. Boulier et al. (Eds.): CASC 2020, LNCS 12291, pp. 188–205, 2020.
https://doi.org/10.1007/978-3-030-60026-6_11

1   Introduction

Speaking in simple generalities, we say that symbolic computation is concerned with mathematical equations that contain *symbols*; symbols are used both for *variables*, which are typically to be solved for, and *parameters*, which are typically carried through and appear in the *solutions*, which are then interpreted as formulae: that is, objects that can be further studied, perhaps by varying the parameters. One prominent early researcher said that the difference between symbolic and numeric computation was merely a matter of *when* numerical values were inserted into the parameters: before the computation meant you were going to do things numerically, and after the computation meant you had done symbolic computation. The words "parameters" and "variables" are therefore not precisely descriptive, and can often be used interchangeably. Indeed as a matter of practice, polynomial equations can often be taken to have one subset of its symbols taken as variables rather than any other subset in quite strategic fashion: it may be better to solve for x as a function of y than to solve for y as a function of x.

In this paper we are concerned with systems of equations containing several symbols, some of which we take to be variables, and all the rest as parameters. More, we restrict our attention to problems in which the *variables* appear only linearly. Parameters are allowed to appear polynomially, of whatever degree.

Parametric linear systems (PLS) arise in many contexts, for instance in the analysis of the stability of equilibria in dynamical systems models such as occur in mathematical biology and other areas. Understanding the different potential kinds of dynamical behavior can be important for model selection as well as analysis. Another important area of interest is the role of parametric linear systems in dealing with the stability of the equilibria of parametric autonomous system of ordinary differential equations (see [25] and [11]). One particularly famous example is the Lotka-Volterra system which arises naturally from predator-prey equations. See also [24] and [23]. Other examples of the use of parametric linear system from science and engineering includes their application in computing the characteristic solutions for differential equations [8], dealing with colored Petri nets [13] and in operations research and engineering [9,17,21,31]. Some problems in robotics [2] and certain modelling problems in mathematical biology, see e.g. [29], also can benefit from the ability to effectively solve PLS.

After some discussion of prior comprehensive solving work in Sect. 2, we proceed with formal problem and solution definitions for parametric linear systems (PLS) in Sect. 3. Our primary tool for solving these is by way of *comprehensive triangular Smith normal form (CTSNF)*, which is introduced in Sect. 4. The following section reduces PLS to CTSNF and Sect. 6 describes the solution of CTSNF problems for the case of up to three parameters.

An application that seems at first to be of only theoretical interest is the computation of the *matrix logarithm*, or indeed any of several other matrix functions such as matrix square root. We briefly discuss this example in more detail with a pair of small matrices in Sect. 7.2. We also give other examples in Sect. 7.

2   Prior Work

Interest in computation of the solution of PLS dates back to the beginning of symbolic computation. For instance, one of the first things users have requested of computer algebra systems is the explicit form of the inverse of a matrix containing only symbolic entries[1]: the user is then typically quite dissatisfied at the complexity of the answer if the dimension is greater than, say, 3. Of course, the determinant itself, which must appear in such an answer, has a factorial number of terms in it, and thus growth in the size of the answer must be more than exponential. Therefore the complexity of any algorithm to solve PLS must be at least exponential in the number of parameters.

An interesting pair of papers addressing the case of only one parameter is [1] and [15]. These papers assume full rank of the linear system—and thus compute the "generic" case when in fact there are isolated values of the parameter for which the rank drops—and use rational interpolation of the numerical solutions of specialized linear systems to recover this generic solution.

Many authors have sought comprehensive solutions—by which is meant complete coverage of all parametric regimes—through various means. One of the first serious methods was the matrix-minor based approach of William Sit [25], which enables practical solution of many problems of interest. Recently, the problem of computing the Jordan form of a parametric matrix once the Frobenius form is known has been attacked by using Regular Chains [4] and this has been moderately successful in practice. Simple methods and heuristics for linear systems containing parameters continue to generate interest, even when Regular Chains are used, such as in [3].

Other authors such as [16,18–20] and [30] have tackled the even more difficult problem of computing the comprehensive solution of systems of *polynomial* equations containing parameters, and of course their methods can be applied to the linear equations being considered here.

By restricting our attention in this paper to linear problems and to those of three parameters or fewer we are able to guarantee better worst case performance (polynomially many solution regimes) and hope to provide better efficiency in many instances than is possible using those general-purpose approaches.

3   Definitions and Notation

Let F be a field and $Y = (y_1, \ldots, y_s)$ a list of parameters. Then $F[Y]$ is the ring of polynomials and $F(Y)$ is the field of rational functions in Y. For each tuple $a = (a_1, \ldots, a_s)$ in F^s, evaluation at a is a mapping $F[Y] \to F$. We will extend this mapping componentwise to polynomials, vectors, matrices, and sets thereof over $F[Y]$. We will use the Householder convention, typesetting matrices in upper case bold, e.g. $\mathbf{A}$, and lower case bold for vectors, e.g. $\mathbf{b}$.

---

[1] This is merely an anecdote, but one of the present authors attests that this really has happened.

For the most part, for such objects over $F[Y]$, we know Y from context and write $\mathbf{A}$ rather than $\mathbf{A}(Y)$, but write $\mathbf{A}(a)$ for the evaluation at $Y = a$.

For a set of polynomials, S, we will denote by $V(S)$ the variety of the ideal generated by S. This is the set of tuples a such that $f(a) = \{0\}$, for all $f \in S$. We will be concerned with pairs N, Z of polynomial sets, $N, Z \subset F[Y]$, defining a semialgebraic set in F^s consisting of those tuples a that evaluate to nonzero on N and to zero on Z. By a slight abuse of notation, we call this semialgebraic set $V(N, Z) = V(Z) \setminus V(N)$. Our inputs are polynomial in the parameters but the output coefficients in general are rational functions. The evaluation mapping extends partially to $F(Y)$: For a rational function $n(Y)/d(Y)$ in lowest terms (n and d relatively prime), the image $n(a)/d(a)$ is well defined so long as $d(a) \neq 0$.

Definition 3.1. *The data for a* **parametric linear system (PLS) problem** *is matrix $\mathbf{A}$ and right hand side vector $\mathbf{b}$ over $F[Y]$, together with a semialgebraic constraint, $V(N, Z)$, with $N, Z \subset F[Y]$. Only of interest are those parameter value tuples in $V(N, Z)$, i.e., on which the polynomials in N are nonzero and the polynomials in Z are zero.*

For the PLS problem $(\mathbf{A}, \mathbf{b}, N, Z)$, a **solution regime** *is a tuple $(\mathbf{u}, \mathbf{B}, N', Z')$ with coefficients of $\mathbf{u}$ and $\mathbf{B}$ in $F(Y)$, such that, for all $a \in V(N', Z')$, $\mathbf{u}(a)$ is a solution vector and $\mathbf{B}(a)$ is a matrix whose columns form a nullspace basis for $\mathbf{A}(a)$.*

A **PLS solution** *is a set of solution regimes that covers $V(N, Z)$, which means, for PLS solution $\{(\mathbf{u}_i, \mathbf{B}_i, N_i, Z_i) \mid i \in 1, \ldots, k\}$, every parameter value assignment that satisfies the problem semialgebraic constraint N, Z also satisfies at least one regime semialgebraic constraint N_i, Z_i. In other words $V(N, Z) \subset \cup_{i=1}^{k} V(N_i, Z_i)$.*

We call entries that *must* occur in any Z in the solution an *intrinsic* restriction, or *singularity*. We call the differing sets $V(N_i, Z_i)$ that may occur in covers of $V(N, Z)$ the *ramifications* of the cover.

We next give an example that illustrates the PLS definition and also sketches the prior approach to PLS given by William Sit in [25]. If, for $\mathbf{M}$ of size $r \times r$, $\mathbf{A}$ is $\begin{bmatrix} \mathbf{M} & \mathbf{B} \\ \mathbf{C} & \mathbf{D} \end{bmatrix}$, and conformally $\mathbf{b} = \begin{bmatrix} \mathbf{c} & \mathbf{d} \end{bmatrix}^T$, then a solution $\mathbf{u} = \begin{bmatrix} \mathbf{v} & \mathbf{w} \end{bmatrix}^T$ satisfies

$$\mathbf{M}\mathbf{v} + \mathbf{B}\mathbf{w} = \mathbf{c} \tag{3.1}$$

and

$$\mathbf{C}\mathbf{v} + \mathbf{D}\mathbf{w} = \mathbf{d}. \tag{3.2}$$

Under the condition that $\det(\mathbf{M})$ is nonzero and all larger minors of $\mathbf{A}$ are zero, equation (3.1) can be solved with specific solution $\mathbf{w} = 0$ and $\mathbf{v} = \mathbf{M}^{-1}\mathbf{c}$. Provided the system is consistent (equation (3.2) holds), we have the regime

$$\left(\begin{bmatrix} \mathbf{v} & \mathbf{w} \end{bmatrix}^T, \begin{bmatrix} -\mathbf{M}^{-1}\mathbf{B} \\ \mathbf{I} \end{bmatrix}, N, Z\right),$$

where $N = \{\det(\mathbf{M})\}$ and $Z = \{$all $(i+1) \times (i+1)$ minors of $\mathbf{A}\})$. Call solution regimes of this type *minor defined* regimes.

Since an $n \times n$ matrix has $\sum_{k=0}^{n} \binom{n}{k}^2 = \binom{2n}{n}$ minors, there are exponentially many minor defined regimes. However, some of these regimes may not be solutions due to inconsistency or it may be possible to combine several regimes into one. For instance if $\det(\mathbf{M})$ is a constant, and $\mathbf{b} = 0$, then all rank r solutions are covered by this one regime. Sit [25] has made a thorough study of minor defined regimes and their simplifications.

Another approach is to base solution regimes on the pivot choices in an LU decomposition. The simplest thing to do is to leave it to the user, although one has to also inform the user through a proviso when this might be necessary [6]. That is, provide the generic answer, but also provide a description of the set N. A more sophisticated approach is developed in [3,4] using the theory of regular chains and its implementation in Maple [18] to manage the algebraic conditions. For example a given matrix entry may be used as a pivot, with validity dependent on adding the polynomial to the non-zero part, N, of the semialgebraic set. For a comprehensive solution the case that entry is zero must also be pursued. In the worst case, this leads to a tree of zero/nonzero choices of depth n and branching factor n.

4 Triangular Smith Forms and Degree Bounds

In this paper we take a different approach, with the solution regimes arising from Hermite normal forms, of which triangular Smith forms are a special case. We give a system of solution regimes of polynomial size in the matrix dimension, n, and polynomial degree, d. Each regime is computed in polynomial time and the regime count is exponential only in the number of parameters. To use Hermite forms we will need to work over a principal ideal domain such as, for parameters x, y, $F(y)[x]$. We will restrict our input matrix to be polynomial in the parameters. This first lemma shows it is not a severe constraint.

Lemma 4.1. *Let $(\mathbf{A}, \mathbf{b}, N, Z)$ be a well defined PLS over field $F(Y)$, for parameter set Y, with $\mathbf{A} \in F(Y)^{m \times n}$ and $\mathbf{b} \in F(Y)^{m}$ with numerator and denominator degrees bounded by d in each parameter of Y. Well defined means that denominators of $\mathbf{A}, \mathbf{B}$ are in N. The problem is equivalent (same solutions) to one in which the entries of the matrix and vector are polynomial in the parameters Y, the dimension is the same, and the degrees are bounded by nd.*

Proof. Because the PLS is well defined, it is specified by N that all denominator factors of $\mathbf{A}(a), \mathbf{b}(a)$ are nonzero for $a \in V(N, Z)$. Let $\mathbf{L}$ be a diagonal matrix with the i-th diagonal entry being the least common multiple (lcm) of the denominators in row i of $\mathbf{A}, \mathbf{b}$. These lcms also evaluate to nonzero on $V(N, Z)$. It follows that $L(a)\mathbf{A}(a)\mathbf{u}(a) = \mathbf{L}(a)\mathbf{b}(a)$ if and only if $\mathbf{A}(a)\mathbf{u}(a) = \mathbf{b}(a)$. Thus the PLS $(\mathbf{LA}, \mathbf{Lb}, V(N, Z))$ is equivalent and its matrix and vector have polynomial entries of degrees bounded by nd.

We will reduce PLS to triangular Smith normal form computations. The rest of this section concerns computation of triangular Smith normal form and bounds for the degrees of the form and its unimodular cofactor.

Definition 4.2. *Given field K and variable x, a matrix $\mathbf{H}$ over $K[x]$ is in (reduced)* **Hermite normal form** *if it is upper triangular, its diagonal entries are monic, and, for each column in which the diagonal entry is nonzero, the off-diagonal entries are of lower degree than the diagonal entry. If each diagonal entry of $\mathbf{H}$ exactly divides all those below and to the right, then $\mathbf{H}$ is column equivalent to a diagonal matrix with the same diagonal entries (its Smith normal form). An equivalent condition is that, for each i, the greatest common divisor of the $i \times i$ minors in the leading i columns equals the greatest common divisor of all $i \times i$ minors. Following Storjohann [27, Section 8, Definition 8.2] we call such a Hermite normal form a* **triangular Smith normal form**. *It will be the central tool in our PLS solution.*

For notational simplicity, we have left out the possibility of echelon structure in a Hermite normal form. We will talk of Hermite normal forms only for matrices having leading columns independent up to the rank of the matrix. Every such matrix over $K[x]$ is row equivalent to a unique matrix in Hermite form as defined above. For given $\mathbf{A}$ we have $\mathbf{UA} = \mathbf{H}$, with $\mathbf{U}$ unimodular, i.e. $\det(\mathbf{U}) \in K^*$, and $\mathbf{H}$ in Hermite form. If $\mathbf{A}$ is nonsingular, the unimodular cofactor $\mathbf{U}$ is unique and has determinant $1/c$, where c is the leading coefficient of $\det(\mathbf{A})$. This follows since $\det(\mathbf{U})\det(\mathbf{A}) = \det(\mathbf{H})$, which is monic.

The next definition and lemma concern assurance that Hermite form computation will yield a triangular Smith form.

Definition 4.3. *Call a matrix* **nice** *if its Hermite form is a triangular Smith form (each diagonal entry exactly divides those below and to the right). In particular, a nice matrix has leading columns independent up to the rank.*

There is always a column transform (unimodular matrix $\mathbf{R}$ applied from the right) such that $\mathbf{AR}$ is nice. The following fact, proven in [14] shows that a random transform over F suffices with high probability.

Fact 4.4. *Let $\mathbf{A}$ be a $m \times n$ matrix over $K[x]$ of degree in x at most d. Let $\mathbf{R}$ be a unit lower triangular matrix with below diagonal elements chosen from subset S of K uniformly at random. Then $\mathbf{AR}$ is nice over $K[x]$ with probability at least $1 - 4n^3d/|S|$.*

Note that $\deg_x(\mathbf{AR}) = \deg_x(\mathbf{A})$ and, for $K = F(y), \mathbf{A} \in F[y,x]^{m \times n}$ and $S \subset F$ we also have $\deg_y(\mathbf{AR}) = \deg_y(\mathbf{A})$.

We continue with analysis of degree bounds for Hermite forms of matrices, particularly degree bounds for triangular Smith forms of nice matrices. The first result needed is the following fact from [10]. Through the remainder of this paper we will employ "soft O" notation, where, for functions $f, g \in \mathbb{R}^k \to \mathbb{R}$ we write $f = O^\sim(g)$ if and only if $f = O(g \cdot \log^c |g|)$ for some constant $c > 0$.

Fact 4.5. *Let F be a field, x, y parameters, and let $\mathbf{A}$ be in $F[y,x]^{n \times n}$, nonsingular, with $\deg_x(\mathbf{A}) \leq d$, $\deg_y(\mathbf{A}) \leq e$. Over $F(y)[x]$, let $\mathbf{H}$ the unique Hermite form row equivalent to $\mathbf{A}$ and $\mathbf{U}$ be the unique unimodular cofactor such that $\mathbf{UA} = \mathbf{H}$. The coefficients of the entries of $\mathbf{H}, \mathbf{U}$ are rational functions of y.*

Let Δ be the least common multiple of the denominators of the coefficients in $\mathbf{H}, \mathbf{U}$, as expressed in lowest terms.

(a) $\deg_x(\mathbf{U}) \le (n-1)d$ and $\deg_x(\mathbf{H}) \le nd$.
(b) $\deg_y(\text{num}(\mathbf{H})), \deg_y(\text{num}(\mathbf{U})) \le n^2 de$ (bounds both numerator and denominator degrees).
(c) $\deg_y(\Delta) \le n^2 de$.
(d) $\mathbf{H}$ and $\mathbf{U}$ can be computed in polynomial time: deterministically in $O^\sim(n^9 d^4 e)$ time and Las Vegas probabilistically (never returns an incorrect result) in $O^\sim(n^7 d^3 e)$ expected time.

Proof. This is [10, Summary Theorem]. The situation there is more abstract, more involved. We offer this tip to the reader: their $\partial, z, \sigma, \delta$ correspond respectively to our x, y, identity, identity.

Item (c) is not stated explicitly in a theorem of [10] but is evident from the proofs of Theorems 5.2 and 5.6 there. The common denominator is the determinant of a matrix over $K[z]$ of dimension $n^2 d$ and with entries of degree in z at most e.

We will generalize this fact to nonsingular and non-square matrices in Theorem 4.6. In that case the unimodular cofactor, $\mathbf{U}$, is not unique and may have arbitrarily large degree entries. The following algorithm is designed to produce a $\mathbf{U}$ with bounded degrees.

Algorithm 1. U, H = HermiteForm(A)

Require: Nice matrix $\mathbf{A} \in F[y, x]^{m \times n}$, for field F and parameters x, y.
Ensure: For $K = F(y)$, Unimodular $\mathbf{U} \in K[x]^{m \times m}$ and $\mathbf{H} \in K[x]^{m \times n}$ in triangular Smith form such that $\mathbf{UA} = \mathbf{H}$. The point of the specific method given here is to be able, in Theorem 4.6, to bound $\deg_x(\mathbf{U}, \mathbf{H})$ and $\deg_y(\mathbf{U}, \mathbf{H})$ (numerators and denominators).

1: Compute $r = \text{rank}(\mathbf{A})$ and nonsingular $\mathbf{U}_0 \in K^{m \times m}$ such that $\mathbf{A} = \mathbf{U}_0 \mathbf{A}$ has nonsingular leading $r \times r$ minor. Because $\mathbf{A}$ is nice the first r columns are independent and such $\mathbf{U}_0$ exists. $\mathbf{U}_0$ could be a permutation found via Gaussian elimination, say, or a random unit upper triangular matrix. In the random case, failure to achieve nonsingular leading minor becomes evident in the next step, so that the randomization is Las Vegas.

2: Let $\mathbf{U}_0 \mathbf{A} = \begin{bmatrix} \mathbf{A}_1 & \mathbf{A}_2 \\ \mathbf{A}_3 & \mathbf{A}_4 \end{bmatrix}$ and $\mathbf{B} = \begin{bmatrix} \mathbf{A}_1 & \mathbf{0}_{r \times m - r} \\ \mathbf{A}_3 & \mathbf{I}_{m-r} \end{bmatrix}$. $\mathbf{B}$ is nonsingular. Compute its unique unimodular cofactor $\mathbf{U}_1$ and Hermite form $\mathbf{T} = \mathbf{U}_1 \mathbf{B} = \begin{bmatrix} \mathbf{H}_1 & * \\ 0 & * \end{bmatrix}$.

If $\mathbf{H}_1$ is in triangular Smith form, let $\mathbf{H} = \mathbf{U}_1 \mathbf{U}_0 \mathbf{A} = \begin{bmatrix} \mathbf{H}_1 & \mathbf{H}_2 \\ 0 & 0 \end{bmatrix}$.

Let $\mathbf{U} = \mathbf{U}_1 \mathbf{U}_0$ and return $\mathbf{U}, \mathbf{H}$.

Otherwise go back to step 1 and choose a better $\mathbf{U}_0$. With high probability this repetition will not be needed; probability of success increases with each iteration.

Theorem 4.6. *Let F be a field, x, y parameters, and let $\mathbf{A}$ be in $F[y, x]^{m \times n}$ of rank r, $\deg_x(\mathbf{A}) \leq d$, and $\deg_y(\mathbf{A}) \leq e$. Then, for the triangular Smith form form $\mathbf{UAR} = H$ computed as $\mathbf{U}, \mathbf{H} = \text{HermiteForm}(\mathbf{AR})$, we have*

(a) Algorithm HermiteForm *is (Las Vegas) correct and runs in expected time $O(m^7 d^3 e)$;*

(b) $\deg_x(\mathbf{U}, \mathbf{H}) \leq md$;

(c) $\deg_y(\mathbf{U}, \mathbf{H}) = O^\sim(m^2 de)$.

Proof. Let $\mathbf{R}$ be as in Fact 4.4 with $K = F(y)$ and $S \subset F$. If the field F is small, an extension field can be used to provide large enough S.

We apply HermiteForm to $\mathbf{AR}$ to obtain $\mathbf{U}, \mathbf{H}$, and use the notation of the algorithm in this proof. We see by construction that $\mathbf{B}$ is nonsingular, from which it follows that $\mathbf{U}_1$ and $\mathbf{T}$ are uniquely determined. $\mathbf{B}$ is nice because $\mathbf{A}$ is nice and all j-minors of $\mathbf{B}$ for $j > r$ are either zero or equal to $\det \mathbf{A}_1$. It follows that the leading r columns of $\mathbf{H}$ must be those of $\mathbf{T}$. The lower left $(m - r) \times (n - r)$ block of $\mathbf{H}$ must be zero because $\text{rank}(\mathbf{H}) = \text{rank}(\mathbf{A})$. The leading r rows are independent, and any nontrivial linear combination of those rows would be nonzero in the lower left block. Then $\mathbf{H}$ is in triangular Smith form and left equivalent to $\mathbf{A}$ as required. The runtime is dominated by computation of $\mathbf{U}_1$ and $\mathbf{T}$ for $\mathbf{B}$, so Fact 4.5 provides the bound in (a).

For the degree in x, applying Fact 4.5, we have $\deg_x(\mathbf{U}_1) \leq (m - 1)d$. Noting that $\mathbf{U}_0$ has degree zero, we have $\deg_x(\mathbf{U}) = \deg_x(\mathbf{U}_1)$ and $\deg_x(\mathbf{H}) = \deg_x(\mathbf{U}) + \deg_x(\mathbf{A}) \leq (m - 1)d + d = md$.

For the degree in y, note first that the bounds d, e for degrees in $\mathbf{A}$ apply as well to $\mathbf{B}$. We have, by Fact 4.5, that $\deg_y(\text{num}(\mathbf{U}_1)) = O^\sim(m^2 de)$ and the same bound for $\deg_y(\text{den}(\mathbf{U}_1))$. For $\mathbf{H}$, note that $\text{num}(\mathbf{H})/\text{den}(\mathbf{H}) = \text{num}(\mathbf{U})\mathbf{A}/\text{den}(\mathbf{U})$ so that and $\deg_y(\text{den}(\mathbf{H})) \leq \deg_y(\mathbf{U}) = O^\sim(m^2 de)$, and $\deg_y(\text{num}(\mathbf{H})) \leq \deg_y(\text{num}(\mathbf{U})A) = O^\sim(m^2 de) + e = O^\sim(m^2 de)$.   □

5   Reduction of PLS to Triangular Smith Forms

In this section we define the Comprehensive Triangular Smith Normal form problem and solution and show that PLS can be reduced to it. The next section addresses the solution of CTSNF itself.

Definition 5.1. *For field F, parameters $Y = (y_1, \ldots, y_s)$, F_Y is a **parameterized extension** of F if $F_Y = F_s$, the top of a tower of extensions $F_0 = F, F_1, \ldots, F_s$ where, for $i \in 1, \ldots, s$, each F_i is either $F_{i-1}(y_i)$ (rational functions) or $F_{i-1}[y_i]/\langle f_i \rangle$, for f_i irreducible in y_i over F_{i-1} (algebraic extension). When a solution regime to a PLS or CTSNF problem is over a parameterized extension F_Y, the irreducible polynomials involved in defining the extension tower for F_Y will be in the constraint set Z of polynomials that must evaluate to zero.*

*A comprehensive triangular Smith normal form problem (***CTSNF** **prob-****
***lem)** is a triple $(\mathbf{A}, N, Z)$ of a matrix $\mathbf{A}$ over $F[Y, x]$ and polynomial sets
$N, Z \subset F[Y, x]$, so that $V(N, Z)$ constrains the range of desired parameter values
as in the PLS problem.*

For CTSNF problem $(\mathbf{A}, N, Z)$ over $F[Y, x]$, a **triangular Smith regime**
*is of the form $(\mathbf{U}, \mathbf{H}, \mathbf{R}, N', Z')$, with $\mathbf{U}, \mathbf{H}$ over $F_Y[x]$, where F_Y is a parame-
terized extension of F and any polynomials defining algebraic extensions in the
tower are in Z', such that on all $a \in V(N', Z')$, $H(a)$ is in triangular Smith
form over $F(a)[x]$, $\mathbf{U}(a)$ is unimodular in x, $\mathbf{R}$ is nonsingular over F, and
$\mathbf{U}(a)\mathbf{A}(a)\mathbf{R} = \mathbf{H}(a)$.*

A **CTSNF solution** *is a list $\{(\mathbf{U}_i, \mathbf{H}_i, \mathbf{R}_i, N_i, Z_i) | i \in 1, \ldots, k\}$, of* **trian-
gular Smith regimes** *that cover $V(N, Z)$, which is to say*

$$V(N, Z) \subset \cup\{V(N_i, Z_i) \mid i \in 1, \ldots, k\}.$$

The goal in this section is to reduce the PLS problem to the CTSNF problem.
The first step is to show it suffices to consider PLS with a matrix already in
triangular Smith form. The second step is to show each CTSNF solution regime
generates a set of PLS solution regimes.

Lemma 5.2. *Given a parameterized field F_Y and matrix $\mathbf{A}$ over $F[Y, x]$, let $\mathbf{H}$
be a triangular Smith form of $\mathbf{A}$ over $F_Y[x]$, with $\mathbf{U}$ unimodular over $F_Y[x]$,
and $\mathbf{R}$ nonsingular over F such that $\mathbf{U}\mathbf{A}\mathbf{R} = \mathbf{H}$. PLS problem $(\mathbf{A}, \mathbf{b}, N, Z)$ over
$F[Y, x]$ has solution regimes $(\mathbf{u}_1, \mathbf{B}_1, N_1, Z_1), \ldots, (\mathbf{u}_s, \mathbf{B}_s, N_s, Z_s)$ if and only if
PLS problem $(\mathbf{H}, \mathbf{Ub}, N, Z)$ has solution regimes $(\mathbf{R}^{-1}\mathbf{u}_1, \mathbf{R}^{-1}\mathbf{B}_1, N_1, Z_1), \ldots,
(\mathbf{R}^{-1}\mathbf{u}_s, \mathbf{R}^{-1}\mathbf{B}_s, N_s, Z_s)$.*

Proof. Under evaluation at any $a \in V(N, Z)$, $\mathbf{U}(a)$ is unimodular and $\mathbf{R}$ is
unchanged and nonsingular. Thus the following are equivalent.

1. $\mathbf{A}(a)\mathbf{u}(a) = \mathbf{b}(a)$.
2. $\mathbf{U}(a)\mathbf{A}(a)\mathbf{u}(a) = \mathbf{U}(a)\mathbf{b}(a)$.
3. $(\mathbf{U}(a)\mathbf{A}(a)\mathbf{R})(\mathbf{R}^{-1}\mathbf{u}(a)) = \mathbf{U}(a)\mathbf{b}(a)$.

$\square$

Then we have the following algorithm to solve a PLS with the matrix already
in triangular Smith form. For simplicity we assume a square matrix, the rectangular case being a straightforward extension.

Algorithm 2. `TriangularSmithPLS`

Require: PLS problem $(\mathbf{H}, \mathbf{b}, N, Z)$, with $\mathbf{H} \in K_Y[x]^{n \times n}$ and $\mathbf{b} \in K_Y[x]^n$, where F_Y
    is a parameterized extension for parameter list Y and x is an additional parameter,
    with $N, Z \subset F[Y]$ and $\mathbf{H}$ in triangular Smith form.
Ensure: S, the corresponding PLS solution (a list of regimes).
  1: For any polynomial $s(x)$ let $\mathrm{sqfr}(s)$ denote the square-free part. Let s_i denote the
     i-th diagonal entry of $\mathbf{H}$, and define $s_0 = 1, s_{n+1} = 0$. Then, for $i \in 0, \ldots, n$, define
     $f_i = \mathrm{sqfr}(s_{i+1})/\mathrm{sqfr}(s_i)$. Let $\mathcal{I}$ be the set of indices such that f_i has positive degree
     or is zero.
  2: For each $r \in \mathcal{I}$ include in the output S the regime $R = (\mathbf{u}, \mathbf{B}, V)$, where $\mathbf{u} = (\mathbf{H}_r^{-1}\mathbf{b}_r, 0_{n-r})$, with $\mathbf{H}_r$ the leading $r \times r$ submatrix of $\mathbf{H}$ and $\mathbf{b}_r = (b_1, \ldots, b_r)$
     and $\mathbf{B} = (\mathbf{e}_{r+1}, \ldots, \mathbf{e}_n)$. Here $\mathbf{e}_i$ denotes the i-th column of the identity matrix.
  3: Return S.

Lemma 5.3. *Algorithm* TriangularSmithPLS *is correct and generates at most*
$\sqrt{2d}$ *regimes, where* $d = \deg(\det(\mathbf{H}))$.

Proof. Note that, for each row k, the diagonal entry s_k divides all other entries
in the row. Then $\mathbf{H}$ has rank r just in case $s_r \neq 0$ and $s_{r+1} = 0$, i.e., in the cases
determined in step 1 of the algorithm. The addition of s_r to N and f_r to Z ensures
rank r and invertibility of $\mathbf{H}_r$. For all evaluation points $a \in V(N, Z)$ satisfying
those two additional conditions, the last $n - r$ rows of $\mathbf{H}$ are zero. Hence the
nullspace $\mathbf{B}$ is correctly the last $n - r$ columns of $\mathbf{I}_n$. For such evaluation points,
the system will be consistent if and only if the corresponding right hand side
entries are zero, hence the addition of $b_{r+1}, \ldots, b_n$ to Z. $\qquad\square$

Algorithm 3. `PLSviaCTSNF`

Require: A PLS problem $(\mathbf{A}, \mathbf{b}, N, Z)$ over $F[Y, x]$, for parameter list Y and additional
    parameter x.
Ensure: A corresponding PLS solution $S = ((\mathbf{u}_i, \mathbf{B}_i, N_i, Z_i) \mid i \in 1, \ldots, s)$.
  1: Over the ring $F(Y)[x]$, Let T solve the CTSNF problem $(\mathbf{A}, N, Z)$. T is a set of
     triangular Smith regimes of form $(\mathbf{U}, \mathbf{H}, R, N', Z')$. Let $S = \emptyset$.
  2: For each Hermite regime $(\mathbf{U}, \mathbf{H}, R, N', Z')$ in T, using algorithm
     `TriangularSmithPLS`, solve the PLS problem $(\mathbf{H}, \mathbf{Ub}, N', Z')$. Adjoin to S
     the solution regimes, adjusted by factor $\mathbf{R}^{-1}$ as in Lemma 5.2.
  3: Return S.

Theorem 5.4. *Algorithm 3 is correct.*

Proof. For every parameter evaluation $a \in V(N, Z)$ at least one triangular Smith
regime of T in step 2 is valid. Then, by Lemmas 5.3 and 5.2, step 3 produces a
PLS regime covering a. $\qquad\square$

6 Solving Comprehensive Triangular Smith Normal Form

In view of the reductions of the preceding section, to solve a parametric linear system it remains only to solve a comprehensive triangular Smith form problem. This is difficult in general but we give a method to give a comprehensive solution with polynomially many regimes in the bivariate and trivariate cases.

Theorem 6.1. *Let* $\mathbf{A} \in F[y, x]^{m \times n}$ *of degree* d *in* x *and degree* e *in* y, *and let* N, Z *be polynomial sets defining a semialgebraic constraint on* y. *Then the CTSNF problem* $(\mathbf{A}, N, Z)$ *has a solution of at most* $O(n^2 de)$ *triangular Smith regimes.*

Proof. If N is nonempty, then at the end of the construction below just adjoin N to the N_* of each solution regime. If Z is nonempty it trivializes the solution to at most one regime: let $z(y)$ be the greatest common divisor of the polynomials in Z. If z is 1 or is reducible, the condition is unsatisfiable, otherwise return the single triangular Smith regime for $\mathbf{A}$ over $F[y]/\langle z(y)\rangle$. Otherwise construct the solution regimes as follows where we will assume the semialgebraic constraints are empty.

First compute triangular Smith form $\mathbf{U}_0, \mathbf{H}_0, \mathbf{R}_0$ over $F(y)[x]$ such that $\mathbf{A} = \mathbf{U}_0 \mathbf{H}_0 \mathbf{R}_0$. This will be valid for evaluations that do not zero the denominators (polynomials in y) of $\mathbf{H}_0, \mathbf{U}_0$. So set $N_0 = \text{den}(\mathbf{U}_0, \mathbf{H}_0)$ (or to be the set of irreducible factors that occur in $\text{den}(\mathbf{U}_0, \mathbf{H}_0)$. Set $Z_0 = \emptyset$ to complete the first regime.

Then for each irreducible polynomial $f(y)$ that occurs as a factor in N_0 adjoin the regime $(\mathbf{U}_f, \mathbf{H}_f, \mathbf{R}_f, N_f = N \setminus \{f\}, Z_f = \{f\})$, that comes from computing the triangular Smith form over $(F[y]/\langle f \rangle)[x]$. From the bounds of Theorem 4.6 we have the specified bound on the number of regimes.   □

We can proceed in a similar way when there are three parameters, but must address an additional complication that arises.

Theorem 6.2. *Let* $\mathbf{A} \in F[z, y, x]^{m \times n}$ *of degree* d *in* x *and degree* e *in* y, z, *and let* N, Z *be polynomial sets defining a semialgebraic constraint on* y *and* z. *Then the CTSNF problem* $(\mathbf{A}, V(N, Z))$ *has a solution of at most* $O(n^4 d^2 e^2)$ *triangular Smith regimes.*

Proof. As in the bivariate case above, we solve the unconstrained case and just adjoin N, Z, if nontrivial, to the semialgebraic condition of each solution regime.

First compute triangular Smith form $\mathbf{U}_0, \mathbf{H}_0, \mathbf{R}_0$ over $F(y, z)[x]$ such that $\mathbf{A} = \mathbf{U}_0 \mathbf{H}_0 \mathbf{R}_0$. This will be valid for evaluations that do not zero the denominators (polynomials in y, z) of $\mathbf{H}_0, \mathbf{U}_0$. Thus we set $(N_0, Z_0) = (\{\text{den}(\mathbf{U}_0, \mathbf{H}_0)\}, \emptyset)$ to complete the first regime.

Then for each irreducible polynomial f that occurs as a factor in N_0, if y occurs in f, adjoin the regime $(\mathbf{U}_f, \mathbf{H}_f, \mathbf{R}_f, N_f = N \setminus \{f\}, Z_f = \{f\})$, that comes from computing the triangular Smith form over $(F(z)[y]/\langle f \rangle)[x]$. If y does not occur in f, interchange the roles of y, z.

In either case we get a solution valid when f is zero and the solution denominator δ_f is nonzero. This denominator is of degree $O(n^2 de)$ in each of y, z by Theorem 4.6. [It is the new complicating factor arising in the trivariate case.] It is relatively prime to f, so Bézout's theorem [7] in the theory of algebraic curves can be applied: there are at most $\deg(f) \deg(\delta)$ points that are common zeroes of f and δ. We can produce a separate regime for each such (y, z)-point by evaluating $\mathbf{A}$ at the point and computing a triangular Smith form over $F[x]$. Summing over the irreducible f dividing the original denominator in N_0 we have $O((n^2 de)^2)$ bounding the number of these denominator curve intersection points. $\qquad\square$

Corollary 6.3. *For a PLS with $m \times n$ matrix $\mathbf{A}$, $\mathbf{b}$ an m-vector, and with $\deg_x(\mathbf{A}, \mathbf{b}) \leq d, \deg_y(\mathbf{A}, \mathbf{b}) \leq e, \deg_z(\mathbf{A}, \mathbf{b}) \leq e$, we have*

1. *$O(m^{1.5} d^{0.5})$ regimes in the PLS solution for the univariate case (domain of $\mathbf{A}$, $\mathbf{b}$ is $F[x]$).*
2. *$O(m^{2.5} d^{1.5} e)$ regimes in the PLS solution for the bivariate case (domain of $\mathbf{A}$, $\mathbf{b}$ is $F[x, y]$).*
3. *$O(m^{4.5} d^{2.5} e^2)$ regimes in the PLS solution for the trivariate case (domain of $\mathbf{A}$, $\mathbf{b}$ is $F[x, y, z]$).*

Proof. By Lemma 5.3, each CTSNF regime expands to at most $\sqrt{md}$ PLS regimes. $\qquad\square$

7 Normal Forms and Eigenproblems

Comprehensive Hermite Normal form and comprehensive Smith Normal form are immediate corollaries of our comprehensive triangular Smith form. For Hermite form, just take the right hand cofactor to be the identity, $\mathbf{R} = \mathbf{I}$, and drop the check for the divisibility condition on the diagonal entries in Algorithm 2. For Smith form one can convert each regime of CTSNF to a Smith regime. Where $\mathbf{UAR} = \mathbf{H}$ with $\mathbf{H}$ a triangular Smith form, perform column operations to obtain $\mathbf{UAV} = \mathbf{S}$ with $\mathbf{S}$ the diagonal of $\mathbf{H}$. In $\mathbf{H}$ the diagonal entries divide the off diagonal entries in the same row. Subtract multiples of the i-th column from the subsequent columns to eliminate the off diagonal entries. Because the diagonal entries are monic, no new denominator factors arise and $\det(\mathbf{V}) = \det(\mathbf{R}) \in F$. Thus when $(\mathbf{U}, \mathbf{H}, \mathbf{R}, N, Z)$ is a valid regime in a CTSNF solution for $\mathbf{A}$, then $(\mathbf{U}, \mathbf{S}, \mathbf{V}, N, Z)$ is a valid regime for Smith normal form.

It is well known that if $\mathbf{A} \in K^{n \times n}$ for field K (that may involve parameters) and λ is an additional variable, then the Smith invariants $s_1, \ldots, s_n$ of $\lambda \mathbf{I} - \mathbf{A}$ are the Frobenius invariants of $\mathbf{A}$ and $\mathbf{A}$ is similar to its Frobenius normal form, $\oplus_{i=1}^n \mathbf{C}_{s_i}$, where $\mathbf{C}_s$ denotes the companion matrix of polynomial s. Thus we have comprehensive Frobenius normal form as a corollary of CTSNF, however it is without the similarity transform. It would be interesting to develop a comprehensive Frobenius form with each regimes including a transform.

Parametric eigenvalue problems for $\mathbf{A}$ correspond to PLS for $\lambda \mathbf{I} - \mathbf{A}$ with zero right hand side. Often eigenvalue multiplicity is the concern. The geometric multiplicity is available from the Smith invariants, as for example on the diagonal of a triangular Smith form. Common roots of 2 or more of the invariants expose geometric multiplicity and square-free factorization of the individual invariants exposes algebraic multiplicity. Note that square-free factorization may impose further restrictions on the parameters. Comprehensive treatment of square-free factorization is considered in [18].

7.1   Eigenvalue Multiplicity Example

The following matrix, due originally to a question on sci.math.num-analysis in 1990 by Kenton K. Yee, is discussed in [5]. We change the notation used there to avoid a clash with other notation used here. The matrix is

$$\mathbf{Y} = \begin{bmatrix} z^{-1} & z^{-1} & z^{-1} & z^{-1} & z^{-1} & z^{-1} & z^{-1} & 0 \\ 1 & 1 & 1 & 1 & 1 & 1 & 0 & 1 \\ 1 & 1 & 1 & 1 & 1 & 0 & 1 & 1 \\ 1 & 1 & 1 & 1 & 0 & 1 & 1 & 1 \\ 1 & 1 & 1 & 0 & 1 & 1 & 1 & 1 \\ 1 & 1 & 0 & 1 & 1 & 1 & 1 & 1 \\ 1 & 0 & 1 & 1 & 1 & 1 & 1 & 1 \\ 0 & z & z & z & z & z & z & z \end{bmatrix}.$$

One of the original questions was to compute its eigenvectors. Since it contains a symbolic parameter z, this is a parametric eigenvalue problem which we can turn into a parametric linear system, namely to present the nullspace regimes for $\lambda\mathbf{I} - \mathbf{Y}$.

Over $F(z)[\lambda]$, after preconditioning, we get as the triangular Smith form diagonal $(1, 1, 1, 1, 1, \lambda^2 - 1, \lambda^2 - 1, (\lambda^2 - 1)f(\lambda))$, where $f(\lambda) = \lambda^2 - (z + 6 + z^{-1})\lambda + 7$.

Remark 7.1. Without preconditioning, the Hermite form diagonal is instead $(1, 1, 1, 1, \lambda - 1, \lambda^2 - 1, (\lambda^2 - 1), (\lambda + 1)f(\lambda))$.

The denominator of $\mathbf{U}$, $\mathbf{H}$ is a power of z, so the only constraint is $z = 0$ which is already a constraint for the input matrix. We get regimes of rank 5 for $\lambda = \pm 1$, rank 7 for λ being a root of f, and rank 8 for all other λ. In terms of the eigenvalue problem, we get eigenspaces of dimension 3 for each of 1, -1 and of dimension 1 for the two roots of $f(\lambda)$.

To explore algebraic multiplicity, we can examine when f has 1 or -1 as a root. When z is a root of $z^2 + 14z + 1$, $f(\lambda)$ factors as $(\lambda - 1)(\lambda - 7)$ and when $z = 1$ we have $f(\lambda) = (\lambda + 1)(\lambda + 7)$. These factorizations may be discovered by taking resultants of f with $\lambda - 1$ or $\lambda + 1$.

7.2   Matrix Logarithm

Theorem 1.28 of [12] states the conditions under which the matrix equation $\exp(\mathbf{X}) = \mathbf{A}$ has so-called *primary matrix logarithm* solutions, and under which conditions there are more. If the number of distinct eigenvalues s of $\mathbf{A}$ is *strictly less* than the number p of distinct Jordan blocks of $\mathbf{A}$ (that is, the matrix $\mathbf{A}$ is *derogatory*), then the equation also has so-called *nonprimary* solutions as well, where the branches of logarithms of an eigenvalue λ may be chosen differently in each instance it occurs.

As a simple example of what this means, consider

$$\mathbf{A} = \begin{bmatrix} a & 1 \\ 0 & a \end{bmatrix}. \tag{7.1}$$

When we compute its matrix logarithm (for instance using the `MatrixFunction` command in Maple), we find

$$\mathbf{X}_A = \begin{bmatrix} \ln(a) & a^{-1} \\ 0 & \ln(a) \end{bmatrix}. \tag{7.2}$$

This is what we expect, and taking the matrix exponential (a single-valued matrix function) gets us back to $\mathbf{A}$, as expected. However, if instead we consider the derogatory matrix

$$\mathbf{B} = \begin{bmatrix} a & 0 \\ 0 & a \end{bmatrix} \tag{7.3}$$

then its matrix logarithm as computed by `MatrixFunction` is also derogatory, namely

$$\mathbf{X}_B = \begin{bmatrix} \ln(a) & 0 \\ 0 & \ln(a) \end{bmatrix}. \tag{7.4}$$

Yet there are other solutions as well: if we add $2\pi i$ to the first entry and $-2\pi i$ to the second logarithm, we unsurprisingly find another matrix $\mathbf{X}_C$ which also satisfies $\exp(\mathbf{X}) = \mathbf{B}$. But adding $2\pi i$ to the first entry of $\mathbf{X}_A$ while adding $-2\pi i$ to its second logarithm, we get another matrix

$$\mathbf{X}_D = \begin{bmatrix} \ln(a) + 2i\pi & a^{-1} \\ 0 & \ln(a) - 2i\pi \end{bmatrix} \tag{7.5}$$

which has the (somewhat surprising) property that $\exp(\mathbf{X}_D) = \mathbf{B}$, not $\mathbf{A}$.

This example demonstrates in a minimal way that the detailed Jordan structure of $\mathbf{A}$ strongly affects the nature of the solutions to the matrix equation $\exp(\mathbf{X}) = \mathbf{A}$. This motivates the ability of code to detect automatically the differing values of the parameters in a matrix that make it derogatory. To explicitly connect this example to CTSNF, consider

$$\mathbf{M} = \begin{bmatrix} a & b \\ 0 & a \end{bmatrix} \tag{7.6}$$

so that $\mathbf{A}$ above is $\mathbf{M}_{b=1}$ and $\mathbf{B} = \mathbf{M}_{b=0}$. The CTSNF applied to $\lambda\mathbf{I} - \mathbf{M}$ produces two regimes, with forms

$$\mathbf{H}_{b\neq0} = \begin{bmatrix} 1 & \lambda/b \\ 0 & (\lambda - a)^2 \end{bmatrix}, \mathbf{H}_{b=0} = \begin{bmatrix} \lambda - a & 0 \\ 0 & \lambda - a \end{bmatrix}, \tag{7.7}$$

exposing when the logarithms will be linked or distinct. Note that in this case the Frobenius structure equals the Jordan structure.

7.3　Model of Infectious Disease Vaccine Effect

Rahman and Zou [22] have made a model of vaccine effect when there are two subpopulations with differing disease susceptibility and vaccination rates. Within this study stability of the model is a function of the eigenvalues of a Jacobian $\mathbf{J}$. Thus we are interested in cases where the following matrix is singular.

$$
\mathbf{A} = \lambda\mathbf{I} - \mathbf{J} = \begin{bmatrix} \lambda - w & 0 & -a & -c \\ 0 & \lambda - x & -b & -d \\ 0 & 0 & \lambda - a - y & c \\ 0 & 0 & b & \lambda - d - z \end{bmatrix}.
$$

Here w, x are vaccination rates for the two populations, y, z are death rates, a, d are within population transmission rates, and b, c are the between population transmission rates. We have simplified somewhat: for instance a, b, c, d are transmission rates multiplied by other parameters concerning population counts. Stability depends on the positivity of the largest real part of an eigenvalue. For the sake of reducing expression sizes in this example we will arbitrarily set $y = z = 1/10$. For the same reason we will skip right multiplication by an R to achieve triangular Smith form. Hermite form $\mathbf{H}$ of $\lambda\mathbf{I} - \mathbf{J}$ will suffice, revealing the eigenvalues that are wanted.

$$
\mathbf{H} = \begin{bmatrix} \lambda + w & 0 & 0 & -(ad - a\lambda - bc - a/10)/c \\ 0 & \lambda + x & 0 & \lambda + 1/10 \\ 0 & 0 & 1 & (d - \lambda - 1/10)/c \\ 0 & 0 & 0 & \lambda^2 + (1/5 - a - d)\lambda + ad - cb - (1/10)d - (1/10)a + 1/100 \end{bmatrix}.
$$

The discriminant of the last entry gives the desired information for the application subject to the denominator validity: $c \neq 0$. When $c = 0$ the matrix is already in Hermite form, so again the desired information is provided.

　　This example illustrates that often more than three parameters can be easily handled. In experiments with this model not reported here, we did encounter cases demanding solution beyond the methods of this paper. On a more positive note, we feel that comprehensive normal form tools could help analyze models like this when larger in scope, for instance modeling 3 or more subpopulations.

7.4　The Kac-Murdock-Szegö Example

In [4] we see reported times for computation of the comprehensive Jordan form for matrices of the following form, taken from [28], of dimensions 2 to about 20:

$$
\mathrm{KMS}_n = \begin{bmatrix} 1 & -\rho & & & \\ -\rho & \rho^2 + 1 & -\rho & & \\ & \ddots & \ddots & \ddots & \\ & & -\rho & \rho^2 + 1 & -\rho \\ & & & -\rho & 1 \end{bmatrix}. \tag{7.8}
$$

This is, apart from the $(1,1)$ entry and the (n,n) entry, a Toeplitz matrix containing one parameter, ρ. The reported times to compute the Jordan form were plotted in [4] on a log scale, and looked as though they were exponentially growing with the dimension, and were reported in that paper as growing exponentially.

The theorem of this paper states instead that polynomial time is possible for this family, because there are only two parameters (ρ and the eigenvalue parameter, say λ). The Hermite forms for these matrices are all (as far as we have computed) trivial, with diagonal all 1 except the final entry which contains the determinant. Thus all the action for the Jordan form must happen with the discriminant of the determinant. Experimentally, the discriminant with respect to λ has degree $n^2 + n - 4$ for KMS matrices of dimension $n \geq 2$ (this formula was deduced experimentally by giving a sequence of these degrees to the Online Encyclopedia of Integer Sequences [26]) and each discriminant has a factor $\rho^{n(n-1)}$, leaving a nontrivial factor of degree $2n - 4$ growing only linearly with dimension. The case $\rho = 0$ does indeed give a derogatory KMS matrix (the identity matrix). The other factor has at most a linearly-growing number of roots for each of which we expect the Jordan form of the corresponding KMS matrix to have one block of size two and the rest of size one. We therefore see only polynomial cost necessary to compute comprehensive Jordan forms for these matrices, in accord with our theorem.

8   Conclusions

We have shown that using the CTNSF to solve parametric linear systems is of cost polynomial in the dimension of the linear system and polynomial in parameter degree, for problems containing up to three parameters. This shows that polynomially many regimes suffice for problems of this type. To the best of our knowledge, this is the first method to achieve this polynomial worst case.

It remains an open question whether, for linear systems with a fixed number of parameters greater than three, a number of regimes suffices that is polynomial in the input matrix dimension and polynomial degree of the parameters, being exponential only in the number of parameters.

Through experiments with random matrices we have indication that the worst case bounds we give are sharp, though we have not proven this point. As the examples indicated, many problems will have fewer regimes, and sometimes substantially fewer regimes. We have not investigated the effects of further restrictions of the type of problem, such as to sparse matrices.

Acknowledgements. This work was supported by the Natural Sciences and Engineering Research Council of Canada and by the Ontario Research Centre for Computer Algebra. The third author, L. Rafiee Sevyeri, would like to thank the Symbolic Computation Group (SCG) at the David R. Cheriton School of Computer Science of the University of Waterloo for their support while she was a visiting researcher there.

References

1. Boyer, B., Kaltofen, E.L.: Numerical linear system solving with parametric entries by error correction. In: Proceedings of the 2014 Symposium on Symbolic-Numeric Computation. pp. 33–38 (2014)
2. Buchberger, B.: Applications of Gröbner bases in non-linear computational geometry. In: Janßen, R. (ed.) Trends in Computer Algebra, pp. 52–80. Springer, Berlin, Heidelberg (1988)
3. Camargos Couto, A.C., Moreno Maza, M., Linder, D., Jeffrey, D.J., Corless, R.M.: Comprehensive LU factors of polynomial matrices. In: Slamanig, D., Tsigaridas, E., Zafeirakopoulos, Z. (eds.) Mathematical Aspects of Computer and Information Sciences, pp. 80–88. Springer International Publishing, Cham (2020)
4. Corless, R.M., Moreno Maza, M., Thornton, S.E.: Jordan canonical form with parameters from Frobenius form with parameters. In: Blömer, J., Kotsireas, I.S., Kutsia, T., Simos, D.E. (eds.) MACIS 2017. LNCS, vol. 10693, pp. 179–194. Springer, Cham (2017). https://doi.org/10.1007/978-3-319-72453-9_13
5. Corless, R.M.: Essential Maple 7: An Introduction For Scientific Programmers. Springer Science & Business Media, Berlin (2002)
6. Corless, R.M., Jeffrey, D.J.: The Turing factorization of a rectangular matrix. SIGSAM Bull. **31**(3), 20–30 (1997). https://doi.org/10.1145/271130.271135
7. Cox, D., Little, J., O'Shea, D.: Ideals, Varieties, and Algorithms: An Introduction to Computational Algebraic Geometry and Commutative Algebra. Springer Science & Business Media, Berlin (2013)
8. Dautray, R., Lions, J.: Mathematical Analysis and Numerical Methods for Science and Technology. Springer-Verlag, Berlin (1988). https://doi.org/10.1007/978-3-642-61566-5
9. Dessombz, O., Thouverez, F., Laîné, J.P., Jézéquel, L.: Analysis of mechanical systems using interval computations applied to finite element methods. J. Sound Vibration **239**(5), 949–968 (2001). https://doi.org/10.1006/jsvi.2000.3191
10. Giesbrecht, M., Kim, M.S.: Computing the Hermite form of a matrix of Ore polynomials. J. Algebra **376**, 341–362 (2013)
11. Goldman, L.: Integrals of multinomial systems of ordinary differential equations. J. Pure Appl. Algebra **45**(3), 225–240 (1987). https://doi.org/10.1016/0022-4049(87)90072-7
12. Higham, N.J.: Functions of Matrices: Theory and Computation. SIAM, Philadelphia (2008)
13. Jensen, K.: Coloured Petri nets. Vol. 1. Monographs in Theoretical Computer Science. An EATCS Series, Springer-Verlag, Berlin (1997). https://doi.org/10.1007/978-3-642-60794-3, https://doi.org/10.1007/978-3-642-60794-3, basic concepts, analysis methods and practical use, Corrected reprint of the second (1996) edition
14. Kaltofen, E., Krishnamoorthy, M., Saunders, B.D.: Fast parallel computation of Hermite and Smith forms of polynomial matrices. SIAM J. Algebraic Discrete Methods **8**(4), 683–690 (1987)
15. Kaltofen, E.L., Pernet, C., Storjohann, A., Waddell, C.: Early termination in parametric linear system solving and rational function vector recovery with error correction. In: Proceedings of the 2017 ACM on International Symposium on Symbolic and Algebraic Computation. pp. 237–244 (2017)
16. Kapur, D., Sun, Y., Wang, D.: An efficient algorithm for computing a comprehensive Gröbner system of a parametric polynomial system. J. Symb. Comput. **49**, 27–44 (2013)

17. Kolev, L.V.: Outer solution of linear systems whose elements are affine functions of interval parameters. Reliab. Comput. **8**(6), 493–501 (2002). https://doi.org/10.1023/A:1021320711392
18. Lemaire, F., Maza, M.M., Xie, Y.: The regularchains library. In: Maple conference. vol. 5, pp. 355–368 (2005)
19. Montes, A.: The Gröbner Cover. ACM, vol. 27. Springer, Cham (2018). https://doi.org/10.1007/978-3-030-03904-2
20. Montes, A., Wibmer, M.: Gröbner bases for polynomial systems with parameters. J. Symb. Comput. **45**(12), 1391–1425 (2010)
21. Muhanna, R.L., Mullen, R.L.: Uncertainty in mechanics problems–interval-based approach. J. Eng. Mech. **127**(6), 557–566 (2001)
22. Rahman, S.A., Zou, X.: Modelling the impact of vaccination on infectious diseases dynamics. J. Biol. Dyn. **9**(sup1), 307–320 (2015)
23. Savageau, M.A., Voit, E.O., Irvine, D.H.: Biochemical systems theory and metabolic control theory. I. Fundamental similarities and differences. Math. Biosci. **86**(2), 127–145 (1987). https://doi.org/10.1016/0025-5564(87)90007-1
24. Savageau, M.A., Voit, E.O., Irvine, D.H.: Biochemical systems theory and metabolic control theory. II. The role of summation and connectivity relationships. Math. Biosci. **86**(2), 147–169 (1987). https://doi.org/10.1016/0025-5564(87)90008-3
25. Sit, W.Y.: An algorithm for solving parametric linear systems. J. Symb. Comput. **13**(4), 353–394 (1992)
26. Sloane, N.: The on-line encyclopedia of integer sequences (2020), http://oeis.org
27. Storjohann, A.: Algorithms for matrix canonical forms. Ph.D. thesis, ETH Zurich (2000)
28. Trench, W.F.: Properties of some generalizations of Kac-Murdock-Szego matrices. Contemp. Math. **281**, 233–246 (2001)
29. Wahl, L.M., Betti, M.I., Dick, D.W., Pattenden, T., Puccini, A.J.: Evolutionary stability of the lysis-lysogeny decision: Why be virulent? Evolution **73**(1), 92–98 (2019)
30. Weispfenning, V.: Comprehensive Gröbner bases. J. Symb. Comput. **14**(1), 1–29 (1992)
31. Winkels, H., Meika, M.: An integration of efficiency projections into the Geoffrion approach for multiobjective linear programming. Eur. J. Oper. Res. **16**(1), 113–127 (1984)

Symbolic-Numeric Algorithm for Computing Orthonormal Basis of O(5) × SU(1,1) Group

Algirdas Deveikis[1], Alexander A. Gusev[2($\boxtimes$)], Vladimir P. Gerdt[2,3],
Sergue I. Vinitsky[2,3], Andrzej Góźdź[4], Aleksandra Pędrak[5], Čestmir Burdik[6],
and George S. Pogosyan[7]

[1] Vytautas Magnus University, Kaunas, Lithuania
[2] Joint Institute for Nuclear Research, Dubna, Russia
gooseff@jinr.ru
[3] RUDN University, 6 Miklukho-Maklaya, 117198 Moscow, Russia
[4] Institute of Physics, Maria Curie-Skłodowska University, Lublin, Poland
[5] National Centre for Nuclear Research, Warsaw, Poland
[6] Czech Technical University, Prague, Czech Republic
[7] Yerevan State University, Yerevan, Armenia

Abstract. We have developed a symbolic-numeric algorithm implemented in Wolfram Mathematica to compute the orthonormal non-canonical bases of symmetric irreducible representations of the O(5) × SU(1,1) and $\overline{O(5)}$ × $\overline{SU(1,1)}$ partner groups in the laboratory and intrinsic frames, respectively. The required orthonormal bases are labelled by the set of the number of bosons N, seniority λ, missing label μ denoting the maximal number of boson triplets coupled to the angular momentum $L = 0$, and the angular momentum (L, M) quantum numbers using the conventional representations of a five-dimensional harmonic oscillator in the laboratory and intrinsic frames. The proposed method uses a new symbolic-numeric orthonormalization procedure based on the Gram–Schmidt orthonormalization algorithm. Efficiency of the elaborated procedures and the code is shown by benchmark calculations of orthogonalization matrix $O(5)$ and $\overline{O(5)}$ bases, and direct product with irreducible representations of SU(1,1) and $\overline{SU(1,1)}$ groups.

Keywords: Orthonormal non-canonical basis · Irreducible representations · Group O(5) × SU(1, 1) · Gram–Schmidt orthonormalization · Wolfram Mathematica

1 Introduction

The Bohr–Mottelson collective model [1, 2] has gained widespread acceptance in calculations of vibrational-rotational spectra and electromagnetic transitions in atomic nuclei [3–5]. For construction of basis functions of this model, different approaches were proposed, for example, [6–9], that lead only to nonorthogonal set of eigenfunctions needed in further orthonormalization, considered only in

© Springer Nature Switzerland AG 2020
F. Boulier et al. (Eds.): CASC 2020, LNCS 12291, pp. 206–227, 2020.
https://doi.org/10.1007/978-3-030-60026-6_12

intrinsic frame [10–15]. However, until now, there are no sufficiently universal algorithms for evaluation of the required orthonormal bases needed for large-scale applied calculations in both intrinsic and laboratory frames used in modern models to revival point symmetries in specified degeneracy spectra [16,17]. Creation of such symbolic-numeric algorithm is a goal of the present paper.

In the present paper, we elaborate an universal effective symbolic-numeric algorithm implemented as the first version of O5SU11 code in Wolfram Mathematica for computing the orthonormal bases of the Bohr–Mottelson(BM) collective model in both intrinsic and laboratory frames. It is done on the base of theoretical investigations for constructing the non-canonical bases for irreducible representations (IRs) of direct product groups $G = O(5) \times SU(1,1)$ in the laboratory frame [8] and $\bar{G} = \overline{O(5)} \times \overline{SU(1,1)}$ in the intrinsic frame [7]. We pay our attention to computing bases in both laboratory and intrinsic frames needed for construction of the algebraic models accounting symmetry group [18,19] based on anti-isomorphism between G and $\bar{G}$ partner groups [16,17], and point symmetries in modern calculations, for example, [20–23]. The required orthonormal bases are labelled by the set of the number of bosons N, seniority λ, missing label μ, denoting the maximal number of boson triplets coupled to the angular momentum $L = 0$, and the angular momentum (L, M) quantum numbers using the conventional representations of a five-dimensional harmonic oscillator in the laboratory and intrinsic frames. In the proposed method, the authors use a symbolic-numeric non-standard recursive and fast orthonormalization procedure based on the Gram–Schmidt (G–S) orthonormalization algorithm. Efficiency of the elaborated procedures and the code is shown by benchmark calculations of orthogonalization matrix $O(5)$ and $\overline{O(5)}$ bases, and IRs of $SU(1,1)$ group.

The structure of the paper is as follows. In the second section, we present characterization of group $G = O(5) \times SU(1,1)$ and characterization of states. In Subsects. 2.5 and 2.6, we give the explicit formulas needed for the construction of symmetric nonorthogonal bases for IRs of the $O(5)$ and $G = O(5) \otimes SU(1,1)$ groups. In the third section, we present the construction of the orthonormal basis of the collective nuclear model in intrinsic frame corresponding IRs of the $\bar{G} = \overline{O(5)} \times \overline{SU(1,1)}$ group. In the fourth section, we present the algorithm and benchmark calculations of overlaps and orthogonalization upper triangular matrices applied for constructing the orthonormal basis vectors in the laboratory and intrinsic frames. In conclusion, we give a resumé and point out some important problems for further applications of proposed algorithms.

2   Characterization of Group O(5) × SU(1,1) and Characterization Of States in the Laboratory Frame

Quantum description of collective motions by using the deformation variables $\hat{\alpha}_m^{(l)}$ needs the Hilbert space $L_2(\hat{\alpha}^{(l)})$, which is the state space of $(2l + 1)$-dimensional harmonic oscillator. The Hamiltonian of this harmonic oscillator has the form

$$H_l = \frac{1}{2} \sum_{\mu} \left(\hat{\pi}_\mu^{(l)} \hat{\pi}^{(l)\mu} + \hat{\alpha}^{(l)\mu} \hat{\alpha}_\mu^{(l)} \right), \tag{1}$$

where

$$\hat{\alpha}_m^{(l)} = \sum_\mu g_{m\mu}\hat{\alpha}^{(l)\mu} = (-1)^m\hat{\alpha}^{(l)-m} \tag{2}$$

denotes the multiplication operator by the variable $\hat{\alpha}_m^{(l)}$ and

$$\hat{\pi}_m^{(l)} = \sum_m g_{m\mu}\hat{\pi}^{(l)\mu} = -i\frac{\partial}{\partial\hat{\alpha}^{(l)m}} \tag{3}$$

denotes the conjugate momentum to the coordinate $\hat{\alpha}_\mu^{(l)}$.

The covariant metric tensor $g_{mm'}$ in the corresponding manifold has the form

$$g_{mm'} = g^{mm'} = (-1)^l\sqrt{2l+1}(lmlm'|00) = (-1)^m\delta_m^{-m'}. \tag{4}$$

The operators $\hat{\alpha}_m^{(l)}, \hat{\pi}_m^{(l)}$ fulfil the standard commutation relations

$$\left[\hat{\alpha}_m^{(l)}, \hat{\pi}^{(l)m'}\right] = i\delta_m^{m'}, \quad \left[\hat{\alpha}_m^{(l)}, \hat{\alpha}_{m'}^{(l)}\right] = 0, \quad \left[\hat{\pi}^{(l)m}, \hat{\pi}^{(l)m'}\right] = 0. \tag{5}$$

By using these operators one can build the creation and annihilation spinless boson operators $\eta_m^{(l)}$ and $\xi_m^{(l)}$ with the angular momentum l

$$\eta_m^{(l)} = \frac{1}{\sqrt{2}}\left(\hat{\alpha}_m^{(l)} - i\hat{\pi}_m^{(l)}\right), \quad \xi_m^{(l)} = \frac{1}{\sqrt{2}}\left(\hat{\alpha}_m^{(l)} + i\hat{\pi}_m^{(l)}\right). \tag{6}$$

Contravariant operators can be built in standard way

$$\eta^m = \sum_\mu g^{m\mu}\eta_\mu, \quad \xi^m = \sum_\mu g^{m\mu}\xi_\mu. \tag{7}$$

They satisfy the following commutation relations

$$[\xi^m, \eta_{m'}] = \delta_{m'}^m, \quad \left[\xi^m, \xi^{m'}\right] = [\eta_m, \eta_{m'}] = 0, \quad (\eta_m)^\dagger = \xi^m \quad (\xi_m)^\dagger = \eta^m. \tag{8}$$

2.1   Characterization of U(2l+1)

It can be shown that the bilinear forms

$$(\eta\otimes\eta)_M^{(L)}, \quad \left(\tilde{\xi}\otimes\tilde{\xi}\right)_M^{(L)}, \quad \left(\eta\otimes\tilde{\xi}\right)_M^{(L)},$$
$$\text{where}\quad L = 0, 1\dots 2l, \quad \tilde{\xi}_m = (-1)^m\xi_{-m} \tag{9}$$

generate the non-compact symplectic group Sp(2(2l+1),R).

Group theory analysis leads to two classifications of boson states:

$$\text{Sp(2(2l+1),R)} \supset \text{U(2l+1)},$$
$$\text{Sp(2(2l+1),R)} \supset \text{O(2l+1)} \times \text{SU(1,1)}. \tag{10}$$

The orthonormal group $O(2l+1)$ and the non-compact unitary group $SU(1,1)$ are complementary in two physical IRs of the symplectic group $Sp(2(2l+1),R)$ (for odd and even number of bosons).

The unitary group $U(2l+1)$ has $(2l+1)^2$ generators $E_{mm'}$ or bosons operators

$$(\eta \otimes \tilde{\xi})_M^{(L)} = \frac{1}{2}(-1)^l \sum_{mm'} (lmlm'|LM)E_{mm'}, \quad \text{where} \quad L = 0, 1, \ldots 2l, \qquad (11)$$

$$E_{mm'} = \frac{1}{2}(N_{mm'} + \Lambda_{mm'}), \quad N_{mm'} = \hat{a}_m \hat{a}_{m'} + \hat{\pi}_m \hat{\pi}_{m'}, \quad \Lambda_{mm'} = i(\hat{a}_m \hat{\pi}_{m'} - \hat{\pi}_m \hat{a}_{m'}).$$

The operators $(\eta \otimes \tilde{\xi})_M^{(L)}$ fulfil the following commutation relations

$$\left[(\eta \otimes \tilde{\xi})_{M_1}^{(L_1)}, (\eta \otimes \tilde{\xi})_{M_2}^{(L_2)}\right] = \sqrt{(2L_1 + 1)(2L_2 + 1)} \sum_{LM} [(-1)^L - (-1)^{L_1 + L_2}]$$

$$\times (L_1 M_1 L_2 M_2 | LM) \left\{ \begin{matrix} L_1 & L_2 & L \\ l & l & l \end{matrix} \right\} (\eta \otimes \tilde{\xi})_M^{(L)}.$$

The second order Casimir invariant of the group $U(2l+1)$ is given by

$$C^2 = \sum_{L=0}^{2l} A_L, \quad A_L = (-1)^L \sqrt{2L + 1} \left[(\eta \otimes \eta)^{(L)} \otimes (\tilde{\xi} \otimes \tilde{\xi})^{(L)}\right]_0^{(0)}. \qquad (12)$$

It can be shown that

$$C^2 = \hat{N}(\hat{N} - 1), \quad \text{where} \quad \hat{N} = \sum_\mu \eta_\mu \xi^\mu = \sqrt{2l + 1}(\eta \otimes \tilde{\xi})_0^{(0)}, \qquad (13)$$

the operator $\hat{N}$ is the boson number operator.

The eigenvalues of C^2 depend only on the number of bosons in a given state. In the state which contains N bosons, the expectation value of C^2 is

$$\langle C^2 \rangle_N = N(N - 1). \qquad (14)$$

At the same time, N uniquely labels symmetric IRs of $U(2l + 1)$.

Arbitrary state of N bosons can be constructed by using the vectors:

$$|n_{-l}, n_{-l+1} \ldots n_l\rangle = \frac{1}{\sqrt{(n_{-l})! (n_{-l+1})! \ldots (n_l)!}} (\eta_{-l})^{n_{-l}} \ldots (\eta_l)^{n_l} |0\rangle. \qquad (15)$$

According to this, to define uniquely the state of bosons, located on a level with angular momentum equal to l, one needs to have a set of $2l+1$ quantum numbers.

2.2   Characteristic of $O(2l+1)$

The orthogonal group $O(2l+1)$ contains one-to-one transformations of linear spaces spanned by the tensors $\alpha^{(l)} = (\alpha_{-l}^{(l)}, \ldots, \alpha_l^{(l)})$ which do not change the quadratic form

$$\beta^2 = \sum_\mu \alpha_\mu^{(l)} \alpha^{(l)\mu}. \qquad (16)$$

Generators of this group are $l(2l+1)$ independent operators $\Lambda_{mm'}$ for $m > m'$. The commutation relation for these generators are

$$[\Lambda_{m_1 m_2}, \Lambda_{m_3 m_4}] = \delta_{m_2 m_3} \Lambda_{m_1 m_4} + \delta_{m_1 m_4} \Lambda_{m_2 m_3} - \delta_{m_1 m_3} \Lambda_{m_3 m_4} - \delta_{m_2 m_4} \Lambda_{m_1 m_3},$$

where $\quad \delta_{mm'} = \sum_\mu g_{m\mu} \delta_{m'}^\mu = (-1)^m \delta_{m'}^{-m}.$ (17)

It is possible to get a more useful form of these generators

$$\Lambda_{mm'} = \eta_m \xi_{m'} - \eta_{m'} \xi_m = (-1)^l \sum_{LM} [1 - (-1)^L](lmlm'|LM)(\eta \otimes \tilde{\xi})_M^{(L)}. \quad (18)$$

This implies that the operators $(\eta \otimes \tilde{\xi})_M^{(L=1, 3, 5, \ldots, 2l+1)}$ are the generators of the group O($2l+1$).

The second-order Casimir invariant of the orthogonal group O($2l+1$) is

$$\Lambda^2 = \sum_{L=0}^{2l} [1 - (-1)^L] A_L. \quad (19)$$

For unique labelling of totally symmetric IRs of O($2l+1$), one needs only one quantum number λ. Eigenvalues of operators Λ^2 are the numbers

$$\langle \Lambda^2 \rangle_\lambda = \lambda(\lambda + 2l - 1). \quad (20)$$

The quantum number λ is called seniority and denotes the number of bosons which are not coupled to pairs with zero angular momentum.

2.3   Characteristic of SU(1,1)

The non-compact unitary group $SU(1,1)$ is the complementary group to the orthogonal group O($2l+1$).

The group SU(1,1) has three generators:

$$S_+ = \frac{\sqrt{2l+1}}{2}(\eta \otimes \eta)_0^{(0)}, \quad S_- = \frac{\sqrt{2l+1}}{2}(\tilde{\xi} \otimes \tilde{\xi})_0^{(0)}, \quad S_0 = \frac{1}{2}\left(\hat{N} + \frac{2l+1}{2}\right). \quad (21)$$

The above generators satisfy the following commutation relations:

$$[S_+, S_-] = -2S_0, \quad [S_0, S_+] = S_+, \quad [S_0, S_-] = -S_-, \quad (22)$$

and the conjugation relation

$$(S_+)^\dagger = S_-. \quad (23)$$

The second-order Casimir invariant of the group SU(1,1) is the following operator

$$S^2 = S_0^2 - S_0 - S_+ S_-. \quad (24)$$

One can show that the following relation is satisfied

$$\Lambda^2 = 4S^2 - \frac{1}{4}(2l - 3)(2l + 1). \quad (25)$$

So, the eigenvalues of S^2 are given by

$$\langle S^2 \rangle = S(S - 1), \quad \text{where } S = \frac{1}{2}\left(\lambda + \frac{2l+1}{2}\right). \quad (26)$$

2.4   Construction of States with $N > \lambda$

Let the state

$$|\lambda, N = \lambda, \chi\rangle = |\lambda\lambda\chi\rangle \tag{27}$$

denote the state having the seniority number λ which is equal to the number of particles N in the system. Then it satisfies the conditions

$$S_-|\lambda\lambda\chi\rangle = 0, \ S_0|\lambda\lambda\chi\rangle = \frac{1}{2}(\lambda + \frac{2l+1}{2})|\lambda\lambda\chi\rangle. \tag{28}$$

In the above equations, χ denotes the set of quantum numbers which are needed for labelling the states of the boson system. One can construct the states having the number of bosons N greater than the seniority number λ ($N > \lambda$) by using the action of creation operators of boson pairs coupled to zero angular momentum S_+:

$$|\lambda N\chi\rangle = \sqrt{\frac{\Gamma\left[\lambda + \frac{1}{2}(2l+1)\right]}{[\frac{1}{2}(N-\lambda)]!\,\Gamma\left[\frac{1}{2}(N+\lambda+2l+1)\right]}}\ (S_+)^{\frac{1}{2}(N-\lambda)}|\lambda\lambda\chi\rangle. \tag{29}$$

Angular momentum is a good quantum number characterizing nuclear states. It implies that the rotation group O(3) generated by the operators

$$L_m^{(1)} = \sqrt{\frac{1}{3}l(l+1)(2l+1)}\ (\eta \otimes \tilde{\xi})_m^{(1)} \tag{30}$$

should be contained in the group chain which classifies these states.

The operator $\hat{L}^2$ of the squared angular momentum (the Casimir operator for SO(3)) can be constructed as follows:

$$\hat{L}^2 = \sum_{m=-1}^{1} (-1)^m L_m^{(1)} L_{-m}^{(1)} = l(l+1)(2l+1)\sum_{L=0}^{2l} \begin{Bmatrix} l & l & 1 \\ l & l & L \end{Bmatrix} A_L + l(l+1)\hat{N}.$$

In conclusion, the quantum boson states for $l = 0, 1, 2, \ldots$ can be classified according to two group chains

$$U(2l{+}1) \supset O(2l{+}1) \supset \cdots \supset O(3) \supset O(2), \tag{31}$$

$$O(2l{+}1) \otimes SU(1{,}1) \supset \cdots \supset, O(3) \otimes U(1) \supset O(2). \tag{32}$$

Unitary subgroup SU(1,1) $\supset$ U(1) is generated by the operator S_0, and the generator of rotation about the z-axis generating the subgroup O(3) $\supset$ O(2) is the operator $L_0^{(1)}$.

The states constructed according to the first group chain (31) will be denoted by

$$|N\lambda\xi LM\rangle, \tag{33}$$

and the states constructed according to the second group chain (32) will be denoted by replacing letters N and λ

$$|\lambda N\xi LM\rangle. \tag{34}$$

The vectors (33) and (34) though constructed in different way can be identified as the same vectors. In the following, we will treat them as identical.

But one has to stress that vectors (33) and (34) span IRs of different groups. The vectors (33) form a basis of IRs of the group U(2l+1), for given N. The vectors (34) span the basis of IRs of the group O(2l+1) $\otimes$ SU(1,1), for given λ. According to the above property, we can construct the states of N bosons by using the easier scheme (32).

Table 1. The set of values of dimensions of IRs O(5) group D_λ^e at even L and D_λ^o at odd L and their sum $D_\lambda = D_\lambda^e + D_\lambda^o$ vs λ

λ	10	20	30	40	50	60	70	80	90	100
D_λ^e	322	1892	5711	12782	24102	40671	63492	93562	131881	179452
D_λ^o	184	1419	4705	11039	21424	36860	58344	86879	123465	169099
D_λ	506	3311	10416	23821	45526	77531	121836	180441	255346	348551

2.5 Construction of the Nonorthogonal Basis for Symmetric IRs of the Group O(5)

As the first step, we start with the construction of a basis for the group O(5) from Subsect. 2.2 at $l = 2$. We start the construction with the state of maximal seniority λ and maximal angular momentum $L_0 = 2\lambda$:

$$|\lambda\rangle = (\eta_2^{(2)})^\lambda |0\rangle \tag{35}$$

generated by the action of the creation spinless boson operator $\eta_2^{(2)} \equiv \eta_2$ from (6) on the vacuum vector $|0\rangle$ in representation (15) of elementary boson basis of symmetric IR group U(5) from Subsect. 2.1 at $l = 2$. Next, we construct the operators $\hat{O}(\lambda, \mu, L, M)$ commuting with the Casimir operator $\hat{\Lambda}^2$ from (19) of group O(5) and with the lowering the angular momentum to the required L

$$\hat{O}(\lambda, \mu, L, M) = \sum_{L \leq m \leq 2\lambda} \beta_m(\lambda, L)(L_-)^{m-M}(L_+)^{m+\lambda-3\mu} \left[(\eta \otimes \tilde{\xi})_{-3}^{(3)} \right]^{\lambda-\mu}, \tag{36}$$

where

$$\beta_m(\lambda, L) = \frac{(-1)^m}{(m-L)!(m+L+1)!}, \quad L_+ = -\frac{1}{\sqrt{2}} L_{+1}^{(1)}, \quad L_- = \frac{1}{\sqrt{2}} L_{-1}^{(1)}, \tag{37}$$

i.e., with commutator $[\hat{L}_i, \hat{L}_j] = +i\varepsilon_{ijk}\hat{L}_k$, where ε_{ijk} is the totally antisymmetric symbol, $\varepsilon_{123} = +1$. The quantum number μ denotes the maximal number of boson triplets coupled to the angular momentum $L = 0$. It can be shown that if

$$(\lambda-3\mu) \leq L \leq 2(\lambda-3\mu)(\text{even } L), \quad (\lambda-3\mu) \leq (L+3) \leq 2(\lambda-3\mu)L(\text{odd } L), \tag{38}$$

where $0 \leq \mu \leq [\lambda/3]$, and $[\frac{\lambda}{3}]$ denotes the integer part of $\frac{\lambda}{3}$, then the vectors $\hat{O}(\lambda, \mu, L, M)|\lambda\rangle$ are linearly independent and they form a basis for IRs of the group O(5), for given λ.

The dimension D_λ of this space is $D_\lambda = \frac{1}{6}(\lambda+1)(\lambda+2)(2\lambda+3)$ at fixed λ is determined by following [6]:

$$D_\lambda = D_\lambda^e + D_\lambda^o = \sum_{\mu=0}^{[\lambda/3]} \sum_{L=2[(\lambda+1-3\mu)/2]}^{[2\lambda-6\mu]'} (2L+1) + \sum_{\mu=0}^{[(\lambda-3)/3]} \sum_{L=2[(\lambda-3\mu)/2]+1}^{[2\lambda-6\mu-3]'} (2L+1), \quad (39)$$

where the prime means summation by step 2 and $[\mu] = \mathtt{Floor}(\mu)$ is the largest integer not greater that μ. For example, see Table 1.

Table 2. The set of accessible values μ of the states $|\lambda\mu LL\rangle$ for $L = 0, \ldots, 17$ and $\lambda = 0, \ldots, 17$ in non empty square depending on accessible values of momentum L and seniority λ. Degeneracy $d_{\lambda L}$ is given by formula $d_{\lambda L} = \mu_{max} - \mu_{min} + 1$.

L,λ	0	1	2	3	4	5	6	7	8	9	10	11	12	13	14	15	16	17
0	0			1			2			3			4			5		
1																		
2		0	0		1	1		2	2		3	3		4	4		5	5
3			0			1			2			3			4			
4		0	0	0	1	1	1	2	2	2	3	3	3	4	4	4	5	
5			0	0		1	1		2	2		3	3		4	4		
6			0	0	0	0,1	1	1	1,2	2	2	2,3	3	3	3,4	4	4	
7				0	0	0	1	1	1	2	2	2	3	3	3	4		
8			0	0	0	0,1	0,1	1	1,2	1,2	2	2,3	2,3	3	3,4	3,4		
9				0	0	0	0,1	1	1	1,2	2	2	2,3	3	3			
10			0	0	0	0,1	0,1	0,1	1,2	1,2	1,2	2,3	2,3	2,3	3,4			
11				0	0	0	0,1	0,1	1	1,2	1,2	2	2,3	2,3				
12			0	0	0	0,1	0,1	0,1	0,1,2	1,2	1,2	1,2,3	2,3	2,3				
13				0	0	0	0,1	0,1	0,1	1,2	1,2	1,2	2,3					
14			0	0	0	0,1	0,1	0,1	0,1,2	0,1,2	1,2	1,2,3	1,2,3					
15				0	0	0	0,1	0,1	0,1	0,1,2	1,2	1,2						
16			0	0	0	0,1	0,1	0,1	0,1,2	0,1,2	0,1,2	1,2,3						
17				0	0	0	0,1	0,1	0,1	0,1,2	0,1,2							

The range of accessible values of μ at given accessible λ and L is determined by inequalities:

$$\mu_{min} = \max\left(0, \mathtt{Ceiling}\left(\frac{\lambda-L}{3}\right)\right), \quad \mu_{max} = \mathtt{Floor}\left(\frac{\lambda-(L+3(L\bmod 2))/2}{3}\right), \quad (40)$$

where $\mathtt{Ceiling}(\mu)$ is the lowest integer not lower that μ and $\mathtt{Floor}(\mu)$ is the largest integer not greater that μ. The multiplicity $d_{\lambda L}$ is given by the value of $d_{\lambda L} = \mu_{\max} - \mu_{\min} + 1$. For example, the set of accessible values μ at the given accessible λ and L of states $|\lambda\mu LL\rangle$ is given in Tables 2 and 3. One can see that there is no degeneracy $d_{vL} = 1$ for the first few angular momenta $L = 0, 2, 3, 4, 5, 7$, but not for $L=6$: $d_{\lambda L} = 2$. The range of angular moment L that corresponds to a given maximum d_{vL}^{max} of μ-degeneracy d_{vL} is [10]

$$6(d_{\lambda L}^{max} - 1) \leq L \leq 6(d_{\lambda L}^{max} - 1) + 5, \quad d_{\lambda L}^{max} = 1, 2, \ldots.$$

For example, see Tables 2 and 3: $0 \leq L \leq 5$, $d_{\lambda L}^{max} = 1$, $6 \leq L \leq 11$, $d_{\lambda L}^{max} = 2$, $12 \leq L \leq 17$, $d_{\lambda L}^{max} = 3$.

Table 3. Continuation of Table 2 for $L = 0, \ldots, 17$ and $\lambda = 18, \ldots, 34$

L,λ	18	19	20	21	22	23	24	25	26	27	28	29	30	31	32	33	34
0	6			7			8			9			10			11	
1																	
2		6	6		7	7		8	8		9	9		10	10		11
3	5			6			7			8			9			10	
4	5	5	6	6	6	7	7	7	8	8	8	9	9	9	10	10	10
5		5	5		6	6		7	7		8	8		9	9		10
6	4,5	5	5	5,6	6	6	6,7	7	7	7,8	8	8	8,9	9	9	9,10	10
7	4	4	5	5	5	6	6	6	7	7	7	8	8	8	9	9	9
8	4	4,5	4,5	5	5,6	5,6	6	6,7	6,7	7	7,8	7,8	8	8,9	8,9	9	9,10
9	3,4	4	4	4,5	5	5	5,6	6	6	6,7	7	7	7,8	8	8	8,9	9
10	3,4	3,4	4,5	4,5	4,5	5,6	5,6	5,6	6,7	6,7	6,7	7,8	7,8	7,8	8,9	8 9	8,9
11	3	3,4	3,4	4	4,5	4,5	5	5,6	5,6	6	6,7	6,7	7	7,8	7,8	8	8,9
12	2,3,4	3,4	3,4	3,4,5	4,5	4,5	4,5,6	5,6	5,6	5,6,7	6,7	6,7	6,7,8	7,8	7,8	7,8,9	8,9
13	2,3	2,3	3,4	3,4	3,4	4,5	4,5	4,5	5,6	5,6	5,6	6,7	6,7	6,7	7,8	7,8	7,8
14	2,3	2,3,4	2,3,4	3,4	3,4,5	3,4,5	4,5	4,5,6	4,5,6	5,6	5,6,7	5,6,7	6,7	6,7,8	6,7,8	7,8	7,8,9
15	1,2,3	2,3	2,3	2,3	2,3,4	3,4	3,4	3,4,5	4,5	4,5	4,5,6	5,6	5,6	5,6,7	6,7	6,7	6,7,8
16	1,2,3	1,2,3	2,3,4	2,3,4	2,3,4	3,4,5	3,4,5	3,4,5	4,5,6	4,5,6	4,5,6	5,6,7	5,6,7	5,6,7	6,7,8	6,7,8	6,7,8
17	1,2	1,2,3	1,2,3	2,3	2,3,4	2,3,4	3,4	3,4,5	3,4,5	4,5	4,5,6	4,5,6	5,6	5,6,7	5,6,7	6,7	6,7,8

As conclusion of this analysis, we get the non-orthogonal basis for the totally symmetric IRs of the group O(5) which is denoted by four quantum numbers λ, μ, L, M

$$|\lambda\mu LM\rangle_{no} = \sum_{L\leq m\leq 2\lambda} \beta_m(\lambda, L)(L_-)^{m-M}(L_+)^{m+\lambda-3\mu}(\eta_{-1})^{\lambda-\mu}(\eta_2)^\mu|0\rangle, \quad (41)$$

where λ denotes the seniority number, μ can be interpreted as the maximal number of boson triplets coupled to the angular momentum $L = 0$.

These results can be rewritten in representation (15) of elementary boson basis of symmetric IRs group U(5) from Subsect. 2.1 at $l = 2$. For this purpose, let us assume that the third component of the angular momentum has its maximal value $M = L$

$$|\lambda\mu L\, M = L\rangle = \sum_{n_{-2}...n_2} \langle n_{-2}\, n_{-1}\ldots n_2|\lambda\mu L\, M = L\rangle_{no}|n_{-2}\ldots n_2\rangle. \quad (42)$$

Here the vectors $\langle n'_{-2} \ldots n'_2 | \lambda \mu' LM = L \rangle_{no}$ in the representation of the five-dimensional harmonic oscillator $\langle n'_{-2} \ldots n'_2 |$ have the form

$$\langle n_{-2} \ldots n_2 | \lambda \mu LM = L \rangle = (2L+1) \sqrt{\frac{6^{n_0} (n_{-2})! (n_{-1})! \ldots (n_2)! (2L)!}{(L + \lambda - 3\mu)! (L - \lambda + 3\mu)!}} \qquad (43)$$

$$\times \sum_{p_1 \ldots p_8 \; q_1 \ldots q_5} (-1)^{p_1 + p_3 + p_5 + p_7 + q_2 + q_4} \; 2^{p_1 + 2p_3 + 2p_6 + p_8 + q_2 + q_4}$$

$$\times \frac{(\lambda - \mu)! \; \mu! \; (p_2 + p_3 + 2p_5 + 2p_6 + 2p_7 + 3p_8)! \; (p_2 + p_6 + p_8 + q_2 + 2q_3 + 3q_4 + 4q_5)!}{(p_1)! (p_2)! \ldots (p_8)! (q_1)! (q_2)! \ldots (q_5)! (L + 2\lambda - p_4 - p_5 + 1)!},$$

where the following conditions are satisfied

$$\sum_i n_i = \lambda, \quad \sum_i i n_i = L, \quad \sum_i p_i = \lambda - \mu, \quad \sum_i q_i = \mu, \qquad (44)$$

$$p_1 + q_1 = n_{-2}, \quad p_3 + p_4 + q_2 = n_{-1}, \quad p_2 + p_7 + q_3 = n_0, \quad p_5 + p_6 + q_4 = n_1, \quad p_8 + q_5 = n_2.$$

Vectors $\langle n_{-2} \ldots n_2 | \lambda \mu LM \rangle$ at $-L \leq M < L$ are calculated from recurrence relations

$$\langle n_{-2} \ldots n_2 | \lambda \mu LM - 1 \rangle = ((L - M + 1)(L + M))^{-1/2} \langle n_{-2} \ldots n_2 | \hat{L}_- | \lambda \mu LM \rangle =$$

$$((L - M + 1)(L + M))^{-1/2} \Big[2\sqrt{n_{-2}(n_{-1}+1)} \langle n_{-2}-1, n_{-1}+1, n_0, n_1, n_2 | \lambda \mu LM \rangle$$

$$+ \sqrt{6n_{-1}(n_0+1)} \langle n_{-2}, n_{-1}-1, n_0+1, n_1, n_2 | \lambda \mu LM \rangle$$

$$+ \sqrt{6n_0(n_1+1)} \langle n_{-2}, n_{-1}, n_0-1, n_1+1, n_2 | \lambda \mu LM \rangle$$

$$+ 2\sqrt{n_1(n_2+1)} \langle n_{-2}, n_{-1}, n_0, n_1-1, n_2+1 | \lambda \mu LM \rangle \Big], \qquad (45)$$

where summation is performed over $n_i \geq 0$ subjected to the following conditions: $\sum_i n_i = \lambda, \sum_i i n_i = M$.

Calculating the above coefficients one gets the vectors of the non-orthogonal basis for the totally symmetric IRs of the group O(5) which is denoted by four quantum numbers λ, μ, L, M for given λ:

$$|\lambda \lambda \mu LM \rangle = \sum_{n_{-2} \ldots n_2} \langle n_{-2} \ldots n_2 | \lambda \lambda \mu LM \rangle | n_{-2} \ldots n_2 \rangle. \qquad (46)$$

2.6   Basis of IRs for Groups O(5) ⊗ SU(1,1)

In this part, we construct the states with an arbitrary number of bosons equal to N, greater than seniority number $N > \lambda$. At this point, we use the construction described in Sect. 2.4. By using Eq. (29) for $l = 2$ one gets

$$|\lambda N \mu LM \rangle = \sqrt{\frac{2^{\frac{N-\lambda}{2}} (2\lambda + 3)!!}{(\frac{N-\lambda}{2})! (N + \lambda + 3)!!}} (S_+)^{\frac{N-\lambda}{2}} |\lambda \mu LM \rangle, \qquad (47)$$

where $(N - \lambda)/2 = 1, 2, \ldots$ is integer. Next we can rewrite operator $(S_+)^{\frac{N-\lambda}{2}}$ in a polynomial form:

$$(S_+)^{\frac{N-\lambda}{2}} = \left(\eta_{-2}\eta_2 - \eta_{-1}\eta_1 + \frac{1}{2}\eta_0\eta_0 \right)^{\frac{N-\lambda}{2}}$$

$$= \sum_{k_1 k_2 k_3} (-1)^{k_2} \left(\frac{1}{2} \right)^{k_3} \left(\frac{\frac{N-\lambda}{2}}{k_1 k_2 k_3} \right) (\eta_{-2}\eta_2)^{k_1} (\eta_{-1}\eta_1)^{k_2} (\eta_0)^{2k_3},$$

$$\left(\begin{array}{c} k \\ k_1 \ldots k_N \end{array} \right) = \delta_{\sum_{i=1}^{N} k_i}^{k} \frac{k!}{k_1! \ldots k_N!}.$$

After easy transformations one gets

$$\langle n_{-2} \ldots n_2 | \lambda N \mu L M \rangle = \sqrt{\frac{2^{\frac{N-\lambda}{2}} (2\lambda + 3)!!}{(\frac{N-\lambda}{2})! (N + \lambda + 3)!!}}$$

$$\times \sum_{k_1 k_2 k_3} \frac{(-1)^{k_2}}{2^{k_3}} \left(\frac{\frac{N-\lambda}{2}}{k_1 k_2 k_3} \right) \sqrt{\frac{n_{-2}! \, n_{-1}! \, n_0! \, n_1! \, n_2!}{(n_{-2}-k_1)!(n_{-1}-k_2)!(n_0-2k_3)!(n_1-k_2)!(n_2-k_1)!}}$$

$$\times \langle n_{-2} - k_1, \, n_{-1} - k_2, \, n_0 - 2k_3, \, n_1 - k_2, \, n_2 - k_1 | \lambda \mu L M \rangle. \tag{48}$$

Calculating the above coefficients one gets the vectors of the non-orthogonal symmetric basis of IRs of the group $O(5) \otimes SU(1,1)$ which is denoted by five quantum numbers λ, N, μ, L, and M for given λ and N:

$$|\lambda N \mu L M \rangle = \sum_{n_{-2} \ldots n_2} \langle n_{-2} \ldots n_2 | \lambda N \mu L M \rangle | n_{-2} \ldots n_2 \rangle, \tag{49}$$

$$\Psi^{\text{lab}}_{\lambda N \mu L M}(\alpha_m) = \sum_{n_{-2} \ldots n_2} \langle \alpha_m | n_{-2} \ldots n_2 \rangle \langle n_{-2} \ldots n_2 | \lambda N \mu L M \rangle, \tag{50}$$

where $\langle \alpha_m | n_{-2} \ldots n_2 \rangle$ is the orthonormal basis from (15) $\langle n_{-2} \ldots n_2 | n'_{-2} \ldots n'_2 \rangle = \delta_{n_{-2}n'_{-2}} \ldots \delta_{n_2 n'_2}$, the following conditions are fulfilled: $\sum_i n_i = N$, $\sum_i i n_i = M$. The effective algorithm for calculation of the required orthonormal basis is given in Sect. 4.

3 Nonorthogonal Basis of the IRs $\overline{O(5)} \times \overline{SU(1,1)}$ Group in the Intrinsic Frame

The collective variables α_m at $m = -2, -1, 0, 1, 2$ in the laboratory frame are expressed through variables $a_{m'} = a_{m'}(\beta, \gamma)$ in the intrinsic frame by the relations

$$\alpha_m = \sum_{m'} D^{2*}_{mm'}(\Omega) a_{m'}, \quad a_{-2} = a_2 = \beta \sin \gamma / \sqrt{2}, \quad a_{-1} = a_1 = 0, \quad a_0 = \beta \cos \gamma, \tag{51}$$

where $D^{2*}_{mm'}(\Omega)$ is the Wigner function of IRs of $\overline{O(3)}$ group in the intrinsic frame [24] (marker * is complex conjugate). The five-dimensional equation of the

B-M collective model in the intrinsic frame $\beta \in R_+^1$ and $\gamma, \Omega \in S^4$ with respect to $\Psi^{int}_{\lambda N \mu L M} \in L_2(R_+^1 \bigotimes S^4)$ with measure $d\tau = \beta^4 \sin(3\gamma) d\beta d\gamma d\Omega$ reads as

$$\{H^{(BM)} - E_n^{BM}\}\Psi_{\lambda N \mu L M} = 0, \quad H^{(BM)} = \frac{1}{2}\left(-\frac{1}{\beta^4}\frac{\partial}{\partial\beta}\beta^4\frac{\partial}{\partial\beta} + \frac{\hat{\Lambda}^2}{\beta^2} + \beta^2\right). \quad (52)$$

Here $E_N^{BM} = (N + \frac{5}{2})$ are eigenvalues, $\hat{\Lambda}^2$ is the quadratic Casimir operator of $\overline{O(5)}$ in $L_2(S^4(\gamma, \Omega))$ at nonnegative integers $N = 2n_\beta + \lambda$, i.e., at even and nonnegative integers $N - \lambda$ determined as

$$(\hat{\Lambda}^2 - \lambda(\lambda+3))\Psi_{\lambda N \mu L M} = 0, \quad \hat{\Lambda}^2 = -\frac{1}{\sin(3\gamma)}\frac{\partial}{\partial\gamma}\sin(3\gamma)\frac{\partial}{\partial\gamma} + \sum_{k-1}^{3}\frac{(\hat{\tilde{L}}_k)^2}{4\sin^2(\gamma - \frac{2}{3}k\pi)}, \quad (53)$$

where the nonnegative integer λ is the so-called seniority and $(\hat{\tilde{L}}_k)^2$ are the angular momentum operators of $\overline{O(3)}$ along the principal axes in intrinsic frame, i.e., with commutator $[\hat{\tilde{L}}_i, \hat{\tilde{L}}_j] = -\imath\varepsilon_{ijk}\hat{\tilde{L}}_k$.

Eigenfunctions $\Psi^{int}_{\lambda N \mu L M}$ of the five-dimensional oscillator have the form

$$\Psi^{int}_{\lambda N \mu L M}(\beta, \gamma, \Omega) = \sum_{K \text{ even}} \Phi^{int}_{\lambda N \mu L K}(\beta, \gamma)\mathcal{D}^{(L)*}_{MK}(\Omega), \quad (54)$$

where $\Phi^{int}_{\lambda N \mu L K}(\beta, \gamma) = F_{N\lambda}(\beta)C_L^{\lambda\mu}\hat{\phi}_K^{\lambda\mu L}(\gamma)$ are the components in the intrinsic frame, $\mathcal{D}^{(L)*}_{MK}(\Omega) = \sqrt{\frac{2L+1}{8\pi^2}}\frac{D^{(L)*}_{MK}(\Omega) + (-1)^L D^{(L)*}_{M,-K}(\Omega)}{1 + \delta_{K0}}$ are the orthonormal Wigner functions with measure $d\Omega$, summation over K runs even values K in range:

$$K = 0, 2, \ldots, L \text{ for even integer } L : 0 \leq L \leq L_{max}, \quad (55)$$
$$K = 2, \ldots, L - 1 \text{ for odd integer } L : 3 \leq L \leq L_{max}.$$

The orthonormal components $F_{N\lambda}(\beta) \in L_2(R_+^1)$ corresponding to reduced functions $\beta^{-2}\mathfrak{F}_{N\lambda}(\beta)$ with measure $d\beta$ of IRs of $\overline{SU(1,1)}$ group [25] are as follows:

$$F_{N\lambda}(\beta) = \sqrt{\frac{2(\frac{1}{2}(N-\lambda)!)}{\Gamma(\frac{1}{2}(N+\lambda+5))}}\beta^\lambda L_{(N-\lambda)/2}^{\lambda+\frac{3}{2}}(\beta^2)\exp\left(-\frac{1}{2}\beta^2\right), \quad (56)$$

where $L_{(N-\lambda)/2}^{\lambda+\frac{3}{2}}(\beta^2)$ is the associated Laguerre polynomial with the number of nodes $n_\beta = (N - \lambda)/2$ [26]. The overlap of the eigenfunctions (54) characterized their nonorthogonality with respect to the missing label μ reads as

$$\langle\Psi^{int}_{\lambda N \mu L M}|\Psi^{int}_{\lambda' N' \mu' L' M'}\rangle = \int d\tau \Psi^{int*}_{\lambda N \mu L M}(\beta, \gamma, \Omega)\Psi^{int}_{\lambda N \mu L M}(\beta, \gamma, \Omega) \quad (57)$$

$$= \delta_{N,N'}\delta_{\lambda,\lambda'}\delta_{L,L'}\delta_{M,M'}\langle\phi^{\lambda\mu L}|\phi^{\lambda\mu' L}\rangle,$$

where $\langle\phi^{\lambda\mu L}|\phi^{\lambda\mu' L}\rangle$ is the reduced overlap: scalar product with integration by γ

$$\langle\phi^{\lambda\mu L}|\phi^{\lambda\mu' L}\rangle = C_L^{\lambda\mu}C_L^{\lambda\mu'}\int_0^\pi d\gamma \sin(3\gamma)\sum_{K \text{ even}}\frac{2(\hat{\phi}_K^{\lambda\mu L}(\gamma)\hat{\phi}_K^{\lambda\mu' L}(\gamma))}{1 + \delta_{K0}}, \quad (58)$$

and $C_L^{\lambda\mu}$ is the corresponding normalization factor of $\phi_K^{\lambda\mu L}(\gamma) = C_L^{\lambda\mu}\hat{\phi}_K^{\lambda\mu L}(\gamma)$

$$(C_L^{\lambda\mu})^{-2} = \int_0^\pi d\gamma \sin(3\gamma) \sum_{K\text{ even}} \frac{2(\hat{\phi}_K^{\lambda\mu L}(\gamma))^2}{1 + \delta_{K0}}. \tag{59}$$

The reduced Wigner coefficients in the chain $\overline{O(5)} \supset \overline{O(3)}$ read as [13]

$$(\lambda\mu L, \lambda'\mu'L', \lambda\mu L'') = \int_0^\pi d\gamma \sin(3\gamma) \sum_{KK'K''} (-1)^{L-L'}(L, L', K, K'|L'', -K'')$$

$$\times \phi_K^{\lambda\mu L}(\gamma)\phi_{K'}^{\lambda'\mu'L'}(\gamma)\phi_{K'}^{\lambda'\mu'L'}(\gamma), \tag{60}$$

where $\phi_K^{\lambda\mu L}(\gamma)$ are the orthonormalized eigenfunctions calculated in the section 4 with respect to the overlap (58)corresponds to the orthonormalized eigenfunctions (54) with respect to the overlap (57) with the set of quantum numbers λ, μ, L, and M.

The components $\hat{\phi}_K^{\lambda\mu L}(\gamma) = (-1)^L\hat{\phi}_{-K}^{\lambda\mu L}(\gamma)$ for even K and $\hat{\phi}_K^{\lambda\mu L}(\gamma) = 0$ for odd L and $K = 0$ as well as for odd K are determined below according to [5–7,12]. It should be noted that for these components, $L \neq 1$, $|K| \leq L$ for $L = $ even and $|K| \leq L - 1$ for $L = $ odd:

$$\hat{\phi}_K^{\lambda\mu L}(\gamma) = \sum_{n=0}^{n_{\max}} F_{n\lambda L}^{\sigma\tau\mu}(\gamma) \left[G_{|K|}^{nL}(\gamma)\delta_{L,\text{even}} + \bar{G}_{|K|}^{nL}(\gamma)\delta_{L,\text{odd}} \right]; \tag{61}$$

$$K = K_{\min}, K_{\min} + 2, \ldots, K_{\max};$$

$$K_{\min} = \begin{cases} 0, L = \text{even}, \\ 2, L = \text{odd}; \end{cases} \quad K_{\max} = \begin{cases} L \quad, L = \text{even}, \\ L-1, L = \text{odd}; \end{cases}$$

$$n_{\max} = \begin{cases} L/2 \quad, L = \text{even}, \\ (L-3)/2, L = \text{odd}; \end{cases}$$

$$\delta_{L,\text{even}} = \begin{cases} 1, L = \text{even}, \\ 0, L = \text{odd}; \end{cases} \quad \delta_{L,\text{odd}} = \begin{cases} 0, L = \text{even}, \\ 1, L = \text{odd}; \end{cases}$$

where $L/2 \leq \lambda - 3\mu \leq L$ for $L = $ even, and $(L+3)/2 \leq \lambda - 3\mu \leq L$ for $L = $ odd;

$$\bar{G}_K^{nL}(\gamma) = \sum_{k=3-L,2}^{L-3} \langle L-3, 3, k, K-k|LK\rangle G_{|k|}^{nL-3}(\gamma) \sin 3\gamma(\delta_{K-k,2} - \delta_{K-k,-2}); \tag{62}$$

$$G_K^{nL}(\gamma) = (-\sqrt{2})^n \sum_{k=2n-L,2}^{L-2n} \langle L-2n, 2n, k, K-k|LK\rangle S_{|k|}^{(L-2n)/2}(\gamma)S_{|K-k|}^n(-2\gamma); \tag{63}$$

$$S_K^r(\gamma) = \left[\frac{(2r+K)!(2r-K)!}{(4r)!}\right]^{1/2} (\sqrt{6})^r r! \sum_{q=K/2}^{[r/2+K/4]} \left(\frac{1}{2\sqrt{3}}\right)^{2q-K/2}$$

$$\times \frac{1}{(r-2q+K/2)!(q-K/2)!q!}(\cos\gamma)^{r+K/2-2q}(\sin\gamma)^{2q-K/2};$$

$$F_{n\lambda L}^{\sigma\tau\mu}(\gamma) = (-1)^{\mu+\tau-n}2^{-n/2} \sum_{r=0}^{[(\mu+\tau-n)/2]} C_{rn\lambda L}^{\sigma\tau\mu}2^{-r}(\cos 3\gamma)^{\mu+\tau-n-2r}; \qquad (64)$$

$$\tau = \begin{cases} \lambda - 3\mu - L/2 & , L = \text{even}, \\ \lambda - 3 - 3\mu - (L-3)/2 & , L = \text{odd}. \end{cases} \quad \sigma = L - \lambda + 3\mu;$$

$$C_{rn\lambda L}^{\sigma\tau\mu} = \frac{3^n \sigma! \lambda! (-1)^r 2^r (2\mu + 2\tau - 2r + \delta_{L,\text{odd}})!(3r)!}{2^{\mu+n} n!(2\lambda+1)! r!(\mu+\tau-r)!(\mu+\tau-n-2r)!} \qquad (65)$$

$$\times \sum_{s=\max(n-\tau,0)}^{\min(\sigma,\lambda,3r-\tau+n)} \frac{(-1)^s 4^s (\tau+s)!(2\lambda+1-2s)!}{s!(\sigma-s)!(\tau-n+s)!(3r-\tau+n-s)!(\lambda-s)!};$$

where $S_K^r(\gamma)$ is taken to be equal 0, if $\sin 3\gamma = 0$ or $\cos 3\gamma = 0$, $F_{n\lambda L}^{\sigma\tau\mu}(\gamma)$ is taken to be equal 0, if $\cos 3\gamma = 0$, $C_{rn\lambda L}^{\sigma\tau\mu}$ is taken to be equal 0, if $\mu+\tau-n-2r < 0$.

For example, at $\lambda = 3\mu$ and $L = 0, M = 0$, and $\lambda = 3\mu+3$ and $L = 3, M = 3$, the eigenfunctions are known:

$$\Psi_{\lambda N\mu LM}(\beta, \gamma, \Omega) = C_\mu^0 \beta^{3\mu} \exp(-\beta^2/2) P_\mu(\cos(3\gamma)), \qquad (66)$$

$$\Psi_{\lambda N\mu LM}(\beta, \gamma, \Omega) = C_\mu^3 \beta^{3\mu+3} \exp(-\beta^2/2) P_{\mu+1}^1(\cos(3\gamma))(D_{32}^{(3)*}(\Omega) - D_{3,-2}^{(3)*}(\Omega)),$$

where $P_{\mu+1}^1(\cos(3\gamma))$ are associated Legendre polynomials [26].

The eigenfunctions $\Psi_{\lambda N\mu LM}(\beta, \gamma, \Omega)$ at $L \le 6$ were calculated in [27,28]. However, for calculation of the required orthogonal basis including large values of λ and L for large-scale calculations of eigenvalue BM problem (52) for Hamiltonian $\mathcal{H} = H^{BM}(\beta, \gamma, \Omega) + V(\beta, \gamma) + \mathcal{K}(\beta, \gamma)$ with potential function $V(\beta, \gamma)$ and additional kinetic function $\mathcal{K}(\beta, \gamma)$ determined in [5,7,10,11], one needs to have a fast algorithm for calculation and orthonormalization of nonorthogonal eigenfunctions $\Psi_{\lambda N\mu LM}(\beta, \gamma, \Omega)$ from (54) at accessible degeneracy characterized by the missing label $\mu_{\min} \le \mu \le \mu_{\max}$ from (40) and also Tables 2 and 3. The effective algorithm for calculation of the required orthonormal basis is given in the Sect. 4.

4   Algorithm and Benchmark Calculations of Overlaps and Orthogonalization Matrices

In the laboratory frame, the overlaps $\langle \hat{u}_\mu | \hat{u}_{\mu'} \rangle \equiv \langle \lambda N\mu LM | \lambda N\mu' LM \rangle$ are calculated by the formula

$$\langle \hat{u}_\mu | \hat{u}_{\mu'} \rangle = \sum_{n'_{-2}...n'_2} \langle \lambda N\mu LM | n'_{-2}...n'_2 \rangle \langle n'_{-2}...n'_2 | \lambda N\mu' LM \rangle. \qquad (67)$$

Here vectors $\langle \alpha_m | \hat{u}_{\mu'} \rangle = \langle \alpha_m | \lambda N\mu' LM \rangle$ in the representation of the orthonormal basis $\langle n'_{-2}...n'_2 |$ of the five-dimensional harmonic oscillator (15) are determined by Eqs. (49) and (50) through the unnormalized and non-orthogonal μ' components $\langle n'_{-2} \ldots n'_2 | \lambda N \mu' L M \rangle$ of the reduced vectors $|\hat{u}'_\mu\rangle$ from Eqs. (43, 45, 48).

In the intrinsic frame, the overlap $\langle \hat{u}_\mu | \hat{u}_{\mu'} \rangle \equiv \langle \lambda N \mu L M | \lambda N \mu' L M \rangle$ reads as:

$$\langle \hat{u}_\mu | \hat{u}_{\mu'} \rangle = \int_0^\pi \sin(3\gamma) d\gamma \sum_{K \geq 0, even} \frac{2 \hat{\phi}_K^{\lambda \mu L}(\gamma) \hat{\phi}_K^{\lambda \mu' L}(\gamma)}{1 + \delta_{K0}}. \tag{68}$$

Here vectors $\langle \beta, \gamma, \Omega | \hat{u}_{\mu'} \rangle = \langle \beta, \gamma, \Omega | \lambda N \mu' L M \rangle$ in the representation of the orthonormal Wigner functions $\mathcal{D}_{MK}^{(L)*}(\Omega)$ and components $F_{N\lambda}(\beta)$ are determined by Eqs. (54)-(59) through the unnormalized and non-orthogonal by μ' components $\hat{\phi}_K^{\lambda \mu' L}(\gamma)$ of the reduced vectors $|\hat{u}_{\mu'}\rangle = \langle \gamma | \hat{\phi}^{\lambda \mu' L} \rangle$ from (61)–(65).

The numerical calculations performed in the program S05U11 use the floating-point arithmetics. In this case, we use instead of the unnormalized nonorthogonal $|\hat{u}_\mu\rangle$ the normalized but nonorthogonal eigenvectors $|u_\mu\rangle$:

$$|u_\mu\rangle = \hat{N}_{\mu\mu}^{-1} |\hat{u}_\mu\rangle, \quad \hat{N}_{\mu\mu} = (\langle \hat{u}_\mu | \hat{u}_\mu \rangle)^{1/2}, \tag{69}$$

where the normalization matrix is equal to $\hat{N}_{\mu\mu'} = \hat{N}_{\mu\mu} \delta_{\mu\mu'}$, and the normalized overlaps are

$$\langle u_\mu | u_{\mu'} \rangle = \langle \hat{u}_\mu | \hat{N}_{\mu\mu}^{-1} \hat{N}_{\mu'\mu'}^{-1} | \hat{u}_{\mu'} \rangle, \quad \langle u_\mu | u_\mu \rangle = 1. \tag{70}$$

We orthonormalize these normalized nonorthogonal BM states $|u_\mu\rangle$:

$$|\phi_\mu\rangle = \sum_{\mu'=\mu_{\min}}^{\mu_{\max}} |u_{\mu'}\rangle A_{\mu',\mu} = \sum_{\mu'=\mu_{\min}}^{\mu_{\max}} |\hat{u}_{\mu'}\rangle \hat{A}_{\mu',\mu}, \quad \mathbf{A} = \hat{\mathbf{N}} \hat{\mathbf{A}}. \tag{71}$$

Below the hat symbol over some vectors and matrices is used to label calculations with unnormalized BM vectors. The symbols $A_{\mu',\mu}$ denote the matrix elements of the upper triangular matrix of the BM basis orthonormalization coefficients. These coefficients satisfy the following condition

$$A_{\mu',\mu} = 0, \quad \text{if } \mu' > \mu, \quad \mu, \mu' = \mu_{\min}, \ldots, \mu_{\max}. \tag{72}$$

The matrix $\mathbf{A}$ is constructed to satisfy the orthonormalization conditions

$$\langle \phi_\mu | \phi_{\mu'} \rangle = \delta_{\mu\mu'}, \quad \sum_{i',k'=1}^{d_{\lambda L}} A_{i',i} \langle u_{i'} | u_{k'} \rangle A_{k',k} = \delta_{i,k}, \quad \langle u_i | u_{k'} \rangle = \sum_{k'=1}^{d_{\lambda L}} A_{k',i}^{-1} A_{k',k}^{-1}. \tag{73}$$

Here the multiplicity index i or internal index $k, k' = 1, \ldots, d_{\lambda L}$ is recalculated by formula $d_{\lambda L} = \mu_{\max} - \mu_{\min} + 1$ to external index $\mu, \mu' = \mu_{\min}, \ldots, \mu_{\max}$ and vice versa was introduced to distinguish the orthonormalized BM states at given values of quantum numbers λ, N, L, M and takes the same number of values as μ. Note the last relation in (73) is a decomposition of the overlap matrix to a product of the low and upper triangular inverse matrices $(\mathbf{A}^{-1})^T \mathbf{A}^{-1}$.

Table 4. *Algorithm* for calculation of elements of the upper triangular matrix $A_{\mu,\mu'}$ which is used for the generation of an orthonormal basis $|\phi_\mu\rangle = \sum_{\mu'=\mu_{\min}}^{\mu_{\max}} |u_{\mu'}\rangle A_{\mu',\mu}$ starting from $|u_{\mu_{\min}}\rangle$ till $|u_{\mu_{\max}}\rangle$, where the external index $\mu, \mu' = \mu_{\min}, \ldots, \mu_{\max}$ is recalculated by formula $k = \mu - \mu_{\min} + 1$ to internal index $k, k' = 1, \ldots, d_{\lambda L}$, $d_{\lambda L} = \mu_{\max} - \mu_{\min} + 1$ and vice versa

Input:	Overlap matrix $\langle u_k | u_{k'}\rangle$;
Output:	Orthogonalization of the upper triangular matrix $A_{k',k}$;
1.1	$A_{k',k} = \delta_{k'k}, \quad k = 1, \ldots, d_{\lambda L}, k' = k, \ldots, d_{\lambda L}$;
1.2	$f_{k,k'} = \langle u_k | u_{k'}\rangle, \quad k = 1, \ldots, d_{\lambda L}, k' = k, \ldots, d_{\lambda L}$;
	for $n = 1$ to $d_{\lambda L}$ do
2.1	$u_k = -f_{k,n}/f_{k,k}, k = 1, \ldots, n-1;$ $\qquad\qquad u_k \equiv u_{k,n};$
2.2	$f_{n,n} = f_{n,n} + \sum_{k=1}^{n-1} u_k^2 f_{k,k} + 2\sum_{k=1}^{n-1} u_k f_{k,n};$
2.3	$f_{n,k} = f_{n,k} + \sum_{k'=1}^{n-1} u_{k'} f_{k',k}, \quad k = n+1, \ldots, d_{vL};$
2.4	$A_{k,n} = \sum_{k'=k}^{n-1} A_{k,k'} u_{k'}, \quad k = 1, \ldots, n-1;$
	end for
3.1	$A_{k',k} = A_{k',k}/\sqrt{f_{kk}}, \quad k = 1, \ldots, n, k' = 1, \ldots, k;$
test:	$\sum_{n',k'=1}^{d_{\lambda L}} A_{n',n} \langle u_{n'} | u_{k'}\rangle A_{k',k} = \delta_{n,k}$

Below we present the analytical orthonormalization algorithm (see Table 4) based on the G-S orthonormalization procedure of a set of non-orthogonal linear independent vectors: $\hat{u}_1, \ldots, \hat{u}_{i_{max}}$ unnormalized or $u_1, \ldots, u_{i_{max}}$ normalized [29]

$$\hat{\phi}_i = u_i - \frac{\langle \hat{\phi}_1 | u_i\rangle}{\langle u_1 | u_1\rangle} - \cdots - \frac{\langle \hat{\phi}_{i-1} | u_i\rangle}{\langle u_{i-1} | u_{i-1}\rangle}, \quad i = 1, \ldots, i_{\max}. \tag{74}$$

Here for the intrinsic frame, the scalar product $\langle \hat{\phi}_i | \hat{\phi}_i\rangle$ is determined by (68) while for laboratory frame $\langle \hat{\phi}_i | \hat{\phi}_i\rangle = \hat{\phi}_i^T \hat{\phi}_i$. After calculation of a set of orthogonal but as yet *unnormalized* vectors $\hat{\phi}_i$ starting from $i = 1$ till $i_{\max}$, one calculates the set of orthogonal and *normalized* vectors ϕ_i: $\phi_i = \hat{\phi}_i / \sqrt{\langle \hat{\phi}_i | \hat{\phi}_i\rangle}$ at $i = 1, \ldots, i_{\max}$. It is important that here the normalization of calculated orthogonal *unnormalized* vectors $\hat{\phi}_i$ is realized after orthogonalization with respect to conventional realization G-S procedure [30]. It gives important possibility to avoid the source of numerical round-off errors in floating-point calculations or if necessary to use the integer arithmetic or symbolic calculations of the recursive algorithm given below. The essential part of the proposed algorithm consists in factorization of the recursive relations (74) by extracting the required orthogonalization upper triangular matrix $A_{\mu'\mu}$ acting on the initial set of non-orthogonal vectors $|u_{\mu'}\rangle$: $|\phi_\mu\rangle = \sum_{\mu'=\mu_{\min}}^{\mu_{\max}} |u_{\mu'}\rangle A_{\mu',\mu}$. It means that the calculated matrix $A_{\mu'\mu} \equiv A_{\mu',\mu}^{lab}(N,M)$ in the laboratory frame is the same on all components $|u_\mu\rangle = \langle n_{-2}, n_{-1}, n_0, n_1, n_2 | u_\mu\rangle$ of the initial set of the non-orthogonal reduced vectors $|u_{\mu'}\rangle$; action of calculated matrix $A_{\mu'\mu} \equiv A_{\mu',\mu}^{int}(\lambda, N, L, M)$ in the intrinsic frame is the same on all components $\phi_K^{\lambda\mu'L}(\gamma) = C_L^{\lambda\mu'} \hat{\phi}_K^{\lambda\mu'L}(\gamma)$ of the initial set of the non-orthogonal reduced vectors $|u_{\mu'}\rangle = \langle \gamma | \phi^{\lambda\mu'L}\rangle$. The accuracy of its calculation is automatically checked by means of orthogonality relations (73) without preliminary calculation of required orthogonal normalized vectors ϕ_i.

Remark. A direct calculation of the overlap of the orthogonal bases $\Psi^{lab}_{\lambda N \mu L M}(\alpha_m)$ in the laboratory frame (50) and $\Phi^{int}_{\lambda N \mu L K}(\beta, \gamma)$ in the intrinsic frame (54) is the tutorial task. Using Eq. (1.16) of Ref. [7] one can check that the following relations hold:

$$\Psi^{lab}_{\lambda N \mu L M}(\alpha_m) = \sum_{K=0,even}^{L} \Phi^{int}_{\lambda N \mu L K}(\beta, \gamma) \mathcal{D}^{L*}_{MK}(\Omega), \qquad (75)$$

$$\Phi^{int}_{\lambda N \mu L K}(\beta, \gamma) = \sum_{M=0}^{L} \Psi^{lab}_{\lambda N \mu L M}(\alpha_m) \mathcal{D}^{L}_{MK}(\Omega), \qquad (76)$$

where the variables α_m in the laboratory frame are expressed through the ones $a_m = a_m(\beta, \gamma)$ in the intrinsic frame by relations (51).

The presented Algorithm (see Table 4) can be realized in any Computer Algebra System. It has been realized here as the function NormOverlapa of the first version of O5SU11 code implemented in Mathematica 11.1 [31].

```
NormOverlapa[Overlap_] :=Module[{},
  For[x = 1, x <= Length[Overlap], x++,
    A[x, x]=1;
    For[xx = x, xx <= Length[Overlap], xx++,
      fover[x, xx] = Part[Overlap, x, xx];
    ]
  ];
  For[n = 1, n <= Length[Overlap], n++,
    For[k = 1, k <= n-1, k++,
      ui[k]=-fover[k, n]/fover[k, k];
    ]
    fover[n, n]=fover[n, n]+Sum[ui[k]*ui[k]*fover[k, k],{k,1,n-1}]
      +Sum[2*ui[k]*fover[k, n],{k,1,n-1}];
    For[k = n+1, k <= Length[Overlap], k++,
      fover[n, k]=fover[n, k]+Sum[ui[kk]*fover[kk, k],{kk,1,n-1}];
    ];
    For[k = 1, k <= n-1, k++,
      A[k, n]=Sum[A[k,kk]*ui[kk],{kk,k,n-1}];
    ]
  ];
  Return[
  Table[
  If[x > xx, 0, A[x, xx]/Sqrt[fover[xx, xx]] ]
    , {x, 1,   Length[Overlap]}, {xx, 1, Length[Overlap]}]]
  ];
(*test: *)
    A=NormOverlapa[Overlap]
    Transpose[A]*Overlap*A    (*gives identity matrix*)
```

Below we present benchmark calculations of the overlap matrices $\langle \hat{u}_\mu | \hat{u}_{\mu'} \rangle$ or $\langle u_\mu | u_{\mu'} \rangle$ and orthogonalization matrices $\hat{A}_{\mu'\mu}$ or $A_{\mu'\mu}$ executed with help of the O5SU11 code.

In the laboratory frame, the unnormalized overlap $\langle \hat{u}_\mu | \hat{u}_{\mu'} \rangle$ from (67) and orthogonalization matrix $\hat{A}_{\mu'\mu}$ from (71) at $\lambda=12, N = 12, \mu=0,1,2, L=12, M=12$ are as follows:

$$\langle \hat{u}_\mu | \hat{u}_{\mu'} \rangle = \begin{pmatrix} 159549545.26713809 & 213803.08882591313 & 57637.968478797638 \\ 213803.08882591313 & 4988824.1342315109 & -422776.94375296634 \\ 57637.968478797638 & -422776.94375296634 & 744945.4277113013 \end{pmatrix},$$

$$\hat{A}_{\mu'\mu} = \begin{pmatrix} 0.0000791684632054189660 & -5.999558704952660 * 10^{-7} & -5.501511852913287 * 10^{-7} \\ 0 & 0.000447714234829464325 & 0.0000982042323533384462 \\ 0 & 0 & 0.00115861132845170549 \end{pmatrix}.$$

The normalized overlap $\langle u_\mu | u_{\mu'} \rangle_{no}$ from (67) and orthogonalization matrix $A_{\mu'\mu}$ from (71) at $\lambda=12, N = 12, \mu=0,1,2, L=12, M=12$:

$$\langle u_\mu | u_{\mu'} \rangle = \begin{pmatrix} 1.0000000000000000000 & 0.00757821796968012826 & 0.00528687022845146168 \\ 0.00757821796968012826 & 1.0000000000000000000 & -0.2193057245440392052 \\ 0.00528687022845146168 & -0.2193057245440392052 & 1.0000000000000000000 \end{pmatrix},$$

$$A_{\mu'\mu} = \begin{pmatrix} 1.0000000000000000000 & -0.00757821796968012826 & -0.00694912043276433934 \\ 0 & 1.0000000000000000000 & 0.2193457895990078230 \\ 0 & 0 & 1.0000000000000000000 \end{pmatrix}.$$

The unnormalized overlap $\langle \hat{u}_\mu | \hat{u}_{\mu'} \rangle$ from (67) and orthogonalization matrix $\hat{A}_{\mu'\mu}$ from (71) at $\lambda=12, N=14, \mu=0,1,2, L=12, M=12$:

$$\langle \hat{u}_\mu | \hat{u}_{\mu'} \rangle = \begin{pmatrix} 4.62693681274700 * 10^9 & 6.200289575951 * 10^6 & 1.671501085885 * 10^6 \\ 6.200289575951 * 10^6 & 1.44675899892714 * 10^8 & -1.22605313688360 * 10^7 \\ 1.671501085885 * 10^6 & -1.22605313688360 * 10^7 & 2.16034174036277 * 10^7 \end{pmatrix},$$

$$\hat{A}_{\mu'\mu} = \begin{pmatrix} 0.0000147012145478876 & -1.1140900826293 * 10^{-7} & -1.021605104012 * 10^{-7} \\ 0 & 0.0000831384462433374 & 0.0000182360681372795 \\ 0 & 0 & 0.0002151487224526028 \end{pmatrix}.$$

The normalized overlap $\langle u_\mu | u_{\mu'} \rangle_{no}$ from (67) and orthogonalization matrix $A_{\mu'\mu}$ from (71) at $\lambda=12, N=14, \mu=0,1,2, L=12, M=12$:

$$\langle u_\mu | u_{\mu'} \rangle = \begin{pmatrix} 1.0000000000000000000 & 0.00757821796968012826 & 0.00528687022845146168 \\ 0.00757821796968012826 & 1.0000000000000000000 & -0.2193057245440392052 \\ 0.00528687022845146168 & -0.2193057245440392052 & 1.0000000000000000000 \end{pmatrix},$$

$$A_{\mu'\mu} = \begin{pmatrix} 1.0000000000000000000 & -0.00757821796968012826 & -0.00694912043276433934 \\ 0 & 1.0000000000000000000 & 0.2193457895990078230 \\ 0 & 0 & 1.0000000000000000000 \end{pmatrix}.$$

Note that the overlaps $\langle \hat{u}_\mu | \hat{u}_{\mu'} \rangle$ calculated in the laboratory frame and defined by Eqs. (43), (45) at fixed values of λ and L are independent of M, i.e., they are equal for different values of the quantum number M. This is due to the Wigner–Eckart theorem for spherical tensors in respect to SO(3) group. It means that the corresponding orthogonalization matrices $A_{\mu',\mu}$ defined by Eq. (73) at fixed values of λ and L with different values of quantum number M are equal too. These facts give essential optimization of the computer resources in the large-scale calculations with increasing seniority number λ determined by eigenvalues of the Casimir operator of the group O(5), see Table 1 and Eq. (39).

One can see also that the overlaps of orthogonalization matrices at fixed L for the case of $N > \lambda$ differ only by an integer multiplier from the case $N = \lambda$, for example, in the above cases, this multiplier is equal to 29.

In intrinsic frame, the unnormalized overlap $\langle \hat{u}_\mu | \hat{u}_{\mu'} \rangle$ is determined by (68) and orthogonalization matrix $\hat{A}_{\mu'\mu}$ from (71) at $\lambda=6$, $N=6$, $\mu=0,1$, $L=6$, with summation over $K=0,2,4,6$:

$$\langle \hat{u}_\mu | \hat{u}_{\mu'} \rangle = \begin{pmatrix} 7572204\pi/385 & -301113\pi/40040 \\ -301113\pi/40040 & 1699\pi/65065 \end{pmatrix} = \begin{pmatrix} 61789.0401503461 & -23.6257339835260 \\ -23.6257339835260 & 0.0820343643809891 \end{pmatrix},$$

$$\hat{A}_{\mu'\mu} = \begin{pmatrix} 0.004022946671779255 & 0.001334983666487966 \\ 0 & 3.491419967151016 \end{pmatrix}.$$

The following normalized overlap $\langle u_\mu | u_{\mu'} \rangle$ and matrix $A_{\mu'\mu}$ are

$$\langle u_\mu | u_{\mu'} \rangle = \begin{pmatrix} 1.000000000000000 & -0.3318422478360953 \\ -0.3318422478360953, & 1.000000000000000 \end{pmatrix},$$

$$A_{\mu'\mu} = \begin{pmatrix} 1.000000000000000 & 0.3318422478360953 \\ 0 & 1.000000000000000 \end{pmatrix}.$$

The unnormalized overlap $\langle \hat{u}_\mu | \hat{u}_{\mu'} \rangle$ from (68) and the orthogonalization matrix $\hat{A}_{\mu'\mu}$ from (71) at $\lambda=12$, $N=12$, $\mu=0,1,2$, $L=12$, with summation over $K=0,2,\ldots,12$:

$$\langle \hat{u}_\mu | \hat{u}_{\mu'} \rangle = \begin{pmatrix} 1116847934437.424 & -48516553.06697824 & 7016.562464786226 \\ -48516553.06697824 & 5966.500516763265 & -0.9790521939740782 \\ 7016.562464786226 & -0.9790521939740782 & 0.0008802758859593768 \end{pmatrix},$$

$$\hat{A}_{\mu'\mu} = \begin{pmatrix} 9.462436483745545e-7 & 5.623880080711865e-7 & -4.629117698040114e-8 \\ 0 & 0.01294613581264879 & 0.003808823740956861 \\ 0 & 0 & 33.70471020973709 \end{pmatrix}.$$

The following normalized overlap $\langle u_\mu | u_{\mu'} \rangle$ and matrix $A_{\mu'\mu}$ are

$$\langle u_\mu | u_{\mu'} \rangle = \begin{pmatrix} 1.000000000000000 & -0.5943374193710569 & 0.2237783001963385 \\ -0.5943374193710569 & 1.000000000000000 & -0.4272052696463735 \\ 0.2237783001963385 & -0.4272052696463735 & 1.000000000000000 \end{pmatrix},$$

$$A_{\mu'\mu} = \begin{pmatrix} 1.000000000000000 & 0.5943374193710569 & -0.04892099097301160 \\ 0 & 1.000000000000000 & 0.2942054521964399 \\ 0 & 0 & 1.000000000000000 \end{pmatrix}.$$

As an example, in Fig. 1 we show the CPU time and MaxMemoryUsed during of calculations of overlap integrals (67) and (68) and execution of the G-S orthonormalization procedure (71)-(72) in the laboratory and intrinsic frames by the above *symbolic algorithm* versus parameter λ with help of the O5SU11 code using the PC Intel Celeron CPU 2.16 GHz 4GB 64bit Windows 8.1. The computations were evaluated numerically to 20-digit precision that have been confirmed by the calculated values of the diagonal matrices from the last arrow test of Algorithm in Table 4 to 20-digit precision. One can see that the CPU time (in logarithmic scale) of execution of the overlap integrals is linearly growing. However, the G-S orthonormalization procedure in the intrinsic frame has reduced the computer resources in comparison with one in the laboratory frame.

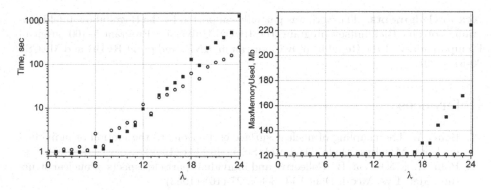

Fig. 1. The CPU time in s. (on the left) and the maximum memory in Mb used to store intermediate data for the current Mathematica session in computation of the overlap integrals and orthogonalization matrices (on the right). Both values are given versus the parameter λ at $L = \lambda$ in the laboratory (marked by squares) and intrinsic (marked by cycles) frames.

5 Conclusion

In present paper, we have elaborated a new universal effective symbolic-numeric algorithm implemented as the first version of O5SU11 code in the Wolfram Mathematica for computing the orthonormal basis of the Bohr–Mottelson(BM) collective model in the both intrinsic and laboratory frames, which can be implemented in any computer algebra system. This kind of basis is widely used for calculating the spectra and electromagnetic transitions in solid, molecular, and nuclear physics. The new symbolic algorithm for orthonormalization of the obtained BM basis based on the Gram–Schmidt orthonormalization procedure has been developed.

The distinct advantage of this method is that it does not involve any square root operation on the expressions coming from the previous steps for computation of the orthonormalization coefficients for this basis. This makes the proposed method very suitable for calculations using computer algebra systems. The symbolic nature of the developed algorithms allows one to avoid the numerical round-off errors in calculation of spectral characteristics (especially close to resonances) of quantum systems under consideration and to study their analytical properties for understanding the dominant symmetries [19].

The program SO5U11 in the Mathematica language for the orthonormalization of the non-canonical basis using the overlap integrals in the laboratory and intrinsic frames (Eqs. (67) and (68)) given by the analytical formula is now prepared and will be published as an open code elsewhere. The great advantage of the program SO5U11 is the possibility to specify an arbitrary precision of calculations which is especially important for large-scale calculations of physical quantities that involve procedures of matrices inversion in eigenvalue problems with degenerated spectra or similar one [32].

Acknowledgments. The work was partially supported by the Bogoliubov–Infeld program, Votruba–Blokhintsev program, the RUDN University Program 5–100, grant of Plenipotentiary of the Republic of Kazakhstan in JINR, and grant RFBR and MECSS 20-51-44001.

References

1. Bohr, A.: The coupling of nuclear surface oscillations to the motion of individual nucleons. Mat Fys. Medd. Dan. Vid. Selsk. 26 (14) (1952)
2. Bohr, A., Mottelson, B.: Collective and individual-particle aspects of nuclear structure. Mat. Fys. Medd. Dan. Vid. Selsk. 27 (16) (1953)
3. Bohr, A., Mottelson, B.R.: Nuclear Structure, vol. 2. W.A. Bejamin Inc., Amsterdam (1970)
4. Eisenberg, J.M., Greiner, W.: Nuclear Theory. Third edition, North-Holland, Vol. 1, (1987)
5. Moshinsky, M., Smirnov, Y.F.: The Harmonic Oscillator in Modern Physics. Harwood Academic Publishers GmbH, Netherlands (1996)
6. Chac'on, E., Moshinsky, M., Sharp, R.T.: $U(5) \supset O(5) \supset O(3)$ and the exact solution for the problem of quadrupole vibrations of the nucleus. J. Math. Phys. **17**, 668–676 (1976)
7. Chac'on, E., Moshinsky, M.: Group theory of the collective model of the nucleus. J. Math. Phys. **18**, 870–880 (1977)
8. Szpikowski, S., Góźdź, A.: The orthonormal basis for symmetric irreducible representations of $O(5) \times SU(1,1)$ and its application to the interacting boson model. Nucl. Phys. A **340**, 76–92 (1980)
9. Góźdź, A., Szpikowski, S.: Complete and orthonormal solution of the five-dimensional spherical harmonic oscillator in Bohr-Mottelson collective internal coordinates. Nucl. Phys. A **349**, 359–364 (1980)
10. Hess, P.O., Seiwert, M., Maruhn, J., Greiner, W.: General Collective Model and its Application to $^{238}_{92}$ UZ. Phys. A **296**, 147–163 (1980)
11. Troltenier, D., Maruhn, J.A., Hess, P.O.: Numerical application of the geometric collective model. In: Langanke, K., Maruhn, J.A., Konin, S.E. (eds.) Computational Nuclear Physics, vol. 1, pp. 116–139. Springer-Verlag, Berlin (1991). https://doi.org/10.1007/978-3-642-76356-4_6
12. Yannouleas, C., Pacheco, J.M.: An algebraic program for the states associated with the $U(5) \supset O(5) \supset O(3)$ chain of groups. Comput. Phys. Commun. **52**, 85–92 (1988)
13. Yannouleas, C., Pacheco, J.M.: Algebraic manipulation of the states associated with the $U(5) \supset O(5) \supset O(3)$ chain of groups: orthonormalization and matrix elements. Comput. Phys. Commun. **54**, 315–328 (1989)
14. Welsh, T.A., Rowe, D.J.: A computer code for calculations in the algebraic collective model of the atomic nucleus. Comput. Phys. Commun. **200**, 220–253 (2016)
15. Ferrari-Ruffino, F., Fortunato, L.: GCM Solver (Ver. 3.0): a mathematica notebook for diagonalization of the geometric collective model (Bohr Hamiltonian) with generalized gneuss-greiner potential. Computation 6, 48 (2018)
16. Chen, J.Q., et al.: Intrinsic Lie group and nuclear collective rotation about intrinsic axes. J. Phys. A: Math. Gen. **16**, 1347–1360 (1983)
17. Chen, J.Q., Pingand, J., Wang, F.: Group Representation Theory for Physicists. World Sci, Singapore (2002)

18. Góźdź, A., et al.: Structure of Bohr type collective spaces - a few symmetry related problems. Nuclear Theor. **32**, 108–122 (2014). Eds. A. Georgiewa, N. Minkov, Heron Press, Sofia (2014)

19. Góźdź, A., Pędrak, A., Gusev, A.A., Vinitsky, S.I.: Point Symmetries in the Nuclear SU(3) Partner Groups Model. Acta Phys. Polonica B Proc. Suppl. **11**, 19–28 (2018)

20. Pr'ochniak, L., Zajac, K.K., Pomorski, K., et al.: Collective quadrupole excitations in the 50<Z, N<82 nuclei with the general Bohr Hamiltonian. Nucl. Phys. A **648**, 181–202 (1999)

21. Pr'ochniak, L., Rohozi'nski, S.G.: Quadrupole collective states within the Bohr collective Hamiltonian. J. Phys. G: Nucl. Part. Phys. **36**, 123101 (2009)

22. Gusev, A.A., Gerdt, V.P., Vinitsky, S.I., Derbov, V.L., Góźdź, A., Pędrak, A.: Symbolic algorithm for generating irreducible bases of point groups in the space of SO(3) group. In: Gerdt, V.P., Koepf, W., Seiler, W.M., Vorozhtsov, E.V. (eds.) CASC 2015. LNCS, vol. 9301, pp. 166–181. Springer, Cham (2015). https://doi.org/10.1007/978-3-319-24021-3_13

23. Gusev, A.A., et al.: Symbolic algorithm for generating irreducible rotational-vibrational bases of point groups. In: Gerdt, V.P., Koepf, W., Seiler, W.M., Vorozhtsov, E.V. (eds.) CASC 2016. LNCS, vol. 9890, pp. 228–242. Springer, Cham (2016). https://doi.org/10.1007/978-3-319-45641-6_15

24. Varshalovitch, D.A., Moskalev, A.N., Hersonsky, V.K.: Quantum Theory of Angular Momentum, Nauka, Leningrad (1975) (also World Scientific, Singapore (1988))

25. Moshinsky, M., Seligman, T.H., Wolf, K.B.: Canonical transformations and the radial oscillator and Coulomb problems. J. Math. Phys. **13**, 901–907 (1972)

26. Abramowitz, M., Stegun, I.A.: Handbook of Mathematical Functions. Dover, New York (1972)

27. Bes, D.R.: The γ-dependent part of the wave functions representing γ-unstable surface vibrations. Nucl. Phys. **10**, 373–385 (1959)

28. Budnik, A.P., Gay, E.V., Rabotnov, N.S., et al.: Basis wave functions and operator matrices of collective nuclear model. Soviet J. Nuclear Phys. **14**(2), 304–314 (1971)

29. Strang, G.: Linear Algebra and its Applications. Academic press, N. Y. (1976)

30. Weisstein, E.W.: Gram-Schmidt Orthonormalization. From MathWorld - A Wolfram WebResource. http://mathworld.wolfram.com/Gram-SchmidtOrthonormalization.html

31. MathWorld - A Wolfram WebResource. http://mathworld.wolfram.com

32. Dvornik, J., Jaguljnjak Lazarevic, A., Lazarevic, D., Uros, M.: Exact arithmetic as a tool for convergence assessment of the IRM-CG method. Heliyon. 6, e03225 (2020). https://doi.org/10.1016/j.heliyon.2020.e03225

Symbolic-Numeric Study of Geometric Properties of Adiabatic Waveguide Modes

Dmitriy V. Divakov(✉)📷, Anastasiia A. Tiutiunnik📷,
and Anton L. Sevastianov📷

Department of Applied Probability and Informatics,
Peoples' Friendship University of Russia (RUDN University),
6 Miklukho-Maklaya St, Moscow 117198, Russia
{divakov-dv,tyutyunnik-aa,sevastianov-al}@rudn.ru

Abstract. The eikonal equation links wave optics to ray optics. In the present work, we show that the eikonal equation is also valid for an approximate description of the phase of vector fields describing guided-wave propagation in inhomogeneous waveguide structures in the adiabatic approximation. The main result of the work was obtained using the model of adiabatic waveguide modes. Highly analytical solution procedure makes it possible to obtain symbolic or symbolic-numerical expressions for vector fields of guided modes. Making use of advanced computer algebra systems, we describe fundamental properties of adiabatic modes in symbolic form. Numerical results are also obtained by means of computer algebra systems.

Keywords: Vector fields · Eikonal equation · Luneburg lens · Focusing · Adiabatic waveguide modes · Symbolic solution of Maxwell equations · Symbolic-numerical method

1 Introduction

Vector problems of electrodynamics usually require significant computational resources and are studied using various numerical methods, such as the finite-difference time-domain (FDTD) or Yee's method, finite element method (FEM), incomplete Galerkin method (IGM) or Kantorovich method, as well as their combinations.

1.1 Purely Numerical Methods

Finite-Difference Methods. Completely numerical methods, e.g., FDTD [1–3] and other finite-difference methods, begin from discretization of the

---

The contribution of D.V. Divakov (investigation – obtaining numerical results) and A.A. Tiutiunnik (investigation – obtaining symbolic results) is supported by the Russian Science Foundation (grant no. 20-11-20257). The contribution of A.L. Sevastianov is conceptualization, formal analysis and writing.

© Springer Nature Switzerland AG 2020
F. Boulier et al. (Eds.): CASC 2020, LNCS 12291, pp. 228–244, 2020.
https://doi.org/10.1007/978-3-030-60026-6_13

continuous variables of the problem and, thereby, offer no possibility of analyzing solutions at the level of symbolic expressions from the very first step.

Finite-difference methods are universal and suitable for the widest class of problems – both linear and nonlinear problems are approximated by finite-difference analogues. However, the price to pay for this versatility is the significant expenditure of computer resources, especially if the object is extended and nonuniform in one or several spatial directions. The success of such methods is directly related to the availability of large computing power.

Finite Element Methods. Finite element methods as well as finite-difference methods are applicable to a wide class of problems [4–6]. Although the solution is represented as a functional dependence, this dependence only ensures smoothness of the solution rather than reflects its physical properties. Due to their versatility, finite element methods are also dependent on computing power.

1.2   Symbolic-Numerical Methods

Galerkin and Kantorovich Methods. In solving electrodynamic problems, there is an "intermediate" class of methods that represent the approximate solution as an expansion in a system of basis functions. The expansion coefficients can be constants (in the classical Galerkin method [12,13]) or functions of one or several spatial variables (in the Kantorovich method [7,8] and in the incomplete Galerkin method [9–11]). The system of functions in which the solution is expanded must be complete in the functional space to which the desired solution should belong, and some additional conditions (e.g., matching, smoothness, etc.).

As a rule, a fortunate choice of basis functions allows solving the problem with sufficient accuracy even keeping a small number of expansion terms.

The main advantages of this "mixed" approach are:

1. Saving computing resources. The initial problem for multidimensional partial differential equations is reduced at a symbolic level to a system of ordinary differential equations (ODE) with initial or boundary conditions. The problem for the ODE system is solved in reasonable time on a personal computer.
2. Representation of results in the form of symbolic expressions allows a more detailed analysis and provides greater clarity of their physical meaning.

Model of Adiabatic Waveguide Modes. We used the symbolic-numerical approach to develop the adiabatic waveguide modes (AWM) method based on the model of adiabatic waveguide modes described in Refs. [14,15].

In Ref. [15], a symbolic form of the adiabatic waveguide modes in an arbitrary homogeneous layer of a multilayer waveguide was derived, basing on which it is possible to construct waveguide modes of multilayer smoothly-irregular waveguide structures. With further use of the symbolic-numerical approach, these modes can serve as a basis for Kantorovich decomposition.

1.3   Formulation of the Problem

The problem of finding the phase of a waveguide mode in a regular waveguide (by the example of a three-layer regular waveguide) was considered and numerically solved in [15]. This problem reduces to finding zeros of the characteristic determinant of the matrix of boundary equations, which yields the phase deceleration coefficients of the guided modes β_j. The phase of each of the guided modes $\varphi_j(z)$ is trivially determined given the phase deceleration coefficient: $\varphi_j(z) = \beta_j(z - z_0) + \varphi^0$, where φ^0 is the initial phase, corresponding to $z = z_0$.

The next stage of the study is to formulate the problem of finding the phase of an adiabatic waveguide mode in an irregular waveguide (by the example of a four-layer waveguide with one irregular layer) and to solve it numerically. In this case the presence of a layer with variable thickness violates the linear behaviour of the phase, so that if the layer irregularity depends on both y and z, the phase will also be a function of y and z.

The formulation of this problem and the development of an approximate method for solving it is the subject of the present paper. As an irregular structure, we consider the Luneburg waveguide lens, which is a three-layer regular waveguide with a fourth layer having variable thickness depending on y and z.

The structure choice was not accidental: it is an object rather complicated for modeling, however, its basic properties are known from physical experiments. So, we will carry out numerical calculations using the example of a Luneburg waveguide lens.

2   Methods and Approaches

2.1   AWM Model, the Form of the Solution

The AWM model approximately describes the guided modes in smoothly-irregular waveguide structures (for details see Section 2 of Ref. [15]). In this study without loss of generality, a four-layered structure will be considered.

The AWM model makes use of the asymptotic method [16], in which electromagnetic fields are presented in the form [15]:

$$\vec{E}(x, y, z, t) = \sum_{s=0}^{\infty} \frac{\vec{E}^s(x; y, z)}{(-i\omega)^{\gamma+s}} \exp\{i\omega t - ik_0\varphi(y, z)\}, \tag{1}$$

$$\vec{H}(x, y, z, t) = \sum_{s=0}^{\infty} \frac{\vec{H}^s(x; y, z)}{(-i\omega)^{\gamma+s}} \exp\{i\omega t - ik_0\varphi(y, z)\}, \tag{2}$$

where k_0 is the wavenumber, $\varphi(y, z)$ is the phase, and $\vec{E}^s(x; y, z)$, $\vec{H}^s(x; y, z)$ determine the amplitude of the s-th order. In the notation of $\vec{E}^s(x; y, z)$, $\vec{H}^s(x; y, z)$ the separation of x by a semicolon means the following assumption: $\partial\vec{E}^s/\partial y$, $\partial\vec{E}^s/\partial z$, $\partial\vec{H}^s/\partial y$, $\partial\vec{H}^s/\partial z$ are small quantities.

In other words, the following expressions for the derivatives are valid:

$$\frac{\partial \vec{E}}{\partial y} = -ik_0 \varphi_y \vec{E},$$

$$\frac{\partial \vec{E}}{\partial z} = -ik_0 \varphi_z \vec{E},$$

and the analogous expressions:

$$\frac{\partial \vec{H}}{\partial y} = -ik_0 \varphi_y \vec{H},$$

$$\frac{\partial \vec{H}}{\partial z} = -ik_0 \varphi_z \vec{H},$$

in which φ_y and φ_z are partial derivatives of $\varphi(y, z)$ in y and z, respectively.

2.2 AWM Model. Reduction of Maxwell Equations

The Maxwell equations in the zero order $(s = 0)$ of the asymptotic expansion reduce to a system of ordinary differential equations of the first order [15]:

$$\frac{\partial \vec{u}}{\partial x} + A(x, y, z)\, \vec{u} = \vec{0}, \tag{3}$$

and two additional relations

$$E_x^0 = \frac{1}{\varepsilon} \left(\varphi_z H_y^0 - \varphi_y H_z^0 \right), \tag{4}$$

$$H_x^0 = -\frac{1}{\mu} \left(\varphi_z E_y^0 - \varphi_y E_z^0 \right), \tag{5}$$

where the desired vector function $\vec{u}(x; y, z)$ consists of the variables

$$\vec{u}(x; y, z) = \left(E_y^0 \; H_z^0 \; H_y^0 \; E_z^0 \right)^T$$

that describe the distribution of the appropriate field components along the x-axis at each point (y, z). Matrix A is defined as follows [15]

$$A(x, y, z) = \begin{pmatrix} 0 & -\frac{ik_0 \varphi_y^2}{\varepsilon} + ik_0 \mu & \frac{ik_0 \varphi_y \varphi_z}{\varepsilon} & 0 \\ -\frac{ik_0 \varphi_z^2}{\mu} + ik_0 \varepsilon & 0 & 0 & \frac{ik_0 \varphi_y \varphi_z}{\mu} \\ -\frac{ik_0 \varphi_y \varphi_z}{\mu} & 0 & 0 & \frac{ik_0 \varphi_y^2}{\mu} - ik_0 \varepsilon \\ 0 & -\frac{ik_0 \varphi_y \varphi_z}{\varepsilon} & \frac{ik_0 \varphi_z^2}{\varepsilon} - ik_0 \mu & 0 \end{pmatrix} \tag{6}$$

where $\varepsilon = \varepsilon(x, y, z)$ and $\mu = \mu(x, y, z)$ are the piecewise constant permittivity and permeability, respectively.

2.3 AWM Model. Reduction of Boundary Conditions for Maxwell Equations

At the discontinuity surfaces of $\varepsilon = \varepsilon(x,y,z)$ and $\mu = \mu(x,y,z)$ the matching conditions must be satisfied that follow from the boundary conditions for Maxwell equations. For planar boundaries $x = c$ ($c = const$) the tangential components $E_y^0, H_z^0, H_y^0, E_z^0$ must be continuous, i.e., in vector notation [15],

$$[\vec{u}]|_{x=c} = \vec{0}, \tag{7}$$

where $[\vec{u}]|_{x=c} = \vec{u}|_{x=c-0} - \vec{u}|_{x=c+0}$ is the jump of vector function $\vec{u}$ at the point $x = c$. For curved boundaries $x = h(y,z)$ the continuity conditions [15]

$$[\vec{u} + V\vec{u}]|_{x=h(y,z)} = \vec{0}, \tag{8}$$

must be fulfilled, where matrix V has the following form:

$$V = \begin{pmatrix} 0 & \dfrac{h_y\varphi_y}{\varepsilon} & -\dfrac{h_y\varphi_z}{\varepsilon} & 0 \\ -\dfrac{h_z\varphi_z}{\mu} & 0 & 0 & \dfrac{h_z\varphi_y}{\mu} \\ \dfrac{h_y\varphi_z}{\mu} & 0 & 0 & -\dfrac{h_y\varphi_y}{\mu} \\ 0 & -\dfrac{h_z\varphi_y}{\varepsilon} & \dfrac{h_z\varphi_z}{\varepsilon} & 0 \end{pmatrix}, \tag{9}$$

where h_y and h_z are partial derivatives of $h(y,z)$ in y and z, respectively.

Guided modes correspond to electromagnetic fields that satisfy the asymptotic conditions [17]

$$\|\vec{u}\| \xrightarrow[x\to\pm\infty]{} 0. \tag{10}$$

2.4 AWM Model. The Approximation of "Horizontal" Boundary Conditions

At first let us restrict ourselves to the approximation of "horizontal" boundary conditions (7), which will play the role of zero-order approximation to the boundary conditions (8) with respect to the small parameter ν at

$$\nu = \|V\| \ll 1, \tag{11}$$

where $\|V\| = \max\limits_{i,j} \{|v_{i,j}|\}$.

Remark. Relation (11) is valid if any of the quantities $h_y\varphi_y$, $h_y\varphi_z$, $h_z\varphi_y$ and $h_z\varphi_z$ is small in absolute value. In the AWM model the smoothly irregular structures are considered, for which h_y, h_z are small. Hence, ν is not small only when the quantities φ_y and φ_z (or at least one of them) are much greater than unity. Therefore, the principal aspect of the considered approximation is the estimation of smallness of ν, which will be performed a posteriori in the course of numerical calculations.

2.5   AWM Model. Setting of the Problem for the Current Study

In Ref. [15], system (3) is solved in the symbolic form for constant ε, μ, which offers a possibility of solving the system (3) for piecewise constant ε, μ. Given the general solution of the system (3) in each domain of constant ε, μ and the matching conditions (7) for the boundaries between these domains, using the conditions (10) for unlimited domains of constant ε, μ, we derive a homogeneous system of equations. The unknowns in this system are coefficients at the functions of the fundamental system of solutions in each domain of constant ε, μ. The determinant of this system should be zero to ensure the existence of a nontrivial solution.

In the layer number α with constant permittivity and permeability $\varepsilon = \varepsilon_\alpha$, $\mu = \mu_\alpha$ the solution of the system of differential Eq. (3) has the form [15]

$$
\vec{u}_\alpha\left(x; y, z\right) = A_\alpha \begin{pmatrix} q_\alpha \\ -i\varepsilon_\alpha \eta_\alpha \\ 0 \\ \varphi_y \varphi_z \end{pmatrix} e^{\gamma_\alpha x} + B_\alpha \begin{pmatrix} -i\mu_\alpha \eta_\alpha \\ p_\alpha \\ \varphi_y \varphi_z \\ 0 \end{pmatrix} e^{\gamma_\alpha x} +
$$

$$
+ C_\alpha \begin{pmatrix} q_\alpha \\ i\varepsilon_\alpha \eta_\alpha \\ 0 \\ \varphi_y \varphi_z \end{pmatrix} e^{-\gamma_\alpha x} + D_\alpha \begin{pmatrix} i\mu_\alpha \eta_\alpha \\ p_\alpha \\ \varphi_y \varphi_z \\ 0 \end{pmatrix} e^{-\gamma_\alpha x},
$$

(12)

where $q_\alpha = \varphi_y^2 - \varepsilon_\alpha \mu_\alpha$, $p_\alpha = \varphi_z^2 - \varepsilon_\alpha \mu_\alpha$, $\eta_\alpha = \sqrt{\varphi_y^2 + \varphi_z^2 - \varepsilon_\alpha \mu_\alpha}$, $\gamma_\alpha = k_0 \eta_\alpha$, and $A_\alpha, B_\alpha, C_\alpha, D_\alpha$ are indefinite constants at each point (y, z).

In this study, using the computer algebra system, we formulate an approximate problem of computing the coefficient of phase deceleration in a general case of a smoothly irregular four-layer structure by an example of the Luneburg waveguide lens.

To present the problem in symbolic form, we consider a four-layer waveguide structure with one layer of variable thickness, which is characterized by the following permittivity and permeability:

$$
\varepsilon = \begin{cases} \varepsilon_c, & x > h_2\left(y, z\right) \\ \varepsilon_l, & h_1 < x < h_2\left(y, z\right) \\ \varepsilon_f, & 0 < x < h_1 \\ \varepsilon_s, & x < 0 \end{cases}, \quad \mu = \begin{cases} \mu_c, & x > h_2\left(y, z\right) \\ \mu_l, & h_1 < x < h_2\left(y, z\right) \\ \mu_f, & 0 < x < h_1 \\ \mu_s, & x < 0 \end{cases}
$$

(13)

The symbolic representation of the solution $\vec{u}_\alpha\left(x; y, z\right)$ in each layer $\alpha = s, f, l, c$ is known (see (12)). This allows writing down a symbolic representation of the characteristic matrix of boundary conditions in the zero-order approximation with respect to ν, i.e., the conditions (7) at the boundaries $x = 0, x = h_1$ and $x = h_2\left(y, z\right)$. Using standard Maple [18] commands subs, expand and simplify, we derive a system of boundary equations based on the solutions $\vec{u}_\alpha\left(x; y, z\right)$ and the boundary conditions. In each of the four layers, the solution comprises four indefinite constants (see (12)), while the boundary conditions at three boundaries $x = 0, x = h_1$ and $x = h_2(y, z)$ yield only 12 equations. The other 4 equations follow from asymptotic conditions (10).

The resulting system of equations at any fixed (y, z) is a system of linear algebraic homogeneous equations having the form

$$M^*(y, z, \varphi_y, \varphi_z) \vec{C} = \vec{0}, \tag{14}$$

where M^* is the matrix of coefficients generally dependent on both y, z and partial derivatives of the sought function φ_y, φ_z; $\vec{C}$ is a vector composed of the indefinite coefficients $A_\alpha, B_\alpha, C_\alpha$ and D_α looked for. System (14) has a nontrivial solution if and only if

$$\det M^*(y, z, \varphi_y, \varphi_z) = 0. \tag{15}$$

Equation (15) is a nonlinear partial differential equation of the first order. It is convenient to solve this equation using the method of characteristics, which reduces the initial nonlinear partial differential equation to a system of ordinary differential equations for the characteristics [19].

Thus, the sought phase of the adiabatic waveguide mode $\varphi(y, z)$ must satisfy the nonlinear Eq. (15), which can be explicitly written only after calculating the determinant in a symbolic form.

Remark. Symbolic calculation of a 12×12 determinant is possible only using the libraries of symbolic transformations. The authors make use of Maple system for this purpose.

Before calculating the determinant (15), we performed symbolic transformations to simplify the elements of matrix M^* specified symbolically. As a result of symbolic simplifications, problem (15) reduces to two problems:

1. Finding zeros of the determinant of the reduced matrix for the considered domain of (y, z), or, in other words, solving the non-linear equation $\det M(y, z, \beta^2(y, z)) = 0$ (where $\beta^2(y, z) = \varphi_y^2 + \varphi_z^2$) and finding desired $\beta^2(y, z)$ for each (y, z) from the considered domain;
2. Subsequent solution of the reduced nonlinear differential equation with the right-hand side calculated at Step 1: $\varphi_y^2 + \varphi_z^2 = \beta^2(y, z)$.

Problem 1 was solved using the function `Determinant` of the Maple package `LinearAlgebra`. The zeros of determinant were approximately found using the classical bisection method [20].

Remark. A specific feature of waveguide problems is that the localizing a zero of the determinant within an interval of 10^{-15} one has to deal with the values of the determinant itself of the order of 10^{30}. Therefore, in the numerical calculations we used the numbers with enlarged mantissa by setting `Digits := 30`.

Problem 2 was solved by the method of characteristics [19] using the command `charstrip` from the library `PDETools` [18], which allows getting a system of ordinary differential equations for characteristics from a nonlinear first-order partial differential equation. This system complemented with the initial conditions was solved numerically using a Fehlberg fourth-fifth-order Runge-Kutta method with degree four interpolant – `rkf45` – with the parameter, determining the relative error `relerr` $= 10^{-12}$ [18]. The method is implemented in Maple in a symbolic-numerical form.

2.6 Numerical Experiment. Verification

To verify the implemented method we consider the waveguide Lunenburg lens of the radius R, designed to focus the waveguide mode TE_0 at the distance $F = 2R$. Within the frameworks of the AWM in the zero-order approximation with respect to $\nu \ll 1$ (approximation of "horizontal" boundary conditions) the problem of finding the phase for different lens radii (from 10^2 to 10^4 wavelengths) was solved. Besides, we estimated a posteriori the order of ν for the same lens radii to determine the range of validity of the approximation of "horizontal" boundary conditions.

As the initial data we took the parameters of the Luneburg lens designed by Konstantin Lovetskiy [21] using the method of cross sections, the initial data for which were provided by the solution of the Morgan equation [22].

3 Results

3.1 Results Obtained in Symbolic Form

We consider the four-layer waveguide structure, formed by a three-layered waveguide on which the fourth layer of variable thickness is deposited, sufficiently extended in the plane yOz to ensure the conditions $|\partial h_2/\partial y| \ll 1$, $|\partial h_2/\partial z| \ll 1$.

Using the Maple toolkit, we write the boundary equations of the AWM model in the zero-order approximation with respect to $\nu \ll 1$ in a symbolic form.

The Main Result. In the zero-order approximation with respect to $\nu \ll 1$ (the approximation of "horizontal" boundary conditions) the phase $\varphi(y, z)$ in the AWM model satisfies the eikonal equation

$$\varphi_y^2 + \varphi_z^2 = \beta^2(y, z), \tag{16}$$

where $\beta^2(y, z)$ is the square of the phase deceleration coefficient.

Appendix. The quantity $\beta^2(y, z)$ is determined as a root of the equation

$$\det M\left(y, z, \beta^2(y, z)\right) = 0, \tag{17}$$

where the 8×8 matrix M is defined as

$$M = \begin{pmatrix} M_{11} & M_{12} \\ M_{21} & M_{22} \end{pmatrix}, \tag{18}$$

236   D. V. Divakov et al.

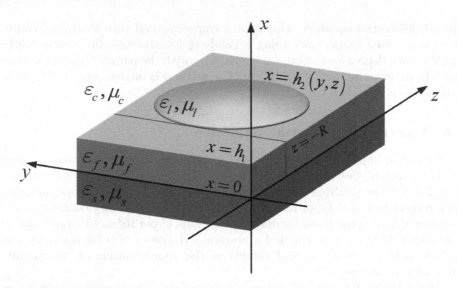

Fig. 1. Structure of the irregular four-layer waveguide (additional waveguide layer has variable thickness $d = h_2(y, z) - h_1$)

where

$$M_{11} = \begin{pmatrix} \theta_s & i\eta_f\mu_f - i\eta_s\mu_s & \theta_s & -i\eta_f\mu_f - i\eta_s\mu_s \\ i\eta_f\varepsilon_f - i\eta_s\varepsilon_s & \theta_s & -i\eta_f\varepsilon_f - i\eta_s\varepsilon_s & \theta_s \\ \theta_f e^{\gamma_f h_1} & -ie^{\gamma_f h_1}\eta_f\mu_f & \theta_f e^{-\gamma_f h_1} & ie^{-\gamma_f h_1}\eta_f\mu_f \\ -ie^{\gamma_f h_1}\eta_f\varepsilon_f & \theta_f e^{\gamma_f h_1} & ie^{-\gamma_f h_1}\eta_f\varepsilon_f & \theta_f e^{-\gamma_f h_1} \end{pmatrix}, \quad (19)$$

$$M_{12} = \begin{pmatrix} 0 & 0 & 0 & 0 \\ 0 & 0 & 0 & 0 \\ 0 & ie^{-\gamma_l d}\eta_l\mu_l & 0 & -ie^{\gamma_l d}\eta_l\mu_l \\ ie^{-\gamma_l d}\eta_l\varepsilon_l & 0 & -ie^{\gamma_l d}\eta_l\varepsilon_l & 0 \end{pmatrix}, \quad (20)$$

$$M_{21} = \begin{pmatrix} e^{\gamma_f h_1} & 0 & e^{-\gamma_f h_1} & 0 \\ 0 & e^{\gamma_f h_1} & 0 & e^{-\gamma_f h_1} \\ 0 & 0 & 0 & 0 \\ 0 & 0 & 0 & 0 \end{pmatrix}, \quad (21)$$

$$M_{22} = \begin{pmatrix} -e^{-\gamma_l d} & 0 & -e^{\gamma_l d} & 0 \\ 0 & -e^{-\gamma_l d} & 0 & -e^{\gamma_l d} \\ \theta_c & i\eta_c\mu_c + i\eta_l\mu_l & \theta_c & i\eta_c\mu_c - i\eta_l\mu_l \\ i\eta_c\varepsilon_c + i\eta_l\varepsilon_l & \theta_c & i\eta_c\varepsilon_c - i\eta_l\varepsilon_l & \theta_c \end{pmatrix} \quad (22)$$

and $\eta_\alpha = \sqrt{\beta^2(y,z) - \varepsilon_\alpha\mu_\alpha}$, $d = h_2(y,z) - h_1$, $\theta_c = \varepsilon_l\mu_l - \varepsilon_c\mu_c$, $\theta_f = \varepsilon_l\mu_l - \varepsilon_f\mu_f$, $\theta_s = \varepsilon_f\mu_f - \varepsilon_s\mu_s$.

3.2 Results Obtained Numerically

To verify the result obtained we consider the Luneburg waveguide lens designed to focus the waveguide mode TE_0 at length $F = 2R$, where R is the waveguide lens radius.

We consider the guided mode TE_0, propagating in a three-layer waveguide from $z = -\infty$ (see Fig. 1) in the positive direction of the z-axis with the phase $\varphi_0(z) = \beta_0(z + R)$, where β_0 is the coefficient of phase deceleration. At $z = -R$ the mode enters the waveguide lens. The phase in this domain satisfies the eikonal Eq. (16), which is to be solved.

Initial Data: The wavelength $\lambda = 0.55\,[\mu m]$; the wavenumber $k_0 = 2\pi/\lambda\,[\mu m^{-1}]$; the waveguide lens radius $R = 10^3\lambda$; the thickness of the main waveguide layer $h_1 = 2\lambda$; the coating and the substrate in the model are semi-infinite; the variable thickness of the additional waveguide layer is defined as $d(y, z) = h_2(y, z) - h_1$. Due to the cylindrical symmetry of the lens, $h_2(y, z) = h(r)|_{r=\sqrt{y^2+z^2}/R}$, the plot of $h(r)$ is shown in Fig. 2; the permittivities of the materials are $\varepsilon_c = 1$, $\varepsilon_f = 2.449225$, $\varepsilon_l = 3.61$, $\varepsilon_s = 2.1609$, and their permeabilities are $\mu_c = \mu_f = \mu_l = \mu_s = 1$; the coefficient of phase deceleration of the mode TE_0 of the three-layer waveguide is $\beta_0 \approx 1.55149273806929012586$.

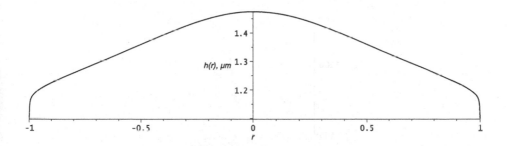

Fig. 2. The upper boundary of the additional waveguide layer

Numerical Results. The variable thickness $d(y, z)$ of the additional waveguide layer corresponds to the function $\beta^2(y, z)$ that determines the square of the phase deceleration coefficient at the point (y, z), presented in Fig. 3.

The characteristics of the eikonal equation with the right-hand side $\beta^2(y, z)$ (shown in Fig. 3) are presented in Fig. 4 by the projections of the characteristics on the plane yOz, which we will refer to as rays, and in Fig. 5 by the integral surface $\varphi(y, z)$ of the eikonal Eq. (16), composed of the family of integral curves.

Remark. The calculations for lenses with radii $R = 10^2\lambda$, $10^3\lambda$, $10^4\lambda$ yield seemingly similar results (like in Fig. 4, 5) differing only in the scale of the considered domain.

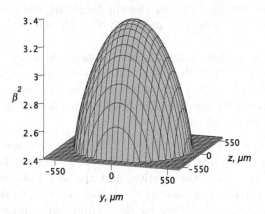

Fig. 3. Plot of $\beta^2(y, z)$

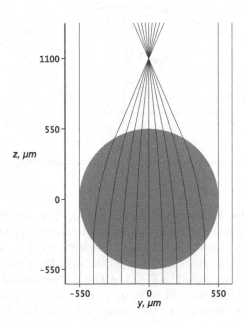

Fig. 4. Projections of characteristics on the yOz plane for the Luneburg lens with $R = 10^3\lambda$, $F = 2R$

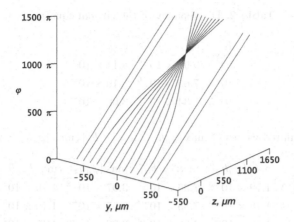

Fig. 5. Integral surface of the phase, composed of the family of integral curves for the Luneburg lens with $R = 10^3\lambda$, $F = 2R$

Table 1. Lengths of the localization interval of the ray crossing point and apertures for different lens radii

R	$|I_F|$, μm	$|I_F|/F$	A/R, %
$10^2\lambda$	1.53×10^{-3}	1.39×10^{-5}	93.8
$10^3\lambda$	1.78×10^{-2}	1.62×10^{-5}	97.9
$10^4\lambda$	1.54×10^{-1}	1.40×10^{-5}	98.8

The crossing points of all calculated rays passed through the lens are localized in the interval $I_F = [z_{\min} \; ; \; z_{\max}]$. In Table 1 for $R = 10^2\lambda$, $10^3\lambda$, $10^4\lambda$ we present the calculated interval lengths $|I_F| = z_{\max} - z_{\min}$, as well as $|I_F|/F$, where $F = 2R$. For each $R = 10^2\lambda$, $10^3\lambda$, $10^4\lambda$ we also present the limit value of A/R in percent, where the rays, coming from the points $-A \leq z \leq A$ cross among themselves, while the rays coming from the points $z > A$ and $z < -A$ are parallel to the z-axis.

We also calculate the discrepancy of the eikonal equation

$$\delta_{abs} = \max \left| \varphi_y^2 + \varphi_z^2 - \beta^2(y,z) \right|,$$

$$\delta_{rel} = \max \left\{ \left| \varphi_y^2 + \varphi_z^2 - \beta^2(y,z) \right| / \left| \beta^2(y,z) \right| \right\}$$

along the rays, where φ_y and φ_z are found approximately using the method of characteristics (Table 2).

The calculated values of $\max|\varphi_y h_y|$, $\max|\varphi_z h_y|$, $\max|\varphi_y h_z|$ and $\max|\varphi_z h_z|$ are summarized in Table 3.

Table 2. Discrepancy of the eikonal equation

R	δ_{abs}	δ_{rel}
$10^2\lambda$	9.95×10^{-6}	4.11×10^{-6}
$10^3\lambda$	7.69×10^{-6}	3.19×10^{-6}
$10^4\lambda$	7.34×10^{-6}	3.00×10^{-6}

Table 3. Calculated values of $\max|\varphi_y h_y|$, $\max|\varphi_z h_y|$, $\max|\varphi_y h_z|$ and $\max|\varphi_z h_z|$

R	$\max|\varphi_y h_y|$	$\max|\varphi_z h_y|$	$\max|\varphi_y h_z|$	$\max|\varphi_z h_z|$
$10^2\lambda$	1.79×10^{-3}	7.36×10^{-2}	3.97×10^{-2}	1.19×10^{0}
$10^3\lambda$	1.79×10^{-4}	7.36×10^{-3}	3.97×10^{-3}	1.19×10^{-1}
$10^4\lambda$	1.79×10^{-5}	7.36×10^{-4}	3.97×10^{-4}	1.19×10^{-2}

4   Discussion

4.1   Symbolic Results

In this paper, we investigate the AWM model [14,15] in the zeroth order of the asymptotic method using the approximation of "horizontal" boundary conditions. The latter is important from a physical point of view, since it allows comparing the AWM model calculations with the results of the cross-sectional method [21,23,24], which also makes use of "horizontal" boundary conditions.

In waveguide problems, the system of boundary equations plays an important role, because its solution determines the phase of the guided modes and the constants for their further numerical construction.

A symbolic-calculation study of the system of boundary equations (in the zeroth approximation with respect to the parameter ν) allowed simplifying the form of the system at the symbolic level and reducing the problem in a form convenient for numerical solution.

Instead of solving equation $\det M^* (y, z, \varphi_y, \varphi_z) = 0$, which is extremely difficult to analyze in the symbolic form (the matrix dimension in the general case is 12×12) we obtain symbolically the eikonal equation $\varphi_y^2 + \varphi_z^2 = \beta^2 (y, z)$, where only the right-hand side is specified numerically. The quantity $\beta^2 (y, z)$ is specified numerically because it is a solution of the equation $\det M (y, z, \beta^2 (y, z)) = 0$, where due to symbolic manipulations the initial system of boundary equations with the 12×12 matrix M^* is reduced to an equivalent system with the 8×8 matrix M (see. (18)–(22)). Moreover, the computer algebra tools allow determination of $\beta^2 (y, z)$ with enhanced accuracy, using the values with extended number of decimal digits.

The eikonal equation links geometric optics to wave optics, and in the present work its explicit derivation in the AWM model for the particular case of small ν is important for geometric interpretation of the guided propagation of adiabatic modes.

Moreover, numerical experiments answer the question about the applicability of the approximation of "horizontal" boundary conditions within the frameworks of the AWM model.

4.2 Numerical Results

In numerical experiments we consider the waveguide Luneburg lens designed to focus the radiation at the distance $F = 2R$, where R is the waveguide lens radius. The calculated rays, passing through the lens, with sufficient accuracy intersect in the focus point (see Fig. 4). The relative error is of the order of 10^{-5} (see Table 1, column 3) for the radii of the lens $10^2\lambda - 10^4\lambda$.

In fact, the rays calculated using the AWM model with high accuracy cross in the lens focus at any considered radii of the lens. It is important that the calculations demonstrate that the greater the lens radius (see Table 1, column 4), the greater is the lens aperture within the given accuracy. Thus, for the lens radius $10^4\lambda$ the aperture amounts to 98.8%.

In other words, the greater the radius of the waveguide lens, the more exactly the behavior of the rays close to the lens edges is described by the AWM model in the approximation of "horizontal" boundary conditions.

The applicability of "horizontal" boundary conditions is largely determined by the smallness of the parameter ν. Commonly ν is considered small if it is by two orders of magnitude smaller than unity, i.e., if $\nu \sim 10^{-2}$. By definition, $\nu = \|V\| = \max\{\max|\varphi_y h_y|, \max|\varphi_z h_y|, \max|\varphi_y h_z|, \max|\varphi_z h_z|\}$. Table 3 presents the values of $\max|\varphi_y h_y|$, $\max|\varphi_z h_y|$, $\max|\varphi_y h_z|$ and $\max|\varphi_z h_z|$ calculated along the rays. Only for the lens with the radius $10^4\lambda$ the parameter ν is of the order of 10^{-2} and can be considered small.

From Table 3 it is also seen that the larger the radius of the waveguide lens, the smaller the parameter ν. Therefore, for extended Luneburg lenses with $R > 10^4\lambda$ the approximation of "horizontal" boundary conditions is likely to be valid.

The AWM model is formulated for smoothly irregular waveguide structures, so that the approximation of "horizontal" boundary conditions is a natural first step. However, this approximation does not describe the complete variety of physical effects, e.g., the effect of mode hybridization. Note, that the AWM model as such can describe vector fields without using the approximation of "horizontal" boundary conditions. In this case it is necessary to solve the problem $\det M^*(y, z, \varphi_y, \varphi_z) = 0$, to which the method of characteristics can be also applied. An additional difficulty will consist in the necessity to calculate partial derivatives of the determinant. This problem is also expected to be solved using the computer algebra system that allows symbolic differentiation of cumbersome expressions like a determinant.

In the present work, we solved only the problem of approximate determination of the phase. We did not set the problems of describing the field in the waveguide lens completely and, what is of primary importance, of constructing the field near the focal point, which is much more difficult.

The present study is focused at developing symbolic-numerical techniques for phase determination. A necessary condition is the use of symbolic manipulations with symbolic expressions, which allows the formulation of the main result.

The Maple option of numerical calculations using extended number of decimal digits appears to be extremely important for solving ill-conditioned problems.

All Maple programs created within the framework of the current study are publicly available at the following link https://bitbucket.org/DmitriyDivakov/waveguide-luneburg-lens/downloads/.

5   Conclusion

In this work, the eikonal equation is symbolically derived, governing the phase of the adiabatic waveguide mode in the approximation of "horizontal" boundary conditions. Based on numerical calculations, it was found that the approximation of "horizontal" boundary conditions is valid for Luneburg waveguide lenses with a radius of more than $10^4 \lambda$.

Potential applicability of the model of adiabatic waveguide modes to describing the electromagnetic field behavior in focusing problems is demonstrated, which is of importance for modeling and design of waveguide lenses.

As the next step, it is planned to consider the same Luneburg waveguide lens without using the approximation of "horizontal" boundary conditions in the frameworks of the AWM model.

Acknowledgments. The authors are grateful to Konstantin Lovetskiy for providing numerical data of the designed Luneburg lens, based on which all numerical calculations were carried out. The authors are grateful to Leonid Sevastianov for useful discussions and assistance provided in writing this article.

References

1. Yee, K.: Numerical solution of initial boundary value problems involving Maxwell's equations in isotropic media. IEEE Trans. Antennas Propag. **14**(3), 302–307 (1966). https://doi.org/10.1109/TAP.1966.1138693
2. Taflove, A.: Application of the finite-difference time-domain method to sinusoidal steady-state electromagnetic-penetration problems. IEEE Trans. Electromagn. Compat. EMC-**22**(3), 191–202 (1980). https://doi.org/10.1109/TEMC.1980.30387
3. Joseph, R., Goorjian, P., Taflove, A.: Direct time integration of Maxwell's equations in two-dimensional dielectric waveguides for propagation and scattering of femtosecond electromagnetic solitons. Opt. Lett. **18**(7), 491–493 (1993). https://doi.org/10.1364/OL.18.000491
4. Bathe, K.J.: Finite Element Procedures in Engineering Analysis. Prentice Hall, Englewood Cliffs (1982)
5. Gusev, A.A., et al.: Symbolic-numerical algorithms for solving the parametric self-adjoint 2D elliptic boundary-value problem using high-accuracy finite element method. In: Gerdt, V.P., Koepf, W., Seiler, W.M., Vorozhtsov, E.V. (eds.) CASC 2017. LNCS, vol. 10490, pp. 151–166. Springer, Cham (2017). https://doi.org/10.1007/978-3-319-66320-3_12

6. Bogolyubov, A.N., Mukhartova, Yu.V., Gao, J., Bogolyubov, N.A.: Mathematical modeling of plane chiral waveguide using mixed finite elements. In: Progress in Electromagnetics Research Symposium, pp. 1216–1219 (2012)

7. Kantorovich, L.V., Krylov, V.I.: Approximate Methods of Higher Analysis. Wiley, New York (1964)

8. Gusev, A.A., Chuluunbaatar, O., Vinitsky, S.I., Derbov, V.L.: Solution of the boundary-value problem for a systems of ODEs of large dimension: benchmark calculations in the framework of Kantorovich method. Discrete Continuous Models Appl. Comput. Sci. **3**, 31–37 (2016)

9. Sveshnikov, A.G.: The incomplete Galerkin method. Dokl. Akad. Nauk SSSR **236**(5), 1076–1079 (1977)

10. Petukhov, A.A.: Joint application of the incomplete Galerkin method and scattering matrix method for modeling multilayer diffraction gratings. Math. Models Comput. Simul. **6**(1), 92–100 (2014). https://doi.org/10.1134/S2070048214010128

11. Divakov, D., Sevastianov, L., Nikolaev, N.: Analysis of the incomplete Galerkin method for modelling of smoothly-irregular transition between planar waveguides. J. Phys: Conf. Ser. **788**, 012010 (2017). https://doi.org/10.1088/1742-6596/788/1/012010

12. Fletcher, C.A.J.: Computational Galerkin Methods. Springer, Heidelberg (1984). https://doi.org/10.1007/978-3-642-85949-6

13. Tiutiunnik, A.A., Divakov, D.V., Malykh, M.D., Sevastianov, L.A.: Symbolic-numeric implementation of the four potential method for calculating normal modes: an example of square electromagnetic waveguide with rectangular insert. In: England, M., Koepf, W., Sadykov, T.M., Seiler, W.M., Vorozhtsov, E.V. (eds.) CASC 2019. LNCS, vol. 11661, pp. 412–429. Springer, Cham (2019). https://doi.org/10.1007/978-3-030-26831-2_27

14. Sevastyanov, L.A., Sevastyanov, A.L., Tyutyunnik, A.A.: Analytical calculations in maple to implement the method of adiabatic modes for modelling smoothly irregular integrated optical waveguide structures. In: Gerdt, V.P., Koepf, W., Seiler, W.M., Vorozhtsov, E.V. (eds.) CASC 2014. LNCS, vol. 8660, pp. 419–431. Springer, Cham (2014). https://doi.org/10.1007/978-3-319-10515-4_30

15. Divakov, D.V., Sevastianov, A.L.: The implementation of the symbolic-numerical method for finding the adiabatic waveguide modes of integrated optical waveguides in CAS maple. In: England, M., Koepf, W., Sadykov, T.M., Seiler, W.M., Vorozhtsov, E.V. (eds.) CASC 2019. LNCS, vol. 11661, pp. 107–121. Springer, Cham (2019). https://doi.org/10.1007/978-3-030-26831-2_8

16. Babich, V.M., Buldyrev, V.S.: Asymptotic Methods in Short-Wave Diffraction Problems. Nauka, Moscow (1972). [English translation: Springer Series on Wave Phenomena 4. Springer, Berlin Heidelberg New York 1991]

17. Adams, M.J.: An Introduction to Optical Waveguides. Wiley, New York (1981)

18. Mathematics-based software and services for education, engineering, and research. https://www.maplesoft.com/

19. Courant, R., Hilbert, D.: Methods of Mathematical Physics, vol. 2. Partial Differential Equations. nterscience, New York (1962)

20. Hamming, R.W.: Numerical Methods for Scientists and Engineers, 2nd Revised edition. Dover Publications (1987)

21. Gevorkyan, M., Kulyabov, D., Lovetskiy, K., Sevastianov, L., Sevastianov, A.: Field calculation for the horn waveguide transition in the single-mode approximation of the cross-sections method. Proc. SPIE **10337**, 103370H (2017). https://doi.org/10.1117/12.2267906

22. Morgan, S.P.: General solution of the Luneburg lens problem. J. Appl. Phys. **29**, 1358–1368 (1958). https://doi.org/10.1063/1.1723441
23. Shevchenko, V.V.: Smooth Transitions in Open Waveguides. Nauka, Moscow (1969). (in Russian)
24. Ivanov, A.A., Shevchenko, V.V.: A planar transversal junction of two planar waveguides. J. Commun. Technol. Electron. **54**(1), 63–72 (2009). https://doi.org/10.1134/S1064226909010057

Intrinsic Complexity for Constructing Zero-Dimensional Gröbner Bases

Amir Hashemi[1,2]($\boxtimes$), Joos Heintz[3], Luis M. Pardo[4], and Pablo Solernó[5]

[1] Department of Mathematical Sciences, Isfahan University of Technology,
84156-83111 Isfahan, Iran
amir.hashemi@iut.ac.ir
[2] School of Mathematics, Institute for Research in Fundamental Sciences (IPM),
19395-5746 Tehran, Iran
[3] Departamento de Computación and ICC, UBA-CONICET,
Facultad de Ciencias Exactas y Naturales, Universidad de Buenos Aires,
Ciudad Universitaria, 1428 Buenos Aires, Argentina
joos@dc.uba.ar
[4] Depto. de Matemáticas, Estadística y Computación. Facultad de Ciencias,
Universidad de Cantabria, Avda. Los Castros s/n, 39071 Santander, Spain
luis.m.pardo@gmail.com
[5] Departamento de Matemática and IMAS, UBA-CONICET,
Facultad de Ciencias Exactas y Naturales, Universidad de Buenos Aires,
Ciudad Universitaria, 1428 Buenos Aires, Argentina
psolerno@dm.uba.ar

Abstract. In this paper, we give a thorough revision of Lakshman's paper by fixing some serious flaws in his approach. Furthermore, following this analysis, an intrinsic complexity bound for the construction of zero-dimensional Gröbner bases is given. Our complexity bound is in terms of the degree of the input ideal as well as the degrees of its generators. Finally, as an application of the presented method, we exhibit and analyze a (Monte Carlo) probabilistic algorithm to compute the degree of an equidimensional ideal.

1 Introduction

Gröbner bases, introduced by Bruno Buchberger in 1965 in his Ph.D. thesis [5] are a powerful tool for constructive problems in polynomial ideal theory. Using the linear algebra method proposed by Lazard [21] to compute Gröbner bases and by having the maximum degree of the intermediate polynomials during the Gröbner basis computation, we are able to give the complexity of this computation. In 1982, Mayr and Meyer [26] proved that, in the worst case, the maximum degree of a reduced Gröbner basis of an ideal may be double exponential in terms of the maximum degree of a generating set of the ideal. However, for the class of zero-dimensional ideals this upper bound becomes single exponential.

Let us recall some of the existing results on the complexity of computing zero-dimensional Gröbner bases. Let R be the polynomial ring $K[x_1, \ldots, x_n]$ and

© Springer Nature Switzerland AG 2020
F. Boulier et al. (Eds.): CASC 2020, LNCS 12291, pp. 245–265, 2020.
https://doi.org/10.1007/978-3-030-60026-6_14

$I \subset R$ a zero-dimensional ideal generated by polynomials of degree at most d. Let us fix a monomial ordering on R. Lazard in [22] showed that if the projective dimension of I is zero then the complexity of computing a Gröbner basis for I is $d^{O(n)}$. Then, Dickenstein et al. in [7] proved the complexity bound $d^{O(n^2)}$ for this problem. Lazard and Lakshman in [20] described a probabilistic algorithm to compute in time $d^{O(n)}$ Gröbner bases of the radical of I as well as of all the irreducible components of $\sqrt{I}$. Finally, based on this algorithm, Lakshman in [19] showed that the reduced Gröbner basis of I and reduced Gröbner bases for all primary components of I can be constructed in a time polynomial in d^n. Unfortunately, the approach presented in [19] has serious flaws which are investigated in this paper. In principle, there are several errors in [19] (as in Lemmata 1, 4 and 5), but the main doubt lies on the statement of [19, Theorem 2]. After fixing these flaws (by a thorough revision of the paper), based on Lakshman's method, we give new upper bounds for the complexity of computing reduced Gröbner bases for I and of the primary components of I. Our bound depends on the maximum degree of a generating set of I, the degree of I and degrees of the primary components of I. Finally, as an application of this approach, we present a probabilistic method to compute the degree of an equi-dimensional ideal and analyze its complexity.

The article is organized as follows. In the next section, we review the basic definitions and notations which will be used throughout. In Sect. 3, we revise the paper [19] in order to fix its flaws and improve its results. In the last section, we illustrate an application of the results obtained in Sect. 3 for computing the degree of an equidimensional ideal.

2   Preliminaries

In this section, we introduce basic notations and preliminaries (related to the Gröbner bases and degree of an ideal) needed in the subsequent sections. Let K be an infinite field, $X = X_1, \ldots, X_n$ be a sequence of variables and $R = K[X]$ be the polynomial ring over K. We consider polynomials $f_1, \ldots, f_k \in R$ and the ideal $\mathfrak{a} = (f_1, \ldots, f_k)$ generated by them (if the f_i's are homogeneous then we will stress it). Furthermore, the dimension of the ideal $\mathfrak{a}$, denoted by $\dim(\mathfrak{a})$, is the (Krull) dimension of the corresponding factor ring $A = R/\mathfrak{a}$.

We denote by $\mathscr{M}(X) = \{X_1^{\alpha_1} \cdots X_n^{\alpha_n} \mid (\alpha_1, \ldots, \alpha_n) \in \mathbb{N}^n\}$ the set of all monomials in R. Let us fix a monomial ordering $\prec$ on $\mathscr{M}(X)$ with $X_1 \prec \cdots \prec X_n$. The *leading monomial* of a non-zero polynomial $f \in R$, denoted by $\mathrm{LM}(f)$, is the greatest monomial w.r.t. $\prec$ appearing in f and its coefficient is the leading coefficient of f, denoted by $\mathrm{LC}(f)$. The leading term of f is the product $\mathrm{LT}(f) = \mathrm{LC}(f)\mathrm{LM}(f)$. We denote by $t(f)$ the set of all monomials appearing in f. We denote by $\sharp t(f)$ the number of terms appearing in f. For a finite set F we write $\sharp t(F) = \max\{\sharp t(f) \mid f \in F\}$.

For every finite set F, we denote by $\mathrm{LM}(F)$ the set $\{\mathrm{LM}(f) \mid f \in F\}$. The *leading monomial monoid* of an ideal $\mathfrak{a} \subset R$ w.r.t. $\prec$ is defined as

$$\mathrm{LM}(\mathfrak{a}) = \{\mathrm{LM}(f) \mid 0 \neq f \in \mathfrak{a}\}.$$

For every non-zero polynomial $f \in R$, we denote by $\langle \mathrm{LM} \rangle(f)$ the set of all $\mathrm{LM}(f)X^{\alpha}$ with $\alpha = (\alpha_1, \ldots, \alpha_n) \in \mathbb{N}^n$. We denote by $\langle \mathrm{LM} \rangle(F)$ the set monomials in the following class:

$$\langle \mathrm{LM} \rangle(F) = \bigcup_{f \in F} \langle \mathrm{LM} \rangle(f) = \bigcup_{f \in F} \mathrm{LM}(f) \cdot \mathscr{M}(X).$$

Recall that a finite subset $G \subset \mathfrak{a}$ is called a *Gröbner basis* for $\mathfrak{a}$ w.r.t. $\prec$ if $\mathrm{LM}(\mathfrak{a}) = \langle \mathrm{LM} \rangle(G)$. A Gröbner basis G is *minimal* if all leading coefficients of all polynomials in G are equal to 1 and that for all $f \in G$ we have $\mathrm{LM}(f) \notin \langle \mathrm{LM} \rangle(G \setminus \{f\})$. Computing a minimal Gröbner basis from a given Gröbner basis does not increase significantly the complexity of the computation. Finally, recall that a *reduced* Gröbner basis G of an ideal $\mathfrak{a}$ is a minimal Gröbner basis such that for every $f \in G$ no term of f lies in $\langle \mathrm{LM} \rangle(G \setminus \{f\})$.

A key ingredient of the computation of Gröbner bases is Hironaka's multi-variate division algorithm, also called computation of a normal form (remainder) w.r.t. a finite set of polynomials (see [6, Theorem 3, page 64], and more precisely, Exercise 11, page 69). Let $F \subset R$ be a finite set of polynomials and $f \in R$. We will denote by $\mathrm{NF}(f, F)$ a remainder of the multi-variate division of f by F. Another relevant ingredient in the computation of Gröbner bases is the *S-polynomial*. For two non-zero polynomials $f, g \in R$, their *S*-polynomial is defined to be

$$S(f, g) = \frac{M}{\mathrm{LT}(f)} f - \frac{M}{\mathrm{LT}(g)} g, \tag{1}$$

where M is the least common multiple of $\mathrm{LM}(f)$ and $\mathrm{LM}(g)$. In addition, $S(f, g, F)$ stands for a remainder of the Hironaka division of $S(f, g)$ by F. We also denote by $S(f, 0, F)$ a normal form of f w.r.t. F. *Buchberger's criterion* asserts that a finite set F is a Gröbner basis if and only if $S(f, g, F) = 0$ for all $f, g \in F$.

The *normal set* of an ideal $\mathfrak{a}$ w.r.t. $\prec$ is the set of all monomials not in $\mathrm{LM}(\mathfrak{a})$. For every finite subset $F \subset R$, we also denote by $N(F)$ the monomials not in $\langle \mathrm{LM} \rangle(F)$. Recall that an ideal $\mathfrak{a} \subset R$ is zero-dimensional if and only if $N(\mathfrak{a})$ is finite. Moreover, given a zero-dimensional ideal $\mathfrak{a} \subset R$ the residue classes $\{f + \mathfrak{a} \mid f \in N(\mathfrak{a})\}$ forms a basis for A as a K-vector space. Hence, if $\mathfrak{a}$ is zero-dimensional, we obviously have $\dim_K(R/\mathfrak{a}) = \sharp N(\mathfrak{a})$. For more details, we refer the reader to [6].

Let E be an $\mathbb{N}$-graded R-module. Given any $s \in \mathbb{N}$, we denote by E_s the union of $\{0\}$ and the set of the elements of E of degree s. Recall that the *Hilbert function* of $\mathfrak{a}$ is defined by $\mathrm{HF}_{\mathfrak{a}}(s) = \dim_K(R_s/\mathfrak{a}_s)$. From a certain degree, this function of s is equal to a (unique) polynomial in s, called the *Hilbert polynomial*, and is denoted by $\mathrm{HP}_{\mathfrak{a}}$. The *Hilbert series* of $\mathfrak{a}$ is the following power series

$$\mathrm{HS}_{\mathfrak{a}}(t) = \sum_{s=0}^{\infty} \mathrm{HF}_{\mathfrak{a}}(s) t^s.$$

By the Hilbert-Serre theorem, we know that the Hilbert series of $\mathfrak{a}$ may be written as $p(t)/(1-t)^r$ where $r = \dim(\mathfrak{a})$ and $p(1) \neq 0$. Let us recall the definitions of the degree of a homogeneous ideal, see e.g. [36, page 43] and [4].

Definition 1. *Suppose that $\mathfrak{a}$ is a homogeneous ideal with $r = \dim(\mathfrak{a})$. If $r > 0$, the (homogeneous) degree of $\mathfrak{a}$, denoted by $\deg(\mathfrak{a})$, is $(r-1)!$ times the leading coefficient of the Hilbert polynomial of $\mathfrak{a}$. If $r = 0$, the degree of $\mathfrak{a}$ is the sum of the coefficients of $\mathrm{HS}_{\mathfrak{a}}(t)$.*

By [18, page 173], we have $\deg(\mathfrak{a}) = p(1)$ and in consequence since $\mathfrak{a}$ and $\mathrm{LM}(\mathfrak{a})$ share the same Hilbert function, $\deg(\mathfrak{a}) = \deg(\langle\mathrm{LM}\rangle(\mathfrak{a}))$. Now, let us turn our attention to the relation between the degrees of an ideal and its primary components. Let $\mathfrak{q}$ be a $\mathfrak{p}$-primary ideal. The *length* of $\mathfrak{q}$, denoted by $\ell(\mathfrak{q})$, is the maximum length ℓ of a chain $\mathfrak{q} = \mathfrak{q}_1 \subset \cdots \subset \mathfrak{q}_\ell = \mathfrak{p}$ of primary ideals. Note that ℓ is the length of the Artinian local ring $(R/\mathfrak{q})_{\mathfrak{p}}$. With these notations, we have $\deg(\mathfrak{q}) = \ell \deg(\mathfrak{p})$. The degree of a homogeneous ideal $\mathfrak{a}$ equals to the sum of the degrees of its primary components of dimension $\dim(\mathfrak{a})$. Let $\mathfrak{m} \subset R$ be a maximal ideal and $\mathfrak{q}$ an $\mathfrak{m}$-primary ideal. Then, $\deg(\mathfrak{q}) = \dim_K(R/\mathfrak{q})$. If $\mathfrak{a} \subset R$ is a zero-dimensional ideal then we have

$$\deg(\mathfrak{a}) = \sum_{i=1}^{m} \deg(\mathfrak{q}_i),$$

where $\mathfrak{a} = \bigcap_{i=1}^{m} \mathfrak{q}_i$ is the irredundant primary decomposition of $\mathfrak{a}$. We conclude this section, by defining the degree of a non-necessarily homogeneous zero-dimensional ideal $\mathfrak{a} \subset R$. Let ${}^h R$ be the ring $K[X_0, X_1, \ldots, X_n]$ where X_0 is a new variable. For any polynomial $f \in R$, we define its *homogenization* to be ${}^h f = X_0^{\deg(f)} f(X_1/X_0, \ldots, X_n/X_0) \in {}^h R$. Furthermore, we define ${}^h \mathfrak{a} = ({}^h f \mid f \in \mathfrak{a}) \subset {}^h R$. It is clear that $\dim({}^h \mathfrak{a}) = 1$. Then, we define $\deg(\mathfrak{a})$ to be $\deg({}^h \mathfrak{a})$. If $\mathfrak{a} = \bigcap_{i=1}^{m} \mathfrak{q}_i$ is the irredundant primary decomposition of a zero-dimensional ideal $\mathfrak{a}$ then ${}^h \mathfrak{q}_1 \cap \cdots \cap {}^h \mathfrak{q}_m$ is an irredundant primary decomposition of ${}^h \mathfrak{a} \subset {}^h R$ and $\dim({}^h \mathfrak{q}_i) = 1$ for each i. In addition, since homogenization of a primary (resp. prime) ideal remains primary (resp. prime) ideal, we conclude that the length of a primary ideal remains stable after homogenization, see e.g. [38, Chapter VII, Theorem 17]. These arguments yield $\deg(\mathfrak{a}) = \sum_{i=1}^{m} \deg(\mathfrak{q}_i) = \sum_{i=1}^{m} \ell_i \deg(\mathfrak{p}_i)$ where $\mathfrak{q}_i$ is $\mathfrak{p}_i$-primary and $\ell_i = \ell(\mathfrak{q}_i)$. Finally, if $\ell(\mathfrak{a})$ stands for $\max\{\ell_1, \ldots, \ell_m\}$ then we have trivially, $\max\{\ell(\mathfrak{a}), m\} \leq \deg(\mathfrak{a})$.

3   A Thorough Revision of Lakshman's Paper

In this section, we make a thorough revision of the paper [19] by Lakshman with special focus on fixing its serious flaws. In addition, using this approach we will present intrinsic complexity bounds for computing Gröbner bases of a zero-dimensional ideal and for all its primary components.

Indeed, there are some flaws in the intermediate statements leading to [19, Theorem 2]. Firstly, in Lemma 1, Lakshman claims that some sequence of ideals is increasing, which is obviously wrong since it is, in fact, a decreasing sequence of ideals. This error may be fixed without any effort. Secondly,

Lemmata 4 and 5 are directly false since he assumes that the index of a primary zero-dimensional ideal agrees with its length and this is not always true (see Corollary 10 below for an obvious counterexample). Lakshman's bound in Lemma 4 must be replaced by what we exhibit in Lemma 15 below. This modifies [19, Lemma 5] making it doubtful and somehow misleading. Not only because of this confusion between index and length, but also because [19] omits a lot of relevant material to bound the arithmetic complexity of the algorithm. The correction of Lemma 5 in [19] now becomes Lemma 16 and none of the bounds we have found agrees with the bounds exhibited by Lakshman in his paper.

Next, [19, Lemma 6] contains the main flaw. The author assumes that the number of field operations required to compute a normal form w.r.t. a finite set F only depends on the number of polynomials in F and the number of non-zero terms involved. As far as we know, no proof of this fact is known. This was the reason to introduce the function $\mathscr{T}_S$ in Definition 5. In addition, this forces us to make a thorough revision of his Lemma 6 and leads to Theorem 17 which summarizes our study of the complexity of Lakshman's algorithm. Finally, [19, Theorems 2 and 3] can be replaced by Corollaries 18 and 22, respectively.

3.1 Complexity of Converting Gröbner Bases

In this subsection, we give the complexity of converting a given Gröbner basis of a zero-dimensional ideal into the reduced Gröbner basis for the same ideal with respect to the same monomial order (see Corollary 7). We first fix a monomial order $\prec$ on R. For a polynomial $f \in R$, we denote by $\deg_{X_i}(f)$ the degree of f w.r.t. the variable X_i.

Lemma 2. *Let $\mathfrak{a} \subset R$ be a zero-dimensional ideal and $F \subset \mathfrak{a}$ be a finite set of generators of $\mathfrak{a}$ (although not necessarily a Gröbner basis for $\mathfrak{a}$). Assume that $N(F)$ is a finite set. Then, for all $S \subseteq \{1, \ldots, n\}$, such that $\sharp S \leq n - 1$, there is some $f \in F$ such that $\deg_{X_i}(\mathrm{LM}(f)) = 0$ for all $i \in S$.*

Proof. Let $\mathscr{M}_S(X) = \{\prod_{k \notin S} X_k^{\alpha_k} \mid \alpha_k \in \mathbb{N}\}$. Since $\sharp S \leq n - 1$ then $\mathscr{M}_S(X)$ is infinite and it is not completely included in $N(F)$, because $N(F)$ is a finite set. Thus, there must be some $m = \prod_{k \notin S} X_k^{\alpha_k} \in \mathscr{M}_S(X)$ such that $m \notin N(F)$. Thus

$$m \in \bigcup_{f \in F} \langle \mathrm{LM} \rangle(f),$$

which implies that exists some $f \in F$ such that $\mathrm{LM}(f)$ divides m. Hence, for any $i \in S$, no non-zero power of X_i may divide $\mathrm{LM}(f)$ and, thus $\deg_{X_i}(\mathrm{LM}(f)) = 0$ for all $i \in S$. $\square$

This lemma implies the following result.

Lemma 3. *With the same hypothesis as in Lemma 2, for every monomial $m \in N(F)$ and for each $1 \leq i \leq n$ we have $\deg_{X_i}(m) < \max\{\deg_{X_i}(f) \mid f \in F\}$.*

Proof. Let $m = X_1^{\alpha_1} \cdots X_n^{\alpha_n} \in N(F)$ and $d_i = \max\{\deg_{X_i}(\mathrm{LM}(f)) \mid f \in F\}$ for each i. Consider the sets $T = \{k \in \{1, \ldots, n\} \mid d_k \le \alpha_k\}$ and $S = \{1, \ldots, n\} \setminus T$. Reasoning by reductio ad absurdum assume that $T \ne \emptyset$. Thus, $\sharp S \le n - 1$ and from Lemma 3, there is $f \in F$ with $\deg_{X_i}(\mathrm{LM}(f)) = 0$, $\forall i \in S$. Thus, we conclude that $\mathrm{LM}(f)$ has the form $\prod_{k \in T} X_k^{\beta_k}$. Then, as d_k is maximal, $\beta_k \le d_k$, for all $k \in T$ and, hence, $\mathrm{LM}(f) = \prod_{k \in T} X_k^{\beta_k} \mid \prod_{k \in T} X_k^{d_k}$. On the other hand $\prod_{k \in T} X_k^{d_k} \mid m$ which implies that $\mathrm{LM}(f)$ divides m, leading to a contradiction. It follows that $T = \emptyset$ and this yields the claim.   □

Next, let us consider $\mathfrak{a} \subset R$ an arbitrary zero-dimensional ideal and let $G = \{f_1, \ldots, f_s\}$ be a minimal Gröbner basis of $\mathfrak{a}$ w.r.t. $\prec$. As G is minimal, we may assume that

$$\mathrm{LM}(f_s) \prec \cdots \prec \mathrm{LM}(f_1). \tag{2}$$

Now, we transform the elements in G as follows. Let $h_i = f_i - \mathrm{LT}(f_i)$ for $1 \le i \le s$. Due to (2), every monomial in h_i is strictly smaller than any leading monomial of f_k for all k satisfying $1 \le k \le i$. Let $\widetilde{h_i} = \mathrm{NF}(h_i, \{f_{i+1}, \ldots, f_s\})$ and $\widetilde{f_i} = \mathrm{LT}(f_i) + \widetilde{h_i}$. We have the following statement.

Lemma 4. *With these notations and assumptions, we have:*

i) $\mathrm{LM}(\widetilde{f_i}) = \mathrm{LM}(f_i)$ *and* $\widetilde{G} = \{\widetilde{f_1}, \ldots, \widetilde{f_s}\}$ *is the reduced Gröbner basis of* $\mathfrak{a}$.
ii) *Any monomial in* $t(\widetilde{h_i})$ *lies in* $N(\mathfrak{a})$ *and* $\sharp t(\widetilde{h_i}) \le \sharp N(\mathfrak{a})$.
iii) *The number of non-zero terms of every element of* $\widetilde{G}$ *is at most* $\sharp N(\mathfrak{a}) + 1$.

Proof. The results are folklore from the theory of Gröbner bases, see e.g. [6]. □

Definition 5. *Let* $\mathscr{T}_S(n, t, k, D)$ *be an upper bound over the number of arithmetic operations of elements in the field* K *required to compute* $S(f, g, F)$ *where*

i) F *generates a zero-dimensional ideal,* $\sharp F \le k$ *and* $N(F)$ *is finite,*
ii) *the number of non-zero terms of any polynomial in* $\{f, g\} \cup F$ *is at most* t,
iii) *the maximum of the degrees of the polynomials in* $\{f, g\} \cup F$ *is at most* D.

Note that $\mathscr{T}_S(n, t, k, D)$ is assumed to be also a bound for the number of arithmetic operations required to compute a normal form $\mathrm{NF}(f, F) = S(f, 0, F)$, provided that f and F satisfy the required conditions.

Remark 6. The bound $\mathscr{T}_S(n, t, k, D)$ is known to be finite, as we have

$$\mathscr{T}_S(n, t, k, D) \in O\left(nt \log(D) \left(tk \binom{D+n}{n} \right)^{\omega} \right)$$

where $\omega < 2.373$ is the exponent of the complexity of matrix multiplication, see [23]. Other bounds under different assumptions can be found in the literature. It is worth noting that van der Hoeven in [34] by using the concepts of *relaxed power series* and *fast sparse polynomial arithmetic* described a fast algorithm for sparse reduction of a polynomial w.r.t. an autoreduced set of polynomials.

Corollary 7. *Let* $\mathfrak{a}$ *be a zero-dimensional ideal. The computation of the reduced Gröbner basis* G_{red} *from a Gröbner basis* G *of* $\mathfrak{a}$ *can be done in a number of arithmetic operations in* K *which is bounded by the following quantity:*

$$O\left(\sharp G\left(n\sharp t(G)+\mathcal{T}_S\left(n,\sharp t(G),\sharp G,D\right)\right)\right)$$

where D *is is an upper bound for the degrees of the polynomials in* G.

Proof. To compute the reduced Gröbner basis, we first compute a minimal Gröbner basis $G_{\min}$ from G. In doing so, let us sort $G = \{f_1, \ldots, f_s\}$ and remove all f_i with $\mathrm{LM}(f_i) = \mathrm{LM}(f_{i-1})$ to obtain a subset $G_1 = \{g_1, \ldots, g_r\} \subseteq G$ such that the $\mathrm{LM}(g_r) \prec \cdots \prec \mathrm{LM}(g_1)$. Then, we eliminate from G_1 all those g_i's such that $\mathrm{LM}(g_i)$ is divisible by some $\mathrm{LM}(g_k)$ with $k > i$. The final result is a minimal Gröbner basis $G_{\min}$. This can be done, obviously in a time linear in n, $\sharp G$ and $\sharp t(G)$ (note that checking the divisibility of two monomials needs n comparisons, and we consider each comparison as a field operation). Now, taking $G_{\min}$ we proceed as explained above to get a reduced Gröbner basis. The total operations performed by this procedure depends polynomially on the number of elements in $G_{\min}$ and the cost of performing the corresponding reductions. But, each normal form computation is performed in a number of field operations bounded by $\mathcal{T}_S\left(n,\sharp t(G),\sharp(G),D\right)$, thus proving the corollary. □

3.2 Complexity of Computing the Primary Decomposition

In this subsection, we present the complexity of computing a Gröbner basis for a primary component of a zero-dimensional ideal under suitable assumptions of genericity (see Corollary 18). Then, we apply this result to prove our main result about the complexity of computing reduced Gröbner bases for zero-dimensional ideals (see Corollary 22). Let us first define the index of a primary ideal.

Definition 8. *Let* $\mathfrak{p} \subset R$ *be a prime ideal and* $\mathfrak{q}$ *a* $\mathfrak{p}$-*primary ideal. We define the index of* $\mathfrak{q}$ *as the minimum positive integer* $\mathrm{Ind}(\mathfrak{q}) = \rho$ *such that* $\mathfrak{p}^\rho \subseteq \mathfrak{q}$.

Observe that in any Noetherian ring (in particular, in R) the index of a primary ideal is always well-defined. We then fix a zero-dimensional ideal $\mathfrak{a} \subset R$. Recall that $\mathfrak{a}$ is zero-dimensional if and only if every associated prime of $\mathfrak{a}$ is maximal in R. The following classical statement may be seen in [35].

Lemma 9. *Let* $\mathfrak{a}$ *be a zero-dimensional ideal as above. Let* $\mathfrak{p} \subset R$ *be an associated prime of* $\mathfrak{a}$ *and* $\mathfrak{q}$ *a* $\mathfrak{p}$-*primary ideal occurring in a minimal primary decomposition of* $\mathfrak{a}$. *Let* ρ *be the index of* $\mathfrak{q}$. *Then, the following properties hold:*

i) if $\sigma < \rho$ *then* $\dim_K(R/(\mathfrak{a}+\mathfrak{p}^{\sigma-1})) < \dim_K(R/(\mathfrak{a}+\mathfrak{p}^\sigma))$,
ii) if $\sigma \geq \rho$ *then* $\mathfrak{q} = \mathfrak{a}+\mathfrak{p}^\rho = \mathfrak{a}+\mathfrak{p}^\sigma$.

Namely, the index of a $\mathfrak{p}$-primary ideal $\mathfrak{q}$ occurring in a minimal primary decomposition of a zero-dimensional ideal $\mathfrak{a}$ is the minimal positive integer ρ such that $\mathfrak{a}+\mathfrak{p}^\rho = \mathfrak{a}+\mathfrak{p}^{\rho+1}$ and, in this case, $\mathfrak{q} = \mathfrak{a}+\mathfrak{p}^\rho$. This is the key ingredient of the Lakshman algorithm in [19].

Corollary 10. *With these notations, let* $\mathfrak{a} \subset R$ *be a zero-dimensional ideal,* $\mathfrak{q}$ *an isolated* $\mathfrak{p}$*-primary component of* $\mathfrak{a}$. *Then,*

i) $\mathrm{Ind}(\mathfrak{q}) \leq \ell(\mathfrak{q})$, *see also [4, Lemma 1]. In particular,* $\mathrm{Ind}(\mathfrak{q}) \deg(\mathfrak{p}) \leq \deg(\mathfrak{q})$.
ii) $\ell(\mathfrak{q})$ *is an upper bound for the minimal positive integer* $\rho \in \mathbb{N}$ *such that*

$$\mathfrak{a} + \mathfrak{p}^\rho = \mathfrak{a} + \mathfrak{p}^{\rho+1}.$$

Proof. Let $\rho = \mathrm{Ind}(\mathfrak{q})$ be the index of $\mathfrak{q}$. Then, according to Claim i) of Lemma 9, for every $\sigma < \rho$, the ideal $\mathfrak{a} + \mathfrak{p}^\sigma$ satisfies $\mathfrak{p} = \sqrt{\mathfrak{q}} = \sqrt{\mathfrak{a} + \mathfrak{p}^\rho} \subseteq \sqrt{\mathfrak{a} + \mathfrak{p}^\sigma} \subseteq \mathfrak{p}$. Hence, $\sqrt{\mathfrak{a} + \mathfrak{p}^\sigma} = \mathfrak{p}$ for every $\sigma < \rho$. Thus, $\mathfrak{a} + \mathfrak{p}^\sigma$ is a $\mathfrak{p}$-primary ideal. Claim i) of Lemma 9 implies that the following is a chain of $\mathfrak{p}$-primary ideals with strict inclusions:

$$\mathfrak{q} = \mathfrak{a} + \mathfrak{p}^\rho \subsetneqq \mathfrak{a} + \mathfrak{p}^{\rho-1} \subsetneqq \cdots \subsetneqq \mathfrak{a} + \mathfrak{p}^2 \subsetneqq \mathfrak{a} + \mathfrak{p} = \mathfrak{p}$$

and, hence, $\mathrm{Ind}(\mathfrak{q}) \leq \ell(\mathfrak{q})$ as claimed. Since $\deg(\mathfrak{q}) = \ell(q)\deg(\mathfrak{p})$, it follows that $\mathrm{Ind}(\mathfrak{q})\deg(\mathfrak{p}) \leq \deg(\mathfrak{q})$.

Finally, Claim ii) of the statement immediately follows from Claim ii) of Lemma 9.   □

Example 11. This example shows that the equality $\mathrm{Ind}(\mathfrak{q}) = \ell(\mathfrak{q})$ in Claim i) of Corollary 10 does not hold in general: let $R = K[x_1, x_2]$, $\mathfrak{p} = (x_1, x_2)$ and $\mathfrak{q}$ the $\mathfrak{p}$-primary ideal given by $\mathfrak{q} = \mathfrak{p}^4$. We can see easily that $\deg(\mathfrak{p}) = 1$ and $\mathrm{Ind}(\mathfrak{q}) = \rho = 4$. It is obvious that a basis of $R/\mathfrak{q}$ as a K-vector space is determined by all the monomials of degree at most 3, hence yielding $\ell(\mathfrak{q}) = 10 > \mathrm{Ind}(\mathfrak{q}) = 4$.

Now, we recall Lakshman's main algorithm to compute a Gröbner basis of an isolated primary component $\mathfrak{q}$ of a zero-dimensional ideal $\mathfrak{a}$, provided that we are given a finite set of generators of $\mathfrak{a}$ and the reduced Gröbner basis of the corresponding associated prime $\mathfrak{p}$.

Algorithm 1. PRIMARYCOMPONENT

Input: A finite generating set F of a zero-dimensional ideal $\mathfrak{a}$ and the reduced Gröbner basis G of the associated prime $\mathfrak{p}$ of $\mathfrak{a}$

Output: The reduced Gröbner basis of the $\mathfrak{p}$-primary component $\mathfrak{q}$ of $\mathfrak{a}$

$C := F$
$B := G$ ▷ *The reduced Gröbner basis of* $\mathfrak{a} + \mathfrak{p} = \mathfrak{p}$.
while $B \neq C$ **do**
   $C := B$ ▷ B *is the reduced Gröbner basis of some* $\mathfrak{a} + \mathfrak{p}^\sigma$, *for* $\sigma < \mathrm{Ind}(\mathfrak{q})$.
   $B :=$ The reduced Gröbner basis of the ideal generated by $F \cup B \cdot G$
      ▷ B *becomes the reduced Gröbner basis of* $\mathfrak{a} + \mathfrak{p} \cdot (\mathfrak{a} + \mathfrak{p}^\sigma) = \mathfrak{a} + \mathfrak{p}^{\sigma+1}$.
end while
return (B)

Lemma 12. *Algorithm 1 computes the reduced Gröbner basis of the isolated primary component* $\mathfrak{q}$. *The number of reduced Gröbner basis calculations (i.e. the number of times the procedure enters in the* **while***-loop) is at most* $\mathrm{Ind}(\mathfrak{q}) \leq \ell(\mathfrak{q})$.

Proof. Both claims are obvious in view of Corollary 10.   □

The complexity analysis then depends on the complexity of computing the reduced Gröbner basis inside each **while**-loop. Let $\mathfrak{a} = \bigcap_{i=1}^{m} \mathfrak{q}_i$ be an irredundant primary decomposition of a zero-dimensional ideal $\mathfrak{a}$ where each $\mathfrak{q}_i$ is $\mathfrak{p}_i$-primary for each i. From now on, we assume that the variables $X_1, \ldots, X_n$ are in generic position w.r.t. $\sqrt{\mathfrak{a}} = \bigcap_{i=1}^{m} \mathfrak{p}_i$. More precisely, we assume that there are $h_1, \ldots, h_n \in K[X_1]$ such that

$$\sqrt{\mathfrak{a}} = (X_n - h_n(X_1), \ldots, X_2 - h_2(X_1), h_1(X_1))$$

where $\deg(h_i) \leq \deg(h_1) - 1$ for each $2 \leq i \leq n$, and $\deg(h_1) = \deg(\sqrt{\mathfrak{a}}) = \sum_{i=1}^{m} \deg(\mathfrak{p}_i)$. We refer to this property as $\sqrt{\mathfrak{a}}$ being in *normal X_1-position*. This may be achieved by a generic linear change of coordinates to transform $\sqrt{\mathfrak{a}}$ into this position, see e.g. [13]. Now, let $\mathfrak{q}$ be a $\mathfrak{p}$-primary component of $\mathfrak{a}$. The generic change of variables is also well suited for each of its associated primes and, hence, we may assume that

$$\mathfrak{p} = (X_n - g_n(X_1), \ldots, X_2 - g_2(X_1), g_1(X_1)), \tag{3}$$

where $\deg(g_i) \leq \deg(\mathfrak{p}) - 1$ for each $2 \leq i \leq n$, and $\deg(g_1) = \deg(\mathfrak{p})$. Observe that g_1 is some irreducible factor of h_1 in $K[X_1]$ and for $2 \leq i \leq n$, we have $g_i = \mathrm{rem}(h_i, g_1)$ is the remainder of the division of h_i by g_1, see [1, Proposition 8.69] for more details. According to the comments introduced in Algorithm 1, the procedure computes reduced Gröbner bases of all $\mathfrak{p}$-primary ideals $\mathfrak{h}_i = \mathfrak{a} + \mathfrak{p}^i$ where $1 \leq i \leq \mathrm{Ind}(\mathfrak{q}) \leq \ell(\mathfrak{q})$. Let us also denote by B_i the reduced Gröbner basis of $\mathfrak{h}_i$ computed in the course of Algorithm 1. The following statement was proved in [19, Lemma 2].

Lemma 13. *With the same notations as above, let B_i be the reduced Gröbner basis of $\mathfrak{h}_i$ computed by Algorithm 1. Let $\delta = \deg(\mathfrak{p})$ be the degree of the prime ideal $\mathfrak{p}$ associated to $\mathfrak{a}$. Then, all leading monomials in B_i are of the form $(X_1^{\delta})^{i_1} X_2^{i_2} \cdots X_n^{i_n}$ with $i_j \geq 0$.*

Let $m \in R$ be a monomial which belongs to the normal set $N(\mathfrak{h}_i) = \mathscr{M}(X) \setminus \mathrm{LM}(\mathfrak{h}_i)$ of some intermediate primary ideal $\mathfrak{h}_i$ computed by Algorithm 1. Define the *class* m of all monomials associated to m w.r.t. $\mathfrak{h}_i$ as:

$$\mathrm{Cl}_{\mathfrak{h}_i}(m) = \{m, X_1 m, \ldots, X_1^{\delta-1} m\}.$$

The following lemma resumes the main properties of these classes.

Lemma 14. *Let $m = (X_1^{\delta})^{j_1} X_2^{j_2} \cdots X_n^{j_n}$ and $m' = (X_1^{\delta})^{j_1'} X_2^{j_2'} \cdots X_n^{j_n'}$ be two monomials. Then, the following statements hold.*

i) If $m \in N(\mathfrak{h}_i)$ then $\mathrm{Cl}_{\mathfrak{h}_i}(m) \subseteq N(\mathfrak{h}_i)$.
ii) Given $m, m' \in N(\mathfrak{h}_i)$ as above, we have:

$$\mathrm{Cl}_{\mathfrak{h}_i}(m) \cap \mathrm{Cl}_{\mathfrak{h}_i}(m') \neq \emptyset \iff \mathrm{Cl}_{\mathfrak{h}_i}(m) = \mathrm{Cl}_{\mathfrak{h}_i}(m') \iff m = m'.$$

iii) The number of classes of equivalence $\mathrm{Cl}_{\mathfrak{h}_i}(m)$ *is at most* $\ell = \ell(\mathfrak{q})$.

In particular, the set of classes $\mathrm{Cl}_{\mathfrak{h}_i}(m)$ *defines a partition of* $N(\mathfrak{h}_i)$. *Namely, there is a finite set of monomials* $m_1, \ldots, m_L \in N(\mathfrak{h}_i)$ *of the form*

$$m_r = (X_1^\delta)^{j_{r,1}} n_r, \qquad (4)$$

where $n_r \in K[X_2, \ldots, X_n]$ *are monomials such that*

$$N(\mathfrak{h}_i) = \bigcup_{r=1}^{L} \mathrm{Cl}_{\mathfrak{h}_i}(m_r) \qquad (5)$$

is a disjoint decomposition of $N(\mathfrak{h}_i)$ *with* $L \leq \ell(\mathfrak{q})$.

Proof. As B_i is the reduced Gröbner basis of $\mathfrak{h}_i$, then by Lemma 13 we may assume that $\mathrm{LM}(B_i)$ has the form

$$\mathrm{LM}(B_i) = \{(X_1^\delta)^{i_1} u_1, \ldots, (X_1^\delta)^{i_t} u_t\},$$

where $t \in \mathbb{N}$ and $u_k \in K[X_2, \ldots, X_n]$ are monomials. Consequently, we have:

$$\mathrm{LM}(\mathfrak{h}_i) = \bigcup_{k=1}^{t} (X_1^\delta)^{i_k} u_k \cdot \{X_1^{\mu_1} \cdots X_n^{\mu_n} \mid (\mu_1, \ldots, \mu_n) \in \mathbb{N}^n\}.$$

Now, to prove Claim i), assume in contrary that for some $0 \leq r \leq \delta - 1$, we have $X_1^r m \in \mathrm{Cl}_{\mathfrak{h}_i}(m) \cap \mathrm{LM}(\mathfrak{h}_i)$. Then, there exist some k and some $a \in \mathbb{N}$ with

$$X_1^{\delta j_1 + r} = X_1^{\delta i_k + a} \quad \text{and} \quad u_k \mid X_2^{j_2} \cdots X_n^{j_n}.$$

Thus, we must have $i_k \leq j_1$ and in turn $(X_1^\delta)^{i_k} u_k \mid (X_1^\delta)^{j_1} X_2^{j_2} \cdots X_n^{j_n} \in N(\mathfrak{h}_i)$ which is impossible and this proves Claim i).

As for Claim ii), assume that there exist r, r' with $0 \leq r, r' \leq \delta - 1$ such that:

$$X_1^r m = X_1^{r'} m' \in \mathrm{Cl}_{\mathfrak{h}_i}(m) \cap \mathrm{Cl}_{\mathfrak{h}_i}(m') \neq \emptyset.$$

Assume that $r' \geq r$, then we have $m = X_1^{r'-r} m' \in \mathrm{Cl}_{\mathfrak{h}_i}(m')$. It follows that $\delta j_1 = \delta j_1' + (r' - r)$ which implies $\delta \mid r' - r$, and $0 \leq r' - r < \delta$. Thus $r = r'$ and, hence, $m = m'$, proving Claim ii).

Assume $m_1, \ldots, m_L$ is a sequence of monomials with $\bigcup_{k=1}^{L} \mathrm{Cl}_{\mathfrak{h}_i}(m_k) \subseteq N(\mathfrak{h}_i)$ where the union is a disjoint union of sets of cardinality δ. Hence, we have

$$\sharp \bigcup_{k=1}^{L} \mathrm{Cl}_{\mathfrak{h}_i}(m_k) = \sum_{k=1}^{L} \sharp \mathrm{Cl}_{\mathfrak{h}_i}(m_k) = L\delta \leq \sharp N(\mathfrak{h}_i) = \deg(\mathfrak{h}_i) \leq \deg(\mathfrak{q}) = \ell(\mathfrak{q}) \deg(\mathfrak{p})$$

and so, $L \leq \ell(\mathfrak{q})$, because $\delta = \deg(\mathfrak{p})$ and this shows Claim iii). Finally, let $u = X_1^{t_1} X_2^{t_2} \cdots X_n^{t_n} \in N(\mathfrak{h}_i)$. Let $i_1, r \in \mathbb{N}$ be the quotient and the remainder of the Euclidean division of t_1 by δ. Let us define $v = (X_1^\delta)^{i_1} X_2^{t_2} \cdots X_n^{t_n}$. Thus, we have $u = X_1^r v \in \mathrm{Cl}_{\mathfrak{h}_i}(v)$: if $v \in \mathrm{LM}(\mathfrak{h}_i)$, then $u \in \mathrm{LM}(\mathfrak{h}_i)$ which is impossible; hence $v \in N(\mathfrak{h}_i)$ and therefore $u \in \mathrm{Cl}_{\mathfrak{h}_i}(v)$. $\square$

The following lemma is the corrected version of [19, Lemma 4] in which we replace the index of a primary component by its length.

Lemma 15. *With the same notations as above, let G be the reduced Gröbner basis of $\mathfrak{p}$ w.r.t $\prec$. Let B_i be the reduced Gröbner basis for $\mathfrak{h}_i$ w.r.t $\prec$ computed by Algorithm 1. Then, $\sharp \mathrm{LM}(G \cdot B_i) \leq n^2 \ell$ and $\sharp N(G \cdot B_i) \leq (n+1)\ell\delta$.*

Proof. Let $m_1, \ldots, m_L \in N(\mathfrak{h}_i)$ be the monomials described in Equality (5). By Claim i) of Lemma 14, for all r and for all k with $0 \leq k \leq \delta - 1$, we have $X_1^k m_r \in N(\mathfrak{h}_i)$. Let $m = (X_1^\delta)^{i_1} X_2^{i_2} \cdots X_n^{i_n} \in \mathrm{LM}(B_i)$ with $i_k \geq 0$. Two cases may happen: if $i_1 \geq 1$, then $m/X_1^\delta = (X_1^\delta)^{i_1-1} X_2^{i_2} \cdots X_n^{i_n} \in N(\mathfrak{q}_i)$. Then, there would exist r and k with $0 \leq k \leq \delta-1$, such that $m/X_1^\delta = X_1^k m_r$. According to Equality (4) of Lemma 14, $m_r = (X_1^\delta)^{j_{r,1}} n_r$, where $n_r \in K[X_2, \ldots, X_n]$. Then,

$$(X_1^\delta)^{i_1} X_2^{i_2} \cdots X_n^{i_n} = m = X_1^k X_1^\delta m_r = X_1^k (X_1^\delta)^{j_{r,1}+1} n_r.$$

We conclude that $k = 0$ and, hence, $m/X_1^\delta = m_r$. Otherwise, if $i_1 = 0$, then $m = X_2^{i_2} \cdots X_n^{i_n} \in \mathrm{LM}(B_i)$ is not a constant. Thus, there exist some k with $2 \leq k \leq n$, such that $M/X_k = X_2^{i_2} \cdots X_k^{i_k-1} \cdots X_n^{i_n} \in N(\mathfrak{h}_i)$. The same argument used in the case $i_1 \geq 1$ applies to conclude that there must be some r with $1 \leq r \leq L$, such that $m/X_k = m_r$. In conclusion, we have proved the inclusion:

$$\mathrm{LM}(B_i) \subseteq \{X_1^\delta m_r \; : \; 1 \leq r \leq L\} \bigcup \left(\bigcup_{k=2}^{n} \{X_k m_r \; : \; 1 \leq r \leq L\} \right). \quad (6)$$

According to Lemma 14, we conclude that $\sharp \mathrm{LM}(B_i) \leq nL \leq n\ell$. Moreover, as B_i is reduced, then $\sharp B_i = \sharp \mathrm{LM}(B_i) \leq n\ell$. From Equality (3) we know that the Gröbner basis G of $\mathfrak{p}$ has n elements. This yields $\sharp \mathrm{LM}(G \cdot B_i) \leq n^2 \ell$.

Let us now study the bound for $\sharp N(G \cdot B_i)$. We have obviously $N(\mathfrak{h}_i) = N(B_i) \subseteq N(G \cdot B_i)$. Next, let us prove that the following inclusion holds:

$$N(G \cdot B_i) \setminus N(\mathfrak{h}_i) \subseteq \bigcup_{t=0}^{\delta-1} \{X_1^t u \mid u \in \mathrm{LM}(B_i)\}. \quad (7)$$

Let $m \in N(G \cdot B_i)$ be a monomial not in $N(\mathfrak{h}_i)$. Then, $m \notin \mathrm{LM}(G \cdot B_i)$ and $m \in \mathrm{LM}(\mathfrak{h}_i) = \mathrm{LM}(B_i)$. Thus, there exists $f \in B_i$ such that $u = \mathrm{LM}(f)$ and $m = X_1^{t_1} \cdots X_n^{t_n} u$. We claim that $t_1 < \delta$ and $t_2 = \cdots = t_n = 0$. If $t_1 \geq \delta$, then

$$m = X_1^{t_1-\delta} X_2^{t_2} \cdots X_n^{t_n} \left(X_1^\delta u \right) = X_1^{t_1-\delta} X_2^{t_2} \cdots X_n^{t_n} \left(\mathrm{LM}(g_1) \mathrm{LM}(f) \right) \in \mathrm{LM}(G \cdot B_i)$$

which contradicts $m \notin \mathrm{LM}(G \cdot B_i)$. Similarly, if $t_k \geq 1$ for some k then

$$m = X_1^{t_1} \cdots X_k^{t_k-1} \cdots X_n^{t_n} \left(X_k u \right) \in \mathrm{LM}(G \cdot B_i),$$

which is also impossible because of the same reason. In conclusion, if m is a monomial in $N(G \cdot B_i) \setminus N(\mathfrak{h}_i)$ there exists $u \in \mathrm{LM}(B_i)$ and t, $0 \leq t \leq \delta - 1$, such that $m = X_1^t u$. This proves the inclusion (7). Finally, we just count the cardinalities of these sets to conclude

$$\sharp N(G \cdot B_i) \le \sharp N(\mathfrak{h}_i) + \delta \sharp \mathrm{LM}(B_i) \le L\delta + \delta nL \le (n+1)\delta \ell.$$

□

Assume that $\mathfrak{a}$ is a zero-dimensional ideal generated by $F = \{f_1, \ldots, f_k\}$ with $\deg(f_i) \le d$. As in Equality (3), assume that $G = \{X_n - g_n(X_1), \ldots, X_2 - g_2(X_1), g_1(X_1)\}$ generates $\mathfrak{p}$. Let B_i be the reduced Gröbner basis of $\mathfrak{h}_i = \mathfrak{a} + \mathfrak{p}^i$ and B_{i+1} the reduced Gröbner basis for $\mathfrak{h}_{i+1} = \mathfrak{a} + \mathfrak{p} \cdot \mathfrak{h}_i = \mathfrak{a} + \mathfrak{p}^{i+1}$ computed by Algorithm 2 from $F \cup G \cdot B_i$. Note that this algorithm is a variant of Buchberger's algorithm applied to our situation. In the next lemma, we bound the size of B_{i+1} and the total number of S-polynomials to construct it, cf. [19, Lemma 5].

Algorithm 2. BUCHBERGER

Input: A finite set $F \cup G \cdot B_i$
Output: A Gröbner basis B_{i+1} for $\mathfrak{h}_{i+1}$
  $\widetilde{F} := \{\mathrm{NF}(f_j, G \cdot B_i) \mid j = 1, \ldots, k\}$
  $T := \widetilde{F} \cup G \cdot B_i$
  $P := \{\{f, g\} \mid f \ne g, f, g \in T\}$
  while $P \ne \emptyset$ **do**
    select $\{f, g\}$ from P
    $r := S(f, g, T)$
    if $r = 0$ **then**
      $P := P \setminus \{\{f, g\}\}$
    else
      $T := T \cup \{r\}$
      $P := P \cup \{\{r, h\} \mid h \in T, h \ne r\}$
    end if
  end while
  $B :=$ The reduced form of T, by performing an autoreduction
  return (B)

Lemma 16. *With the above notations, the following statements hold.*

i) $\sharp B_{i+1} \le k + (n+1)^2 \deg(\mathfrak{q})$.
ii) $\sharp N(\mathfrak{h}_{i+1}) = \sharp N(B_{i+1}) \le \deg(\mathfrak{q})$.
iii) *The total number of treated S-polynomials to compute B_{i+1} is at most*

$$\left(k + (n+1)^2 \deg(\mathfrak{q})\right)^3.$$

iv) *For every $S(f, g, H)$ constructed in the course of the algorithm, we have:*

$$\max\left(\{\deg_{X_1}(f), \deg_{X_1}(g)\} \cup \{\deg_{X_1}(h) \mid h \in H\}\right) \le \max\{d, \deg(\mathfrak{q})\},$$

and for all k, $2 \le k \le n$,

$$\max\left(\{\deg_{X_k}(f), \deg_{X_k}(g)\} \cup \{\deg_{X_k}(h) \mid h \in H\}\right) \le \max\{d, \ell(\mathfrak{q})\}.$$

Proof. To bound the size of B_{i+1}, let us first count the number of S-polynomials leading to a non-zero normal form computed in this algorithm. Let $\mathscr{R}$ be this number. For this, we shall need to introduce some subindices to determine the size of the intermediate sets computed in the course of this algorithm. We initialize with $T_0 := T$ and $P_0 := P$. In addition, one step is one iteration of the algorithm leading to a non-zero normal form. Thus, $T_{\mathscr{R}}$ is a Gröbner basis of the ideal $\mathfrak{h}_{i+1}$. Whereas the sequence T_s is an increasing sequence, P_{s+1} does not necessarily contain P_s, since some of the elements of P_s may have been removed.

An upper bound for the number of the treated S-polynomials is given by

$$\sharp \bigcup_{s=0}^{\mathscr{R}} P_j + k + \sharp T_{\mathscr{R}} \leq (\mathscr{R} + 1)\max\{\sharp P_s \mid 0 \leq s \leq \mathscr{R}\} + k + \sharp T_{\mathscr{R}}. \tag{8}$$

Note that, in the above bound, we considered the fact that the computation of $\widetilde{F}$ from F requires k additional S-polynomials. Here we took also into account the last step of the algorithm to compute the reduced Gröbner basis B. For this purpose, the number of S-polynomials that we need to perform is at most $\sharp T_{\mathscr{R}}$. Now, one observes that for each s we have

$$\begin{cases} \sharp T_{s+1} = \sharp T_s + 1, \\ \sharp P_{s+1} \leq \sharp P_s + \sharp T_s. \end{cases} \tag{9}$$

Let us denote by $\mathscr{S}_0$ the cardinality of the initial set T_0. According to (6) we have $\sharp B_i = \sharp \mathrm{LM}(B_i) \leq n\ell$. As G contains only n polynomials, we obtain

$$\begin{cases} \mathscr{S}_0 = \sharp T_0 = \sharp(F \cup G \cdot B_i) \leq \sharp F + \sharp(G \cdot B_i) \leq k + n^2\ell, \\ \sharp P_0 = \binom{\mathscr{S}_0}{2}. \end{cases} \tag{10}$$

Next, observe that $T_s \supseteq F \cup G \cdot B_i \supseteq G \cdot B_i$ and if $r = S(f, g, T_s) \neq 0$, the leading monomial of r belongs to the normal set $N(T_s)$, which satisfies $N(T_s) \subseteq N(G \cdot B_i)$. Thus, according to Lemma 15, we conclude that

$$\mathscr{R} \leq \sharp N(G \cdot B_i) \leq (n+1)\ell\delta = (n+1)\deg(\mathfrak{q})$$

and in turn $\sharp B_{i+1} \leq \sharp T_{\mathscr{R}} \leq k + n^2\ell + (n+1)\ell\delta \leq k + (n+1)^2\deg(\mathfrak{q})$, proving the first claim. We notice that $\mathfrak{h}_{i+1} \subset \mathfrak{q}$ and therefore $N(\mathfrak{h}_{i+1}) \subset N(\mathfrak{q})$. This implies that $\sharp N(\mathfrak{h}_{i+1}) \leq \sharp N(\mathfrak{q}) = \deg(\mathfrak{q})$ and the second claim now easily follows. To prove Claim iii), proceeding by induction and using (9) we can show that for all s with $0 \leq s \leq \mathscr{R}$, it holds

$$\sharp P_s \leq \binom{\mathscr{S}_0}{2} + s\mathscr{S}_0 + \sum_{t=1}^{s} t. \tag{11}$$

It follows that for all s, we have

$$\sharp P_s \leq \binom{\mathscr{S}_0}{2} + \mathscr{R}\mathscr{S}_0 + \sum_{t=1}^{\mathscr{R}} t \leq \frac{1}{2}\left(\mathscr{S}_0^2 + \mathscr{R}^2\right) + \mathscr{R}\mathscr{S}_0 = \frac{1}{2}(\mathscr{S}_0 + \mathscr{R})^2$$

and hence

$$\sharp P_s \leq \frac{1}{2}\left(k + n^2\ell + (n+1)\deg(\mathfrak{q})\right)^2 \leq \frac{1}{2}\left(k + (n+1)^2\deg(\mathfrak{q})\right)^2.$$

Thus, according to Inequality (8), the total number of S-polynomials calculated by our version of Buchberger's algorithm is bounded by

$$(\mathscr{R}+1)\max\{\sharp P_s \mid 0 \leq s \leq \mathscr{R}\} + k + \sharp T_{\mathscr{R}} \leq \left(k + (n+1)^2\deg(\mathfrak{q})\right)^3.$$

In order to prove Claim iv), we proceed by induction on i. For every finite set Q and for all k, $1 \leq k \leq n$, we denote by $\delta_k(Q) = \max\{\deg_{X_k}(h) \mid h \in Q\}$. As $B_1 = \{X_n - g_n(X_1), \ldots, X_2 - g_2(X_1), g_1(X_1)\}$, we then conclude that

$$\delta_1(B_1) = \max\{\deg_{X_1}(h) \mid h \in B_1\} = \deg(\mathfrak{p}) = \delta$$

and, for all k with $2 \leq k \leq n$, $\delta_k(B_1) = \max\{\deg_{X_k}(h) \mid h \in B_1\} = 1$. Hence the first step of the induction holds. Now observe that B_{i+1} is computed from F and $G \cdot B_i$ using a sequence of intermediate sets of polynomials $T_0, T_1, \ldots, T_{\mathscr{R}}$. Every polynomial f in T_0 satisfies one of the following two conditions:

– If $f \in G \cdot B_i$, for any k with $2 \leq k \leq n$, we have

$$\deg_{X_1}(f) \leq \delta + \delta_1(B_i), \qquad \deg_{X_k}(f) \leq 1 + \delta_k(B_i). \tag{12}$$

– If $f \in \widetilde{F}$ then f is a sum of non-zero terms in $N(G \cdot B_i)$, then, from Lemma 3 we conclude that for all k with $1 \leq k \leq n$, it holds

$$\deg_{X_k}(f) \leq \max\{\deg_{X_k}(h) \mid h \in G \cdot B_i\},$$

and the degree bounds of (12) also apply to $\deg_{X_1}(f)$ and $\deg_{X_k}(f)$.

We know that $T_{j+1} = T_j \cup \{S(f, g, T_j)\}$ for some polynomials $f, g \in T_j$. However, as $G \cdot B_i \subseteq T_j$, we conclude $N(T_j) \subseteq N(G \cdot B_i)$. Hence, every polynomial added to T_j to build up T_{j+1} is a linear combination of monomials in $N(G \cdot B_i)$. Applying once more Lemma 3, we conclude that, for all k with $2 \leq k \leq n$ it holds

$$\begin{aligned}\delta_1(T_{j+1}) &\leq \delta + \delta_1(B_i),\\\delta_k(T_{j+1}) &\leq 1 + \delta_k(B_i).\end{aligned} \tag{13}$$

The reduced Gröbner basis B_{i+1} is obtained from $T_{\mathscr{R}}$ by applying some computations of normal forms of polynomials whose degrees are bounded by those of $T_{\mathscr{R}}$. All in all, for all k with $2 \leq k \leq n$ we immediately see that

$$\begin{aligned}\delta_1(B_{i+1}) &\leq \delta + \delta_1(B_i) \leq (i+1)\delta_1(B_1) = (i+1)\delta,\\\delta_k(B_{i+1}) &\leq 1 + \delta_k(B_i) \leq i+1.\end{aligned}$$

Finally, we note that the reduced Gröbner basis of $\mathfrak{q}$ is obtained by the sequence of intermediate reduced Gröbner bases $B_1, \ldots, B_L$ with $L = \ell$. We then conclude that for all i and k, we have $\delta_1(B_i) \leq \ell\delta = \ell(\mathfrak{q})\deg(\mathfrak{p}) = \deg(\mathfrak{q})$ and

$\delta_k(B_i) \leq \ell = \ell(\mathfrak{q})$. Taking into account that we have considered some preliminary computations of normal forms to obtain $\widetilde{F}$ from F and that polynomials in F have total degrees at most d, we get $\delta_1(B_L) \leq \max\{d, \deg(\mathfrak{q})\}$ and $\delta_k(B_L) \leq \max\{d, \ell(\mathfrak{q})\}$. The same bounds apply for every term $S(f, g, H)$ involved in the computation of S-polynomials along this process. $\square$

Our first partial complexity estimate is the content of the following result, cf. the corollary on page 231 of [19].

Theorem 17. *Let $\mathfrak{q}$ be a zero-dimensional ideal such that $\sqrt{\mathfrak{a}}$ is in normal X_1-position. The number of field operations required by Algorithm 1 is at most*

$$O\left(\ell(\mathfrak{q}) \left(k + (n+1)^2 \deg(\mathfrak{q})\right)^3\right) \times \mathscr{T}_S(n, \tau, \sharp\Theta, D)$$

where $\mathscr{T}_S$ is the function that measures the number of arithmetic operations required to compute an S-polynomial with the following parameters:

- *$\tau := \max\{\sharp t(F), (n+1)(\deg(\mathfrak{q})+1)^2\}$ being an upper bound for the number of non-zero terms involved,*
- *$\sharp\Theta := k + (n+1)^2 \deg(\mathfrak{q})$ being an upper bound for the maximum cardinalities of the sets of polynomials involved,*
- *$D := \max\{2d, \deg(\mathfrak{q}) + (n-1)\ell(\mathfrak{q})\}$ being an upper bound for the maximum of the degrees of the polynomials involved.*

Proof. From Lemma 12, we know that the number of times that Algorithm 1 enters the **while**-loop is at most $\ell(\mathfrak{q}) \leq \deg(\mathfrak{q})$. On the other hand, as already mentioned, to compute the reduced Gröbner basis for $\mathfrak{q}$, we shall compute the intermediate reduced Gröbner bases $B_1, \ldots, B_L$ with $L = \ell(\mathfrak{q})$ by starting from the reduced Gröbner basis $B_0 := G$ for $\mathfrak{p}$. Thus, it suffices to prove that the claimed bound without $\ell(\mathfrak{q})$ holds for the number of arithmetic operations to construct B_{i+1} from B_i for each i. For this purpose, by Lemma 16, the number of computations of S-polynomials and normal forms is bounded by

$$O\left(\left(k + (n+1)^2 \deg(\mathfrak{q})\right)^3\right). \tag{14}$$

In addition, at each step, we shall need to compare whether B_i and B_{i+1} are equal, before doing anything. As both of them are reduced Gröbner basis, we just have to compare them element-by-element. This can be done in time

$$O(\sharp B_i \sharp B_{i+1} \max\{\sharp t(B_i), \sharp t(B_{i+1})\}).$$

Observe that in Lemma 16, it was shown that for each i, $\sharp B_{i+1} \leq k + (n+1)^2 \deg(\mathfrak{q})$. Moreover, in Lemma 15, we have proved that $\sharp t(B_{i+1}) \leq (n+1)\deg(\mathfrak{q})$. These arguments confirm the upper bound (14) for the number of operations required in Algorithm 1.

To complete the proof, it is enough to show that the number of arithmetic operations in the field K to calculate an S-polynomial is bounded by

$\mathscr{T}_S(n, \tau, \sharp\Theta, D)$. Following Definition 5, we first prove that the number of non-zero terms in any involved polynomial is at most $\max\{\sharp t(F), (n+1)(\deg(\mathfrak{q})+1)^2\}$. Remark that to compute the Gröbner basis B_{i+1} using Algorithm 2, we construct an increasing sequence of sets of generators of $\mathfrak{h}_{i+1}$, say $T_0 \subset T_1 \subset \cdots \subset T_{\mathscr{R}}$. We want to bound the maximum number of terms occurring in any of the T_s's. As B_i is a reduced Gröbner basis of $\mathfrak{h}_i$, from Lemmata 4 and 16, the maximum number of non-zero terms of any polynomial in B_i is bounded by $\sharp(N(B_i))+1 \leq \deg(\mathfrak{q})+1$. Now, we consider an arbitrary polynomial $h \in G \cdot B_i$ that generates the ideal $\mathfrak{p}\mathfrak{h}_i$. The number of terms in h is bounded by

$$\sharp t(G \cdot B_i) \leq \sharp t(G) \times \sharp t(B_i) \leq (\delta+1)(\sharp N(B_i)+1) \leq (\deg(\mathfrak{p})+1)(\deg(\mathfrak{q})+1).$$

Recall that in Algorithm 2 we defined $T_0 := \widetilde{F} \cup G \cdot B_i$. Every term of $\widetilde{f}_j$ belongs to $N(G \cdot B_i)$ and, hence, from Lemma 15 we have

$$\sharp t(\widetilde{F}) \leq \sharp N(G \cdot B_i) \leq (n+1)\deg(\mathfrak{q}).$$

We conclude that the number of terms of any polynomial in T_0 is at most

$$\max\{\sharp N(G \cdot B_i), (\deg(\mathfrak{p})+1)(\deg(\mathfrak{q})+1)\} \leq (n+1)(\deg(\mathfrak{q})+1)^2.$$

It can be shown similarly, by induction and using the fact that $T_0 \subseteq T_s$ and $G \cdot B_i \subseteq T_s$ for all s, that the same bound holds for T_s as well. Thus, by considering the polynomials in F, the number of non-zero terms involved in the algorithm is bounded by $\max\{\sharp t(F), (n+1)(\deg(\mathfrak{q})+1)^2\}$. Moreover, in Lemma 16, we proved that $k+(n+1)^2\deg(\mathfrak{q})$ is an upper bound for the maximum cardinalities of the sets of polynomials. Finally, from Claim iv) of Lemma 16, we conclude that the maximum degree of $S(f, g, H)$ constructed in the Algorithm 2 is at most $\max\{2d, \deg(\mathfrak{q})+(n-1)\ell(\mathfrak{q})\}$ and this finishes the proof. □

Corollary 18. *Let $\mathfrak{a}$ be a zero-dimensional ideal such that $\sqrt{\mathfrak{a}}$ is in normal X_1-position and K a field with efficient factorization of univariate polynomials. Assume that the radical $\sqrt{\mathfrak{a}}$ is given by the Kronecker description of the form $\sqrt{\mathfrak{a}} = (X_n - g_n(X_1), X_{n-1} - g_{n-1}(X_1), \ldots, X_2 - g_2(X_1), g_1(X_1))$ where $g_1, \ldots, g_n \in K[X_1]$ with $\deg(g_i) < \deg(g_1)$, for all i, $2 \leq i \leq n$. Then, we can compute the reduced Gröbner bases for all isolated primary components of $\mathfrak{a}$ in a number of field operations which is bounded by $PFC(\mathfrak{a}) + Q(\mathfrak{a})$ where*

* *the quantity $PFC(\mathfrak{a})$ is the Polynomial Factorization Cost, i.e. the number of arithmetic operations required in K to factorize a univariate polynomial over K of total degree bounded by the degree of $\mathfrak{a}$ and coefficients of bit length bounded by the logarithmic height of $V(\mathfrak{a})$ viewed as an arithmetic variety.*
* $Q(\mathfrak{a})$ *is* $O\left(\ell(\mathfrak{a})\deg(\sqrt{\mathfrak{a}})\left((n+1)^2\deg(\mathfrak{a})+k\right)^3\right) \times \mathscr{T}_S(n, \tau, \sharp\Theta, D)$.

Proof. From Theorem 17, we know that the reduced Gröbner basis of the primary component $\mathfrak{q}$ can be computed in a number of arithmetic operations in K bounded by

$$O\left(\ell(\mathfrak{q})\left((n+1)^2\deg(\mathfrak{q})+k\right)^3\right) \times \mathscr{T}_S(n, \tau, \sharp\Theta, D).$$

On the other hand, the number of primary components of $\mathfrak{a}$ is bounded by the number of associated primes of $\mathfrak{a}$ and this number is at most $\deg(\sqrt{\mathfrak{a}})$.   □

Remark 19. The quantity $PFC(\mathfrak{a})$ aims to capture the complexity of factoring a univariate polynomial which can be obtained as the instantiation of an elimination polynomial related to $\mathfrak{a}$.

In the case K is a number field, since the old times of LLL, Trager's method and Landau-Miller norm method approach, we know that $PFC(\mathfrak{a})$ is bounded by a polynomial in the degree of the field extension $[K : \mathbb{Q}]$, $\deg(\mathfrak{a})$ and the logarithmic height of $V(\mathfrak{a})$. In our conditions, the less precise quantity could be the logarithmic height of $V(\mathfrak{a})$. Bézout Inequality (see [17,36]) and the Arithmetic Bézout Inequality (see e.g., [3,25,27,29–32], or the references in [12,28]) imply the following upper bounds for $\deg(\mathfrak{a})$ and the logarithmic height of $V(\mathfrak{a})$:

$$\deg(\mathfrak{a}) \leq d^{n-\dim(\mathfrak{a})},\ \ \mathrm{ht}(V(\mathfrak{a})) \leq O(d^{n-\dim(\mathfrak{a})}(h + n\log(d)))$$

where $\mathrm{ht}(V(\mathfrak{a}))$ is equal to $\mathrm{ht}(\mathfrak{a})$ and h is a bound for the bit length of coefficients of the f_i's. Hence, in this case $PFC(\mathfrak{a})$ is bounded by a polynomial in $[K : \mathbb{Q}], d^{n-\dim(\mathfrak{a})}, n$ and h Note that in the case $K = \mathbb{Q}$, in [2, Corollary 5.3] an algorithm is presented for factoring a polynomial $f \in \mathbb{Q}[X]$ within the bit complexity $O(r^8 + r^6p^2)$ where $r = \deg(f)$ and $p = \log(\|f\|_2)$. Hence, in this case and under our hypothesis, $PFC(\mathfrak{a})$ is bounded by a quantity which is a polynomial of order:

$$PFC(\mathfrak{a}) \leq \left(d^{n-\dim(\mathfrak{a})}nh\right)^{O(1)}.$$

Remark 20. The same type of statement holds for a perfect field with an efficient univariate factorization algorithm. In the case of finite fields we have to add the phrase "with enough elements", in order to have efficient probabilistic non-zero polynomial identity tests (as those in [16,33,39]), for polynomials with degrees bounded by $\deg(\mathfrak{a})$. Note that in the case K is a finite field, as it is known in the literature of the topic, the number of arithmetic operations in the ground field K of deterministic factorization of univariate polynomials (and, hence, $PFC(\mathfrak{a})$) is bounded by a polynomial in $\deg(\mathfrak{a})$ and in the characteristic of the field. If we admit probabilistic algorithms, the quantity $PFC(\mathfrak{a})$ would be bounded by a polynomial in $\deg(\mathfrak{a}) \leq d^{n-\dim(\mathfrak{a})}$ and the logarithm of the field cardinality. See the survey [37] for more detailed references.

Remark 21. From the estimation of $\mathscr{T}_S(n, \tau, \sharp\Theta, D)$ in Remark 6, it holds $Q(\mathfrak{a}) \leq k^3 d^{O(n^2)}$. Our insistence to exhibit an accurate bound for $Q(\mathfrak{a})$ comes from our interest to analyze whether $Q(\mathfrak{a})$ can be of order $\left(d^{n-\dim(\mathfrak{a})}\right)^{O(1)} = d^{O(n)}$. From our study, we conclude that this bound can be achieved provided that there exist methods to compute S-polynomials faster than those cited in Remark 6.

Corollary 22. *The same bound presented in Corollary 18 holds for the number of arithmetic operations in K to compute the reduced Gröbner basis for $\mathfrak{a}$.*

Proof. In [19, page 232], it has been shown that by computing the reduced Gröbner bases of the primary components of $\mathfrak{a}$ and by applying a variant of the FGLM algorithm [11] one is able to construct the reduced Gröbner basis for $\mathfrak{a}$. Note that, in this variant of the FGLM algorithm, we shall consider the vectors of size at most $\deg(\mathfrak{a})$ and therefore the FGLM cost would be $O(n \deg(\mathfrak{a})^3) \leq Q(\mathfrak{a})$, see [11, Proposition 4.1]. □

4   Complexity of Computing the Degree of an Ideal

In this section, we give an algorithm to compute the degree of an ideal. Then, after stating some required preliminaries, we provide the complexity analysis of this algorithm (see Theorem 23).

Let K be an algebraically closed field with efficient factorization of univariate polynomials. Let $\mathfrak{a} = (f_1, \ldots, f_k) \subset K[X]$ be an equidimensional[1] ideal of dimension r. In [15], it has been shown that one is able to choose generic linear polynomials $\ell_1, \ldots, \ell_r$ so that the ideal $\mathfrak{a} + (\ell_1, \ldots, \ell_r)$ is zero-dimensional and the sum of the degrees of all primary components of this ideal is defined to be the degree of $\mathfrak{a}$. On the other hand, in [13], it has been shown that by a generic linear change of coordinates $\sqrt{\mathfrak{a}}$ is transformed to normal X_1-position. Thus, we may choose generic linear polynomials $\ell_1, \ldots, \ell_n$ such that $\mathfrak{b} = \mathfrak{a} + (\ell_1, \ldots, \ell_n)$ is zero-dimensional and $\sqrt{\mathfrak{b}}$ is in normal X_1-position. Based on the results presented in Section 3 we give the next probabilistic algorithm to compute $\deg(\mathfrak{a})$.

Algorithm 3. DEGREE

Input: A finite generating set F of an equidimensional $\mathfrak{a}$ of dimension r
Output: $\deg(\mathfrak{a})$
    choose generic linear polynomials $\ell_1, \ldots, \ell_n \in K[X_1, \ldots, X_n]$
    compute a Kronecker description of $V(\sqrt{\mathfrak{b}})$ where $\mathfrak{b} = \mathfrak{a} + (\ell_1, \ldots, \ell_n)$
    factor the univariate minimal equation of the chosen primitive element of the residue ring $K[V(\mathfrak{b})] := K[X_1, \ldots, X_n]/\sqrt{\mathfrak{b}}$.
    compute a Kronecker description of each of the associated prime $\mathfrak{p}$ of $\mathfrak{b}$.
    compute $\deg(\mathfrak{q})$, for every primary component $\mathfrak{q}$ of $\mathfrak{b}$.
    return $\sum_{\mathfrak{q}} \deg(\mathfrak{q})$ where $\mathfrak{q}$ is a primary component of the ideal $\mathfrak{b}$

It is worth noting that different approaches have been developed in literature to compute primary components of a zero-dimensional ideal generated by $g_1, \ldots, g_s$. For example, we can point out the method of *Kronecker solver* introduced by Giusti et al. [14] (see also [8, 10, 24]). In this approach, we shall require several assumptions: g_i for $1 < i \leq n$ forms a non-zero divisor in $K[X]/(g_1, \ldots, g_{i-1})$. In addition, we shall perform a linear change of variables

---

[1] That is all isolated primes of $\mathfrak{a}$ share the same dimension. Note that, in general, an equidimensional is not unmixed. A proper ideal is said to be unmixed if its dimension is equal to the dimension of every associated prime of the ideal.

so that the 12 conditions mentioned in [10, page 127] are satisfied. Under these assumptions, an efficient method, without the use of Gröbner bases, is given to compute Kronecker representations for all the primary components of the ideal. Now, let K be of characteristic zero, d_i denote the degree of g_i so that $d_1 \geq \cdots \geq d_s$ and let $d = d_1 \cdots d_n$. Then, in [9] an algorithm has been proposed for these computations within the arithmetic complexity $\tilde{O}(d^{11} + (L + ns)d^6)$ where $g_1, \ldots, g_s$ are given by a straight-line program of size L. As a consequence of Corollary 22, we obtain the main result of this section.

Theorem 23. *Algorithm 3 is a (Monte Carlo) probabilistic algorithm such that for every input equidimensional ideal $\mathfrak{a} \subset R$, given by a system of generators $F = \{f_1, \ldots, f_k\}$ of degree at most d and coefficients of bit length at most h, outputs the degree $\deg(\mathfrak{a})$. The total number of arithmetic field operations performed by this algorithm is bounded by the sum of two quantities depending on the input ideal $Kron(\mathfrak{a}) + PFC(\mathfrak{a}) + Q(\mathfrak{a})$ where $Kron(\mathfrak{a})$ denotes the number of arithmetic operations in K to compute the Kronecker description for $\sqrt{\mathfrak{a}}$.*

Acknowledgments. The research was partially supported by the following Iranian, Argentinian and Spanish Grants:
- IPM Grant No. 98550413 (Amir Hashemi)
- UBACyT 20020170100309BA and PICT-2014-3260 (Joos Heintz)
- Spanish Grant MTM2014-55262-P (Luis M. Pardo)

References

1. Becker, T., Weispfenning, V.: Gröbner Bases: A Computational Approach to Commutative Algebra. In cooperation with Heinz Kredel. Springer, New York (1993). https://doi.org/10.1007/978-1-4612-0913-3
2. Belabas, K., van Hoeij, M., Klüners, J., Steel, A.: Factoring polynomials over global fields. J. Théor. Nombres Bordx. **21**(1), 15–39 (2009)
3. Bost, J.B., Gillet, H., Soulé, C.: Un analogue arithmétique du théorème de Bézout. C. R. Acad. Sci. Paris Sér. I **312**(11), 845–848 (1991)
4. Brownawell, W.D., Masser, D.W.: Multiplicity estimates for analytic functions. II. Duke Math. J. **47**, 273–295 (1980)
5. Buchberger, B.: Bruno Buchberger's PhD thesis 1965: an algorithm for finding the basis elements of the residue class ring of a zero dimensional polynomial ideal. Translation from the German. J. Symb. Comput. **41**(3–4), 475–511 (2006). https://doi.org/10.1016/j.jsc.2005.09.007
6. Cox, D.A., Little, J., O'Shea, D.: Ideals, Varieties, and Algorithms. An Introduction to Computational Algebraic Geometry and Commutative Algebra, 4th edn. Springer, Cham (2015). https://doi.org/10.1007/978-3-319-16721-3
7. Dickenstein, A., Fitchas, N., Giusti, M., Sessa, C.: The membership problem for unmixed polynomial ideals is solvable in single exponential time. Discrete Appl. Math. **33**(1–3), 73–94 (1991)
8. Durvye, C.: Algorithmes pour la décomposition primaire des idéaux polynomiaux de dimension nulle donnés en évaluation. Ph.D. thesis, University of Versailles-St Quentin en Yvelines (2008)

9. Durvye, C.: Evaluation techniques for zero-dimensional primary decomposition. J. Symb. Comput. **44**(9), 1089–1113 (2009)
10. Durvye, C., Lecerf, G.: A concise proof of the Kronecker polynomial system solver from scratch. Expo. Math. **26**(2), 101–139 (2008)
11. Faugère, J.C., Gianni, P., Lazard, D., Mora, T.: Efficient computation of zero-dimensional Gröbner bases by change of ordering. J. Symb. Comput. **16**(4), 329–344 (1993)
12. Fernández, M., Pardo, L.M.: An arithmetic Poisson formula for the multi-variate resultant. J. Complexity **29**(5), 323–350 (2013)
13. Gianni, P., Mora, T.: Algebrric solution of systems of polynomirl equations using Groebher bases. In: Huguet, L., Poli, A. (eds.) AAECC 1987. LNCS, vol. 356, pp. 247–257. Springer, Heidelberg (1989). https://doi.org/10.1007/3-540-51082-6_83
14. Giusti, M., Lecerf, G., Salvy, B.: A Gröbner free alternative for polynomial system solving. J. Complexity **17**(1), 154–211 (2001)
15. Hashemi, A., Heintz, J., Pardo, L.M., Solernó, P.: On Bézout inequalities for non-homogeneous polynomial ideals. arXiv:1701.04341 (2017)
16. Heintz, J., Schnorr, C.P.: Testing polynomials which are easy to compute. In: International Symposium on Logic and Algorithmic, Zürich 1980, Monographs of L'Enseignement Mathematique, vol. 30, pp. 237–254 (1982)
17. Heintz, J.: Definability and fast quantifier elimination in algebraically closed fields. Theor. Comput. Sci. **24**, 239–277 (1983)
18. Kemper, G.: A Course in Commutative Algebra, vol. 256. Springer, Heidelberg (2011). https://doi.org/10.1007/978-3-642-03545-6
19. Lakshman, Y.N.: A single exponential bound on the complexity of computing Gröbner bases of zero dimensional ideals. In: Mora, T., Traverso, C. (eds.) Effective Methods in Algebraic Geometry. Progress in Mathematics, vol. 94, pp. 227–234. Birkhäuser, Boston (1991). https://doi.org/10.1007/978-1-4612-0441-1_15
20. Lakshman, Y.N., Lazard, D.: On the complexity of zero-dimensional algebraic systems. In: Mora, T., Traverso, C. (eds.) Effective Methods in Algebraic Geometry. Progress in Mathematics, vol. 94, pp. 217–225. Birkhäuser, Boston (1991)
21. Lazard, D.: Gröbner bases, Gaussian elimination and resolution of systems of algebraic equations. In: van Hulzen, J.A. (ed.) EUROCAL 1983. LNCS, vol. 162, pp. 146–156. Springer, Heidelberg (1983). https://doi.org/10.1007/3-540-12868-9_99
22. Lazard, D.: Résolution des systèmes d'équations algébriques. Theor. Comput. Sci. **15**, 77–110 (1981). https://doi.org/10.1016/0304-3975(81)90064-5
23. Le Gall, F.: Powers of tensors and fast matrix multiplication. In: Proceedings of the 39th International Symposium on Symbolic and Algebraic Computation, ISSAC 2014, pp. 296–303. Association for Computing Machinery (ACM), New York (2014)
24. Lecerf, G.: Computing the equidimensional decomposition of an algebraic closed set by means of lifting fibers. J. Complexity **19**(4), 564–596 (2003)
25. Lelong, P.: Mesure de Mahler et calcul de constantes universelles pour les polynômes de n variables. Math. Ann. **299**(4), 673–695 (1994)
26. Mayr, E.W., Meyer, A.R.: The complexity of the word problems for commutative semigroups and polynomial ideals. Adv. Math. **46**, 305–329 (1982). https://doi.org/10.1016/0001-8708(82)90048-2
27. McKinnon, D.: An arithmetic analogue of Bezout's theorem. Compos. Math. **126**(2), 147–155 (2001)
28. Pardo, L.M., Pardo, M.: On the zeta Mahler measure function of the Jacobian determinant, condition numbers and the height of the generic discriminant. Appl. Algebra Eng. Commun. Comput. **27**(4), 303–358 (2016). https://doi.org/10.1007/s00200-016-0284-9

29. Philippon, P.: Critères pour l'indépendance algébrique. Publ. Math. Inst. Hautes Étud. Sci. **64**, 5–52 (1986)
30. Philippon, P.: Sur des hauteurs alternatives. I. (On alternative heights. I). Math. Ann. **289**(2), 255–283 (1991)
31. Philippon, P.: Sur des hauteurs alternatives. II. (On alternative heights. II). Ann. Inst. Fourier **44**(4), 1043–1065 (1994)
32. Philippon, P.: Sur des hauteurs alternatives. III. J. Math. Pures Appl. (9) **74**(4), 345–365 (1995)
33. Schwartz, J.T.: Fast probabilistic algorithms for verification of polynomial identities. J. Assoc. Comput. Mach. **27**, 701–717 (1980)
34. Hoeven, J.: On the complexity of multivariate polynomial division. In: Kotsireas, I.S., Martínez-Moro, E. (eds.) ACA 2015. SPMS, vol. 198, pp. 447–458. Springer, Cham (2017). https://doi.org/10.1007/978-3-319-56932-1_28
35. Van der Waerden, B.L.: Algebra. Volume II. Based in part on lectures by. E. Artin and E. Noether. Transl. from the German 5th ed. by John R. Schulenberger. Springer, New York (1991)
36. Vogel, W.: Lectures on results on Bezout's theorem. Notes by D. P. Patil. Lectures on Mathematics and Physics. Mathematics, 74. Tata Institute of Fundamental Research. Springer, Berlin, ix, 132 p. (1984)
37. von zur Gathen, J., Panario, D.: Factoring polynomials over finite fields: a survey. J. Symb. Comput. **31**(1 2), 3 17 (2001)
38. Zariski, O., Samuel, P.: Commutative algebra. Vol. II. The University Series in Higher Mathematics. Princeton, N.J.-Toronto-London-New York: D. Van Nostrand Company, Inc. x, 414 p. (1960)
39. Zippel, R.: Interpolating polynomials from their values. J. Symb. Comput. **9**(3), 375–403 (1990)

On the Study of the Motion of a System of Two Connected Rigid Bodies by Computer Algebra Methods

Valentin Irtegov$^{(\boxtimes)}$ and Tatiana Titorenko

Matrosov Institute for System Dynamics and Control Theory SB RAS,
134, Lermontov Str., Irkutsk 664033, Russia
irteg@icc.ru

Abstract. Stationary motions of a system of two connected rigid bodies in a constant gravity field are studied. With the help of computer algebra tools, we construct the symbolic characteristic function of the system and use it as a starting point for deriving the equations of motion. Stationary solutions of the equations which correspond to permanent rotations of the system are found and their stability in the sense of Lyapunov is investigated. The computer algebra system *Mathematica* and the software package developed on its basis are applied to solve the problems under study.

1 Introduction

Systems of connected rigid bodies are widely used in designing complicated technical devices and instruments as their models to study the dynamical properties of these objects. In particular, concerns spacecraft, platform mechanisms, industrial robots, and etc. By now there are many publications devoted to the modelling and analysis of the dynamics of connected rigid bodies, because this area of research has been actively developing since the middle of the last century (see, e.g., [1–3]). From recent works, we mention [4–6]. Nevertheless, in the dynamics of connected rigid bodies there exist unresolved problems so far. Particularly it concerns the qualitative analysis of such systems. As a rule, the systems are multiparametric, depend on many variables, and their symbolic analysis requires bulky computations. Trying to simplify the systems for the purpose of their qualitative analysis, one imposes constraints on the geometry of masses of the bodies, the points of their connection, etc. We consider the problem in more general formulation. The motion of a system of two rigid asymmetric bodies linked together by an ideal spherical hinge in a constant gravity field is studied. The attachment point does not lie on the principal inertia axes of one of the bodies. For the given mechanical system, we find and analyze stationary motions by means of computer algebra methods as well as the software package *LinModel.m* written in the language of the computer algebra system (CAS) *Mathematica* [7]. This package is used for obtaining the symbolic equations of motion of the system.

© Springer Nature Switzerland AG 2020
F. Boulier et al. (Eds.): CASC 2020, LNCS 12291, pp. 266–281, 2020.
https://doi.org/10.1007/978-3-030-60026-6_15

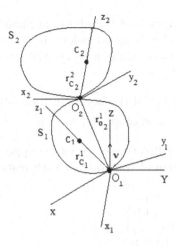

Fig. 1. Bodies system.

Constructing the equations of motion is based on the Lagrangian formalism. First, using a geometrical description of a mechanical system, the package computes its symbolic characteristic function – the Lagrange function. At the next step, this function is used as a starting point for deriving the equations of motion. Then the equations are analyzed. We find their solutions possessing the following property: the first integrals of the problem take stationary values on them. Such solutions are called stationary. In our work, we restrict ourselves to the solutions corresponding to permanent rotations of the system. The Lyapunov stability of the solutions is investigated. The energy integral and other first integrals of the equations of motion are used to obtain sufficient conditions of their stability.

The paper is organized as follows. In Sect. 2 and 3, deriving the Lagrange function and the equations of motion are described. In Sect. 4–6, we find the stationary solutions and invariant manifolds (IMs) for the equations. In Sect. 7, the stability of the stationary solutions is analyzed. Finally, we give a conclusion in Sect. 8.

2   Constructing the Lagrange Function

The motion of the system of two rigid bodies S_1 and S_2 (Fig. 1) in a constant gravity field is considered. The first of them has the fixed point O_1. The bodies are linked together by the ideal spherical hinge O_2.

The following coordinate systems are introduced to describe the motion of the mechanical system: the inertial O_1XYZ (its Z axis with the unit vector ν is directed vertically upwards), the moving frames $O_1x_1y_1z_1$ and $O_2x_2y_2z_2$ attached rigidly to the bodies S_1 and S_2, respectively. The x_i, y_i, z_i ($i = 1, 2$) axes are directed along the principal inertia axes of the bodies. The centers of masses c_1 and c_2 correspondingly lie on the O_1z_1 and O_2z_2 axes. The positions of $O_1x_1y_1z_1$ with respect to O_1XYZ and $O_2x_2y_2z_2$ with respect to $O_1x_1y_1z_1$ are defined by Euler's angles $\psi_1, \theta_1, \varphi_1$ and $\psi_2, \theta_2, \varphi_2$.

The geometrical description of the above mechanical system for the *Mathematica* package *LinModel.m* is given below. It is used as input data by the package to construct the Lagrange function of the system.

1. The general system data:

$$\text{the number of the bodies in the system} = 2,$$
$$\text{the direction of gravity} = -Z.$$

The latter means that the direction of gravity is opposite to the $O_1 Z$ axis.

2. The description of motion of the body S_1 with respect to $O_1 XYZ$:

$$mass = m_1,$$
$$\mathbf{r}^0_{O_1} = \{0, 0, 0\},$$
$$\mathbf{r}^1_{c_1} = \{0, 0, c_1\},$$
$$\mathbf{v}^0_{O_1} = \{0, 0, 0\},$$
$$\text{rotation axes of the body} = \{3, 1, 3\},$$
$$\text{rotation angles} = \{\psi_1, \theta_1, \varphi_1\},$$
$$J^{O_1} = \{\{A_1, 0, 0\}, \{0, B_1, 0\}, \{0, 0, C_1\}\}.$$

3. The description of motion of the body S_2 with respect to $O_1 x_1 y_1 z_1$:

$$mass = m_2,$$
$$\mathbf{r}^1_{O_2} = \{s_1, s_2, s_3\},$$
$$\mathbf{r}^2_{c_2} = \{0, 0, c_2\},$$
$$\mathbf{v}^1_{O_2} = \{0, 0, 0\},$$
$$\text{rotation axes of the body} = \{3, 1, 3\},$$
$$\text{rotation angles} = \{\psi_2, \theta_2, \varphi_2\},$$
$$J^{O_2} = \{\{A_2, 0, 0\}, \{0, B_2, 0\}, \{0, 0, C_2\}\}.$$

Here m_1, m_2 are the masses of the bodies, $\mathbf{r}^0_{O_1}$, $\mathbf{r}^1_{O_2}$ are the radius vectors of the points O_1, O_2 in the coordinate systems $O_1 XYZ$ and $O_1 x_1 y_1 z_1$, respectively; $\mathbf{v}^0_{O_1}$, $\mathbf{v}^1_{O_2}$ are the vectors of linear velocities of the points O_1, O_2 in the projections onto the axes of $O_1 XYZ$ and $O_1 x_1 y_1 z_1$; A_i, B_i, C_i are the principal inertia moments of the bodies (from now and further $i = 1, 2$); $\mathbf{r}^i_{c_i}$ are the radius vectors of the point c_i in the frame $O_i x_i y_i z_i$; J^{O_i} is the inertia tensor of the body S_i with respect to the point O_i in the frame $O_i x_i y_i z_i$.

The above description encoded in the language of the CAS *Mathematica* is placed by the user in a file. *LinModel* reads it from the file and computes for the body S_i its absolute angular velocity $\boldsymbol{\omega}_i$, the absolute linear velocity $\mathbf{v}_{O_i}$ of the point O_i, and the kinetic energy of the body by the formula:

$$2T_i = m_i \mathbf{v}^2_{O_i} + \boldsymbol{\omega}_i \cdot \mathbf{J}^{O_i} \cdot \boldsymbol{\omega}_i + 2m_i [\mathbf{v}_{O_i} \times \boldsymbol{\omega}_i] \cdot \mathbf{r}^i_{c_i}.$$

The kinetic energy of the system is defined as follows: $T = \sum_{i=1}^{2} T_i$.

The force function of the system and each body S_i in a constant gravity field are calculated by the formulae:

$$U = \sum_{i=1}^{2} U_i, \quad U_i = -m_i g(\boldsymbol{\nu} \cdot \mathbf{r}^0_{c_i}) + \text{const},$$

where g is the acceleration of gravity, $\mathbf{r}^0_{c_1} = \mathbf{r}^0_{O_1} + \mathbf{r}^1_{c_1}$, $\mathbf{r}^0_{c_2} = \mathbf{r}^0_{O_1} + \mathbf{r}^1_{O_2} + \mathbf{r}^2_{c_2}$.

As a result, *LinModel* returns the kinetic energy

$$
\begin{aligned}
2T = {} & (A_1 p_1^2 + B_1 q_1^2 + C_1 r_1^2) + A_2(b_{11}p_1 + p_2 + b_{12}q_1 + b_{13}r_1)^2 \\
& + B_2(b_{21}p_1 + b_{22}q_1 + q_2 + b_{23}r_1)^2 + C_2(b_{31}p_1 + b_{32}q_1 + b_{33}r_1 + r_2)^2 \\
& + m_2 \left[(s_2 p_1 - s_1 q_1)^2 + (s_1 r_1 - s_3 p_1)^2 + (s_3 q_1 - s_2 r_1)^2 \right] \\
& + c_2 m_2 \big[(b_{21}p_1 + b_{22}q_1 + q_2 + b_{23}r_1) [b_{13}(s_2 p_1 - s_1 q_1) + b_{12}(s_1 r_1 - s_3 p_1) \\
& + b_{11}(s_3 q_1 - s_2 r_1)] - (b_{11}p_1 + p_2 + b_{12}q_1 + b_{13}r_1)(b_{23}(s_2 p_1 - s_1 q_1) \\
& + b_{22}(s_1 r_1 - s_3 p_1) + b_{21}(s_3 q_1 - s_2 r_1)),
\end{aligned}
$$

the force function

$$U = -g \left[a_{33} c_1 m_1 + m_2 [a_{13}(c_2 b_{31} + s_1) + a_{23}(c_2 b_{32} + s_2) + a_{33}(c_2 b_{33} + s_3)] \right],$$

and the Lagrange function $L = T + U$ of the mechanical system.

Here p_i, q_i, r_i are the projections of the angular velocity vector of the body S_i onto the axes of $O_i x_i y_i z_i$:

$$
\begin{aligned}
p_i &= \dot{\psi}_i \sin \varphi_i \sin \theta_i + \dot{\theta}_i \cos \varphi_i, \\
q_i &= \dot{\psi}_i \cos \varphi_i \sin \theta_i - \dot{\theta}_i \sin \varphi_i, \\
r_i &= \dot{\varphi}_i + \dot{\psi}_i \cos \theta_i,
\end{aligned}
$$

a_{kl}, b_{kl} are the elements of cosine matrices of angles between the axes of $O_1 XYZ$ and $O_1 x_1 y_1 z_1$, and $O_1 x_1 y_1 z_1$ and $O_2 x_2 y_2 z_2$, respectively. These are related to Euler's angles ψ_i, θ_i, and φ_i as follows:

$$a_{kl} = \zeta^{(1)}_{kl}, \quad b_{kl} = \zeta^{(2)}_{kl} \quad (k, l = 1, \ldots, 3), \quad \text{where}$$

$$\zeta^{(i)}_{11} = \cos \varphi_i \cos \psi_i - \cos \theta_i \sin \varphi_i \sin \psi_i,$$

$$\zeta^{(i)}_{12} = \cos \psi_i \cos \theta_i \sin \varphi_i + \cos \varphi_i \sin \psi_i, \quad \zeta^{(i)}_{13} = \sin \varphi_i \sin \theta_i,$$

$$\zeta^{(i)}_{21} = -\cos \psi_i \sin \varphi_i - \cos \varphi_i \cos \theta_i \sin \psi_i,$$

$$\zeta^{(i)}_{22} = \cos \varphi_i \cos \psi_i \cos \theta_i - \sin \varphi_i \sin \psi_i,$$

$$\zeta^{(i)}_{23} = \cos \varphi_i \sin \theta_i, \quad \zeta^{(i)}_{31} = \sin \psi_i \sin \theta_i, \quad \zeta^{(i)}_{32} = -\cos \psi_i \sin \theta_i,$$

$$\zeta^{(i)}_{33} = \cos \theta_i \quad (i = 1, 2).$$

As one can see, the kinetic energy and the force function are represented in algebraic form. Such representation is more suitable from the viewpoint of symbolic computations.

3 Deriving the Equations of Motion

Using the Lagrange function L as input data, *LinModel* produces the equations of motion of the system under study by the formulae [8]:

$$\frac{d}{dt}\left(\frac{\partial L}{\partial \boldsymbol{\omega}^{(i)}}\right) = \frac{\partial L}{\partial \boldsymbol{\omega}^{(i)}} \times \boldsymbol{\omega}^{(i)} + \frac{\partial L}{\partial \boldsymbol{\alpha}^{(i)}} \times \boldsymbol{\alpha}^{(i)} + \frac{\partial L}{\partial \boldsymbol{\beta}^{(i)}} \times \boldsymbol{\beta}^{(i)} + \frac{\partial L}{\partial \boldsymbol{\gamma}^{(i)}} \times \boldsymbol{\gamma}^{(i)},$$

$$\dot{\boldsymbol{\alpha}}^{(i)} = \boldsymbol{\alpha}^{(i)} \times \boldsymbol{\omega}^{(i)}, \ \dot{\boldsymbol{\beta}}^{(i)} = \boldsymbol{\beta}^{(i)} \times \boldsymbol{\omega}^{(i)}, \ \dot{\boldsymbol{\gamma}}^{(i)} = \boldsymbol{\gamma}^{(i)} \times \boldsymbol{\omega}^{(i)} (i = 1, 2).$$

Here $\boldsymbol{\omega}^{(i)} = (p_i, q_i, r_i)$, $\boldsymbol{\alpha}^{T(1)} = (a_{11}, a_{21}, a_{31})$, $\boldsymbol{\beta}^{T(1)} = (a_{12}, a_{22}, a_{32})$, $\boldsymbol{\gamma}^{T(1)} = (a_{13}, a_{23}, a_{33})$, $\boldsymbol{\alpha}^{T(2)} = (b_{11}, b_{21}, b_{31})$, $\boldsymbol{\beta}^{T(2)} = (b_{12}, b_{22}, b_{32})$, $\boldsymbol{\gamma}^{T(2)} = (b_{13}, b_{23}, b_{33})$.

The resulting equations are written as

$$
\begin{aligned}
&(A_1 + m_2(s_2^2 + s_3^2))\,\dot{p}_1 + (A_2 b_{11} + c_2 m_2(b_{22} s_3 - b_{23} s_2))\,\dot{p}_2 \\
&+(B_2 b_{21} + c_2 m_2(b_{13} s_2 - b_{12} s_3))\,\dot{q}_2 - (A_2 - B_2) b_{21}(b_{21} p_1 + b_{22} q_1 + b_{23} r_1) \\
&-b_{31}(A_2 - C_2)(b_{31} p_1 + b_{32} q_1 + b_{33} r_1) - m_2 s_1(s_2 \dot{q}_1 + s_3 \dot{r}_1) + A_2(V_7 p_1 + V_{10} q_1 \\
&+V_{11} \dot{r}_1) + c_2 m_2[(b_{13} b_{21} - b_{11} b_{23})(2 s_2 \dot{p}_1 - s_1 \dot{q}_1) + (b_{11} b_{22} - b_{12} b_{21})(2 s_3 \dot{p}_1 - s_1 \dot{r}_1) \\
&+(b_{13} b_{22} - b_{12} b_{23})(s_2 \dot{q}_1 + s_3 \dot{r}_1)] + b_{31} C_2 \dot{r}_2 + A_2[\,q_1\,(V_{11} p_1 - V_8 r_1) \\
&+r_1(V_9 q_1 - V_{10} p_1) + V_{12}(q_1^2 - r_1^2)] + A_2 p_2(b_{13} q_1 - b_{12} r_1) + B_2 q_2(b_{23} q_1 - b_{22} r_1) \\
&+C_2 r_2(b_{33} q_1 - b_{32} r_1) + (C_1 - B_1)\,q_1 r_1 - (A_2 - C_2)[(b_{33} q_1 + b_{11} q_2 - b_{32} r_1) \\
&\times(b_{32} q_1 + b_{33} r_1) + b_{31}(q_2(b_{11} p_1 + b_{12} q_1) + p_1(b_{33} q_1 + b_{11} q_2) + r_1(b_{13} q_2 - b_{32} p_1)) \\
&-b_{21} p_2 r_2] - (A_2 - B_2)[(b_{23} q_1 - b_{22} r_1)(b_{21} p_1 + b_{22} q_1 + b_{23} r_1) - r_2(b_{21}(b_{11} p_1 \\
&+b_{12} q_1) + b_{11}(b_{21} p_1 + b_{22} q_1) + r_1(b_{13} b_{21} + b_{11} b_{23})) + b_{31} p_2 q_2] \\
&+(B_2 - C_2)[b_{31} p_2(b_{21} p_1 + b_{22} q_1 + b_{23} r_1) + b_{21} p_2(b_{31} p_1 + b_{32} q_1 + b_{33} r_1) \\
&-b_{11} q_2 r_2] + c_2 m_2\Big[(b_{31}(b_{12} p_2 + b_{22} q_2) - b_{32}(b_{11} p_2 + b_{21} q_2))(r_1 s_1 - 2 p_1 s_3) \\
&-(b_{31}(b_{13} p_2 + b_{23} q_2) - b_{33}(b_{11} p_2 + b_{21} q_2))(q_1 s_1 - 2 p_1 s_2) + (b_{32}(b_{13} p_2 + b_{23} q_2) \\
&-b_{33}(b_{12} p_2 + b_{22} q_2) + b_{21} p_2)(q_1 s_2 + r_1 s_3) \\
&+(b_{13} b_{21} - b_{11} b_{23})(r_1(p_1 s_1 + 2 q_1 s_2) + (r_1^2 - q_1^2) s_3) \\
&+(b_{12} b_{21} - b_{11} b_{22})(q_1(p_1 s_1 + 2 r_1 s_3) + (q_1^2 - r_1^2) s_2) \\
&-(b_{11} q_2(q_1 s_2 + r_1 s_3) + q_1 s_1(b_{22} p_2 - b_{12} q_2) + r_2 s_3(b_{12} p_2 + b_{22} q_2)) \\
&+(b_{12} b_{23} - b_{13} b_{22}) p_1(r_1 s_2 - q_1 s_3) - (b_{33} s_2 - b_{32} s_3)(p_2^2 + q_2^2) \\
&+((b_{13} q_2 - b_{23} p_2)\, r_1 s_1 + (b_{13} p_2 + b_{23} q_2)\, r_2 s_2)\Big] + m_2(r_1 s_2 - q_1 s_3)(p_1 s_1 + q_1 s_2 \\
&+r_1 s_3) - g[a_{23} c_1 m_1 + m_2(c_2(a_{33} b_{32} - a_{23} b_{33}) + (a_{33} s_2 - a_{23} s_3))] = 0, \\
&(B_1 + m_2(s_1^2 + s_3^2))\,\dot{q}_1 + (A_2 b_{12} + c_2 m_2(b_{23} s_1 - b_{21} s_3))\,\dot{p}_2 \\
&+(B_2 b_{22} + c_2 m_2(b_{11} s_3 - b_{13} s_1))\,\dot{q}_2 - (A_2 - B_2) b_{22}(b_{21} \dot{p}_1 + b_{22} \dot{q}_1 + b_{23} \dot{r}_1) \\
&-b_{32}(A_2 - C_2)(b_{31} \dot{p}_1 + b_{32} \dot{q}_1 + b_{33} \dot{r}_1) - m_2 s_2(s_1 \dot{p}_1 + s_3 \dot{r}_1) + A_2(V_{10} \dot{p}_1 + V_8 \dot{q}_1 \\
&+V_{12} \dot{r}_1) + c_2 m_2[(b_{13} b_{22} - b_{12} b_{23})(s_1 \dot{p}_1 - 2 s_1 \dot{q}_1) + (b_{11} b_{22} - b_{12} b_{21})(2 s_3 \dot{q}_1 - s_2 \dot{r}_1) \\
&-(b_{13} b_{21} - b_{11} b_{23})(s_1 \dot{p}_1 + s_3 \dot{r}_1)] + b_{32} C_2 \dot{r}_2 + A_2[\,p_1(V_7 r_1 - V_{12} q_1) + r_1(V_{10} q_1 \\
&-V_9 p_1) + V_{11}(r_1^2 - p_1^2)] + A_2 p_2(b_{11} r_1 - b_{13} p_1) + B_2 q_2(b_{21} r_1 - b_{23} p_1) \\
&+C_2 r_2(b_{31} r_1 - b_{33} p_1) + (A_1 - C_1) p_1 r_1 + (B_2 - C_2)[b_{22} p_2(b_{31} p_1 + b_{32} q_1 + b_{33} r_1) \\
&+b_{32} p_2(b_{21} p_1 + b_{22} q_1 + b_{23} r_1) - b_{12} q_2 r_2] - (A_2 - C_2)[(b_{12} q_2 + b_{31} r_1 - b_{33} p_1) \\
&\times(b_{31} p_1 + b_{32} q_1 + b_{33} r_1) + b_{32} q_2(b_{11} p_1 + b_{13} r_1) + b_{12} q_2((b_{32} q_1 + b_{33} r_1) - b_{33} r_1) \\
&-b_{22} p_2 r_2] + (A_2 - B_2)[(b_{23} p_1 - b_{21} r_1)(b_{21} p_1 + b_{22} q_1 + b_{23} r_1) + r_2(b_{12}(b_{21} p_1 \\
&+b_{22} q_1) + b_{22}(b_{11} p_1 + b_{13} r_1) + b_{12}(b_{22} q_1 + b_{23} r_1)) - b_{32} p_2 q_2]
\end{aligned}
$$

$$\tag{1}$$

$$+c_2 m_2 \Big[[b_{31}(b_{12}p_2 + b_{22}q_2) - b_{32}(b_{11}p_2 + b_{21}q_2)](r_1 s_2 - 2q_1 s_3)$$
$$-[b_{32}(b_{13}p_2 + b_{23}q_2) - b_{33}(b_{12}p_2 + b_{22}q_2)](2q_1 s_1 - p_1 s_2)$$
$$-[(b_{13}p_2 + b_{23}q_2)r_2 s_1 + (b_{23}p_2 - b_{13}q_2)r_1 s_2]$$
$$+(b_{11}b_{23} - b_{13}b_{21})q_1(r_1 s_1 - p_1 s_3) + (p_2^2 + q_2^2)(b_{33}s_1 - b_{31}s_3)$$
$$+[b_{33}(b_{11}p_2 + b_{21}q_2) - b_{31}(b_{13}p_2 + b_{23}q_2) + b_{22}p_2 - b_{12}q_2](p_1 s_1 + r_1 s_3)$$
$$+[(b_{11}q_2 - b_{21}p_2)p_1 s_2 + (b_{11}p_2 + b_{21}q_2)r_2 s_3]$$
$$+(b_{12}b_{23} - b_{13}b_{22})[p_1(p_1 s_3 - 2r_1 s_1) - r_1(q_1 s_2 + r_1 s_3)]$$
$$-(b_{12}b_{21} - b_{11}b_{22})[(p_1^2 - r_1^2)s_1 + p_1(q_1 s_2 + 2r_1 s_3)]\Big]$$
$$+m_2[p_1 s_3(q_1 s_2 + r_1 s_3) - r_1 s_1(p_1 s_1 + q_1 s_2) + (p_1^2 - r_1^2)s_1 s_3]$$
$$+g[m_2(a_{13}s_3 - a_{33}s_1) + c_1 m_1 a_{13} + c_2 m_2(a_{13}b_{33} - a_{33}b_{31})] = 0,$$
$$(C_1 + m_2(s_1^2 + s_2^2))\dot{r}_1 + (A_2 b_{13} + c_2 m_2(b_{21}s_2 - b_{22}s_1))\dot{p}_2$$
$$+(B_2 b_{23} + c_2 m_2(b_{12}s_1 - b_{11}s_2))\dot{q}_2 - (A_2 - B_2)b_{23}(b_{21}\dot{p}_1 + b_{22}\dot{q}_1 + b_{23}\dot{r}_1)$$
$$-(A_2 - C_2)b_{33}(b_{31}\dot{p}_1 + b_{32}\dot{q}_1 + b_{33}\dot{r}_1) - m_2 s_3(s_1\dot{p}_1 + s_2\dot{q}_1) + A_2(V_{11}\dot{p}_1 + V_{12}\dot{q}_1$$
$$+V_9\dot{r}_1) + c_2 m_2[(b_{12}b_{21} - b_{11}b_{22})(s_1\dot{p}_1 + s_2\dot{q}_1) + (b_{13}b_{22} - b_{12}b_{23})(s_3\dot{p}_1 - 2s_1\dot{r}_1)$$
$$+(b_{11}b_{23} - b_{13}b_{21})(s_3\dot{q}_1 - 2s_2\dot{r}_1)] + b_{33}C_2\dot{r}_2 + A_2[p_1(V_{12}r_1 - V_7 q_1)$$
$$+q_1(V_8 p_1 - V_{11}r_1) + V_{10}(p_1^2 - q_1^2)] - (A_1 - B_1)p_1 q_1$$
$$+A_2(b_{12}p_1 - b_{11}q_1)p_2 + B_2(b_{22}p_1 - b_{21}q_1)q_2 + C_2(b_{32}p_1 - b_{31}q_1)r_2$$
$$-(A_2 - C_2)[(b_{31}p_1 + b_{32}q_1)(b_{32}p_1 - b_{31}q_1 + b_{13}q_2) + b_{33}((b_{32}p_1 - b_{31}q_1$$
$$+b_{13}q_2)r_1 + (b_{11}p_1 + b_{12}q_1 + b_{13}r_1)q_2) - b_{23}p_2 r_2]$$
$$+(B_2 - C_2)[b_{33}p_2(b_{21}p_1 + b_{22}q_1 + b_{23}r_1) + b_{23}p_2(b_{31}p_1 + b_{32}q_1 + b_{33}r_1)$$
$$-b_{13}q_2 r_2] + (A_2 - B_2)[(b_{21}q_1 - b_{22}p_1)(b_{21}p_1 + b_{22}q_1 + b_{23}r_1)$$
$$+r_2(b_{23}(b_{11}p_1 + b_{12}q_1) + b_{13}(b_{21}p_1 + b_{23}r_1) + b_{13}(b_{22}q_1 + b_{23}r_1)) - b_{33}p_2 q_2]$$
$$-c_2 m_2 \Big[(p_2^2 + q_2^2)(b_{32}s_1 - b_{31}s_2) + (b_{11}b_{22} - b_{12}b_{21})r_1(q_1 s_1 - p_1 s_2)$$
$$+[(b_{23} + b_{12}b_{31} - b_{11}b_{32})p_2 - (b_{13} - b_{22}b_{31} + b_{21}b_{32})q_2](p_1 s_1 + q_1 s_2)$$
$$+[(b_{12}b_{33} - b_{13}b_{32})p_2 + (b_{22}b_{33} - b_{23}b_{32})q_2](2r_1 s_1 - p_1 s_3)$$
$$+[(b_{13}b_{31} - b_{33}b_{11})p_2 + (b_{23}b_{31} - b_{33}b_{21})q_2](2r_1 s_2 - q_1 s_3)$$
$$-[(b_{21}s_2 - b_{22}s_1)q_2 r_2 + (b_{21}p_1 + b_{22}q_1)p_2 s_3]$$
$$+[(b_{12}s_1 - b_{11}s_2)p_2 r_2 + (b_{11}p_1 + b_{12}q_1)q_2 s_3]$$
$$+(b_{13}b_{22} - b_{12}b_{23})[p_1(p_1 s_2 - 2q_1 s_1) - q_1(q_1 s_2 + r_1 s_3)]$$
$$-(b_{13}b_{21} - b_{11}b_{23})[(p_1^2 - q_1^2)s_1 + p_1(2q_1 s_2 + r_1 s_3)]\Big]$$
$$+m_2[(q_1 s_1 - p_1 s_2)(p_1 s_1 + q_1 s_2 + r_1 s_3) + g(c_2(a_{23}b_{31} - a_{13}b_{32})$$
$$+(a_{23}s_1 - a_{13}s_2))] = 0,$$
$$A_2(b_{11}\dot{p}_1 + \dot{p}_2 + b_{12}\dot{q}_1 + b_{13}\dot{r}_1) + c_2 m_2[b_{23}(s_1\dot{q}_1 - s_2\dot{p}_1) + b_{22}(s_3\dot{p}_1 - s_1\dot{r}_1)$$
$$+b_{21}(s_2\dot{r}_1 - s_3\dot{q}_1)] - (A_2 + B_2 - C_2)(b_{31}p_1 q_2 + b_{32}q_1 q_2 + b_{33}q_2 r_1)$$
$$+(C_2 - B_2)[(b_{21}p_1 + b_{22}q_1 + b_{23}r_1)(b_{31}p_1 + b_{32}q_1 + b_{33}r_1) + q_2 r_2]$$
$$+(A_2 - B_2 + C_2)(b_{21}p_1 r_2 + b_{22}q_1 r_2 + b_{23}r_1 r_2)$$
$$+c_2 m_2 \Big[(b_{12}b_{31} - b_{11}b_{32})((p_1^2 + q_1^2)s_3 - r_1(p_1 s_1 + q_1 s_2)) + (b_{13}b_{32} - b_{12}b_{33})$$
$$\times(q_1(q_1 s_1 - p_1 s_2) + r_1(r_1 s_1 - p_1 s_3)) - (b_{13}b_{31} - b_{11}b_{33})(p_1(p_1 s_2 - q_1 s_1)$$
$$+r_1(r_1 s_2 - q_1 s_3)) - g(a_{13}b_{21} + a_{23}b_{22} + a_{33}b_{23})\Big] = 0,$$
$$B_2(b_{21}\dot{p}_1 + b_{22}\dot{q}_1 + \dot{q}_2 + b_{23}\dot{r}_1) + c_2 m_2[b_{13}(s_2\dot{p}_1 - s_1\dot{q}_1) + b_{12}(s_1\dot{r}_1 - s_3\dot{p}_1)$$
$$+b_{11}(s_3\dot{q}_1 - s_2\dot{r}_1)] + (A_2 + B_2 - C_2)p_2(b_{31}p_1 + b_{32}q_1 + b_{33}r_1)$$
$$+(A_2 - C_2)[(b_{11}p_1 + b_{12}q_1 + b_{13}r_1)(b_{31}p_1 + b_{32}q_1 + b_{33}r_1) + p_2 r_2]$$
$$+(A_2 - B_2 - C_2)r_2(b_{11}p_1 + b_{12}q_1 + b_{13}r_1)$$
$$+c_2 m_2 \Big[(b_{22}b_{31} - b_{21}b_{32})((p_1^2 + q_1^2)s_3 - r_1(p_1 s_1 + q_1 s_2)) - (b_{23}b_{31} - b_{21}b_{33})$$
$$\times(p_1(p_1 s_2 - q_1 s_1) + r_1(r_1 s_2 - q_1 s_3)) + (b_{23}b_{32} - b_{22}b_{33})((q_1^2 + r_1^2)s_1$$
$$-p_1(q_1 s_2 + r_1 s_3)) + g(a_{13}b_{11} + a_{23}b_{12} + a_{33}b_{13})\Big] = 0,$$

$$C_2(b_{31}\dot{p}_1 + b_{32}\dot{q}_1 + b_{33}\dot{r}_1 + \dot{r}_2) - (A_2 - B_2 + C_2)p_2(b_{21}p_1 + b_{22}q_1$$
$$+b_{23}r_1) - (A_2 - B_2 - C_2)\,q_2(b_{11}p_1 + b_{12}q_1 + b_{13}r_1)$$
$$-(A_2 - B_2)[\,(b_{11}p_1 + b_{12}q_1 + b_{13}r_1)(b_{21}p_1 + b_{22}q_1 + b_{23}r_1) + p_2q_2] = 0,$$

$$\dot{a}_{11} = a_{21}r_1 - a_{31}q_1, \ \ \dot{a}_{12} = a_{22}r_1 - a_{32}q_1, \ \ \dot{a}_{13} = a_{23}r_1 - a_{33}q_1,$$
$$\dot{a}_{21} = a_{31}p_1 - a_{11}r_1, \ \ \dot{a}_{22} = a_{32}p_1 - a_{12}r_1, \ \ \dot{a}_{23} = a_{33}p_1 - a_{13}r_1, \qquad (2)$$
$$\dot{a}_{31} = a_{11}q_1 - a_{21}p_1, \ \ \dot{a}_{32} = a_{12}q_1 - a_{22}p_1, \ \ \dot{a}_{33} = a_{13}q_1 - a_{23}p_1,$$

$$\dot{b}_{11} = b_{21}r_1 - b_{31}q_1, \ \ \dot{b}_{12} = b_{22}r_1 - b_{32}q_1, \ \ \dot{b}_{13} = b_{23}r_1 - b_{33}q_1,$$
$$\dot{b}_{21} = b_{31}p_1 - b_{11}r_1, \ \ \dot{b}_{22} = b_{32}p_1 - b_{12}r_1, \ \ \dot{b}_{23} = b_{33}p_1 - b_{13}r_1, \qquad (3)$$
$$\dot{b}_{31} = b_{11}q_1 - b_{21}p_1, \ \ \dot{b}_{32} = b_{12}q_1 - b_{22}p_1, \ \ \dot{b}_{33} = b_{13}q_1 - b_{23}p_1.$$

Equations (1)–(3) have the following first integrals. The integrals of energy and kinetic moment:

$$H = T - U = h,$$
$$V = \frac{\partial L}{\partial \boldsymbol{\omega}^{(1)}} \cdot \boldsymbol{\gamma}^{(1)} = a_{13}\Big[A_1p_1 + A_2b_{11}p_2 + B_2b_{21}q_2 + b_{31}C_2r_2$$
$$-(A_2 - B_2)b_{21}(b_{21}p_1 + b_{22}q_1 + b_{23}r_1) - b_{31}(A_2 - C_2)(b_{31}p_1 + b_{32}q_1 + b_{33}r_1)$$
$$+c_2m_2[(b_{11}b_{23} - b_{13}b_{21})(q_1s_1 - 2p_1s_2) + (b_{12}b_{21} - b_{11}b_{22})(r_1s_1 - 2p_1s_3)$$
$$+((b_{13}q_2 - b_{23}p_2)s_2 + (b_{22}p_2 - b_{12}q_2)s_3) + (b_{13}b_{22} - b_{12}b_{23})(q_1s_2 + r_1s_3)]$$
$$+m_2(p_1(s_2^2 + s_3^2) - s_1(q_1s_2 + r_1s_3)) + A_2(V_7p_1 + V_{10}q_1 + V_{11}r_1)\Big]$$

$$+a_{23}\Big[A_2b_{12}p_2 + B_1q_1 + B_2b_{22}q_2 + b_{32}C_2r_2$$
$$-(A_2 - B_2)b_{22}(b_{21}p_1 + b_{22}q_1 + b_{23}r_1) - b_{32}(A_2 - C_2)(b_{31}p_1 + b_{32}q_1 + b_{33}r_1)$$
$$+c_2m_2[(b_{12}b_{23} - b_{13}b_{22})(2q_1s_1 - p_1s_2) + (b_{12}b_{21} - b_{11}b_{22})(r_1s_2 - 2q_1s_3)$$
$$+((b_{23}p_2 - b_{13}q_2)s_1 - b_{21}p_2s_3 + b_{11}q_2s_3) - (b_{13}b_{21} - b_{11}b_{23})(p_1s_1 + r_1s_3)]$$
$$+m_2(q_1(s_1^2 + s_3^2) - s_2(p_1s_1 + r_1s_3)) + A_2[V_{10}p_1 + V_8q_1 + V_{12}r_1]\Big]$$

$$+a_{33}\Big[A_2b_{13}p_2 + B_2b_{23}q_2 + b_{33}C_2r_2 + C_1r_1 \qquad\qquad (4)$$
$$-(A_2 - B_2)b_{23}(b_{21}p_1 + b_{22}q_1 + b_{23}r_1) - b_{33}(A_2 - C_2)(b_{31}p_1 + b_{32}q_1 + b_{33}r_1)$$
$$+c_2m_2[(b_{12}b_{21} - b_{11}b_{22})(p_1s_1 + q_1s_2) + ((b_{12}q_2 - b_{22}p_2)s_1 + (b_{21}p_2 - b_{11}q_2)s_2)$$
$$+(b_{12}b_{23} - b_{13}b_{22})(2r_1s_1 - p_1s_3) + (b_{13}b_{21} - b_{11}b_{23})(2r_1s_2 - q_1s_3)]$$
$$+m_2(r_1(s_1^2 + s_2^2) - (p_1s_1 + q_1s_2)s_3) + A_2(V_{11}p_1 + V_{12}q_1 + V_9r_1)\Big] = c,$$

where h and c are arbitrary constants.

The geometric integrals:

$$V_1 = a_{11}^2 + a_{21}^2 + a_{31}^2 = 1, V_7 = b_{11}^2 + b_{21}^2 + b_{31}^2 = 1,$$
$$V_2 = a_{12}^2 + a_{22}^2 + a_{32}^2 = 1, V_8 = b_{12}^2 + b_{22}^2 + b_{32}^2 = 1,$$
$$V_3 = a_{13}^2 + a_{23}^2 + a_{33}^2 = 1, V_9 = b_{13}^2 + b_{23}^2 + b_{33}^2 = 1,$$
$$V_4 = a_{11}a_{12} + a_{21}a_{22} + a_{31}a_{32} = 0, \ V_{10} = b_{11}b_{12} + b_{21}b_{22} + b_{31}b_{32} = 0, \qquad (5)$$
$$V_5 = a_{11}a_{13} + a_{21}a_{23} + a_{31}a_{33} = 0, \ V_{11} = b_{11}b_{13} + b_{21}b_{23} + b_{31}b_{33} = 0,$$
$$V_6 = a_{12}a_{13} + a_{22}a_{23} + a_{32}a_{33} = 0, \ V_{12} = b_{12}b_{13} + b_{22}b_{23} + b_{32}b_{33} = 0.$$

For brevity, the denotations V_j $(j = 7, \ldots, 12)$ are used in (1), (4) instead of the corresponding expressions.

4   Invariant Manifolds of the Equations of Motion

Using geometric integrals (5), we can eliminate some variables from the equations of motion and the rest of the integrals.

The relations

$$V_1 - 1 = 0, \ V_2 - 1 = 0, \ V_3 - 1 = 0, \ V_4 = 0, \ V_5 = 0, \ V_6 = 0 \qquad (6)$$

define the IM of codimension 6 for differential equations (2).

Neither equations (1) nor integrals (4) contain the variables $a_{11}, a_{21}, a_{31}, a_{12}, a_{22}, a_{32}$. So, these equations can be considered as differential equations on IM (6).

The relations

$$V_{10} = 0, \ V_{11} = 0, \ V_{12} = 0 \qquad (7)$$

define the IM of codimension 3 for differential equations (3).

Eliminating b_{12}, b_{23}, and b_{33} from (3) with the help of (7), we obtain the differential equations on IM (7). These have two IMs of codimension 1:

$$b_{13} \mp b_{22}b_{31} \pm b_{21}b_{32} = 0.$$

Two IMs of codimension 4 correspond to them in the original phase space:

$$b_{11}b_{12} + b_{21}b_{22} + b_{31}b_{32} = 0, \ b_{11}b_{23} - b_{21}b_{22}b_{31} - (b_{11}^2 + b_{31}^2)b_{32} = 0,$$
$$(b_{11}^2 + b_{21}^2)b_{22} + b_{21}b_{31}b_{32} + b_{11}b_{33} = 0, \ b_{13} - b_{22}b_{31} + b_{21}b_{32} = 0 \qquad (8)$$

and

$$b_{11}b_{12} + b_{21}b_{22} + b_{31}b_{32} = 0, \ b_{11}b_{23} - b_{21}b_{22}b_{31} - (b_{11}^2 + b_{31}^2)b_{32} = 0,$$
$$(b_{11}^2 + b_{21}^2)b_{22} + b_{21}b_{31}b_{32} + b_{11}b_{33} = 0, \ b_{13} + b_{22}b_{31} - b_{21}b_{32} = 0. \qquad (9)$$

Equations (6), (8) and (6), (9) determine two IMs of codimension 10 of differential equations (2) and (3). The equations of motion (1)–(3) are further analyzed on IM (6), (8).

5   Stationary Solutions of Differential Equations on the IM

Differential equations (2), (3) and their first integrals on IM (6), (8) are given by

$$\dot{a}_{13} = a_{23}r_1 - a_{33}q_1, \ \dot{a}_{23} = a_{33}p_1 - a_{13}r_1, \ \dot{a}_{33} = a_{13}q_1 - a_{23}p_1,$$
$$\dot{b}_{11} = -b_{31}q_2 + b_{21}r_2, \ \dot{b}_{21} = b_{31}p_2 - b_{11}r_2, \ \dot{b}_{31} = -b_{21}p_2 + b_{11}q_2,$$
$$\dot{b}_{22} = b_{32}p_2 + \frac{1}{b_{11}}[(b_{21}b_{22} + b_{31}b_{32})\,r_2],$$
$$\dot{b}_{32} = -(b_{22}p_2 + \frac{1}{b_{11}}[(b_{21}b_{22} + b_{31}b_{32})\,q_2]). \qquad (10)$$

$$\tilde{V}_1 = a_{13}^2 + a_{23}^2 + a_{33}^2 = 1, \ \tilde{V}_2 = b_{11}^2 + b_{21}^2 + b_{31}^2 = 1,$$

$$\tilde{V}_3 = b_{22}^2 + b_{32}^2 + \frac{1}{b_{11}^2}(b_{21}b_{22} + b_{31}b_{32})^2 = 1. \tag{11}$$

These relations have been derived from equations (2), (3) and integrals (5) by elimination of $a_{11}, a_{21}, a_{31}, a_{12}, a_{22}, a_{32}, b_{12}, b_{13}, b_{23}, b_{33}$ from them with the help of (6), (8).

Having eliminated $b_{12}, b_{13}, b_{23}, b_{33}$ from differential equations (1) and integrals (4) with the help of (8), we obtain the equations written on IM (8).

Let us consider the problem of the existence of the solutions like

$$p_i = p_i^0, \ q_i = q_i^0, \ r_i = r_i^0, \ a_{k3} = a_{k3}^0, \ b_{k1} = b_{k1}^0, \ b_{22} = b_{22}^0, \ b_{32} = b_{32}^0 \tag{12}$$
$$(i = 1, 2; k = 1, 2, 3)$$

for the differential equations on IM (6), (8). Here $p_i^0, q_i^0, r_i^0, a_{k3}^0, b_{k1}^0, b_{22}^0, b_{32}^0$ are some constants.

To solve this problem, we substitute (12) into equations (10) and relations $\tilde{V}_2 = 1, \tilde{V}_3 = 1$ (11). These are satisfied, e.g., under the following values of the variables:

$$p_1 = p_1^0, \ q_1 = q_1^0, \ r_1 = r_1^0, \ p_2 = q_2 = r_2 = 0, \ a_{13} = \frac{a_{33}p_1^0}{r_1^0},$$

$$a_{23} = \frac{a_{33}q_1^0}{r_1^0}, \ b_{11} = 1, \ b_{21} = 0, \ b_{22} = 0, \ b_{31} = 0, \ b_{32} = 1.$$

Next the above values are substituted into differential equations (1) written on the IM under consideration. The equations take the form:

$$(B_2 - B_1 + C_1 - C_2) q_1^0 r_1^{0^2} + m_2 r_1^0 (s_2 r_1^0 - s_3 q_1^0)(s_1 p_1^0 + s_2 q_1^0 + s_3 r_1^0)$$
$$-c_2 m_2 r_1^0 [(s_1 p_1^0 + 2s_2 q_1^0) r_1^0 + s_3 (r_1^{0^2} - q_1^{0^2})] + a_{33} g [m_2 ((c_2 + s_2) r_1^0 - s_3 q_1^0)$$
$$-c_1 m_1 q_1^0] = 0,$$
$$(A_1 + A_2 - B_2 - C_1) p_1^0 r_1^{0^2} + m_2 r_1^0 (s_3 p_1^0 - s_1 r_1^0)(s_1 p_1^0 + s_2 q_1^0 + s_3 r_1^0)$$
$$+c_2 m_2 q_1^0 r_1^0 (s_1 r_1^0 - s_3 p_1^0) + a_{33} g [c_1 m_1 p_1^0 - m_2 (s_1 r_1^0 + s_3 p_1^0)] = 0,$$
$$(B_1 + C_2 - A_1 - A_2) p_1^0 q_1^0 r_1^0 + m_2 r_1^0 (s_1 q_1^0 - s_2 p_1^0)(s_1 p_1^0 + s_2 q_1^0 + s_3 r_1^0) \tag{13}$$
$$+c_2 m_2 r_1^0 [s_1 (p_1^{0^2} - q_1^{0^2}) + (2s_2 q_1^0 + s_3 r_1^0) p_1^0] - a_{33} m_2 g [(c_2 + s_2) p_1^0 - s_1 q_1^0] = 0,$$
$$(C_2 - B_2) q_1^0 r_1^0 + c_2 m_2 [r_1^0 (s_1 p_1^0 + s_2 q_1^0) - s_3 (p_1^{0^2} + q_1^{0^2})] - a_{33} c_2 m_2 g = 0,$$
$$(A_2 - C_2) p_1^0 q_1^0 r_1^0 + c_2 m_2 r_1^0 [s_1 q_1^0 (q_1^0 - s_2 p_1^0) + (s_1 r_1^0 - s_3 p_1^0) r_1^0]$$
$$+a_{33} c_2 m_2 g p_1^0 = 0, \ \ -(A_2 - B_2) p_1^0 r_1^0 = 0.$$

Now, using the *Mathematica* built-in function *GroebnerBasis*, we compute a lexicographical basis with respect to, e.g., s_1, s_2, c_1, p_1^0 for the polynomials of system (13). It enables us to obtain both the values for p_1^0, q_1^0, r_1^0 and the conditions of existence of the desired solution.

As a result, we have the following system:

$$p_1^0 = 0,$$
$$((B_2 - C_2)^2 - (B_1 + B_2 - C_1)\, c_2^2 m_2)\, q_1^{0^2} r_1^{0^2} + c_2 m_2 q_1^0 r_1^0\, [c_2 C_2 q_1^0 r_1^0$$
$$+ s_3 (B_2 - C_2 - c_2^2 m_2)(q_1^{0^2} + r_1^{0^2})] + a_{33} c_2 m_2 g\, [2 c_2 m_2 g a_{33}$$
$$- q_1^0\, (c_1 c_2 m_1 q_1^0 + (3\,(C_2 - B_2) + c_2^2 m_2)\, r_1^0) + c_2 m_2 s_3 (q_1^{0^2} + r_1^{0^2})] = 0,$$
$$((C_2 - B_2)\, r_1^0 + c_2 m_2\, (s_2 r_1^0 - s_3 q_1^0))\, q_1^0 - c_2 m_2 g\, a_{33} = 0,$$
$$s_1 = 0,$$

whence it follows:

$$p_1^0 = 0,\ s_1 = 0,\ s_2 = \frac{c_2 m_2 g\, a_{33} + q_1^0 [(B_2 - C_2)\, r_1^0 + c_2 m_2 s_3 q_1^0]}{c_2 m_2\, q_1^0 r_1^0},$$

$$c_1 = \frac{1}{c_2^2 m_1 m_2 g\, a_{33} q_1^{0^2}} \Big[c_2 m_2 g a_{33}\, [2 c_2 m_2 g\, a_{33} + [(3(B_2 - C_2) - c_2^2 m_2)\, q_1^0 r_1^0$$
$$+ c_2 m_2 s_3\, (q_1^{0^2} + r_1^{0^2})]] + q_1^0 r_1^0\, [((B_2 - C_2)^2 + c_2^2 m_2 (C_1 + C_2 - B_1$$
$$- B_2))\, q_1^0 r_1^0 - c_2 m_2 s_3 (C_2 - B_2 + c_2^2 m_2)(q_1^{0^2} + r_1^{0^2})] \Big]. \tag{14}$$

Using the 3rd relation of (14) and setting $q_1^0 = r_1^0$, we find r_1^0:

$$r_1^0 = \pm \sqrt{\frac{c_2 m_2 g\, a_{33}}{z}}, \tag{15}$$

where $z = C_2 - B_2 + c_2 m_2 (s_2 - s_3)$.

From $\tilde{V}_1 = 1$ (11), taking into account the values for $a_{13}, a_{23}, p_1^0, q_1^0, r_1^0$, we have $a_{33} = \pm 1/\sqrt{2}$.

Thus, the following solutions for the differential equations on IM (6), (8) are obtained:

$$p_1 = 0,\ q_1 = \pm 2^{-1/4} \sqrt{\frac{c_2 m_2 g}{z_1}},\ r_1 = \pm 2^{-1/4} \sqrt{\frac{c_2 m_2 g}{z_1}},\ p_2 = q_2 = r_2 = 0,$$

$$a_{13} = 0,\ a_{23} = -\frac{1}{\sqrt{2}},\ a_{33} = -\frac{1}{\sqrt{2}},\ b_{11} = 1,\ b_{21} = 0,\ b_{22} = 0,\ b_{31} = 0,$$

$$b_{32} = 1; \tag{16}$$

$$p_1 = 0,\ q_1 = \pm 2^{-1/4} \sqrt{\frac{c_2 m_2 g}{z_2}},\ r_1 = \pm 2^{-1/4} \sqrt{\frac{c_2 m_2 g}{z_2}},\ p_2 = q_2 = r_2 = 0,$$

$$a_{13} = 0,\ a_{23} = \frac{1}{\sqrt{2}},\ a_{33} = \frac{1}{\sqrt{2}},\ b_{11} = 1,\ b_{21} = 0,\ b_{22} = 0,\ b_{31} = 0,$$

$$b_{32} = 1. \tag{17}$$

Here $z_1 = B_2 - C_2 - c_2 m_2 (s_2 - s_3)$, $z_2 = -z_1$. The conditions for the solutions to be real are $z_1 > 0\,(z_2 > 0)$.

On substituting the above solutions into the differential equations under consideration, these are satisfied under the following constraints on the parameters:

$$c_1 = \frac{1}{m_1(B_2 - C_2 + c_2 m_2(s_3 - s_2))} \Big[m_2((B_2 - C_2)(s_2 - s_3) + c_2(B_1 - C_1$$
$$+ 2m_2 s_2(s_3 - s_2) + c_2 m_2(s_2 + s_3))) \Big], \; s_1 = 0. \tag{18}$$

Thus, the differential equations on IM (6), (8) have the solutions of type (12) when their parameters satisfy conditions (18).

From a mechanical viewpoint, solutions (16), (17) correspond to the following motion of the mechanical system under study: the body S_1 rotates about an immobile axis positioned in the body (in the plane $O_1 y_1 z_1$) with the angular velocity $\omega^2 = \pm(\sqrt{2} c_2 m_2\, g)/(B_2 - C_2 + c_2 m_2(s_3 - s_2))$, and the body S_2 is at rest relative to S_1. As can be seen from (16) and (17), the body S_1 has opposite positions with respect to the coordinate system $O_1 XYZ$ in these cases. The motions of similar type for the symmetric bodies were studied in [3].

6   On Families of the Integrals Assuming Stationary Values

Now we consider the problem of obtaining the families of integrals assuming stationary values on solutions (16), (17). To this end, the linear combination from the first integrals of the differential equations on IM (6), (8) is constructed:

$$2K = \lambda_0 \tilde{H} - \lambda_1 \tilde{V} - \lambda_2 \tilde{V}_1 - \lambda_3 \tilde{V}_2 - \lambda_4 \tilde{V}_3 \; (\lambda_i = \text{const}). \tag{19}$$

Here $\tilde{H}$ and $\tilde{V}$ are the integrals H and V (4) from which b_{12}, b_{13}, b_{23}, and b_{33} have been eliminated with the help of (8); λ_i $(i = 0, 1, \ldots, 4)$ are the parameters of the family of integrals K.

The necessary extremum conditions for K with respect to the phase variables are written as

$$\frac{\partial K}{\partial p_i} = 0, \; \frac{\partial K}{\partial q_i} = 0, \; \frac{\partial K}{\partial r_i} = 0, \; \frac{\partial K}{\partial a_{k3}} = 0, \; \frac{\partial K}{\partial b_{k1}} = 0,$$
$$\frac{\partial K}{\partial b_{22}} = 0, \; \frac{\partial K}{\partial b_{32}} = 0 (i = 1, 2; k = 1, 2, 3).$$

Next, taking into account conditions (18), we substitute solutions (16) into the above equations and find the constraints on λ_i under which the solutions satisfy them:

$$\lambda_2 = -2^{-1/2} \alpha^2 [(B_1 c_2 + B_2(c_2 + s_2) - s_2(C_2 + c_2 m_2(c_2 + s_2))$$
$$+ c_2 m_2 s_3(2c_2 + s_3))] \lambda_0, \; \lambda_1 = \mp 2^{1/4} \sqrt{c_2}\, \alpha \lambda_0,$$
$$\lambda_3 = -2^{-3/2} c_2 \alpha^2 (B_2 + 2c_2 m_2(s_3 - s_2)) \lambda_0,$$
$$\lambda_4 = -2^{-1/2} c_2 \alpha^2 (B_2 + c_2 m_2(s_3 - s_2)) \lambda_0,$$

where $\alpha = \sqrt{m_2 g / z_1}$.

Having substituted the latter expressions into (19), we have:

$$2K_{1,2} = \tilde{H} \pm 2^{1/4}\alpha\sqrt{c_2}\,\tilde{V} + 2^{-1/2}\alpha^2(B_1 c_2 + B_2(c_2 + s_2))$$
$$-s_2(C_2 + c_2 m_2(c_2 + s_2)) + c_2 m_2 s_3(2c_2 m_2 + s_3))\tilde{V}_1$$
$$+2^{-3/2}\alpha^2 c_2(B_2(\tilde{V}_2 + 2\tilde{V}_3) - 2c_2 m_2(s_2 - s_3)(\tilde{V}_2 + \tilde{V}_3)). \tag{20}$$

The integrals K_1 and K_2 assume stationary values, respectively, on the 1st and 2nd solutions of (16). It is easy to verify by direct computations.

One can derive similarly the integrals K_3 and K_4 taking stationary values on the 1st and 2nd solutions of (17):

$$2K_{3,4} = \tilde{H} \pm 2^{1/4}\beta\sqrt{c_2}\,\tilde{V} + 2^{-1/2}\beta^2(B_1 c_2 + B_2(c_2 + s_2))$$
$$-s_2(C_2 + c_2 m_2(c_2 + s_2)) + c_2 m_2 s_3(2c_2 m_2 + s_3))\tilde{V}_1$$
$$+2^{-3/2}\beta^2 c_2(B_2(\tilde{V}_2 + 2\tilde{V}_3) - 2c_2 m_2(s_2 - s_3)(\tilde{V}_2 + \tilde{V}_3)),$$

where $\beta = \sqrt{m_2 g / z_2}$.

7   On the Stability of Stationary Solutions

Let us investigate the stability of the solutions (16) by the Routh–Lyapunov method [9]. To this end, the integrals K_1 and K_2 are used. We solve this problem under the condition $s_2 = c_2$ to obtain the observable form of stability conditions.

Under the above restriction, the integrals K_1 and K_2 take the form

$$4\tilde{K}_{1,2} = 2\tilde{H} \pm 2\sqrt{2}\,\bar{z}\tilde{V} + 2\,\bar{z}^2[B_1 + 2B_2 - C_2 + m_2(s_3^2 - 2c_2(c_2 + s_3))]\tilde{V}_1$$
$$+\bar{z}^2[B_2(\tilde{V}_2 + 2\tilde{V}_3) - 2c_2 m_2(c_2 - s_3)(\tilde{V}_2 + \tilde{V}_3)], \tag{21}$$

$\bar{z} = \sqrt{c_2 m_2 g}/(2^{1/4}\sqrt{B_2 - C_2 + c_2 m_2(s_3 - c_2)})$.

For the equations of perturbed motion, the second variation of the integral $\tilde{K}_1$ $(\tilde{K}_2)$ obtained in the neighborhood of the solution under study on the linear manifold

$$\delta\tilde{V}_1 = -\sqrt{2}(y_2 + y_3) = 0, \; \delta\tilde{V}_2 = 2y_4 = 0, \; \delta\tilde{V}_3 = 2y_8 = 0$$

can be written as $\delta^2 \tilde{K}_{1,2} = Q_1 + Q_2$, where

$$4Q_1 = A_2 y_{10}^2 + 2(A_2 - c_2^2 m_2)\,y_{10}y_9 + (A_1 + A_2 - m_2(c_2^2 - s_3^2))\,y_9^2$$
$$\pm 2\sqrt{2}(A_2 - c_2^2 m_2)\,\bar{z}y_1 y_{10} \pm 2\sqrt{2}[A_1 + A_2 - m_2(c_2^2 - s_3^2)]\,\bar{z}y_1 y_9$$
$$+2[B_1 + 2B_2 - C_2 + m_2(2c_2(s_3 - c_2) + s_3^2)]\,\bar{z}^2 y_1^2 - 2\sqrt{2}(A_2 - B_2)$$
$$\times \bar{z}^2 y_1 y_5 - (A_2 - B_2)\bar{z}^2 y_5^2 - [A_2 - B_2 + 2c_2 m_2(c_2 - s_3)]\,\bar{z}^2 y_7^2$$
$$+[\sqrt{2}c_2 g m_2 - 2\bar{z}^2\,(A_2 - C_2 - c_2 m_2(c_2 - s_3))]\,y_5 y_7 + 2[\sqrt{2}(C_2 - A_2)$$
$$+c_2 m_2(g + \sqrt{2}(c_2 + s_3)\bar{z}^2)]\,\bar{z}^2 y_1 y_7, \tag{22}$$

$$4Q_2 = (B_1 + C_2 + m_2 s_3^2)\, y_{11}^2 + 2c_2 m_2 s_3\, y_{11} y_{12} + B_2\, y_{12}^2$$
$$+ 2(B_2 - c_2^2 m_2)\, y_{12} y_{13} + (B_2 + C_1 - c_2^2 m_2) y_{13}^2 + 2C_2 y_{11} y_{14} + C_2 y_{14}^2$$
$$\pm 2\sqrt{2}(B_1 + C_2 + m_2 s_3^2)\, \bar{z} y_2 y_{11} \mp 2\sqrt{2}[B_2 - c_2 m_2(c_2 + s_3)]\, \bar{z} y_2 y_{12}$$
$$\mp 2\sqrt{2}(B_2 + C_1 - c_2^2 m_2)\, \bar{z} y_2 y_{13} \pm 2\sqrt{2} C_2\, \bar{z} y_2 y_{14} + 4(B_1 + 2B_2$$
$$- C_2 + m_2(2c_2(s_3 - c_2) + s_3^2))\, \bar{z}^2 y_2^2 + (B_2 - C_2 + 2c_2 m_2(s_3 - c_2))\, \bar{z}^2 y_6^2$$
$$+ 2c_2 m_2(2\sqrt{2} s_3\, \bar{z}^2 + g) y_2 y_6. \tag{23}$$

Here $y_1 = a_{13}$, $y_2 = a_{23} + \frac{1}{\sqrt{2}}$, $y_3 = a_{33} + \frac{1}{\sqrt{2}}$, $y_4 = b_{11} - 1$, $y_5 = b_{21}$, $y_6 = b_{22}$, $y_7 = b_{31}$, $y_8 = b_{32} - 1$, $y_9 = p_1$, $y_{10} = p_2$, $y_{11} = q_1 \mp \bar{z}$, $y_{12} = q_2$, $y_{13} = r_1 \mp \bar{z}$, $y_{14} = r_2$ are the deviations from the unperturbed motion.

The conditions for the quadratic forms Q_1 and Q_2 to be positive definite are sufficient for the stability of the solutions under consideration. According to the Sylvester criterion, the conditions of positive definiteness for the quadratic form Q_1 are given by

$$\Delta_1 = A_2 > 0,$$
$$\Delta_2 = A_2(B_2 - A_2)\, \bar{z}^2 > 0,$$
$$\Delta_3 = (B_2 - A_2)[A_1 A_2 + m_2(A_2(c_2^2 + s_3^2) - c_2^4 m_2)]\, \bar{z}^2 > 0,$$
$$\Delta_4 = c_2 m_2(c_2 - s_3)(A_2 - B_2)[A_1 A_2 + m_2(A_2(c_2^2 + s_3^2) - c_2^4 m_2)]\, \bar{z}^4 > 0,$$
$$\Delta_5 = c_2 m_2(A_2 - B_2)[(A_1 - B_1 - B_2 + C_2 + 3c_2^2 m_2)s_3 - c_2(A_1 - B_1$$
$$- B_2 + C_2 + c_2^2 m_2)]\,[A_1 A_2 + m_2(A_2(c_2^2 + s_3^2) - c_2^4 m_2)]\bar{z}^6 > 0.$$

These are reduced to the equivalent system of inequalities by elementary transformations:

$$A_1 A_2 + m_2(A_2(c_2^2 + s_3^2) - c_2^4 m_2) > 0,$$

$$\frac{c_2 - s_3}{(A_2 - B_2)\, c_2} > 0, \qquad \frac{A_1 - B_1 - B_2 + C_2 + \frac{c_2^2 m_2(c_2 - 3s_3)}{c_2 - s_3}}{A_2 - B_2} > 0. \tag{24}$$

We obtain in a similar way the conditions for the quadratic form Q_2 to be positive definite:

$$B_1 B_2 + m_2(B_2 - c_2^2 m_2)s_3^2 > 0,$$
$$B_1(B_2(C_1 + c_2^2 m_2) - c_2^4 m_2^2) + C_1 m_2(B_2 - c_2^2 m_2)s_3^2 > 0,$$
$$B_2 - C_2 - 2c_2^2 m_2 + 2c_2 m_2 s_3 > 0, \tag{25}$$
$$(B_2 - C_2 - 2c_2^2 m_2 + 2c_2 m_2 s_3)(B_1 + 2B_2 - C_1 - 2C_2 - 3c_2^2 m_2 + m_2 s_3^2)$$
$$- c_2^2 m_2^2(c_2 + s_3)^2 > 0.$$

Taking into account the condition for solutions (16) to be real, we find that inequalities (24) and (25) are consistent when the following constraints hold:

$$c_2 > 0 \text{ and } s_3 \geq 5c_2 \text{ and } \Big[B_1 + 2B_2 + m_2 s_3^2 > C_1 + 2C_2 + c_2 m_2 (4c_2 + s_3)$$

$$\text{and } B_1 + B_2 + \frac{2c_2^3 m_2}{c_2 - s_3} > A_1 + C_2 + 3c_2^2 m_2 \text{ and } C_2 < B_2 + c_2 m_2 (s_3 - 3c_2)$$

$$\text{and } \frac{B_2(c_2 - s_3)}{c_2^2(c_2 - 3s_3)} \leq m_2 \leq \frac{B_2}{c_2^2}$$

$$\text{and } \Big[\Big(B_1 + B_2 > C_2 + m_2 \Big(\frac{c_2^4 m_2}{A_2} + s_3 \Big(\frac{2c_2^2}{s_3 - c_2} - s_3 \Big) \Big)$$

$$\text{and } \Big(A_2 < \frac{c_2^4 m_2}{c_2^2 + s_3^2} \text{ and } A_1 > \frac{c_2^4 m_2^2}{A_2} - m_2(c_2^2 + s_3^2) \Big) \Big)$$

$$\text{or } \Big(\frac{c_2^4 m_2}{c_2^2 + s_3^2} < A_2 < B_2 \text{ and } B_1 + B_2 > C_2 + \frac{c_2^2 m_2 (c_2 - 3s_3)}{c_2 - s_3} \Big) \Big] \Big]. \tag{26}$$

The above constraints have been derived with *Mathematica* built-in function *Reduce*. This function has produced a great deal of variants for the consistent conditions. Only part of them is represented here for space reasons. From the presented conditions, it follows that the motions under study are stable when the attachment point O_2 lies above the center of mass of the body S_2, and the parameters of the bodies satisfy constraints (26).

Similar sufficient conditions have been obtained for solutions (17): $c_2 < 0$ and $s_3 \leq 5c_2$ and [...] (here the expressions in the square brackets are the same as (26)). Whence we conclude: the motions under study are stable when the attachment point of the bodies lies below the center of mass of the body S_2.

Sufficient stability conditions are usually compared with the corresponding necessary ones. The sufficient conditions are considered good enough when they correspond closely to the necessary ones. In the case of the mechanical system under study, it is another problem for our future work. Here we restrict ourselves to a graphical illustration of our result. To do this, we put

$$B_1 = 2A_1, \; C_1 = \frac{3}{2}A_1, \; B_2 = \frac{3}{2}A_2, \; C_2 = 2A_2, \; m_2 = \frac{3A_2}{2c_2^2}.$$

Under the above values, solutions (16) and inequalities (26) correspondingly take the form:

$$p_1 = 0, \; q_1 = \pm \frac{\sqrt{3g}}{2^{1/4}\sqrt{3s_3 - 4c_2}}, \; r_1 = \pm \frac{\sqrt{3g}}{2^{1/4}\sqrt{3s_3 - 4c_2}}, \; p_2 = q_2 = r_2 = 0,$$

$$a_{13} = 0, \; a_{23} = -\frac{1}{\sqrt{2}}, \; a_{33} = -\frac{1}{\sqrt{2}}, \; b_{11} = 1, \; b_{21} = 0, \; b_{22} = 0, \; b_{31} = 0, \; b_{32} = 1.$$

$$s_3 > 0 \text{ and } 0 < c_2 \leq \frac{s_3}{5} \text{ and } n > \frac{2c_2 - 5s_3}{c_2 - s_3}, \text{ where } n = A_1/A_2.$$

The stability region defined by the latter inequalities and plotted with *Mathematica* built-in function *RegionPlot3D* is shown in Fig. 2 (the dark region). So, when s_3, c_2, and n assume the values from this region, and the parameters B_i, C_i, and m_2 have the above values, the solutions under investigation are stable.

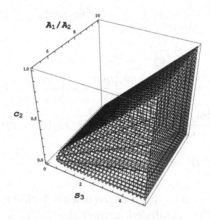

Fig. 2. Stability area.

8   Conclusion

Using computer algebra methods and the software package developed on the basis of the CAS *Mathematica*, we have constructed the symbolic characteristic function and the equations of motion for the system of two connected rigid bodies in a constant gravity field. New particular solutions for the equations have been found. These correspond to the following motions of the mechanical system: the first body rotates permanently and the second is at rest relative to the first. The linear combinations of the first integrals of the problem, which take stationary values on the solutions, have been derived. These combinations were used to investigate the stability of the solutions. It was shown that the solutions are stable when the parameters of the bodies of the system obey corresponding restrictions.

References

1. Ishlinsky, A.Y.: Mechanics of Gyroscopic Systems. USSR Academy Science, Moscow (1963)
2. Wittenburg, I.S.: Dynamics of Systems of Rigid Bodies. Teubner, Stuttgart (1977)
3. Savchenko, A.Ya., Bolgrabskaya, I.A., Kononyhin, G.A.: Stability of systems of connected rigid bodies. Naukova Dumka, Kyiv (1991)
4. Gutnik, S.A., Sarychev, V.A.: Application of computer algebra methods to investigate the dynamics of the system of two connected bodies moving along a circular orbit. Program. Comput. Softw. **45**(2), 51–57 (2019). https://doi.org/10.1134/S0361768819020051

5. Leutin, A.P.: Modeling the motion of connected bodies. Mech. Solids. **47**(1), 19–33 (2012)
6. Pogorelov, D.Y.: Contemporary algorithms for computer synthesis of equations of motion of multibody systems. Comp. Syst. Sci. **44**(4), 503–512 (2005)
7. Banshchikov, A.V., Burlakova, L.A., Irtegov, V.D., Titorenko, T.N.: Software Package LinModel for the Analysis of the Dynamics of Large Dimensional Mechanical Systems. Certificate of State Registration of Software Programs. FGU-FIPS. 2008610622 (2008)
8. Borisov, A.V., Mamaev, I.S.: Mechanics of a Rigid Body. Chaos. Institute of Computer Science, Moscow-Izhevsk, Hamiltonian Methods, Integrability (2005)
9. Collected Works On permanent helical motions of a rigid body in fluid. **1**, 276–319 (1954). USSR Acad. Sci., Moscow-Leningrad

Complexity Estimates for Fourier-Motzkin Elimination

Rui-Juan Jing[1], Marc Moreno-Maza[2], and Delaram Talaashrafi[2(✉)]

[1] Jiangsu University, Zhenjiang, China
rjing@ujs.edu.cn
[2] Western University, London, ON, Canada
moreno@csd.uwo.ca, dtalaash@uwo.ca

Abstract. In this paper, we propose an efficient method for removing all redundant inequalities generated by Fourier-Motzkin Elimination. This method is based on an improved version of Balas' work and can also be used to remove all redundant inequalities in the input system. Moreover, our method only uses arithmetic operations on matrices and avoids resorting to linear programming techniques. Algebraic complexity estimates and experimental results show that our method outperforms alternative approaches, in particular those based on linear programming and the simplex algorithm.

Keywords: Polyhedral set · Fourier-Motzkin Elimination · Algebraic complexity · Efficient implementation

1 Introduction

Polyhedral sets play an important role in computational sciences. For instance, they are used to model, analyze, transform and schedule for-loops of computer programs; we refer to [3,4,6,15,16,21,38]. Of prime importance are the following operations on polyhedral sets: (i) conversion between H-representation and V-representation (performed, for instance, by the double description method); and (ii) projection, as performed by Fourier-Motzkin Elimination.

Although the double description (DD) method and Fourier-Motzkin Elimination (FME) have a lot in common, and, they are considered as the same algorithm in the paper [8] of Winfried Bruns and Bogdan Ichim, they are not totally similar. Quoting Komei Fukuda and Alain Prodon from [18]: "The FME algorithm is more general than the DD method, but often considered as the same method partly because it can be used to solve the extreme ray enumeration problem".

Fourier-Motzkin Elimination is an algorithmic tool for projecting a polyhedral set onto a linear subspace. It was proposed independently by Joseph Fourier and Theodore Motzkin, respectively in 1827 and 1936. See the paper [14] of George Danzing and Section 12.2 of the book [35] of Alexander Schrijver, for a presentation of Fourier-Motzkin Elimination. The original version of this

© Springer Nature Switzerland AG 2020
F. Boulier et al. (Eds.): CASC 2020, LNCS 12291, pp. 282–306, 2020.
https://doi.org/10.1007/978-3-030-60026-6_16

algorithm produces large amounts of redundant inequalities and has a double exponential algebraic complexity. Removing all these redundancies is equivalent to giving the so-called *minimal representation* of the projection of a polyhedron. Leonid Khachiyan explained in [28] how linear programming (LP) could be used to remove all redundant inequalities, thereby reducing the cost of Fourier-Motzkin Elimination to a number of machine word operations singly exponential in the dimension of the ambient space. However, Khachiyan did not state a more precise running time estimate taking into account the characteristics of the polyhedron being projected, such as the number of its facets.

As we shall prove in this paper, rather than using linear programming one may use only matrix arithmetic, increasing the theoretical and practical efficiency of Fourier-Motzkin Elimination while still producing an irredundant representation of the projected polyhedron.

Other algorithms for projecting polyhedral sets remove some (but not all) redundant inequalities with the help of extreme rays: see the work of David A. Kohler [29]. As observed by Jean-Louis Imbert in [24], the method he proposed in that paper and that of Sergei N. Chernikov in [11] are equivalent. On the topic of finding extreme rays of a polyhedral set in H-representation, see Natalja V. Chernikova [12], Hervé Le Verge [30] and Komei Fukuda [18]. These methods are very effective in practice, but none of them can remove all redundant inequalities generated by Fourier-Motzkin Elimination.

Fourier-Motzkin Elimination is well suited for projecting a polyhedron, described by its facets (given by linear inequalities), onto different sub-spaces. And our paper is about projecting polyhedral sets to lower dimensions, eliminating one variable after another, thanks to the Fourier-Motzkin Elimination algorithm as described in Schrijver's book [35]. In fact, our goal is to find the minimal representations of all of the successive projections of a given polyhedron (in H-representation, thus given by linear inequalities), by eliminating variables one after another, using the Fourier-Motzkin Elimination algorithm. Computing these successive projections has applications in the analysis, scheduling and transformation of for-loop nests of computer programs. For instance, after applying a uni-modular transformation to the loop counters of a for-loop nest, the loop bounds of the new for-loop nest are derived from the successive projections of a well-chosen polyhedron.

In this paper, we show how to remove all the redundant inequalities generated by Fourier-Motzkin Elimination, considering a non-empty, full-dimensional, and pointed polyhedron as the input. Our approach is based on an improved version of a method proposed by Egon Balas in [2]. To be more specific, we first compute a so-called *initial redundancy test cone* from which we can derive the so-called *redundancy test cone*, which is used to detect the redundant inequalities generated after each elimination of a variable.

Consider a non-empty, full-dimensional, and pointed polyhedron $Q \subseteq \mathbb{Q}^n$ as input, given by a system of m linear inequalities of height h. We show, see Theorem 5, that eliminating the variables from that system, one after another (thus performing Fourier-Motzkin Elimination) can be done within

$O(m^{\frac{5n}{2}}n^{\theta+1+\epsilon}h^{1+\epsilon})$ bit operations, for any $\epsilon > 0$, where θ is the exponent of linear algebra, as defined in the landmark book [19].

Therefore, we obtain a more favourable estimate than the one presented in [25,26] for Fourier-Motzkin Elimination with a removal of the redundant inequalities via linear programming. Indeed, in those papers, the estimate is $O(n^2 m^{2n} \mathsf{LP}(n, 2^n hn^2 m^n))$ bit operations, where $\mathsf{LP}(d, H)$ is an upper bound for the number of bit operations required for solving a linear program in n variables and with total bit size H. For instance, in the case of Karmarkar's algorithm [27], we have $\mathsf{LP}(d, H) \in O(d^{3.5} H^2 \cdot \log H \cdot \log \log H)$. Then, comparing the exponents of m, n and h, we have $\frac{5n}{2}$, $\theta + 1 + \epsilon$, $1 + \epsilon$ respectively with the method proposed in the present paper and $4n + \epsilon$, $6 + \epsilon$, $2 + \epsilon$ respectively with the estimate of [25,26].

Our algorithm is stated in Sect. 4 and follows the revisited version of Balas' algorithm presented in Sect. 3. Since the maximum number of facets of any standard projection of Q is $O(m^{\lfloor n/2 \rfloor})$, our running time for Fourier-Motzkin Elimination is satisfactory; the other factors in our estimate come from the cost of linear algebra operations for testing redundancy.

We have implemented the algorithms proposed in Sect. 4 using the BPAS library [10] publicly available at www.bpaslib.org. We have compared our code against other implementations of Fourier-Motzkin Elimination including the CDD library [17]. Our experimental results, reported in Sect. 6, show that our proposed method can solve more test-cases (actually all) that we used, while the counterpart software failed to solve some of them.

Section 2 provides background materials about polyhedral sets and polyhedral cones together with the original version of Fourier-Motzkin Elimination. As mentioned above, Sect. 3 contains our revisited version of Balas' method and detailed proofs of its correctness. Based on this, Sect. 4 presents a new algorithm producing a *minimal projected representation* for a given full-dimensional pointed polyhedron. Complexity results are established in Sect. 5. In Sect. 6 we report on our experimentation and in Sect. 7 we discuss related works.

To summarize, our contributions are: (i) making Balas' algorithm to be practical, by devising a method for finding the initial redundancy test cone efficiently and using it in the Fourier-Motzkin Elimination, (ii) exhibiting the theoretical efficiency of the proposed algorithm by analyzing its bit complexity, and, (iii) demonstrating its practical effectiveness (implemented as part of the BPAS library) compared to other available related software.

2   Background

In this section, we review the basics of polyhedral geometry. Section 2.1 is dedicated to the notions of polyhedral sets and polyhedral cones. Sections 2.2 and 2.3 review the double description method and Fourier-Motzkin elimination. We conclude this section with the cost model that we shall use for complexity analysis, see Sect. 2.4. As we omit most proofs, for more details please refer to [18,35,37]. For the sake of simplicity in the complexity analysis of the presented algorithms, we constraint our coefficient field to the field of rational numbers $\mathbb{Q}$. However, all of the algorithms presented in this paper apply to polyhedral sets with coefficients in the field $\mathbb{R}$ of real numbers.

Throughout this paper, we use bold letters, e.g. $\mathbf{v}$, to denote vectors and we use capital letters, e.g. A, to denote matrices. Also, we assume that vectors are column vectors. For row vectors, we use the transposition notation, as in A^t for the transposition of matrix A. The concatenation of two column vectors $\mathbf{v}$ and $\mathbf{w}$ is denoted $(\mathbf{v}, \mathbf{w})$, thus using parentheses, while the concatenation of two row vector $\mathbf{v}^t$ and $\mathbf{w}^t$ is denothed $[\mathbf{v}^t, \mathbf{w}^t]$, thus using square brackets. For a matrix A and an integer k, we denote by A_k is the row of index k in A. More generally, if K is a set of integers, we denote by A_K the sub-matrix of A with row indices in K.

2.1 Polyhedral Cones and Polyhedra

Polyhedral Cone. A set of points $C \subseteq \mathbb{Q}^n$ is called a *cone* if for each $\mathbf{x} \in C$ and each real number $\lambda \geq 0$ we have $\lambda \mathbf{x} \in C$. A cone $C \subseteq \mathbb{Q}^n$ is called *convex* if for all $\mathbf{x}, \mathbf{y} \in C$, we have $\mathbf{x} + \mathbf{y} \in C$. If $C \subseteq \mathbb{Q}^n$ is a convex cone, then its elements are called the *rays* of C. For two rays $\mathbf{r}$ and $\mathbf{r}'$ of C, we write $\mathbf{r}' \simeq \mathbf{r}$ whenever there exists $\lambda \geq 0$ such that we have $\mathbf{r}' = \lambda \mathbf{r}$. A cone $C \subseteq \mathbb{Q}^n$ is a *polyhedral cone* if it is the intersection of finitely many half-spaces, that is, $C = \{x \in \mathbb{Q}^n \mid Ax \leq \mathbf{0}\}$ for some matrix $A \in \mathbb{Q}^{m \times n}$. Let $\{\mathbf{x}_1, \ldots, \mathbf{x}_m\}$ be a set of vectors in $\mathbb{Q}^n$. The *cone generated* by $\{\mathbf{x}_1, \ldots, \mathbf{x}_m\}$, denoted by $\mathsf{Cone}(\mathbf{x}_1, \cdots, \mathbf{x}_m)$, is the smallest convex cone containing those vectors. In other words, we have $\mathsf{Cone}(\mathbf{x}_1, \ldots, \mathbf{x}_m) = \{\lambda_1 \mathbf{x}_1 + \cdots + \lambda_m \mathbf{x}_m \mid \lambda_1 \geq 0, \ldots, \lambda_m \geq 0\}$. A cone obtained in this way is called a *finitely generated cone*.

Polyhedron. A set of vectors $P \subseteq \mathbb{Q}^n$ is called a *convex polyhedron* if $P = \{\mathbf{x} \mid Ax \leq \mathbf{b}\}$ holds, for a matrix $A \in \mathbb{Q}^{m \times n}$ and a vector $\mathbf{b} \in \mathbb{Q}^m$, for some positive integer m. Moreover, the polyhedron P is called a *polytope* if P is bounded. From now on, we always use the notation $P = \{\mathbf{x} \mid Ax \leq \mathbf{b}\}$ to represent a polyhedron in $\mathbb{Q}^n$. The system of linear inequalities $\{Ax \leq \mathbf{b}\}$ is called a *representation* of P. We say an inequality $\mathbf{c}^t \mathbf{x} \leq c_0$ is *redundant* w.r.t. a polyhedron representation $Ax \leq \mathbf{b}$ if this inequality is implied by $Ax \leq \mathbf{b}$. A representation of a polyhedron is *minimal* if no inequality of that representation is implied by the other inequalities of that representation. To obtain a minimal representation for the polyhedron P, we need to remove all the redundant inequalities in its representation. This requires the famous Farkas' lemma. Since this lemma has different variants, we simply mention here the variant from [35] which we use in this paper.

Lemma 1 (Farkas' lemma). *Let $A \in \mathbb{Q}^{m \times n}$ be a matrix and $\mathbf{b} \in \mathbb{Q}^m$ be a vector. Then, there exists a vector $\mathbf{t} \in \mathbb{Q}^n$ with $\mathbf{t} \geq \mathbf{0}$ satisfying $A\mathbf{t} = \mathbf{b}$ if and if $\mathbf{y}^t \mathbf{b} \geq 0$ holds for every vector $\mathbf{y} \in \mathbb{Q}^m$ satisfying $\mathbf{y}^t A \geq 0$.*

A consequence of Farkas' lemma is the following criterion for testing whether an inequality $\mathbf{c}^t \mathbf{x} \leq c_0$ is redundant w.r.t. a polyhedron representation $Ax \leq \mathbf{b}$.

Lemma 2 (Redundancy test criterion). *Let* $\mathbf{c} \in \mathbb{Q}^n$, $c_0 \in \mathbb{Q}$, $A \in \mathbb{Q}^{m \times n}$ *and* $\mathbf{b} \in \mathbb{Q}^m$. *Assume* $A\mathbf{x} \leq \mathbf{b}$ *is a consistent linear inequality system. Then, the inequality* $\mathbf{c}^t\mathbf{x} \leq c_0$ *is redundant w.r.t.* $A\mathbf{x} \leq \mathbf{b}$ *if and only if there exists a vector* $\mathbf{t} \geq \mathbf{0}$ *and a number* $\lambda \geq 0$ *satisfying* $\mathbf{c}^t = \mathbf{t}^t A$ *and* $c_0 = \mathbf{t}^t\mathbf{b} + \lambda$.

Characteristic Cone and Pointed Polyhedron. The *characteristic cone* of P is the polyhedral cone denoted by $\mathsf{CharCone}(P)$ and defined by $\mathsf{CharCone}(P) = \{\mathbf{y} \in \mathbb{Q}^n \mid \mathbf{x} + \mathbf{y} \in P, \forall \mathbf{x} \in P\} = \{\mathbf{y} \mid A\mathbf{y} \leq \mathbf{0}\}$. The *linearity space* of the polyhedron P is the linear space denoted by $\mathsf{LinearSpace}(P)$ and defined as $\mathsf{CharCone}(P) \cap -\mathsf{CharCone}(P) = \{\mathbf{y} \mid A\mathbf{y} = \mathbf{0}\}$, where $-\mathsf{CharCone}(P)$ is the set of the $-\mathbf{y}$ for $\mathbf{y} \in \mathsf{CharCone}(P)$. The polyhedron P is *pointed* if its linearity space is $\{\mathbf{0}\}$.

Lemma 3 (Pointed polyhedron criterion). *The polyhedron P is pointed if and only if the matrix A is full column rank.*

Extreme Point and Extreme Ray. The *dimension* of the polyhedron P, denoted by $\dim(P)$, is the maximum number of linearly independent vectors in P. We say that P is *full-dimensional* whenever $\dim(P) = n$ holds. An inequality $\mathbf{a}^t\mathbf{x} \leq b$ (with $\mathbf{a} \in \mathbb{Q}^n$ and $b \in \mathbb{Q}$) is an *implicit equation* of the inequality system $A\mathbf{x} \leq \mathbf{b}$ if $\mathbf{a}^t\mathbf{x} = b$ holds for all $\mathbf{x} \in P$. Then, P is full-dimensional if and only if it does not have any implicit equation. A subset F of the polyhedron P is called a *face* of P if F equals $\{\mathbf{x} \in P \mid A_{\mathrm{sub}}\mathbf{x} = \mathbf{b}_{\mathrm{sub}}\}$ for a sub-matrix A_{sub} of A and a sub-vector $\mathbf{b}_{\mathrm{sub}}$ of $\mathbf{b}$. A face of P, distinct from P and of maximum dimension is called a *facet* of P. A non-empty face that does not contain any other face of a polyhedron is called a *minimal face* of that polyhedron. Specifically, if the polyhedron P is pointed, each minimal face of P is just a point and is called an *extreme point* or *vertex* of P. Let C be a cone such that $\dim(\mathsf{LinearSpace}(C)) = t$. Then, a face of C of dimension $t + 1$ is called a minimal proper face of C. In the special case of a pointed cone, that is, whenever $t = 0$ holds, the dimension of a minimal proper face is 1 and such a face is called an *extreme ray*. We call an *extreme ray* of the polyhedron P any extreme ray of its characteristic cone $\mathsf{CharCone}(P)$. We say that two extreme rays $\mathbf{r}$ and $\mathbf{r}'$ of the polyhedron P are *equivalent*, and denote it by $\mathbf{r} \simeq \mathbf{r}'$, if one is a positive multiple of the other. When we consider the set of all extreme rays of the polyhedron P (or the polyhedral cone C) we will only consider one ray from each equivalence class. A pointed cone C can be generated by its extreme rays, that is, we have $C = \{\mathbf{x} \in \mathbb{Q}^n \mid (\exists \mathbf{c} \geq \mathbf{0}) \ \mathbf{x} = R\mathbf{c}\}$, where the columns of R are the extreme rays of C. We denote by $\mathsf{ExtremeRays}(C)$ the set of extreme rays of the cone C. Recall that all cones considered here are polyhedral. The following, see [32,37], is helpful in the analysis of algorithms manipulating extreme rays of cones and polyhedra. Let $E(C)$ be the number of extreme rays of a polyhedral cone $C \in \mathbb{Q}^n$ with m facets. Then, we have:

$$E(C) \leq \binom{m - \lfloor \frac{n+1}{2} \rfloor}{m - 1} + \binom{m - \lfloor \frac{n+2}{2} \rfloor}{m - n} \leq m^{\lfloor \frac{n}{2} \rfloor}. \tag{1}$$

Algebraic Test of (Adjacent) Extreme Rays. Given $\mathbf{t} \in C$ and a cone $C = \{\mathbf{x} \in \mathbb{Q}^n \mid A\mathbf{x} \leq \mathbf{0}\}$, we define the *zero set* $\zeta_A(\mathbf{t})$ as the set of row indices i such that $A_i \mathbf{t} = 0$, where A_i is the i-th row of A. For simplicity, we use $\zeta(\mathbf{t})$ instead of $\zeta_A(\mathbf{t})$ when there is no ambiguity. The proof of the following, which we call the *algebraic test*, can be found in [18]: Let $\mathbf{r} \in C$. Then, the ray $\mathbf{r}$ is an extreme ray of C if and only if we have $\mathrm{rank}(A_{\zeta(\mathbf{r})}) = n - 1$. Two distinct extreme rays $\mathbf{r}$ and $\mathbf{r}'$ of the polyhedral cone C are called *adjacent* if they span a 2-dimensional face of C. From [18], we have: Two distinct extreme rays, $\mathbf{r}$ and $\mathbf{r}'$, of C are adjacent if and only if $\mathrm{rank}(A_{\zeta(\mathbf{r}) \cap \zeta(\mathbf{r}')}) = n - 2$ holds.

Polar Cone. Given a polyhedral cone $C \subseteq \mathbb{Q}^n$, the *polar cone* induced by C, denoted by C^*, is defined as: $C^* = \{\mathbf{y} \in \mathbb{Q}^n \mid \mathbf{y}^t \mathbf{x} \leq 0, \forall \mathbf{x} \in C\}$. The proof of the following property can be found in [35]: For a given cone $C \in \mathbb{Q}^n$, there is a one-to-one correspondence between the faces of C of dimension k and the faces of C^* of dimension $n - k$. In particular, there is a one-to-one correspondence between the facets of C and the extreme rays of C^*.

Homogenized Cone. The *homogenized cone* of the polyhedron $P = \{\mathbf{x} \in \mathbb{Q}^n \mid A\mathbf{x} \leq \mathbf{b}\}$ is denoted by $\mathsf{HomCone}(P)$ and defined by: $\mathsf{HomCone}(P) = \{(\mathbf{x}, x_{\mathrm{last}}) \in \mathbb{Q}^{n+1} \mid A\mathbf{x} - \mathbf{b}x_{\mathrm{last}} \leq \mathbf{0}, x_{\mathrm{last}} \geq 0\}$.

Lemma 4 (H-representation correspondence). *An inequality $A_i \mathbf{x} \leq b_i$ is redundant in P if and only if the corresponding inequality $A_i \mathbf{x} - b_i x_{\mathrm{last}} \leq 0$ is redundant in $\mathsf{HomCone}(P)$.*

Theorem 1 (Extreme rays of the homogenized cone). *Every extreme ray of the homogenized cone $\mathsf{HomCone}(P)$ associated with the polyhedron P is either of the form $(\mathbf{x}, 0)$ where $\mathbf{x}$ is an extreme ray of P, or $(\mathbf{x}, 1)$ where $\mathbf{x}$ is an extreme point of P.*

2.2 The Double Description Method

It follows from Sect. 2.1 that any pointed polyhedral cone C can be represented either as the intersection of finitely many half-spaces (given as a system of linear inequalities $A\mathbf{x} \leq \mathbf{0}$ and called *H-representation* of C) or as $\mathsf{Cone}(R)$, where R is a matrix, the columns of which are the extreme rays of C (called *V-representation* of C). The pseudo-code of the *double description method*, as presented in [18], and implemented in the CDD library [17] is shown in Algorithm 1. This algorithm calls partition and AdjacencyTest functions. Given a set of vectors J and an inequality A_i, the partition function places each member j of J into one of the sets J^+, J^0, J^-, according to the sign (positive, null or negative) of A_{ij}. Moreover, the AdjacencyTest determines adjacency of the input extreme rays. This algorithm produces the V-representation of a pointed polyhedral cone given by its H-representation. Some of the results presented in our paper depend on algebraic complexity estimates for the double description method. In [18], one can find an estimate in terms of arithmetic operations on the coefficients of the input H-representation. Since we need a bit complexity estimate, we provide one as Lemma 9.

Algorithm 1. DDmethod

1: **Input:** a matrix $A \in \mathbb{Q}^{m \times n}$ defining the H-representation of a pointed cone C
2: **Output:** a matrix R defining the V-representation of C
3: let K be the set of the indices of A's independent rows
4: $R' := (A_K)^{-1}$
5: let J be the set of the columns of R'
6: **while** $K \neq \{1, \ldots, m\}$ **do**
7:     select a A-row index $i \notin K$
8:     set R' to be an empty matrix
9:     J^+, J^0, $J^- :=$ partition(J, A_i)
10:     add the vectors in J^+ and J^0 as columns to R'
11:     **for** $p \in J^+$ **do**
12:         **for** $n \in J^-$ **do**
13:             **if** AdjacencyTest$(A_K, \mathbf{r}_p, \mathbf{r}_n) =$ true **then**
14:                 $\mathbf{r}_{new} := (A_i \mathbf{r}_p)\mathbf{r}_n - (A_i \mathbf{r}_n)\mathbf{r}_p$
15:                 add $\mathbf{r}_{new}$ as a column to R'
16:             **end if**
17:         **end for**
18:     **end for**
19:     let J be the set of the columns of R'
20:     $K := K \cup \{i\}$
21: **end while**
22: let R be the matrix created by the vectors in J as its columns
23: **return** (R)

2.3 Fourier-Motzkin Elimination

Let $A \in \mathbb{Q}^{m \times p}$ and $B \in \mathbb{Q}^{m \times q}$ be matrices. Let $\mathbf{c} \in \mathbb{Q}^m$ be a vector. Consider the polyhedron $P = \{(\mathbf{u}, \mathbf{x}) \in \mathbb{Q}^{p+q} \mid A\mathbf{u} + B\mathbf{x} \leq \mathbf{c}\}$. We denote by proj$(P; \mathbf{x})$ the *projection of P on $\mathbf{x}$*, that is, the subset of $\mathbb{Q}^q$ defined by proj$(P; \mathbf{x}) = \{\mathbf{x} \in \mathbb{Q}^q \mid \exists\, \mathbf{u} \in \mathbb{Q}^p, (\mathbf{u}, \mathbf{x}) \in P\}$.

Fourier-Motzkin elimination (FME for short) is an algorithm computing the projection proj$(P; \mathbf{x})$ of the polyhedron of P by successively eliminating the $\mathbf{u}$-variables from the inequality system $A\mathbf{u} + B\mathbf{x} \leq \mathbf{c}$. This process shows that proj$(P; \mathbf{x})$ is also a polyhedron.

Let ℓ_1, ℓ_2 be two inequalities: $a_1 x_1 + \cdots + a_n x_n \leq c_1$ and $b_1 x_1 + \cdots + b_n x_n \leq c_2$. Let $1 \leq i \leq n$ such that the coefficients a_i and b_i of x_i in ℓ_1 and ℓ_2 are positive and negative, respectively. The *combination* of ℓ_1 and ℓ_2 w.r.t. x_i, denoted by Combine(ℓ_1, ℓ_2, x_i), is:

$$-b_i(a_1 x_1 + \cdots + a_n x_n) + a_i(b_1 x_1 + \cdots + b_n x_n) \leq -b_i c_1 + a_i c_2.$$

Theorem 2 shows how to compute proj$(P; \mathbf{x})$ when $\mathbf{u}$ consists of a single variable x_i. When $\mathbf{u}$ consists of several variables, FME obtains the projection proj$(P; \mathbf{x})$ by repeated applications of Theorem 2.

Theorem 2 (Fourier-Motzkin theorem [29]). *Let $A \in \mathbb{Q}^{m \times n}$ be a matrix and let $\mathbf{c} \in \mathbb{Q}^m$ be a vector. Consider the polyhedron $P = \{\mathbf{x} \in \mathbb{Q}^n \mid A\mathbf{x} \leq \mathbf{c}\}$.*

Let S be the set of inequalities defined by $\mathbf{Ax} \leq \mathbf{c}$. Also, let $1 \leq i \leq n$. We partition S according to the sign of the coefficient of x_i:

$$S^+ = \{\ell \in S \mid \mathrm{coeff}(\ell, x_i) > 0\},$$
$$S^- = \{\ell \in S \mid \mathrm{coeff}(\ell, x_i) < 0\},$$
$$S^0 = \{\ell \in S \mid \mathrm{coeff}(\ell, x_i) = 0\}.$$

We construct the following system of linear inequalities:

$$S' = \{\mathsf{Combine}(s_p, s_n, x_i) \mid (s_p, s_n) \in S^+ \times S^-\} \cup S^0.$$

Then, S' is a representation of $\mathsf{proj}(P; \{\mathbf{x} \setminus \{x_i\}\})$.

With the notations of Theorem 2, assume that each of S^+ and S^- counts $\frac{m}{2}$ inequalities. Then, the set S' counts $(\frac{m}{2})^2$ inequalities. After eliminating p variables, the projection would be given by $O((\frac{m}{2})^{2^p})$ inequalities. Thus, FME is *double exponential* in p.

On the other hand, from [33] and [26], we know that the maximum number of facets of the projection on $\mathbb{Q}^{n-p}$ of a polyhedron in $\mathbb{Q}^n$ with m facets is $O(m^{\lfloor n/2 \rfloor})$. Hence, it can be concluded that most of the generated inequalities by FME are redundant. Eliminating these redundancies is the main subject of the subsequent sections.

2.4 Cost Model

For any rational number $\frac{a}{b}$, thus with $b \neq 0$, we define the *height* of $\frac{a}{b}$, denoted as $\mathsf{height}(\frac{a}{b})$, as $\log \max(|a|, |b|)$. For a given matrix $A \in \mathbb{Q}^{m \times n}$, let $\|A\|$ denote the infinity norm of A, that is, the maximum absolute value of a coefficient in A. We define the height of A, denoted by $\mathsf{height}(A) := \mathsf{height}(\|A\|)$, as the maximal height of a coefficient in A. For the rest of this section, our main reference is the PhD thesis of Arne Storjohann [36]. Let k be a non-negative integer. We denote by $\mathcal{M}(k)$ an upper bound for the number of bit operations required for performing any of the basic operations (addition, multiplication, and division with remainder) on input $a, b \in \mathbb{Z}$ with $|a|, |b| < 2^k$. Using the multiplication algorithm of Arnold Schönhage and Volker Strassen [34] one can choose $\mathcal{M}(k) \in O(k \log k \log \log k)$.

We also need complexity estimates for some matrix operations. For positive integers a, b, c, let us denote by $\mathcal{MM}(a, b, c)$ an upper bound for the number of arithmetic operations (on the coefficients) required for multiplying an $(a \times b)$-matrix by an $(b \times c)$-matrix. In the case of square matrices of order n, we simply write $\mathcal{MM}(n)$ instead of $\mathcal{MM}(n, n, n)$. We denote by θ the exponent of linear algebra, that is, the smallest real positive number such that $\mathcal{MM}(n) \in O(n^\theta)$.

We now give the complexity estimates in terms of $\mathcal{M}(k) \in O(k \log k \log \log k)$ and $\mathcal{B}(k) = \mathcal{M}(k) \log k \in O(k(\log k)^2 \log \log k)$. We replace every term of the form $(\log k)^p (\log \log k)^q (\log \log \log k)^r$, (where p, q, r are positive real numbers) with $O(k^\epsilon)$ where ϵ is a (positive) infinitesimal. Furthermore, in the complexity estimates of algorithms operating on matrices and vectors over $\mathbb{Z}$, we use a parameter β, which is a bound on the magnitude of the integers occurring during the algorithm. Our complexity estimates are measured in terms of *machine word operations*. Let $A \in \mathbb{Z}^{m \times n}$ and $B \in \mathbb{Z}^{n \times p}$. Then, the product of A by B can be computed within $O(\mathcal{MM}(m, n, p)(\log \beta) + (mn + np + mp)\mathcal{B}(\log \beta))$ word operations, where $\beta = n \|A\| \|B\|$ and $\|A\|$ (resp. $\|B\|$) denotes the maximum absolute value of a coefficient in A (resp. B). Neglecting logarithmic factors, this estimate becomes $O(\max(m, n, p)^{\theta} \max(h_A, h_b))$ where $h_A = \mathsf{height}(A)$ and $h_B = \mathsf{height}(B)$. For a matrix $A \in \mathbb{Z}^{m \times n}$, a cost estimate of Gauss-Jordan transform is $O(nmr^{\theta-2}(\log \beta) + nm(\log r)\mathcal{B}(\log \beta))$ word operations, where r is the rank of the input matrix A and $\beta = (\sqrt{r}\|A\|)^r$. Let h be the height of A, for a matrix $A \in \mathbb{Z}^{m \times n}$, with height h, the rank of A is computed within $O(mn^{\theta+\epsilon}h^{1+\epsilon})$ word operations, and the inverse of A (when this matrix is invertible over $\mathbb{Q}$ and $m = n$) is computed within $O(m^{\theta+1+\epsilon}h^{1+\epsilon})$ word operations. Let $A \in \mathbb{Z}^{n \times n}$ be an integer matrix, which is invertible over $\mathbb{Q}$. Then, the absolute value of any coefficient in A^{-1} (inverse of A) can be bounded up to $(\sqrt{n-1}\|A\|^{(n-1)})$.

3   Revisiting Balas' Method

As recalled in Sect. 2, FME produces a representation of the projection of a polyhedron by eliminating one variable at a time. However, this procedure generates lots of redundant inequalities limiting its use in practice to polyhedral sets with a handful of variables only. In this section, we propose an efficient algorithm which generates the minimal representation of a full-dimensional pointed polyhedron, as well as its projections. Throughout this section, we use Q to denote a full-dimensional pointed polyhedron in $\mathbb{Q}^n$, where

$$Q = \{(\mathbf{u}, \mathbf{x}) \in \mathbb{Q}^p \times \mathbb{Q}^q \mid A\mathbf{u} + B\mathbf{x} \leq \mathbf{c}\}, \tag{2}$$

with $A \in \mathbb{Q}^{m \times p}$, $B \in \mathbb{Q}^{m \times q}$ and $\mathbf{c} \in \mathbb{Q}^m$. Thus, Q has no implicit equations in its representation and the coefficient matrix $[A, B]$ has full column rank. Our goal in this section is to compute the minimal representation of the projection $\mathsf{proj}(Q; \mathbf{x})$ given by $\mathsf{proj}(Q; \mathbf{x}) := \{\mathbf{x} \mid \exists \mathbf{u}, s.t.(\mathbf{u}, \mathbf{x}) \in Q\}$. We call the cone $C := \{\mathbf{y} \in \mathbb{Q}^m \mid \mathbf{y}^t A = 0 \text{ and } \mathbf{y} \geq 0\}$ the *projection cone* of Q w.r.t.$\mathbf{u}$. When there is no ambiguity, we simply call C the projection cone of Q. Using the following so-called *projection lemma*, we can compute a representation for the projection $\mathsf{proj}(Q; \mathbf{x})$:

Lemma 5 ([11]). *The projection* $\mathsf{proj}(Q; \mathbf{x})$ *of the polyhedron* Q *can be represented by*

$$S := \{\mathbf{y}^t B\mathbf{x} \leq \mathbf{y}^t\mathbf{c}, \forall \mathbf{y} \in \mathsf{ExtremeRays}(C)\},$$

where C *is the projection cone of* Q *defined above.*

Lemm 5 provides the main idea of the block elimination method. However, the represention produced in this way may have redundant inequalities. In [2], Balas observed that if the matrix B is invertible, then we can find a cone such that its extreme rays are in one-to-one correspondence with the facets of the projection of the polyhedron (the proof of this fact is similar to the proof of our Theorem 3). Using this fact, Balas developed an algorithm to find all redundant inequalities for all cases, including the cases where B is singular.

In this section, we will explain Balas' algorithm[1] in detail. To achieve this, we lift the polyhedron Q to a space in higher dimension by constructing the following objects.

Construction of B_0. Assume that the first q rows of B, denoted as B_1, are independent. Denote the last $m - q$ rows of B as B_2. Add $m - q$ columns, $e_{q+1}, \ldots, e_m$, to B, where e_i is the i-th vector in the canonical basis of $\mathbb{Q}^m$, thus with 1 in the i-th position and 0's anywhere else. The matrix B_0 has the following form:

$$B_0 = \begin{bmatrix} B_1 & \mathbf{0} \\ B_2 & I_{m-q} \end{bmatrix}.$$

To maintain consistency in the notation, let $A_0 = A$ and $c_0 = c$.

Construction of Q^0. We define:

$$Q^0 := \{(\mathbf{u}, \mathbf{x}') \in \mathbb{Q}^p \times \mathbb{Q}^m \mid A_0\mathbf{u} + B_0\mathbf{x}' \leq c_0 , \ x_{q+1} = \cdots = x_m = 0\}.$$

From now on, we use $\mathbf{x}'$ to represent the vector $\mathbf{x} \in \mathbb{Q}^q$, augmented with $m - q$ variables $(x_{q+1}, \ldots, x_m)$. Since the extra variables $(x_{q+1}, \ldots, x_m)$ are assigned to zero, we note that $\mathsf{proj}(Q; \mathbf{x})$ and $\mathsf{proj}(Q^0; \mathbf{x}')$ are "isomorphic" by means of the bijection Φ:

$$\Phi : \quad \begin{array}{l} \mathsf{proj}(Q; \mathbf{x}) \to \mathsf{proj}(Q^0; \mathbf{x}') \\ (x_1, \ldots, x_q) \mapsto (x_1, \ldots, x_q, 0, \ldots, 0) \end{array}$$

In the following, we will treat $\mathsf{proj}(Q; \mathbf{x})$ and $\mathsf{proj}(Q^0; \mathbf{x}')$ as the same polyhedron when there is no ambiguity.

Construction of W^0. Define W^0 to be the set of all $(\mathbf{v}, \mathbf{w}, v_0) \in \mathbb{Q}^q \times \mathbb{Q}^{m-q} \times \mathbb{Q}$ satisfying

$$W^0 = \{(\mathbf{v}, \mathbf{w}, v_0) \mid [\mathbf{v}^t, \mathbf{w}^t]B_0^{-1}A_0 = 0, [\mathbf{v}^t, \mathbf{w}^t]B_0^{-1} \geq 0, \\ -[\mathbf{v}^t, \mathbf{w}^t]B_0^{-1}c_0 + v_0 \geq 0\}. \tag{3}$$

Similar to the discussion in the work of Balas, the extreme rays of the cone $\mathsf{proj}(W^0; \{\mathbf{v}, v_0\})$ are used to construct the minimal representation of the projection $\mathsf{proj}(Q; \mathbf{x})$.

Theorem 3 shows that extreme rays of the cone $\overline{\mathsf{proj}(W^0; \{\mathbf{v}, v_0\})}$, which is defined as

$$\overline{\mathsf{proj}(W^0; \{\mathbf{v}, v_0\})} := \{(\mathbf{v}, -v_0) \mid (\mathbf{v}, v_0) \in \mathsf{proj}(W^0; \{\mathbf{v}, v_0\})\},$$

---

[1] It should be noted that, although we are using his idea, we have found a flaw in Balas' paper. In fact, the last inequality in representation of W^0 is written as equality that paper.

are in one-to-one correspondence with the facets of the homogenized cone of proj(Q; $\mathbf{x}$). As a result its extreme rays can be used to find the minimal representation of HomCone(proj(Q; $\mathbf{x}$)).

Lemma 6. *The operations "computing the characteristic cone" and "computing projections" commute. To be precise, we have:*

$$\mathsf{CharCone}(\mathsf{proj}(Q; \mathbf{x})) = \mathsf{proj}(\mathsf{CharCone}(Q); \mathbf{x}).$$

Proof. By the definition of the characteristic cone, we have CharCone(Q) = $\{(\mathbf{u}, \mathbf{x}) \mid A\mathbf{u} + B\mathbf{x} \le \mathbf{0}\}$, whose representation has the same left-hand side as the one of Q. The lemma is valid if we can show that the representation of proj(CharCone(Q); $\mathbf{x}$) has the same left-hand side as proj(Q; $\mathbf{x}$). This is obvious with the Fourier-Motzkin Elimination procedure.

<u>Theorem 3.</u> *The polar cone of* HomCone(proj(Q; $\mathbf{x}$)) *equals to* $\overline{\mathsf{proj}(W^0; \{\mathbf{v}, v_0\})}$.

Proof. By definition, the polar cone (HomCone(proj(Q; $\mathbf{x}$))* is equal to

$$\{(\mathbf{y}, y_0) \mid [\mathbf{y}^t, y_0][\mathbf{x}^t, x_{\text{last}}]^t \le 0, \forall \ (\mathbf{x}, x_{\text{last}}) \in \mathsf{HomCone}(\mathsf{proj}(Q; \mathbf{x}))\}.$$

This claim follows immediately from (HomCone(proj(Q; $\mathbf{x}$))* = $\overline{\mathsf{proj}(W^0; \{\mathbf{v}, v_0\})}$. We prove this latter equality in two steps.

($\supseteq$) For any $(\overline{\mathbf{v}}, -\overline{v}_0) \in \overline{\mathsf{proj}(W^0; \{\mathbf{v}, v_0\})}$, we need to show that

$$[\overline{\mathbf{v}}^t, -\overline{v}_0][\mathbf{x}^t, x_{\text{last}}]^t \le 0$$

holds when $(\mathbf{x}, x_{\text{last}}) \in$ HomCone(proj(Q; $\mathbf{x}$)). Remember that Q is pointed. As a result, HomCone(proj(Q; $\mathbf{x}$)) is also pointed. Therefore, we only need to verify the desired property for the extreme rays of HomCone(proj(Q; $\mathbf{x}$)), which either have the form $(\mathbf{s}, 1)$ or $(\mathbf{s}, 0)$ (Theorem 1). Before continuing, we should notice that since $(\overline{\mathbf{v}}, \overline{v}_0) \in$ proj(W^0; $\{\mathbf{v}, v_0\}$), there exists $\overline{\mathbf{w}}$ such that $[\overline{\mathbf{v}}^t, \overline{\mathbf{w}}^t, \overline{v}_0] \in W^0$. Cases 1 and 2 below conclude that $(\overline{\mathbf{v}}, -\overline{v}_0) \in$ HomCone(proj(Q; $\mathbf{x}$))* holds.

Case 1: for the form $(\mathbf{s}, 1)$, we have $\mathbf{s} \in$ proj(Q; $\mathbf{x}$). Indeed, $\mathbf{s}$ is an extreme point of proj(Q; $\mathbf{x}$). Hence, there exists $\overline{\mathbf{u}} \in \mathbb{Q}^p$, such that we have $A\overline{\mathbf{u}} + B\mathbf{s} \le \mathbf{c}$. By construction of Q^0, we have $A_0\overline{\mathbf{u}} + B_0\mathbf{s}' \le \mathbf{c}_0$, where $\mathbf{s}' = [\mathbf{s}^t, s_{q+1}, \dots, s_m]^t$ with $s_{q+1} = \dots = s_m = 0$. Therefore, we have:

$$[\overline{\mathbf{v}}^t, \overline{\mathbf{w}}^t]B_0^{-1}A_0\overline{\mathbf{u}} + [\overline{\mathbf{v}}^t, \overline{\mathbf{w}}^t]B_0^{-1}B_0\mathbf{s}' \le [\overline{\mathbf{v}}^t, \overline{\mathbf{w}}^t]B_0^{-1}\mathbf{c}_0.$$

This leads us to $\overline{\mathbf{v}}^t\mathbf{s} = [\overline{\mathbf{v}}^t, \overline{\mathbf{w}}^t]\mathbf{s}' \le [\overline{\mathbf{v}}^t, \overline{\mathbf{w}}^t]B_0^{-1}\mathbf{c}_0 \le \overline{v}_0$. Therefore, we have $[\overline{\mathbf{v}}^t, -\overline{v}_0][\mathbf{s}^t, x_{\text{last}}]^t \le 0$, as desired.

Case 2: for the form $(\mathbf{s}, 0)$, we have

$$\mathbf{s} \in \mathsf{CharCone}(\mathsf{proj}(Q; \mathbf{x})) = \mathsf{proj}(\mathsf{CharCone}(Q); \mathbf{x}).$$

Thus, there exists $\overline{\mathbf{u}} \in \mathbb{Q}^p$ such that $A\overline{\mathbf{u}} + B\mathbf{s} \le \mathbf{0}$. Similarly to Case 1, we have $[\overline{\mathbf{v}}^t, \overline{\mathbf{w}}^t]B_0^{-1}A_0\overline{\mathbf{u}} + [\overline{\mathbf{v}}^t, \overline{\mathbf{w}}^t]B_0^{-1}B_0\mathbf{s}' \le [\overline{\mathbf{v}}^t, \overline{\mathbf{w}}^t]B_0^{-1}\mathbf{0}$. Therefore, we have

$\overline{\mathbf{v}}^t \mathbf{s} = [\overline{\mathbf{v}}^t, \overline{\mathbf{w}}^t] \mathbf{s}' \leq [\overline{\mathbf{v}}^t, \overline{\mathbf{w}}^t] B_0^{-1} \mathbf{0} = 0$, and thus, we have $[\overline{\mathbf{v}}^t, -\overline{v}_0][\mathbf{s}^t, x_{\text{last}}]^t \leq 0$, as desired.

($\subseteq$) For any $(\overline{\mathbf{y}}, \overline{y}_0) \in \mathsf{HomCone}(\mathsf{proj}(Q; \mathbf{x}))^*$, we have $[\overline{\mathbf{y}}^t, \overline{y}_0][\mathbf{x}^t, x_{\text{last}}]^t \leq 0$ for all $(\mathbf{x}, x_{\text{last}}) \in \mathsf{HomCone}(\mathsf{proj}(Q; \mathbf{x}))$. For any $\overline{\mathbf{x}} \in \mathsf{proj}(Q; \mathbf{x})$, we have $\overline{\mathbf{y}}^t \overline{\mathbf{x}} \leq -\overline{y}_0$ since $(\overline{\mathbf{x}}, 1) \in \mathsf{HomCone}(\mathsf{proj}(Q; \mathbf{x}))$. Therefore, we have $\overline{\mathbf{y}}^t \mathbf{x} \leq -\overline{y}_0$, for all $\mathbf{x} \in \mathsf{proj}(Q; \mathbf{x})$, which makes the inequality $\overline{\mathbf{y}}^t \mathbf{x} \leq -\overline{y}_0$ redundant in the system $\{A\mathbf{u} + B\mathbf{x} \leq \mathbf{c}\}$. By Farkas' Lemma (see Lemma 2), there exists $\mathbf{p} \geq \mathbf{0}, \mathbf{p} \in \mathbb{Q}^m$ and $\lambda \geq 0$ such that $\mathbf{p}^t A = \mathbf{0}$, $\overline{\mathbf{y}} = \mathbf{p}^t B$, $\overline{y}_0 = \mathbf{p}^t \mathbf{c} + \lambda$. Remember that $A_0 = A$, $B_0 = [B, B']$, $\mathbf{c}_0 = \mathbf{c}$. Here B' is the last $m - q$ columns of B_0 consisting of $\mathbf{e}_{q+1}, \ldots, \mathbf{e}_m$. Let $\overline{\mathbf{w}} = \mathbf{p}^t B'$. We then have

$$\{\mathbf{p}^t A_0 = \mathbf{0}, \ [\overline{\mathbf{y}}^t, \overline{\mathbf{w}}^t] = \mathbf{p}^t B_0, -\overline{y}_0 \geq \mathbf{p}^t \mathbf{c}_0, \mathbf{p} \geq \mathbf{0}\},$$

which is equivalent to

$$\{\mathbf{p}^t = [\overline{\mathbf{y}}^t, \overline{\mathbf{w}}^t] B_0^{-1}, [\overline{\mathbf{y}}^t, \overline{\mathbf{w}}^t] B_0^{-1} A_0 = \mathbf{0},$$
$$-\overline{y}_0 \geq [\overline{\mathbf{y}}^t, \overline{\mathbf{w}}^t] B_0^{-1} \mathbf{c}_0, [\overline{\mathbf{y}}^t, \overline{\mathbf{w}}^t] B_0^{-1} \geq \mathbf{0}\}.$$

Therefore, $(\overline{\mathbf{y}}, \overline{\mathbf{w}}, -\overline{y}_0) \in W^0$, and $(\overline{\mathbf{y}}, -\overline{y}_0) \in \mathsf{proj}(W^0; \{\mathbf{v}, v_0\})$. From this, we deduce that $(\overline{\mathbf{y}}, \overline{y}_0) \in \mathsf{proj}(W^0; \{\mathbf{v}, v_0\})$ holds.

Theorem 4. *The minimal representation of* $\mathsf{proj}(Q; \mathbf{x})$ *is given exactly by*

$$\{\mathbf{v}^t \mathbf{x} \leq v_0 \mid (\mathbf{v}, v_0) \in \mathsf{ExtremeRays}(\mathsf{proj}(W^0; (\mathbf{v}, v_0))) \setminus \{(\mathbf{0}, 1)\}\}.$$

Proof. By Theorem 3, the minimal representation of the homogenized cone $\mathsf{HomCone}(\mathsf{proj}(Q; \mathbf{x}))$ is given exactly by

$$\{\mathbf{v}\mathbf{x} - v_0 x_{\text{last}} \leq 0 \mid (\mathbf{v}, v_0) \in \mathsf{ExtremeRays}(\mathsf{proj}(W^0; (\mathbf{v}, v_0)))\}.$$

Using Lemma 4, any minimal representation of $\mathsf{HomCone}(\mathsf{proj}(Q; \mathbf{x}))$ has at most one more inequality than any minimal representation of $\mathsf{proj}(Q; \mathbf{x})$. This extra inequality is $x_{\text{last}} \geq 0$ and, in this case, $\mathsf{proj}(W^0; (\mathbf{v}, v_0))$ will have the extreme ray $(\mathbf{0}, 1)$, which can be detected easily. Therefore, the minimal representation of $\mathsf{proj}(Q; \mathbf{x})$ is given by

$$\{\mathbf{v}^t \mathbf{x} \leq v_0 \mid (\mathbf{v}, v_0) \in \mathsf{ExtremeRays}(\mathsf{proj}(W^0; (\mathbf{v}, v_0))) \setminus \{(\mathbf{0}, 1)\}\}.$$

For simplicity, we call the cone $\mathsf{proj}(W^0; \{\mathbf{v}, v_0\})$ the redundancy test cone of Q w.r.t. $\mathbf{u}$ and denote it by $\mathcal{P}_{\mathbf{u}}(Q)$. When $\mathbf{u}$ is empty, we define $\mathcal{P}(Q) := \mathcal{P}_{\mathbf{u}}(Q)$ and we call it the initial redundancy test cone. It should be noted that $\mathcal{P}(Q)$ can be used to detect redundant inequalities in the input system, as it is shown in Steps 3 to 8 of Algorithm 4.

4 Minimal Representation of the Projected Polyhedron

In this section, we present our algorithm for removing all the redundant inequalities generated during Fourier-Motzkin elimination. Our algorithm detects and eliminates redundant inequalities, right after their generation, using the redundancy test cone introduced in Sect. 3. Intuitively, we need to construct the cone W^0 and obtain a representation of the redundancy test cone, $\mathcal{P}_{\mathbf{u}}(Q)$, where $\mathbf{u}$ is the vector of eliminated variables, each time we eliminate a variable during FME. This method is time consuming because it requires to compute the projection of W^0 onto $\{\mathbf{v}, v_0\}$ space at each step. However, as we prove in Lemma 7, we only need to compute the initial redundancy test cone, using Algorithm 2, and the redundancy test cones, used in the subsequent variable eliminations, can be found incrementally without any extra cost. After generating the redundancy test cone, the algorithm, using Algorithm 3, keeps the newly generated inequality only if it is an extreme ray of the redundancy test cone.

Note that a byproduct of this process is a *minimal projected representation* of the input system, according to the specified variable ordering. This representation is useful for finding solutions of linear inequality systems. The notion of projected representation was introduced in [25,26] and will be reviewed in Definition 1.

For convenience, we rewrite the input polyhedron Q defined in Eq. (2) as: $Q = \{\mathbf{y} \in \mathbb{Q}^n \mid \mathbf{A}\mathbf{y} \leq \mathbf{c}\}$, where $\mathbf{A} = [A, B] \in \mathbb{Q}^{m \times n}$, $n = p + q$ and $\mathbf{y} = [\mathbf{u}^t, \mathbf{x}^t]^t \in \mathbb{Q}^n$. We assume the first n rows of $\mathbf{A}$ are linearly independent.

Algorithm 2. Generate initial redundancy test cone

Input: $S = \{\mathbf{A}\mathbf{y} \leq \mathbf{c}\}$, a representation of the input polyhedron Q;
Output: $\mathcal{P}$, a representation of the initial redundancy test cone;
1: Construct $\mathbf{A}_0$ in the same way we constructed B_0, that is, $\mathbf{A}_0 := [\mathbf{A}, \mathbf{A}']$, where
   $\mathbf{A}' = [\mathbf{e}_{n+1}, \ldots, \mathbf{e}_m]$ with $\mathbf{e}_i$ being the i-th vector of the canonical basis of $\mathbb{Q}^m$;
2: Let $W := \{(\mathbf{v}, \mathbf{w}, v_0) \in \mathbb{Q}^n \times \mathbb{Q}^{m-n} \times \mathbb{Q} \mid -[\mathbf{v}^t, \mathbf{w}^t]\mathbf{A}_0^{-1}\mathbf{c} + v_0 \geq 0, [\mathbf{v}^t, \mathbf{w}^t]\mathbf{A}_0^{-1} \geq \mathbf{0}\}$;

3: $\mathcal{P} = \mathsf{proj}(W; \{\mathbf{v}, v_0\})$;
4: **return** $(\mathcal{P})$;

Remark 1. There are two important points about Algorithm 2. First, we only need a representation of the initial redundancy test cone. This representation does not need to be minimal. Therefore, calling Algorithm 2 in Algorithm 4 (which computes a minimal projected representation of a polyhedron) does not lead to a recursive call to Algorithm 4. Second, to compute the projection $\mathsf{proj}(W; \{\mathbf{v}, v_0\})$, we need to eliminate $m - n$ variables from $m + 1$ inequalities. The block elimination method is applied to achieve this. As it is shown in Lemma 5, the block elimination method will require to compute the extreme rays of the projection cone (denoted by C), which contains $m + 1$ inequalities and $m + 1$ variables. However, considering the structural properties of the coefficient matrix of the representation of C, we found out that computing the extreme rays of C is equivalent to computing the extreme rays of another simpler cone, which still has $m + 1$ inequalities but only $n + 1$ variables.

Lemma 7. *A representation of the redundancy test cone $\mathcal{P}_{\mathbf{u}}(Q)$ can be obtained from $\mathcal{P}(Q)$ by setting coefficients of the corresponding p eliminated variables to 0 in the representation of $\mathcal{P}(Q)$.*

Proof. To distinguish from the construction of $\mathcal{P}(Q)$, we rename the variables $\mathbf{v}, \mathbf{w}, v_0$ as $\mathbf{v_u}, \mathbf{w_u}, v_u$, when constructing W^0 and computing the test cone $\mathcal{P}_{\mathbf{u}}(Q)$.

That is, we have $\mathcal{P}_{\mathbf{u}}(Q) = \mathrm{proj}(W^0; \{\mathbf{v_u}, v_u\})$, where W^0 is the set of all $(\mathbf{v_u}, \mathbf{w_u}, v_u) \in \mathbb{Q}^q \times \mathbb{Q}^{m-q} \times \mathbb{Q}$ satisfying

$$\{(\mathbf{v_u}, \mathbf{w_u}, v_u) \mid [\mathbf{v_u^t}, \mathbf{w_u^t}]B_0^{-1}A = \mathbf{0}, -[\mathbf{v_u^t}, \mathbf{w_u^t}]B_0^{-1}\mathbf{c} + v_u \geq 0, [\mathbf{v_u^t}, \mathbf{w_u^t}]B_0^{-1} \geq \mathbf{0}\},$$

while we have $\mathcal{P}(Q) = \mathrm{proj}(W; \{\mathbf{v}, v_0\})$ where W is the set of all $(\mathbf{v}, \mathbf{w}, v_0) \in \mathbb{Q}^n \times \mathbb{Q}^{m-n} \times \mathbb{Q}$ satisfying $\{(\mathbf{v}, \mathbf{w}, v_0) \mid -[\mathbf{v^t}, \mathbf{w^t}]A_0^{-1}\mathbf{c} \mid v_0 \geq 0, [\mathbf{v^t}, \mathbf{w^t}]A_0^{-1} \geq \mathbf{0}\}$.

By Step 1 of Algorithm 2, $[\mathbf{v^t}, \mathbf{w^t}]A_0^{-1}A = \mathbf{v}^t$ holds for all $(\mathbf{v}, \mathbf{w}, v_0) \in W$. We can rewrite $\mathbf{v}$ as $\mathbf{v}^t = [\mathbf{v_1^t}, \mathbf{v_2^t}]$, where $\mathbf{v_1}$ and $\mathbf{v_2}$ are the first p and last $n-p$ variables of $\mathbf{v}$. Then, we have $[\mathbf{v^t}, \mathbf{w^t}]A_0^{-1}A = \mathbf{v_1^t}$ and $[\mathbf{v^t}, \mathbf{w^t}]A_0^{-1}B = \mathbf{v_2^t}$. Similarly, we have $[\mathbf{v_u^t}, \mathbf{w_u^t}]B_0^{-1}A = \mathbf{0}$ and $[\mathbf{v_u^t}, \mathbf{w_u^t}]B_0^{-1}B = \mathbf{v_u^t}$ for all $(\mathbf{v_u}, \mathbf{w_u}, v_u) \in W^0$. This lemma holds if we can show $\mathcal{P}_{\mathbf{u}} = \mathcal{P}|_{\mathbf{v_1}=\mathbf{0}}$. We prove this in two steps:

($\subseteq$) For any $(\overline{\mathbf{v}}_{\mathbf{u}}, \overline{v}_{\mathbf{u}}) \in \mathcal{P}_{\mathbf{u}}(Q)$, there exists $\overline{\mathbf{w}}_{\mathbf{u}} \in \mathbb{Q}^{m-q}$, such that

$$(\overline{\mathbf{v}}_{\mathbf{u}}, \overline{\mathbf{w}}_{\mathbf{u}}, \overline{v}_{\mathbf{u}}) \in W^0.$$

Let $[\overline{\mathbf{v}}^t, \overline{\mathbf{w}}^t] := [\overline{\mathbf{v}}_{\mathbf{u}}^t, \overline{\mathbf{w}}_{\mathbf{u}}^t]B_0^{-1}A_0$, where $\overline{\mathbf{v}}^t = [\overline{\mathbf{v}}_1^t, \overline{\mathbf{v}}_2^t]$ ($\overline{\mathbf{v}}_1 \in \mathbb{Q}^p, \overline{\mathbf{v}}_2 \in \mathbb{Q}^{n-p}$, and $\overline{\mathbf{w}} \in \mathbb{Q}^{m-n}$). Then, because $(\overline{\mathbf{v}}_{\mathbf{u}}, \overline{\mathbf{w}}_{\mathbf{u}}, \overline{v}_{\mathbf{u}}) \in W^0$, we have $\overline{\mathbf{v}}_1^t = [\overline{\mathbf{v}}_{\mathbf{u}}^t, \overline{\mathbf{w}}_{\mathbf{u}}^t]B_0^{-1}A = \mathbf{0}$ and $\overline{\mathbf{v}}_2^t = [\overline{\mathbf{v}}_{\mathbf{u}}^t, \overline{\mathbf{w}}_{\mathbf{u}}^t]B_0^{-1}B = \overline{\mathbf{v}}_{\mathbf{u}}$. Let $\overline{v}_0 = \overline{v}_{\mathbf{u}}$, it is easy to verify that $(\overline{\mathbf{v}}, \overline{\mathbf{w}}, \overline{v}_0) \in W$. Therefore, $(\mathbf{0}, \overline{\mathbf{v}}_{\mathbf{u}}, \overline{v}_{\mathbf{u}}) = (\overline{\mathbf{v}}, \overline{v}_0) \in \mathcal{P}(Q)$.

($\supseteq$) For any $(\mathbf{0}, \overline{\mathbf{v}}_2, \overline{v}_0) \in \mathcal{P}(Q)$, there exists $\overline{\mathbf{w}} \in \mathbb{Q}^{m-n}$, such that

$$(\mathbf{0}, \overline{\mathbf{v}}_2, \overline{\mathbf{w}}, \overline{v}_0) \in W.$$

Let $[\overline{\mathbf{v}}_{\mathbf{u}}^t, \overline{\mathbf{w}}_{\mathbf{u}}^t] := [\mathbf{0}, \overline{\mathbf{v}}_2^t, \overline{\mathbf{w}}^t]A_0^{-1}B_0$. We have $\overline{\mathbf{v}}_{\mathbf{u}} - [\mathbf{0}, \overline{\mathbf{v}}_2^t, \overline{\mathbf{w}}^t]A_0^{-1}B = \overline{\mathbf{v}}_2$. Let $\overline{v}_{\mathbf{u}} = \overline{v}_0$, it is easy to verify that $(\overline{\mathbf{v}}_{\mathbf{u}}, \overline{\mathbf{w}}_{\mathbf{u}}, \overline{v}_{\mathbf{u}}) \in W^0$. Therefore, $(\overline{\mathbf{v}}_2, \overline{v}_0) = (\overline{\mathbf{v}}_{\mathbf{u}}, \overline{v}_{\mathbf{u}}) \in \mathcal{P}_{\mathbf{u}}(Q)$.

Consider again the polyhedron $Q = \{\mathbf{y} \in \mathbb{Q}^n \mid A\mathbf{y} \leq \mathbf{c}\}$, where $A = [A, B] \in \mathbb{Q}^{m \times n}$, $n = p + q$ and $\mathbf{y} = [\mathbf{u}^t, \mathbf{x}^t]^t \in \mathbb{Q}^n$. Fix a variable ordering, say $y_1 > \cdots > y_n$, For $1 \leq i \leq n$, we denote by $A^{(y_i)}$ the inequalities in the representation $A\mathbf{y} \leq \mathbf{c}$ of Q whose largest variable is y_i. We denote by $\mathrm{ProjRep}(Q; y_1 > \cdots > y_n)$ the linear system $A^{(y_1)}$ if $n = 1$ and the conjunction of $A^{(y_1)}$ and $\mathrm{ProjRep}(\mathrm{proj}(Q; \mathbf{y}_2); y_2 > \cdots > y_n)$ otherwise, where $\mathbf{y}_2 = (y_2, \ldots, y_n)$. Of course, $\mathrm{ProjRep}(Q; y_1 > \cdots > y_n)$ depends on the representation which is used of Q.

Definition 1 (Projected representation). *For the polyhedron $Q \subseteq \mathbb{Q}^n$, we call projected representation of Q w.r.t. the variable order $y_1 > \cdots > y_n$ any linear system of the form $\mathrm{ProjRep}(Q; y_1 > \cdots > y_n)$. We say that such a linear system P is a minimal projected representation of Q if, for all $1 \leq k \leq n$, every inequality of P, with y_k as largest variable, is not redundant among all the inequalities of P with variables among $y_k, \ldots, y_n$.*

We can generate a *minimal projected representation* of a polyhedron, w.r.t. an specific variable ordering by Algorithm 4.

Algorithm 3. Extreme ray test

Input: $(\mathcal{P}, \ell)$, where (i) $\mathcal{P} := \{(\mathbf{v}, v_0) \in \mathbb{Q}^n \times \mathbb{Q} \mid M[\mathbf{v}^t, v_0]^t \leq \mathbf{0}\}$ with $M \in \mathbb{Q}^{m \times (n+1)}$,
    (ii) $\ell : \mathbf{a}^t \mathbf{y} \leq c$ with $\mathbf{a} \in \mathbb{Q}^n$ and $c \in \mathbb{Q}$;
Output: true if $[\mathbf{a}^t, c]^t$ is an extreme ray of $\mathcal{P}$, false otherwise;
  1: Let $\mathbf{s} := M[\mathbf{a}^t, c]^t$;
  2: Let $\zeta(\mathbf{s})$ be the index set of the zero coefficients of $\mathbf{s}$;
  3: **if** rank$(M_{\zeta(\mathbf{s})}) = n$ **then**
  4:     return (true);
  5: **else**
  6:     return (false);
  7: **end if**

---

5 Complexity Estimates

In this section, we analyze the computational complexity of Algorithm 4, which computes a minimal projected representation of a given polyhedron. This computation is equivalent to eliminating all variables, one after another, in Fourier-Motzkin elimination. We prove that using our algorithm, finding a minimal projected representation of a polyhedron is singly exponential in the dimension n of the ambient space. The most consuming procedure in Algorithm 4 is finding the initial redundancy test cone. This operation requires another polyhedron projection in higher dimension. As it is shown in Remark 1, we can use block elimination method to perform this task efficiently. This requires the computation of the extreme rays of the projection cone. The double description method is an efficient way to solve this problem. We begin this section by computing the bit complexity of the double description algorithm.

Lemma 8 (Coefficient bound of extreme rays). *Let*

$$S = \{\mathbf{x} \in \mathbb{Q}^n \mid A\mathbf{x} \leq \mathbf{0}\}$$

be a minimal representation of a cone $C \subseteq \mathbb{Q}^n$, where $A \in \mathbb{Q}^{m \times n}$. Then, the absolute value of a coefficient in any extreme ray of C is bounded over by $(n-1)^n \|A\|^{2(n-1)}$.

Proof. From the properties of extreme rays, see Sect. 2.1, we know that when $\mathbf{r}$ is an extreme ray, there exists a sub-matrix $A' \in \mathbb{Q}^{(n-1) \times n}$ of A, such that $A'\mathbf{r} = 0$. This means that $\mathbf{r}$ is in the null-space of A'. Thus, the claim follows by proposition 6.6 of [36].

Algorithm 4. Minimal Projected Representation of Q

Input: $S = \{\mathbf{A}\mathbf{y} \leq \mathbf{c}\}$: a representation of the input polyhedron Q;
Output: A minimal projected representation of Q;
1: Generate the initial redundancy test cone $\mathcal{P}$ by Algorithm 2;
2: $S_0 := \{\ \}$;
3: **for** i from 1 to m **do**
4:     Let f be the result of applying Algorithm 3 with the inputs $\mathcal{P}$ and $\mathbf{A}_i\mathbf{y} \leq \mathbf{c}_i$;
5:     **if** $f = $ true **then**
6:         $S_0 := S_0 \cup \{\mathbf{A}_i\mathbf{y} \leq \mathbf{c}_i\}$;
7:     **end if**
8: **end for**
9: $\mathcal{P} := \mathcal{P}|_{v_1=0}$;
10: **for** i from 0 to $n-1$ **do**
11:     $S_{i+1} := \{\ \}$;
12:     **for** $\ell_{\text{pos}} \in S_i$ with positive coefficient of y_{i+1} **do**
13:         **for** $\ell_{\text{neg}} \in S_i$ with negative coefficient of y_{i+1} **do**
14:             $\ell_{\text{new}} := \textsf{Combine}(\ell_{\text{pos}}, \ell_{\text{neg}}, y_{i+1})$;
15:             Let f be the result of applying Algorithm 3 with the inputs $\mathcal{P}$ and ℓ_{new};
16:             **if** $f = $ true **then**
17:                 $S_{i+1} := S_{i+1} \cup \{\ell_{\text{new}}\}$;
18:             **end if**
19:         **end for**
20:     **end for**
21:     **for** $\ell \in S_i$ with zero coefficient of y_{i+1} **do**
22:         Let f be the result of applying Algorithm 3 with the inputs $\mathcal{P}$ and ℓ ;
23:         **if** $f = $ true **then**
24:             $S_{i+1} := S_{i+1} \cup \{\ell\}$;
25:         **end if**
26:     **end for**
27:     $\mathcal{P} := \mathcal{P}|_{v_{i+1}=0}$;
28: **end for**
29: **return** $(S_0 \cup S_1 \cup \cdots \cup S_n)$;

Lemma 9. *Let $S = \{\mathbf{x} \in \mathbb{Q}^n \mid A\mathbf{x} \leq \mathbf{0}\}$ be the minimal representation of a cone $C \subseteq \mathbb{Q}^n$, where $A \in \mathbb{Q}^{m \times n}$. The double description method requires $O(m^{n+2}n^{\theta+\epsilon}h^{1+\epsilon})$ bit operations, where h is the height of the matrix A.*

Proof. The cost of Algorithm 1 during the processing of the first n inequalities (Line 4) is negligible (in comparison to the subsequent computations) since it is equivalent to find the inverse of an $n \times n$ matrix. Therefore, to analyze the complexity of the DD method, we focus on the while-loop located at Line 6 of the Algorithm 1. After adding t inequalities, with $n \leq t \leq m$, the first step is to partition the extreme rays at the $t-1$-iteration, with respect to the newly added inequality (Line 9 of Algorithm 1). Note that we have at most $(t-1)^{\lfloor \frac{n}{2} \rfloor}$ extreme rays (Eq. (1)) whose coefficients can be bounded over by $(n-1)^n \|A\|^{2(n-1)}$ (Lemma 8) at the $t-1$-iteration. Hence, this step needs at most $C_1 := (t-1)^{\lfloor \frac{n}{2} \rfloor} \times n \times \mathcal{M}(\log((n-1)^n \|A\|^{2(n-1)})) \leq O(t^{\lfloor \frac{n}{2} \rfloor} n^{2+\epsilon} h^{1+\epsilon})$

bit operations. After partitioning the vectors, the next step is to check adjacency for each pair of vectors (Line 13 of of Algorithm 1). The cost of this step is equivalent to computing the rank of a sub-matrix $A' \in \mathbb{Q}^{(t-1)\times n}$ of A. This should be done for $\frac{t^n}{4}$ pairs of vectors. This step needs at most $C_2 := \frac{t^n}{4} \times O((t-1)n^{\theta+\epsilon}h^{1+\epsilon}) \leq O(t^{n+1}n^{\theta+\epsilon}h^{1+\epsilon})$ bit operations. We know there are at most $t^{\lfloor\frac{n}{2}\rfloor}$ pairs of adjacent extreme rays. The next step is to combine every pair of adjacent vectors in order to obtain a new extreme ray (Line 14 of Algorithm 1). This step consists of n multiplications in $\mathbb{Q}$ of coefficients with absolute value bounded over by $(n-1)^n\|A\|^{2(n-1)}$ (Lemma 8) and this should be done for at most $t^{\lfloor\frac{n}{2}\rfloor}$ vectors. Therefore, the bit complexity of this step, is no more than $C_3 := t^{\lfloor\frac{n}{2}\rfloor} \times n \times \mathcal{M}(\log((n-1)^n\|A\|^{2(n-1)})) \leq O(t^{\lfloor\frac{n}{2}\rfloor}n^{2+\epsilon}h^{1+\epsilon})$. Finally, the complexity of iteration t of the while loop is $C := C_1 + C_2 + C_3$. The claim follows after simplifying $m \times C$.

Lemma 10 (Complexity of constructing the initial redundancy test cone). *Let h be the maximum height of A and $\mathbf{c}$ in the input system, then generating the initial redundancy test cone (Algorithm 2) requires at most*

$$O(m^{n+3+\epsilon}(n+1)^{\theta+\epsilon}h^{1+\epsilon})$$

bit operations. Moreover, $\mathsf{proj}(W; \{\mathbf{v}, v_0\})$ can be represented by $O(m^{\lfloor\frac{n+1}{2}\rfloor})$ inequalities, each with a height bound of $O(m^\epsilon n^{2+\epsilon}h)$.

Proof. We analyze Algorithm 2 step by step.

Step 1: Construction of A_0 from A. The cost of this step can be neglected. However, it should be noticed that the matrix A_0 has a special structure. Without loss of generality, we can assume that the first n rows of A are linearly independent. The matrix A_0 has the following structure:

$$A_0 = \begin{pmatrix} A_1 & \mathbf{0} \\ A_2 & I_{m-n} \end{pmatrix},$$

where A_1 is a full rank matrix in $\mathbb{Q}^{n\times n}$ and $A_2 \in \mathbb{Q}^{(m-n)\times n}$.

Step 2: Construction of the Cone W. Using the structure of the matrix A_0, its inverse can be expressed as

$$A_0^{-1} = \begin{pmatrix} A_1^{-1} & \mathbf{0} \\ -A_2A_1^{-1} & I_{m-n} \end{pmatrix}.$$

Also, from Sect. 2.4 we have $\|A_1^{-1}\| \leq (\sqrt{n-1}\|A_1\|)^{n-1}$. Therefore, $\|A_0^{-1}\| \leq n^{\frac{n+1}{2}}\|A\|^n$, and $\|A_0^{-1}\mathbf{c}\| \leq n^{\frac{n+3}{2}}\|A\|^n\|\mathbf{c}\| + (m-n)\|\mathbf{c}\|$. That is, $\mathrm{height}(A_0^{-1}) \in O(n^{1+\epsilon}h)$ and $\mathrm{height}(A_0^{-1}\mathbf{c}) \in O(m^\epsilon + n^{1+\epsilon}h)$. As a result, height of coefficients of W can be bounded over by $O(m^\epsilon + n^{1+\epsilon}h)$.

To estimate the bit complexity, we need the following consecutive steps:

- Computing A_0^{-1}, which requires

$$O(n^{\theta+1+\epsilon}h^{1+\epsilon}) + O((m-n)n^2\mathcal{M}(\max(\mathsf{height}(A_2), \mathsf{height}(A_1^{-1}))))$$
$$\leq O(mn^{\theta+1+\epsilon}h^{1+\epsilon}) \text{ bit operations;}$$

- Constructing $W := \{(\mathbf{v}, \mathbf{w}, v_0) \mid -[\mathbf{v}^t, \mathbf{w}^t]A_0^{-1}\mathbf{c} + v_0 \geq 0, [\mathbf{v}^t, \mathbf{w}^t]A_0^{-1} \geq \mathbf{0}\}$ requires at most

$$C_1 := O(m^{1+\epsilon}n^{\theta+1+\epsilon}h^{1+\epsilon}) + O(mn\mathcal{M}(\mathsf{height}(A_0^{-1}, \mathbf{c})))$$
$$+ O((m-n)h) \leq O(m^{1+\epsilon}n^{\theta+\epsilon+1}h^{1+\epsilon}) \text{ bit operations.}$$

Step 3: Projecting W and Finding the Initial Redundancy Test Cone.
Following Lemma 5, we obtain a representation of $\mathsf{proj}(W; \{\mathbf{v}, v_0\})$ through finding extreme rays of the corresponding projection cone.
Let $E = (-A_2A_1^{-1})^t \in \mathbb{Q}^{n\times(m-n)}$ and $\mathbf{g}^t$ be the last $m-n$ elements of $(A_0^{-1}\mathbf{c})^t$. Then, the projection cone can be represented by:

$$C = \{\mathbf{y} \in \mathbb{Q}^{m+1} \mid \mathbf{y}^t \begin{pmatrix} E \\ \mathbf{g}^t \\ I_{m-n} \end{pmatrix} = \mathbf{0}, \mathbf{y} \geq \mathbf{0}\}.$$

Note that $y_{n+2}, \ldots, y_{m+1}$ can be solved from the system of equations in the representation of C. We substitute them in the inequalities and obtain a representation of the cone C', given by:

$$C' = \{\mathbf{y}' \in \mathbb{Q}^{n+1} \mid \mathbf{y}'^t \begin{pmatrix} E \\ \mathbf{g}^t \end{pmatrix} \leq \mathbf{0}, \mathbf{y}' \geq \mathbf{0}\}$$

In order to find the extreme rays of the cone C, we can find the extreme rays of the cone C' and then back-substitute them into the equations to find the extreme rays of C. Applying the double description algorithm to C', we can obtain all extreme rays of C', and subsequently, the extreme rays of C. The cost estimate of this step is bounded over by the complexity of the double description algorithm with C' as input. This operation requires at most $C_2 := O(m^{n+3}(n+1)^{\theta+\epsilon}\max(\mathsf{height}(E, \mathbf{g}^t))^{1+\epsilon}) \leq O(m^{n+3+\epsilon}(n+1)^{\theta+\epsilon}h^{1+\epsilon})$ bit operations. The overall complexity of the algorithm can be bounded over by: $C_1 + C_2 \leq O(m^{n+3+\epsilon}(n+1)^{\theta+\epsilon}h^{1+\epsilon})$. Also, by Lemma 8 and Lemma 9, we know that the cone C has at most $O(m^{\lfloor\frac{n+1}{2}\rfloor})$ distinct extreme rays, each with height no more than $O(m^\epsilon n^{2+\epsilon}h)$. That is, $\mathsf{proj}(W^0; \{\mathbf{v}, v_0\})$ can be represented by at most $O(m^{\lfloor\frac{n+1}{2}\rfloor})$ inequalities, each with a height bound of $O(m^\epsilon n^{2+\epsilon}h)$.

Lemma 11. *Algorithm 3 runs within $O(m^{\frac{n}{2}}n^{\theta+\epsilon}h^{1+\epsilon})$ bit operations.*

Proof. The first step is to multiply the matrix M and the vector $(\mathbf{t}, t_0)$. Let d_M and c_M be the number of rows and columns of M, respectively.

Thus, $M \in \mathbb{Q}^{d_M \times c_M}$. We know that M is the coefficient matrix of $\mathsf{proj}(W^0, \{\mathbf{v}, v_0\})$. Therefore, after eliminating p variables $c_M = q + 1$, where $q = n - p$ and $d_M \leq m^{\frac{n}{2}}$. Also, we have $\mathsf{height}(M) \in O(m^\epsilon n^{2+\epsilon} h)$. With these specifications, the multiplication step and the rank computation step need $O(m^{\frac{n}{2}} n^{2+\epsilon} h^{1+\epsilon})$ and $O(m^{\frac{n}{2}} (q + 1)^{\theta+\epsilon} h^{1+\epsilon})$ bit operations, respectively. The claim follows after simplification.

Using Algorithms 2 and 3, we can find the minimal projected representation of a polyhedron in singly exponential time w.r.t. the number of variables n.

Theorem 5. *Algorithm 4 is correct. Moreover, a minimal projected representation of Q can be produced within $O(m^{\frac{5n}{2}} n^{\theta+1+\epsilon} h^{1+\epsilon})$ bit operations.*

Proof. The correctness of the algorithm follows from Theorem 4 and Lemma 7.

By [24,29], we know that after eliminating p variables, the projection of the polyhedron has at most m^{p+1} facets. For eliminating the next variable, there will be at most $(\frac{m^{p+1}}{2})^2$ pairs of inequalities to be considered and each of the pairs generate a new inequality which should be checked for redundancy. Therefore, the overall complexity of the algorithm is:

$$O(m^{n+3+\epsilon} (n + 1)^{\theta+\epsilon} h^{1+\epsilon}) + \sum_{p=0}^{n} m^{2p+2} O(m^{\frac{n}{2}} n^{\theta+\epsilon} h^{1+\epsilon}) = O(m^{\frac{5n}{2}} n^{\theta+1+\epsilon} h^{1+\epsilon}).$$

6 Experimentation

In this section we report on our software implementation of the algorithms presented in the previous sections. Our implementation as well as our test cases are part of the **BPAS** library, available at http://www.bpaslib.org/.

We report on serial and parallel implementation of the Minimal Projected Representation (MPR) algorithm. Comparing with the **Project** command of the **PolyhedralSets** package of Maple 2017 and the famous **CDD** library (version 2018), we have been able to solve our test cases more efficiently. We believe that this is the result of using a more effective algorithm and an efficient implementation in C.

As test cases we use 16 consistent linear inequality systems. The first 9 test cases, (t1 to t9) are linear inequality systems that are randomly generated. The systems S24 and S35 are 24-simplex and 35-simplex polytopes. The systems C56 and C510 are cyclic polytopes in dimension five with six and ten vertices, The system C68 is a cyclic polytope in dimension six with eight vertices, C1011 is cyclic polytope in dimension ten with eleven vertices, and, Cro6 is the cross polytope in 6 dimension [22]. The **test** column of Table 1 shows these systems along with the number of variables and the number of inequalities for each of them.

We implemented the MPR algorithm with two different approaches: one iterative following closely Algorithm 4, and the other reorganizing that algorithm by means of a divide and conquer scheme. In both implementations, we use a dense representation for the linear inequalities. In the first approach, we use *unrolled linked lists* to encode linear inequality systems. Indeed, using this data structure, we are able to store an array of inequalities in each node of a linked list and we can improve data locality. However, we use simple linked lists in the divide and conquer version to save time on dividing and joining lists. Although both these approaches have shown quite similar and promising results in terms of running time, we anticipate to get better results if we combine unrolled linked lists with the divide and conquer scheme while using a varying threshold for recursion as the algorithm goes on.

Columns MPR-itr and MPR-rec of the Table 1 give the running time (in milliseconds) of these implementations on a configuration with an Intel-i7-7700T CPU (4 cores, 8 threads, clocking at 3.8 GHz). Also, columns CDD, Maple, and Maple-MPR are corresponding to running times of the Fourier algorithm in the CDD library, which uses LP for redundancy elimination, the function PolyhedralSets:-Project of Maple, and, an implementation of our algorithm in the Maple programming language, on the same system, respectively.

Table 1. Running time (in milliseconds) table for a set of examples, varying in the number of variables and inequalities, collected on a system with Intel-i7-7700T 4-core processor, clocking at 3.8 GHz.

Test (var, ineq)	MPR-itr	MPR-rec	CDD	Maple	Maple-MPR
S24 (24,25)	46	41	411	6485	3040
S35 (35,36)	205	177	2169	57992	9840
Cro6 (6,64)	28	29	329	246750	8610
C56 (5,6)	1	1	13	825	140
C68 (6,16)	4	4	866	20154	650
C1011 (10,11)	95	92	>1h	>1h	>1h
C510 (5,42)	23	22	7674	6173	6070
T1 (5,10)	7	7	142	7974	1400
T2 (10,12)	109	112	122245	3321217	13330
T3 (7,10)	26	26	8207	117021	2900
T4 (10,12)	368	370	1177807	>1h	26650
T5 (5,11)	7	7	75	8229	1650
T6 (10,20)	26591	26156	>1h	>1h	>1h
T7 (9,19)	162628	158569	>1h	>1h	>1h
T8 (8,19)	21411	20915	>1h	>1h	>1h
T9 (6,18)	1281	1263	77372	>1h	267920

Using the divide and conquer scheme, we have been able to parallelize our program, with Cilk [5]. We call this algorithm Parallel Minimal Projected Representation (PMPR). Table 2 presents the running time (in milliseconds) and speedup of the multi-core version of the algorithm. The columns PMPR-1, PMPR-4, PMPR-8, and, PMPR-12 demonstrate the running time of the multi-core program on a system with Intel-Xeon-X5650 (12 cores, 24 threads, clocking at 2.6 GHz), using 1, 4, 8, and 12 Cilk workers, respectively. The numbers in brackets show the speedup we gain using multi-threading.

Table 2. Running time (in milliseconds) table for our set of examples, with different number of Cilk workers, collected on a system Intel-Xeon-X5650 and 12 CPU cores, clocking at 2.6GHz.

Test	PMPR-1	PMPR-4		PMPR-8		PMPR-12	
S24	67	71	(0.9 x)	73	(0.9 x)	83	(0.8 x)
S35	291	308	(0.9 x)	310	(0.9 x)	375	(0.7 x)
Cro6	54	45	(1.2 x)	36	(1.5 x)	34	(1.5 x)
C56	2	3	(0.6 x)	3	(0.6 x)	12	(0.1 x)
C68	8	7	(1.1 x)	7	(1.1 x)	19	(0.4 x)
C1011	176	62	(2.8 x)	47	(3.7 x)	53	(3.3 x)
C510	38	33	(1.1 x)	34	(1.1 x)	40	(0.9 x)
T1	13	8	(1.6 x)	9	(1.4 x)	17	(0.7 x)
T2	205	67	(3.0 x)	55	(3.7 x)	57	(3.5 x)
T3	48	20	(2.4 x)	18	(2.6 x)	20	(2.4 x)
T4	685	207	(3.3 x)	141	(4.8 x)	126	(5.4 x)
T5	14	9	(1.5 x)	10	(1.3 x)	11	(1.2 x)
T6	44262	12995	(3.4 x)	6785	(6.5 x)	5163	(8.5 x)
T7	282721	78176	(3.6 x)	48048	(5.8 x)	35901	(7.8 x)
T8	41067	10669	(3.8 x)	5689	(7.2 x)	4471	(9.1 x)
T9	2407	742	(3.2 x)	491	(4.8 x)	448	(5.3 x)

7   Related Works and Concluding Remarks

As we previously discussed, removing redundant inequalities during the execution of Fourier-Motzkin Elimination is the central issue towards efficiency. Different algorithms have been developed to solve this problem. They also have been implemented in the various software libraries, including but not limited to: CDD[17], VPL[7], PPL[1], Normaliz[9], PORTA[13], and Polymake[20] In this section, we briefly review *some* of these works.

In [11], Chernikov proposed a redundancy test with little added work, which greatly improves the practical efficiency of Fourier-Motzkin Elimination. Kohler proposed a method in [29] which only uses matrix arithmetic operations to test the redundancy of inequalities. As observed by Imbert in his work [24], the method he proposed in his paper as well as those of Chernikov and Kohler are essentially equivalent. Even though these works are effective in practice, none of them can remove all redundant inequalities generated by Fourier-Motzkin Elimination.

Besides Fourier-Motzkin Elimination, block elimination is another algorithmic tool to project polyhedra on a lower dimensional subspace. This method relies on the extreme rays of the so-called projection cone. Although there exist efficient methods to enumerate the extreme rays of this projection cone, like the *double description method* [18] (also known as Chernikova's algorithm [12,30]), this method can not remove all the redundant inequalities.

In [2], Balas shows that if certain *inconvertibility conditions* are satisfied, then the extreme rays of the redundancy test cone exactly defines a minimal representation of the projection of a polyhedron. As Balas mentioned in his paper, this method can be extended to any polyhedron.

A drawback of Balas' work is the necessity of enumerating the extreme rays of the redundancy test cone (so as to produce a minimal representation of the projection $\mathsf{proj}(Q; \mathbf{x})$) which is time consuming. Our algorithm tests the redundancy of the inequality $\mathbf{a}\mathbf{x} \leq c$ by checking whether $(\mathbf{a}, c)$ is an extreme ray of the redundancy test cone or not.

Another related topic to our work is the concept of subsumption cone, as defined in [23]. Consider the polyhedron Q given in Eq. (2), define $T := \{(\lambda, \alpha, \beta) \mid \lambda^t \mathbf{A} = \alpha^t, \lambda^t \mathbf{c} \leq \beta, \lambda \geq \mathbf{0}\}$, where λ and α are vectors of dimension m and n respectively, and β is a variable. The *subsumption cone* of Q is obtained by eliminating λ in T, that is, $\mathsf{proj}(T; \{\alpha, \beta\})$. We proved that considering a full-dimensional, pointed polyhedron, where the first n rows of the coefficient matrix are linearly independent, the initial redundancy test cone and the subsumption cone are equivalent.

Given a V-representation of a polyhedron P, one can obtain the V-representation of any projection of P^2. The double description method turns the V-representation of the projection to its H-representation. Most existing software libraries dealing with polyhedral sets store a polyhedron with these two representations, like the *Parma Polyhedra Library (PPL)* [1]. In this case, it is convenient to compute the projection using the block elimination method. When we are only given the H-representation, the first thing is to compute the V-representation, which is equivalent to the procedure of computing the initial test cone in our method. When we need to perform successive projections, it is well-known that Fourier-Motzkin Elimination performs better than repeated applications of the double description method.

Recently, the verified polyhedron library (VPL) [7] takes advantage of parametric linear programming to project a polyhedron. Like PPL, VPL may not beat Fourier-Motzkin Elimination when we need to perform successive projections. In VPL, the authors rely on *raytracing* to remove redundant inequalities. This is an efficient way of removing redundancies, but this cannot remove

---

[2] For example, P is generated by $\{(1,2,3,4)^t, (2,3,4,5)^t, (2,3,7,9)^t\}$, the projection of P onto the last two coordinates is generated by $\{(3,4)^t, (4,5)^t, (7,9)^t\}$.

them all, thus Linear Programming (LP) is still needed. As pointed out in [31], raytracing is effective when there are not many redundancies; unfortunately, Fourier-Motzkin Elimination typically generates lots of redundancies.

Another modern library dealing with polyhedral sets computation is the Normaliz library [9]. In this library, Fourier-Motzkin Elimination is used for conversion between different descriptions of polyhedral sets. This is a different strategy than the one of our paper. As discussed in the introduction, we are motivated here by performing successive projections as required in the analysis, scheduling and transformation of for loop nests of computer programs.

References

1. Bagnara, R., Hill, P.M., Zaffanella, E., Bagnara, A.: The parma polyhedra library user's manual (2002)
2. Balas, E.: Projection with a minimal system of inequalities. Comput. Optim. Appl. **10**(2), 189–193 (1998)
3. Bastoul, C.: Code generation in the polyhedral model is easier than you think. In: Proceedings of the 13th International Conference on Parallel Architectures and Compilation Techniques, PACT 2004, pp. 7–16. IEEE Computer Society, Washington, DC (2004). https://doi.org/10.1109/PACT.2004.11
4. Benabderrahmane, M.-W., Pouchet, L.-N., Cohen, A., Bastoul, C.: The polyhedral model is more widely applicable than you think. In: Gupta, R. (ed.) CC 2010. LNCS, vol. 6011, pp. 283–303. Springer, Heidelberg (2010). https://doi.org/10.1007/978-3-642-11970-5_16
5. Blumofe, R.D., Joerg, C.F., Kuszmaul, B.C., Leiserson, C.E., Randall, K.H., Zhou, Y.: Cilk: an efficient multithreaded runtime system. SIGPLAN Not. **30**(8), 207–216 (1995). https://doi.org/10.1145/209937.209958
6. Bondhugula, U., Hartono, A., Ramanujam, J., Sadayappan, P.: A practical automatic polyhedral parallelizer and locality optimizer. SIGPLAN Not. **43**(6), 101–113 (2008). https://doi.org/10.1145/1379022.1375595
7. Boulme, S., Marechaly, A., Monniaux, D., Perin, M., Yu, H.: The verified polyhedron library: an overview, pp. 9–17 (2018)
8. Bruns, W., Ichim, B.: Normaliz: algorithms for affine monoids and rational cones. J. Algebra **324**(5), 1098–1113 (2010)
9. Bruns, W., Römer, T., Sieg, R., Söger, C.: Normaliz 3.0 (2008)
10. Chen, C., Covanov, S., Mansouri, F., Maza, M.M., Xie, N., Xie, Y.: The basic polynomial algebra subprograms. In: Hong, H., Yap, C. (eds.) ICMS 2014. LNCS, vol. 8592, pp. 669–676. Springer, Heidelberg (2014). https://doi.org/10.1007/978-3-662-44199-2_100
11. Chernikov, S.N.: Contraction of systems of linear inequalities. Dokl. Akad. Nauk SSSR **131**(3), 518–521 (1960)
12. Chernikova, N.V.: Algorithm for finding a general formula for the non-negative solutions of a system of linear inequalities. Zhurnal Vychislitel'noi Mat. Mat. Fiziki **5**(2), 334–337 (1965)
13. Christof, T., Löbel, A.: PORTA: polyhedron representation transformation algorithm, version 1.3. 2. Konrad-Zuse-Zentrum für Informationstechnik Berlin (2000)
14. Dantzig, G.B.: Fourier-Motzkin elimination and its dual. Technical report (1972)

15. Feautrier, P.: Dataflow analysis of array and scalar references. Int. J. Parallel Program. **20**, 23–53 (1991). http://citeseerx.ist.psu.edu/viewdoc/summary? doi=10.1.1.31.1342

16. Feautrier, P.: Automatic parallelization in the polytope model. In: Perrin, G.-R., Darte, A. (eds.) The Data Parallel Programming Model. LNCS, vol. 1132, pp. 79–103. Springer, Heidelberg (1996). https://doi.org/10.1007/3-540-61736-1_44

17. Fukuda, K.: The CDD and CDDplus homepage. https://www.inf.ethz.ch/personal/fukudak/cdd_home/

18. Fukuda, K., Prodon, A.: Double description method revisited. In: Deza, M., Euler, R., Manoussakis, I. (eds.) CCS 1995. LNCS, vol. 1120, pp. 91–111. Springer, Heidelberg (1996). https://doi.org/10.1007/3-540-61576-8_77

19. von zur Gathen, J., Gerhard, J.: Modern Computer Algebra, 2nd edn. Cambridge University Press, New York (2003)

20. Gawrilow, E., Joswig, M.: Polymake: a framework for analyzing convex polytopes. In: Kalai, G., Ziegler, G.M. (eds.) Polytopes-Combinatorics and Computation, pp. 43–73. Springer, Heidelberg (2000). https://doi.org/10.1007/978-3-0348-8438-9_2

21. Grosser, T., Zheng, H., Aloor, R., Simbürger, A., Größlinger, A., Pouchet, L.: Polly - polyhedral optimization in LLVM. In: First International Workshop on Polyhedral Compilation Techniques (IMPACT 2011), Chamonix, France, April 2011

22. Henk, M., Richter-Gebert, J., Ziegler, G.M.: 16 basic properties of convex polytopes. Handb. Discrete Comput. Geometry 255–382 (2004)

23. Huynh, T., Lassez, C., Lassez, J.L.: Practical issues on the projection of polyhedral sets. Ann. Math. Artif. Intell. **6**(4), 295–315 (1992)

24. Imbert, J.L.: Fourier's elimination: which to choose? In: PPCP, pp. 117–129 (1993)

25. Jing, R.-J., Moreno Maza, M.: Computing the integer points of a polyhedron, I: algorithm. In: Gerdt, V.P., Koepf, W., Seiler, W.M., Vorozhtsov, E.V. (eds.) CASC 2017. LNCS, vol. 10490, pp. 225–241. Springer, Cham (2017). https://doi.org/10.1007/978-3-319-66320-3_17

26. Jing, R.-J., Moreno Maza, M.: Computing the integer points of a polyhedron, II: complexity estimates. In: Gerdt, V.P., Koepf, W., Seiler, W.M., Vorozhtsov, E.V. (eds.) CASC 2017. LNCS, vol. 10490, pp. 242–256. Springer, Cham (2017). https://doi.org/10.1007/978-3-319-66320-3_18

27. Karmarkar, N.: A new polynomial-time algorithm for linear programming. In: Proceedings of the Sixteenth Annual ACM Symposium on Theory of Computing, STOC 1984, pp. 302–311. ACM, New York (1984). https://doi.org/10.1145/800057.808695

28. Khachiyan, L.: Fourier-Motzkin elimination method. In: Floudas, C.A., Pardalos, P.M. (eds.) Encyclopedia of Optimization, 2nd edn., pp. 1074–1077. Springer, Heidelberg (2009). https://doi.org/10.1007/978-0-387-74759-0_187

29. Kohler, D.A.: Projections of convex polyhedral sets. Technical report, California University at Berkeley, Operations Research Center (1967)

30. Le Verge, H.: A note on Chernikova's algorithm. Ph.D. thesis, INRIA (1992)

31. Maréchal, A., Périn, M.: Efficient elimination of redundancies in polyhedra by ray-tracing. In: Bouajjani, A., Monniaux, D. (eds.) VMCAI 2017. LNCS, vol. 10145, pp. 367–385. Springer, Cham (2017). https://doi.org/10.1007/978-3-319-52234-0_20

32. McMullen, P.: The maximum numbers of faces of a convex polytope. Mathematika **17**(2), 179–184 (1970)

33. Monniaux, D.: Quantifier elimination by lazy model enumeration. In: Touili, T., Cook, B., Jackson, P. (eds.) CAV 2010. LNCS, vol. 6174, pp. 585–599. Springer, Heidelberg (2010). https://doi.org/10.1007/978-3-642-14295-6_51

34. Schönhage, A., Strassen, V.: Schnelle multiplikation großer zahlen. Computing **7**(3–4), 281–292 (1971). https://doi.org/10.1007/BF02242355
35. Schrijver, A.: Theory of Linear and Integer Programming. Wiley, New York (1986)
36. Storjohann, A.: Algorithms for matrix canonical forms. Ph.D. thesis, Swiss Federal Institute of Technology Zurich (2000)
37. Terzer, M.: Large scale methods to enumerate extreme rays and elementary modes. Ph.D. thesis, ETH Zurich (2009)
38. Verdoolaege, S., Carlos Juega, J., Cohen, A., Ignacio Gómez, J., Tenllado, C., Catthoor, F.: Polyhedral parallel code generation for CUDA. TACO **9**(4), 54 (2013)

Progress Report on the Scala Algebra System

Raphaël Jolly[(✉)]

Databeans, Vélizy-Villacoublay, France
raphael.jolly@free.fr

Abstract. The new extension methods in Dotty (codename for Scala 3) allow us to solve a long-standing problem with type classes and coercion. It enables the realization of a new Scala adapter for the Java Algebra System (JAS). Our design is compared to the existing Scala DSL for the Rings computer algebra library, and shown to be interchangeable with it. The question whether Scala can be used for library development is raised, and it turns out to be adequately efficient, taking advantage of the new language inline feature for code specialization.

Keywords: Type classes · Implicit conversion · Operator overloading · Domain specific language

1 Introduction

In a previous incarnation of this work [2], we have shown that it is possible to use type classes in Scala to model categories as an alternative to f-bounded polymorphism, used in the Java Algebra System [3]. It has several benefits: it allows post-facto extensions [7], making it possible to reuse existing classes without wrappers. It also allows for generic numeric-symbolic implementations with unboxed primitive types for improved efficiency. There was however a problem with coercion and its interaction with type classes. In consequence, we had to devise a hybrid scheme: as type classes make it possible to work with values of any type, why not exercise them on f-bounded wrappers, which are coercion-friendly?

The downside of this approach was that we could not use it to implement a Scala DSL to existing libraries (namely JAS) like is currently possible with Jython or JRuby. For this, we had to wait for improvements in the Scala language itself, which are now beginning to emerge in Scala 3 (codenamed "Dotty").

2 Outline

The paper is organized as follows. Section 3 presents the new extension methods introduced in Dotty and their suitability for our purpose. In Sect. 4 we discuss the problematic interaction of type classes with coercion, and propose a solution.

© Springer Nature Switzerland AG 2020
F. Boulier et al. (Eds.): CASC 2020, LNCS 12291, pp. 307–315, 2020.
https://doi.org/10.1007/978-3-030-60026-6_17

In Sect. 5, a new Scala adapter for JAS is introduced, based on this solution. Section 6 reviews a similar implementation in the Rings computer algebra system. Lastly, Sect. 7 outlines the benefits of the language for library in addition to interface development.

3   New in Dotty: Extension Methods

There are two important features a computer algebra language must support to allow a proper mathematical notation: operator overloading and implicit conversion. In [2], we have investigated how to implement operator overloading with type classes. In Dotty, these are now enhanced to support extension methods [4], which allows to define infix operators, with their parameters on both sides:

```
trait Ring[T]:
  def (x: T) + (y: T): T
  def zero: T
```

This simple definition allows to endow Java's `BigInteger` with an arithmetic operator, defined as a method with a + symbol as its name, and also to provide the 0 of the ring:

```
type BigInteger = java.math.BigInteger

given BigInteger as Ring[BigInteger]:
  def (x: BigInteger) + (y: BigInteger) = x.add(y)
  def zero = java.math.BigInteger.valueOf(0)
```

For reference, in former Scala versions the typeclass definition was much more complicated:

```
trait Ring[T] {
  def plus(x: T, y: T): T
  def zero: T
}
object Ring {
  trait ExtraImplicits {
    implicit def infixRingOps[T: Ring](lhs: T): Ops[T] =
      new OpsImpl(lhs)
  }
  trait Ops[T] {
    def lhs: T
    def factory: Ring[T]
    def +(rhs: T) = factory.plus(lhs, rhs)
  }
  class OpsImpl[T: Ring](val lhs: T) extends Ops[T] {
    val factory = implicitly[Ring[T]]
  }
}
```

With this mechanism, we can add two `BigIntegers` with a symbolic + despite the fact that it is not present in the original class. This is how operator overloading is addressed in our design. There is however a problem when implicit conversion is added into the mix, as explained in the next section.

4   Type Classes and Implicit Conversion: A Difficult Marriage

A vast majority of computer languages support some form of implicit conversion (or "coercion"). For instance, Scala can automatically convert not only from `Int` to `Long`, but also to custom classes like `BigInt`, which makes it an extensible mechanism.

```
scala> 11 + 1
scala> 1 + 11
// res1: Long = 2

scala> BigInt(1) + 1
scala> 1 + BigInt(1)
// res3: scala.math.BigInt = 2
```

However, even with such simple example, one quickly runs into difficulties, as outlined below, regarding the `divideAndRemainder` operation. This is in the case of older Scala versions:

```
scala> BigInt(1) /% 1
res4: (scala.math.BigInt, scala.math.BigInt) = (1,0)
scala> 1 /% BigInt(1)
<console>:8: error: value /% is not a member of Int
           1 /% BigInt(1)
             ^
```

In Dotty, there is progress, as the error message gives us a hint on how to solve the problem.

```
scala> 1 /% BigInt(1)
       ^^^^
value /% is not a member of Int, but could be made
    available as an extension method.
```

Here, we should note that `BigInt`, which is part of the Scala standard library, is implemented following the "wrapper" approach, that is to say, it is an adapter to Java's `BigInteger`, with appropriate symbolic methods in front of the class' arithmetic methods.

This is in contrast with the typeclass approach, exposed in the previous section, which does not play well with coercion. We have to lift by hand the values to the desired ring, and we can only add two BigIntegers but not mix a simple Int. Likewise for polynomials: we must lift coefficients to the polynomial ring using a constructor (r in the example).

```
BigInteger(1) + 1 => BigInteger(1) + BigInteger(1)
x + 1            => x + r(1)
```

This is solved in Dotty, thanks to extension methods, but only for operations to the right. For operations to the left, we need to import the operators explicitly, as exposed in more detail in the next section. Alternatively, we could also restrict ourselves to a "big endian notation", that is, in descending monomial order from left to right, leaving "naked" coefficients only to the right of the expression. But this is not an ideal solution.

5   Scala Adapter for JAS

Let us now introduce our Scala adapter for JAS. We declare a class Ring as a subtype of the abstract Ring structure of the ScAS hierarchy. Then, we define our extension methods to call the non-symbolic methods of JAS. First we have the ring operations, and a compare method, inherited from the fact that it is an ordered ring. Then we have some ring specific methods isUnit, characteristic, and the 0 and the 1 of the ring. Lastly, there are some prettyprint methods to display the value as a String or in MathML.

```
class Ring[T <: RingElem[T] : RingFactory]
    extends scas.structure.ordered.Ring[T] {
  def factory = summon[RingFactory[T]]
  def (x: T) + (y: T) = x.sum(y)
  def (x: T) - (y: T) = x.subtract(y)
  def (x: T) * (y: T) = x.multiply(y)
  def compare(x: T, y: T) = x.compareTo(y)
  def (x: T).isUnit = x.isUnit
  def characteristic = factory.characteristic
  def zero = factory.getZERO()
  def one = factory.getONE()
  def (x: T).toCode(level: Level) = x.toString
  def (x: T).toMathML = ???
  def toMathML = ???
}
```

Here, we have a perfect instance of an object-functional implementation, with a subtyping relation in the type parameter of Ring. This is because, as mentionned previously, JAS uses f-bounded polymorphism (a mechanism by which a type is allowed to have itself as type parameter bound) and we have to take it into account. But we also have a RingFactory context bound, which is a functional programming concept. So we have both an object-oriented and a functional programming concept in the same program, which is quite an achievement of Scala.

Here is a use case for the JAS adapter, with the "Polypower" benchmark.

```
import jas.{ZZ, BigInteger, poly2scas, coef2poly,
    int2bigInt, bigInt2scas}


given r as GenPolynomialRing[BigInteger](ZZ,
    Array("x", "y", "z"), TermOrderByName.INVLEX)
val Array(one, x, y, z) = r.gens
val s = poly2scas(r)
import s.{+, *}

val p = 1 + x + y + z
val q = p \ 20
val q1 = q + 1
val q2 = q * q1
q2.length
// 12341
```

The import section contains all the required items for clarity, but it could be replaced with a wildcard import. Next, we define a typeclass instance for polynomials over the integers, with the variables and term order. On the next line, we get the generators into variables with suitable names. Our main result is that with the import of the typeclass' operators +, * it is possible for the design to work as expected. But our second result is that, contrarily to what is exposed in the documentation [4], it is not sufficient to have the typeclass instance in scope to make the extension methods available. This is however necessary when the coefficient is on the left, and we might still expect some improvement in this regard in a future version of the language.

Here is the equivalent Python code for comparison.

```
from jas import PolyRing, ZZ
# sparse polynomial powers

r = PolyRing( ZZ(), "(x, y, z)", PolyRing.lex );
# [one, x, y, z] = r.gens()

p = 1 + x + y + z;
q = p ** 20;
q1 = q + 1;
q2 = q * q1;
len(q2)
// 12341
```

The code is a bit more concise, but Scala does not look bad in comparison. In Python, we do not have to assign the generators because they are injected automatically in the scripting interface.

6   Related Work : Scala DSL for Rings

The Rings project [6] has opted for a similar, typeclass-based design with its Scala DSL interface. To address the coercion problem, as far as we can tell the retained solution looks as follows in the new typeclass syntax.

```
trait Ring[E]:
  def (x: E) + (y: Int): E
  def (x: E) + (y: E): E
  def (x: Int) + (y: E): E

trait IPolynomialRing[Poly <: IPolynomial[Poly], E]
    extends Ring[Poly]:
  def (x: Poly) + (y: E): Poly
  def (x: E) + (poly: Poly): Poly
```

Here, in addition to the + operator acting on the polynomial ring elements, two additional extension methods are provided to add a coefficient on both sides. Hence, there is no need to lift coefficients to the polynomial ring, and no implicit conversion is involved.

```
import cc.redberry.rings

import rings.poly.PolynomialMethods._
import rings.scaladsl._
import syntax._

implicit val ring = UnivariateRing(UnivariateRing(Z, "x"), "y")
val x = ring("x")
val y = ring("y")
ring.show(x+y)
// x+y
```

However, the drawback of this solution occurs when there are several levels of recursion, as in the case of nested polynomial rings. In that case, the design will fail, as shown in the code excerpt below. To mitigate this, all generators are first lifted to the topmost ring by means of parsing, which the scripting principle is especially meant to avoid, for type-safety.

```
implicit val r = UnivariateRing(Z, "x")
implicit val s = UnivariateRing(r, "y")
val x = r("x")
val y = s("y")
s.show(1+x)

javax.script.ScriptException:
    overloaded method value + with alternatives:
```

```
(x: Int)Int <and>
...
```

```
cannot be applied to (UnivariatePolynomial[BigInteger])
  in s.show(1+x)
```

Thanks to its ability to reuse existing classes, our own design can be adjusted to the Rings library with a similar adapter as to JAS, allowing to solve this nested polynomial problem. The adapter is shown below.

```
abstract class Ring[T] extends scas.structure.ordered.Ring[T]:
  def ring: cc.redberry.rings.Ring[T]
  def coder = Coder.mkCoder(ring)
  def (x: T) + (y: T) = ring.add(x, y)
  def (x: T) - (y: T) = ring.subtract(x, y)
  def (x: T) * (y: T) = ring.multiply(x, y)
  def compare(x: T, y: T) = ring.compare(x, y)
  def (x: T).isUnit = ring.isUnit(x)
  def characteristic = ring.characteristic
  def zero = ring.getZero()
  def one = ring.getOne()
  def (x: T).toCode(level: Level) = coder.stringify(x)
  def (x: T).toMathML = ???
  def toMathML = ???
```

7   A Language for Computer Algebra Libraries

The Rings as well as the JAS libraries both have to make some trade-offs to safeguard efficiency versus genericity. For instance, in JAS the exponent vectors are specialized by hand with Long exponents. In Rings, some liberties are taken with type safety, as a wildcard is used in the definition of `cc.redberry.rings.poly.MultivariateRing` and `asInstanceOf` in the definition of `cc.redberry.rings.scaladsl.Rings.MultivariateRing`. As far as we can tell, this is all to allow unboxed primitive polynomial term coefficients.

On the other hand, Scala offers some mechanisms for automatic code specialization. In the former versions, there was a @specialized annotation, but it is abandonned in Dotty. Fortunately, there is a replacement mechanism with the new Inline feature [5], which is exemplified in the code excerpt below.

```
abstract class MathLib[N : Numeric]:
  def dotProduct(xs: Array[N], ys: Array[N]): N

object MathLib:
  inline def apply[N : Numeric] = new MathLib[N]:
    def dotProduct(xs: Array[N], ys: Array[N]) =
      require(xs.length == ys.length)
      var i = 0
```

```
var s: N = Numeric[N].zero
while (i < xs.length)
  s = s + xs(i) * ys(i)
  i += 1
s
```

Here, the `inline` modifier in front of `apply` means that the code will be copied at every use site, replacing the type parameter by some concrete type. Below, an example is given that computes a dot product which is specialized in `Double`.

```
val mlib = MathLib[Double]

val xs = Array(1.0, 1.0)
val ys = Array(2.0, -3.0)
mlib.dotProduct(xs, ys)
// -1.0
```

We have used this principle to specialize a generic implementation of power product exponents in the Scala Algebra System [1]. To assess the gain in efficiency of inline specialized code, we have used the Polypower benchmark that we have seen previously. Computations were made on a Intel Atom x5-Z8300 processor at 1.44GHz with 2GB of RAM on Linux debian 4.19.0-8-amd64 and OpenJDK 64-Bit Server VM version "11.0.6". The result is shown in Fig. 1.

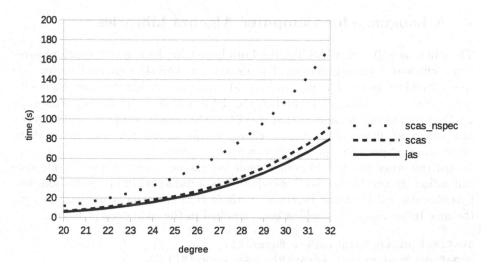

Fig. 1. Polypower benchmark. Execution times in seconds versus degree of polynomial multiplication with non-specialized and specialized generic exponent vectors, respectively, together with originally specialized JAS code for comparison. As we can see, there is considerable cost for boxing, which is removed by specialization. (In the legend, scas_nspec stands for non-specialized vectors, scas is for specialized vectors, and jas for the original JAS data)

8   Conclusion

The new extension methods in Dotty allowed us to solve a long-standing problem with type classes and coercion. It enabled the realization of a new Scala adapter for JAS. Our design compares favorably to the existing Scala DSL for Rings, as it allows arithmetic operations in nested polynomial rings without parsing. Thanks to type classes, it is even possible to use our interface as a front-end replacement for Rings. The question whether Scala could be used for library development is raised, and it turns out to be adequately efficient, taking advantage of the new language inline feature for code specialization.

Acknowledgments. The author would like to thank the anonymous referees for their useful suggestions.

References

1. Jolly, R.: ScAS - Scala Algebra System. Technical report (2010–2020). https://github.com/rjolly/scas
2. Jolly, R.: Categories as type classes in the scala algebra system. In: Gerdt, V.P., Koepf, W., Mayr, E.W., Vorozhtsov, E.V. (eds.) CASC 2013. LNCS, vol. 8136, pp. 209–218. Springer, Cham (2013). https://doi.org/10.1007/978-3-319-02297-0_18
3. Kredel, H.: Parametric solvable polynomial rings and applications. In: Gerdt, V.P., Koepf, W., Seiler, W.M., Vorozhtsov, E.V. (eds.) CASC 2015. LNCS, vol. 9301, pp. 275–291. Springer, Cham (2015). https://doi.org/10.1007/978-3-319-24021-3_21
4. Odersky, M.: Extension methods - Dotty documentation. Technical report (2019–2020). https://dotty.epfl.ch/docs/reference/contextual/extension-methods.html
5. Odersky, M.: Inline - Dotty documentation. Technical report (2019–2020). https://dotty.epfl.ch/docs/reference/metaprogramming/inline.html
6. Poslavsky, S.: Rings: an efficient JVM library for commutative algebra (*invited talk*). In: England, M., Koepf, W., Sadykov, T.M., Seiler, W.M., Vorozhtsov, E.V. (eds.) CASC 2019. LNCS, vol. 11661, pp. 1–11. Springer, Cham (2019). https://doi.org/10.1007/978-3-030-26831-2_1
7. Watt, S.: Post facto type extensions for mathematical programming. In: Proceedings Domain-Specific Aspect languages (SIGPLAN/SIGSOFT DSAL 2006). ACM (2006). http://www.csd.uwo.ca/~watt/home/courses/2006-07/cs888b/materials/2006-dsal-postfacto.pdf

Routh – Hurwitz Stability of a Polynomial Matrix Family. Real Perturbations

Elizaveta A. Kalinina[✉][iD], Yuri A. Smol'kin, and Alexei Yu. Uteshev[iD]

Faculty of Applied Mathematics, St. Petersburg State University,
7–9 Universitetskaya Nab., St. Petersburg, 199034 Saint Petersburg, Russia
{e.kalinina,a.uteshev}@spbu.ru, st040343@student.spbu.ru
http://www.apmath.spbu.ru/ru/

Abstract. The problem of Routh–Hurwitz stability of a polynomial matrix family is considered as that of discovering the structure of the stability domain in the parameter space. Algorithms for finding the spectral abscissa and the distance to instability from any internal point of the stability domain to its boundary for the case of real perturbations are proposed. The treatment is performed in the ideology of analytical algorithm for elimination of variables and localization of zeros of algebraic systems. Some examples are given.

Keywords: Matrix polynomials · Robust Routh – Hurwitz stability · Parameters · Frobenius norm

1 Introduction

Matrix $A \in \mathbb{R}^{n \times n}$ is called **stable** (**Routh – Hurwitz stable**) if all its eigenvalues are situated in the open left half plane of the complex plane. For a stable matrix A, some perturbation $E \in \mathbb{R}^{n \times n}$ may lead to that eigenvalues of $A + E$ cross the imaginary axis, i.e., to the loss of stability.

The smallest real perturbation E that makes $A + E$ unstable is called the **destabilizing real perturbation**. It is connected with the notion of the **distance to instability (stability radius) under real perturbations**.

Denote by $\Lambda(A)$ the spectrum of a matrix A. The distance of a stable matrix A to instability under real perturbations is formally defined as

$$\beta_{\mathbb{R}}(A) = \min\{||E|| \, | \, \eta(A + E) \geq 0, E \in \mathbb{R}^{n \times n}\}, \tag{1}$$

where $|| \cdot ||$ denotes some norm in $\mathbb{R}^{n \times n}$ while

$$\eta(A) = \max\{\Re(\lambda) \, | \, \lambda \in \Lambda(A)\} \tag{2}$$

is the **spectral abscissa** of the matrix A.

Supported by RFBR according to the project No 17-29-04288.

© Springer Nature Switzerland AG 2020
F. Boulier et al. (Eds.): CASC 2020, LNCS 12291, pp. 316–334, 2020.
https://doi.org/10.1007/978-3-030-60026-6_18

The problem of finding the distance to instability has a rich history. There are many different approaches and results concerning bounds and characterizations of complex and real stability radii [14]. The real stability radius problem is known as essentially more difficult than its complex counterpart [9]. We will tackle exactly this case: hereinafter, the notion "distance to instability" should be understood in the meaning "distance to instability under real perturbations". The next narrowing of the object of the present investigation relates to the choice of the norm in (1): we will treat this to be Frobenius norm (contrary to the complex case, the spectral and Frobenius norms define different real stability radii [6]). The Frobenius norm is considered to be more applicable to the problems arising in Control Theory, admitting natural extensions to infinite-dimensional systems [2]. However, it should be mentioned that while the 2-norm variant of the problem has been explored intensively [5,10,17], there are just a few studies [3,8,14] on the Frobenius norm counterpart. The treatment of the latter is considered as far more complex than the former due to the fundamental difference between the spectral and Frobenius norms. To confirm this statement, one can notice that the orders of the matrices from the examples presented in the cited papers (and in the works [1,18,20]) never exceed 4. The present paper should be viewed as the continuation of the two preceding papers on the Schur stability [11] and on the Routh – Hurwitz stability under complex perturbations [12].

The stated problem can be treated as a particular case of the general problem of finding the distance to instability in the parameter space for a matrix with entries polynomially depending on a parameter vector $\mu = (\mu_1, \mu_2, \ldots, \mu_k)$, $k \geq 1$. Indeed, one can pose the problem of finding the **domain of stability** in the parameter space $\mathbb{R}^k$, i.e., the domain

$$\mathbb{P} = \{\mu \in \mathbb{R}^k \mid A(\mu) \text{ is stable}\}. \tag{3}$$

If $\mathbb{P} \neq \emptyset$, the next problem is its estimation with the aid of simple domains like boxes or discs, or finding the distance to its boundary from a particular specialization $\mu_{[0]} \in \mathbb{R}^k$.

In Sect. 2, we outline the algebraic backgrounds of the suggested approach that is based on the resultant and discriminant computation using their representation in the form of the appropriate Hankel determinant. This toolkit provides an algebraic method for finding the spectral abscissa of a polynomial matrix family.

In Sect. 3, we detail the structure of the boundary of set (3). We also propose an algebraic approach to the problem of finding **the distance to instability** in the parameter space, i.e., the Euclidean distance $d_*(\mu_{[0]})$ from a given point $\mu_{[0]} \in \mathbb{R}^k$ corresponding to a stable matrix $A(\mu_{[0]})$ to the nearest point $\mu_* \in \mathbb{R}^k$ at the boundary of domain (3). Reduction of the problem to that one of the univariate algebraic equation solving allows one to utilize symbolic procedures for zero approximation within the desired accuracy [4,23]. In Sect. 4, we present a new algorithm for finding the distance to instability in the matrix space (or **stability radius**) and corresponding destabilizing perturbation for a matrix family $A(\mu)$ of the order 3.

In Sect. 5, some numerical examples are presented illuminating the efficiency of the suggested algorithms.

Remark. All the numerical computations were made in CAS Mathematica 12.0 with the help of functions `Discriminant` and `Resultant`. We present the results of the approximate computations with the 10^{-6} accuracy.

2   Algebraic Preliminaries

Here we give some auxiliary results regarding the properties of the zero sets of polynomials.

Given a polynomial

$$f(z) = a_0 z^n + a_1 z^{n-1} + a_2 z^{n-2} + \ldots + a_n \in \mathbb{C}[z], \ a_0 \neq 0, \ n \geq 2$$

with zeroes $\alpha_1, \alpha_2, \ldots, \alpha_n$ and a polynomial

$$g(z) = b_0 z^m + b_1 z^{m-1} + \cdots + b_{m-1} z + b_m \in \mathbb{C}[z], \ b_0 \neq 0$$

with zeros $\beta_1, \ldots, \beta_m$, the **resultant** of these polynomials is formally defined by the formula

$$\mathcal{R}(f, g) := a_0^m b_0^n \prod_{\ell=1}^{n} \prod_{j=1}^{m} (\alpha_\ell - \beta_j), \tag{4}$$

while practically it can be expressed as a polynomial in the coefficients of $f(z)$ and $g(z)$ using several determinantal representations. Here the following one [15] is used:

Theorem 1 (Kronecker). *Let* $\deg f > \deg g$. *Expand* $g(z)/f(z)$ *in Laurent series in powers of* z^{-1}:

$$\frac{g(z)}{f(z)} = \frac{c_0}{z} + \frac{c_1}{z^2} + \cdots + \frac{c_j}{z^{j+1}} + \cdots \tag{5}$$

and compose the Hankel matrix

$$C = [c_{j+k}]_{j,k=0}^{n-1} = \begin{bmatrix} c_0 & c_1 & c_2 & \cdots & c_{n-1} \\ c_1 & c_2 & c_3 & \cdots & c_n \\ c_2 & c_3 & c_4 & \cdots & c_{n+1} \\ \cdots & & & & \cdots \\ c_{n-1} & c_n & c_{n+1} & \cdots & c_{2n-2} \end{bmatrix}. \tag{6}$$

Denote by C_j *its* jth *leading principal minor. One has*

$$\mathcal{R}(f, g) = a_0^{n+m} C_n .$$

Polynomials $f(z)$ and $g(z)$ possess a unique common zero iff $C_n = 0, C_{n-1} \neq 0$. This zero can be expressed by the formula

$$\lambda = -\frac{a_1}{a_0} - \frac{1}{C_{n-1}} \begin{vmatrix} c_0 & c_1 & c_2 & \cdots & c_{n-2} \\ c_1 & c_2 & c_3 & \cdots & c_{n-1} \\ c_2 & c_3 & c_4 & \cdots & c_n \\ \cdots & & & & \cdots \\ c_{n-3} & c_n & c_{n-2} & \cdots & c_{2n-5} \\ c_{n-1} & c_n & c_{n+1} & \cdots & c_{2n-3} \end{vmatrix}. \tag{7}$$

For the particular case $g(z) \equiv f'(z)$, the coefficients of expansion (5) are known as the Newton sums of the polynomial $f(z)$. They can be computed recursively by the formulas

$$s_k = \begin{cases} -(a_1 s_{k-1} + a_2 s_{k-2} + \cdots + a_{k-1} s_1 + a_k k)/a_0, & \text{if } k \leq n; \\ -(a_1 s_{k-1} + a_2 s_{k-2} + \cdots + a_n s_{k-n})/a_0, & \text{if } k > n. \end{cases}$$

Matrix (6) is transformed into the matrix

$$S = [s_{j+k}]_{j,k=0}^{n-1} \tag{8}$$

that relates to the **discriminant** of the polynomial $f(z)$, i.e., to

$$\mathcal{D}(f) := a_0^{2n-2} \prod_{1 \leq j < k \leq n} (\alpha_k - \alpha_j)^2.$$

Corollary 1. *Denote by S_j the jth leading principal minor of matrix (8). Then one has*

$$\mathcal{D}(f) = a_0^{2n-2} S_n .$$

The polynomial $f(z)$ possesses a single multiple zero of the multiplicity 2 iff $S_n = 0, S_{n-1} \neq 0$. This zero can be represented as a rational function of the coefficients of $f(z)$ via formula (7) where substitution $\{c_j := s_j\}_{j=0}^{2n-3}$ is made.

Remark. In some further formulas involving resultant and discriminant of the multivariate polynomials, we will occasionally specify with subscripts the variable of the considered polynomials like $\mathcal{R}_z$ or $\mathcal{D}_z$.

The resultant can effectively solve several problems connected with computation of symmetric functions of the zeros of a polynomial like the one in the following result [13].

For the polynomial $f(z) \in \mathbb{R}[z]$, find the real and imaginary part of $f(x+iy)$ ($\{x, y\} \subset \mathbb{R}$):

$$f(z) = f(x+iy) = \Phi(x, y^2) + iy\Psi(x, y^2),$$

where

$$\Phi(x, Y) = f(x) - \frac{1}{2!}f''(x)Y + \frac{1}{4!}f^{(4)}(x)Y^2 - \cdots,$$

$$\Psi(x, Y) = f'(x) - \frac{1}{3!}f^{(3)}(x)Y + \frac{1}{5!}f^{(5)}(x)Y^2 - \cdots.$$

Theorem 2. *Polynomial*

$$\mathcal{X}(x) = \mathcal{R}_Y(\Phi(x, Y), \Psi(x, Y))$$

is of the degree $N = n(n-1)/2$ and possesses the set of zeros coinciding with

$$\left\{ \frac{1}{2}(\alpha_j + \alpha_k) \,\middle|\, 1 \le j < k \le n \right\}. \tag{9}$$

The leading coefficient of $\mathcal{X}(x)$ equals $(-1)^{\lfloor n/4 \rfloor} 2^N a_0^{n-1}$, while its free term equals

$$K(f) = \mathcal{R}_Y(\Phi(0, Y), \Psi(0, Y))$$
$$= \mathcal{R}_Y(a_n - a_{n-2}Y + a_{n-4}Y^2 + \dots, a_{n-1} - a_{n-3}Y + a_{n-5}Y^2 + \dots). \tag{10}$$

For a polynomial $f(z)$, the following criterion of stability is valid [19]:

Theorem 3 (Routh). *Polynomial $f(z)$ is stable iff*
(a) all the coefficients $a_0, \dots, a_n$ are of the same sign;
(b) all the coefficients of $\mathcal{X}(x)$ are of the same sign.

It is evident that set (9) contains the real parts of the complex-conjugate pairs of the zeros of the polynomial $f(z)$ (if any). Therefore, one can apply Theorem 2 for finding spectral abscissa (2) of a stable matrix A.

Theorem 4. *Let $f(z) = \det(zI - A)$. The value $\eta(A)$ equals the maximum of the real zeros of the polynomials $f(z)$ and $\mathcal{X}(x)$.*

Example 1. Find the spectral abscissa for the matrix

$$A(\mu) = \begin{bmatrix} -1 & -\mu^2 & -1 \\ \mu & -\mu-1 & \mu \\ \mu^2 & 1 & -\mu^2-1 \end{bmatrix}. \tag{11}$$

Solution. We have

$$f(z; \mu) = z^3 + (\mu^2 + \mu + 3)z^2 + (2\mu^3 + 3\mu^2 + \mu + 3)z + 2\mu^5 + 3\mu^3 + 2\mu^2 + \mu + 1;$$
$$\mathcal{X}(x; \mu) = -8x^3 - (8\mu^2 + 8\mu + 24)x^2 - (2\mu^4 + 8\mu^3 + 20\mu^2 + 14\mu + 24)x$$
$$- (5\mu^4 + 7\mu^3 + 11\mu^2 + 5\mu + 8).$$

Domain of stability (3) in the parameter line is defined as $\mathbb{P} = (\mu^{(0)}, +\infty)$, where $\mu^{(0)}$ stands for the zero of $f(0, \mu) = 0$, namely $\mu^{(0)} \approx -0.682328$. For these values of parameters, each of the equations $f(z, \mu) = 0$ and $\mathcal{X}(x; \mu) = 0$ has a single real zero, the corresponding branches of these implicit functions compose the plot of the spectral abscissa $\eta(A)$ as the function of μ (Fig. 1).   □

Among the conditions of stability given in Theorem 3, those imposed on the free terms of the polynomials $f(z)$ and $\mathcal{X}(x)$ are somehow the mostly critical for keeping the stability property. This statement can be clarified when treating the problem of stability of matrix families. Let $A(\mu)$ be a matrix with the entries polynomially depending on a parameter vector $\mu = (\mu_1, \dots, \mu_k)$; then its characteristic polynomial $f(z, \mu) = \det(zI - A(\mu))$ is a polynomial in z and μ.

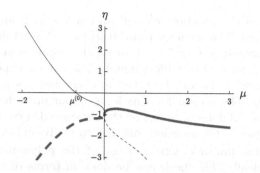

Fig. 1. Example 1. Plot of $\eta(\mu)$; branch of $f(z, \mu) = 0$ (thin blue), branch of $\mathcal{X}(x; \mu) = 0$ (red boldface) (Color figure online)

Theorem 5. *If $A(\mu)$ is stable for at least one specialization of the parameter vector, then the equations*

$$a_n(\mu) = 0 \text{ and } K(f; \mu) = 0 \tag{12}$$

define implicit manifolds in $\mathbb{R}^k$ that form the boundary for stability domain (3) in the parameter space.

The domain of stability $\mathbb{P}$ can be estimated via testing the inclusions of simpler subdomains like, for instance, the box:

$$\left\{ \mu_1^- \leq \mu_1 \leq \mu_1^+, \mu_2^- \leq \mu_2 \leq \mu_2^+, \ldots, \mu_k^- \leq \mu_k \leq \mu_k^+ \right\}.$$

In [12], this approach is tackled via reduction to that of localization of real zeros of a univariate polynomial. Other types of approximations can be done with the aid of balls inscribed in $\mathbb{P}$: if $\mu_{[0]} \in \mathbb{P}$, then the distance to instability in the parameter space gives the maximal radius for this ball. Since this distance is that from the point to algebraic manifolds (12), one can try to investigate this problem using algebraic approach.

3   Distance to Instability in the Parameter Space

The problem of finding the Euclidean distance from a point X_0 to an algebraic manifold defined implicitly by the equation

$$G(X) = 0 \tag{13}$$

in $\mathbb{R}^k, k \in \{2, 3\}, \deg G > 1$ can evidently be reduced to the constrained optimization problem for the objective function $\|X - X_0\|^2$. The critical values of this function are the zeroes of the univariate algebraic equation that is further referred to as the **distance equation** [11, 21].

First consider the case of $\mathbb{R}^2$. Let $X_0 = (0, 0)$ and the polynomial $G(x, y)$ be even in y and such that $G(0, 0) \neq 0$. Denote $\widetilde{G}(x, y^2) \equiv G(x, y)$. Let

$$\mathcal{F}(z) := \mathcal{D}_x(\widetilde{G}(x, z - x^2)) = 0. \tag{14}$$

Any critical value of the function $x^2 + y^2$ on curve (13) is among the positive zeros of this equation. The distance from $(0,0)$ to (13) is attained either on the real zero of the polynomial $G(x,0) = 0$ or on the square root of some positive zero of distance equation (14). Unfortunately, one cannot expect that only the minimal positive zero z_* should be taken into account (as it was erroneously claimed in [11,21]). The reason for this is an opportunity for this zero to be generated by the pair of nonreal points on the considered curve $(\alpha_*, \pm \imath \beta_*)$, where $\{\alpha_*, \beta_*\} \subset \mathbb{R}$. To block this occasion, one has to verify an additional condition $z_* - x_*^2 > 0$ for the multiple zero $x = x_*$ of the polynomial $\widetilde{G}(x, z_* - x^2)$. Fortunately, generically this check can be done in terms of the minors of the determinantal representation of the discriminant, namely using Corollary 1.

With this in mind, one can generalize the suggested approach first to the case of arbitrary polynomial $G(x,y)$, not necessarily even in any of its variables. Indeed, the problem can be *evenized*. Let us take in (14)

$$\widetilde{G}(x, y^2) := G(x,y)G(x,-y) \equiv G_1^2(x, y^2) - y^2 G_2^2(x, y^2)$$

where the polynomials $G_1(x, y^2)$ and $G_2(x, y^2)$ are the even and odd terms in y of the expansion

$$G(x,y) \equiv G_1(x, y^2) + y G_2(x, y^2), \quad \{G_1, G_2\} \subset \mathbb{R}[x, y^2].$$

This trick results in some extraneous factors in the expression for $\mathcal{F}(z)$:

$$\mathcal{F}(z) \equiv \mathcal{F}_1(z)\mathcal{F}_2^2(z) \text{ with } \mathcal{F}_2(z) := \mathcal{R}_x(G_1(x, z - x^2), G_2(x, z - x^2)).$$

The true distance equation is then $\mathcal{F}_1(z) = 0$.

Example 2. For the matrix

$$A(\mu_1, \mu_2) = \begin{bmatrix} -3\mu_1 - 6 & -\mu_1 + 3\mu_2 - 1 & -4\mu_1 + 1 \\ -3\mu_1 + 3\mu_2 + 4 & 3\mu_1 - 6 & -5\mu_1 + 5\mu_2 + 1 \\ -4\mu_1 + 2\mu_2 - 2 & -\mu_1 - 2\mu_2 - 2 & 2\mu_1 - 2\mu_2 - 7 \end{bmatrix},$$

find the distance to instability in the parameter plane from $(\mu_1, \mu_2) = (0,0)$.

Solution. Curves (12) are as follows

$$\det A = -89\mu_1^3 + 165\mu_2\mu_1^2 + 196\mu_1^2 - 148\mu_2^2\mu_1 - 49\mu_2\mu_1 + 181\mu_1 + 48\mu_2^3$$
$$-19\mu_2^2 - 89\mu_2 - 310;$$
$$K(f, \mu) = -23\mu_1^3 + 69\mu_2\mu_1^2 - 359\mu_1^2 - 120\mu_2^2\mu_1 + 114\mu_2\mu_1 - 759\mu_1 + 50\mu_2^3$$
$$+50\mu_2^2 + 642\mu_2 + 2122.$$

For the curve $\det A(\mu_1, \mu_2) = 0$ and the point $(\mu_1, \mu_2) = (0,0)$, the distance equation is as follows:

$$
\begin{aligned}
\mathcal{F}_1(z) = {} & 10624349317642400000z^9 - 170280038894384762568 0z^8 \\
& + 88798497635238525972984z^7 - 168284332643351232907352 1z^6 \\
& + 1361969711002693284493533 7z^5 - 69025641522462473951213248z^4 \\
& + 18505011930218286551599186 8z^3 - 32270437413033755135495711 2z^2 \\
& + 3531997943049534307786114 08z - 1635749217256831992076424 00 = 0.
\end{aligned}
$$

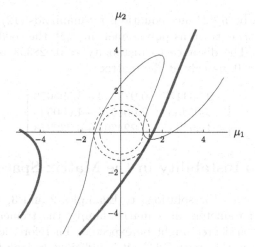

Fig. 2. Example 2. Boundaries for the stability domain: $\det A = 0$ (thin blue) and $K(f, \mu) = 0$ (red boldface) (Color figure online)

The distance equals ≈ 1.067503.

Analogously, the distance equation for the curve $K(f, \mu) = 0$ is

$$\mathcal{F}_1(z) = 4078209762893700000z^9 - \cdots - 301912001786643991081146553533316 = 0$$

and the distance equals ≈ 1.378284 (Fig. 2).   □

Distance evaluation for an arbitrary point (x_0, y_0) to curve (13) can be reduced to the case just considered via linear substitution.

The treatment of the problem for the manifolds in $\mathbb{R}^3$ is carried out in a similar manner with the aid of the notion of the discriminant of a bivariate polynomial $f(x, y)$. Using the result [16,22], one can utilize the following iterative formula

$$\mathcal{D}_{x,y}(f(x, y)) := \gcd(\mathcal{D}_x(\mathcal{D}_y(f(x, y))), \mathcal{D}_y(\mathcal{D}_x(f(x, y))))$$

for its computation. On computing every internal discriminant in the right-hand side, one should get rid of extraneous square factors.

Then for manifold (13) where $G(x_1, x_2, x_3)$ is an even polynomial in x_3, the distance equation can be found as

$$\mathcal{F}(z) := \mathcal{D}_{x_1, x_2}(\widetilde{G}(x_1, x_2, z - x_1^2 - x_2^2)) = 0 \text{ where } \widetilde{G}(x_1, x_2, x_3^2) \equiv G(x_1, x_2, x_3).$$

Extension of the result to the case of arbitrary polynomial G is carried out in a manner similar to that utilized for the bivariate case.

Example 3. For the matrix

$$A = \begin{bmatrix} -3\mu_1 + 2\mu_3 - 6 & -\mu_1 + 3\mu_2 - 2\mu_3 - 1 & -4\mu_1 + 3\mu_3 + 1 \\ -3\mu_1 + 3\mu_2 + 4 & 3\mu_1 - 2\mu_3 - 6 & -5\mu_1 + 5\mu_2 + 2\mu_3 + 1 \\ -4\mu_1 + 2\mu_2 - 2\mu_3 - 2 & -\mu_1 - 2\mu_2 + \mu_3 - 2 & 2\mu_1 - 2\mu_2 + 2\mu_3 - 7 \end{bmatrix},$$

find the distance to instability in the parameter space from $(\mu_1, \mu_2, \mu_3) = (0, 0, 0)$.

Solution. Here both distance equations for manifolds (12) are of the degree 21, and on representing them as polynomials in $\mathbb{Z}[z]$, the coefficients are of the orders up to 10^{92}. The distance to instability ≈ 0.928353 is achieved in the manifold $\det A(\mu) = 0$, namely at the matrix

$$\approx \begin{bmatrix} -9.334477 & -0.972349 & -3.595076 \\ 1.506453 & -2.665523 & -4.050548 \\ -4.394287 & -3.224794 & -6.232273 \end{bmatrix}.$$

4   Distance to Instability in the Matrix Space

As one can notice from the solutions to Examples 2 and 3, with the increase of the number of parameters in a matrix family, the tremendous increase of the distance equation degree might be expected. Therefore, for the distance to instability problem in the space $\mathbb{R}^{n \times n}$, it is judicious to look for an alternative approach that reduces somehow the number of parameters involved. Indeed, this can be done via reducing the problem to that of finding the destabilizing perturbation. It turns out that this matrix E_* should be of a low rank, namely either 1 or 2. We recall first the relative result [24].

Theorem 6. *Suppose $A \in \mathbb{R}^{n \times n}$ is stable. The distance to instability $\beta_{\mathbb{R}}(A)$ equals the minimal of the two values: either*

$$\sigma_{\min}(A) \tag{15}$$

or

$$\sqrt{\min_{\substack{\{X,Y\} \in \mathbb{R}^n, \|X\|=1, \|Y\|=1, X^T Y=0 \\ (X^T AY)(Y^T AX) \leq 0}} F(X,Y)}, \tag{16}$$

where $\sigma_{\min}(A)$ stands for the smallest singular value of the matrix A,

$$F(X,Y) = \|AX\|^2 + \|AY\|^2 - (X^T AY)^2 - (Y^T AX)^2$$

and all vector norms here are 2-norms.

This result can be interpreted in the ideology of Sect. 3. We precede this interpretation with the following statement known as the Principle of the Irrelevance of Algebraic Inequalities:

Theorem 7 (H. Weyl[25]). *Let $\mathfrak{R}$ be an infinite integral domain with n independent indeterminates $x_1, \ldots, x_n$. Let $\mathbf{P}$ and $\mathbf{Q}$ be polynomials in $\mathfrak{R}[x_1, \ldots x_n]$ such that if $\mathbf{P}(\eta_1, \ldots, \eta_n) \neq 0$, for some η_i in $\mathfrak{R}$, then $\mathbf{Q}(\eta_1, \ldots, \eta_n) = 0$. Then $\mathbf{Q} \equiv 0$.*

Theorem 8. *For the nonsingular matrix A and for the manifold $\det B = 0$ in $\mathbb{R}^{n \times n}$, the distance to instability equation coincides with the characteristic equation*

$$\det(zI - A^T A) = 0.$$

Proof. Let V be an eigenvector of the matrix $A^T A$ corresponding to its eigenvalue σ^2 with $\sigma > 0$:

$$A^T A V = \sigma^2 V. \tag{17}$$

We first prove that the matrix

$$E = -\frac{A V V^T}{V^T V} \tag{18}$$

satisfies the conditions $\det(A + E) = 0$. Indeed, by (18) one has:

$$[\det(A + E)]^2 = \det(A + E)^T (A + E)$$

$$= \det\left(A^T A - \frac{A^T A V V^T}{V^T V} - \frac{V V^T A^T A}{V^T V} + \frac{V V^T A^T A V V^T}{(V^T V)^2} \right)$$

and, due to (17),

$$= \det\left(A^T A - 2\sigma^2 \frac{V V^T}{V^T V} + \sigma^2 \frac{V V^T V V^T}{(V^T V)^2} \right) = \det\left(A^T A - \sigma^2 \frac{V V^T}{V^T V} \right).$$

Since the column $V/\sqrt{V^T V}$ is an eigenvector of the symmetric matrix $A^T A$ corresponding to the eigenvalue σ^2, the last determinant vanishes.

In a similar way, the relation $\|E\| = \sigma$ can be established.

Finally, let us prove that matrix (18) is orthogonal to the manifold $\det B = 0$ at the "point" $B = A + E$. For simplicity, we will assume $\|V\| = 1$ in the rest of the proof. Let us compute the gradient of $\det B$:

$$\nabla(\det B) = [B_{11}, B_{12}, \ldots, B_{1n}, B_{21}, B_{22}, \ldots, B_{nn}]$$

where B_{jk} stands for the algebraic complement to the entry b_{jk} in $\det B$. This row is just a vectorization of the matrix $[\mathrm{adj}(B)]^T$ where $\mathrm{adj}(B)$ stands for the adjugate matrix of B. Let us find the expression for $\mathrm{adj}(B)$ for the choice $B = A + E$, i.e., for $\det B = 0$. For this aim, the Sherman – Morrison formula can be utilized:

$$(A + u v^T)^{-1} = A^{-1} - \frac{1}{1 + v^T A^{-1} u} A^{-1} u v^T A^{-1}, \tag{19}$$

Here $\{u, v\} \in \mathbb{R}^n$ are column vectors and it is assumed that

$$1 + v^T A^{-1} u \neq 0. \tag{20}$$

Note that

$$\det(A + u v^T) = (1 + v^T A^{-1} u) \det A.$$

The last equality is just a consequence of the Schur complement formula for the determinant of the order $(n + 1)$:

$$-\det(A + u v^T) = \begin{vmatrix} A & u \\ v^T & -1 \end{vmatrix} = (-1 - v^T A^{-1} u) \det A.$$

Under condition (20), formula (19) is equivalent to the equality

$$\text{adj}(A + uv^T) = \det A \left\{ (1 + v^T A^{-1} u) A^{-1} + A^{-1} u v^T A^{-1} \right\}. \qquad (21)$$

This equality can be treated as an algebraic identity with respect to the entries of the matrix A and the columns u, v. Being valid under restriction (20) that is also algebraic w.r.t. the just mentioned entries, it should be also valid for the case $1 + v^T A^{-1} u = 0$. The last claim is a consequence of Theorem 7.

Therefore, for the matrix $E = -AVV^T$, formula (21) yields the relationship that is sufficient for the collinearity of E and $\nabla(\det B)$ computed at $B = A + E$:

$$E = \tau \left\{ (-\det A)(VV^T A^{-1})^T \right\} \quad \text{for } \tau := \frac{\sigma^2}{\det A}.$$

This is equivalent to the equality

$$AVV^T = \sigma^2 (A^{-1})^T VV^T$$

that is valid due to (17).

$\square$

Formula (18) yields the destabilizing perturbation E_* for the case $\beta_{\mathbb{R}}(A) = \sigma_{\min}(A)$; this matrix has rank 1. If $\beta_{\mathbb{R}}(A)$ equals the value (16) that is attained at X_* and Y_*, then

$$E_* = (aX_* - AY_*)Y_*^T + (bY_* - AX_*)X_*^T \quad \text{where } a := X_*^T AY_*, b := Y_*^T AX_*. \qquad (22)$$

It is known [7] that matrix (22) has rank 2. We can now recognize its nonzero eigenvalues.

Theorem 9. *If $a \neq -b$, then matrix (22) has a unique nonzero eigenvalue $\lambda = -X_*^T AX_* = -Y_*^T AY_*$ of the multiplicity 2.*

Proof. Evidently, the vectors X_* and Y_* are the eigenvectors of the matrix E_*^T:

$$X_*^T E_* = X_*^T (aX_* - AY_*)Y_*^T + X_*^T (bY_* - AX_*)X_*^T$$

$$= aY_*^T - (X_*^T AY_*)Y_*^T - (X_*^T AX_*)X_*^T = -(X_*^T AX_*)X_*^T.$$

Similarly, one has: $Y_*^T E_* = -(Y_*^T AY_*)Y_*^T$.

To prove that the values $X_*^T AX_*$ and $Y_*^T AY_*$ are identical, let us utilize the Lagrange multiplier method for the constrained minimization problem involved in (16). According to Theorem 6, the Lagrangian function is given by

$$F_1(X, Y, p, q, r) = X^T A^T AX + Y^T A^T AY - (X^T AY)^2 - (Y^T AX)^2$$
$$+ p(X^T X - 1) + q(Y^T Y - 1) + r(X^T Y),$$

where p, q, and r are the Lagrange multipliers.

Differentiating F_1 w.r.t. X and Y and setting to zero vector $\mathbb{O} \in \mathbb{R}^n$, we get

$$2A^T AX - 2(X^T AY)AY - 2(Y^T AX)A^T Y + 2pX + rY = \mathbb{O}, \qquad (23)$$
$$2A^T AY - 2(Y^T AX)AX - 2(X^T AY)A^T X + 2qY + rX = \mathbb{O}. \qquad (24)$$

Pre-multiplying (23) by Y^T and equation (24) by X^T, we get

$$2Y^T A^T AX - 2(X^T AY)(Y^T AY) - 2(Y^T AX)(Y^T AY) + r = 0,$$
$$2X^T A^T AY - 2(Y^T AX)(X^T AX) - 2(X^T AY)(X^T AX) + r = 0,$$

whence it follows that

$$(X^T AY + Y^T AX)(Y^T AY - X^T AX) = 0.$$

$\square$

Theorem 6 reduces essentially the number of variables in the distance to instability evaluation problem (roughly, from n^2 to $2n$). However, the new constrained optimization problem is still complicated in treatment even for the low order matrices. Below we suggest a new algorithm for finding the distance to instability for a polynomial matrix of the order 3.

In what follows, we will consider the most general case $a \neq -b$, that is, $X_*^T AY_* \neq -Y_*^T AX_*$.

Lemma 1. *Let the nonzero vectors X, Y be orthogonal and $X^T AY \neq -Y^T AX$. Then there exists $\varphi \in [0, 2\pi)$ such that the vectors*

$$X' = X \cos\varphi - Y \sin\varphi, \; Y' = X \sin\varphi + Y \cos\varphi \qquad (25)$$

satisfy the condition

$$X'^T AX' = Y'^T AY' = \frac{1}{2}\left(X^T AX + Y^T AY\right). \qquad (26)$$

Proof. Condition $X'^T AX' = Y'^T AY'$ will be satisfied iff

$$X^T AX \cos^2\varphi - X^T AY \cos\varphi \sin\varphi - Y^T AX \cos\varphi \sin\varphi + Y^T AY \sin^2\varphi$$

$$= X^T AX \sin^2\varphi + X^T AY \sin\varphi \cos\varphi + Y^T AX \sin\varphi \cos\varphi + Y^T AY \cos^2\varphi.$$

This yields the following relation for φ:

$$\tan 2\varphi = \frac{X^T AX - Y^T AY}{X^T AY + Y^T AX}. \qquad (27)$$

For this value, one has:

$$X'^T AX' = X^T AX \cos^2\varphi + Y^T AY \sin^2\varphi - \frac{1}{2}(X^T AY + Y^T AX)\sin 2\varphi$$

$$= X^T AX \cos^2\varphi + Y^T AY \sin^2\varphi - \frac{1}{2}(X^T AX - Y^T AY)\cos 2\varphi$$

$$= \frac{1}{2}(X^T AX + Y^T AY).$$

$\square$

It is evident that four vectors X, Y, AX, and AY are linearly dependent.

Lemma 2. *Suppose that*

$$AX = \alpha_1 X + \beta_1 Y + U, \quad AY = \alpha_2 X + \beta_2 Y + cU, \tag{28}$$

for some vector U orthogonal to the plane $\mathrm{span}(X, Y)$*, some scalars α_1, α_2, β_1, β_2, c and*

$$\|X\| = \|Y\| = 1, U^T X = U^T Y = X^T Y = 0. \tag{29}$$

Then for any rotation of the plane $\mathrm{span}(X, Y)$ *around the origin, the sum of squared norms of vector rejections of AX and AY from the plane* $\mathrm{span}(X, Y)$ *equals* $(1 + c^2)\|U\|^2$.

Proof. Rotation through an angle φ maps vectors X and Y to vectors (25). Then we get

$$AX' = (\alpha_1 \cos\varphi - \alpha_2 \sin\varphi)X + (\beta_1 \cos\varphi - \beta_2 \sin\varphi)Y + (\cos\varphi - c\sin\varphi)U,$$
$$AY' = (\alpha_1 \sin\varphi + \alpha_2 \cos\varphi)X + (\beta_1 \sin\varphi + \beta_2 \cos\varphi)Y + (\sin\varphi + c\cos\varphi)U.$$

Hence, the vector rejections of AX' and AY' from the plane $\mathrm{span}(X, Y)$ are $(\cos\varphi - c\sin\varphi)U$ and $(\sin\varphi + c\cos\varphi)U$.

$\square$

Lemma 3. *For vectors X, Y, U satisfying conditions (28) and (29), we have*

$$F(X, Y) = (X^T AX)^2 + (Y^T AY)^2 + (1 + c^2)\|U\|^2.$$

Proof. By direct calculation.

$\square$

Now we can prove the main result.

Theorem 10. *Let $A \in \mathbb{R}^{3 \times 3}$ be stable. The distance to instability $\beta_{\mathbb{R}}(A)$ equals the minimal of the two values: either (15) or*

$$\sqrt{\min_{Z \in \mathbb{R}^3} H(Z)} \ \text{where} \ H(Z) := \frac{F(Z)}{2G^2(Z)} \tag{30}$$

with

$$F(Z) := (Z^T Z)\mathfrak{G}(Z, AZ, A^2 Z) + \mathfrak{G}(Z, AZ)\mathfrak{G}(Z, A^2 Z), \tag{31}$$
$$G(Z) = \mathfrak{G}(Z, AZ). \tag{32}$$

Here $\mathfrak{G}(Z_1, Z_2, \ldots)$ stands for the Gram determinant of the columns $\{Z_1, Z_2, \ldots\} \subset \mathbb{R}^3$ *(i.e., $\mathfrak{G}(Z_1, Z_2, \ldots) = \det([Z_1, Z_2, \ldots]^T [Z_1, Z_2, \ldots]))$.*

Proof. We want to find vectors X_* and Y_* in $\mathbb{R}^3$ minimizing the function $F(X,Y)$ from Theorem 6. Since vectors X_*, Y_*, AX_* and AY_* are linearly dependent, it follows that the planes $\mathrm{span}(X_*, Y_*)$ and $\mathrm{span}(AX_*, AY_*)$ have nonzero intersection and there exists a vector Z in the plane $\mathrm{span}(X_*, Y_*)$ such that the vector AZ belongs to the same plane. Let us take this vector and, with the help of the Gram–Schmidt process, form an orthonormal basis of the plane $\mathrm{span}(Z, AZ)$:

$$X' = \frac{1}{\sqrt{Z^T Z - \frac{(Z^T AZ)^2}{Z^T A^T AZ}}} \left(Z - \frac{Z^T AZ}{Z^T A^T AZ} AZ \right) \text{ and } Y' = \frac{AZ}{\sqrt{Z^T A^T AZ}} \tag{33}$$

Since coordinates of a vector w.r.t. an orthonormal basis can be found as inner products, we have

$$\begin{cases} AX' = (X'^T AX')X' + (Y'^T AX')Y' + X'_\perp, \\ AY' = (X'^T AY')X' + (Y'^T AY')Y' + Y'_\perp, \end{cases} \tag{34}$$

where the vectors $X'_\perp$ and $Y'_\perp$ are orthogonal to the plane $\mathrm{span}(X', Y') = \mathrm{span}(Z, AZ)$. Now, by formulas (33), we obtain

$$\begin{cases} X'^T AX' = \dfrac{1}{\Delta(z)} \left(\dfrac{(Z^T AZ)^2}{Z^T A^T AZ} Z^T A^T A^2 Z - (Z^T AZ)(Z^T A^2 Z) \right), \\[2ex] Y'^T AX' = \dfrac{1}{\sqrt{\Delta(z)}} \left(Z^T A^T AZ - \dfrac{Z^T AZ}{Z^T A^T AZ} Z^T A^T A^2 Z \right), \\[2ex] X'^T AY' = \dfrac{1}{\sqrt{\Delta(z)}} \left(Z^T A^2 Z - \dfrac{Z^T AZ}{Z^T A^T AZ} Z^T A^T A^2 Z \right), \\[2ex] Y'^T AY' = \dfrac{Z^T A^T A^2 Z}{Z^T A^T AZ}, \end{cases} \tag{35}$$

where

$$\Delta(z) := Z^T Z (Z^T A^T AZ) - (Z^T AZ)^2 \overset{(32)}{=} G(Z).$$

From formulas (34) and (35), one can express $X'_\perp$ and $Y'_\perp$ in terms of Z.

Hence, taking into account formula (26) and Lemma 3, we reduce constrained optimization problem (16) to that one with the objective function

$$\frac{1}{2}(X'^T AX' + Y'^T AY')^2 + X'^T_\perp X'_\perp + Y'^T_\perp Y'_\perp \equiv H(Z),$$

where $H(Z)$ is introduced by (30).

$\square$

Now let us compute the destabilizing perturbation. For vector Z_* providing the minimum of the function $H(Z)$, calculate the orthonormal basis X', Y' of the plane $\mathrm{span}(Z_*, AZ_*)$ by formulas (33). Formula (27) then gives the angle of rotation φ. Formulas (25) yield then the new basis

$$X_* = X' \cos\varphi - Y' \sin\varphi, \ Y_* = X' \sin\varphi + Y' \cos\varphi, \tag{36}$$

and E_* is evaluated via substitution of these vectors into (22).

5   Numerical Examples

Example 4. Find the distance to instability for the Frobenius matrix

$$\begin{bmatrix} 0 & 1 & 0 \\ 0 & 0 & 1 \\ -91 & -55 & -13 \end{bmatrix}.$$

Solution. According to Theorem 10, we first compute the smallest singular value of the matrix: $\sigma_{\min}(A) = \frac{1}{2}(\sqrt{11658} - \sqrt{11294}) \approx 0.849493$.

To find value (30), we get

$$F(z_1, z_2, z_3) = 11524253650z_1^6 z_2 z_3 + 425696830z_1^4 z_2 z_3^3 + 5064605728z_1^3 z_2^3 z_3^2$$
$$+ 22232496040z_1^3 z_2^4 z_3 + \ldots + 2861716z_2^3 z_3^5 + 2248z_1 z_3^7 + 8866z_2 z_3^7,$$

$$G(z_1, z_2, z_3) = 8281z_1^4 + 10010z_1^3 z_2 + 2366z_1^3 z_3 + 11306z_1^2 z_2^2 + 1612z_1^2 z_2 z_3$$
$$+ 170z_1^2 z_3^2 + \ldots + 280z_2^2 z_3^2 + 208z_1 z_2 z_3^2 + 26z_2 z_3^3.$$

Consider polynomials

$$f(x_1, x_2) := F(x_1, x_2, 1) \text{ and } g(x_1, x_2) := G(x_1, x_2, 1).$$

To find the required minimum, we have to solve the following system of equations

$$P(x_1, x_2) := \frac{\partial f}{\partial x_1} g - 2\frac{\partial g}{\partial x_1} f = 0, \ Q(x_1, x_2) := \frac{\partial f}{\partial x_2} g - 2\frac{\partial g}{\partial x_2} f = 0. \quad (37)$$

The resultant $\mathcal{R}_{x_2}(P, Q)$ is a polynomial of the degree 111 in x_1 and it is reducible over $\mathbb{Z}$. The factor that we are interested in is as follows:

$$9419067565404001702089 1x_1^9 - 2268129372077996450568 1381x_1^8 + \ldots$$
$$-2731954511313476401720 23654739x_1 + 19252634670894308375 3790294889.$$

Its real zeros are:

$$x_{1*} \approx -6.678096, \ 0.975914, \ 238.671049.$$

The distance equation is

$$\mathcal{F}_1(z) := 264608725603218736465972352\, z^9$$
$$-2482778534463182978702672498368\, z^8$$
$$-60414043532276844722433124139834 16\, z^7$$
$$-18775588060618420451517932796910938 79\, z^6$$
$$+34445391711550097171504986029660547611 44\, z^5$$
$$-118654280158217887185584184480139888631 2992\, z^4$$
$$+181692433990859305990739054557502710597775512\, z^3$$
$$-1319875747217437472761622582226443869581406 3040\, z^2$$
$$+36828691764231816963867758330996666202586585 5312\, z$$
$$-7666642328922100939978493312753417857840062 2592 = 0$$

and its real zeros are

$$z_* \approx 0.209742, \ 80.144078, \ 11426.774168.$$

The distance to instability $\sqrt{z_*} \approx 0.457976$ is achieved at $x_{1*} \approx -6.678096$, $x_{2*} \approx 12.477326$.

Finally let us find the destabilizing perturbation E_* via formula (22). Vectors (33) are as follows:

$$X' \approx \begin{bmatrix} -0.457503 \\ 0.887647 \\ -0.052659 \end{bmatrix}, Y' \approx \begin{bmatrix} 0.135037 \\ 0.010822 \\ -0.990773 \end{bmatrix}.$$

Hence, we get $\tan 2\varphi \approx -0.018430$ and basis (36) is:

$$X_* \approx \begin{bmatrix} -0.456238 \\ 0.887709 \\ -0.061793 \end{bmatrix}, Y_* \approx \begin{bmatrix} -0.139251 \\ -0.002636 \\ 0.990245 \end{bmatrix}.$$

Therefore, $a \approx 0.883746, b \approx -6.563069$ and

$$E_* \approx \begin{bmatrix} 0.043826 & 0.024309 & -0.398268 \\ -0.007436 & 0.070758 & -0.208628 \\ -0.002061 & 0.003541 & 0.001451 \end{bmatrix}.$$

$\square$

Example 5. For matrix (11), find the dependence of the distance to instability on the parameter μ.

Remark. For this matrix, a plot of distance to instability for complex perturbations was presented in [12].

Solution. For matrix (11), we have

$$F(Z) = 4 + 18\mu^2 + 4\mu^3 + 22\mu^4 + 4\mu^5 + 12\mu^6 + 3\mu^8 - 4\mu z_1 + 4\mu^2 z_1 + \ldots,$$
$$G(Z) = 1 + \mu^2 + 4\mu^2 z_1^2 + \mu^4 z_1^2 + 2\mu^2 z_1^3 - 2\mu^4 z_1^3 + \mu^2 z_1^4 + \mu^4 z_1^4 + \ldots.$$

Then polynomials (37) have the following degrees:

$$\deg_{z_1} P = 10, \deg_{z_2} P = 11, \deg_\mu P = 16, \deg_{z_1} Q = 11, \deg_{z_2} Q = 10, \deg_\mu Q = 16.$$

The resultant

$$\mathcal{R}_{z_2}(P, Q) = 704482010726400\mu^{330} z_1^{111} - 13526054605946880\mu^{329} z_1^{111} + \ldots$$

is a polynomial in μ and z_1, $\deg_\mu \mathcal{R}_{z_2} = 332, \deg_{z_1} \mathcal{R}_{z_2} = 111$. It happens to be reducible over $\mathbb{Z}$, and on elimination of the extraneous factors, one gets the equation

$$(12\mu^{29} + 152\mu^{28} + 719\mu^{27} + \ldots)z_1^9 + \ldots + 490\mu^4 - 98\mu^3 - 151\mu^2 - 54\mu - 8 = 0.$$

For any specialization of the parameter μ, the z_1 value can be evaluated with any predetermined accuracy. Then we can find the corresponding values z_2 and (30). In Fig. 3, the dependence of $\beta_{\mathbb{R}}$ on the parameter μ is presented.

When $\mu \leq \mu^{(0)} \approx -0.682328$, the matrix $A(\mu)$ is unstable, i.e., $\beta_{\mathbb{R}}(A) = 0$. The distance to instability over the interval $\mu^{(0)} \leq \mu \leq \mu^{(1)} \approx 0.772543$ is equal to the minimal singular value of the matrix (displayed as thin blue plot in Fig. 3). The red boldface plot demonstrates the values of $\beta_{\mathbb{R}}$ found by (30).

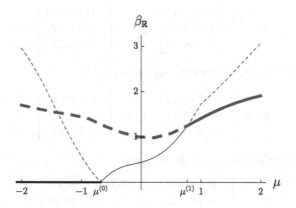

Fig. 3. Example 4. Distance to instability in the matrix space

The switching value $\mu = \mu^{(1)}$ corresponds to the matrix $A(\mu)$ that is equidistant from the two boundaries of the stability domain, i.e., it can be destabilized by two matrices of the identical norm:

$$E_{*1} \approx \begin{bmatrix} -0.042420 & 0.023236 & 0.031892 \\ -0.623514 & 0.341538 & 0.458770 \\ -0.670580 & 0.367319 & 0.504151 \end{bmatrix}, E_{*2} \approx \begin{bmatrix} 0.871781 & 0.073653 & -0.066321 \\ 0.084323 & 0.406861 & 0.430345 \\ -0.077178 & 0.437408 & 0.490915 \end{bmatrix}.$$

$\square$

Remark. In all the examples of the matrices of the order 3 that have been done by the authors, the degree of the distance equation in the matrix space equals 9. Thus, the fear of the permanent degree growth with that of the number of parameters of the family expressed at the beginning of Sect. 4, is not confirmed. Therefore, it is of interest to clarify this dependence.

6   Conclusions

We have investigated the Routh – Hurwitz stability property for matrices with the entries polynomially depending on parameters in the case of real perturbations. We have concerned with the following issues

- finding the spectral abscissa of a matrix;
- description and estimation of the domain of stability in the parameter space;
- computation of the distance to instability in the matrix space.

The purely algebraic procedures based on symbolic algorithms for the elimination of variables and localization of the real zeros for algebraic equation systems have been suggested for solving the stated problems. This provides one with precise information on the obtained solution, i.e., the results do not depend on the precision of calculations and round-off errors. For further investigation, it remains to optimize the computational efficiency of the suggested algorithms for their application to the matrices of arbitrary order.

Acknowledgments. The authors are grateful to Prof Evgenii Vorozhtzov and to the anonimous referees for valuable suggestions that helped to improve the quality of the paper.

References

1. Ackermann, J., Muench, R.: Robustness analysis in a plant parameter plane. IFAC Proc. vol. **20**(5), part 8, 205–209 (1987)
2. Bobylev, N.A., Bulatov, A.V.: A bound on the real stability radius of continuous-time linear infinite-dimensional systems. Comput. Math. Model. **12**(4), 359–368 (2001)
3. Bobylev, N.A., Bulatov, A.V., Diamond, P.H.: Estimates of the real structured radius of stability of linear dynamic systems. Autom. Remote Control **62**, 505–512 (2001)
4. Emiris, I., Mourrain, B., Tsigaridas, E.: Separation bounds for polynomial systems. J. Symb. Comput. **101**, 128–151 (2020)
5. Freitag, M.A., Spence, A.: A Newton-based method for the calculation of the distance to instability. Linear Algebra Appl. **435**, 3189–3205 (2011)
6. Guglielmi, N., Gürbüzbalaban, M., Mitchell, T., Overton, M.: Approximating the real structured stability radius with Frobenius norm bounded perturbations. SIAM J. Matrix Anal. Appl. **38**(4), 1323–1353 (2017)
7. Guglielmi, N., Lubich, C.: Low-rank dynamics for computing extremal points of real pseudospectra. SIAM J. Matrix Anal. Appl. **34**, 40–66 (2013)
8. Guglielmi, N., Manetta, M.: Approximating real stability radii. IMA J. Numer. Anal. **35**(3), 1401–1425 (2014)
9. Hinrichsen, D., Pritchard, A.J.: Stability radii of linear systems. Syst. Control Lett. **7**(1), 1–10 (1986)
10. Hinrichsen, D., Pritchard, A.J.: Mathematical Systems Theory I: ModellingState Space analysis. Stability and Robustness. Springer-Verlag, Berlin Heidelberg (2005)
11. Kalinina, E., Smol'kin, Y., Uteshev, A.: Robust schur stability of a polynomial matrix family. In: England, M., Koepf, W., Sadykov, T.M., Seiler, W.M., Vorozhtsov, E.V. (eds.) CASC 2019. LNCS, vol. 11661, pp. 262–279. Springer, Cham (2019). https://doi.org/10.1007/978-3-030-26831-2_18
12. Kalinina, E.A., Smol'kin, Yu.A. Uteshev, A.Yu.: Stability and distance to instability for polynomial matrix families. complex perturbations. Linear Multilinear Algebra (2020) https://doi.org/10.1080/03081087.2020.1759500

13. Kalinina, E.A., Uteshev, A.Y.: Determination of the number of roots of a polynomial lying in a given algebraic domain. Linear Algebra Appl. **185**, 61–81 (1993)
14. Katewa, V., Pasqualetti, F.: On the real stability radius of sparse systems. Automatica **113**, 108685 (2020)
15. Kronecker, L.: Zur Theorie der Elimination einer Variablen aus zwei algebraischen Gleichungen. Werke. Bd. 2, pp. 113–192. Teubner, Leipzig (1897)
16. Lazard, D., McCallum, S.: Iterated discriminants. J. Symb. Comput. **44**, 1176–1193 (2009)
17. Qiu, L., Davison, E.J.: The stability robustness determination of state space models with real unstructured perturbations. Math. Control Sig. Syst. **4**(3), 247–267 (1991)
18. Rem, S., Kabamba, P.T., Bemstein, D.S.: Guardian map approach to robust stability of linear systems with constant real parameter uncertainty. IEEE Trans. Autom. Control **39**(1), 162–164 (1994)
19. Routh, E.J.: A Treatise on the Stability of a Given State of Motion: Particularly Steady Motion. Macmillan and Co., London (1877)
20. Savov, S., Popchev, I.: Robust stability analysis for a perturbed single-area power system model. Cybern. Inf. Technol. **15**(4), 42–49 (2015)
21. Uteshev, A.Yu., Goncharova, M.V.: Metric problems for algebraic manifolds: analytical approach. In: Constructive Nonsmooth Analysis and Related Topics – CNSA 2017 Proc., 7974027 (2017)
22. Uteshev, A.Yu., Goncharova, M.V.: Approximation of the distance from a point to an algebraic manifold. In: Proceeding of 8th International Conference on Pattern Recognition Applications and Methods. vol. 1, pp. 715–720 (2019)
23. Uteshev, A.Y., Shulyak, S.G.: Hermite's method of separation of solutions of systems of algebraic equations and its applications. Linear Algebra Appl. **177**, 49–88 (1992)
24. Van Loan, C.F.: How near is a stable matrix to an unstable matrix? In: Datta, B.N. et al. (eds.) Linear Algebra and its Role in Systems Theory 1984, Contemporary Math., vol. 47, pp. 465–478. Amer. Math. Soc., Providence, Rhode Island (1985) https://doi.org/10.1090/conm/047
25. Weyl, H.: Classical Groups: Their Invariants and Representations. Princeton Univ. Press, Princeton, New Jork (1953)

Hermite Rational Function Interpolation with Error Correction

Erich L. Kaltofen[1,2] ⓘ, Clément Pernet[3] ⓘ, and Zhi-Hong Yang[1,2(✉)] ⓘ

[1] Department of Mathematics, North Carolina State University, Raleigh, NC 27695-8205, USA
{kaltofen,zyang28}@ncsu.edu
[2] Department of Computer Science, Duke University, Durham, NC 27708-0129, USA
{kaltofen,zy99}@cs.duke.edu
[3] Laboratoire Jean Kuntzmann, Univ. Grenoble Alpes, CNRS, 38058 Grenoble Cedex 09, France
clement.pernet@univ-grenoble-alpes.fr
https://users.cs.duke.edu/~elk27,
http://ljk.imag.fr/membres/Clement.Pernet/

Abstract. We generalize Hermite interpolation with error correction, which is the methodology for multiplicity algebraic error correction codes, to Hermite interpolation of a rational function over a field K from function and function derivative values.

We present an interpolation algorithm that can locate and correct $\leq E$ errors at distinct arguments $\xi \in \mathsf{K}$ where at least one of the values or values of a derivative is incorrect. The upper bound E for the number of such ξ is input. Our algorithm sufficiently oversamples the rational function to guarantee a unique interpolant. We sample $(f/g)^{(j)}(\xi_i)$ for $0 \leq j \leq \ell_i, 1 \leq i \leq n, \xi_i$ distinct, where $(f/g)^{(j)}$ is the j-th derivative of the rational function f/g, $f, g \in \mathsf{K}[x]$, $\mathrm{GCD}(f, g) = 1$, $g \neq 0$, and where $N = \sum_{i=1}^{n}(\ell_i + 1) \geq D_f + D_g + 1 + 2E + 2\sum_{k=1}^{E} \ell_k$; D_f is an upper bound for $\deg(f)$ and D_g an upper bound for $\deg(g)$, which are input to our algorithm. The arguments ξ_i can be poles, which is truly or falsely indicated by a function value ∞ with the corresponding $\ell_i = 0$. Our results remain valid for fields K of characteristic $\geq 1 + \max_i \ell_i$. Our algorithm has the same asymptotic arithmetic complexity as that for classical Hermite interpolation, namely $N(\log N)^{O(1)}$.

For polynomials, that is, $g = 1$, and a uniform derivative profile $\ell_1 = \cdots = \ell_n$, our algorithm specializes to the univariate multiplicity code decoder that is based on the 1986 Welch-Berlekamp algorithm.

Keywords: Hermite interpolation · Cauchy interpolation · Error correction codes · Multiplicity codes · List decoding

This research was supported by the National Science Foundation under Grant CCF-1717100 (Kaltofen and Yang).

© Springer Nature Switzerland AG 2020
F. Boulier et al. (Eds.): CASC 2020, LNCS 12291, pp. 335–357, 2020.
https://doi.org/10.1007/978-3-030-60026-6_19

1   Introduction

Algebraic error correction codes are based on interpolating a polynomial f from its values $a_i = f(\xi_i)$ at distinct argument scalars ξ_i, when some of the inputs $\hat{a}_\lambda$ for the evaluations are incorrect, namely $\hat{a}_\lambda \neq a_\lambda$. The coefficients of f are from a field K, as are the arguments ξ_i and the list of correct and incorrect evaluations $\hat{a}_i$. The 1960 algorithm by Irving Reed and Gustave Solomon [16] reconstructs a polynomial f of degree $\leq D$ from $n = D + 1 + 2E$ values $\hat{a}_i$ when $\leq E$ of the values are incorrect, namely $|\{\lambda \mid \hat{a}_\lambda \neq f(\xi_\lambda)\}| \leq E$. The number of evaluations is optimal: for $n = D + 2E$ there may exist two polynomials that interpolate with $\leq E$ errors. The Reed-Solomon decoder generalizes to rational functions $f/g \in \mathsf{K}(x)$ with $n = D_f + D_g + 1 + 2E$, where $D_f \geq \deg(f)$, $D_g \geq \deg(g)$ [1]. Decoding can be performed by the extended Euclidean Algorithm [19,21] or by solving a linear system [8]. Lemma 3.2 in [8] shows that for $D_f = \deg(f)$, $D_g = \deg(g)$, $n - 1$ evaluations are always insufficient to correct E errors.

Multiplicity codes [3–5,11,13,15,17] generalize the Reed-Solomon problem to the Hermite interpolation problem with error correction. In classical (error-free) Hermite interpolation one reconstructs a polynomial (or rational function) from the values of the polynomial and its derivatives. The classical algorithm of divided differences can reconstruct f from $a_{i,j} = f^{(j)}(\xi_i)$, where $\xi_1, \ldots, \xi_n \in \mathsf{K}$ are distinct scalars, $f^{(j)}$ is the j-th derivative of f, and $0 \leq j \leq \ell_i$ with $(\ell_1 + 1) + \cdots + (\ell_n + 1) = D + 1$. We shall assume that the characteristic of the field K is either 0 or $\geq 1 + \max \ell_i$. For the case that $n = 1$, one has $f(x) = \sum_{0 \leq j \leq \ell_1} f^{(j)}(\xi_1)/j! \, (x - \xi_1)^j$. As in the Reed-Solomon decoding problem, one assumes that some inputs $\hat{a}_{i,j} \neq f^{(j)}(\xi_i)$. It is clear that not all evaluation profiles $(\ell_1, \ldots, \ell_n)$ with $\sum_{i=1}^{n}(\ell_i + 1) = D + 1 + 2E$ are decodable when there are $\leq E$ errors. For example, if $n = 1$ the oversampled derivatives cannot reveal all errors, since $f^{(j)}(x) = 0$ for $j \geq \deg(f) + 1$. Furthermore, if all $\hat{a}_{i,0}$ are erroneous, the constant coefficient of f is unrecoverable. Example 1 below shows that if one has $D = \deg(f)$, $n = D + 2E$ and $\ell_1 = \cdots = \ell_{2E} = 1$ and $\ell_{2E+1} = \cdots = \ell_{D+2E} = 0$, that is $N = D + 4E$, there may be a second polynomial of degree $\leq D$ that fits all but E evaluations.

The reason why, in the Hermite case, one may have to match every bad value with more than one additional good value, in contrast to the Reed-Solomon decoder, is apparent from the Birkhoff generalization of the Hermite interpolation problem. In Birkhoff interpolation one does not have all consecutive derivatives at a scalar ξ_i. Schoenberg [18] uses a matrix $\Theta = [\theta_{i,j}]_{1 \leq i \leq n, 0 \leq j \leq D} \in \{0, 1\}^{n \times (D+1)}$ with exactly $(D + 1)$ 1-entries. If $\theta_{i,j} = 1$ then the evaluation $a_{i,j} = f^{(j)}(\xi_i)$ is input. For $\mathsf{K} = \mathbb{R}$ one asks for which Θ's one always gets a unique interpolant. The Pólya-Schoenberg Theorem states that for $n = 2$ there is a unique interpolant if and only if $\forall j, 0 \leq j \leq D-1$: $\sum_{0 \leq \mu \leq j}(\theta_{1,\mu} + \theta_{2,\mu}) \geq j+1$. If those Pólya conditions are violated, then either there are more than one solution or there is no solution. In the error correction setting, for example, in erasure codes, the errors invalidate evaluations, that is, set $\theta_{i,j} = 0$ whenever $\hat{a}_{i,j}$ is an error. Thus the remaining good points may constitute an (oversampled) Birkhoff problem that does not have one unique solution. For example, for $f(x) = (x^2 - 1^2)(x^2 - 7^2)$ we have $f(x)' = 4x(x^2 - 5^2)$, so

$f(-1) = f(1) = f(-7) = f(7) = 0$, $f'(0) = f'(-5) = f'(5) = 0$ is interpolated at 7 good values by the polynomials 0 and f (see also Example 2). Multiplicity code decoders also need to locate the erroneous locations. Our problem is more difficult: the correct values are those of a rational function, not of a polynomial.

The algorithms for error-correcting Hermite interpolation of polynomials and rational functions, in analogy to the Pólya conditions, interpolate a unique polynomial or rational function from its upper bounds for the degrees and the number of errors by sufficient oversampling. Suppose the profile of derivatives at each distinct argument ξ_i ($1 \le i \le n$) is sorted: $\ell_1 \ge \cdots \ge \ell_n \ge 0$. Again, we assume that the characteristic of K is either 0 or $\ge \ell_1 + 1$. For a rational function $f/g \in \mathsf{K}(x)$ we input $D_f \ge \deg(f)$, $D_g \ge \deg(g)$, $\hat{a}_{i,j} \in \mathsf{K}$ for $1 \le i \le n$ and $0 \le j \le \ell_i$, and E such that for $\le E$ of all arguments ξ_i there is an error at least one j:

$$E \ge |\{i \mid 1 \le i \le n \text{ and } \exists j, 0 \le j \le \ell_i \colon \hat{a}_{i,j} \ne (f/g)^{(j)}(\xi_i)\}|.$$

We use the fresh symbol $\infty = (f/g)(\xi_i)$ if $g(\xi_i) = 0$ and allow both false (non-pole) scalars $\hat{a}_{i,j} \in \mathsf{K}$ at such poles, as well as false poles $\hat{a}_{i,j} = \infty$ when $g(\xi_i) \ne 0$. We shall assume that the only $\hat{a}_{i,j} = \infty$ are at evaluations $j = 0$ and that then no derivatives are present, that is, $\ell_i = 0$. If $\hat{a}_{i,j} = \infty \ne \hat{a}_{i,k}$ for some $j \ne k$ then one of the values is erroneous, unless the characteristic of K is positive and $\le D_g$. In that case, the list of values $\hat{a}_{i,0}, \hat{a}_{i,1}, \ldots$ at ξ_i is pre-processed: see Remark 2. Note that without errors, f/g cannot be interpolated at a single argument ξ_1 when all derivative values are ∞.

Our algorithm recovers f/g if the number of evaluations, N, at n distinct ξ_i satisfies

$$N \overset{\text{def}}{=} \sum_{i=1}^{n} (\ell_i + 1) = D_f + D_g + 1 + 2 \sum_{i=1}^{E} (\ell_i + 1) = D_f + D_g + 1 + 2E + 2 \sum_{i=1}^{E} \ell_i \quad (1)$$

(see Theorem 1). The equation (1) implies $2E + 1 \le n$ (see (6)). Note that if $N <$ the right-side of (1), one needs to increase either n or $\ell_{E+1}, \ldots, \ell_n$ and sample more values. If $N >$ the right-side of (1), one can decrease $\ell_n, \ldots, \ell_{E+1}$ and/or n. If equality in (1) is achieved, further reduction of oversampling may be possible while preserving (1); see Remark 4. For polynomial interpolation we can set $D_g = 0$. In relation to Example 1: with $D = \deg(f), g = 1$, if $\ell_1 = \cdots = \ell_{2E+1} = 1$ and $\ell_{2E+2} = \cdots = \ell_{D+2E} = 0$ then f is recovered uniquely from $N = D + 4E + 1$ evaluations with $\le E$ errors. For $\ell_i = 0$ for all i, our algorithm specializes to rational function recovery with errors with $n = N = D_f + D_g + 1 + 2E$.

1.1 Comparison to Multiplicity Code Decoders

Multiplicity codes are based on Hermite polynomial interpolation with error correction, that is, $D_f = D$, $D_g = 0$. In [11] the following parameter settings are used: $n = q$ and the field of scalars is $\mathsf{K} = \mathbb{F}_q$, a finite field of q elements. The number of derivatives is uniformly $\ell_1 = \cdots = \ell_q = s - 1$. There are $\le E = (sq - D - 1)/(2s)$ indices λ_κ where at least one of the s derivative values $\hat{a}_{\lambda_\kappa, j}$

$(0 \le j \le s-1)$ is an error. At each error index λ_κ, there can be as many as s errors, for a total of $(sq - D - 1)/2$ errors, the latter of which is the degree of the error locator polynomial in [11, Section 3.1]. Multiplicity codes then recover the code polynomial from the $N = sq$ values $\hat{a}_{i,j}$ for $1 \le i \le q$ and $0 \le j \le s-1$, which agrees with the right-side of (1): $D + 1 + 2Es = sq$. Our decoders here allow for unequal ℓ_i.

Our main contribution is the generalization to Hermite interpolation of rational functions from such partially erroneous values, including the handling of arguments at roots of the denominator, that is, poles. An important idea behind Algorithm 5.1 is from the algorithm in [20] (as cited in ([6])) for Hermite rational function interpolation, which in turn is based on Cauchy interpolation via the extended Euclidean algorithm. Our algorithm essentially performs Warner's algorithm, now on an unreduced fraction of polynomials, where both numerator and denominator are multiplied with the error locator polynomial, which the Cauchy interpolation algorithm computes (see Lemma 1). The Welch-Berlekamp decoder for Reed-Solomon codes [21] and its generalization to multiplicity code decoders [11, Section 3.1.1] also has our interpretation of solving such a Cauchy problem.

Because of derivatives, in the Hermite setting the roots of the error locator polynomial have multiplicities. With our assumption that $\hat{a}_{i,j} = \infty$ only if $j = \ell_i = 0$, we can prove that the N values in (1) are sufficient for unique recovery if there are $\le E$ arguments ξ_i with some $\hat{a}_{i,j}$ being an error (see Theorem 1).

The half-GCD algorithm [14] and fast Hermite interpolation algorithms [2] then yield an arithmetic complexity of $N(\log N)^{O(1)}$. We note that the uniqueness of the interpolant for the error-free Hermite rational function problem implies uniqueness with errors when oversampled at N points (1), which yields a linear system for the coefficients of the unreduced numerator and denominator polynomials. Our approach computes a solution via the extended Euclidean algorithm and additionally optimizes the required polynomial division: see Remark 3. Our Algorithm 5.1 also diagnoses if no valid rational function interpolant exists, which can be used to perform list-decoding: see Remark 6.

2   Polynomial Hermite Interpolation

Let $n \ge 1$, $\xi_i \in \mathsf{K}$ for $1 \le i \le n$ be distinct values, $\ell_1 \ge \ell_2 \ge \cdots \ge \ell_n \ge 0$, $a_{i,j} \in \mathsf{K}$ for $1 \le i \le n$ and $0 \le j \le \ell_i$. Suppose that the characteristic of K is either 0 or $\ge \ell_1 + 1$. For $d = (\ell_1 + 1) + \cdots + (\ell_n + 1) - 1$ there exists a unique $f \in \mathsf{K}[x]$ with $\deg(f) \le d$ such that $a_{i,j} = f^{(j)}(\xi_i)$ for $1 \le i \le n$ and $0 \le j \le \ell_i$ where $f^{(j)}(x)$ is the j-th derivative of $f(x) = c_d x^d + \cdots + c_1 x + c_0$ defined by

$$f^{(j)}(x) = \Big(\sum_{\delta=0}^{d} c_\delta x^\delta\Big)^{(j)} = \sum_{\delta=j}^{d} c_\delta\, \delta(\delta-1)\cdots(\delta-j+1)\, x^{\delta-j}. \qquad (2)$$

We note that if K has any characteristic, the rational function field $\mathsf{K}(x)$ is a differential field with the derivative $'$ being a function satisfying $c' = 0$ for all

$c \in \mathsf{K}$, $x' = 1$ and $(F+G)' = F'+G'$ and $(FG)' = F'G+FG'$ for all $F, G \in \mathsf{K}(x)$, which yields (2). See also Remark 2 below.

The algorithm of divided differences, which goes back to at least Guo Shoujing (1231–1316), computes the coefficients $\bar{c}_{i,j}$ forming the decomposition of $f(x)$ in mixed-shifted-basis representation

$$f(x) = \sum_{\nu=1}^{n} \sum_{\mu=0}^{\ell_\nu} \bar{c}_{\nu-1,\mu} \Big(\prod_{\kappa=1}^{\nu-1} (x - \xi_\kappa)^{\ell_\kappa+1} \Big) (x - \xi_\nu)^\mu.$$

For $1 \le i \le n$ and $0 \le j \le \ell_i$ the interpolant

$$H_{i,j}(x) = \sum_{\nu=1}^{i} \sum_{\mu=0}^{\ell'_\nu} \bar{c}_{\nu-1,\mu} \Big(\prod_{\kappa=1}^{\nu-1} (x - \xi_\kappa)^{\ell_\kappa+1} \Big) (x - \xi_\nu)^\mu$$

$$\text{with } \ell'_\nu = \ell_\nu \text{ for } \nu < i \text{ and } \ell'_i = j,$$

fits the values $a_{\nu,\mu}$ for $1 \le \nu \le i$ and $0 \le \mu \le \ell'_\nu$. We compute the next $H_{i,j+1}$ ($j < \ell_i$) or $H_{i+1,0}$ ($j = \ell_i$) to fit $a_{i,j+1}$ or $a_{i+1,0}$, respectively. For $j < \ell_i$ we have

$$H_{i,j+1}(x) = H_{i,j}(x) + \bar{c}_{i-1,j+1} G_{i,j+1}(x),$$

$$G_{i,j+1}(x) = \Big(\prod_{\kappa=1}^{i-1} (x - \xi_\kappa)^{\ell_\kappa+1} \Big) (x - \xi_i)^{j+1}.$$

Note that $G_{i,j+1}^{(\mu)}(\xi_\nu) = 0$ for $1 \le \nu \le i$ and $0 \le \mu \le \ell'_\nu$, so $H_{i,j+1}(x)$ interpolates all of $H_{i,j}$'s values. Finally, $G_{i,j+1}^{(j+1)}(\xi_i) = \big(\prod_{\kappa=1}^{i-1} (\xi_i - \xi_\kappa)^{\ell_\kappa+1} \big) (j+1)!$, which is $\neq 0$ by our assumption that the characteristic of K is 0 or $\ge \ell_1 + 1$. Therefore $H_{i,j+1}^{(j+1)}(\xi_i) = a_{i,j+1}$ has a unique solution $\bar{c}_{i-1,j+1}$. The case $H_{i+1,0}$ is similar.

Algorithms for computing the Hermite interpolant f in soft-linear arithmetic complexity go back to [2].

3   Rational Function Recovery

Our algorithms are a generalization of the Cauchy interpolation algorithm for rational functions, which is based on the extended Euclidean algorithm. We now state the key lemma, which goes back to Leopold Kronecker's algorithm for computing Padé approximants.

Lemma 1. *Let d and e be non-negative integers, and let $H(x) \in \mathsf{K}[x]$, K an arbitrary field, $\deg(H) \le d + e$; furthermore, let ξ_i, $1 \le i \le d + e + 1$, be not necessarily distinct elements in K.*

1. *Define $r_0 = \prod_{i=1}^{d+e+1} (x - \xi_i)$ and $r_1(x) = H(x)$. Now let $r_\rho(x), q_\rho(x) \in \mathsf{K}[x]$ be the ρ-th remainder and quotient respectively, in the Euclidean polynomial remainder sequence*

$$r_{\rho-2}(x) = q_\rho(x) r_{\rho-1}(x) + r_\rho(x), \quad \deg(r_\rho) < \deg(r_{\rho-1}) \text{ for } \rho \ge 2.$$

In the exceptional case $H = 0$ the sequence is defined to be empty. Finally, let $s_\rho(x), t_\rho(x) \in \mathsf{K}[x]$ be the multipliers in the extended Euclidean scheme $s_\rho r_1 + t_\rho r_0 = r_\rho$, namely,

$$s_0 = t_1 = 0, \quad t_0 = s_1 = 1,$$

$$s_\rho = s_{\rho-2} - q_\rho\, s_{\rho-1}, \quad t_\rho = t_{\rho-2} - q_\rho\, t_{\rho-1} \quad for\ \rho \geq 2.$$

Then there exists an index $\gamma \geq 1$, such that $\deg(r_\gamma) \leq d < \deg(r_{\gamma-1})$ and

$$r_\gamma \equiv s_\gamma H \pmod{r_0} \quad and \quad \deg(s_\gamma) \leq e. \tag{3}$$

2. Let $R(x), S(x) \in \mathsf{K}[x]$ be another solution of (3), namely

$$R \equiv S H \pmod{r_0} \quad d \geq \deg(R)\ and\ e \geq \deg(S). \tag{4}$$

Then $s_\gamma R = r_\gamma S$. If furthermore $\mathrm{GCD}(R, S) = 1$ then $R = c\,r_\gamma$, $S = c\,s_\gamma$ for some $c \in \mathsf{K} \setminus \{0\}$.

Proof. See [7, Lemma 1].

4   Error-Correcting Hermite Interpolation

Let $f(x) \in \mathsf{K}[x]$ be a univariate polynomial and D be an upper bound of $\deg(f)$. One is given a set of n distinct arguments $\xi_1, \ldots, \xi_n \in \mathsf{K}$, and for each argument ξ_i, one is given a row vector

$$\widehat{A}_{i,*} = [\hat{a}_{i,0}, \ldots, \hat{a}_{i,\ell_i}] \in \mathsf{K}^{1 \times (\ell_i+1)}.$$

We call $\hat{a}_{i,j}$ an error if $\hat{a}_{i,j} \neq f^{(j)}(\xi_i)$, and we call $\widehat{A}_{i,*}$ error-free if $\hat{a}_{i,j} = f^{(j)}(\xi_i)$ for all $j = 0, \ldots, \ell_i$. Let $\{\lambda_1, \ldots, \lambda_k\} \subset \{1, \ldots, n\}$ be the set of indices where every row vector $\widehat{A}_{\lambda_1,*}, \ldots, \widehat{A}_{\lambda_k,*}$ has at least one error, and let $E \geq k$ (if all row vectors $\widehat{A}_{1,*}, \ldots, \widehat{A}_{n,*}$ are error-free, then let $E = k = 0$). As in Section 2 we assume that K is a field of characteristic 0 or $\geq 1 + \max_i \ell_i$. To uniquely recover $f(x)$, a condition $n \geq 2E + 1$ is necessary: if $n = 2E$, one can have for $f \in \mathsf{K}[x]$ with E errors $\hat{a}_{i,0} = f(\xi_i) + 1 \neq f(\xi_i)$ where $1 \leq i \leq E$, and for $f + 1$ with E errors $\hat{a}_{i,0} = f(\xi_i) \neq f(\xi_i) + 1$ where $E + 1 \leq i \leq 2E$, that is both f and $f + 1$ are valid interpolants with E errors. Without loss of generality, we assume that $\ell_1 \geq \cdots \geq \ell_n \geq 0$, and let

$$\widehat{A} = \begin{bmatrix} \widehat{A}_{1,*} \\ \vdots \\ \widehat{A}_{n,*} \end{bmatrix} \in \left(\mathsf{K}^{1 \times (\ell_1+1)} \cup \cdots \cup \mathsf{K}^{1 \times (\ell_n+1)}\right)^n$$

be the vector of those value row vectors. The total number of values in $\widehat{A}$ is $N = \sum_{i=1}^{n}(\ell_i + 1)$. We will show how to recover the polynomial $f(x)$ from the points $\xi_1, \ldots, \xi_n$ and the values in $\widehat{A}$ by the extended Euclidean algorithm if

$$n + \sum_{i=E+1}^{n} \ell_i = D + 1 + 2E + \sum_{i=1}^{E} \ell_i. \tag{5}$$

Note that the equality (5) is equivalent to $\sum_{i=1}^{n}(\ell_i+1) = D+1+2E+2\sum_{i=1}^{E}\ell_i$, which means $\widehat{A}$ has $N = D + 1 + 2E + 2\sum_{i=1}^{E}\ell_i$ values. Furthermore, the equality (5) implies that $n \geq 2E+1$, because recovery is unique and for $n \leq 2E$ we would have the ambiguous solution above; more explicitly, for $n \leq 2E$ we have the contradiction

$$n + \sum_{i=E+1}^{n} \ell_i \leq 2E + \ell_{E+1}E < D+1+2E+ \sum_{i=1}^{E}\ell_i. \qquad (6)$$

We remark that the condition (5) can be relaxed to $n + \sum_{i=E+1}^{n}\ell_i \geq D + 1 + 2E + \sum_{i=1}^{E}\ell_i$, because in that case, one can decrease $\ell_n, \ell_{n-1}, \ldots, \ell_{E+1}$ successively to achieve the equality (5). In fact, even if the equality is satisfied, one may still be able to reduce the ℓ_i's on both sides so that the algorithm can recover $f(x)$ with fewer values (see Remark 4).

4.1 Error-Correcting Polynomial Hermite Interpolation

Input: ‣ A field K, nonnegative integers $D, E \in \mathbb{Z}_{\geq 0}$;
  ‣ A set of distinct points $\{\xi_1, \ldots, \xi_n\} \subseteq \mathsf{K}$;
  ‣ A list of n row vectors $\widehat{A} = [\widehat{A}_{i,*}]_{1\leq i\leq n}$ where
   ‣ $\ell_1 \geq \cdots \geq \ell_n \geq 0$; the characteristic of K is either 0 or $\geq \ell_1 + 1$;
   ‣ $\widehat{A}_{i,*} = [\hat{a}_{i,0}, \ldots, \hat{a}_{i,\ell_i}]$;
   ‣ $n + \sum_{i=E+1}^{n}\ell_i = D + 1 + 2E + \sum_{i=1}^{E}\ell_i \ (\Longrightarrow n \geq 2E + 1)$.
Output: ‣ The interpolant $f(x) \in \mathsf{K}[x]$ and the error locator polynomial $\Lambda(x)$ in $\mathsf{K}[x]$ which satisfy
   ‣ $\deg(f) \leq D$;
   ‣ $\Lambda(x) = 1$ or $\Lambda(x) = \prod_{\kappa=1}^{k}(x-\xi_{\lambda_\kappa})^{\delta_\kappa}$ where $\xi_{\lambda_1}, \ldots, \xi_{\lambda_k}$ are distinct and $k \leq E$;
   ‣ row vector $\widehat{A}_{i,*}$ is error free if and only if $i \notin \{\lambda_1, \ldots, \lambda_k\}$;
   ‣ $\delta_\kappa = \ell_{\lambda_\kappa} + 1 - \min\{j \mid f^{(j)}(\xi_{\lambda_\kappa}) \neq \hat{a}_{\lambda_\kappa,j}\}$.
  ‣ Or a message indicating there is no such interpolant.

1. *If $\hat{a}_{i,j} = 0$ for all $i = 1, \ldots, n$ and $j = 0, \ldots, \ell_i$, then return $f = 0$ and $\Lambda = 1$.*
2. *Compute the Hermite interpolant $H(x) \in \mathsf{K}[x]$ of the data set $\{(\xi_i; \hat{a}_{i,0}, \ldots, \hat{a}_{i,\ell_i}) \mid i = 1, \ldots, n\}$, namely compute a polynomial $H(x) \in \mathsf{K}[x]$ such that $H^{(j)}(\xi_i) = \hat{a}_{i,j}$.*
   If $E = 0$ and $\deg(H) \leq D$, then return $f = H$ and $\Lambda = 1$. If $E = 0$ and $\deg(H) > D$, then return a message indicating there is no such interpolant.
3. *Let $r_0 = (x - \xi_1)^{\ell_1+1} \cdots (x - \xi_n)^{\ell_n+1}$, $r_1 = H$, $s_0 = 0$, $s_1 = 1$, and $\rho = 2$.*
   3a. Compute the ρ-th Euclidean polynomial remainder r_ρ and the multiplier s_ρ in the extended Euclidean scheme $s_\rho r_1 + t_\rho r_0 = r_\rho$, namely

$$r_\rho(x) = r_{\rho-2}(x) - q_\rho(x)r_{\rho-1}(x), \quad \deg(r_\rho) < \deg(r_{\rho-1}),$$
$$s_\rho(x) = s_{\rho-2}(x) - q_\rho(x)s_{\rho-1}(x).$$

3b. If $\deg(r_\rho) \leq D + E + \sum_{i=1}^{E} \ell_i$, then let $\gamma = \rho$ and go to Step 4.

3c. Otherwise, let $\rho = \rho + 1$ and go to Step 3a.

By the half-GCD algorithm, Step 3 can be performed in soft-linear arithmetic complexity.

4. If s_γ divides r_γ, then factorize s_γ over K; if s_γ has $\leq E$ distinct factors, then go to Step 5. Otherwise return a message indicating there is no such interpolant.

5. Compute $f = r_\gamma/s_\gamma$. If $\deg(f) \leq D$, then return f and $\Lambda = s_\gamma/\text{lc}(s_\gamma)$, where $\text{lc}(s_\gamma)$ is the leading coefficent of s_γ. Otherwise return a message indicating there is no such interpolant.

Step 3 computes (r_γ, s_γ) as in Lemma 1 with $d = D + E + \sum_{i=1}^{E} \ell_i$ and $e = E + \sum_{i=1}^{E} \ell_i$. We will prove that if there is an interpolant $f(x) \in \mathsf{K}[x]$ which satisfies the output specifications, then $r_\gamma/s_\gamma = (f\Lambda)/\Lambda = f$ (see Lemma 2) and $\Lambda = s_\gamma/\text{lc}(s_\gamma)$ (see Lemma 3). Here $s_\gamma \neq 0$ because $\text{GCD}(s_\gamma, t_\gamma) = 1$ for $\gamma \geq 2$.

On the other hand, if the polynomial r_γ/s_γ computed in Step 5 has degree $\leq D$, then it satisfies the output specifications, which we will prove as a special case in Lemma 5. Therefore, we can check the validity of r_γ/s_γ without computing all the values $(r_\gamma/s_\gamma)^{(j)}(\xi_i)$ for $i = 1, \dots, n$ and $j = 0, \dots, \ell_i$.

Lemma 2. With the notation as in Algorithm 4.1, if there is a polynomial $f \in \mathsf{K}[x]$ which satisfies the output specifications, then

$$f\Lambda \equiv H\Lambda \pmod{r_0}. \tag{7}$$

Moreover, $r_\gamma/s_\gamma = (f\Lambda)/\Lambda = f$, which implies the interpolant f is unique.

Proof. Recall that $\Lambda(x) = (x - \xi_{\lambda_1})^{\delta_1} \cdots (x - \xi_{\lambda_k})^{\delta_k}$ is the error locator polynomial where

(i) $\xi_{\lambda_1}, \dots, \xi_{\lambda_k}$ are the arguments with erroneous values, that is, for indices $\lambda_\kappa \in \{\lambda_1, \dots, \lambda_k\}$, there exists $j \in \{0, \dots, \ell_{\lambda_\kappa}\}$ such that $f^{(j)}(\xi_{\lambda_\kappa}) \neq \hat{a}_{\lambda_\kappa, j}$;

(ii) $\delta_\kappa = \ell_{\lambda_\kappa} + 1 - \min\{j \mid f^{(j)}(\xi_{\lambda_\kappa}) \neq \hat{a}_{\lambda_\kappa, j}\}$, $\kappa = 1, \dots, k$.

Since $r_0 = (x - \xi_1)^{\ell_1+1} \cdots (x - \xi_n)^{\ell_n+1}$, proving the equality (7) is equivalent to proving $(x - \xi_i)^{\ell_i+1}$ divides $(f\Lambda - H\Lambda)$ for all $i = 1, \dots, n$, which is again equivalent to proving the following equality:

$$(f\Lambda)^{(j)}(\xi_i) = (H\Lambda)^{(j)}(\xi_i) \text{ for all } i = 1, \dots, n \text{ and } j = 0, \dots, \ell_i. \tag{8}$$

If $i \notin \{\lambda_1, \dots, \lambda_k\}$ then $f^{(j)}(\xi_i) = \hat{a}_{i,j} = H^{(j)}(\xi_i)$ for all $j = 0, \dots, \ell_i$ and (8) follows immediately. For $\xi_{\lambda_\kappa} (1 \leq \kappa \leq k)$ and $j = 0, \dots, \ell_{\lambda_\kappa}$,

$$(f\Lambda)^{(j)}(\xi_{\lambda_\kappa}) = \sum_{\tau=0}^{j} \binom{j}{\tau} f^{(j-\tau)}(\xi_{\lambda_\kappa}) \Lambda^{(\tau)}(\xi_{\lambda_\kappa}),$$

$$(H\Lambda)^{(j)}(\xi_{\lambda_\kappa}) = \sum_{\tau=0}^{j} \binom{j}{\tau} H^{(j-\tau)}(\xi_{\lambda_\kappa}) \Lambda^{(\tau)}(\xi_{\lambda_\kappa}).$$

Moreover,

$$\Lambda^{(\tau)}(\xi_{\lambda_\kappa}) = 0 \text{ if } \tau < \delta_\kappa, \tag{9}$$

$$f^{(j-\tau)}(\xi_{\lambda_\kappa}) = \hat{a}_{\lambda_\kappa, j-\tau} = H^{(j-\tau)}(\xi_{\lambda_\kappa}) \text{ if } \tau \geq \delta_\kappa. \tag{10}$$

The equality (9) holds because $\Lambda(x)$ has a factor $(x - \xi_{\lambda_\kappa})^{\delta_\kappa}$; the equality (10) follows from

$$j - \tau \leq \ell_{\lambda_\kappa} - \tau < \ell_{\lambda_\kappa} + 1 - \delta_\kappa = \min\{j \mid f^{(j)}(\xi_{\lambda_\kappa}) \neq \hat{a}_{\lambda_\kappa, j}\}.$$

Therefore, $f^{(j-\tau)}(\xi_{\lambda_\kappa})\Lambda^{(\tau)}(\xi_{\lambda_\kappa}) = H^{(j-\tau)}(\xi_{\lambda_\kappa})\Lambda^{(\tau)}(\xi_{\lambda_\kappa})$ for all $\tau = 0, \ldots, j$, and (8) is proved.

Let $d = D + E + \sum_{i=1}^{E} \ell_i$, $e = E + \sum_{i=1}^{E} \ell_i$, $R = f\Lambda$, and $S = \Lambda$. Then $\deg(r_0) = d + e + 1$, $\deg(H) \leq d + e$, $\deg(R) \leq d$, and $\deg(S) \leq e$. By Lemma 1, $r_\gamma/s_\gamma = R/S = f\Lambda/\Lambda = f$.

Lemma 3. *With the notation as in Algorithm 4.1, we have* $\Lambda = s_\gamma/\text{lc}(s_\gamma)$.

Proof. By Lemma 2, $r_\gamma = s_\gamma f$. On the other hand, $r_\gamma \equiv s_\gamma H \pmod{r_0}$. Therefore

$$s_\gamma f \equiv s_\gamma H \pmod{r_0}. \tag{11}$$

Let $(x - \xi_{\lambda_\kappa})^{\delta_\kappa}$ be a factor of Λ and denote $\epsilon_\kappa = \min\{j \mid f^{(j)}(\xi_{\lambda_\kappa}) \neq \hat{a}_{\lambda_\kappa, j}\}$, then

$$\delta_\kappa + \epsilon_\kappa = \ell_{\lambda_\kappa} + 1.$$

Since $(x - \xi_{\lambda_\kappa})^{\ell_{\lambda_\kappa}+1}$ is a factor of r_0, it follows from (11) that $(x - \xi_{\lambda_\kappa})^{\delta_\kappa+\epsilon_\kappa}$ divides $(f - H)s_\gamma$. In addition, $\epsilon_\kappa = \min\{j \mid f^{(j)}(\xi_{\lambda_\kappa}) \neq \hat{a}_{\lambda_\kappa, j} = H^{(j)}(\xi_{\lambda_\kappa})\}$ implies that

$$\text{GCD}((x - \xi_{\lambda_\kappa})^{\delta_\kappa+\epsilon_\kappa}, f - H) = (x - \xi_{\lambda_\kappa})^{\epsilon_\kappa}.$$

Therefore $(x - \xi_{\lambda_\kappa})^{\delta_\kappa}$ divides s_γ, and so Λ divides s_γ.

Assume that $s_\gamma = \Lambda w$ for some $w \in K[x]$, then $r_\gamma = s_\gamma f = f\Lambda w$, so the extended Euclidean scheme $s_\gamma r_1 + t_\gamma r_0 = r_\gamma$ becomes

$$f\Lambda w = H\Lambda w + t_\gamma r_0.$$

However, from Lemma 2, we know that $f\Lambda \equiv H\Lambda \pmod{r_0}$, which means there is $\tilde{t} \in K[x]$ such that

$$f\Lambda = H\Lambda + \tilde{t} r_0.$$

Therefore $t_\gamma = \tilde{t} w$, and this leads to $w \in K$ because $\text{GCD}(s_\gamma, t_\gamma) = 1$. Since the leading coefficient of Λ is 1, we have $s_\gamma = \text{lc}(s_\gamma)\Lambda$.

Note that Reed-Solomon decoding is a special case of our setting where $\ell_1 = \cdots = \ell_n = 0$ and $n = N = D+1+2E$. When $\ell_1 \geq 1$ and $n \leq D+2E$, our method requires $N = D + 1 + 2E + 2\sum_{i=1}^{E} \ell_i$ which is more than the values required by Reed-Solomon decoding. However, in some cases, the number of values we required is necessary for computing a unique interpolant $f(x)$, that is, there can be two valid interpolants if fewer values are given. We show this by the following example.

Example 1. Let $K = \text{algclo}(\mathbb{Q}) \cap \mathbb{R}$ be the real algebraic closure of $\mathbb{Q}$, and assume that $2E \leq D - 1$. Let $\xi_{2E+1}, \ldots, \xi_{D+2E}$ be D distinct points in K and

$$f(x) = \prod_{i=2E+1}^{D+2E} (x - \xi_i).$$

By Rolle's theorem, $f(x)'$ has $D - 1$ distinct roots in K, which allows us to choose $2E$ distinct points $\xi_1, \ldots, \xi_{2E}$ from these roots. Moreover, all the points $\xi_1, \ldots, \xi_{2E}, \xi_{2E+1}, \ldots, \xi_{D+2E}$ are distinct. Now we have $f'(\xi_1) = \cdots = f'(\xi_{2E}) = 0$ and $f(\xi_{2E+1}) = \cdots = f(\xi_{D+2E}) = 0$. Let $\ell_1 = \cdots = \ell_{2E} = 1$, $\ell_{2E+1} = \cdots = \ell_{D+2E} = 0$ and $n = D + 2E$, then $N = \sum_{i=1}^{n}(\ell_i + 1) = D + 4E$. Suppose the N values are given as follows:

$$\left.\begin{array}{ll} \hat{a}_{i,0} = f(\xi_i) & \text{for } i = 1, \ldots, E, \\ \hat{a}_{i,0} = 0 & \text{for } i = E+1, \ldots, D+2E, \\ \hat{a}_{i,1} = 0 & \text{for } i = 1, \ldots, 2E. \end{array}\right\} \tag{12}$$

If the E errors are $\hat{a}_{1,0}, \ldots, \hat{a}_{E,0}$, then 0 is a valid interpolant; if the E errors are $\hat{a}_{E+1,0}, \ldots, \hat{a}_{2E,0}$ then f is a valid interpolant. Thus for the points $\xi_1, \ldots, \xi_{D+2E}$ and the $D + 4E$ values in (12), there are ≥ 2 valid interpolants. $\square$

Example 2. As we have shown in Sect. 1, from Birkhoff problems with multiple solutions one obtains Hermite interpolation problems with errors that have multiple solutions. For instance, for the polynomial $f(x) = (x^2 - 1^2)^3(x^2 - 7^2)^3$ we have $f^{(j)}(\xi) = 0$ for $\xi = \pm 1$, $\xi = \pm 7$ and $j = 0, 1, 2$, and $f'(\xi) = 0$ for $\xi = 0$ and $\xi = \pm 5$. Therefore with those $n = 7$ arguments ξ and $\ell_i = 2$ for $1 \leq i \leq 7$, one has both f and the zero polynomial as a solution with $E = 3$ errors at $N = 21 = \deg(f) + 1 + 2E + 2$ values. $\square$

Example 3. If the field of scalars K has finite characteristic $\geq \ell_1 + 1$, our count (5) is optimal for higher derivatives. Let $n = 2E + 1$ and let $\ell_1 = \cdots = \ell_{2E+1} = p - 1$ for a prime number p which is the characteristic of the field of scalars K, whose cardinality is $|K| \geq 2E + 2$, so that there exist $n + 1$ distinct elements ξ_i in K. Let $f(x) = (x - \xi_1)^p$. Then $f(\xi_1) = 0$ and $f^{(j)}(\xi_i) = 0$ for all $1 \leq i \leq n$ and $1 \leq j \leq \ell_i$. Therefore f and the zero polynomial interpolate all $(2E + 1)p - 2E$ zero values, and E errors cannot be unambiguously corrected from $N = (2E + 1)\deg(f)$ values. If one adds an $(N + 1)$'st value $f(\xi_{n+1})$ then $N + 1 = \deg(f) + 1 + 2E + 2E(p - 1) = (2E + 1)p + 1$ (cf. (5)) and Algorithm 4.1 interpolates a unique polynomial with $\leq E$ erroneous values. $\square$

Remark 1. If all $\ell_i \leq 1$, we can prove that $N = 2D + 2E$ is the optimal count in the case that $n \geq 2E + 1$ which is necessary, and that $2E \geq D - 1$ and that the characteristic of K is either 0 or $\geq D + 1$. For $D = 0$ we have $N = n = 2E + 1$. We first show that the zero polynomial is the only interpolant of evaluations that yield 0 at any of $N - 2E$ of the evaluations. If $E_0 \leq 2E$ values $f(\xi_i)$ are removed, a non-zero polynomial of degree D can be zero at the remaining $n - E_0$ values only if $n - E_0 \leq D \iff E_0 \geq n - D$. There are $N - n \geq 0$

values of f', of which one removes $2E - E_0 \leq 2E - (n - D)$ values. There remain $\geq N - n - (2E - n + D) = D$ values of f' at distinct arguments, which are zero, which means $f' = 0$ and, by our assumption on the characteristic, $\deg(f) = 0$. Because $f(\xi_i) = 0$ at one of the $n \geq 2E + 1$ arguments ξ_i, $f = 0$.

If there are $N = 2D + 2E - 1$ values, we choose $n = 2E + D$. We know from Example 1 that there exists a non-zero polynomial f and argument values ξ_i for $1 \leq i \leq n = D + 2E$, such that $f(\xi_i) = 0$ for $i = 2E + 1, \ldots, 2E + D$ and $f'(\xi_i) = 0$ for $i = 1, \ldots D - 1 \leq 2E$. $\square$

5   The Rational Function Case

Let $f(x), g(x) \in \mathsf{K}[x]$, $g \neq 0$, $\deg(f) \leq D_f$, $\deg(g) \leq D_g$, $\mathrm{GCD}(f, g) = 1$. One is given a set of n distinct arguments $\xi_1, \ldots, \xi_n \in \mathsf{K}$, and for each argument ξ_i, one is given a row vector

$$\widehat{A}_{i,*} = [\hat{a}_{i,0}, \ldots, \hat{a}_{i,\ell_i}] \in (\mathsf{K} \cup \{\infty\})^{1 \times (\ell_i + 1)}.$$

We call $\hat{a}_{i,j}$ an error if one of the two cases happens: ξ_i is not a pole of $(f/g)^{(j)}$ and $\hat{a}_{i,j} \neq (f/g)^{(j)}(\xi_i)$, or, ξ_i is a pole of $(f/g)^{(j)}$ and $\hat{a}_{i,j} \neq \infty$. Let $\{\lambda_1, \ldots, \lambda_k\} \subset \{1, \ldots, n\}$ be the set of indices where every row vector $\widehat{A}_{\lambda_1,*}, \ldots, \widehat{A}_{\lambda_k,*}$ has at least one error, and let $E \geq k$ (if all row vectors are error-free then let $E = k = 0$). Let $\widehat{A}$ be the list of these row vectors:

$$\widehat{A} = \begin{bmatrix} \widehat{A}_{1,*} \\ \vdots \\ \widehat{A}_{n,*} \end{bmatrix} \in \left((\mathsf{K} \cup \{\infty\})^{1 \times (\ell_1 + 1)} \cup \cdots \cup (\mathsf{K} \cup \{\infty\})^{1 \times (\ell_n + 1)} \right)^n.$$

We assume that $\ell_1 \geq \cdots \geq \ell_n \geq 0$ and the last n_∞ (n_∞ can be zero) rows of $\widehat{A}$ only have one value ∞ and all other rows of $\widehat{A}$ have values in K, that is, we have the input specifications:

▸ $\ell_1 \geq \cdots \geq \ell_{n-n_\infty} \geq 0$; if the characteristic p of K is > 0, then $p \geq \ell_1 + 1$ is required.
▸ $\hat{a}_{i,j} \neq \infty$ for all $i = 1, \ldots, n - n_\infty$ and $0 \leq j \leq \ell_i$;
▸ $\hat{a}_{i,0} = \infty$ and $\ell_i = 0$ for all $i = n - n_\infty + 1, \ldots, n$.

For an arbitrary $\widehat{A}$, we process the inputs as is discussed in the following remark.

Remark 2. If for a location i one has $\hat{a}_{i,j} = \infty$ for all j, then a pole is indicated either truly or falsely. In this case we compress the list to a single value $\hat{a}_{i,0} = \infty$ and reset $\ell_i = 0$. For a true pole, and for characteristic of K either 0 or $\geq \deg(g) + 1$, all values are correct, but the additional $\hat{a}_{i,j} = \infty$ for $j \geq 1$ yield no additional information. In fact, $f(x) = 1/x^D$ cannot be interpolated from the values $f^{(j)}(0) = \infty$ for all $0 \leq j \leq D$ without errors. Our handling of poles is not a restriction of our algorithm, but is in the nature of the Hermite interpolation problem.

If for a location i, the list of values $\hat{a}_{i,0}, \hat{a}_{i,1}, \ldots$ is a mix of both elements $\in \mathsf{K}$ and ∞'s, we remove or truncate the list depending on the characteristic of K.

1. For characteristic of K either 0 or $\geq D_g + 1$, we have for $g(x) = (x - \alpha_1)^{\mu_1} \cdots (x - \alpha_\nu)^{\mu_\nu}$, α_i distinct $\in$ algclo(K), that

$$\left(\frac{f(x)}{g(x)}\right)' = \frac{f(x)'}{g(x)} - \frac{f(x)g(x)'}{g(x)^2}$$
$$= \frac{f(x)' \prod_i (x - \alpha_i) - f(x) \sum_i \mu_i \prod_{j \neq i}(x - \alpha_j)}{g(x) \prod_i (x - \alpha_i)}, \qquad (13)$$

where the right-side of (13) is a reduced rational function because no α_i is a root of the numerator: $f(\alpha_i) \neq 0$ because f/g is reduced, and $\mu_i \prod_{j \neq i}(\alpha_i - \alpha_j) \neq 0$ in K for all i by our assumption on the characteristic of K. Therefore, if the list $\hat{a}_{i,0}, \hat{a}_{i,1}, \ldots$ is a mix of both elements $\in \mathsf{K}$ and ∞'s, then some values in the list must be errors, so we remove the argument ξ_i and the list of values altogether. We also reduce the number of errors accordingly.

Note that our algorithms do not account for error distributions and assume the worst case. For instance, if in a list of $\ell_i = 20$ values there is a single ∞, we do not treat ∞ as a likely error. If fact, if there is a burst of errors, that ∞ may be the correct value.

2. For positive characteristic $p \leq \deg(g)$, a mix of ∞'s and field element values may not indicate an error: for $\xi_1 = 0$ and $f/g = (cx^{p+1} + 1)/x^p$, $(f/g)' = c$ and has no pole at 0. For such a field, if $\hat{a}_{i,0} \neq \infty$ and $\hat{a}_{i,j} = \infty$ for some $j \geq 1$, then either $\hat{a}_{i,0}$ or $\hat{a}_{i,j}$ is an error, so we remove the argument ξ_i and the list of values altogether and adjust the number of errors. Otherwise, we truncate the list to a single value $\hat{a}_{i,0} = \infty$ and reset $\ell_i = 0$. $\square$

Now we show how to recover the rational function f/g by the extended Euclidean algorithm with the following condition:

$$n + \sum_{i=E+1}^{n} \ell_i = n + \sum_{i=E+1}^{n-n_\infty} \ell_i = D_f + D_g + 1 + 2E + \sum_{i=1}^{E} \ell_i. \qquad (14)$$

Let E_∞ be the number of false poles, i. e., $E_\infty = |\{i \mid \hat{a}_{i,0} = \infty, g(\xi_i) \neq 0\}|$. Note that $n_\infty \leq D_g + E_\infty$. The condition (14) implies that $n - n_\infty \geq 2(E - E_\infty) + 1$, since otherwise we have the contradiction:

$$n_\infty + (n - n_\infty) + \sum_{i=E+1}^{n-n_\infty} \ell_i \leq (D_g + E_\infty) + 2(E - E_\infty) + (E - 2E_\infty)\ell_{E+1}$$

$$< D_f + D_g + 1 + 2E + \sum_{i=1}^{E} \ell_i.$$

The condition (14) can also be relaxed to $n + \sum_{i=E+1}^{n-n_\infty} \ell_i \geq D_f + D_g + 1 + 2E + \sum_{i=1}^{E} \ell_i$, because in that case, one can always adjust the ℓ_i's to achieve (14) (see Remark 4).

5.1   Error-Correcting Rational Function Hermite Interpolation

Input: ▸ A field K, nonnegative integers $D_f, D_g, E \in \mathbb{Z}_{\geq 0}$;
　　　▸ A set of distinct points $\{\xi_1, \ldots, \xi_n\} \subset$ K.
　　　▸ A list of n row vectors $\widehat{A} = [\widehat{A}_{i,*}]_{1 \leq i \leq n}$ and $n_\infty \in \mathbb{Z}_{\geq 0}$ where
　　　　　▸ $\ell_1 \geq \cdots \geq \ell_{n-n_\infty} \geq 0$, $\ell_{n-n_\infty+1} = \cdots = \ell_n = 0$;
　　　　　▸ the characteristic of K is either 0 or $\geq \ell_1 + 1$;
　　　　　▸ $\widehat{A}_{i,*} = [\hat{a}_{i,0}, \ldots, \hat{a}_{i,\ell_i}]$ and $\hat{a}_{i,j} \in$ K for all $i = 1, \ldots, n - n_\infty$ and
　　　　　　$j = 0, \ldots, \ell_i$;
　　　　　▸ $\hat{a}_{i,0} = \infty$ for all $i = n - n_\infty + 1, \ldots, n$;
　　　　　▸ $n + \sum_{i=E+1}^{n-n_\infty} \ell_i = D_f + D_g + 1 + 2E + \sum_{i=1}^{E} \ell_i$.
Output: ▸ The rational function $f/g \in$ K(x) such that
　　　　　▸ $f, g \in$ K$[x]$, $g \neq 0$, GCD$(f, g) = 1$;
　　　　　▸ $\deg(f) \leq D_f$ and $\deg(g) \leq D_g$;
　　　　　▸ f/g produces errors in $\leq E$ row vectors of $\widehat{A}$.
　　　▸ Or a message indicating there is no such function.

1. *If $\hat{a}_{i,j} = 0$ for all $i = 1, \ldots, n$ and $j = 0, \ldots, \ell_i$, then return $f/g = 0$.*
2. *Let $I_\infty = \{n - n_\infty + 1, \ldots, n\}$ and $P_\infty(x) = \prod_{i \in I_\infty}(x - \xi_i)$.*
3. *For $i = 1, \ldots, n - n_\infty$ and $j = 1, \ldots, \ell_i$, compute*

$$\hat{b}_{i,j} \stackrel{\text{def}}{=} \sum_{\tau=0}^{j} \binom{j}{\tau} \hat{a}_{i,\tau} P_\infty^{(j-\tau)}(\xi_i).$$

4. *Compute the polynomial Hermite interpolant $\bar{H}(x)$ of the data set $\{(\xi_i; \hat{b}_{i,0}, \ldots, \hat{b}_{i,\ell_i}) \mid i = 1, \ldots, n - n_\infty\}$ (namely $\bar{H}^{(j)}(\xi_i) = \hat{b}_{i,j}$, see Section 2). Let $H(x) = \bar{H}(x) P_\infty(x)$.*
5. *Let $r_0(x) = P_\infty(x) \prod_{i=1}^{n-n_\infty}(x - \xi_i)^{\ell_i+1}$, $r_1 = H$, $s_0 = 0$, $s_1 = 1$ and $\rho = 2$.*
   5a. *Compute the ρ-th Euclidean polynomial remainder r_ρ and the multiplier s_ρ in the extended Euclidean scheme $s_\rho r_1 + t_\rho r_0 = r_\rho$, namely*

$$r_\rho(x) = r_{\rho-2}(x) - q_\rho(x) r_{\rho-1}(x), \quad \deg(r_\rho) < \deg(r_{\rho-1}),$$
$$s_\rho(x) = s_{\rho-2}(x) - q_\rho(x) s_{\rho-1}(x).$$

   5b. *If $\deg(r_\rho) \leq D_f + n_\infty + E + \sum_{i=1}^{E} \ell_i$, then let $\gamma = \rho$ and go to step 6.*
   5c. *Otherwise, let $\rho = \rho + 1$ and go to Step 5a.*
   Step 5 computes (r_γ, s_γ) as in Lemma 1 with $d = D_f + E + \sum_{i=1}^{E} \ell_i + n_\infty$ and $e = D_g + E + \sum_{i=1}^{E} \ell_i - n_\infty$. We will prove in Lemma 4 that if there are $f, g \in$ K$[x]$ satisfy the output specifications, then $r_\gamma/(s_\gamma P_\infty^2) = f/g$. Here we also have $s_\gamma \neq 0$ because GCD$(s_\gamma, t_\gamma) = 1$ when $\gamma \geq 2$ and $s_1 = 1$.
6. *Compute $\Gamma = $ GCD(r_γ, s_γ) and $f/g = r_\gamma/(s_\gamma P_\infty^2)$ with GCD$(f, g) = 1$.*
   6a. *If $\deg(f) \leq D_f$ and $\deg(g) \leq D_g$, compute $k_1 = |\{i \mid 1 \leq i \leq n - n_\infty, \Gamma(\xi_i) = 0\}|$ and $k_2 = |\{i \mid n - n_\infty + 1 \leq i \leq n, g(\xi_i) \neq 0\}|$; if $k_1 + k_2 \leq E$ then return f/g.*
   6b. *If $\deg(f) > D_f$, or $\deg(g) > D_g$, or $k_1 + k_2 > E$, return a message indicating there is no such function.*

We will prove in Lemma 5 that the rational function f/g returned by Step 6a satisfies the output specifications. Therefore, we can check the validity of f/g without computing all the values $f^{(j)}(\xi_i)$ and $g^{(j)}(\xi_i)$ for $i = 1, \ldots, n$ and $j = 0, \ldots, \ell_i$.

We will define the error locator polynomial $\Lambda(x)$ in Lemma 4, and then based on Lemma 6, we show how to compute f/g and $\Lambda(x)$ more efficiently other than reducing the fraction $r_\gamma/(s_\gamma P_\infty^2)$ and evaluating Γ and g (see Remark 3).

Lemma 4. *We use the notation of Algorithm 5.1 and assume that there exists a rational function $f/g \in \mathsf{K}(x)$ which satisfies the output specifications. Let $\xi_{\lambda_1}, \ldots, \xi_{\lambda_k}$ be the arguments with erroneous values, that is, for indices $\lambda_\kappa \in \{\lambda_1, \ldots, \lambda_k\}$, there exists $j \in \{0, \ldots, \ell_{\lambda_\kappa}\}$ such that $\hat{a}_{i,j}$ is an error. For $\lambda_\kappa \notin I_\infty$, let $\delta_\kappa = \ell_{\lambda_\kappa} + 1 - \min\{j \mid \hat{a}_{\lambda_\kappa,j} \text{ is an error }\}$. Let*

$$\bar{\Lambda}(x) = \prod_{\lambda_\kappa \in \{\lambda_1, \ldots, \lambda_k\}\setminus I_\infty} (x - \xi_{\lambda_\kappa})^{\delta_\kappa},$$

$$\Lambda_\infty(x) = \prod_{\lambda_\kappa \in \{\lambda_1, \ldots, \lambda_k\}\cap I_\infty} (x - \xi_{\lambda_\kappa}),$$

$$g_\infty(x) = \prod_{1 \le \nu \le n,\, \nu \in I_\infty \setminus \{\lambda_1, \ldots, \lambda_k\}} (x - \xi_\nu). \tag{15}$$

Let $\Lambda(x) = \bar{\Lambda}(x)\Lambda_\infty(x)$ and $\bar{g} = g/g_\infty$. Then

$$f P_\infty \Lambda \equiv H \bar{g} \bar{\Lambda} \pmod{r_0}. \tag{16}$$

Moreover, $f/g = r_\gamma/(s_\gamma P_\infty^2)$, which implies the interpolant f/g is unique.

Proof. Note that $P_\infty = \Lambda_\infty g_\infty$, we have $H \bar{g} \bar{\Lambda} = \bar{H} P_\infty \bar{g} \bar{\Lambda} = \bar{H} g \Lambda$, hence (16) is equivalent to

$$f P_\infty \Lambda \equiv \bar{H} g \Lambda \pmod{r_0}. \tag{17}$$

By the same argument as in the proof of (7) in Lemma 2, proving (17) is equivalent to proving the following two equalities:

$$(f P_\infty \Lambda)(\xi_i) = (\bar{H} g \Lambda)(\xi_i) \text{ for } i \in I_\infty, \tag{18}$$

$$(f P_\infty \Lambda)^{(j)}(\xi_i) = (\bar{H} g \Lambda)^{(j)}(\xi_i) \text{ for } i \notin I_\infty,\ j = 0, \ldots, \ell_i. \tag{19}$$

Note that $\bar{H} g \Lambda = (\bar{H} \bar{g} \bar{\Lambda}) P_\infty$, therefore both sides of the equation in (18) are equal to zero because $P_\infty(\xi_i) = 0$ for $i \in I_\infty$.

It remains to prove (19). For $i \notin I_\infty$ and $j = 0, \ldots, \ell_i$,

$$(f P_\infty \Lambda)^{(j)}(\xi_i) = \sum_{\tau=0}^{j} \binom{j}{\tau} (f P_\infty)^{(j-\tau)}(\xi_i) \Lambda^{(\tau)}(\xi_i)$$

$$(\bar{H} g \Lambda)^{(j)}(\xi_i) = \sum_{\tau=0}^{j} \binom{j}{\tau} (\bar{H} g)^{(j-\tau)}(\xi_i) \Lambda^{(\tau)}(\xi_i),$$

we show that either $\Lambda^{(\tau)}(\xi_i) = 0$ or $(fP_\infty)^{(j-\tau)}(\xi_i) = (\bar{H}g)^{(j-\tau)}(\xi_i)$ by considering the following three cases.

Case 1. $\xi_i \notin \{\xi_{\lambda_1}, \ldots, \xi_{\lambda_k}\}$, then for $j = 0, \ldots, \ell_i$,

$$(fP_\infty)^{(j)}(\xi_i) = \sum_{\sigma=0}^{j} \binom{j}{\sigma} f^{(\sigma)}(\xi_i) P_\infty^{(j-\sigma)}(\xi_i) \tag{20}$$

$$= \sum_{\sigma=0}^{j} \binom{j}{\sigma} \sum_{\mu=0}^{\sigma} \binom{\sigma}{\mu} \hat{a}_{i,\sigma-\mu} g^{(\mu)}(\xi_i) P_\infty^{(j-\sigma)}(\xi_i) \tag{21}$$

$$= (\bar{H}g)^{(j)}(\xi_i). \tag{22}$$

The equality (21) follows from

$$f^{(\sigma)} = \left((f/g)\, g\right)^{(\sigma)} = \sum_{\mu=0}^{\sigma} \binom{\sigma}{\mu}(f/g)^{(\sigma-\mu)} g^{(\mu)}.$$

Case 2. $\xi_i = \xi_{\lambda_\kappa}$ for some $\kappa \in \{1, \ldots, k\}$ and $\tau < \delta_\kappa$, then $\Lambda^{(\tau)}(\xi_{\lambda_\kappa}) = 0$.
Case 3. $\xi_i = \xi_{\lambda_\kappa}$ for some $\kappa \in \{1, \ldots, k\}$ and $\tau \geq \delta_\kappa$, then $j - \tau < \min\{j \mid (f/g)^{(j)}(\xi_{\lambda_\kappa}) \neq \hat{a}_{\lambda_\kappa, j}\}$, and one can prove that $(fP_\infty)^{(j-\tau)}(\xi_{\lambda_\kappa}) = (\bar{H}g)^{(j-\tau)}(\xi_{\lambda_\kappa})$ as in (22).

Now (19) is proved, which completes the proof of (16). Let $R = fP_\infty\Lambda$ and $S = \bar{g}\bar{\Lambda}$, we rewrite (16) as

$$R \equiv SH \pmod{r_0}.$$

Let $d = D_f + E + \sum_{i=1}^{E} \ell_i + n_\infty$ and $e = D_g + E + \sum_{i=1}^{E} \ell_i - n_\infty$ we have $\deg(r_0) = d + e + 1$ by the input specifications of the Algorithm 5.1 (or the condition (14)). Moreover, $\deg(H) \leq d \mid c$, $\deg(R) \leq d$ and $\deg(S) \leq e$, by Lemma 1, we have $R/S = r_\gamma/s_\gamma$. Thus $f/g = R/(SP_\infty^2) = r_\gamma/(s_\gamma P_\infty^2)$.

Lemma 5. *Let $\Gamma = \mathrm{GCD}(r_\gamma, s_\gamma)$ and $f/g = r_\gamma/(s_\gamma P_\infty^2)$ with $\mathrm{GCD}(f, g) = 1$ be as in Step 6 of Algorithm 5.1. If $\deg(f) \leq D_f$, $\deg(g) \leq D_g$ and $k_1 + k_2 \leq E$, then f/g satisfies the output specifications of Algorithm 5.1.*

Proof. It is sufficient to prove that f/g produces errors in $\leq k_1$ row vectors of the list $[\widehat{A}_{1,*}, \ldots, \widehat{A}_{n-n_\infty,*}]$. By the extended Euclidean scheme $s_\gamma r_1 + t_\gamma r_0 = r_\gamma$,

$$r_\gamma \equiv s_\gamma H \pmod{P}, \tag{23}$$

where $H = r_1$ and $P = r_0$. Since $fP_\infty^2/g = r_\gamma/s_\gamma$ and $\Gamma = \mathrm{GCD}(r_\gamma, s_\gamma)$, (23) leads to

$$fP_\infty^2\Gamma \equiv gH\Gamma \pmod{P}. \tag{24}$$

By dividing P_∞, we get

$$fP_\infty\Gamma \equiv g\bar{H}\Gamma \pmod{\prod_{i=1}^{n-n_\infty} (x - \xi_i)^{\ell_i+1}}. \tag{25}$$

Therefore, if $\Gamma(\xi_i) \neq 0$, then $(fP_\infty)^{(j)}(\xi_i) = (g\bar{H})^{(j)}(\xi_i)$ for all $j = 0, \ldots, \ell_i$, and this equality expands to (20), (21), and (22). Because $P_\infty(\xi_i) \neq 0$ for all $i = 1, \ldots, n - n_\infty$, it follows from (20, 21) that $f^{(j)}(\xi_i) = \sum_{\mu=0}^{j} \binom{j}{\mu} \hat{a}_{i,j-\mu} g^{(\mu)}(\xi_i)$ if $i \in \{1, \ldots, n - n_\infty\}$ and $\Gamma(\xi_i) \neq 0$. This means for f/g, the list $[\hat{A}_{1,*}, \ldots, \hat{A}_{n-n_\infty,*}]$ has at least $n - n_\infty - k_1$ error-free row vectors.

From Lemma 4 and Lemma 5, we conclude the correctness of the Algorithm 5.1 in the following theorem.

Theorem 1. *Let D_f, D_g, E and $\ell_1 \geq \cdots \geq \ell_n$ be nonnegative integers, and let K be a field of characteristic $\geq \ell_1 + 1$. For a set of n distinct points $\{\xi_1, \ldots, \xi_n\} \subset \mathsf{K}$ and a list of n row vectors $\hat{A} = [\hat{A}_{i,*}]_{1 \leq i \leq n}$ with $\hat{A}_{i,*} = [\hat{a}_{i,0}, \ldots, \hat{a}_{i,\ell_i}] \in (\mathsf{K} \cup \{\infty\})^{1 \times (\ell_i + 1)}$, if $\hat{A}$ satisfies the input specifications of the Algorithm 5.1, then either there is a unique rational function interpolant f/g satisfying the output specifications and the Algorithm 5.1 will return it, or there is no such rational function interpolant and the Algorithm 5.1 will report the nonexistence.*

Lemma 6. *With the notation as in Algorithm 5.1 and Lemma 4, we have*

$$\mathrm{GCD}(r_\gamma, s_\gamma) = \bar{A} \cdot \mathrm{GCD}(\bar{g}, g_\infty),$$

and Λ_∞^2 divides r_γ.

Proof. We first prove that $\bar{A}$ divides s_γ and r_γ. Since $r_\gamma \equiv s_\gamma H \pmod{r_0}$, we have

$$r_\gamma \bar{g} \equiv s_\gamma H \bar{g} = s_\gamma (\bar{H}g) \Lambda_\infty \pmod{r_0}. \tag{26}$$

On the other hand, let $R = fP_\infty \Lambda$ and $S = \bar{g}\bar{A}$, as it is shown in the proof of Lemma 4 that

$$r_\gamma S = s_\gamma R, \tag{27}$$

which is $r_\gamma(\bar{g}\bar{A}) = s_\gamma fP_\infty \Lambda$. Because $\bar{A} \neq 0$, dividing $\bar{A}$ on both sides results in

$$r_\gamma \bar{g} \equiv s_\gamma(fP_\infty) \Lambda_\infty \pmod{r_0}. \tag{28}$$

Combining (26) and (28) leads to

$$s_\gamma(\bar{H}g)\Lambda_\infty \equiv s_\gamma(fP_\infty)\Lambda_\infty \pmod{r_0}. \tag{29}$$

Since $\bar{A}$ is a factor of r_0 and $\mathrm{GCD}(\bar{A}, \Lambda_\infty) = 1$, $\bar{A}$ divides $s_\gamma(\bar{H}g - fP_\infty)$. Let $(x - \xi_{\lambda_\kappa})^{\delta_\kappa}$ be a factor of $\bar{A}$, and let $\epsilon_\kappa = \min\{j \mid (f/g)^{(j)}(\xi_{\lambda_\kappa}) \neq \hat{a}_{\lambda_\kappa, j}\}$, using the same argument as in the proof of Lemma 3, one can prove that $(x - \xi_{\lambda_\kappa})^{\delta_\kappa}$ divides s_γ, and so $\bar{A}$ divides s_γ. Because $r_\gamma \equiv s_\gamma H \pmod{r_0}$, $\bar{A}$ also divides r_γ.

Now assume that $\mathrm{GCD}(r_\gamma, s_\gamma) = \bar{A}w$ for some $w \in \mathsf{K}[x]$. Let $v = \mathrm{GCD}(\bar{g}, g_\infty)$, then $\mathrm{GCD}(R, S) = \bar{A}v$. From (27), we have the reduced fractions:

$$\frac{r_\gamma/(\bar{A}w)}{s_\gamma/(\bar{A}w)} = \frac{R/(\bar{A}v)}{S/(\bar{A}v)}, \tag{30}$$

which implies that

$$r_\gamma/w = c(R/v), \ s_\gamma/w = c(S/v) \text{ for some } c \in \mathsf{K} \setminus \{0\}. \tag{31}$$

Combining the Euclidean scheme $s_\gamma r_1 + t_\gamma r_0 = r_\gamma$, we have

$$\frac{R - SH}{v} = \frac{r_\gamma - s_\gamma H}{cw} = \frac{t_\gamma r_0}{cw}, \tag{32}$$

and so $(R - SH)/r_0 = (t_\gamma v)/(cw)$. By Lemma 4, $(R - SH)/r_0$ is a polynomial, which implies that w divides $(t_\gamma v)$. But $\mathrm{GCD}(t_\gamma, w) = 1$ because s_γ and t_γ are relatively prime, therefore w divides v.

We now prove that $v = w$: suppose $v = ww^*$ with $\deg(w^*) \geq 1$. Since v divides g_∞, there exists a ξ_i, with $1 \leq i \leq n$ and $i \notin \{\lambda_1, \ldots, \lambda_k\}$, such that $g_\infty(\xi_i) = w^*(\xi_i) = 0$. We have the following contradiction:

$$0 = r_\gamma(\xi_i) - H(\xi_i)s_\gamma(\xi_i) \tag{33}$$
$$= cf(\xi_i)\Lambda(\xi_i)\Lambda_\infty(\xi_i)(g_\infty/w^*)(\xi_i) \tag{34}$$
$$\neq 0. \tag{35}$$

The Eq. (33) follows from $r_\gamma = s_\gamma r_1 + t_\gamma r_0$; (34) is a consequence of (31) and $P_\infty(\xi_i) = 0$; since g_∞ in (15) has single roots, $(g_\infty/w^*)(\xi_i) \neq 0$, which leads to (35).

Finally, $f/g = r_\gamma/(s_\gamma P_\infty^2) = r_\gamma/(s_\gamma \Lambda_\infty^2 g_\infty^2)$ and $\mathrm{GCD}(g, \Lambda_\infty) = 1$, so Λ_∞^2 must be a factor of the numerator r_γ.

Remark 3. Instead of computing f/g by reducing the fraction $r_\gamma/(s_\gamma P_\infty^2)$ as in Step 6 of the Algorithm 5.1, we can compute $\bar{\Lambda}$ and Λ_∞ first, and then compute f/g. In other words, for computing Λ and f/g, we can replace the Step 6 of the Algorithm 5.1 with the following steps:

6a. *Compute* $\Gamma = \mathrm{GCD}(r_\gamma, s_\gamma)$, $\tilde{r} = r_\gamma/\Gamma$, *and* $\tilde{s} = s_\gamma/\Gamma$.
6b. *Compute* $w = \mathrm{GCD}(\Gamma, P_\infty)$, $u = P_\infty/w$ *and* $\bar{\Lambda} = \Gamma/w$.
6c. *Compute* $\tilde{u} = \tilde{r}/u$, $\Lambda_\infty = \mathrm{GCD}(\tilde{u}, P_\infty)$, $\tilde{f} = \tilde{u}/\Lambda_\infty$, $g_\infty = P_\infty/\Lambda_\infty$, *and* $\tilde{g} = \tilde{s} w g_\infty$.
6d. *Let* k_1 *and* k_2 *be the number of distinct factors of* $\bar{\Lambda}$ *and* Λ_∞ *respectively.*
6d(i) *If* $\deg(\tilde{f}) \leq D_f$, $\deg(\tilde{g}) \leq D_g$, *and* $k_1 + k_2 \leq E$, *return* $\tilde{f}/\tilde{g}$, $\bar{\Lambda}$, *and* Λ_∞.
6d(ii) *Else, return a message indicating there are no* $f, g \in \mathsf{K}[x]$ *such that* $\deg(f) \leq D_f$, $\deg(g) \leq D_g$ *and* f/g *produces errors in* $\leq E$ *row vectors of* $\widehat{A}$.

Proof. By Lemma 6, $\Gamma = \bar{\Lambda} \cdot \mathrm{GCD}(\tilde{g}, g_\infty)$. Since $g_\infty(\xi_{\lambda_\kappa}) \neq 0$ (see (15)) for all $\lambda_\kappa \in \{\lambda_1, \ldots, \lambda_k\}$, we have $\mathrm{GCD}(\Lambda, g_\infty) = 1$, thus $\mathrm{GCD}(\Gamma, P_\infty) = \mathrm{GCD}(\tilde{g}, g_\infty)$ and $\Gamma = \bar{\Lambda} w$. From (31), we have

$$\tilde{r} = r_\gamma/\Gamma = c(R/\Gamma) \text{ and } \tilde{s} = s_\gamma/\Gamma = c(S/\Gamma) \text{ for some } c \in \mathsf{K} \setminus \{0\}.$$

Using the substitutions $R = f \Lambda P_\infty$, $S = \bar{g}\bar{\Lambda}$, $\Gamma = \bar{\Lambda} w$, and $u = P_\infty/w$, one can verify that $\tilde{u} = cf\Lambda_\infty$ and $\tilde{g} = cg$. Because f and g are relatively prime, we have $\mathrm{GCD}(f, g_\infty) = 1$, so $\mathrm{GCD}(\tilde{u}, P_\infty) = \Lambda_\infty$.

6   Further Remarks

Remark 4. As stated in the introduction, the sufficient conditions (1, 5, 14) for an interpolation profile of orders of derivatives $\ell_1 \geq \cdots \geq \ell_n$ at distinct arguments may oversample, because the ℓ_i's are on both sides and could be reduced simultaneously while preserving the conditions. Therefore, one can add the following "pre-processing data" Step 0 at the beginning of the Algorithm 4.1, which may reduce the number of values for recovering f and improve the efficiency of the algorithm.

0. *For every* $j = 0, 1, \ldots, \ell_1$, *let* $m_j = \max\{ i \mid \hat{a}_{i,j}$ *is given as input*$\}$ *(the dimension of the j-th column of $\widehat{A}$, the number of inputs for the j-th derivative). Furthermore, let* $M_j = \sum_{\mu=0}^{j} m_\mu$, *which is the number of inputs up to the j-th derivative. Compute the minimal β such that* $M_\beta \geq D + 1 + 2(\beta + 1)E$. *Let*

$$N^{[\text{new}]} = D + 1 + 2(\beta + 1)E \tag{36}$$

and

$$\ell_i^{[\text{new}]} = \begin{cases} \beta & \text{for } 1 \leq i \leq N^{[\text{new}]} - M_{\beta-1}, \\ \beta - 1 & \text{for } N^{[\text{new}]} - M_{\beta-1} + 1 \leq i \leq m_\beta, \\ \ell_i & \text{for } i > m_\beta. \end{cases} \tag{37}$$

Now $m_\beta^{[\text{new}]} = N^{[\text{new}]} - M_{\beta-1}$ *and* $\sum_{i=1}^{n}(\ell_i^{[\text{new}]} + 1) = N^{[\text{new}]}$.

One can also add Step 0 at the beginning of Algorithm 5.1 by replacing D with $D_f + D_g$. Fig. 1 shows how Step 0 removes redundant values. Recall we are given n distinct points $\xi_1, \ldots, \xi_n$, and for each point ξ_i, we are given a row vector of values: $\widehat{A}_{i,*} = [\hat{a}_{i,0}, \ldots, \hat{a}_{i,\ell_i}]$ with $\ell_1 \geq \cdots \geq \ell_n \geq 0$, and $\widehat{A}$ is the list of these row vectors

$$\widehat{A} = \begin{bmatrix} \widehat{A}_{1,*} \\ \vdots \\ \widehat{A}_{n,*} \end{bmatrix}.$$

$\widehat{A}$ is shown as the "staircase" in Figure 1, which has $D = 15$, $E = 2$, $n = 8$, $\ell_1 = 11$, $\ell_2 = 10$, $\ell_3 = \ell_4 = 8$, $\ell_5 = \ell_6 = 7$, $\ell_7 = 3$, $\ell_8 = 0$, $N = 62$, $\beta = 5$, $N^{[\text{new}]} = 40$, $\ell_1^{[\text{new}]} = \cdots = \ell_5^{[\text{new}]} = 5$, $\ell_6^{[\text{new}]} = 4$. Intuitively, Step 0 cuts $\widehat{A}$ by the red line and removes the right part, and the left part has $N^{[\text{new}]}$ values.

Lemma 7. *The β computed in the Step 0 above is no more than D.*

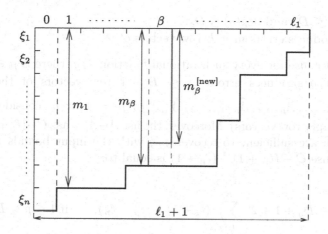

Fig. 1. Truncation by β

Proof. By the minimality of β, we have $M_{\beta-1} \leq D + 2\beta E$, and so

$$m_\beta = M_\beta - M_{\beta-1} \geq N^{[\text{new}]} - M_{\beta-1} \geq 2E + 1.$$

Moreover, $m_0 \geq \cdots \geq m_{\beta-1} \geq m_\beta \geq 2E + 1$, thus

$$(2E + 1)\beta \leq \sum_{j=0}^{\beta-1} m_j = M_{\beta-1} \leq D + 2\beta E,$$

which concludes that $\beta \leq D$.

Remark 5. If we are given an error rate $1/q$ ($q \in \mathbb{Z}_{\geq 3}$) instead of an upper bound E on the number of errors, and we are also given bounds

(i) $D_f \geq \deg(f)$, $D_g \geq \deg(g)$
(ii) $\beta = \max\{j \mid (f/g)^{(j)} \text{ is available for evaluation}\}$,

then the Algorithm 5.1 can recover f/g for $q - 2(\beta+1) = \eta > 0$ with $n = \lceil \frac{q\delta}{(\beta+1)\eta} \rceil$ distinct arguments and $N = \delta + 2(\beta+1)\lfloor \frac{\delta}{\eta} \rfloor$ values where $\delta = D_f + D_g(\beta+1) + 1$. Cf. [9, Remark 1.1] and [10, Remark 1.1].

Remark 6. In the input specifications of the Algorithm 5.1, the number of values in $\widehat{A}$ is $C = D_f + D_g + 1 + 2\sum_{i=1}^{E}(\ell_i + 1)$ (see also (1)), which guarantees that the Algorithm 5.1 either returns a unique valid interpolant f/g or determines no such interpolant exists. If $E \geq 1$ and $\widehat{A}$ has $C - (\ell_n + 1)$ values, we can use Algorithm 5.1 on every $n - 1$ row vectors of $\widehat{A}$ and with input $D_f, D_g, E - 1$, to compute all possible rational functions f/g which satisfy:

▸ $f, g \in \mathsf{K}[x]$, $g \neq 0$, $\mathrm{GCD}(f, g) = 1$;

- $\deg(f) \leq D_f$ and $\deg(g) \leq D_g$;
- f/g produces errors in $\leq E$ row vectors of $\widehat{A}$.

This is because for every such rational function f/g, there is $\mu \in \{1, \ldots, n\}$ for which f/g produces errors in $\leq E - 1$ row vectors of the list $\widehat{A} - \widehat{A}_{\mu,*} \stackrel{\text{def}}{=} [\widehat{A}_{1,*}, \ldots, \widehat{A}_{\mu-1,*}, \widehat{A}_{\mu+1,*}, \ldots, \widehat{A}_{n,*}]$ (if $\mu = 1$ or n, consider $\widehat{A}_{0,*}$ and $\widehat{A}_{n+1,*}$ as empty row vectors). Moreover, the list $\widehat{A} - \widehat{A}_{\mu,*}$ has $C - (\ell_n + 1) - (\ell_\mu + 1)$ values which are sufficient to recover f/g with the input bounds D_f, D_g and $E - 1$, because $C - (\ell_n + 1) - (\ell_\mu + 1)$ is equal to:

$$
\begin{cases}
D_f + D_g + 1 + 2 \sum_{i=1, i \neq \mu}^{E} (\ell_i + 1) + (\ell_\mu - \ell_n), & \text{if } 1 \leq \mu \leq E, \\
D_f + D_g + 1 + 2 \sum_{i=1}^{E-1} (\ell_i + 1) + (2\ell_E - \ell_\mu - \ell_n), & \text{if } E + 1 \leq \mu \leq n
\end{cases}
$$

This method can be generalized to situations where $\widehat{A}$ has $C - \sum_{i=n}^{n-n_0}(\ell_i + 1)$ values and n_0 is a small constant compared to E. For the polynomial case with a uniform derivative profile, that is, $D_g = 0$ and $\ell_1 = \cdots = \ell_n$, [4] and [12] give algorithms to list-decode derivative (or multiplicity) codes by solving differential equations.

7   Conclusion

Interpolation algorithms go back to ancient Chinese mathematicians. Algorithms that also can tolerate errors in the evaluations appeared as error correction algebraic codes in the early1960 s. Table 1 gives a brief history. Our paper completes the second column by giving an error correction interpolation algorithm of nearly linear arithmetic complexity for a rational function from values at its derivatives.

Table 1. A brief history of univariate interpolation.

	Polynomial	*Rational Function*
at values	Sun-Tsu/Lagrange, Guo Shoujing/Newton	Cauchy
at values of derivatives	Hermite, Birkhoff	Warner [20]
at values with errors	Reed and Solomon [16]	Beelen, Høholdt, Nielsen, Wu [1]
at values of derivatives with errors	Multiplicity codes: Rosenbloom and Tsfasman [17]	This paper

A   Appendix

Notation (in alphabetic order):

$\hat{a}_{i,j}$	the input value for the j-th derivative of f, or an error, at the i-th point
$\widehat{A}_{i,*}$	$= [\hat{a}_{i,0}, \ldots, \hat{a}_{i,\ell_i}]$, the row vector of values for the i-th point ξ_i
$\widehat{A}$	$= [\widehat{A}_{1,*}, \ldots, \widehat{A}_{n,*}]^T$, the collection of all input values
$\hat{b}_{i,j}$	$= \sum_{\tau=0}^{j} \binom{j}{\tau} \hat{a}_{i,\tau} P_{\infty}^{(j-\tau)}(\xi_i)$ the value for the j-th derivative of H at the i-th point
β	the minimal integer such that there are $\geq D + 1 + 2E + 2\beta E$ values for derivatives of order $\leq \beta$
c_j	the coefficient of x^j in f
D	an upper bound of the degree of the polynomial interpolant
D_f	an upper bound of the degree of the numerator of the rational interpolant
D_g	an upper bound of the degree of the denominator interpolant
δ_κ	$= \ell_{\lambda_\kappa} + 1 - \min\{ j \mid \hat{a}_{\lambda_\kappa,j} \text{ is an error}\}$
E	an upper bound on the number of errors in the input values to the algorithm
ξ_i	the i-th interpolation point
ξ_{λ_κ}	$1 \leq \kappa \leq k$, are the points with erroneous values, namely, $\exists j$ s.t. $\hat{a}_{\lambda_\kappa,j}$ is an error
ϵ_κ	$= \min\{ j \mid \hat{a}_{\lambda_\kappa,j} \text{ is an error}\} = \ell_{\lambda_\kappa} + 1 - \delta_\kappa$
f	polynomial interpolant or numerator of the rational interpolant for the correct values
g	the denominator of the rational interpolant for the correct values
$\bar{g}$	a factor of g indicating true non-poles
g_∞	a factor of g indicating true poles
H	the polynomial Hermite interpolant for all input values (including $\leq E$ errors)
I_∞	$= \{i \mid \exists j \text{ s.t. } \hat{a}_{i,j} = \infty\}$
k	the actual number of points with erroneous input values
K	a field
ℓ_i	the highest derivative order at the i-th point
Λ	the error locator polynomial
$\bar{\Lambda}$	$= \prod_{\kappa \in \{1,\ldots,k\}, \lambda_\kappa \notin I_\infty} (x - \xi_{\lambda_\kappa})^{\delta_\kappa}$
Λ_∞	$= \prod_{\kappa \in \{1,\ldots,k\}, \lambda_\kappa \in I_\infty} (x - \xi_{\lambda_\kappa})$
m_j	the number of input values for the j-th derivative of f
M_j	the number of input values for up to the j-th derivative of f
n	the number of distinct points
n_∞	degree of P_∞
N	the number of the input values
P_∞	$= \prod_{\exists j \text{ s.t. } \hat{a}_{i,j}=\infty} (x - \xi_i)$, the polynomial indicating all poles
r_0	$= (x - \xi_1)^{\ell_1+1} \cdots (x - \xi_n)^{\ell_n+1}$
r_γ	the γ-th remainder of the Euclidean polynomial remainder sequence $r_0, r_1, \ldots$
s_γ	the Bézout coefficient of r_1 in the γ-th extended Euclidean scheme: $s_\gamma r_1 + t_\gamma r_0 = r_\gamma$
t_γ	the Bézout coefficient of r_0 in the γ-th extended Euclidean scheme: $s_\gamma r_1 + t_\gamma r_0 = r_\gamma$

References

1. Beelen, P., Høholdt, T., Nielsen, J.S.R., Wu, Y.: On rational interpolation-based list-decoding and list-decoding binary Goppa codes. IEEE Trans. Inf. Theory shape it. **59**(6), 3269–3281 (2013)
2. Chin, F.Y.: A generalized asymptotic upper bound for fast polynomial evaluation and interpolation. SIAM J. Comput. 5(4), 682–690 (1976)

3. Coxon, N.: Fast systematic encoding of multiplicity codes. J. Symbolic Comput. **94**, 234–254 (2019)
4. Guruswami, V., Wang, C.: Optimal rate list decoding via derivative codes. In: Goldberg, L.A., Jansen, K., Ravi, R., Rolim, J.D.P. (eds.) APPROX/RANDOM -2011. LNCS, vol. 6845, pp. 593–604. Springer, Heidelberg (2011). https://doi.org/10.1007/978-3-642-22935-0_50
5. Guruswami, V., Wang, C.: Linear-algebraic list decoding for variants of Reed-Solomon codes. IEEE Trans. Inf. Theory **59**(6), 3257–3268 (2013)
6. Gustavson, F.G., Yun, D.Y.Y.: Fast computation of the rational Hermite interpolant and solving Toeplitz systems of equations via the extended Euclidean algorithm. In: Ng, E.W. (ed.) Proceedings of the International Symposium on Symbolic and Algebraic Computation. pp. 58–64. EUROSAM '79, Springer, Berlin, Heidelberg (1979)
7. Kaltofen, E., Trager, B.: Computing with polynomials given by black boxes for their evaluations: Greatest common divisors, factorization, separation of numerators and denominators. J. Symbolic Comput. **9**(3), 301–320 (1990)
8. Kaltofen, E., Pernet, C., Storjohann, A., Waddell, C.A.: Early termination in parametric linear system solving and rational function vector recovery with error correction. In: Burr, M. (ed.) ISSAC '17 Proceedings of 2017 ACM International Symposium Symbolic Algebraic Computer. pp. 237–244. Association for Computing Machinery, New York (2017). http://users.cs.duke.edu/~elk27/bibliography/17/KPSW17.pdf
9. Kaltofen, E., Yang, Z.: Sparse multivariate function recovery from values with noise and outlier errors. In: Kauers, M. (ed.) ISSAC 2013 Proceedings of 38th International Symposium Symbolic Algebraic Computer. pp. 219–226. Association for Computing Machinery, New York (2013). http://users.cs.duke.edu/~elk27/bibliography/13/KaYa13.pdf EKbib/13/KaYa13.pdf
10. Kaltofen, E., Yang, Z.: Sparse multivariate function recovery with a high error rate in evaluations. In: Nabeshima, K. (ed.) ISSAC 2014 Proceedings of 39th International Symposium Symbolic Algebraic Computer pp. 280–287. Association for Computing Machinery, New York (2014) http://users.cs.duke.edu/~elk27/bibliography/14/KaYa14.pdf
11. Kopparty, S.: Some remarks on multiplicity codes. In: Barg, A., Musin, O.R. (eds.) Discrete Geometry and Algebraic Combinatorics: AMS Spec. Session. Contemporary Mathematics, vol. 625, pp. 155–176 (2014)
12. Kopparty, S.: List-decoding multiplicity codes. Theor. Comput. **11**(1), 149–182 (2015)
13. Kopparty, S., Saraf, S., Yekhanin, S.: High-rate codes with sublinear-time decoding. J. ACM (JACM) **61**(5), 1–20 (2014)
14. Moenck, R.T.: Fast computation of GCDs. In: Proceedings of 5th ACM Symposium Theory Computing pp. 142–151 (1973)
15. Nielsen, R.R.: List decoding of linear block codes. Ph.D. thesis, Technical University of Denmark (2001)
16. Reed, I.S., Solomon, G.: Polynomial codes over certain finite fields. J. Soc. Ind. Appl. Math. **8**(2), 300–304 (1960)
17. Rosenbloom, M.Y., Tsfasman, M.A.: Codes for the m-metric. Problemy Peredachi Informatsii **33**(1), 55–63 (1997)
18. Schoenberg, I.J.: On Hermite-Birkhoff interpolation. J. Math. Anal. Appl. **16**, 538–543 (1967). https://doi.org/10.1016/0022-247X(66)90160-0
19. Sugiyama, Y., Kasahara, M., Hirasawa, S., Namekawa, T.: A method for solving key equation for decoding Goppa codes. Inf. Control **27**(1), 87–99 (1975)

20. Warner, D.D.: Hermite interpolation with rational functions. Ph.D. thesis, University of California, San Diego (1974)
21. Welch, L.R., Berlekamp, E.R.: Error correction of algebraic block codes. US Patent 4,633,470 (1986). http://patft.uspto.gov/

Good Pivots for Small Sparse Matrices

Manuel Kauers[ID] and Jakob Moosbauer[(✉)][ID]

Institute for Algebra, Johannes Kepler University, Linz, Austria
{manuel.kauers,jakob.moosbauer}@jku.at

Abstract. For sparse matrices up to size 8×8, we determine optimal choices for pivot selection in Gaussian elimination. It turns out that they are slightly better than the pivots chosen by a popular pivot selection strategy, so there is some room for improvement. We then create a pivot selection strategy using machine learning and find that it indeed leads to a small improvement compared to the classical strategy.

1 Introduction

It can be cumbersome to solve a sparse linear system with Gaussian elimination because a poor choice of a pivot can have a dramatic effect on the sparsity. In the worst case, we start with a fairly sparse matrix, and already after a small number of elimination steps, we are faced with a dense matrix, for which continuing the elimination procedure may be too costly. The principal goal of a linear system solver for sparse matrices is therefore to maintain as much of the sparsity as possible, for as long as possible. To achieve this goal, we can pay special attention to the pivot selection. A popular pivot selection strategy which aims at maintaining the sparsity of a matrix is attributed to Markowitz [3,10]. It is based on the notion of "fill-in", which is defined as the number of matrix entries that get affected when a particular element is chosen as pivot. More precisely, for a matrix $A = ((a_{i,j}))_{i,j=1}^{n,m}$, let r_i be the number of nonzero entries of the ith row $(i = 1, \ldots, n)$ and c_j be the number of nonzero entries of the jth column $(j = 1, \ldots, m)$. Then the fill-in associated to the entry at position (i,j) is defined as $(r_i - 1)(c_j - 1)$. Note that this is exactly the number of cells into which something gets added when the entry at (i,j) is chosen as pivot:

M.K. was supported by the Austrian FWF grants F50-04, W1214-13, and P31571.
J.M. was supported by the Land Oberüsterreich through the LIT-AI Lab.

© Springer Nature Switzerland AG 2020
F. Boulier et al. (Eds.): CASC 2020, LNCS 12291, pp. 358–367, 2020.
https://doi.org/10.1007/978-3-030-60026-6_20

The selection strategy of Markowitz is to choose among the eligible candidates a pivot for which the fill-in is minimized. The strategy thus neglects that touching a cell that already is nonzero does not decrease the sparsity (it may in fact increase if we are lucky enough).

It is known that finding a pivot that minimizes the number of new entries introduced during the whole elimination process has been shown to be NP-complete by Yannakakis [16]. But how much does this matter? In other words: how close does the Markowitz pivot selection strategy get to the theoretical optimum? This is the first question we address in this paper. By an exhaustive search through all square matrices up to size 8×8 in an idealized setting, we have determined the pivot choices that minimize the total number of operations. As expected, it turns out that with the optimal pivot choice, the number of operations is indeed smaller than with the Markowitz strategy, albeit just by a small amount. This confirms the common experience that the Markowitz strategy is a good approach, especially since it also has the feature that it can be easily implemented and does not cost much. Nevertheless, there is some room for improvement, and the second question we address in this paper is how this room for improvement could possibly be exploited. We tried to do so by training a pivot selection strategy using machine learning. It turns out that the resulting neural network performs indeed a bit better than the Markowitz strategy, at least in the setting under consideration.

Our study is limited to small matrices because determining the optimal pivots by exhaustive search is prohibitively expensive for larger matrices. It is clear that sparsity optimization for matrices of this size does not have any practical relevance. In fact, it may be argued that there are no nonzero sparse matrices of size 8×8 at all, because each such matrix has at least 12.5% nonzero entries, which is far more than the sparsity of matrices arising in many numerical applications. However, our results indicate that the gap between Markowitz criterion and the optimal choice can possibly be narrowed by adequate use of machine learning, and it may have some relevance for sparse matrices with symbolic entries for which the cost of arithmetic is so high that spending some additional time on searching for a better pivot may be justified. In future work, we will investigate whether the machine learning approach can also be used to construct a selection strategy for symbolic matrices of more realistic sizes.

2   Algebraic Setting

We will distinguish two kinds of matrix entries: 0 (zero) and $*$ (nonzero), and we ignore the possibility of accidental cancellations, so we adopt the simplifying assumption that the sum of two nonzero elements is always nonzero. As coefficient domain, we therefore take the set $S = \{0, *\}$ together with addition and multiplication defined as follows:

$$
\begin{array}{c|cc}
+ & 0 & * \\
\hline
0 & 0 & * \\
* & * & *
\end{array}
\qquad
\begin{array}{c|cc}
\cdot & 0 & * \\
\hline
0 & 0 & 0 \\
* & 0 & *
\end{array}.
$$

The operations we count are $*+*$ and $*\cdot*$, i.e., operations not involving zero. As we have primarily applications with symbolic matrices in mind (originating, e.g., from applications in symbolic summation [9], Gröbner bases computation [5], or experimental mathematics [1,2]), we do not worry about stability issues.

The exact number of operations depends not only on the choice of the pivot but also on how the elimination is performed. We consider two variants. In the first variant, we add a suitable multiple of the pivot row to all rows which have a nonzero entry in the pivot column:

$$\begin{pmatrix} a & b & 0 \\ 0 & c & d \\ e & f & g \end{pmatrix} \begin{array}{c} {}^{-e/a} \\ \rightharpoondown \\ \hookleftarrow_{+} \end{array} \rightsquigarrow \begin{pmatrix} a & b & 0 \\ 0 & c & d \\ 0 & f - eb/a & g \end{pmatrix}.$$

In this case, we count the following operations. First, there are $c-1$ divisions to compute the factors corresponding to e/a in the above sketch, where c is the number of nonzero elements in the pivot column. Secondly, there are $(r-1)(c-1)$ multiplications to compute all the numbers corresponding to eb/a, where r is the number of nonzero elements in the pivot row. Finally, for each clash of two nonzero elements in the submatrix, like f and eb/a in the sketch above, we count one addition.

The second variant is inspired by fraction free elimination [6]. Here we do not compute a multiplicative inverse of the pivot. Instead, the affected rows in the submatrix are multiplied by the pivot:

$$\begin{pmatrix} a & b & 0 \\ 0 & c & d \\ e & f & g \end{pmatrix} \begin{array}{c} {}^{-e} \\ \rightharpoondown \\ \hookleftarrow_{+} \end{array} \rightsquigarrow \begin{pmatrix} a & b & 0 \\ 0 & c & d \\ 0 & af - eb & ag \end{pmatrix}.$$

In this case, we count the following operations. First, the number of multiplications by a is given by $\sum_i (r_i - 1)$ where r_i is the number of nonzero entries in the ith row and the summation ranges over the rows which have a nonzero entry in the pivot column, excluding the pivot row. Secondly, there are again $(r-1)(c-1)$ multiplications compute all the numbers corresponding to eb in the above sketch, where r is is the number of nonzero elements in the pivot row and c the number of nonzero elements in the pivot column. Finally, for each clash of two nonzero elements in the submatrix, like af and eb in the sketch above, we count one addition.

In the following, we refer to the first variant as the "field case" (because it involves a division) and to the second variant as the "ring case" (because it is fraction free).

3   How Many Matrices Are There?

It is clear that when we distinguish two kinds of entries, 0 and $*$, then there are 2^{n^2} different matrices of size $n \times n$. However, for the problem under consideration, we do not need to consider all of them. Pivot search will not be affected by permuting the rows of a matrix. For two $n \times n$ matrices A, B, write $A \approx B$ if a suitable permutation of rows turns A into B. Then $\approx$ is an equivalence relation, and we get a normal form with respect to $\approx$ by simply sorting the rows. It suffices to consider these normal forms, which reduces the problem from 2^{n^2} matrices to $\binom{2^n+n-1}{n}$ equivalence classes. The number of equivalence classes is the number of sorted n-tuples of binary numbers less than 2^n.

As we consider a pivot search that also allows column exchanges, we can go a step further. Write $A \sim B$ if applying a suitable permutation to the rows and a suitable permutation to the columns turns A into B. This is also an equivalence relation, but it is less obvious how to get a normal form. It was observed by Zivkovic [17] that deciding $\sim$ on binary matrices is equivalent to deciding the graph isomorphism problem for bipartite graphs. The idea is to interpret the matrix as adjacency matrix where row indices correspond to vertices of the first kind and column indices correspond to vertices of the second kind of a bipartite graph. Permuting rows or columns then amounts to applying permutations to each of the two kinds of vertices. Graph isomorphism is a difficult problem, but it is not a big deal for the matrix sizes we consider here. We used the nauty library [11] to compute normal forms with respect to $\sim$, and only analyzed matrices in normal form.

There does not appear to be a simple formula for the number of equivalence classes with respect to $\sim$, but for small n, the numbers can be determined by Polya enumeration theory, as explained for example in [7], and they are available as A002724 in the OEIS [14]. Here are the counts for $n = 1, \ldots, 9$.

n	$|S^{n \times n}|$	$|S^{n \times n}/{\approx}|$	$|S^{n \times n}/{\sim}|$
1	2	2	2
2	16	10	7
3	512	120	36
4	65 536	3 876	317
5	33 554 432	376 992	5 624
6	68 719 476 736	119 877 472	251 610
7	562 949 953 421 312	131 254 487 936	33 642 660
8	18 446 744 073 709 551 616	509 850 594 887 712	14 685 630 688
9	2 417 851 639 229 258 349 412 352	7 145 544 812 472 168 960	21 467 043 671 008

An 8×8 binary matrix can be conveniently represented in a 64bit word, and for storing some information about the elimination cost for various pivot choices (best, worst, average, Markowitz), we spend altogether 56 bytes per matrix. With this encoding, a database for all 6×6 matrices consumes about 3.8 TB of space, a database for all 7×7 matrices up to row permutations consumes about 7.3 TB, and a database for all 8×8 matrices up to row and column permutations consumes 784 GB. The number of equivalence classes for 9×9 matrices is so much larger that we have not considered them. Not only would a database for 9×9

cost an absurd amount of disk space, it would also make the programming more cumbersome and the analysis less efficient because we need more than one word per matrix.

4 Exhaustive Search

We use exhaustive search to create databases with the following information: the matrix itself, the cost of elimination over a field and over a ring with the best pivot and what is the best pivot, the cost with the worst pivot and what is the worst pivot, the median of the elimination costs, and the same considering only the pivots which produce minimum fill-in. For matrices of size $n \times n$ we only do one elimination step and then use the data from the database of matrices of size $n-1$. We include a simplified pseudocode to show how the data is produced. By $M(n)$ we denote a set of representatives of $S^{n \times n}$ with respect to $\sim$. The function $\texttt{Eliminate}$ takes a matrix and a row and column index and returns the result of performing one step of Gaussian elimination. The function $\texttt{Cost}$ takes the same input and returns the number of operations needed in the elimination step. The mappings $costmin$, $costmax$ and $costmed$ are the database entries.

> **input** : A size n, a set of $n \times n$ matrices $M(n)$ and three mappings
> $costmin$, $costmax$ and $costmed$ from $M(n-1)$ to the reals
> **output**: three mappings $costmin$, $costmax$ and $costmed$ from $M(n)$ to
> the reals

```
1  for m ∈ M(n) do
2      if there exist i, j ∈ {1, . . . , n} with Cost(m, i, j) = 0 then
3          m̄ ← Eliminate(m, i, j)
4          costmin(m) ← costmin(m̄)
5          costmax(m) ← costmax(m̄)
6          costmed(m) ← costmed(m̄)
7      else
8          costmin(m) ← ∞; costmax(m) ← −∞
9          for all (i, j) ∈ {1, . . . , n}² with m_{ij} = ∗ do
10             m̄ ← Eliminate(m, i, j)
11             o ← Cost(m, i, j)
12             costmin(m) ← Min(costmin(m̄) + o, costmin(m))
13             costmax(m) ← Max(costmax(m̄) + o, costmax(m))
14             cost_{ij} ← costmed(m̄) + o
15         costmed(m) ← Median(cost)
```

The median in line 15 is taken only over those pivots which have been considered. For the minimum fill-in strategy we adjust the **if** statement in line 9 such that only pivots which produce minimum fill-in are considered. In this case also the database entries are taken from the minimum fill-in strategy.

Note that not only a pivot with no other elements in its column results in zero elimination cost (since there is nothing to eliminate), but also a pivot with no other elements in its row results in no elimination cost, since it does not change any matrix entries apart from those which become 0. In line 2 we ensure that regardless of the strategy we are analyzing a free pivot is always chosen. Also note that if there are several such pivots, then the order in which we chose them does not affect the total cost.

We use the function *genbg* from the nauty package to produce the list of n × n matrices and we use the nauty package main function to compute a canonical form after the elimination step.

The full source code and the databases up to size 6 × 6 can be found at https://github.com/jakobmoosbauer/pivots. The larger databases can be provided by the authors upon request.

5 Machine Learning

There have been some recent advances in applying machine learning for improving selection heuristics in symbolic computation. As it is the case for Gaussian elimination, many algorithms allow different choices which do not affect correctness of the result, however may have a large impact on the performance. Machine learning models have for example been applied in cylindrical algebraic decomposition [8] and Buchberger's algorithm [13]. For more applications see [4].

Without going into too much detail on the background of machine learning, we give here a summary of our approach, mainly in order to document the computations we performed in order to facilitate a proper interpretation of the experimental results reported later. For explanations of the technical terms used in this section, we refer to the literature on machine learning.

We train a reinforcement learning agent [15] to select a good pivot in Gaussian elimination. In reinforcement learning an agent interacts with an environment by choosing an action in a given state. In every time step the agent is presented with a state s_t, chooses an action a_t and the environment returns the new state s_{t+1} and a reward r_{t+1}. In our case the state is the matrix and the action is a row and a column index. The new state is the matrix we get after performing the elimination with the chosen pivot and the reward is minus the number of operations needed in the elimination step. An episode is a sequence of steps that transform a matrix to row echelon form. The agent tries to maximize the return $G_t = \sum_{k=0}^{T-t} \gamma^k r_{t+k+1}$ with the discount factor $0 < \gamma \leq 1$. The return measures the expected future rewards and the discount factor makes the agent prefer earlier rewards over later rewards. Since we have a bound on the length of the episode we can choose $\gamma = 1$, in this case the return equals the total number of operations needed.

A difference to the more common concept of supervised learning is that we do not need a training set with training data. Instead we can sample from all possible inputs and the reward signal replaces the training data. This reduces the problem that an agent can produce bad answers outside of the training pool. We sample matrices equally distributed from all binary matrices. During the learning process actions are chosen using an ϵ-greedy policy. This means that with a probability of ϵ we choose a random action, with probability $1-\epsilon$ we choose a greedy action, i.e., the action the agent would choose in the given situation. This policy ensures accurate predictions for the actions the agent eventually chooses and also that we do not miss good choices because we never try them.

The main component of the reinforcement learning agent is the deep Q-network which approximates the Q-value function [12,15]. The Q-value function maps a state and an action to the expected return value. We can use this information to pick a pivot that has the lowest expected cost. Since we need to fix the network size in advance we can only consider matrices of bounded size. For the present paper we stick to small matrix sizes since we can evaluate the overall performance using the database and the network stays of easy manageable size.

We encode the row and column of the pivot by a one-hot vector. This means we use $2n$ inputs nodes which each correspond to a row or a column and set those of the pivot to 1 and the others to 0. We also tried to use the fill-in as additional input feature to see whether it improves the performance. After experimenting with different network structures we settled with a fully connected network with $n(n+2)$, respectively $n(n+2)+1$ input nodes and two hidden layers with $n(n+2)$ nodes and a relu activation function. We chose this architecture to fit the task at hand, for larger matrices training and evaluating this neural network would get prohibitively expensive. Selecting appropriate features or decomposing the problem in smaller parts are possible ways around this.

The network is trained using deep Q-learning with experience replay and a target network [12]. Experience replay means that we keep a replay buffer with state-action-reward pairs we created and in each training step we sample a batch of training data for the network to learn from. This helps the network not to "forget" what it already learned.

For the learning process there are different hyperparameters which control the learning process. They can have a huge impact on the performance and the convergence of the learning process. We did not invest a lot of time in tuning network architecture and hyperparameters, since our goal was to evaluate the general applicability of machine learning to this problem rather than finding best-possible results. The ϵ-greedy policy starts at $\epsilon = 0.5$ and slowly decays to $\epsilon = 0.1$. We use a discount factor of 1, learning rate 0.001, and a batch size of 50. The learning rate describes the step size of the parameter adjustment for the neural network. If the learning rate is too small, then the convergence is very slow, if the learning rate is too big the process does not converge at all.

The goal of reinforcement learning to maximize a reward function directly applies to our problem, which is to minimize the number of operations needed. Another advantage is that we have a totally observable deterministic environment. A difficulty we are faced with is that we can choose among a large number of different actions, which depends on the current state. So for each possible action we need an extra call to the neural network. This could be avoided by only considering pivots with small fill-in instead of all possible pivots. Neural networks are quite good at pattern recognition, which seems to be useful in our context. However, since the computational cost is invariant under row and column permutation, there is no locality of features in the pivot selection problem. Therefore it is not reasonable to use convolutional layers, which proved very useful in other tasks involving pattern recognition.

6   Results

In this section we analyze the results of the exhaustive search and the performance of machine learning. In the first graph we show the savings achieved by using the minimum fill-in strategy compared to the median cost. We observe that for matrices of size 8×8 the Markowitz criterion saves on average 42% of the operations needed in the field case and 37% in the ring case (left graph). Moreover, these numbers tend to grow as the matrices grow larger. So it is reasonable to expect that even larger savings are achieved for very large matrices.

In the graph on the right we compare the median cost when choosing pivots with minimal fill-in to the optimal pivot. For matrices of size 6×6 to 8×8 there are possible savings of about 5% in the field case and about 7% in the ring case. These numbers are increasing up to size 7×7, but there is a decrease from 7×7 to 8×8. In view of this decrease, it is hard to predict how the graph continues for matrix sizes that are currently out of the reach of exhaustive search.

For matrices of size 6×6 we also analyzed how the improvement potential depends on the sparsity of the matrix. For matrices with less than 30% nonzero entries, almost all choices of pivots perform equally. We see that the highest savings compared to the minimum fill-in strategy can be made for matrices with 40 to 60% nonzero entries, in the ring case up to 80% nonzero entries. Although for rather dense matrices there is still room for sparsity improvements, for these matrices the Markowitz strategy is almost optimal. During the elimination matrices become denser every step, as we introduce new nonzero entries. By minimizing the fill-in we look ahead only one step in the elimination process. For matrices which are almost dense this seems to be sufficient, whereas for sparser matrices it might be helpful to look ahead further.

Since several different pivots can have minimal fill-in, the question arises whether we just need a refined criterion to pick an optimal pivot among those that have minimal fill-in, or if we need to do something completely different. In order to address this question, we test if the best pivot which produces minimal fill-in is already optimal. We observe that the percentage of matrices where Markowitz's strategy is optimal throughout the whole computation drops to 46% in the field case and 30% in the ring case for 8×8 matrices. It seems reasonable to assume that for large matrices there is almost always a better strategy, even if you could choose the best pivot among those which have minimal fill-in. Even though pivots that have minimal fill-in are not always optimal, it seems reasonable that optimal pivots still have rather small fill-in. Although we did not analyze this with our database, we did not observe any examples where the optimal pivot had very large fill-in compared to other choices.

The table below shows the improvement achieved by the machine learning model compared to the fill-in strategy. The machine learning model is able to surpass the Markowitz strategy by a very small amount. The network was trained on 40000 to 100000 matrices to a point where additional training did not result in further improvement. While for 4×4 and 5×5 matrices this means that the model was presented every matrix multiple times, for the larger sizes the model was only trained on a small part of all matrices. This indicates that the model can generalize well from a small amount of samples. The results achieved using the fill-in as additional input to the network did not noticeably differ from those where we did not provide it. However, using the fill-in as feature we needed fewer training episodes to achieve similar results. So the machine learning model was able to find a better strategy knowing only the matrix entries, but providing additional features helps to speed up training. Since neither further training nor using a deeper network did improve the results it is likely that the model gets stuck in a local maximum.

n	4	5	6	7
field	0.75%	1.31%	1.72%	1.92%
ring	1.19%	2.19%	3.27%	3.99%

Let us consider an example where we can see why the minimum fill-in strategy does sometimes not perform very well. This is the 6×6 matrix (actually 6×5, since the last row consists of zeros) with the largest difference between the best pivot and the best pivot that produces minimum fill-in both over a ring and over a field:

$$\begin{pmatrix} * & * & * & * & * & * \\ * & * & * & * & * & * \\ 0 & 0 & * & * & * & * \\ 0 & 0 & 0 & * & * & * \\ 0 & 0 & 0 & 0 & * & * \end{pmatrix}.$$

When the rows and columns are sorted like this it is easy to spot that if we choose a pivot in the first column, we get a row echelon form after one elimination step with 11 operations. However, the pivots in the first column produce fill-in 5 and those in the last row produce only fill-in 4. Choosing the pivot with the minimal fill-in results in four elimination steps and in each step we have to eliminate every row, resulting in a total of 32 field operations. Over a ring the difference becomes even larger, 15 operations are needed with the best pivot and 55 if we always pick what produces minimal fill-in. There are two takeaways from this example. First we notice that the fill-in for the better pivot is produced by elements in its row and the fill-in for the other pivot is produced by the column. Especially in the ring case this leads to a large amount of extra computations since for every element we want to eliminate all the elements in the corresponding row have to be multiplied by the pivot. This suggests to use some kind of weighted fill-in where the number of elements in the column is weighted higher than the number of elements in the row. Another heuristic criterion motivated by this example is to do a two-step lookahead. In the first two columns there is a block of four nonzero elements and all other elements are 0. If we choose one of these four elements as pivot, then we only have to eliminate one row and the elimination is completed for both rows, since the second column contains no other elements. In this particular example the neural network finds the optimal pivot with the higher fill-in.

References

1. Brualdi, R.A., et al.: Combinatorial Matrix Theory. Springer, New York (1991). https://doi.org/10.1007/978-3-319-70953-6
2. Corless, R.M., Thornton, S.E.: The bohemian eigenvalue project. ACM Commun. Comput. Algebra **50**(4), 158–160 (2017). https://doi.org/10.1145/3055282.3055289
3. Duff, I.S., Erisman, A.M., Reid, J.K.: Direct Methods for Sparse Matrices. Clarendon Press, Oxford (1986)
4. England, M.: Machine learning for mathematical software. In: Davenport, J.H., Kauers, M., Labahn, G., Urban, J. (eds.) ICMS 2018. LNCS, vol. 10931, pp. 165–174. Springer, Cham (2018). https://doi.org/10.1007/978-3-319-96418-8_20
5. Faugère, J.C.: A new efficient algorithm for computing Gröbner bases. J. Pure Appl. Algebra **139**(1–3), 61–88 (1999)
6. Geddes, K.O., Czapor, S.R., Labahn, G.: Algorithms for Computer Algebra. Kluwer, Dordrecht (1992)
7. Harary, F., Palmer, E.M.: Graphical Enumeration. Academic Press, New York (1973)
8. Huang, Z., England, M., Wilson, D., Davenport, J.H., Paulson, L.C., Bridge, J.: Applying machine learning to the problem of choosing a heuristic to select the variable ordering for cylindrical algebraic decomposition. In: Watt, S.M., Davenport, J.H., Sexton, A.P., Sojka, P., Urban, J. (eds.) CICM 2014. LNCS (LNAI), vol. 8543, pp. 92–107. Springer, Cham (2014). https://doi.org/10.1007/978-3-319-08434-3_8
9. Koutschan, C.: Creative telescoping for holonomic functions. In: Computer Algebra in Quantum Field Theory: Integration, Summation and Special Functions. Texts and Monographs in Symbolic Computation, pp. 171–194. Springer (2013). https://doi.org/10.1007/978-3-7091-1616-6_7
10. Markowitz, H.M.: The elimination form of the inverse and its application to linear programming. Manage. Sci. **3**(3), 255–269 (1957). http://www.jstor.org/stable/2627454
11. McKay, B.D., Piperno, A.: Practical graph isomorphism, ii. J. Symbolic Comput. **60**, 94–112 (2014). https://doi.org/10.1016/j.jsc.2013.09.003
12. Mnih, V., et al.: Human-level control through deep reinforcement learning. Nature **518**(7540), 529–533 (2015)
13. Peifer, D., Stillman, M., Halpern-Leistner, D.: Learning selection strategies in buchberger's algorithm (2020). arXiv preprint arXiv:2005.01917
14. Sloane, N.J.A.: The on-line encyclopedia of integer sequences (2020). https://oeis.org/
15. Sutton, R.S., Barto, A.G.: Reinforcement Learning: An Introduction. MIT Press, Cambridge (1998)
16. Yannakakis, M.: Computing the minimum fill-in is NP-complete. SIAM J. Algebraic Discrete Methods **2**(1), 77–79 (1981). https://doi.org/10.1137/0602010
17. Živković, M.: Classification of small (0,1) matrices. Linear Algebra Appl. **414**(1), 310–346 (2006). https://doi.org/10.1016/j.laa.2005.10.010

Nullstellensatz-Proofs for Multiplier Verification

Daniela Kaufmann[(⊠)] and Armin Biere

Johannes Kepler University, Linz, Austria
daniela.kaufmann@jku.at

Abstract. Automated reasoning techniques based on computer algebra are an essential ingredient in formal verification of gate-level multiplier circuits. Generating and independently checking proof certificates helps to validate the verification results. Two algebraic proof systems, Nullstellensatz and polynomial calculus, are well-known in proof complexity. The practical application of the polynomial calculus has been studied recently. However, producing and checking Nullstellensatz certificates for multiplier verification has not been considered so far. In this paper we show how Nullstellensatz proofs can be generated as a by-product of multiplier verification and present our Nullstellensatz proof checker NUSS-CHECKER. Additionally, we prove quadratic upper bounds on the proof size for simple array multipliers.

1 Introduction

Formal verification aims to prove or disprove the correctness of a given system with respect to a certain specification. Nonetheless, the verification process might not be correct and contain errors. Thus it is common to produce proof certificates, which can be checked by stand-alone proof checkers in order to increase the confidence in the results of the verification process.

For example, many applications of formal verification use satisfiability (SAT) solving and various resolution or clausal proof formats [17], such as DRUP [13,14], DRAT [18], and LRAT [11] are available to validate the verification results. In the annual SAT competition it is even required to provide certificates since 2013.

However, in certain applications SAT solving cannot be applied successfully. For instance formal verification of arithmetic circuits, more precisely multiplier circuits is considered to be hard for SAT solving. The current state of the art in verifying multiplier circuits relies on computer algebra [9,24,31,32]. In this approach the circuit is modeled as a set of polynomials and it is shown that the specification, also encoded as a polynomial, is implied by the polynomials that are induced by the circuit. That is, for each gate in the circuit a polynomial is defined that captures the relations of the inputs and output of the gate. These gate polynomials generate a Gröbner basis [7]. Preprocessing techniques based

This work is supported by the LIT AI Lab funded by the State of Upper Austria.

© Springer Nature Switzerland AG 2020

F. Boulier et al. (Eds.): CASC 2020, LNCS 12291, pp. 368–389, 2020.
https://doi.org/10.1007/978-3-030-60026-6_21

on variable elimination are applied to rewrite and thus simplify the Gröbner basis [24,31]. After preprocessing the specification polynomial is reduced by the rewritten gate polynomials until no further reduction is possible. The given multiplier is correct if and only if the final result is zero.

Besides circuit verification, algebraic reasoning in combination with SAT solving [6] is successfully used to solve complex combinatorial problems, e.g., finding faster ways for matrix multiplication [19,20], computing small unit-distance graphs with chromatic number 5 [16], or solving the Williamson conjecture [5], and has possible future applications in cryptanalysis [8,40]. All these applications raise the need to invoke algebraic proof systems for proof validation.

Two algebraic proof systems are commonly known in the proof complexity community, polynomial calculus (PC) [10] and Nullstellensatz (NSS) [3]. Both systems are well-studied, with the main focus on deriving complexity measures, such as degree and proof size, e.g., [2,22,33,34]. Proofs in PC allow us to dynamically capture that a polynomial can be derived from a given set of polynomials using algebraic ideal theory. However, PC as defined in [10], is not suitable for practical proof checking [23], thus we introduced the practical algebraic calculus (PAC) in [37] that can be checked efficiently.

Proofs in NSS capture whether a polynomial can be represented as a linear combination from a given set of polynomials. Since NSS proofs are more static we made the following conjecture for the application of multiplier circuit verification in [23]: "In a correct NSS proof we would also need to express the rewritten polynomials as a linear combination of the given set of polynomials and thus lose the optimized representation, which will most likely lead to an exponential blow-up of monomials in the NSS proof."

In this paper we show that this conjecture has to be rejected, at least for those multiplier architectures considered in this paper. We introduce how NSS proofs can be produced in our verification tool AMULET [24,26] and our experimental results demonstrate that we are able to generate concise NSS proofs. For simple array multipliers, which consist only of full- and half-adders that are arranged in a grid-like structure, we prove quadratic bounds for the proof size. Furthermore, we present our NSS proof checker NUSS-CHECKER and discuss important design decisions that help to improve the checking time and memory usage.

2 Preliminaries

We describe our state-of-the-art approach in gate-level multiplier verification using computer algebra [24], and give an introduction to the algebraic proof systems PC, PAC, and NSS.

2.1 Multiplier Verification

Digital circuits are used in computers and digital systems and compute binary digital values for the logical function they implement, given binary values at the input. The computation is usually realized by logic gates, representing simple

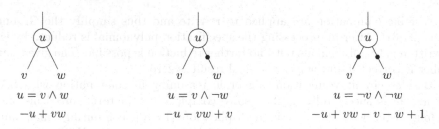

$$u = v \wedge w$$
$$-u + vw$$

$$u = v \wedge \neg w$$
$$-u - vw + v$$

$$u = \neg v \wedge \neg w$$
$$-u + vw - v - w + 1$$

Fig. 1. All polynomial encodings covered by AIG nodes

Boolean functions, such as NOT, AND, OR. The specification of a circuit is a desired relation between its inputs and outputs and the goal of verification is to formally prove that the circuit fulfills its specification, i.e., for all inputs the outputs of the circuit match the specification.

In our setting, we consider gate-level integer multiplier circuits C with $2n$ input bits $a_0, \ldots, a_{n-1}$, $b_0, \ldots, b_{n-1} \in \{0,1\}$ and $2n$ output bits $s_0, \ldots, s_{2n-1} \in \{0,1\}$. The internal gates are denoted by $g_1, \ldots, g_k \in \{0,1\}$. Let R be a commutative ring with unity and let $R[a_0, \ldots, a_{n-1}, b_0, \ldots, b_{n-1}, g_1, \ldots, g_k, s_0, \ldots, s_{2n-1}] = R[X]$. Since we consider integer multipliers, we will later set the ring $R = \mathbb{Z}$, but for now let us keep the more general ring R. The multiplier C is correct iff for all possible inputs a_i, $b_i \in \{0,1\}$ the following specification $\mathcal{L} = 0$ holds:

$$\mathcal{L} = - \sum_{i=0}^{2n-1} 2^i s_i + \left(\sum_{i=0}^{n-1} 2^i a_i \right) \left(\sum_{i=0}^{n-1} 2^i b_i \right) \tag{1}$$

A common representation of circuits are And-Inverter-Graphs (AIG) [28], which are directed acyclic graphs consisting of two-input nodes that represent logical conjunction. The edges may contain a marking that indicates logical negation. The semantics of each node implies a polynomial relation, cf. Fig. 1.

Let $G(C) \subseteq R[X]$ be the set of polynomials that contains for each gate of the given circuit C the corresponding polynomial of Fig. 1, with u, v, and w replaced by corresponding variables $x \in X$. We call these polynomials *gate constraints*.

All variables $x \in X$ are Boolean and we enforce this property by adding for each variable a *Boolean value constraint* $x(x - 1) = 0$. Let $B(Y) = \{y(1 - y) \mid y \in Y\} \subseteq R[X]$ for $Y \subseteq X$, be the set of Boolean value constraints for Y.

On the set of terms we fix an order $\leq$ such that for all terms τ, σ_1, σ_2 it holds that $1 \leq \tau$ and $\sigma_1 \leq \sigma_2 \Rightarrow \tau\sigma_1 \leq \tau\sigma_2$. An order is called a *lexicographic term order* if for all terms $\sigma_1 = x_1^{d_1} \cdots x_r^{d_r}$, $\sigma_2 = x_1^{e_1} \cdots x_r^{e_r}$ we have $\sigma_1 < \sigma_2$ iff $\exists i \in \mathbb{N}$ with $d_j = e_j$ for all $j < i$, and $d_i < e_i$. For a polynomial $p = c\tau + \cdots$ the largest term τ (w.r.t. $\leq$) is called the *leading term* $< (p) = \tau$. Furthermore $\mathrm{lc}(p) = c$ is called the *leading coefficient* and $\mathrm{lm}(p) = c\tau$ is called the *leading monomial* of p.

Definition 1 ([24]). *Let* $P \subseteq R[X]$. *If for a term order, all leading terms of* P *only consist of a single variable with exponent 1, are unique, and further* $\mathrm{lc}(p) \in R^{\times}$ *for all* $p \in P$, *we say* P *has* unique monic leading terms *(UMLT)*.

We order the polynomials in $G(C)$ according to a lexicographic term order, such that the output variable of a gate is always greater than the inputs of the gate. Such an order is also called *reverse topological term order* [29]. It immediately follows that $G(C)$ has UMLT. Let $X_0 \subseteq X$ be the set of all variables that do not occur as leading terms in $G(C)$ and let $J(C) = \langle G(C) \cup B(X_0) \rangle \subseteq R[X]$. The circuit fulfills its specification if and only if we can derive that $\mathcal{L} \in J(C)$ [24].

For the remainder of this section let $R = \mathbb{Z}$. Because of the UMLT property of the gate polynomials, $G(C) \cup B(X_0)$ defines a D-Gröbner basis [4] for $J(C) \subseteq \mathbb{Z}[X]$ [24]. We further showed in [24] that $J(C) = \langle G(C) \cup B(X) \rangle \subseteq \mathbb{Z}[X]$, i.e., $J(C)$ contains all Boolean value constraints for $x \in X$. Thus the correctness of the circuit can be established by reducing $\mathcal{L}$ by the gate polynomials and all Boolean value constraints and checking whether the result is zero.

It was shown in [30] that simply reducing the specification by $G(C) \cup B(X)$ leads to large intermediate reduction results. Thus, we developed preprocessing techniques based on variable elimination [24]. Typical components in multipliers are full- and half-adders, which are used to add three resp. two bits and produce a two-bit output c, s. The specification is $-2c - s + x + y + z = 0$ for a full-adder and $-2c - s + x + y = 0$ for a half-adder, with x, y, z representing the inputs. We include these specifications in the D-Gröbner basis by eliminating the internal variables of the full- and half-adders in C. After preprocessing $\mathcal{L}$ is reduced by the rewritten D-Gröbner basis $G(C)'$ until completion.

However, parts of the multiplier, more precisely final stage adders that are generate-and-propagate (GP) adders [36], are hard to verify using computer algebra. Contrarily, equivalence checking of adder circuits is easy for SAT solving. Hence, we combine SAT solving and computer algebra and our verification tool AMULET automatically replaces the complex GP adders by simple ripple-carry adders [24]. The correctness of the replacement is verified by SAT solvers and the rewritten multiplier is verified using computer algebra techniques. We generate DRUP proofs in SAT solvers and PAC proofs in AMULET. These proofs can be merged into one single PAC proof [25].

2.2   Algebraic Proof Systems

In the following we introduce algebraic proof formats, which are able to generate proof certificates using algebraic reasoning methods. Algebraic proof systems typically reason over polynomials in $\mathbb{K}[X]$, where $\mathbb{K}$ is a field and the variables X represent Boolean values. The aim of an algebraic proof is to derive whether a polynomial f can be derived from a given set of polynomials $G = \{g_1, \ldots, g_l\} \subseteq \mathbb{K}[X]$ together with the Boolean value constraints $B(X) = \{x_i^2 - x_i \mid x_i \in X\}$. In algebraic terms this means to show that the polynomial $f \in \langle G \cup B(X) \rangle$.

The first proof system we consider is the *polynomial calculus* (PC) [10]. A proof in PC is a sequence of proof rules $P = (p_1, \ldots, p_m)$, with $p_i \in \mathbb{K}[X]$ and $p_m = f$. Each rule has the following form that model the properties of an ideal:

Axiom $\qquad \dfrac{}{p} \qquad p \in G \cup B(X)$

Addition $\qquad \dfrac{p \quad q}{p + q} \qquad p, q$ both appear in P

Multiplication $\qquad \dfrac{p}{qp} \qquad p$ appears earlier in P, $q \in \mathbb{K}[X]$

The following metrics for PC are common in proof complexity, e.g. in [22,34]:

Definition 2. *Let* $\deg(p)$ *be the degree of a polynomial* p. *The degree of a PC proof* P *is the maximum degree of any proof rule* p_i, *i.e.,* $\deg(P) = \max\{\deg(p_i)\}$.

Definition 3. *The length of a PC proof* P *is defined as the maximum number of proof rules, i.e.* $\text{length}(P) = m$.

Definition 4. *Let* $\text{msize}(p)$ *denote the number of monomials in a polynomial* p. *The size of a PC proof* P *is the number of monomials in all proof rules* p_i, *i.e.,*

$$\text{size}(P) = \sum_{i=1}^{m} \text{msize}(p_i).$$

However, PC proofs cannot be checked efficiently, as the sequence of proof rules only contains the conclusion polynomials of each proof rule. Thus we modified PC in [23,37] and extended PC by adding information on the derivation of each p_i, yielding the practical polynomial algebraic calculus (PAC).

Furthermore, in our application with $G = G(C)$, all polynomials in G have UMLT. Thus we were able to generalize the soundness and completeness arguments of PC to polynomial rings $R[X]$ over commutative rings R with unity [24], thus also to $\mathbb{Z}[X]$. Additionally, we treat the Boolean value constraints implicitly, i.e., we consider proofs in the ring $\mathbb{Z}[X]/\langle B(X)\rangle$ to admit shorter proofs [23,27].

The metrics degree, length, and size can be directly applied to PAC proofs. PAC proofs can be checked using our proof checkers PACHECK or PASTÈQUE [23,27]. The proof checkers read the given set of polynomials $G \cup B(X)$ and verify the correctness of each proof line by checking whether the necessary conditions are fulfilled. We furthermore check whether it holds for one proof rule that $p_i = f$.

The *Nullstellensatz proof system* [3] derives whether a polynomial $f \in \mathbb{K}[X]$ can be represented as a linear combination from a given set of polynomials $G = \{g_1, \ldots, g_l\} \subseteq \mathbb{K}[X]$ and the Boolean value constraints $B(X)$. That is, an NSS proof for a given polynomial f and a set of polynomials G is an equality

$$\sum_{i=1}^{l} h_i g_i + \sum_{x_j \in X} r_j(x_j^2 - x_j) = f, \text{ for } h_i, r_j \in \mathbb{K}[X]. \tag{2}$$

By the same arguments given for PAC [24], we are able to generalize the soundness and completeness arguments of NSS proofs to rings $R[X]$ for our application where $G = G(C)$ has UMLT. We consider $R = \mathbb{Z}$ and again treat the Boolean value constraints implicitly to yield shorter proofs. Thus, the NSS proof we consider for a given polynomial $f \in \mathbb{Z}[X]/\langle B(X)\rangle$ and a set of polynomials $G = \{g_1, \ldots, g_l\} \subseteq \mathbb{Z}[X]/\langle B(X)\rangle$ is an equality P, such that

$$\sum_{i=1}^{l} h_i g_i = f \in \mathbb{Z}[X]/\langle B(X)\rangle, \tag{3}$$

with $h_i \in \mathbb{Z}[X]/\langle B(X)\rangle$. We call g_i the *base of the NSS proof* and h_i *co-factors*. The following metrics for NSS are common in proof complexity, e.g. in [1, 15]:

Definition 5. *The degree of an NSS proof P is* $\max\{\deg(h_i g_i)\}$.

Definition 6. *The size of an NSS proof P is given as*

$$\text{size}(P) = \sum_{i=0}^{l} \text{msize}(h_i)\,\text{msize}(g_i).$$

A further metric is the representation size that measures the total number of monomials in the polynomials g_i and the co-factors h_i. As the name indicates, it estimates the number of monomials needed to write down an NSS proof.

Definition 7. *The representation size of an NSS proof P is given as*

$$\text{repsize}(P) = \sum_{i=0}^{l} (\text{msize}(h_i) + \text{msize}(g_i)).$$

Checking NSS proofs seems straightforward as we simply need to expand the products $h_i g_i$, calculate the sum, and compare the derived polynomial to the given target polynomial f. However, we discuss practical issues of proof checking in Sect. 5, where we introduce our proof checker NUSS-CHECKER.

3   Proof Generation

In this section we discuss how NSS proofs can be generated in our verification tool AMULET [24]. We introduced in Sect. 2 that we distinguish two phases during verification of multipliers. In the preprocessing step we eliminate variables from the induced D-Gröbner basis $G(C)$ to gain a simpler polynomial representation $G(C)'$. In the second step the specification is reduced by the rewritten D-Gröbner basis $G(C)'$ to determine whether the given circuit is correct. Both phases have to be included in the NSS proof to yield a representation of the specification $\mathcal{L}$ as a linear combination of the original gate constraints $G(C) \in \mathbb{Z}[X]/\langle B(X)\rangle$.

AMULET reads the given AIG, determines a reverse topological term ordering and encodes each AIG node by a corresponding polynomial to derive the set of gate constraints $G(C)$. All polynomials from $G(C)$ are kept in the memory even if they are removed from the D-Gröbner basis during preprocessing.

In the preprocessing step, we repeatedly eliminate all variables $v \in X \setminus X_0$ from $G(C)$ that occur in the tail of only one polynomial, cf. Sect. 4.2. in [26]. Let $p_v \in G(C)$ such that $< (p_v) = v$. Since $G(C)$ has UMLT and $v \notin X_0$, such a p_v exists. All polynomials $p \in G(C) \setminus \{p\}$, with $v \in p$ are reduced by p_v to remove v from p. The reduction algorithm is depicted in Algorithm 1 and returns polynomials $h, r \in \mathbb{Z}[X]/\langle B(X) \rangle$ such that $p + hp_v = r \in \mathbb{Z}[X]/\langle B(X) \rangle$. In contrast to more general polynomial division/reduction algorithms we use the fact in Algorithm 1 that $\mathrm{lm}(p_v) = -v$.

Algorithm 1: Reduction(p, p_v, v)

Input   : Polynomials p, $p_v \in \mathbb{Z}[X]/\langle B(X) \rangle$, $\mathrm{lm}(p_v) = -v$
Output: Polynomials $h, r \in \mathbb{Z}[X]/\langle B(X) \rangle$ such that $p + hp_v = r$
1 $t \leftarrow p, r \leftarrow p, h \leftarrow 0$;
2 **while** $t \neq 0$ **do**
3 | **if** $v \in< (t)$ **then**
4 | | $h = h + \mathrm{lm}(t)/v$;
5 | | $r = r + p_v \, \mathrm{lm}(t)/v \mod \langle B(X) \rangle$;
6 | $t = t - \mathrm{lm}(t)$;
7 **return** h, r

Algorithm 2: Add-to-basis-representation$(p_v, h, \mathrm{base}(r))$

Input   : Polynomials p_v, $h \in \mathbb{Z}[X]/\langle B(X) \rangle$, basis representation $\mathrm{base}(r)$
Output: Updated basis representation $\mathrm{base}(r)$ such that (p_v, h) is included
1 **if** $p_v \rightarrow orig$ **then**
2 | **if** $(p_v, h_i) \in \mathrm{base}(r)$ for any h_i **then**
3 | | $\mathrm{base}(r) \leftarrow (\mathrm{base}(r) \setminus \{(p_v, h_i)\}) \cup \{(p_v, h_i + h)\}$;
4 | **else**
5 | | $\mathrm{base}(r) \leftarrow \mathrm{base}(r) \cup \{(p_v, h)\}$;
6 **else**
7 | **foreach** $(p_i', h_i') \in \mathrm{base}(p_v)$ **do** Add-to-basis-representation(p_i', hh_i')
8 **return** $\mathrm{base}(r)$

We replace the polynomial p by the calculated remainder r, and remove p_v from the D-Gröbner basis [24]. To keep track of the rewriting steps we want to store information on the derivation of the rewritten polynomials r.

Definition 8. *We call* $\mathrm{base}(r) = \{(p_i, q_i) \mid p_i \in G(C), \ q_i \in \mathbb{Z}[X]/\langle B(X) \rangle\}$ *the basis representation of* $r \in \mathbb{Z}[X]/\langle B(X) \rangle$, *such that* $r = \sum_{(p_i, q_i) \in \mathrm{base}(r)} q_i p_i$.

For the rewritten polynomial r that is derived by Algorithm 1, we have to include the tuples $(p, 1)$, (p_v, h) in the basis representation $\text{base}(r)$, cf. Algorithm 2. However, we want to represent r in terms of the original gate constraints $G(C)$ only, thus we need to take into account whether the polynomial p resp. p_v are original gate constraints or whether they are rewritten, that is $\text{base}(p) \neq \{\}$.

If p_v is an original gate constraint we include the tuple (p_v, h) in $\text{base}(r)$. If p_v does not occur in any tuple in $\text{base}(r)$, we simply add (p_v, h) to $\text{base}(r)$. Otherwise $\text{base}(r)$ contains a tuple (p_v, h_i) that has to be updated to $(p_v, h_i + h)$, which corresponds to merging common factors in $\text{base}(r)$.

If the polynomial p_v is not an original gate constraint, $\text{base}(p_v) \neq \{\}$, i.e., p_v can be written as a linear combination $p_v = h'_1 p_1 + \cdots + h'_l p_l$ for some original constraints p_i and $h'_i \in \mathbb{Z}[X]/\langle B(X) \rangle$. Thus the tuple (p_v, h) corresponds to $h p_v = h h'_1 p_1 + \cdots + h h'_l p_l$. We traverse through the tuples $(p_i, h'_i) \in \text{base}(p_v)$, multiply each of the co-factors h'_i by h and add the corresponding tuple $(p_i, h h'_i)$ to $\text{base}(r)$. Multiplying and expanding the product $h h_i$ may lead to an exponential blow-up in the size of the NSS proof as the following example shows.

Algorithm 3: Spec-Reduction$(\mathcal{L}, G(C)')$

Input : Circuit specification $\mathcal{L} \in \mathbb{Z}[X]$, D-Gröbner basis $G(C)'$
Output: Remainder r, Basis representation $\text{base}(\mathcal{L})$
1 $r \leftarrow \mathcal{L}, \text{base}(\mathcal{L}) \leftarrow \{\}$;
2 **foreach** $g \in G(C)'$ **do**
3 $\quad$ $r, h \leftarrow \text{Reduction}(r, g, < (g))$;
4 $\quad$ $\text{base}(\mathcal{L}) \leftarrow \text{Add-to-basis-representation}(r, g, h, \text{base}(\mathcal{L}))$;
5 **return** $r, \text{base}(\mathcal{L})$

Example 1. Consider a set of polynomials $G = \{-y_1 + (1 \mid x_0) y_0, -y_2 + (1 + x_1) y_1, \ldots, -y_k + (1 + x_{k-1}) y_{k-1}\} \subseteq \mathbb{Z}[y_0, \ldots y_k, x_0, \ldots x_k]$ and assume we eliminate $y_1, \ldots, y_{k-1}$, yielding $-y_k + (1 + x_0)(1 + x_1) \ldots (1 + x_{k-1}) y_0$. The expanded form of the co-factor of y_0 contains 2^k monomials.

Surprisingly our experiments, cf. Sect. 6, show that this blow-up does not occur in arithmetic circuit verification, rejecting our conjecture of [23].

Example 2. We demonstrate a sample run of Algorithm 2. Let $G(C) = \{p_1, p_2, p_3\} \subseteq \mathbb{Z}[X]/\langle B(X) \rangle$ and $x, y, z \in \mathbb{Z}[X]/\langle B(X) \rangle$. Assume $q_1 = p_1 + x p_2$, and $q_2 = p_3 + y p_2$. Thus $\text{base}(q_1) = \{(p_1, 1), (p_2, x)\}$ and $\text{base}(q_2) = \{(p_2, y), (p_3, 1)\}$. Let $p = q_1 + z q_2$. We receive $\text{base}(p)$ by adding $(q_1, 1)$ and (q_2, z) to $\text{base}(p) = \{\}$.

$(q_1, 1)$: Since $q_1 \notin G(C)$, we and add each tuple of $\text{base}(q_1) = \{(p_1, 1), (p_2, x)\}$ with co-factors multiplied by 1 to $\text{base}(p)$. We gain $\text{base}(p) = \{(p_1, 1), (p_2, x)\}$.

(q_2, z): We consider $\text{base}(q_2) = \{(p_2, y), (p_3, 1)\}$ and add (p_2, yz) and (p_3, z) to $\text{base}(p)$. Since p_3 is not yet contained in the ancestors of p, we directly add (p_3, z) to $\text{base}(p)$. The polynomial p_2 is already contained in $\text{base}(p)$, thus we add yz to the co-factor x of p_2 and we derive $\text{base}(p) = \{(p_1, 1), (p_2, x + yz), (p_3, z)\}$.

After preprocessing is completed, we repeatedly apply Algorithm 1 and reduce the specification polynomial $\mathcal{L}$ by the rewritten D-Gröbner basis $G(C)'$. We consider the polynomials $g \in G(C)'$ in reverse topological order, such that each polynomial in $G(C)'$ has to be considered exactly once for reduction. We generate the final NSS proof by deriving a basis representation for $\mathcal{L}$. Therefore we add after each reduction step the tuple (g, h), where h is the corresponding co-factor of polynomial g, to the base representation base($\mathcal{L}$) using Algorithm 2. Algorithm 3 shows the complete reduction process.

We check whether the final remainder r is zero. If so, base($\mathcal{L}$) represents an NSS proof and is printed to a file. If r is not zero, r contains only input variables $a_i, b_i \in X_0$ and can be used to generate counter-examples [23].

4   Proof Size

In this section we examine the proof complexity of the induced NSS proofs in AMULET for certain multiplier architectures. In particular we are interested in the degree and proof (representation) size. First, we examine these proof metrics for *btor-multipliers* that are generated by Boolector [35]. In this architecture AND-gates are used to produce the partial products, which are accumulated in an *array structure* using full- and half-adders. The final-stage adder is a ripple-carry adder. These multipliers are considered as "simple" multipliers, because they can be fully decomposed into full- and half-adders, cf. Fig. 2 for input bit-width 4. The AIG representation of full- and half-adders is shown in Figs. 3 and 4.

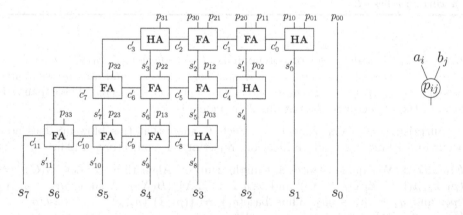

Fig. 2. The architecture of btor-multipliers for input bit-width 4

In previous work [37] we studied the proof complexity of PAC proofs and empirically demonstrated that checking commutativity of btor-multipliers induces PAC proofs of quadratic length and cubic size. However these proofs were produced using existing computer algebra systems [41] that are not targeted for multiplier verification. In more recent work [26] we investigated the proof metrics for PAC proofs that are generated in our verification tool AMULET [24]. We formally derived that n-bit btor-multipliers generate PAC proofs with degree 3 that have a length of $16n^2 - 20n - 1$ and a proof size (cf. Definition 4) in $\mathcal{O}(n^2 \log(n))$.

In the following we will investigate the complexity of NSS proofs that are generated by AMULET for btor-multipliers. We split the gate constraints in $G(C)$ into three categories: the *output polynomials* that link an output s_i to an output of a full- or half-adder, that is $-s_i + s'_k$ or $-s_i - s'_k + 1$ depending on the sign of s'_k. For example, the multiplier in Fig. 2 induces the output polynomial $-s_3 + s'_8$. Furthermore, we consider *polynomials representing partial products*, i.e., $-p_{ij} + a_i b_j$. All remaining polynomials in $G(C)$ are induced from the full- and half-adders in the circuit, i.e., *the internal adder polynomials*.

We are able to express the specification of each full- and half-adder, cf. Sect. 2 as a linear combination of the internal adder polynomials. Figure 3 shows the AIG representing a full-adder, as it occurs in btor-multipliers. Depending on the position of the full-adder in the multiplier, the sign of the inputs x, y, and z may be inverted and thus internal variables of the full-adder are negated, which affects the proof size. The full-adder in Fig. 3 represents the full-adder in btor-multipliers that yields the largest NSS proofs (input x and output c are inverted). We use the proof size of these full-adders to estimate an upper bound of the proof size. The corresponding gate polynomials can be seen on the right side of Fig. 3 together with the co-factors that are induced in AMULET. Expanding the linear combination yields the specification $-2(1 - c) + s + (1 - x) - y - z$. Figure 4 shows the same result for a half-adder resulting in $-2c - s + x + y$. From the polynomials in Figs. 3 and 4 we are able to derive the following lemmas.

Lemma 1. *The NSS proof generated in* AMULET *for a half-adder has maximum size 61 and maximum representation size 45. The NSS proof for a half-adder has maximum size 23 and maximum representation size 19.*

Proof. Figures 3 and 4 show the representation of the full-adder and half-adders that occur in btor-multipliers that maximize the NSS proof size. Furthermore the

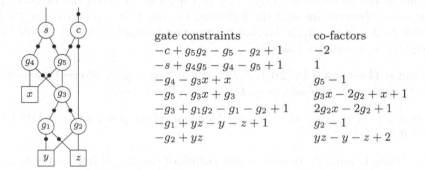

gate constraints	co-factors
$-c + g_5 g_2 - g_5 - g_2 + 1$	-2
$-s + g_4 g_5 - g_4 - g_5 + 1$	1
$-g_4 - g_3 x + x$	$g_5 - 1$
$-g_5 - g_3 x + g_3$	$g_3 x - 2g_2 + x + 1$
$-g_3 + g_1 g_2 - g_1 - g_2 + 1$	$2g_2 x - 2g_2 + 1$
$-g_1 + yz - y - z + 1$	$g_2 - 1$
$-g_2 + yz$	$yz - y - z + 2$

Fig. 3. Full-adder architecture in btor-multipliers

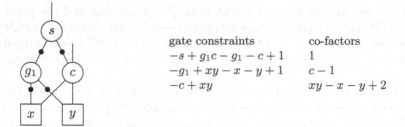

gate constraints	co-factors
$-s + g_1 c - g_1 - c + 1$	1
$-g_1 + xy - x - y + 1$	$c - 1$
$-c + xy$	$xy - x - y + 2$

Fig. 4. Half-adder architecture in btor-multipliers

induced co-factors in AMULET are shown. We simply count the number of monomials in the polynomials and use the definition of proof (representation) size, cf. Definitions 6 and 7 to yield the desired results.

Lemma 2. *The degree of an NSS proof for a full- or half-adder is 3.*

Proof. It can be seen in Figs. 3 and 4 that multiplying each of the gate polynomials by the corresponding co-factor yields degree at most 3 in $\mathbb{Z}[X]/\langle B(X)\rangle$.

We use the full- and half-adder specifications to derive a concise NSS proof. That is, we want to find co-factors, such that we are able to express the specification $\mathcal{L}$ cf. Eq. 1 as a linear combination of the output polynomials, adder specifications and the polynomials that represent partial products.

It is easy to see that all the output polynomials, i.e., $-s_i + s'_k$ or $-s_i - s'_k + 1$ need to be multiplied by the corresponding constant 2^i, because neither the internal adder polynomials nor the polynomials representing partial products contain any output variable s_i of the multiplier. Furthermore, since all adder specifications are linear, we multiply these polynomials by constants to cancel output variables of an adder that are input to another adder. For example, the multiplier of Fig. 2 induces the polynomials $-2c'_{11} - s'_{11} + c'_7 + p_{33} + c'_{10}$, $-2c'_{10} - s'_{10} + s'_7 + p_{23} + c'_9$. We multiply the first polynomial by two to cancel the monomials containing c'_{10}. It follows by the same arguments that we only need to multiply the polynomials $-p_{ij} + a_i b_j$ by constants to cancel the variables p_{ij}. Using these observations and the following lemmas that are derived in [26] we are able to derive quadratic bounds for the proof (representation) size of btor-multipliers in Theorems 1 and 2.

Lemma 3 (Lemma 2 in [26]). *Let C be a btor-multiplier of input bit-width n. Then C contains n half-adders and $n^2 - 2n$ full-adders.*

Theorem 1. *The proof size of n-bit btor-multipliers produced in AMULET is bounded by $63n^2 - 93n$.*

Proof. Using Lemma 3, we derive that the proof size for all full- and half-adder specifications is at most $23n + 61n^2 - 122n = 61n^2 - 99n$. These specifications are only multiplied by constants during reduction, thus reduction has no effect on the proof size. The $2n$ polynomials representing the circuit outputs are multiplied by constants, thus each polynomial contributes at most 3 monomials. Each of the n^2 polynomials representing partial products is also multiplied by a constant, adding 2 monomials to the proof size. Collecting the results leads to a proof size of $61n^2 - 99n + 6n + 2n^2 = 63n^2 - 93n$.

Theorem 2. *The proof representation size of n-bit btor-multipliers produced in* AMULET *is bounded by* $48n^2 - 63n$.

Proof. Using Lemma 3 and multiplying the co-factors by appropriate constants we derive that the proof representation size for all full- and half-adder specifications is at most $19n + 45n^2 - 90n = 45n^2 - 71n$. The $2n$ polynomials representing the circuit outputs are multiplied by constants. Thus for each of the $2n$ products we derive a representation size 4. Each of the n^2 polynomials representing partial products is also multiplied by a constant, adding 3 monomials to the proof representation size. Collecting the results leads to a proof representation size of $45n^2 - 71n + 8n + 3n^2 = 48n^2 - 63n$.

Theorem 3. *The degree of the NSS proof of n-bit btor-multipliers is* 3.

Proof. It follows from Lemma 2 that the degree of the NSS proof for an adder specification is 3. This linear adder specification is only multiplied by constants in the NSS proof for btor-multipliers. Furthermore, the degree of the output polynomials is 1 and the degree of the polynomials representing the partial products is 2, and both are multiplied only by constant factors in the NSS proof. Thus the maximum degree of a polynomial product in the NSS is 3.

Figure 5 shows the proof (representation) size together with the derived bounds of Theorems 1 and 2 for btor-multipliers with an input bit-width n in $[4, 128]$. The absolute error of the bounds can be seen in Fig. 6, which empirically indicates that the difference between the upper bound and the real proof size is in $\mathcal{O}(n)$, giving us a precise bound on the coefficient of the quadratic terms.

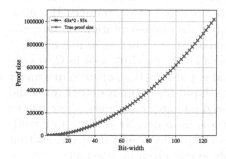 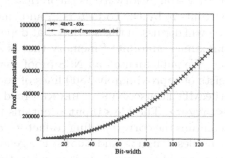

Fig. 5. Proof size (left) and proof representation size (right) for btor-multipliers

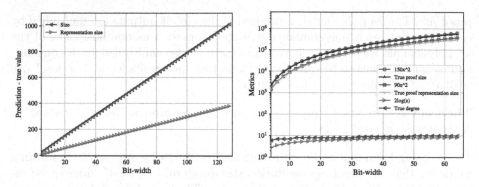

Fig. 6. Absolute error of the estimated bounds for proof (representation) size (left). Empirical evaluation of proof metrics for bp-wt-rc–multipliers (right)

The second multiplier architecture we consider are the complex *bp-wt-rc*–multipliers that are part of the AOKI benchmarks [21]. These benchmarks only scale up to input bit-width 64. The bp-wt-rc–multipliers use a Booth encoding [36] to generate the partial products, which are then accumulated using a Wallace-tree. The final-stage adder is a ripple-carry adder. The abbreviations of these components "Booth encoding" – "Wallace-tree" – "ripple-carry adder" give this architecture its name. Due to their irregular structure we only give empirical evidence for the proof metrics, which can be seen in the right side of Fig. 6.

Proposition 1. *Let C be a bp-wt-rc–multiplier of input bit-width n. The degree of the NSS proof is in $\mathcal{O}(\log(n))$. The proof (representation) size is in $\mathcal{O}(n^2)$.*

5   Proof Checking

We validate the correctness of the generated NSS proofs by checking whether $\sum_{i=1}^{l} q_i p_i = \mathcal{L} \in \mathbb{Z}[X]/\langle B(X)\rangle$ for $p_i \in G(C)$, $q_i \in \mathbb{Z}[X]/\langle B(X)\rangle$. This sounds rather straightforward as theoretically we only need to multiply the original constraints p_i by the co-factors q_i and calculate the sum of the products. However, we will discuss in this section that depending on the implementation the time and maximum amount of memory that is allocated varies by orders of magnitude.

We implemented an NSS proof checker, called NUSS-CHECKER in C. It consists of approximately 1800 lines of code and is published[1] as open source under the MIT license. NUSS-CHECKER reads three input files `<input>`, `<cofact>`, and `<target>`. The file `<input>` contains the original gate constraints p_i, `<cofact>` contains the corresponding co-factors q_i in the same order. NUSS-CHECKER reads the files `<input>` and `<cofact>`, generates the products and then verifies that the sum of the products is equal to the polynomial given in `<target>`.

The polynomials in NUSS-CHECKER are internally stored as ordered linked lists of monomials. The coefficients are represented using the GMP library and

---

[1] http://fmv.jku.at/nussproofs.

the terms are ordered linked lists of variables. All internally allocated terms are shared using a hash table. We already discussed in [23] that the variable ordering has an enormous effect on the memory usage of the tool, since different variable orderings induce different terms. In the default mode NUSS-CHECKER orders the variables by their name using the function strcmp, as this minimized memory usage for our application [23]. NUSS-CHECKER furthermore supports to use the same variable ordering as in the given files. That is, whenever a new variable is parsed we assign an increasing numerical level value and sort according to this value. Both orderings strcmp and level can be applied in reverse order too.

NUSS-CHECKER generates the products on the fly. That is, we parse both files <input> and <cofact> simultaneously, read two polynomials q_i and p_i from each file and calculate the product $q_i p_i$.

The polynomial arithmetic needed for multiplication and addition is implemented from scratch, because in the default setting we always calculate modulo the ideal $\langle B(X) \rangle$. General algorithms for polynomial arithmetic need to take exponent arithmetic over $\mathbb{Z}$ into account [38], which is not the case in our setting. Furthermore, in our previous work on PAC [37] we used modern computer algebra systems, Mathematica [41] and Singular [12], for proof checking, which turned out to be much slower than our own implemented algorithms.

Addition of two polynomials is implemented by pushing the monomials of both polynomials on a stack, which is then sorted (using Quicksort) according to the fixed term ordering and monomials with equal terms are merged to yield the final sum. Multiplication is implemented in a similar way.

Since addition of polynomials in $\mathbb{Z}[X]$ is associative, we are able to derive different addition schemes. We experimented with four different addition patterns, which are depicted in Fig. 7 for adding six polynomials. The subscript i of "$+_i$" shows the order of the addition operation.

If we *sum up all polynomials at once*, we do not generate the intermediate addition results. Instead we push all monomials of the l products $p_i q_i$ onto one big stack. Afterwards, the monomials on the stack are sorted and merged, which corresponds to one big addition. In this addition scheme we do not compute any intermediate summands, which makes the algorithm very fast, because we sort the stack only once. However, all occurring monomials of the products are pushed on the stack and stored until the final sorting and merging, which increases the memory usage of NUSS-CHECKER.

If we *add up in sequence*, we only store one polynomial in the memory, and always add the lastest product $p_i q_i$. This allows for monomials to cancel, which helps to reduce the memory usage. On the other hand, in our application the target polynomial $\mathcal{L}$ contains n^2 partial products that lead to intermediate summands of quadratic size, which slows down the checking time.

If we add up in a *tree structure with breadth first*, we add two consecutive products of the NSS proof and store the resulting sum. After parsing the proof, we have $\frac{l}{2}$ polynomials on a stack. We repeatedly iterate over the stack and always sum up two consecutive polynomials, until only one polynomial is left. This has the effect that we do not collect and carry along the n^2 partial

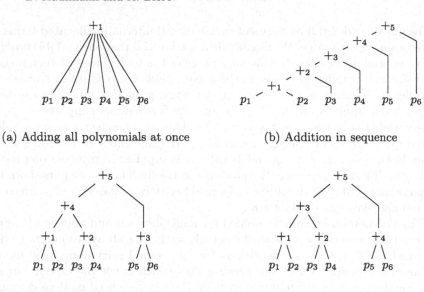

(a) Adding all polynomials at once   (b) Addition in sequence

(c) Tree structure, breadth first   (d) Tree structure, depth first

Fig. 7. Addition schemes of 6 polynomials

products. However, the memory usage increases, because we store $\frac{l}{2}$ polynomials simultaneously.

In the addition scheme, where we use a *tree structure and sum up depth first*, we develop the tree on-the-fly by always adding two polynomials of the same layer as soon as possible. It may be necessary to sum up remaining intermediate polynomials that are elements of different layers, as can be seen in Fig. 7. Similar to using a tree structure with breadth first addition, we do not collect and carry along the partial products. Furthermore, we always store at most $\lceil \log(l) \rceil$ polynomials in the memory, as a binary tree with l leafs has height $\lceil \log(l) \rceil$ and we never have more polynomials than layers in the memory.

We apply the presented addition schemes on btor-multipliers, cf. Sect. 4 and it can be seen in Fig. 8 that the results compare favorably to our conjectures of checking time and memory usage for each addition scheme. However, NUSS-CHECKER supports all presented options for addition, with *adding up in binary tree, depth first* set as default, because for different applications, using other addition schemes may be more beneficial.

For example, we shuffled the order of the polynomials in the NSS proof of 128-bit btor-multipliers 200 times and report the box-plots of the checking time and memory usage in Fig. 9. Since "adding up in sequence" always exceeded the time limit of 300 s, we omit its box-plot. It can be seen that the fastest addition scheme is now "all at once". However, the "tree based, depth first" approach still has the smallest memory usage.

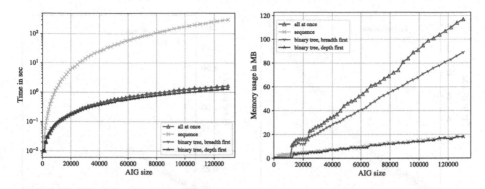

Fig. 8. Time (left) and memory usage (right) of addition schemes

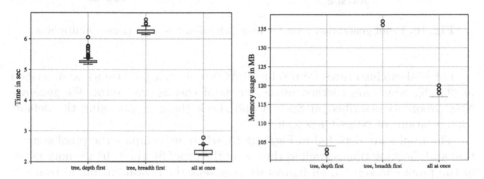

Fig. 9. Checking time (left) and memory usage (right) of shuffled NSS proofs

6   Evaluation

In this section we provide experimental results for generating and checking NSS proofs for multiplier verification and we aim to provide a comprehensive comparison between PAC and NSS proofs for the selected multiplier architectures.

In our experiments we use an Intel Xeon E5-2620 v4 CPU at 2.10 GHz (with turbo-mode disabled) with a memory limit of 128 GB. The time is listed in rounded seconds (wall-clock time). The wall-clock time is measured from starting the tools until they are finished. Source code, benchmarks and experimental data are available at http://fmv.jku.at/nussproofs.

In our experiments we consider the simple btor-multipliers with an input bit-width n in $[4, 128]$ and the complex bp-wt-rc–multipliers with an input bit-width n in $[4, 64]$. These architectures are already discussed in detail in Sect. 4.

More complex multipliers include GP adders [36]. However, in our verification approach [24] these GP adders are replaced by ripple-carry adders and only the rewritten multiplier is verified using computer algebra. Thus it suffices to consider complex multipliers that include a ripple-carry adder in this paper.

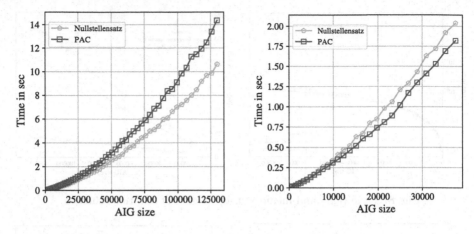

Fig. 10. Proof generation time for btor (left) and bp-wt-rc (right) multipliers

For both architectures we produce PAC proofs using AMULET as described in [23,26], which are checked using our proof checker PACHECK. We generate NSS proofs as described in Sect. 3 and check these proofs using the default configurations of NUSS-CHECKER.

The results are depicted in Figs. 10–13, where we compare the proof generation and checking time as well as the size of the proof files and the memory usage of the proof checkers. In all figures we represent the measurements in terms of the size of the input AIG, i.e., the number of circuit constraints, because the number of gates in these multipliers is quadratic in the bit-width n.

Figure 10 shows the time needed to generate the NSS and PAC proofs in AMULET. It can be seen that for btor-multipliers the generation time of PAC proofs is around 30% slower than for NSS proofs. For bp-wt-rc–multipliers PAC proofs are produced slightly faster than NSS proofs.

The size of the proof files (in megabyte) is shown in Fig. 11. Depending on the multiplier architecture the size of the NSS proof file is 5–10 times smaller than the size of the PAC proof. This result is actually expected as the PAC proof includes all intermediate steps and results of generating and adding the products. In the NSS proof file we only store the co-factors without any intermediate steps.

Figure 12 depicts that NSS proofs can be checked faster than the corresponding PAC proofs. In fact, even for a btor-multiplier with input bit-width 128, where the AIG contains more than 129 000 nodes, checking the NSS proof takes around 1 s and is four times faster than checking the PAC proof. We observed that the proportion between multiplication and addition in NUSS-CHECKER is around 1:1.7, e.g. for 128-bit btor-multipliers 0.25 s are used by the multiplication function and 0.4 s are used by the addition operation.

Last, we compare the memory usage of PACHECK and NUSS-CHECKER, i.e., the maximum amount of memory that is allocated during proof checking and it can be seen in Fig. 13 that NSS proofs need less than a third of the memory.

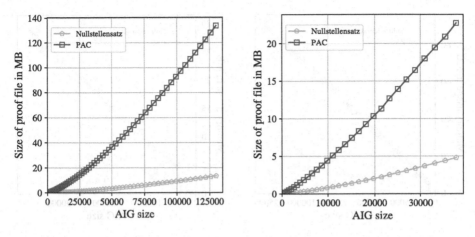

Fig. 11. Size of the proof files for btor (left) and bp-wt-rc (right) multipliers

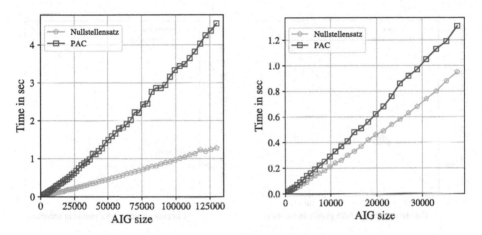

Fig. 12. Proof checking time for btor (left) and bp-wt-rc (right) multipliers

The AOKI benchmarks contain 192 different multiplier architectures, 168 of which can be successfully verified using AMULET. We compare the proof generation and checking time of PAC and NSS proofs for these 168 multipliers in Fig. 14. We fixed the input bit-width of all multipliers to 64. It can be seen that for multipliers that use Booth encoding to generate the partial products the generation time of NSS proofs is slightly slower than for PAC proofs. However, checking the NSS proof is almost always faster than checking PAC proofs.

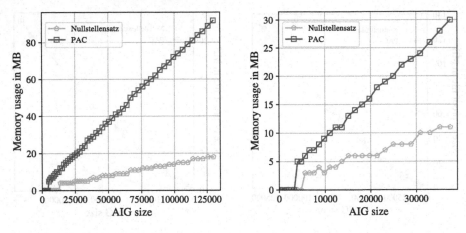

Fig. 13. Memory usage of checkers for btor (left) and bp-wt-rc (right) multipliers

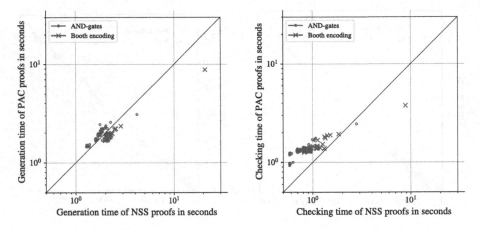

Fig. 14. Generation (left) and checking time (right) for 64-bit multipliers

7   Conclusion

In this paper we elaborated whether concise Nullstellensatz proofs can be generated to validate the results of multiplier verification using computer algebra. We discussed how Nullstellensatz proofs are developed as by-product in our verification tool AMULET. Our experiments showed that we are able to produce compact Nullstellensatz proofs that are faster to check than proof certificates based on the polynomial calculus. For simple array multipliers we formally derived quadratic bounds on the proof size for Nullstellensatz proofs. Furthermore, we presented our Nullstellensatz proof checker NUSS-CHECKER and discussed several design decisions that allow efficient proof checking.

In the future we want to further investigate the connection between polynomial calculus and Nullstellensatz for multiplier verification and want to derive possibilities to convert DRUP proofs to Nullstellensatzproofs, similar to converting DRUP proofs into PAC proofs as in [25]. Another intriguing research direction is to develop techniques that allow production of smaller Nullstellensatz proofs and connect it to SAT solving [24]. More general problems beyond the Boolean case may be also of interest [39].

References

1. Atserias, A., Ochremiak, J.: Proof complexity meets algebra. ACM Trans. Comput. Log. **20**(1), 1:1–1:46 (2019)
2. Beame, P., Cook, S.A., Edmonds, J., Impagliazzo, R., Pitassi, T.: The relative complexity of NP search problems. J. Comput. Syst. Sci. **57**(1), 3–19 (1998)
3. Beame, P., Impagliazzo, R., Krajícek, J., Pitassi, T., Pudlák, P.: Lower bounds on Hilbert's Nullstellensatz and propositional proofs. In: Proceedings of the London Mathematical Society, vol. s3-73, pp. 1–26 (1996)
4. Becker, T., Weispfenning, V., Kredel, H.: Gröbner Bases. Graduate texts in mathematics, vol. 141. Springer, Heidelberg (1993)
5. Bright, C., Kotsireas, I., Ganesh, V.: Applying computer algebra systems and SAT solvers to the Williamson conjecture. J. Symb. Comput. (2019, in press)
6. Bright, C., Kotsireas, I., Ganesh, V.: SAT solvers and computer algebra systems: a powerful combination for mathematics. CoRR abs/1907.04408 (2019)
7. Buchberger, B.: Ein Algorithmus zum Auffinden der Basiselemente des Restklassenringes nach einem nulldimensionalen Polynomideal. Ph.D. thesis, University of Innsbruck (1965)
8. Choo, D., Soos, M., Chai, K.M.A., Meel, K.S.: Bosphorus: bridging ANF and CNF solvers. In: DATE 2019, pp. 468–473. IEEE (2019). https://doi.org/10.23919/DATE.2019.8715061
9. Ciesielski, M.J., Su, T., Yasin, A., Yu, C.: Understanding algebraic rewriting for arithmetic circuit verification: a bit-flow model. IEEE TCAD 1 (2019). Early acces
10. Clegg, M., Edmonds, J., Impagliazzo, R.: Using the Groebner basis algorithm to find proofs of unsatisfiability. In: STOC 1996, pp. 174–183. ACM (1996)
11. Cruz-Filipe, L., Heule, M.J.H., Hunt, W.A., Kaufmann, M., Schneider-Kamp, P.: Efficient certified RAT verification. In: de Moura, L. (ed.) CADE 2017. LNCS (LNAI), vol. 10395, pp. 220–236. Springer, Cham (2017). https://doi.org/10.1007/978-3-319-63046-5_14
12. Decker, W., Greuel, G.M., Pfister, G., Schönemann, H.: SINGULAR 4-1-0 – a computer algebra system for polynomial computations (2016). http://www.singular.uni-kl.de
13. Gelder, A.V.: Verifying RUP proofs of propositional unsatisfiability. In: ISAIM 2008 (2008)
14. Gelder, A.V.: Producing and verifying extremely large propositional refutations - have your cake and eat it too. Ann. Math. Artif. Intell. **65**(4), 329–372 (2012)
15. Grigoriev, D., Hirsch, E.A., Pasechnik, D.V.: Exponential lower bound for static semi-algebraic proofs. In: Widmayer, P., Eidenbenz, S., Triguero, F., Morales, R., Conejo, R., Hennessy, M. (eds.) ICALP 2002. LNCS, vol. 2380, pp. 257–268. Springer, Heidelberg (2002). https://doi.org/10.1007/3-540-45465-9_23
16. Heule, M.J.H.: Computing small unit-distance graphs with chromatic number 5. CoRR abs/1805.12181 (2018)

17. Heule, M.J.H., Biere, A.: Proofs for satisfiability problems. In: All about Proofs, Proofs for All Workshop, APPA 2014. vol. 55, pp. 1–22. College Publications (2015)
18. Heule, M.J.H., Jr., W.A.H., Wetzler, N.: Trimming while checking clausal proofs. In: FMCAD 2013, pp. 181–188. IEEE (2013)
19. Heule, M.J.H., Kauers, M., Seidl, M.: Local search for fast matrix multiplication. In: Janota, M., Lynce, I. (eds.) SAT 2019. LNCS, vol. 11628, pp. 155–163. Springer, Cham (2019). https://doi.org/10.1007/978-3-030-24258-9_10
20. Heule, M.J.H., Kauers, M., Seidl, M.: New ways to multiply 3×3-matrices. CoRR abs/1905.10192 (2019)
21. Homma, N., Watanabe, Y., Aoki, T., Higuchi, T.: Formal design of arithmetic circuits based on arithmetic description language. IEICE Trans. **89-A**(12), 3500–3509 (2006)
22. Impagliazzo, R., Pudlák, P., Sgall, J.: Lower bounds for the polynomial calculus and the Gröbner basis algorithm. Comput. Complex. **8**(2), 127–144 (1999)
23. Kaufmann, D.: Formal verification of multiplier circuits using computer algebra. Ph.D. thesis, Informatik, Johannes Kepler University Linz (2020)
24. Kaufmann, D., Biere, A., Kauers, M.: Verifying large multipliers by combining SAT and computer algebra. In: FMCAD 2019, pp. 28–36. IEEE (2019)
25. Kaufmann, D., Biere, A., Kauers, M.: From DRUP to PAC and back. In: DATE 2020, pp. 654–657. IEEE (2020)
26. Kaufmann, D., Biere, A., Kauers, M.: SAT, computer algebra, multipliers. In: Vampire 2018 and Vampire 2019. EPiC Series in Computing, vol. 71, pp. 1–18. EasyChair (2020)
27. Kaufmann, D., Fleury, M., Biere, A.: Pacheck and Pastèque, checking practical algebraic calculus proofs. In: FMCAD 2020. IEEE (2020, to appear). http://fmv.jku.at/pacheck_pasteque/,
28. Kuehlmann, A., Paruthi, V., Krohm, F., Ganai, M.: Robust Boolean reasoning for equivalence checking and functional property verification. IEEE TCAD **21**(12), 1377–1394 (2002)
29. Lv, J., Kalla, P., Enescu, F.: Efficient Gröbner basis reductions for formal verification of Galois field arithmetic circuits. IEEE TCAD **32**(9), 1409–1420 (2013)
30. Mahzoon, A., Große, D., Drechsler, R.: PolyCleaner: clean your polynomials before backward rewriting to verify million-gate multipliers. In: ICCAD 2018, pp. 129:1–129:8. ACM (2018)
31. Mahzoon, A., Große, D., Drechsler, R.: RevSCA: using reverse engineering to bring light into backward rewriting for big and dirty multipliers. In: DAC 2019, pp. 185:1–185:6. ACM (2019)
32. Mahzoon, A., Große, D., Scholl, C., Drechsler, R.: Towards formal verification of optimized and industrial multipliers. In: DATE 2020, pp. 544–549. IEEE (2020)
33. Meir, O., Nordström, J., Robere, R., de Rezende, S.F.: Nullstellensatz size-degree trade-offs from reversible pebbling. ECCC **137**, 18:1–18:16 (2019)
34. Miksa, M., Nordström, J.: A generalized method for proving polynomial calculus degree lower bounds. In: Conference on Computational Complexity, CCC 2015. LIPIcs, vol. 33, pp. 467–487. Schloss Dagstuhl (2015)
35. Niemetz, A., Preiner, M., Wolf, C., Biere, A.: BTOR2, BtorMC and Boolector 3.0. In: Chockler, H., Weissenbacher, G. (eds.) CAV 2018. LNCS, vol. 10981, pp. 587–595. Springer, Cham (2018). https://doi.org/10.1007/978-3-319-96145-3_32
36. Parhami, B.: Computer Arithmetic - Algorithms and Hardware designs. Oxford University Press, New York (2000)
37. Ritirc, D., Biere, A., Kauers, M.: A practical polynomial calculus for arithmetic circuit verification. In: SC2 2018, pp. 61–76. CEUR-WS (2018)

38. Roche, D.S.: What can (and can't) we do with sparse polynomials? In: ISSAC, pp. 25–30. ACM (2018)
39. Saraf, S., Volkovich, I.: Black-box identity testing of depth-4 multilinear circuits. Combinatorica **38**(5), 1205–1238 (2018). https://doi.org/10.1007/s00493-016-3460-4
40. Soos, M., Meel, K.S.: BIRD: engineering an efficient CNF-XOR SAT solver and its applications to approximate model counting. In: AAAI 2019, pp. 1592–1599. AAAI Press (2019). https://doi.org/10.1609/aaai.v33i01.33011592
41. Wolfram Research Inc: Mathematica (2016), version 10.4

"Mathemachines" via LEGO, GeoGebra and CindyJS

Zoltán Kovács[✉]

The Private University College of Education of the Diocese of Linz,
Salesianumweg 3, 4020 Linz, Austria
zoltan@geogebra.org

Abstract. The embedding of computer algebra technology within some software and hardware environments is a part of some recent technological improvements for mathematics education, in particular for gifted students. Computer algebra methods for simulation and modeling can be effectively used to connect symbolic computation and dynamic geometry in the popular dynamic mathematics software GeoGebra, based on fast calculation of Gröbner bases. As a result, teaching of algebraic curves via linkages and LEGO constructions can be approached with the help of a combination of novel tools.

In our contribution, we describe a set of three tools: the software tool LEGO Digital Designer, the program GeoGebra and the use of a web-camera through the CindyJS system for the introduction of basic issues concerning algebraic curves and geometric loci.

Keywords: Linkages · Algebraic geometry · LEGO · GeoGebra · CindyJS

1   From Watt's Steam Machine to STEAM Education

In the era of modern machining tools like computer-numerically-controlled (CNC) drills, boring tools and lathes, there seems to be no challenge on producing straight line movements anymore. Historically, as Kempe writes,

> "...until 1874 no-one in England knew of a method for drawing a straight line that was, in principle, perfect. The first solution was found by a French army officer called Peaucellier and was brought to England by Professor Sylvester in a lecture at the Royal Institution in January 1874." [8]

Peaucellier's method was independently discovered by Lipkin in 1871, a young Russian mathematician, Chebyshev's student. He received a substantial reward from the Russian government for this discovery [9]. Also Peaucellier was awarded by the French government by winning the Prix Montyon in 1875 for this invention, in the same year when Lipkin died. All of these events were about 90 years after James Watt's invention of the parallel linkage that was described in his patent specification of 1784 for the steam engine.

© Springer Nature Switzerland AG 2020
F. Boulier et al. (Eds.): CASC 2020, LNCS 12291, pp. 390–401, 2020.
https://doi.org/10.1007/978-3-030-60026-6_22

Watt's invention was extremely important for the industry, but the motion it provided was just almost-straight. It is nowadays, however, still in use, for example, in vehicle suspensions (Fig. 1)—it prevents relative sideways motion between the axle and body of the car.

Fig. 1. Watt's linkage in a 1998 Ford Ranger EV suspension [15]

These revolutionary results may deserve new interest according to the STEAM approach in mathematical education. Precise study of the motions these linkages produce is still a challenge from the scientific point of view (see e.g. [7] or [1]), and, without the freshly available digital tools they cannot even be well described at schools. Luckily, recent improvements in *GeoGebra*, and also in some other software like *CindyJS* [5], and, in addition, the free availability of *LEGO Digital Designer*, can already provide an achievable way to explain the most part of the underlying theory at schools as well. For STEAM educators, the connection here between science (S), technology (T), engineering (E), and mathematics (M) should be evident. In addition, arts (A) can also be affected when the beauty of motions or even the linkages themselves are admired (see e.g. Fig. 2 on Theo Jansen's artistic linkages, and [4] for their mathematical description).

This paper presents a possible way to offer digital resources in this topic aiming to highlight the didactical use of linkages in a STEAM-driven methodology supported by effective computer algebra techniques. Some feedback on the proposals was already given by a set of very gifted students after a summer camp in the United States in Colorado Springs in 2019 [11]. To emphasize the connection between mathematics and these mechanisms, we can simply call them "mathemachines". (See, for example, http://www.macchinematematiche.org, to read more about their long history.)

Fig. 2. Mathematical artist Theo Jansen with one of his beach-beasts

In this paper, the computer algebra part of the novel technology is high-lighted, but, as a harmonic extension of it, two other technical perspectives are mentioned.

2 Mathematical Background

It is surprisingly easy to describe a four-bar linkage by mathematical means. In Fig. 3, 4, and 5, Watt's linkage is shown as LEGO constructions, and also a mathematical sketch. In the latter figure, points A and C are fixed in the coordinate system, while points (m, n) and (f, g) have a circular motion around them, respectively. Point (x, y) is the midpoint of the connecting bar between (m, n) and (f, g). The movement of point (x, y) is to be determined—we assume that a pen refill is inserted at this position in the LEGO construction, and a curve is drawn by the constrained motion. This system can be described by 5 algebraic equations, and the only preliminary knowledge is the Pythagorean theorem (or, equivalently, the equation of a circle, or computing the distance between two points), and the two co-ordinate equations of a midpoint. Figure 6 shows the result of manual drawing with the help of the LEGO linkage.

Finding the geometric positions of (x, y) mathematically, that is, describing them in an algebraic way requires only simple operations of the input equations, namely additions, subtractions and multiplications. The theoretical basis and a practical algorithm that is based on it is well-known since Buchberger's 1965 PhD thesis [3] (namely Gröbner basis and elimination). The method is, however, not part of the secondary curriculum, simply because of the required theoretical

Fig. 3. Watt's parallel linkage given by a LEGO construction drawn in LDD

Fig. 4. Watt's parallel linkage built by LEGO elements in real life

knowledge in algebra (see [13] for more details), and the extremely high amount of the required operations—from the practical point of view. As a result, some parts of these concepts need to be used as black-box in the education at secondary level.

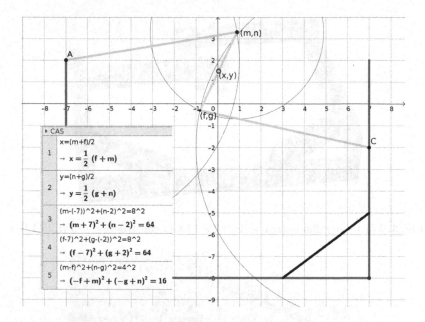

Fig. 5. Description of Watt's parallel linkage by mathematical means

Fig. 6. Watt's curve, constructed with a LEGO linkage and a G2 type pen refill

3   Digital Tools

3.1   LEGO Digital Designer

LEGO Digital Designer (LDD) is a free tool for Windows and Mac operating systems[1], developed by LEGO System A/S. It is available for download at https://www.lego.com/en-us/ldd#full-section-2. LDD offers constructing, saving, and re-opening LEGO constructions in a handy way. Young learners can observe pre-constructed digital assemblies in all details, and can move and rotate them to have a full understanding on how the parts are connected (see Fig. 3 and 7). This supports individual work. (See [10] for a complete set of .lxf files for immediate use with LDD.)

In fact, LDD has advanced capabilities to try to find the exact alignment of parts. This is performed internally with numerical means when using the Hinge

---

[1] Linux users can also use the software freely with the WINE emulator. See www.winehq.org for more details.

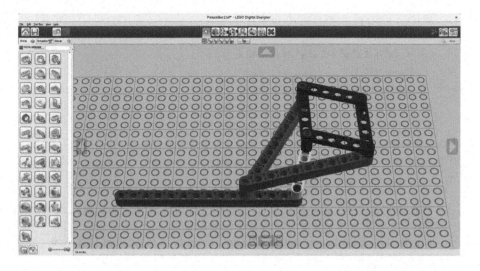

Fig. 7. LDD version 4.3.10 displays the Peaucellier-Lipkin linkage

Align Tool. LDD can also export the assembly plan in various formats, including a web page or a single figure.

3.2 GeoGebra

Technology plays an important role when manipulating equation systems. Several computer programs, and GeoGebra [6] is included on this list, have an effective way to compute the algebraic equation for the positions of the moving point (x, y) of a setup given in Watt's linkage. This process is called elimination—we actually get rid of all variables in an equation system but x and y.

Here we give a simple example of elimination. A two-meter-long ladder is sliding down the wall from its vertical position, until it reaches its horizontal stage. A cat is sitting in its middle. One needs to determine the movement of the cat. By solving this question, we assume that we put the edge of the wall in the origin and the wall corresponds to a part of the y-axis. Now an equation system can be given which contains the end points of the ladder with their coordinates $(a, 0)$, $(0, b)$, the length of the ladder (here 2), and the position (x, y) for the cat. In this case, the equation system $(a - 0)^2 + (0 - b)^2 = 2^2$, $\frac{a+0}{2} = x$, $\frac{0+b}{2} = y$ describes the movement. After eliminating all variables but x and y we obtain $x^2 + y^2 = 1$ that clearly corresponds to a circle with its center in the origin and radius 1. But, in fact, the cat does not move on the full circle, only on a quarter of it. Actually, by elimination we obtain the Zariski closure of the expected solution—it is the algebraic closure of the quadrant.

GeoGebra's built-in command **Eliminate** can be directly used to obtain the equation as seen in Fig. 8. Its syntax is the form **Eliminate**(*<List of Polynomials>*, *<List of Variables>*).

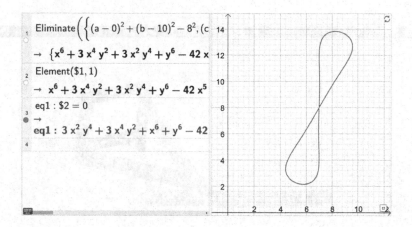

Fig. 8. The **Eliminate** command and its result in GeoGebra Classic 6

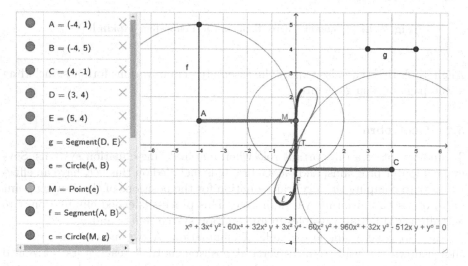

Fig. 9. A dynamic construction that shows the movement of a configurable linkage and computes its algebraic counterpart automatically

On the other hand, elimination is computed so effectively in GeoGebra that it is possible to compute several different eliminations with different setups for the linkage, multiple times in one second. Practically, this means that the learners can do real-time experiments by dragging the fixed points A and C (see Fig. 5) or change the lengths of the bars. To achieve this, the user has to construct a geometrical model of the linkage by means of a strict Euclidean construction, that is, using GeoGebra tools that can be internally translated to straightedge and compass construction steps. Figure 9 shows a possible way of assembling such an online activity, available at https://www.geogebra.org/m/sDCTVGrg# material/DpPyzRdx.

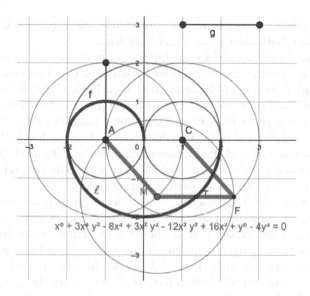

$$x^6 + 3x^4 y^2 - 8x^4 + 3x^2 y^4 - 12x^2 y^2 + 16x^2 + y^6 - 4y^4 = 0$$

Fig. 10. A setup that produces the union of three circles for the algebraic closure

In Fig. 9, the learner has options to drag A or C, or change the length of f or g. This covers all possible setups of a four-bar linkage. The thick (in the online version: green) curve shows all possible positions of point (x, y) denoted by T, while the thin (online: red) curve is the so-called algebraic closure of this set, also described by a formula at the bottom. From the user's perspective, after fixing A, C, f and g the movement of point T (it is draggable) is constrained on the thick curve that is a subset of the thin curve (namely, the Zariski closure of the thick curve).

Digital resources like the one shown in Fig. 9 can be produced by using the command **LocusEquation** in GeoGebra.

Further commands like **Factor** help the learners to understand some elements of the theory of factorization of polynomials at a basic level. The produced curve of a four-bar linkage is sometimes a union of two or more curves and their corresponding polynomials are presented only as their product. In Fig. 10, we can see a setup of A, C, f, and g that produces a union of three curves. Their algebraic counterpart is a circle for each case, but their product is not automatically factorized by GeoGebra. The learners should be motivated to find the factorization of a sextic polynomial, first by hand, and later to use computer algebra to do this step automatically.

GeoGebra fully supports online applets that include all the above mentioned commands. Technically, GeoGebra has a built-in version of the *Giac* computer algebra system [12], and this combination can be used on several platforms including desktop computers, web environments or even mobile devices. For presentations, therefore, a web-based activity can be preferred (as a GeoGebra book, for instance). For do-it-yourself activities, a downloadable version of GeoGebra can be recommended, both the Classic 5 and 6 versions.

On the other hand, GeoGebra has further support that allows understanding the mathematics behind the Peaucellier–Lipkin linkage. The basic idea of the linkage mechanism is a circle inverter, that is, a linkage that have two moving points P and P' and they are the corresponding images of each other according to a map of a circle inversion with a fixed center point O and radius r. Mathematically speaking, for a certain circle inversion I the equation $I(P) = P'$ holds. To connect the basic idea of the linkage mechanism and its mathematical counterpart we use GeoGebra's built-in tool that supports reflecting any geometrical object about a circle, namely, to perform an inversion.

Figure 11 shows how GeoGebra is capable of presenting several objects and their inverted images at the same time (see https://www.geogebra.org/m/kqta2rwa for an online version). Here the reference circle of the inversion is the unit circle (drawn in magenta in the online version), and the set of parallel input lines (drawn in red) are mapped into circles (drawn in blue) that are going through the origin O and all of them have their centers on the y-axis. The input object is technically a single (red) line. Its trace is always shown as the user drags it, while the output object has the same behavior.

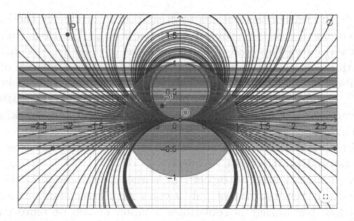

Fig. 11. Observing some fundamental properties of circle inversion with GeoGebra (Color figure online)

3.3   CindyJS

CindyJS is a programming language designed for web based experiments. Among many other features, it successfully connects the user's web camera and the machine's GPU computations with its WebGL plugin [14].

There are several mathematical experiments that can be supported with CindyJS and a web camera. Among others, the input picture can be considered point-by-point and transformed by using algebraic functions, eventually multiple times. In such a way, the properties of complex functions can be visually demonstrated, or self-similar geometric objects can be investigated, including fractals.

In fact, CindyJS can also be used to introduce some fundamental properties of circle inversion by inverting rectangle formed bitmaps, pixel-by-pixel. Effective processing of the high amount of numerical computations allows that the learner or the teacher can use the built-in web camera of the laptop as the input bitmap, and for each pixel, the output can be quickly computed. In such a way it is very easy and quick to verify that a set of parallel lines of the plane is mapped into a set of circles (and eventually into an invariant line).

For instance, in Fig. 12, the author is shown in a striped shirt with the reference circle, before and after inversion. The points outside the input image are shown as black points in the middle of the output image. The black points form a "black snowman": its border is a union of four circle arcs that are joining perpendicularly to their neighbors—this geometrical object is actually the map of a rectangle after inversion. (Note that the consecutive sides of a rectangle are perpendicular to each other, and inversion preserves angles.) Also, a set of parallel lines are mapped into a set of circles (except the one that goes through the center of the reference circle).

Fig. 12. Observing some fundamental properties of circle inversion with CindyJS

4   Finalizing the Story

As it is well-known today, the Peaucellier–Lipkin linkage is a combination of a circle inversor and a constrained circular motion. The assembly provides an exact linear motion, and the main part of explanation is that a circle inversion maps each circle that goes through the center of the reference circle, to an exact straight line. This follows immediately from the converse of property seen in Fig. 11 and 12. (See also [2] for a more detailed explanation.)

By combining LDD to help constructing linkages manually, and using GeoGebra to compute the movements of the "mathemachines" in an exact way, and finally, to learn some basic properties of circle inversion (with the extensive help of CindyJS' WebGL capabilities), we have an effective set of digital tools to help learners in finding Peaucellier's revolutionary idea themselves. Learners need a set of LEGO parts with 27 elements of 9 kinds to build several types of linkages (see Table 1), and, optionally[2], their own laptops with Internet connection to access the freely available software tools.

Table 1. Shopping list of the LEGO bricks with suggested colors to minimize costs

Part	Beam 15	Beam 11	Beam 9	Half beam 6	Beam 5	Beam 3	Beam 1×1	Long pin	Pin	Total
Color	red	gray	yellow	gray	white	gray	gray	blue	black	
Code	32278	32525	40490	32063	32316	32523	18654	6558	2780	

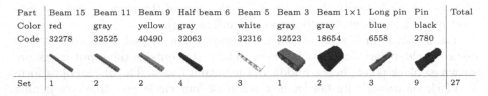

Set	1	2	2	4	3	1	2	3	9	27

The concept is not just promoting understanding but allowing the students to achieve finding a *mathemachine* that produces an exact straight movement, by themselves, with some hints from the teacher. According to the author's experience, this is possible not just among very gifted 11 years old learners, but also with students between 15 and 18, with more average skills. Processing of students' feedback on a mathematics camp in Upper Austria, held in Schloß Weinberg at Kefermarkt in February 2020 (see http://www.projektwoche.jku.at/2020/projekt2020_proj05.shtml), is still under evaluation and planned as input for further research.

5   Conclusion

In this paper, we have demonstrated heavy use of computer algebra in teaching algebraic curves via LEGO linkages. Computer algebra was triggered by Gröbner basis computations to obtain elimination of a set of variables from a polynomial ideal. Effective computation of Gröbner bases was crucial to enable fast visualization of the Zariski closure of the movement of a four-bar linkage, but also to allow deciding if a given LEGO linkage produces a linear motion or not.

However, some theoretical details had to be skipped at the presented educational level. On the other hand, the complexity of the algorithms could still be communicated. The achieved results, namely, the geometry of the set of outputs, were possible to discuss and interpret among young learners as well. As a conclusion, recent computer algebra methods can fruitfully improve students' understanding on mathematical concepts by a careful collection of novel technical means.

---

[2] Computer use for children at camp in Colorado Springs was very limited. Instead of using their own computers, the lecturer's computer was used during the classroom work. Tutorials on the assembly of LEGO linkages were printed instead of using them electronically. This made building the linkages a bit more difficult but still achievable.

Computer algebra was just one part of the applied technology. LDD and CindyJS completed the educational scenario in a harmonic combination.

Acknowledgments. The author is thankful to Tomás Recio for his suggestions on finalizing the paper. The author was partially supported by a grant MTM2017-88796-P from the Spanish MINECO (Ministerio de Economia y Competitividad) and the ERDF (European Regional Development Fund).

References

1. Bai, S., Angeles, J.: Coupler-curve synthesis of four-bar linkages via a novel formulation. Mech. Mach. Theor. **94**, 177–187 (2015)
2. Bryant, J., Sangwin, C.: How Round is your Circle? Where Engineering and Mathematics Meet. Princeton University Press, Princeton, New Jersey (2008)
3. Buchberger, B.: Bruno Buchberger's PhD thesis 1965: an algorithm for finding the basis elements of the residue class ring of a zero dimensional polynomial ideal. J. Symb. Comput. **41**, 475–511 (2006)
4. Dankert, J.: Theo Jansens Strandbeest-Mechanismus. Internet Service für die Auflagen 5 bis 7 des Lehrbuchs, Dankert/Dankert: Technische Mechanik (2013). http://www.tm-aktuell.de/TM5/Viergelenkketten/Strandbeest.html
5. von Gagern, M., Kortenkamp, U., Richter-Gebert, J., Strobel, M.: CindyJS. In: Greuel, G.-M., Koch, T., Paule, P., Sommese, A. (eds.) ICMS 2016. LNCS, vol. 9725, pp. 319–326. Springer, Cham (2016). https://doi.org/10.1007/978-3-319-42432-3_39
6. Hohenwarter, M., et al.: GeoGebra 5, September 2014. http://www.geogebra.org
7. Kapovich, M., Millson, J.J.: Universality theorems for configuration spaces of planar linkages. Topology **41**(6), 1051–1107 (2002)
8. Kempe, A.B.: On a general method of describing plane curves of the n-th degree by linkwork. Proc. London Math. Soc. **7**, 213–216 (1875)
9. Kempe, A.B.: How to Draw a Straight Line; A Lecture on Linkages. MacMillan and Co., London (1877)
10. Kovács, Z.: Lego-linkages. A GitHub project, 6 2019. https://github.com/kovzol/lego-linkages
11. Kovács, Z.: Teaching algebraic curves for gifted learners at age 11 by using LEGO linkages and GeoGebra. CoRR abs/1909.04964 (2019). https://arxiv.org/abs/1909.04964
12. Kovács, Z., Parisse, B.: Giac and GeoGebra – improved Gröbner basis computations. In: Gutierrez, J., Schicho, J., Weimann, M. (eds.) Computer Algebra and Polynomials, pp. 126–138. Lecture Notes in Computer Science, Springer (2015). https://doi.org/10.1007/978-3-319-15081-9_7, http://dx.doi.org/10.1007/978-3-319-15081-9_7
13. Kovács, Z., Recio, T., Vélez, M.P.: Reasoning about linkages with dynamic geometry. J. Symb. Comput. **97**, 16–30 (2020). https://doi.org/10.1016/j.jsc.2018.12.003
14. Montag, A., Richter-Gebert, J.: Bringing together dynamic geometry software and the graphics processing unit. CoRR abs/1808.04579 (2018). https://arxiv.org/abs/1808.04579
15. Wikipedia: Watt's linkage – Wikipedia, The Free Encyclopedia (2020). https://en.wikipedia.org/w/index.php?title=&oldid=952952034, Accessed 8 Jun 2020

Balanced NUCOMP

Sebastian Lindner[1]($\boxtimes$), Laurent Imbert[2], and Michael J. Jacobson Jr.[1]

[1] Department of Computer Science, University of Calgary, Calgary, Canada
{salindne,jacobs}@ucalgary.ca
[2] CNRS, LIRMM, Université de Montpellier, Montpellier, France
laurent.imbert@lirmm.fr

Abstract. Arithmetic in the divisor class group of a hyperelliptic curve is a fundamental component of algebraic geometry packages implemented in computer algebra systems such as Magma and Sage. In this paper, we present an adaptation of Shanks' NUCOMP algorithm for split model hyperelliptic curves of arbitrary genus that uses balanced divisors and includes a number of enhancements to optimize its efficiency in that setting. Our version of NUCOMP offers better performance than Cantor's algorithm in the balanced divisor setting. Compared with Magma's built-in arithmetic, our Magma implementation shows significant speed-ups for curves of all but the smallest genera, with the improvement increasing as the genus grows.

1 Introduction

The divisor class group of a hyperelliptic curve defined over a finite field is a finite abelian group at the center of a number of important open questions in algebraic geometry and number theory. Sutherland [14] surveys some of these, including the computation of the associated L-functions and zeta functions used in his investigation of Sato-Tate distributions [13]. Many of these problems lend themselves to numerical investigation, and as emphasized by Sutherland, fast arithmetic in the divisor class group is crucial for their efficiency. Indeed, implementations of these fundamental operations are at the core of the algebraic geometry packages of widely-used computer algebra systems such as Magma and Sage.

All hyperelliptic curves are represented as models that are categorized as either ramified (imaginary), split (real), or inert according to their number of points at infinity defined over the base field. Ramified curves have one point at infinity, whereas split curves have two. Inert (also called unusual) curves have no infinite points defined over the base field and are usually avoided in practice as they have cumbersome divisor class group arithmetic and can be transformed to a split model over at most a quadratic extension of the base field.

Divisor class group arithmetic differs on ramified and split models. The split scenario is more complicated. As a result, optimizing divisor arithmetic on split

The third author is supported in part by NSERC of Canada.

© Springer Nature Switzerland AG 2020
F. Boulier et al. (Eds.): CASC 2020, LNCS 12291, pp. 402–420, 2020.
https://doi.org/10.1007/978-3-030-60026-6_23

hyperelliptic curves has received less attention from the research community. However, split models have many interesting properties; most importantly, there exists a large array of hyperelliptic curves that cannot be represented with a ramified model and require a split model representation. Thus, exhaustive computations such as those in [13] conduct the bulk of their work on split models by necessity.

Arithmetic in the divisor class group of a hyperelliptic curve can be described algebraically using an algorithm due to Cantor [1], and expressed in terms of polynomial arithmetic. Various improvements and extensions to Cantor's algorithm have been proposed for ramified model curves, including an adaptation of Shank's NUCOMP algorithm [11] for composing binary quadratic forms [6]. The main idea behind NUCOMP is that instead of composing two divisors directly and then reducing to find an equivalent reduced divisor, a type of reduction is applied part way through the composition, so that when the composition is finished the result is almost always reduced. The effect is that the sizes of the intermediate operands are reduced, resulting in better performance in most cases. Improvements to NUCOMP have been proposed, most recently the work of [5], where best practices for computing Cantor's algorithm and NUCOMP are empirically investigated.

NUCOMP has also been proposed for arithmetic in the so-called infrastructure of a split model curve [8]. However, as shown by Galbraith et al. [4,9], arithmetic on split model hyperelliptic curves is most efficiently realized via a divisor arithmetic framework referred to as balanced. Although the balanced and the infrastructure frameworks are similar, NUCOMP had yet to be applied explicitly to the former.

In this paper, we present an adaptation of NUCOMP for divisor class group arithmetic on split model hyperelliptic curves in the balanced divisor framework. We incorporate optimizations from previous works in the ramified model setting and introduce new balanced setting-specific improvements that further enhance practical performance. Specifically, our version of NUCOMP includes various improvements over its infrastructure counterpart [8]:

- it describes for the first time exactly how to use NUCOMP in the framework of balanced divisors, including explicit computations of the required balancing coefficients;
- it introduces a novel normalization of divisors in order to eliminate the extra adjustment step required in [8] for typical inputs when the genus of the hyperelliptic curve is odd, so that in all cases typical inputs require no extra reduction nor adjustment steps;
- it uses certain aspects of NUCOMP to compute one adjustment step almost for free in some cases.

We present empirical results that demonstrate the efficiency gains realized from our new version of NUCOMP as compared with the previous best balanced divisor class group arithmetic based on Cantor's algorithm and the arithmetic implemented in Magma, showing that NUCOMP is the method of choice for all but the smallest genera. With our improvements, NUCOMP is more efficient than Cantor's algorithm for genus as low as 5, compared to 7 using the version in [8], both of which are within the possible range of applications related to numerical investigations of number-theoretic conjectures. Our implementation is faster than Magma's built-in arithmetic for $g \geq 7$, and the gap increases with the genus; we assume that a more fair comparison using, for example, optimized C implementations would further narrow this gap in performance.

The rest of the paper is organized as follows. In Sect. 2 we provide background information on split model hyperelliptic curves. Balanced divisor arithmetic using Cantor's algorithm is presented in Sect. 3. In Sect. 4 we present our version of NUCOMP for the balanced divisor setting, as well as details of our improvements. In Sect. 5 we present empirical results comparing our version of NUCOMP to Cantor's algorithm and Magma's built-in arithmetic. Finally, we give some conclusions and directions for future work in Sect. 6.

2   Background

In this section we recall the essential relevant notions related to divisor classes of hyperelliptic curves and their arithmetic. For more details and background, the reader is referred to [10, § 12.4] for Sect. 2.1 and [3, Chapter 7] for Sect. 2.2.

2.1   Split Model Hyperelliptic Curves

As described in [10, Definition 12.4.1], a *hyperelliptic curve* C of genus g defined over a finite field k is given by a hyperelliptic equation

$$y^2 + h(x)y = f(x), \quad \text{with } h, f \in k[x],$$

that is absolutely irreducible and non-singular. A *split model* for a hyperelliptic curve C of genus g over k is given by a hyperelliptic equation satisfying $\deg(f) = 2g + 2$ and $\deg(h) \leq g + 1$. In addition, the leading coefficient of f is a square except over fields of characteristic 2 where it is of the form $s^2 + s$ for some $s \in k^*$.

Let $C(\overline{k})$ be the set of $\overline{k}$-rational points of C. The *hyperelliptic involution* of C is the map $\iota : C(\overline{k}) \to C(\overline{k})$ that sends a finite point $P = (x, y)$ on C to the point $\overline{P} = \iota(P) = (x, -y - h(x))$ on C. A point P on C is *ramified* if $\iota(P) = P$, and unramified otherwise.

The model used to represent a hyperelliptic curve determines the number and type of points at infinity. A split model representation has two unramified k-rational points at infinity denoted ∞^+ and ∞^-, where $\iota(\infty^+) = \infty^-$. Ramified models have a single ramified k-rational point at infinity, and inert models have none. It is sometimes possible to change the model of a curve C without modifying the field of definition k by translating other points to infinity. If C has a ramified k-rational point, one can obtain a ramified model for C by translating this point to infinity, by [10, Theorem 12.4.12]. If C does not have a ramified k-rational point, but has an unramified k-rational point, then similarly, that point can be translated to infinity, providing two points at infinity ∞^+, ∞^-

and thus C can be represented with a split model. If no k-rational points exist, including at infinity, then the hyperelliptic curve C can only be represented by an inert model; no alternative ramified or split models are possible over k. However, hyperelliptic curves that have neither a ramified nor an unramified k-rational point are rare and only exist over fields whose cardinality is small relative to the genus. If k is a finite field of cardinality q, the Weil bound $\#C(k) \geq q+1-2g\sqrt{q}$ guarantees that a genus g hyperelliptic curve C over k has a k-rational point whenever $q > 4g^2$, and an unramified k-rational point when $q > 4g^2 + 2g + g$.

Split models therefore are more general than ramified, as ramified models are only obtainable when the curve has a ramified point. Inert models of curves can easily be avoided in practice by translating to a split or ramified model when q is sufficiently large to guarantee a k-rational point, or by considering the curve as a split model over a quadratic extension of k otherwise. Thus, in this work we only consider improvements for hyperelliptic curves given by a split model, with performance comparisons to ramified models given in Sect. 5.

2.2 Divisor Class Groups of Split Model Hyperelliptic Curves

A *divisor* on a hyperelliptic curve C defined over k is a formal sum $D = \sum n_P P$ of points $P \in C(\overline{k})$ with only finitely many $n_P \neq 0$. The *support* of D, denoted $\mathrm{supp}(D)$, is the set of points $P \in C(\overline{k})$ occurring in D with $n_P \neq 0$. The *degree* of D is $\deg(D) = \sum n_p$. A divisor D is said to be *defined over* k if $\sigma D = \sum n_P \sigma P = D$ for all $\sigma \in \mathrm{Gal}(\overline{k}/k)$. The set of all degree zero divisors on C defined over k, denoted $\mathrm{Div}_k^0(C)$, is an Abelian group under component-wise addition. A divisor is *principal* if it is of the form $\mathrm{div}(\alpha) = \sum_P \mathrm{ord}_P(\alpha)P$ for some function $\alpha \in k(C)^*$ where $k(C) = k(x, y)$ is the function field of C. Principal divisors have degree zero and the set of all principal divisors $\mathrm{Prin}_k^0(C) = \{\mathrm{div}(\alpha) \mid \alpha \in k(C)^*\}$ is a subgroup of $\mathrm{Div}_k^0(C)$. The *divisor class group* of C defined over k is the quotient group $\mathrm{Pic}_k^0(C) = \mathrm{Div}_k^0(C)/\mathrm{Prin}_k^0(C)$. The principal divisor corresponding to $\infty^+ - \infty^-$ (resp. $\infty^- - \infty^+$) is denoted D_{∞^+} (resp. D_{∞^-}).

A divisor $D = \sum_P n_P P$ is *affine* if $n_{P_\infty} = 0$ for all k-rational points at infinity P_∞ on C. The divisor D is *effective* if $n_P \geq 0$ for all points P. An effective divisor can be written as $\sum P_i$, where the P_i need not be distinct. An affine effective divisor $D = \sum P_i$ is *semi-reduced* if for any $P_i \in \mathrm{supp}(D)$, $\iota(P_i) \notin \mathrm{supp}(D)$, unless $P_i = \iota(P_i)$. A semi-reduced divisor D is *reduced* if $\deg(D) \leq g$.

A semi-reduced divisor D has a compact *Mumford representation* $D = (u, v)$ such that $u, v \in k[x]$, $\deg(v) < \deg(u)$, u is monic, and $u \mid (v^2 + vh - f)$. Explicitly, u is defined as the polynomial whose roots are the x-coordinates of every affine point in the support of $D = \sum n_P P$ accounting for multiplicity, i.e. $u = \prod_i (x - x_i)^{n_{P_i}}$ for all $P_i = (x_i, y_i) \in \mathrm{supp}(D)$. The polynomial v is the interpolating polynomial that passes through the points P_i. The Mumford representation of D is said to be reduced if $\deg(u) \leq g$. The *degree* of a semi-reduced divisor $D = (u, v)$ in Mumford representation is given by $\deg(D) = \deg(u)$.

Every rational divisor class in $\mathrm{Pic}_k^0(C)$ can be represented by a degree zero divisor $[D]$ that has a semi-reduced affine portion, but this representation is not necessarily unique. As described in [4], let D_∞ be an effective divisor of degree g supported on k-rational points at infinity. Over split model curves, let $D_\infty = \lceil g/2 \rceil \infty^+ + \lfloor g/2 \rfloor \infty^-$. Then, every divisor class $[D]$ over genus g hyperelliptic curves described with a split model can be uniquely written as $[D_0 - D_\infty]$ where $D_0 = D_a + D_i$ is a k-rational divisor of degree g with the affine portion D_a reduced, and D_i is a specially-chosen divisor supported on the infinite points, for example as described in the next section. Note that as is standard practice, we refer to the degree of such a divisor class representative as the degree of the affine part D_a, although this is a slight abuse of notation as the divisor D technically has degree zero.

3   Balanced Divisor Arithmetic Using Cantor's Algorithm

Over split model curves, divisor classes in $\mathrm{Pic}^0_k(C)$ do not have a unique, reduced Mumford representation. The Mumford representation only utilizes information about the affine portion D_a of $D_0 = D_a + D_i$ for $D = D_0 - D_\infty$; uniqueness is lost because the same affine portion of D_0 could be combined with different multiples of ∞^+ and ∞^- in D_i to represent different divisor classes. Galbraith et al. [4] defined a *reduced balanced divisor representation* for split model curves which appends to the polynomials u, v a balancing coefficient n, the number of copies of ∞^+ in D_0, hence $[D] = [u, v, n]$. In order for this representation to be unique and reduced, n is kept small, in the range $[0, g - \deg(u)]$ and $\deg u \leq g$. A divisor class $[D] = [u, v, n]$ therefore corresponds to

$$[u, v, n] = [u, v] + n\infty^+ + (g - \deg(u) - n)\infty^- - D_\infty.$$

In this notation, for example,

$$[1, 0, \lceil g/2 \rceil] = [1, 0] + \lceil g/2 \rceil \infty^+ + \lfloor g/2 \rfloor \infty^- - D_\infty$$

is the unique representative of the neutral divisor class in $\mathrm{Pic}^0_k(C)$.

Addition of divisor classes represented as reduced balanced divisors, as described in [4], is done via a two-step process. First, the affine parts of the divisors are added and reduced using Cantor's algorithm, while computing the new balancing coefficient n of the result. At this point it is possible that the resulting divisor class is neither reduced nor balanced, so a series of adjustment steps is applied, *up-adjustments* if n needs to be increased and *down-adjustments* if it needs to decrease, until the n value satisfies $0 \leq n \leq g - \deg u$ and is thus balanced. The main advantage of using balanced divisor representatives is that in the generic case, where both divisors have degree g and $n = 0$, the number of adjustment steps required is zero for even genus and one for odd genus.[1]

The two algorithms that we present in the following sections for addition and reduction (Algorithm 1) and for adjustment (Algorithm 2) follow this strategy

---

[1] The required adjustment step over odd genus reduces the degree of the intermediate divisor, similar to a reduction step.

with a variety of practical improvements. We adopt an alternative normalization of the v polynomial from the Mumford representation, as well as well-known algorithmic improvements to Cantor's algorithm described, for example, in [5], as described next. In all algorithms presented, let $\text{lc}(a)$ denote the leading coefficient of polynomial a and $\text{monic}(a) = a/\text{lc}(a)$, i.e. the polynomial a made monic.

3.1 Extended Mumford Representation and Tenner's Algorithm

One standard optimization for arithmetic with ideals of quadratic number fields is to represent the ideal as a binary quadratic form, a representation that includes a third redundant coefficient that is useful computationally. In the context of divisor arithmetic, this means adding the polynomial $w = (f - v(v+h))/u$ to the Mumford representation, so that balanced divisor classes in our implementation have four coordinates, $[u, v, w, n]$.

This polynomial must be computed in every application of Cantor's algorithm as well as in reduction steps and adjustment steps. Having it available as part of the divisor representation results in some savings in the divisor addition part, and allows for the use of Tenner's algorithm for reduction and adjustment steps, a standard optimization for computing continued fraction expansions of quadratic irrationalities (see, for example, [7, §3.4]).

3.2 Divisor Representation Using Reduced Bases

The standard Mumford representation of a divisor $[u, v]$ has v reduced modulo u, but any other polynomial congruent to v modulo u can also be used. In split model curves, an alternate representation called the *reduced basis* turns out to be computationally superior in practice. Reduced bases are defined in terms of the unique polynomial V^+, the principal (polynomial) part of the root y of $y(y + h(x)) - f(x) = 0$ for which $\deg(f - V^+(V^+ + h)) \leq g$, or the other root $V^- = -V^+ - h$. Note that such V^+ and V^- only exist for split models.

We say that a representation of the affine divisor $[u, v]$ given by $[u, \tilde{v}]$ is in *reduced basis* or *positive reduced basis* if $\tilde{v} = V^+ - [(V^+ - v) \pmod{u}]$ and in *negative reduced basis* if $\tilde{v} = V^- - [(V^- - v)) \pmod{u}]$. To convert a divisor $[u, v, w, n]$ into negative reduced basis $[u, v', w', n]$, first compute $(q, r) = \text{DivRem}(V^- - v, u)$, where we define $\text{DivRem}(a, b)$ as q, r, the quotient and remainder, respectively, obtained when dividing a by b, i.e. $a = qb + r$. For uniqueness, we take the remainder r satisfying $\deg(r) < \deg(b)$. Then $\hat{v} = V^- - r$, $w' = w - q(v + h + \hat{v})$, and let $v' = \hat{v}$. To convert back to positive reduced basis, first compute $q \neq 0$, r such that $v' = qu + r$, then $w = w' - q(v' + h + r)$, and let $v = r$.

In both types of reduced basis, cancellations cause the degree of $f - \tilde{v}(\tilde{v} + h)$ to be two less than that obtained using $v \bmod u$ instead of $\tilde{v}$, resulting in more efficient divisor addition. Although divisor class composition and reduction are not affected by this representation, a negative reduced basis is computed in an up adjustment, and positive in a down adjustment. By working with divisors that are already in a reduced basis, we avoid having to change basis when the corresponding type of adjustments are required.

In our implementation, we use the negative reduced basis to represent our balanced divisors. For even genus curves, no adjustments are required for typical-case inputs, so either type of reduced basis will do. However, for odd genus one up adjustment is always required for typical inputs. Having divisors always represented via a negative reduced basis ensures that base changes are not required before computing this adjustment step.

3.3 Balanced Add

Balanced Add, described in Algorithm 1, for adding divisor classes over split model curves, closely follows an optimized version of Cantor's algorithm (Algorithm 1 of [5]), with the addition of keeping v in negative reduced basis, keeping track of the balancing coefficient n, and applying adjustment steps at the end as described in [4]. The algorithm is optimized for the frequently-occurring situation where $\gcd(u_1, u_2) = 1$, based on a description due to Shanks of Gauss's composition formulas for binary quadratic forms. A more efficient doubling algorithm can be obtained by specializing to the case that $D_2 = D_1$ and simplifying.

Balanced Add, and indeed all the divisor class addition algorithms presented here, require applications of the extended Euclidean algorithm for polynomials. Throughout, we use the notation $(d, s, t) = \mathrm{XGCD}(a, b)$ to denote the output of this algorithm, specifically $d = \gcd(a, b) = as + bt$ with s, t normalized so that $\deg(s) < \deg(b) - \deg(d)$ and $\deg(t) < \deg(a) - \deg(d)$.

In the balanced setting, addition and reduction are similar to that over ramified curves, the only difference being the threshold for applying a reduction step is $\deg(u) > g + 1$ instead of $\deg(u) > g$; adjustment steps are applied when $\deg(u) = g + 1$. Reduction steps decrease the degree of the affine portion of the divisor class by at least two, so at most $\lfloor g/2 \rfloor$ steps are required to reduce the output of the composition portion to a linearly equivalent divisor whose affine part has degree at most $g + 1$.

3.4 Balanced Adjust

Balanced Adjust, described in Algorithm 2, is called after partially reducing a divisor $D = [u, v, w, n]$, for a final reduction from degree $g+1$ if necessary, and for balancing if n is outside the required range $0 \le n \le g - \deg(u)$. Balanced Adjust can be viewed as a composition of the affine portion with D_{∞^+}, when n is above the threshold (down adjustments) or D_{∞^-} when n is below (up adjustments). This can be thought of as transferring a symbolic copy of the point ∞^+ (for down) or $\infty^- = -\infty^+$ (for up) into the affine portion, keeping the divisor class the same. The number of adjustment steps required is at most $\lceil g/2 \rceil$.

4 Balanced NUCOMP

Cantor's algorithm [1] is closely related to Gauss' composition and reduction of binary quadratic forms. In 1988, Shanks [11] described an alternative to Gauss'

Algorithm 1. Balanced Add

Input: $[u_1, v_1, w_1, n_1]$, $[u_2, v_2, w_2, n_2]$, f, h, V^-.
Output: $[u, v, w, n] = [u_1, v_1, w_1, n_1] + [u_2, v_2, w_2, n_2]$.

1: $t_1 = v_1 + h$.
2: Compute $(S, a_1, b_1) = \mathrm{XGCD}(u_1, u_2)$.
3: $K = a_1(v_2 - v_1) \pmod{u_2}$.
4: **if** $S \neq 1$ **then**
5:     Compute $(S', a_2, b_2) = \mathrm{XGCD}(S, v_2 + t_1)$.
6:     $K = a_2 K + b_2 w_1$.
7:     **if** $S' \neq 1$ **then**
8:         $u_1 = u_1/S'$, $u_2 = u_2/S'$, $w_1 = w_1 S'$.
9:     $K = K \pmod{u_2}$.
10:    $S = S'$.
11: $T = u_1 K$, $u = u_1 u_2$, $v = v_1 + T$.
12: $w = (w_1 - K(t_1 + v))/u_2$.
13: $n = n_1 + n_2 + \deg(S) - \lceil g/2 \rceil$.
14: **if** $\deg(u) \leq g$ **then**
15:     **if** $\deg(v) \geq \deg(u)$ **then**
16:         $(q, r) = \mathrm{DivRem}(V^- - v, u)$.
17:         $tv = V^- - r$, $w = w - q(v + h + tv)$, $v = tv$.
18: **else**
19:     **while** $\deg(u) > g + 1$ **do**
20:         **if** $\deg(v) = g + 1$ and $\mathrm{lc}(v) = \mathrm{lc}(-V^- - h)$ **then**
21:             $n = n + \deg(u) - g - 1$.
22:         **else if** $\deg(v) = g + 1$ and $\mathrm{lc}(v) = \mathrm{lc}(V^-)$ **then**
23:             $n = n + g + 1 - \deg(w)$.
24:         **else** $n + (\deg(u) - \deg(w))/2$.
25:         $u_o = u$, $u = w$.
26:         $(q, r) = \mathrm{DivRem}(V^- + v + h, u)$.
27:         $v_t := V^- - r$, $w = u_o - q(v_t - v)$, $v = v_t$.
28:     $w = \mathrm{lc}(u)w$, $u = \mathrm{monic}(u)$.
29: **return** Balanced Adjust$([u, v, w, n], f, h, V^-)$.

method called NUCOMP. Instead of composing and then reducing, which results in a non-reduced intermediate quadratic form with comparatively large coefficients, the idea of NUCOMP is to start the composition process and to apply an intermediate reduction of the operands using a simple continued fraction expansion before completing the composition. The result is that the intermediate operands are smaller, and at the end the resulting quadratic form is in most cases reduced without having to apply any additional reduction steps. Jacobson and van der Poorten [6] showed how to apply the ideas of NUCOMP to divisor class group arithmetic, obtaining analogous reductions in the degrees of the intermediate polynomial operands.

Applied to our setting, the main idea is that the element $(v + y)/u \in k(C)$ is approximated by the rational function u_2/K with u_2, K from Algorithm 1.

Algorithm 2. Balanced Adjust

Input: $[u_a, v_a, w_a, n_a]$, f, h, V^+, where $\deg(u_a) \leq g + 1$.
Output: $[u, v, w, n] = [u_a, v_a, w_a, n_a]$, where $\deg(u) \leq g$ and $0 \leq n \leq \deg(u) - g$.
 1: $u = u_a$, $v = v_a$, $w = w_a$, $n = n_a$,
 2: **if** $n < 0$ **then**
 3:    **while** $n < 0$ **do**
 4:        $u_o = u$, $u = w$.
 5:        $(q, r) = \text{DivRem}(V^- + v + h, u)$.
 6:        $v_t := V^- - r$, $w = u_o - q(v_t - v)$, $v = v_t$.
 7:        $n = n + g + 1 - \deg(u)$.
 8:    $w = \text{lc}(u)w$, $u = \text{monic}(u)$.
 9: **else if** $n > g - \deg(u)$ **then**
10:    $t = -V^- - h$.
11:    $(q, r) = \text{DivRem}(t - v, u)$.
12:    $v_t = t - r$, $w = w - q(v + h + v_t)$, $v = v_t$.
13:    **while** $n > g - \deg(u) + 1$ **do**
14:        $n = n + \deg(u) - g - 1$, $u_o = u$, $u = w$.
15:        $(q, r) = \text{DivRem}(t + v + h, u)$.
16:        $v_t := t - r$, $w = u_o - q(v_t - v)$, $v = v_t$.
17:    **if** $n > g - \deg(u)$ **then**
18:        $n = n + \deg(u) - g - 1$, $u_o = u$, $u = w$.
19:        $(q, r) = \text{DivRem}(V^- + v + h, u)$.
20:        $v_t := V^- - r$, $w = u_o - q(v_t - v)$, $v = v_t$.
21:    **else**
22:        $t = V^- - V^+$, $(q, r) = \text{DivRem}(t, u)$.
23:        $v_t = v + t - r$, $w = w - q(v + v_t)$, $v = v_t$.
24:    $w = \text{lc}(u)w$, $u = \text{monic}(u)$.
25: **return** $[u, v, w, n]$.

Cantor's Algorithm first computes the non-reduced divisor, and subsequently applies a reduction algorithm that can be expressed in terms of expanding the continued fraction of the quadratic irrationality $(v + y)/u$. The first several partial quotients of the simple continued fraction expansion of the rational approximation u_2/K are the same as that of $(v+y)/u$, and these can be computed without having to first compute the non-reduced divisor (u, v). Given those partial quotients, the final reduced divisor can be computed via expressions involving them and other low-degree operands, again without having to first compute the non-reduced divisor (u, v). For more details on the theory behind NUCOMP, see [8].

The most recent work on NUCOMP [5] provides further optimizations and empirical results demonstrating that it outperforms Cantor's algorithm for hyperelliptic curves of genus as small as 7, and that the relative performance improves as the genus increases. An enhanced version of NUCOMP for adding and reducing divisors without balancing, that works for curves defined over arbitrary fields and incorporates all the optimizations described in the previous section, is presented in Algorithm 3.

Algorithm 3. NUCOMP

Input: $[u_1, v_1, w_1]$, $[u_2, v_2, w_2]$, f, h.
Output: $[u, v, w]$ with $[u, v, w] = [u_1, v_1, w_1] + [u_2, v_2, w_2]$.

1: **if** $\deg(u_1) < \deg(u_2)$ **then**
2:     $[u_t, v_t, w_t] = [u_2, v_2, w_2]$, $[u_2, v_2, w_2] = [u_1, v_1, w_1]$.
3:     $[u_1, v_1, w_1] = [u_t, v_t, w_t]$.
4: $t_1 = v_1 + h$, $t_2 = v_2 - v_1$.
5: Compute $(S, a_1, b_1) = \text{XGCD}(u_1, u_2)$.
6: $K = a_1 t_2 \pmod{u_2}$.
7: **if** $S \neq 1$ **then**
8:     Compute $(S', a_2, b_2) = \text{XGCD}(S, v_2 + t_1)$.
9:     **if** $S' \neq 1$ **then**
10:         $u_1 = u_1/S'$, $u_2 = u_2/S'$.   (exact divisions)
11:         $w_1 = w_1 S'$.
12:         $K = K \pmod{u_2}$.
13:         $S = S'$.
14: **if** $\deg(u_2) + \deg(u_1) \leq g$ **then**
15:     $u = u_2 u_1$, $v = v_1 + u_1 K$.
16:     $w = (w_1 - K(t_1 + v))/u_2$.   (exact division)
17:     **if** $\deg(v) \geq \deg(u)$ **then**
18:         $(q, r) = \text{DivRem}(v, u)$.
19:         $w = w + q(v + h + r)$, $v = r$.
20: **else**
21:     Set $r = K$, $r' = u_2$, $c' = 0$, $c = -1$, $l = -1$.
22:     **while** $\deg(r) > (\deg(u_2) - \deg(u_1) + g)/2$ **do**
23:         $(q, r_n) = \text{DivRem}(r', r)$.
24:         Set $r' = r$, $r = r_n$, $c_n = c' - qc$, $c' = c$, $c = c_n$, $l = -l$.
25:     $t_3 = u_1 r$.
26:     $M_1 = (t_3 + t_2 c)/u_2$.   (exact division)
27:     $M_2 = (r(v_2 + t_1) + w_1 c)/u_2$.   (exact division)
28:     $u' = l(r M_1 - c M_2)$.
29:     $z = (t_3 + c'u')/c$.   (exact division)
30:     $v = (z - t_1) \pmod{u'}$.
31:     $u = \text{monic}(u')$.
32:     $w = (f - v(v + h))/u$.
33: **return** $[u, v, w]$.

Although this version of NUCOMP is intended for divisor class group addition on ramified model hyperelliptic curves, it also works for adding reduced affine divisors and producing a reduced output over split model curves. In [8], the authors also describe how to use this to perform arithmetic in the infrastructure of a split model hyperelliptic curve, but not in the divisor class group. It was shown that for split model curves the output of NUCOMP is always reduced for even genus curves, but that for odd genus at least one extra reduction step is required.

In the following (Algorithm 4), we present an adaptation of NUCOMP, denoted Balanced NUCOMP, for performing divisor class group arithmetic on a split model hyperelliptic curve using balanced divisor arithmetic. A more efficient doubling algorithm optimized for the case that the input divisors are equal, denoted Balanced NUDUPL, is used for our testing in Sect. 5 and presented as Algorithm 5 for the reader's convenience. Our additions and improvements to Algorithm 3 include the following:

4.1 using divisors normalized with the negative reduced basis so that in *both* even and odd genus, divisor additions generically require no further reduction nor adjustment steps after NUCOMP;

4.2 adapting NUCOMP to the balanced setting by tracking and updating the balancing coefficient n appropriately, including determining how to update the balancing coefficient n after the simple continued fraction steps;

4.3 using simple continued fraction steps of NUCOMP to eliminate an adjustment step for certain non-generic cases where the degree of the output divisor is small.

In the following subsections, we provide more details and justification for each of these modifications.

4.1 Normalization with Negative Reduced Basis

Let $[u_1, v_1, w_1, n_1]$ and $[u_2, v_2, w_2, n_2]$ be the input for Balanced NUCOMP. For split model curves, the simple continued fraction portion of NUCOMP can absorb at most one adjustment step while still ensuring that the output divisor is reduced, by setting the bound for the simple continued fraction expansion in line 22 appropriately. If the bound is set any lower than in Algorithm 4, then the resulting u polynomial ends up having degree greater than g, meaning that the divisor is not reduced.

We make the choice to normalize all our divisor class representatives using the negative reduced basis for the following reasons:

1. For odd genus, the generic case for divisor class arithmetic requires an up adjustment. Ensuring that our divisors are normalized using negative reduced basis allows us to perform this adjustment step via an extra step in the NUCOMP simple continued fraction part, so that after NUCOMP the output for the generic case is both reduced and balanced without any further steps.

2. For even genus, the generic case requires no adjustments, so either positive reduced or negative reduced basis works equally well.

3. Non-generic cases of divisor class addition over even and odd genus require either up adjustments, down adjustments or no adjustment. Over even genus, out of all cases that require adjustments, exactly half are down and half are up. Over odd genus, far more non-generic cases require an up adjustment than down. This can be seen by analyzing the computation of the balancing coefficient n in the composition portion of Algorithm 1. Line 12 states $n = n_1 + n_2 + \deg(S) - \lceil g/2 \rceil$, where the ceiling function increases the cases for which $n < 0$ for odd genus curves.

Algorithm 4. Balanced NUCOMP

Input: $[u_1, v_1, w_1, n_1]$, $[u_2, v_2, w_2, n_2]$, f, h, V^-.
Output: $[u, v, w, n]$ with $[u, v, w, n] = [u_1, v_1, w_1, n_1] + [u_2, v_2, w_2, n_2]$.

1: **if** $\deg(u_1) < \deg(u_2)$ **then**
2: $\quad [u_t, v_t, w_t, n_t] = [u_2, v_2, w_2, n_2]$, $[u_2, v_2, w_2, n_2] = [u_1, v_1, w_1, n_1]$.
3: $\quad [u_1, v_1, w_1, n_1] = [u_t, v_t, w_t, n_t]$.
4: $t_1 = v_1 + h$, $t_2 = v_2 - v_1$.
5: Compute $(S, a_1, b_1) = \mathrm{XGCD}(u_1, u_2)$.
6: $K = a_1 t_2 \pmod{u_2}$.
7: **if** $S \neq 1$ **then**
8: $\quad$ Compute $(S', a_2, b_2) = \mathrm{XGCD}(S, v_2 + t_1)$.
9: $\quad K = a_2 K + b_2 w_1$.
10: $\quad$ **if** $S' \neq 1$ **then**
11: $\quad\quad u_1 = u_1/S'$, $u_2 = u_2/S'$. (exact divisions)
12: $\quad\quad w_1 = w_1 S'$.
13: $\quad\quad K = K \pmod{u_2}$.
14: $\quad\quad S = S'$.
15: $D = \deg(u_2) + \deg(u_1)$.
16: $n = n_1 + n_2 + \deg(S) - \lceil g/2 \rceil$.
17: **if** $D \leq g$ and $((n \geq 0$ and $n \leq g - D)$ or $\deg(w_1) - \deg(u_2) > g))$ **then**
18: $\quad T = u_1 K$, $u = u_2 u_1$, $v = v_1 + T$.
19: $\quad w = (w_1 - K(t_1 + v))/u_2$. (exact division)
20: $\quad$ **if** $\deg(v) \geq \deg(u)$ **then**
21: $\quad\quad (q, r) = \mathrm{DivRem}(V^- - v, u)$.
22: $\quad\quad tv = V^- - r$, $w = w - q(v + h + tv)$, $v = tv$.
23: **else**
24: $\quad$ Set $r = K$, $r' = u_2$, $c' = 0$, $c = -1$, $l = -1$.
25: $\quad$ **while** $\deg(r) \geq (\deg(u_2) - \deg(u_1) + g + 1)/2$ **do**
26: $\quad\quad (q, r_n) = \mathrm{DivRem}(r', r)$.
27: $\quad\quad$ Set $r' = r$, $r = r_n$, $c_n = c' - qc$, $c = c_n$, $c' = c$, $l = -l$.
28: $\quad t_3 = u_1 r$.
29: $\quad M_1 = (t_3 + ct_2)/u_2$, $M_2 = (r(v_2 + t_1) + w_1 c)/u_2$. (exact divisions)
30: $\quad u = l(rM_1 - cM_2)$.
31: $\quad z = (t_3 + c'u)/c$. (exact division)
32: $\quad v = V^- - [(t_1 - z + V^-) \pmod{u}]$.
33: $\quad u = \mathrm{monic}(u)$.
34: $\quad w = (f - v(v + h))/u$. (exact division)
35: $\quad$ **if** $\deg(z) < g + 1$ **then**
36: $\quad\quad n = n + \deg(u_2) - \deg(r') + g + 1 - \deg(u)$.
37: $\quad$ **else**
38: $\quad\quad n = n + \deg(u_2) + \deg(r)$.
39: **return** Balanced Adjust$([u, v, w, n], f, h, V^-)$.

Note that it is possible to identify some non-generic cases that require a down adjustment directly from the input divisors. One could then consider using this information to change the basis to positive reduced at the beginning of the

---

Algorithm 5. Balanced NUDUPL

Input: $[u_1, v_1, w_1, n_1]$, f, h, V^-.
Output: $[u, v, w, n]$ with $[u, v, w, n] = 2[u_1, v_1, w_1, n_1]$.

1: $t_1 = v_1 + h$, $t_2 = t_1 + v_1$.
2: Compute $(S, a_1, b_1) = \text{XGCD}(u_1, t_2)$.
3: $K = b_1 w_1$.
4: **if** $S \neq 1$ **then**
5:    $u_1 = u_1/S$.   (exact division)
6:    $w_1 = w_1 S$.
7: $K = K \pmod{u_1}$.
8: $D = 2 \deg(u_1)$.
9: $n = 2n_1 + \deg(S) - \lceil g/2 \rceil$.
10: **if** $D \leq g$ and $((n \geq 0$ and $n \leq g - D)$ or $\deg(w_1) - \deg(u_1) > g))$ **then**
11:    $T = u_1 K$, $u = u_1^2$, $v = v_1 + T$.
12:    $w = (w_1 - K(t_2 + T))/u_1$.   (exact division)
13:    **if** $\deg(v) \geq \deg(u)$ **then**
14:       $(q, r) = \text{DivRem}(V^- - v, u)$.
15:       $tv = V^- - r$, $w = w - q(v + h + tv)$, $v = tv$.
16: **else**
17:    Set $r = K$, $r' = u_1$, $c' = 0$, $c = -1$, $l = -1$.
18:    **while** $\deg(r) \geq (g + 1)/2$ **do**
19:       $(q, r_n) = \text{DivRem}(r', r)$.
20:       Set $r' = r$, $r = r_n$, $c_n = c' - qc$, $c = c_n$, $c' = c$, $l = -l$.
21:    $M_2 = (rt_2 + w_1 c)/u_1$.   (exact division)
22:    $u = l(r^2 - cM_2)$.
23:    $z = (u_1 r + c'u)/c$.   (exact division)
24:    $v = V^- - [(V^- - z + t_1) \pmod{u}]$.
25:    $u = \text{monic}(u)$.
26:    $w = (f - v(v + h))/u$.   (exact division)
27:    **if** $\deg(z) < g + 1$ **then**
28:       $n = n + \deg(u_1) - \deg(r') + g + 1 - \deg(u)$.
29:    **else**
30:       $n = n + \deg(u_1) + \deg(r)$.
31: **return** Balanced Adjust$([u, v, w, n], f, h, V^-)$.

---

algorithm, so that the adjustment saved via NUCOMP's simple continued fraction steps is a down adjustment. However, applying the change of basis requires roughly the same amount of computation as one adjustment step in the right direction relative to the basis. We found that in practice any savings obtained were negligible, as adjustments in the wrong direction rarely occur, so we chose not to include this functionality in our algorithm.

4.2 Adapting NUCOMP to the Balanced Setting

Most of the logic for updating the balancing coefficient n is the same as in Cantor's algorithm as presented above (Algorithm 1). The main difference is that NUCOMP does not require reduction steps, as the output is already reduced due to the simple continued fraction reduction of coefficients in lines 24–26 of Balanced NUCOMP. However, it is necessary to determine how these NUCOMP reduction steps affect the resulting balancing coefficient n.

The computation of the simple continued fraction expansion in NUCOMP implicitly keeps track of a principal divisor D_δ, such that for input divisors D_1, D_2 and the reduced output divisor D_3, $D_1 + D_2 = D_3 + D_\delta$, and knowledge of D_δ gives us the information needed to update n. Some of this is described in the version of NUCOMP from [8], but this version does not account for special cases of the last reduction step (where the leading coefficient of input v is the same as the leading coefficient of V^+ or V^-) nor the use of negative reduced basis. In our analysis we account for both, aligning with the special cases from the reduction portion of Balanced Addition (Algorithm 1) and from Balanced Adjust (Algorithm 2).

The last continued fraction step may either be a normal reduction step, a special reduction step or an adjustment step. Special reductions steps can be viewed as reductions that encounter cancellation with either ∞^+ or ∞^-. The cancellation effectively mimics a composition with ∞^+ or ∞^-, thus requiring the same accounting of the balancing coefficient n as an adjustment step. If the last step is an adjustment step, Balanced NUCOMP attempts a reduction, but a reduced basis effectively already applies composition at infinity, so the attempted reduction completes the adjustment. In both cases, the choice of either positive or negative reduced basis solely dictates the direction of the adjustment. We refer to the last simple continued fraction step as *special* if either an adjustment or special reduction step is computed; otherwise we refer to it as *normal*.

There are four possible cases for the computation of n dictated by the choice of positive or negative reduced basis and either normal or special last steps. Note that we do not include cases that arise with positive reduced basis in Algorithm 4, due to our choice of working exclusively with negative reduced basis, but we do describe the computation of n for this case below, too, as we implement and compare both versions in the next Section.

First we describe how to test for special last steps, then how the n value is computed depending on the type of basis used and whether the last step is special or normal. The last step is special exactly when $\deg(z) < g + 1$ as in line 34 of the Balanced NUCOMP algorithm, where z is given in line 30. To see this, we first recall that, as described in [8], each continued fraction step of NUCOMP corresponds to a divisor equivalent to the sum of the input divisors D_1 and D_2. Let $[u', v']$ denote the Mumford representation of the divisor corresponding to the second-last continued fraction step. The last continued fraction step is a special step whenever $\deg(v') = g + 1$ and the leading coefficient of v' is the same as that of V^- (or V^+ if positive reduced basis is being using), because in that case cancellations in the leading coefficients of $V^- - v'$ (or $V^+ - v'$) in the computation of v cause the degree of u to be less than g, implying that the last step is special.

Comparing the computation of v in line 31 of balanced NUCOMP with the computation of v in the reduction step of Balanced Addition (Algorithm 1), and also in any case of Balanced Adjust (Algorithm 2), we see that $v' = v_1 - z$. Thus, the conditions for the last continued fraction step being special are satisfied when $\deg(z) < g + 1$, because this implies that the degree and leading coefficients of v' and v_1 are the same. Note that $\deg(v_1) = g + 1$ and the leading coefficient of v_1 is the same as that of V^+ (or V^-) because the input divisor $[u_1, v_1]$ is given in negative (or positive) reduced basis. As stated earlier, the simple continued fraction steps of Algorithm 4 (lines 24–26) can only incorporate up to one adjustment step in addition to all the required reduction steps. Thus, a final call to Algorithm 2 is required in order to ensure that the output divisor is both reduced and balanced.

4.3   Eliminating an Adjustment for Some Non-generic Cases

The non-balanced version of NUCOMP presented at the beginning of this section (Algorithm 3, lines 14–19) makes use of the observation that if $D = \deg(u_1/S) + \deg(u_2/S) \leq g$, then completing the composition using Cantor's algorithm will produce a divisor that is reduced without having to do any subsequent reduction steps. In the balanced setting, the corresponding balancing coefficient is $n = n_1 + n_2 + \deg(S) - \lceil g/2 \rceil$. If this divisor is not balanced, i.e. $n < 0$ or $n > g - D$, then one may apply NUCOMP's simple continued fraction-based reduction in order to compute one adjustment step, saving one of the more expensive standard adjustment steps. However, this is only beneficial if $\deg(w_1) - \deg(w_2) \leq g$, because otherwise the resulting output divisor will not be reduced due to the fact that $\deg(u)$ depends on the degree of $w_1 S/(u_2/S) = w_1/u_2$. Thus, we only finish the composition with Cantor's algorithm (lines 16–21 of Algorithm 4) if the resulting divisor is reduced and balanced, or if it is reduced and not balanced but performing a NUCOMP reduction step would result in an non-reduced divisor.

5   Empirical Results

In this section we provide empirical data to illustrate the relative performance of the composition algorithms presented above over both ramified curves and split model curves using positive and negative reduced basis representations. We implemented all the algorithms for addition and doubling in Magma as a proof of concept[2]. Therefore, the absolute timings are not of great importance. The reader should rather focus on the relative cost between the various algorithms and models. See https://github.com/salindne/divisorArithmetic/tree/master/generic for raw data, auxiliary graphs and Magma scripts of implementations used in this section.

---

[2] The experiments were performed on a workstation with an Intel Xeon 7550 processor that has 64 cores, each of which is 64-bit and runs at 2.00 GHz.

As a preliminary benchmark, we compared addition using the versions of Cantor's algorithm described earlier and our version of NUCOMP over ramified and split model curves by computing a Fibonacci-like sequence of divisors using $D_{i+1} = D_i + D_{i-1}$, starting from two random divisors. We collected timings for all genus ranging from 2 to 50 and prime fields of sizes 2, 4, 8, 16, 32, 64, 128, 256, 512 and 1024 bits. All timings were run over random hyperelliptic curves with $h = 0$, using implementations of our algorithms that were specialized to exclude any computations with h.

We also performed similar experiments for our doubling algorithms (Cantor's algorithm and NUDUPL, which is NUCOMP specialized to doubling a divisor) over ramified and split model curves, by computing series of thousands of additions of a divisor class with itself. The data for doubling is omitted below, as the relative performance between the various algorithms considered was the same as for addition.

For ramified model curves, the Cantor-based algorithms we used are the same as Algorithm 1 but with the steps dealing with the balancing coefficient n removed and with divisors normalized via $v \bmod u$ as opposed to a reduced basis. We used Algorithm 3 for NUCOMP. For split model curves, the positive reduced basis algorithms are based on Algorithms 1, 2 and 4, but with divisors normalized via $V^+ - [(V^+ - v) \pmod{u}]$ as opposed to a negative reduced basis. We also include timings using Magma's built-in arithmetic for ramified and split model curves.

Apart from the absolute timings, we observed that the relative performances of the various algorithms do not depend on the field size. In the next figures, we illustrate our comparison for 32-bit fields only, as these results are also representative of the other field sizes. From these plots, we can draw the following conclusions:

- For split model curves, as illustrated in the first graph, negative reduced and positive reduced basis perform about the same for even genus. As expected, negative reduced basis is slightly better for odd genus due to the fact that generic cases require no adjustments steps in negative reduced basis as opposed to one adjustment step in positive reduced basis.
- The second graph shows that, for split model curves, our implementation of balanced NUCOMP rapidly becomes faster than Cantor as g grows. It also shows that, when using balanced NUCOMP, the difference between the best algorithms for split and ramified model curves is negligible for all genus. Furthermore, all of our implementations are considerably faster than Magma's built in arithmetic as the genus grows. The graph does not include timings for Magma for $g > 32$ so that the comparisons between the other algorithms are easier to see. We note that our best split model algorithm is about five times faster at genus 50.

– Not surprisingly, as shown in the last graph, for small genus ($g < 5$), Cantor's algorithms are slightly faster than the NUCOMP algorithms. Magma's built-in arithmetic is also faster for $g < 7$. We suspect that this is due to Magma's implementation having access to faster internal primitives, while our implementation has to use the generic polynomial ring setting. Even so, our implementation of NUCOMP is the fastest option for $g \geq 7$.

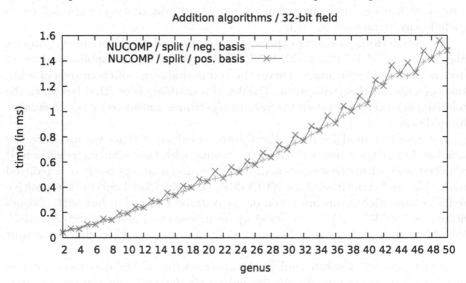

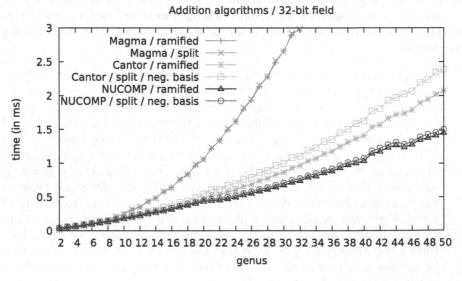

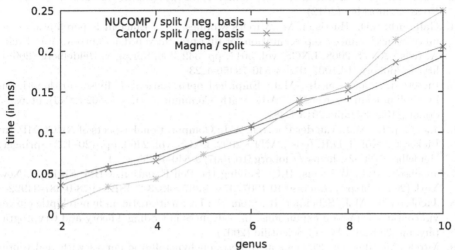

6   Conclusions and Future Work

Our results indicate that Balanced NUCOMP provides an improvement for computing balanced divisor class arithmetic in split model hyperelliptic curves with a cross-over as low as genus 5. As expected, our choice of normalizing v in negative reduced basis and therefore incorporating up-adjustments into NUCOMP performs equally well when compared to positive reduced over even genus, and slightly better over odd. Furthermore, our algorithm performs almost as well and sometimes better than ramified curve NUCOMP and closes the performance gap between ramified model and split model divisor arithmetic.

Integrating our algorithms directly into Magma's built in arithmetic might reduce the relative performance, either lowering or elimination any cross over points between our algorithms and Magma's arithmetic. It would be of interest to adapt NUCOMP for divisor arithmetic over non-hyperelliptic $C_{a,b}$ curves as this setting also plays a role in computational number theoretic applications [13]. Adapting NUCOMP for addition in the divisor class group of superelliptic curves based on [12] may also yield favourable results. It would also be interesting to see if explicit formulas for divisor class group arithmetic based on Balanced NUCOMP applied to arithmetic in split model hyperelliptic curves of low genus, can improve on current best [2,13].

References

1. Cantor, D.: Computing in the Jacobian of a hyperelliptic curve. Math. Comput. **48**(177), 95–101 (1987)
2. Erickson, S., Jacobson Jr., M.J., Stein, A.: Explicit formulas for real hyperelliptic curves of genus 2 in affine representation. Adv. Math. Commun. **5**(4), 623–666 (2011)

3. Galbraith, S.D.: Mathematics of Public Key Cryptography. Cambridge University Press, Cambridge (2012)
4. Galbraith, S.D., Harrison, M., Mireles Morales, D.J.: Efficient hyperelliptic arithmetic using balanced representation for divisors. In: van der Poorten, A.J., Stein, A. (eds.) ANTS 2008. LNCS, vol. 5011, pp. 342–356. Springer, Heidelberg (2008). https://doi.org/10.1007/978-3-540-79456-1_23
5. Imbert, L., Jacobson Jr., M.J.: Empirical optimization of divisor arithmetic on hyperelliptic curves over $\mathbb{F}_{2^m}$. Adv. Math. Commun. **7**(4), 485–502 (2013). https://doi.org/10.3934/amc.2013.7.485
6. Jacobson Jr., M.J., van der Poorten, A.J.: Computational aspects of NUCOMP. In: Fieker, C., Kohel, D.R. (eds.) ANTS 2002. LNCS, vol. 2369, pp. 120–133. Springer, Heidelberg (2002). https://doi.org/10.1007/3-540-45455-1_10
7. Jacobson, M.J., Williams, H.C.: Solving the Pell Equation. CBM. Springer, New York (2009). https://doi.org/10.1007/978-0-387-84923-2. ISBN: 978-0-387-84922-5
8. Jacobson Jr., M.J., Scheidler, R., Stein, A.: Fast arithmetic on hyperelliptic curves via continued fraction expansions. In: Advances in Coding Theory and Cryptography, pp. 200–243. World Scientific (2007)
9. Mireles-Morales, D.: Efficient arithmetic on hyperelliptic curves with real representation. Ph.D. thesis, University of London (2009)
10. Mullen, G., Panario, D.: Handbook of Finite Fields. Chapman and Hall/CRC, New York (2013)
11. Shanks, D.: On Gauss and composition I, II. In: Mollin, R. (ed.) Proceedings of NATO ASI on Number Theory and Applications, pp. 163–179. Kluwer Academic Press (1989)
12. Shashka, T., Kopeliovich, Y.: The addition on Jacobian varieties from a geometric viewpoint. arXiv e-prints arXiv:1907.11070v2, October 2019
13. Sutherland, A.: Sato-Tate distributions. arXiv e-prints arXiv:1604.01256, April 2016
14. Sutherland, A.: Fast Jacobian arithmetic for hyperelliptic curves of genus 3. The Open Book Series **2**(1), 425–442 (2018)

Contact Linearizability of Scalar Ordinary Differential Equations of Arbitrary Order

Yang Liu, Dmitry Lyakhov$^{(\boxtimes)}$, and Dominik L. Michels

Visual Computing Center, King Abdullah University of Science and Technology,
Al-Khawarizmi Bldg 1, Thuwal 23955-6900, Kingdom of Saudi Arabia
{yang.liu.4,dmitry.lyakhov,dominik.michels}@kaust.edu.sa

Abstract. We consider the problem of the exact linearization of scalar nonlinear ordinary differential equations by contact transformations. This contribution is extending the previous work by Lyakhov, Gerdt, and Michels addressing linearizability by means of point transformations. We have restricted ourselves to quasi-linear equations solved for the highest derivative with a rational dependence on the occurring variables. As in the case of point transformations, our algorithm is based on simple operations on Lie algebras such as computing the derived algebra and the dimension of the symmetry algebra. The linearization test is an efficient algorithmic procedure while finding the linearization transformation requires the computation of at least one solution of the corresponding system of the Bluman-Kumei equation.

Keywords: Contact symmetry · Differential Thomas decomposition ·
Exact linearization · Nonlinear ordinary differential equations ·
Symbolic computation

1 Introduction

Symmetry analysis as a systematic method was discovered by Sophus Lie more than 150 years ago and then rediscovered by Ovsyannikov and his colleagues in the 20 century. Sophus Lie himself considered groups of point and contact transformations to integrate systems of partial differential equations. His key idea was to obtain first infinitesimal generators of one-parameter symmetry subgroups and then to construct the full symmetry group. The study of symmetries of differential equations allows one to gain insights into the structure of the problem they describe. Existence of symmetry group allows to decrease the order of differential equation, reduce from partial to ordinary differential equations, construct particular exact solutions or sometimes even general solutions.

In contrast to Lie, recently Lyakhov, Gerdt, and Michels discovered [7,8] that such kind of properties like exact linearizations could be detected completely algorithmically without solving the determining system. It relies strongly on differential algebra and symbolic manipulations with differential equations. Their work in this field was inspired by Ibragimov and Meleshko [2,18]. We want to

© Springer Nature Switzerland AG 2020
F. Boulier et al. (Eds.): CASC 2020, LNCS 12291, pp. 421–430, 2020.
https://doi.org/10.1007/978-3-030-60026-6_24

exclude obtaining of explicit expressions (as they are really large and not really meaningful) and instead of it obtain an algorithm to test the exact linearization property.

This paper is organized as follows. In Sect. 2, we briefly describe the mathematical objects we deal with and the former result on linearization by point transformation [7]. In Sect. 3, we introduce contact symmetry and prove the main theorem of our paper. The implementation of algorithms and its application is illustrated in Sect. 4 by several examples. Finally, we provide a conclusion in Sect. 5.

2   Point Symmetry

We consider an arbitrary order ordinary differential equation (ODE) of the form

$$y^{(n)} = f(x, y, y', \dots, y^{(n-1)}), \quad y^{(k)} := \frac{d^k y}{dx^k} \tag{1}$$

with a rational right-hand side which is solved with respect to the highest derivative.

If an ODE of the form (1) admits transformation into a linear nth order homogeneous equation

$$u^{(n)}(t) + \sum_{k=0}^{n-1} a_k(t)\, u^{(k)}(t) = 0, \quad u^{(k)} := \frac{d^k u}{dt^k}$$

by means of functions[1]

$$u = \phi(x, y), \quad t = \psi(x, y), \tag{2}$$

then we say that (1) admits exact linearization or is linearizable by point transformation.

The invertibility of (2) is provided by the local differential condition

$$J := \phi_x \psi_y - \phi_y \psi_x \neq 0.$$

Our way to check the linearizability of Eq. (1) is based on Lie's approach [4]. We study the symmetry properties of (1) under the *infinitesimal* transformation

$$\tilde{x} = x + \varepsilon\, \xi(x, y) + \mathcal{O}(\varepsilon^2), \quad \tilde{y} = y + \varepsilon\, \eta(x, y) + \mathcal{O}(\varepsilon^2). \tag{3}$$

The *invariance condition* for (1) under the transformation (3) is given by the equality

$$\mathcal{X}(y^{(n)} - f(x, y, \dots, y^{(n-1)}))\big|_{y^{(n)} = f(x, y, \dots, y^{(n-1)})} = 0,$$

---

[1] Please note, that we assumed analytical homeomorphisms.

where the *symmetry operator* reads

$$\mathcal{X} := \xi \partial_x + \sum_{k=0}^{n} \eta^{(k)} \partial_{y^{(k)}}, \quad \eta^{(k)} := D_x \eta^{(k-1)} - y^{(k)} D_x \xi,$$

$\eta^{(0)} := \eta$, and $D_x := \partial_x + \sum_{k \geq 0} y^{(k+1)} \partial_{y^{(k)}}$ is the total derivative operator with respect to x.

This set of symmetry operators forms a basis of the *Lie symmetry algebra*

$$[\mathcal{X}_i, \mathcal{X}_j] = \sum_{k=1}^{m} C_{i,j}^k \mathcal{X}_k, \quad 1 \leq i < j \leq m. \tag{4}$$

Let L denote the Lie symmetry algebra and $m = \dim(L)$. An important role for the analysis plays the *derived algebra $L' \subset L$* which by definition is a subalgebra that consists of all commutators of pairs of elements in L.

Lie showed ([5], Ch. 12, p. 298, "Satz" 3) that the Lie point symmetry algebra of an n-order ODE has a dimension m satisfying

$$n = 1, \ m = \infty; \quad n = 2, \ m \leq 8; \quad n \geq 3, \ m \leq n + 4.$$

Interrelations between n and m ensure the linearizability of the differential equation (1) by point transformation. Here we present the two theorems that describe such interrelations and form the basis of our exact linearization test.

Theorem 1. ([9], Thm. 1) *A necessary and sufficient condition for the linearization of (1) with $n \geq 3$ via a point transformation is the existence of an n-dimensional abelian subalgebra of (4).*

The proof is based on the following lemma which is important for further discussions of contact symmetries.

Lemma 1. *Let us suppose three linear independent operators $X_i = f_i(x, y) \frac{\partial}{\partial x} + g_i(x, y) \frac{\partial}{\partial y}, i = 1, 2, 3$ commuting each other. Then, there exists an appropriate point transformation which maps X_i onto $\bar{X}_i = \bar{f}_i(t) \frac{\partial}{\partial u}$.*

Proof. By rectification the theorem for the non-singular point, we can also find a point transformation to map one operator (e.g. X_1) to shift $\frac{\partial}{\partial u}$. Then,

$$\bar{X}_i = \bar{g}_i(t) \frac{\partial}{\partial t} + \bar{f}_i(t) \frac{\partial}{\partial u}, i = 2, 3.$$

Since $[X_2, X_3] = 0$, direct calculations show that

$$\bar{g}_2(t) \bar{f}_3'(t) - \bar{g}_3(t) \bar{f}_2'(t) = 0, \quad \bar{g}_2(t) \bar{g}_3'(t) - \bar{g}_3(t) \bar{g}_2'(t) = 0.$$

One of two possibilities may hold: either $f_2' g_3' - f_3' g_2' = 0$ or both $g_2 = g_3 = 0$. The first case is not possible without $g_2 = g_3 = 0$. Otherwise it contradicts the linear independence of the operators.

Corollary. Lemma 1 could be easily generalized to an arbitrary number of operators more than 3.

The main result for point symmetry is based on the following theorem, which forms a basis for the algebraic test linearization.

Theorem 2. ([7]) *Eq. (1) with $n \geq 2$ is linearizable by a point transformation if and only if one of the following conditions is fulfilled:*

1. $n = 2$, $m = 8$;
2. $n \geq 3$, $m = n + 4$;
3. $n \geq 3$, $m \in \{n+1, n+2\}$ and the derived algebra of (4) is abelian of dimension n.

This theorem shows that the verification of linearizability requires only checking of dimensions and also finding the derived algebra, which is simple from a computational point of view and abstract theory of finite-dimensional Lie algebra.

3   Contact Symmetry

The most general smooth invertible transformation of variables for an ODE is a contact transformation. It is a local diffeomorphism of the jet bundle $J^1\pi$ into itself defined in standard coordinates by the formulas

$$X = X(x, y, p), \ Y = Y(x, y, p), \ P = Y_p/X_p.$$

Here, we use the standard notation $y' = p$ and $Y' = P$. Also, as a contact transformation, $X_p(Y_x + pY_y) = Y_p(X_x + pX_y)$ is required. One should be aware that the third formula is only valid for nontrivial contact transformations (i.e. $X_p \neq 0$). We use $P = (Y_x + pY_y)/(X_x + pX_y)$ instead for point transformations. The notion of contact transformation was introduced in Lie's doctoral dissertation first. One-parameter group of a contact symmetry is a flow of contact transformation[2]

$$\bar{x} = \bar{x}(x, y, p, a), \ \bar{y} = \bar{y}(x, y, p, a), \ \bar{p} = \bar{p}(x, y, p, a).$$

It defines in a similar way as for point transformation the infinitesimal operator

$$\mathcal{X} := \xi(x, y, p)\,\partial_x + \eta(x, y, p)\,\partial_y + \eta^{[1]}(x, y, p)\partial_p,$$

which is an appropriate derivation of the one-parameter group at $a = 0$. But contact transformation comes with additional constraints on the components of the generator $\mathcal{X}$:

$$\eta_p - p\xi_p = 0, \ \eta^{[1]} = \eta_x + p(\eta_y - \xi_x) - p^2\eta_y.$$

---

[2] It defines an identity transformation if $a = 0$.

Definition 1. *A Lie algebra of contact vector fields is reducible if there exists a local contact transformation around a non-singular point which maps these vector fields onto the first prolongations of point vector fields. Otherwise, it is irreducible.*

The beautiful property of contact symmetries is that except the three specific Lie algebras on plane, all other ones are reducible. Moreover, the following theorem clarifies it.

Theorem 3. ([13], page 134; [17]) *Finite-dimensional irreducible Lie algebras of contact transformations in the complex plane (x, y), where x and y are in general complex numbers, belong to one of the following three classes modulo local contact transformations: L_6, L_7, and L_{10}, which dimensions are 6, 7 and 10.*

The direct computation of the derived algebra shows that

$$[L_6, L_6] = L_6, \ [L_7, L_7] = L_6, \ [L_{10}, L_{10}] = L_{10}.$$

This leads to an interesting observation. Any abelian contact Lie algebra possesses the zero derived algebra by definition, thus it is reducible. Then, transforming to basis when it is merely a prolongation of the point Lie algebra, it is possible to apply Lemma 1 and the corollary, which immediately leads to the following theorem.

Theorem 4. *A necessary and sufficient condition for the linearizability of (1) with $n \geq 3$ via a contact transformation is the existence of an n-dimensional abelian subalgebra in the contact symmetry algebra.*

Proof. By reducibility, this subalgebra can be taken as a point Lie algebra. In the light of Lemma 1, its n generators under new variables (t, u) imply that the n-dimensional symmetry group acts on some solution $u_0(t)$ by rule

$$u(t) = u_0 + \sum_{i=1}^{n} C_i f_i(t),$$

where C_i are group parameters. Without loss of generality, we can assume that $u_0 = 0$, otherwise we apply one more transformation of variables by the rule $U = u - u_0, T = t$. Every solution $u(t)$ of an nth order scalar ODE is defined completely by its initial conditions $u(t_0), u'(t_0), \ldots, u^{(n-1)}(t_0)$ at some point t_0. Varying C_i, it is possible to get any set of initial conditions from the list, because it has a non-zero Wronskian determinant. Finally, if $v(t)$ and $w(t)$ are solutions, then $u(t) = v(t) + w(t)$ is also a solution. This concludes the proof.

Important results for contact symmetries of linear ODE were obtained by Svirshchevskii and Yumaguzhin described by the two following theorems.

Theorem 5. ([19]) *A linear ODE of kth order with $k \geq 4$ does not possess nontrivial (non-point) contact symmetries.*

Theorem 6. ([20]) *The dimension of the contact symmetry algebra of any third order linear ODE is equal to one of the following numbers: 4, 5, and 10. Moreover,*

1. *any third order linear ODE with a 10-dimensional contact symmetry algebra is equivalent to the trivial equation $y''' = 0$,*
2. *any third order linear ODE with 5-dimensional contact symmetry algebra is equivalent to one of the equations of the form $y''' = Ky' + y, K = $ const.*

Theorem 6, together with Theorem 2 and Theorem 5, shows that except for $y''' = 0$, all other linear cases do not posses nontrivial contact symmetries. Thus, the dimensions are

1. $n = 3$, $m = 10$,
2. $n \geq 4$, $m = n + 4$,
3. $n \geq 3$, $m \in \{n + 1, n + 2\}$.

The first two items characterize the case of a maximal symmetry dimension. A remarkable point is that it implies linearizability like it was shown by Lie.

Theorem 7. ([4,5]) *Let Eq. (1) be an nth order scalar ODE.*

1. *If $n = 3$, then Eq. (1) admits at most a ten-dimensional symmetry group of contact transformations. Moreover, the symmetry group is ten-dimensional if and only if Eq. (1) is equivalent (up to a local contact transformation) to $u^{(3)}(t) = 0$.*
2. *If $n \geq 4$, then Eq. (1) admits at most an $(n + 4)$-parameter symmetry group of contact transformations. In addition, the symmetry group is $(n+4)$-dimensional if and only if Eq. (1) is equivalent (up to a local contact transformation) to $u^{(n)}(t) = 0$.*

According to Theorem 5 and Theorem 6, a linear equation which is not trivializable should correspond to the third case (i.e., $n \geq 3$ and $m \in \{n+1, n+2\}$). Thus, its derived algebra is abelian and has the dimension n. Vice versa, following Theorem 4, this is also a sufficient condition.

Theorem 8. *Equation (1) with $n \geq 3$ is linearizable by a contact transformation if and only if one of the following conditions is fulfilled:*

1. *$n = 3$, $m = 10$ or $n \geq 4$, $m = n + 4$ (maximal dimension),*
2. *$n \geq 3$, $m = n + 1$ or $n + 2$ and the derived algebra of contact symmetry is abelian of dimension n.*

4   Algorithm and Examples

The main result of this paper is Theorem 8, which serves as the foundation for the algebraic test for exact linearizability by contact transformations. Once a system of determining equations for symmetry generators is given, we can complete them to involution [15,16], and then by computing the differential Hilbert polynomial find the dimension of the symmetry algebra [3]. In this regard, we prefer to use the differential Thomas decomposition [1] which already showed its convenience and superiority for such kind of tasks. There are a lot of packages for finding particular solutions of determining systems based on some heuristics. It is of great relevance in geometry and physics. Unfortunately, there is no algorithm to solve completely any determining system of symmetries for scalar ODEs, because the existence of this algorithm would immediately imply the ability to solve any linear ODE.

A beautiful property of the finite-dimensional Lie symmetry algebra is that the structure constants could be found exactly without any heuristics. We follow here an approach proposed by Reid [14]. Any N linear independent solutions of the determining system span an N-dimensional Lie symmetry algebra. They could be expressed via a power series solution for the determining system in involution. Substitution of these expressions into (4) leads to an infinite system of linear equations for a finite number of structure constants $C_{i,j}^k$. This system is always equivalent to some truncated version, which leads to an efficient procedure for obtaining structure constants given by Algorithm 1[3].

Algorithm 1. *Contact Linearization Test*

Input: q, a nonlinear differential equation of form (1).
Output: True, if q is linearizable, and False, otherwise.

1: $n :=$ ***DifferentialOrder*** (q);
2: $DS :=$ ***DeterminingSystem*** (q);
3: $IDS :=$ ***InvolutiveDeterminingSystem*** (DS);
4: $m := \dim($***LieSymmetryAlgebra*** $(IDS))$;
5: **if** $(n = 3 \wedge m = 10) \vee (n > 3 \wedge m = n + 4)$ **then**
6:      **return** True;
7: **else if** $n \geq 3 \wedge (m = n + 1 \vee m = n + 2)$ **then**
8:      $SC :=$ ***StructureConstants*** (IDS);
9:      $DA :=$ ***DerivedAlgebra*** (SC);
10:     **if** DA is abelian and $\dim(DA) = n$ **then**
11:         **return** True;
12:     **end if**
13: **end if**
14: **return** False;

We illustrate our theory by presenting the following two examples.

Example 1. [18] We start with a classical example. Equation

$$y''' = \frac{3y''^2}{2y'}$$

---

[3] A modern package for calculations of determining systems is discussed in the literature [6].

describes the family of hyperbolas. As it was shown by Lie, it could be transformed into the simplest equation $y''' = 0$ using a Legendre transformation. Computation of the Hilbert dimension polynomial for the determining system for contact symmetries shows that the dimension is 10, which by Theorem 8 immediately implies trivialization. Let us compute the same for the point transformation. The dimension of symmetry group then is 6, which corresponds to the case in which the linearization is not possible. Thus, it is essentially a contact transformation.

Example 2. [18] Let us consider

$$- 16y'^2 y'' y^{(4)} + 48y'^2 y'''^2 + y' y''^5 x - 48y' y''^2 y''' - y''^5 y + 12y''^4 = 0. \qquad (5)$$

This example also passes our linearization test with dimension $m = 6$. It requires also the computation of the derived algebra which is 4-dimensional and abelian.

In order to recover the linearizing mapping, we will use an analog of the method described in [10–12]. We will briefly discuss it here. Our analysis is based on the Bluman-Kumei equations:

$$\xi(x,y,p)\frac{\partial X}{\partial x} + \eta(x,y,p)\frac{\partial X}{\partial y} + \eta^{[1]}(x,y,p)\frac{\partial X}{\partial p} = \bar{\xi}(x,y,p),$$

$$\xi(x,y,p)\frac{\partial Y}{\partial x} + \eta(x,y,p)\frac{\partial Y}{\partial y} + \eta^{[1]}(x,y,p)\frac{\partial Y}{\partial p} = \bar{\eta}(x,y,p),$$

$$\xi(x,y,p)\frac{\partial P}{\partial x} + \eta(x,y,p)\frac{\partial P}{\partial y} + \eta^{[1]}(x,y,p)\frac{\partial P}{\partial p} = \bar{\eta}^{[1]}(x,y,p),$$

where $\bar{\xi}(x,y,p), \bar{\eta}(x,y,p), \bar{\eta}^{[1]}(x,y,p)$ are generators mapped by contact transformation (X, Y, P) being expressed in old coordinates. So, in Example 1, we

1. solve the system of contact symmetry of $y^{(3)} = 0$ (i.e. $\bar{\xi}, \bar{\eta}, \bar{\eta}^{[1]}$ are expressed in polynomials of X, Y, P);
2. solve ξ, η from Bluman-Kumei equations (as a linear system);
3. substitute the solution from the previous step into the system of ξ, η (i.e., the determining system of contact symmetry algebra of $y''' = \frac{3y''^2}{2y'}$).

Completion to involution forms giant nonlinear determining system[4]. It has one particular solution $X = \sqrt{p}, Y = xp - y$ which coincides with [18].

Example 2 corresponds to the constant coefficient case, and requires different consideration. Since every linear constant coefficient ODE admits only trivial contact symmetries (point symmetries), by Lemma 1, all elements in derived algebra (DA) have the form $f(X)\frac{\partial}{\partial Y} + f'(X)\frac{\partial}{\partial P}$ (being expressed in new coordinates), thus we

1. set $\bar{\xi} = 0$, $\frac{\bar{\eta}_x}{X_x} = \frac{\bar{\eta}_y}{X_y} = \frac{\bar{\eta}_p}{X_p}$;
2. reduce the equations in step 1 with the system of DA;

---

[4] We will not write it down for brevity.

3. vanish all the coefficients of parametric derivatives in step 2.

Completion to involution gives the system of differential equations and in equations

$$\{X_x = 0, X_y = 0, Y_{xx} = 0, Y_{xy} = 0, Y_x + pY_y = 0\}, \{X_p \neq 0, Y_p \neq 0, Y_x \neq 0\}$$

which forms basis of linearizing mappings of (5).

5   Conclusion

We constructed a new algebraic linearization test for scalar ordinary differential equations by contact transformation. It indicates that this approach could be applied to large classes of differential equations including systems of ordinary and partial differential equations. Of course, in the case of partial differential equations the main problem is the infinite-dimensionality of their symmetry algebras. Thus, devising a general scheme for the detection of infinite-dimensional abelian symmetry subalgebras is still a challenge.

Acknowledgments. This work has been funded by the King Abdullah University of Science and Technology (KAUST baseline funding). The authors are grateful to Peter Olver for helpful discussions and to the anonymous reviewers for comments that led to improvement of the paper.

References

1. Bächler, T., Gerdt, V., Lange-Hegermann, M., Robertz, D.: Algorithmic Thomas decomposition of algebraic and differential systems. J. Symb. Comput. **47**(10), 1233–1266 (2012)
2. Ibragimov, N., Sergey Meleshko, S.: Linearization of third-order ordinary differential equations by point and contact transformations. J. Math. Anal. Appl. **308**, 266–289 (2005)
3. Lange-Hegermann, M.: The differential counting polynomial. Found. Comput. Math. **18**, 291–308 (2018)
4. Lie, S.: Klassifikation und Integration von gewöhnlichen Differentialgleichungen zwischen x, y, die eine Gruppe von Transformationen gestatten. III. Archiv for Matematik og Naturvidenskab, vol. 8, 4, pp. 371–458 (1883). Reprinted in Lie's Gesammelte Abhandlungen, paper XIY, vol. 5, pp. 362–427 (1924)
5. Lie, S.: Vorlesungen über kontinuierliche Gruppen mit geometrischen und anderen Anwendungen. Bearbeitet und herausgegeben von Dr. G. Schefferes, Teubner, Leipzig (1883)
6. Lisle, I., Tracy Huang, T.: Algorithmic calculus for Lie determining systems. J. Symb. Comput. **79**, 482–498 (2017)
7. Lyakhov, D.A., Gerdt, V.P., Michels, D.L.: Algorithmic verification of linearizability for ordinary Ddifferential equations. In: Burr, M. (ed.) Proceedings 42nd International Symposium on Symbolic and Algebraic Computation. ISSAC'2017, New York, pp. 285–292. ACM (2017)

8. Lyakhov, D.A., Gerdt, V.P., Michels, D.L.: On the algorithmic linearizability of nonlinear ordinary differential equations. J. Symb. Comput. **98**, 3–22 (2019)
9. Mahomed, F., Peter Leach, P.: Symmetry Lie algebra of nth order ordinary differential equations. J. Math. Anal. Appl. **151**(1), 80–107 (1990)
10. Mohammadi, Z., Reid, G., Tracy Huang, T.: Introduction of the MapDE algorithm for determination of mappings relating differential equations. In: Proceedings 2019 International Symposium on Symbolic and Algebraic Computation, New York, pp. 331–338. ACM (2019)
11. Mohammadi, Z., Reid, G., Huang, T.: Extensions of the MapDE algorithm for mappings relating differential equations (2019). arXiv preprint arXiv:1903.03727
12. Mohammadi, Z., Reid, G.J., Huang, S.-L.T.: The Lie algebra of vector fields package with applications to mappings of differential equations. In: Gerhard, J., Kotsireas, I. (eds.) MC 2019. CCIS, vol. 1125, pp. 337–340. Springer, Cham (2020). https://doi.org/10.1007/978-3-030-41258-6_27
13. Olver, P.: Equivalence, Invariance and Symmetry. Cambridge University Press, Cambridge (1995)
14. Reid, G.: Finding abstract Lie symmetry algebras of differential equations without integrating determining equations. Eur. J. Appl. Math. **2**(04), 319–340 (1991)
15. Robertz, D.: Formal Algorithmic Elimination for PDEs. LNM, vol. 2121. Springer, Cham (2014). https://doi.org/10.1007/978-3-319-11445-3
16. Seiler, W.: Involution: The Formal Theory of Differential Equations and its Applications in Computer Algebra. In: Algorithms and Computation in Mathematics, vol. 24. Springer, Heidelberg (2010)
17. Soh, C., Mahomed, F.: Contact symmetry algebras of scalar ordinary differential equations. Nonlinear Dyn. **28**, 213–230 (2002)
18. Suksern, S., Ibragimov, N., Meleshko, S.: Criteria for fourth-order ordinary differential equations to be linearizable by contact transformations. Commun. Nonlinear Sci. Numer. Simul. **14**, 266–289 (2009)
19. Svirshchevskii, S.: Lie-Bäcklund symmetries of linear ODEs and generalized separation of variables in nonlinear equations. Phys. Lett. A **199**, 344–348 (1995)
20. Yumaguzhin, V.: Classification of 3rd order linear ODE up to equivalence. Differ. Geom. Appl. **6**(4), 343–350 (1996)

Faster Numerical Univariate Polynomial Root-Finding by Means of Subdivision Iterations

Qi Luan[1], Victor Y. Pan[1,2(✉)], Wongeun Kim[3], and Vitaly Zaderman[1]

[1] Ph.D. Programs in Mathematics and Computer Science, The Graduate Center
of the City University of New York (CUNY), New York, NY 10036, USA
qi_luan@yahoo.com, vza52@aol.com
[2] Department of Computer Science, Lehman College of CUNY, Bronx,
NY 10468, USA
victor.pan@lehman.cuny.edu
[3] Department of Mathematics, The Lander College for Men, Touro College,
Kew Garden, NY 11367, USA
won-geun.kim2@touro.edu
http://comet.lehman.cuny.edu/vpan/

Abstract. Root-finding for a univariate polynomial is four millennia old and still highly important for Computer Algebra and various other fields. Subdivision root-finders for a complex univariate polynomial are known to be highly efficient and practically promising. The recent one by Becker et al. [2] competes for user's choice and is nearly optimal for dense polynomials represented in monomial basis, but [18] proposes and analyzes further significant acceleration, which becomes dramatic for polynomials admitting their fast evaluation (e.g., sparse ones). Here and in the companion paper [19], we present some of these results and algorithms.

Keywords: Polynomial roots · Subdivision · Sparse polynomials · Real polynomial root-finding

1 Introduction

1.1 State of the Art

Root-finding for univariate polynomials has been the central subject of Mathematics and Computational Mathematics for four millennia since Sumerian times and until the middle of 19th century A.D and began its new life with the advent of computers. Presently this subject is highly important for Computer Algebra and many other computational areas. Since 2000, the root-finder of user's choice has been the package MPSolve (Multiprecision Polynomial Solver) [3,5], which implements Ehrlich's, aka Aberth's iterations, but recent progress in subdivision iterations has made them potentially competitive. Due to [23], advanced in [2,8,14,21], and known in Computational Geometry as Quad-tree Construction,

© Springer Nature Switzerland AG 2020
F. Boulier et al. (Eds.): CASC 2020, LNCS 12291, pp. 431–446, 2020.
https://doi.org/10.1007/978-3-030-60026-6_25

they extend bisection of a line segment to root-finding in the complex plane. Their advanced version of 2016–2018 by Becker et al. [2] is the second known nearly optimal root-finder[1] for a polynomial represented in monomial basis – by its coefficients:

$$p = p(x) = \sum_{i=0}^{d} p_i x^i = p_d \prod_{j=1}^{d} (x - x_j), \ p_d \neq 0. \tag{1}$$

The implementation in [11] has slightly outperformed MPSolve for root-finding in a region containing a small number of roots,[2] while the implementation in [12] is user's current choice for the highly important task of real polynomial root-finding, where the input and output are real.

The main and bottleneck block of the known subdivision iterations, including those of [2,14,21], is the application of an *exclusion test*, which either certifies or does not certify that a fixed disc on the complex plane contains no roots of an input polynomial p. This test is a special case of *root-counting* in the disc, which is another basic block of subdivision algorithms. According to [2] its main algorithmic novelty versus its predecessors is its root-counting by means of pairwise comparison of the absolute values of the coefficients of $p(x)$ and invoking Pellet's classical theorem.

1.2   Our Progress

A new significant acceleration in [18] relies on another novel approach to root-counting and exclusion tests. Unlike the case of [2] these blocks and the whole root-finder work for a *black box input polynomial* – defined by a black box for its evaluation at a given point. This class includes polynomials represented in Bernstein and Chebyshev bases – admitting numerically stable evaluation, as well as sparse, Mandelbrot's, and various other polynomials, which admit dramatically faster evaluation. [2] takes no advantage of this huge benefit, but [18] fully exploits it and thus dramatically accelerates [2] for the latter input class. In particular, Taylor's shift of the variable and Dandelin–Lobachevsky–Gräffe's recursive root-squaring [9], being two well-known drawbacks of the subdivision root-finder [2], are avoided in [18].

Work [18] also accelerates polynomial root-finding based on Ehrlich's, Newton's, and other functional iterations, as well as numerical multipoint polynomial evaluation, extensively involved in polynomial root-finding but also having independent importance. Because of the size limitation, however, we skip these subjects, omit many details, and leave to [18] formal support for our algorithms,

---

[1] The first such root-finder, of [16], is nearly optimal also for the task of numerical factorization of a polynomial into the product of its linear factors, having independent importance.

[2] Throughout the paper we count m times a root of multiplicity m and handle it as a cluster of m roots whose diameter is smaller than the tolerance to the output approximation errors.

including correctness proof and Boolean cost estimates. We occasionally estimate arithmetic complexity where we can control the precision of computing.
Already an initial implementation of our algorithms in [10] demonstrates 3-fold
acceleration of the previous best implementation of subdivision root-finding,
even though we have not yet incorporated many promising directions for further
progress specified in [18]. Our present paper and its companion [19] together
cover only a fraction of the results of [18], focusing on new exclusion test and
root-counting in subdivision iterations and extension to real root-finding.

1.3   Power Sums and Cauchy Sums

We adopt rather than counter the subdivision iterations of [2,14,21] but enrich
them with performing their main two blocks by means of the approximation of
the power sums of the roots of $p(x)$ that lie in a fixed disc on the complex plain:
(i) the sum of their 0th powers is precisely the number of the roots in the disc,
and (ii) such a disc contains no roots if and only if all the power sums vanish.
We only approximate integers (0 or the number of roots), perform computations
with a low precision, and use just order of $\log(d)$ arithmetic operations in an
exclusion test and root-counting for a degree d input polynomial.

A technical point of our departure was the study of the power sums in the
extensive advanced work on the Boolean complexity of polynomial root-finding
by Schönhage in [22, Sects. 12 and 13]. He has approximated the power sums s_h
of the roots lying in the unit disc $D(0,1) = \{x : |x| \leq 1\}$ by means of *Cauchy
sums* s_h^*, being discretizations of Cauchy's contour integral:[3]

$$s_h := \sum_{x_j \in D(c,\rho)} x_j^h = \int_{C(c,\rho)} \frac{p'(x)}{p(x)} x^h \, dx, \text{ for } h = 0, 1, \ldots, \tag{2}$$

$$s_h^* := \frac{1}{q} \sum_{g=0}^{q-1} \zeta^{(h+1)g} \frac{p'(c+\rho\zeta^g)}{p(c+\rho\zeta^g)} \text{ for } h = 0, 1, \ldots, q-1, \ \zeta := \exp\left(\frac{2\pi i}{q}\right), \tag{3}$$

where ζ denotes a primitive qth root of unity, for a fixed $q > 1$.

1.4   Real Root-Finding

Real root-finding is highly important because in many applications, e.g., to geometric and algebraic-geometric optimization, only real roots of a polynomial are
of interest and because they are typically much less numerous than all d complex
roots. In particular, under a random coefficient model, a polynomial of degree d
is expected to have $O(\log(d))$ real roots (cf. [6]).

Real roots of a polynomial defined numerically, with rounding errors, turn
into nearly real roots, whose approximation has rarely if at all been addressed
properly in the known algorithms, while we handle this issue by following [18].

---

[3] Schönhage was seeking a factor of $p(x)$ with root set made up of the roots of $p(x)$
lying in that disc; he only approximated the power sums s_h for positive h.

Namely [18] proposes, elaborates upon and analyzes efficient root-counting and deflation techniques for the roots of a polynomial lying on and near a circle on the complex plane, and in Sect. 4, we extend these techniques to the roots lying on and near a fixed segment of the real axis by applying Zhukovsky's function and its inverse. Given a black box polynomial $p(x)$, we deflate its factor f whose root set is precisely the root set of $p(x)$ lying on or near a fixed segment of the real axis. Since $\deg(f)$ tends to be much smaller than d, deflation of the factor f and its subsequent root-finding are performed at a low Boolean cost [18].

A preliminary version of this algorithm appeared in [17, Section 7], but presently we simplify it substantially.[4] We perform its stage 1 by means of evaluation and interpolation without involving more advanced algorithm of [4]. At its stage 3, we use more efficient [18, Algorithm 46] instead of [18, Algorithm 45], and we simplify root-finding stage 4 by first applying our new Algorithm 4, which decreases by twice the degree of the factor f of p and still keeps in its root set the images of all real roots of p.

Moreover we propose an alternative extension of our study of root-counting from the complex plain to real interval. This achieves less than deflation but at a much lower cost, and is still a major stage of real and nearly real root-finding. Our non-costly root-counter in and near a line segment provides more information than the customary ones – based on the Descartes rule of signs or Budan–Fourier theorem, involves no costly computation used in Sturm sequences and, unlike Budan–Fourier theorem, can be applied to black box polynomials. And as we said already, unlike the known real root-counters we output the overall number of roots lying in and near a fixed segment of the real axis.

[18, Section 6] and the paper [20] approximate pairwise well-isolated real roots fast by narrowing the range for their search.

1.5 Organization of the Paper

We recall some background material in the next section, cover Cauchy sum computation, root-counting and exclusion tests in Sect. 3, and devote Sect. 4 to real polynomial root-finding. The Boolean complexity of the new algorithms is estimated in [18] in some detail; we do not include this study because of size limitation.

2 Background

2.1 Definitions and Auxiliary Results

– $S(c,\rho)$, $D(c,\rho)$, $C(c,\rho)$, and $A(c,\rho_1,\rho_2)$ denote square, disc, circle (circumference), and annulus on the complex plain, respectively:

$$S(c,\rho) := \{x : |\Re(c-x)| \le \rho,\ |\Im(c-x)| \le \rho\},$$
$$D(c,\rho) := \{x : |x-c| \le \rho\}, \tag{4}$$

---

[4] Otherwise [17] focuses on deflation, and only half-page [17, Section 6.3] overlaps with us.

$$C(c, \rho) := \{x : |x - c| = \rho\}, \ A(c, \rho_1, \rho_2) := \{x : \rho_1 \leq |x - c| \leq \rho_2|\}. \quad (5)$$

- An annulus $A(c, \rho_1, \rho_2)$ has *relative width* $\frac{\rho_2}{\rho_1}$.
- We freely denote polynomials $p(x)$, $t(x) = \sum_i t_i x^i$, $u(x) = \sum_i u_i x^i$ etc. by p, t, u, etc. unless this can cause confusion.
- $|u| = \sum_{i=0}^d |u_i|$ denotes the norm of a polynomial $u(x) = \sum_{i=0}^d u_i x^i$.
- IND($\mathcal{R}$), the index of a region $\mathcal{R}$ of the complex plain (e.g., a square, a disc, an annulus, or a circle), is the number of roots of p contained in it.
- A disc $D(c, \rho)$ and circle $C(c, \rho)$ have an *isolation ratio* θ or equivalently are *θ-isolated* for a polynomial p, real $\theta \geq 1$, and complex c if no roots of p lie in the open annulus $A(c, \rho/\theta, \rho\theta)$, of relative width θ^2, or equivalently if IND($D(c, \rho/\theta)$) =IND($D(c, \rho\theta)$). (See Fig. 1.) A disc and a circle are *well-isolated* if they are θ-isolated for $\theta - 1$ exceeding a positive constant.
- Define the reverse polynomial of $p(x)$:

$$p_{\mathrm{rev}}(x) := x^d p\left(\frac{1}{x}\right) = \sum_{i=0}^d p_i x^{d-i}, \ p_{\mathrm{rev}}(x) = p_0 \prod_{j=1}^d \left(x - \frac{1}{x_j}\right) \text{ if } p_0 \neq 0. \quad (6)$$

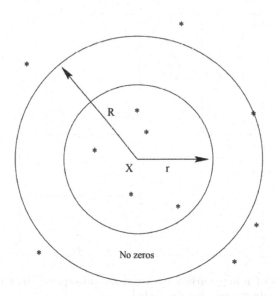

Fig. 1. The internal disc $D(X, r)$ (cf. (4)) is R/r-isolated

Equation (6) implies that the roots of p_{rev} are the reciprocals of the roots of p, which leads to the following results:

$$r_j(0, p) r_{d+1-j}(0, p_{\mathrm{rev}}) = 1 \text{ for } j = 1, \ldots, d. \quad (7)$$

Proposition 1. *The unit disc $D(0,1)$ is θ-isolated for p if and only if it is θ-isolated for p_{rev}.*

The proof of the following theorem of [13] and [1, Theorem 2] is constructive.

Theorem 1. *An algorithm that evaluates at x_0 a black box polynomial $p(x)$ over a field K of constants by using A additions/subtractions, S multiplications by elements from the field K, and M other multiplications/divisions can be extended to evaluate both $p(x_0)$ and $p'(x_0)$ at the cost $2A + M$, $2S$, and $3M$.*

2.2   Subdivision Iterations

Suppose that we seek all roots of p in a fixed square on the complex plane well-isolated from the external roots of p; call this square *suspect*. One can readily compute such a square centered at the origin and containing all roots of p (cf. [18, Section 6.2]). A subdivision iteration divides every suspect square into four congruent sub-squares and to each of them applies *an exclusion test:* a sub-square is discarded if the test proves that it contains no roots of p; otherwise the sub-square is called suspect and is processed in the next iteration (see Fig. 2).

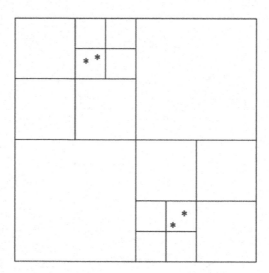

Fig. 2. Four roots of p are marked by asterisks; sub-squares that contain them are suspect; the other sub-squares are discarded

There are at most kd suspect squares at every iteration for a constant k. A root of p can make at most four squares suspect if an exclusion test enabled us to discard every square that contains no roots of p, and then we would have $k \leq 4$. Realistically the subdivision processes have been made less expensive overall by means of incorporation of *soft exclusion tests*, which keep a tested square $S(c, \rho)$

suspect if a disc $D(c, u\rho)$ contains a root of p for some u exceeding $\sqrt{2}$. Then the constant k grows above 4, but the cost of performing exclusion test and the overall cost of subdivision root-finding decrease.

A subdivision iteration begins with approximation of every root of p by the center of some suspect square with an error of at most one half of the diameter of the square and ends with decreasing this bound by twice; the papers [2, 14, 21] accelerate such a linear convergence to a root to quadratic based on Newton's or QIR iterations applied where one knows: (i) a component formed by suspect squares containing a root of a polynomial p and covered with a well-isolated disc and (ii) a number of the roots of p in that component.

3 Cauchy Root-Counting and Soft Exclusion Test

Algorithm 1. Cauchy sum computation.
Input: An integer $q > 1$, a disc $D(c, \rho)$, and a black box polynomial p of degree d satisfying the following inequalities (cf. Remark 1):

$$p(c + \rho\zeta^g) \neq 0 \text{ for } g = 0, 1, \ldots, q - 1. \tag{8}$$

Output:[5] The vector $\mathbf{s}_* = (s_{q-1}^*, s_0^*, \ldots, s_{q-2}^*)^T$ for $s_0^*, \ldots, s_{q-1}^*$ of (3).
Computations: Successively compute the values

1. $p(c + \rho\zeta^g)$ and $p'(c + \rho\zeta^g)$ for $g = 0, 1, \ldots, q - 1$.
2. $r_g := \frac{p'(c+\rho\zeta^g)}{p(c+\rho\zeta^g)}$ for $g = 0, 1, \ldots, q - 1$,
3. $\tilde{s}_h := \sum_{g=0}^{q-1} \zeta^{(h+1)g} r_g$ for $h = q - 1, 0, 1, \ldots, q - 2$, and
4. $s_h^* = \tilde{s}_h/q$ for $h = q - 1, 0, 1, \ldots, q - 2$.

We evaluate the polynomials $p(x)$ and $p'(x)$ at the q points $c + \rho\zeta^g$ for $g = 0, 1, \ldots, q - 1$, perform q divisions at each of stages 2 and 4, and perform DFT on q points at stage 3.

Remark 1. We can ensure (8) with probability 1 if we randomly rotate an input disc $D(c, \rho)$: $p(x) \leftarrow t(x)$ for $t(x - c) = p(\alpha \cdot (x - c))$ and a random α sampled under the uniform probability distribution on $C(0, 1)$. At stage 1, we can detect if $p(c + \rho\zeta^g) \approx 0$ and then recursively reapply random rotation. The rotation can be generalized to other maps [18, Remark 8].

Hereafter **Algorithm 1a** denotes Algorithm 1 restricted to the computation of just the first Cauchy sum s_0^* and its closest integer $\bar{s}_0$ provided that it breaks ties by assigning $\bar{s}_0 = \lceil s_0^* \rceil$. In transition to Algorithm 1a both stages 1 (dominant) and 2 of Algorithm 1 stay unchanged, but stages 3 and 4 are simplified and involve just $2q$ arithmetic operations.

---

[5] With this order of its components the vector $\mathbf{s}_*$ turns into the vector of discrete Fourier transform (*DFT*) at q points (upon a reviewer request we recall its celebrated fast solution FFT in the Appendix). Here and hereafter we assume that $\mathbf{v}$ denotes a column vector, while $\mathbf{v}^T$ denotes its transpose.

In the special case where $D(c, \rho) = D(0, 1)$, expressions (3) are simplified and if a polynomial p is represented in the monomial basis, then the computation at the bottleneck stage 1 of Algorithm 1a can be reduced to performing DFT at q points twice [18, Sub-algorithm 7.1].

Cauchy sum s_h^* is a weighted power sum s_h, with the weights $\frac{1}{1-x_j^q}$ (see Theorem 2); as a simple Corollary 1, we obtain the bounds of [22] on $|s_h^* - s_h|$.

Theorem 2 *[18]. For the roots x_j of $p(x)$ and all h, the Cauchy sums s_h^* (3) satisfy $s_h^* = \sum_{j=1}^{d} \frac{x_j^h}{1-x_j^q}$ unless $x_j^q = 1$ for some j.*

Corollary 1 *[22].[6] Let the disc $D(0, 1)$ be θ-isolated and let d_{in} and $d_{\text{out}} = d - d_{\text{in}}$ denote the numbers of the roots of p lying in and outside that disc, respectively. Write $\eta := 1/\theta$. Then*

$$|s_h^* - s_h| \leq \frac{d_{\text{in}}\eta^{q+h} + d_{\text{out}}\eta^{q-h}}{1 - \eta^q} \text{ for } h = 0, 1, \ldots, q-1. \tag{9}$$

In particular[7]

$$s_h = 0 \text{ and } |s_h^*| \leq \frac{d\eta^{q+h}}{1 - \eta^q} \text{ for } h = 0, 1, \ldots, q-1 \text{ if } d_{\text{in}} = 0. \tag{10}$$

$$\mu := |s_0^* - s_0| \leq \frac{d}{\theta^q - 1}, \text{ and so } \mu < 1/2 \text{ if } q > \frac{\log(2d+1)}{\log(\theta)}, \tag{11}$$

$$\theta \leq \left(\frac{\mu + d}{\mu}\right)^{1/q}, \text{ and so } \theta \leq (d+1)^{1/q} \text{ if } \mu = |s_0^* - s_0| \geq 1. \tag{12}$$

Corollary 2. *Suppose that Algorithm 1a, applied to the unit disc $D(0, 1)$ for $q \geq b\log_2(2d + 1)$ and $b > 0$, outputs $s_0^* > 1/2$ and consequently outputs a positive integer $\bar{s}_0$. Then the disc $D(0, \theta)$ contains a root of p for $\theta = 2^{1/b}$.*

Proof. Suppose that the disc $D(0, \theta)$ contains no roots of p. Then the unit disc $D(0, 1)$ contains no roots of p as well and is θ-isolated. Apply bound (11) for $\theta = 2^{1/b}$ and obtain $|s_0^* - s_0| \leq \frac{d}{2^{q/b}-1}$ and $q \geq b\log_2(2d + 1)$. Conclude that $2^{q/b} \geq 2d + 1$ and hence $|s_0^* - s_0| \leq 1/2$, while we assumed that $s_0^* > 1/2$.

Remark 2. (i) By applying equations (3) to the reverse polynomial $p_{\text{rev}}(x) = x^d p(\frac{1}{x})$ rather than $p(x)$ extend Corollaries 1 and 2 to the approximation of the power sums of the roots of $p(x)$ lying outside the unit disc $D(0, 1)$, whose isolation

---

[6] Unlike paper [22], this result is deduced in [18] from Theorem 2, which is also the basis for probabilistic support of correctness of Cauchy root-counter in [18].

[7] Clearly, we can only improve our approximation of the integer s_0 by the Cauchy sum s_0^* if we drop its imaginary part $\Im(s_0^*)$. The power sum s_0 of the roots in a well-isolated disc is only slightly closer to $\Re(s_0^*)$ than to s_0^* but can be dramatically closer when some or all roots lie on the boundary circle of an input disc (see [18, Section 3.7]).

ratio is invariant in the transition from p to p_{rev}, by virtue of Proposition 1.
(ii) Extend Corollaries 1 and 2 and part (i) of this remark to the case of any
disc $D(c, \rho)$ by means of shifting and scaling the variable $y \leftarrow \frac{x-c}{\rho}$ and observing
that this does not change the isolation ratio of the disc (see the definition of the
isolation ratio and Proposition 1).

We obtain a root-counter s_0 in a disc by means of rounding the 0th Cauchy
sum s_0^* if $\tau = |s_0 - s_0^*| < 0.5$, e.g., if $q > \log_\theta(2d + 1)$ for $\theta > 1$ by virtue of (11),
and if $\theta = 2$, then we can choose any $q \geq 21$ for $d \leq 1,000,000$.

Seeking correct output of a Cauchy root-counter or exclusion test without
unnecessary increase of the parameter q, one can first apply Algorithm 1a for a
small integer q and then recursively double it, reusing the results of the previous
computations, until the computed values of the Cauchy sum s_0^* stabilize near an
integer or just until they approximate an integer closely enough. [18, Section 5]
proves that such an integer is s_0 with a high probability (hereafter *whp*) under
random root models.

This result supports root-finding computations in [18, Section 6.4], but not
in the subdivision processes of [2, 14, 21], where root-counting is applied only
where an input disc is well-isolated, and then Algorithm 1a yields non-costly
solution s_0 by virtue of Corollary 1. For correctness of our exclusion test, we
seek stronger support because in the subdivision iterations of [2, 14, 21], such a
test is applied to the discs for whose isolation ratios no estimates are known.
By virtue of Corollary 2 Algorithm 1a applied to such a disc certifies that its
controlled dilation contains a root of p unless the algorithm outputs $\bar{s}_0 = 0$. The
following algorithm completes an exclusion test in the latter case.

Algorithm 2. Completion of a Cauchy soft exclusion test.
Input: A black box polynomial $p(x)$ of degree d such that Algorithm 1a, applied
to the or equivalently to the disc $D(0, 2)$ and[8] the polynomial $p(x)$, has output $\bar{s}_0 = 0$.
Output: Certification that (i) the disc $D(0, 2)$ contains a root of p definitely if
$q > d$ or whp otherwise or (ii) the unit disc $D(0, 1)$ definitely contains no roots
of p, where cases (i) and (ii) are compatible.
Initialization: Choose an integer q such that

$$q_0 < q \leq 2q_0 \text{ for } q_0 \geq \max\{1, \ \log_2(\frac{d}{q_0 \alpha_d \sqrt{3}})\} \text{ and } \alpha_d = \sqrt{d + \sqrt{d}}. \tag{13}$$

Computations: Apply Algorithm 1 to the unit disc $D(0, 1)$ for the selected q.
Let

$$\mathbf{s}_* := (s_{q-1}^*, s_0^*, s_1^*, \ldots, s_{q-2}^*)^T \tag{14}$$

denote the vector of the values s_h^* of the Cauchy sums output by the algorithm
and let $|\mathbf{s}_*|$ denote the Euclidean norm $(\sum_{h=0}^{q-1} |s_h^*|^2)^{1/2}$. If $|\mathbf{s}_*| \ q_0 \ \alpha_d \geq 1$, con-
clude that the disc $D(0, \theta)$ definitely contains a root of p. Otherwise conclude
that the disc $D(0, 1)$ contains no roots of p definitely if $q > d$ or whp otherwise.

---

[8] One can extend the algorithm by applying Algorithm 1a to a disc $D(0, \theta)$ for smaller
$\theta > 1$ and modifying bound (13) accordingly.

Correctness proof. If the disc $D(0,\theta)$ contains no roots of p, then the unit disc $D(0,1)$ is θ-isolated, and we can apply bound (10) to the Cauchy sums s_h^* output in the above application of Algorithm 1. This would imply that $|s_h^*| \leq \frac{d}{(\theta^q-1)\theta^h}$, and then we would deduce that $|s_h^*|^2 \leq \frac{d^2}{(\theta^q-1)^2\theta^{2h}}$, and so $|\mathbf{s}_*|^2 \leq \frac{d^2}{(\theta^q-1)^2(\theta^2-1)}$. Under the assumed choice of $\theta = 2$ it follows that

$$|\mathbf{s}_*| \leq \frac{d}{(2^q-1)\sqrt{3}} < \frac{d}{(2^{q_0}-1)\sqrt{3}} \quad \text{for } q_0 < q,$$

and then (13) would imply that $|\mathbf{s}_*|\, q_0\, \alpha_d \leq 1$ and hence $|\mathbf{s}_*|\, q\, \alpha_d < 1$. Therefore, the disc $D(0,\theta)$ contains a root of p unless the latter bound holds, as claimed. Correctness of the algorithm in the case where $|\mathbf{s}_*|\, q\, \alpha_d < 1$ follows from [18, Corollaries 4.2 and 4.3].

Remark 3. For the computation of Cauchy sums for q of order of d we should evaluate p and p' at order of d points; by applying our reduction of multi-point polynomial evaluation (MPE) to fast multipole method (FMM) (see [18, Appendix E]) we can do this by using order of $d\log^2(d)$ arithmetic operations, performed numerically with the precision of order $\log(d)$ bits. It outputs the vector of the first q Cauchy sums $s_0, \ldots, s_{q-1}$ within a relative error of order $\log(d)$. This should be sufficient in order to verify the bounds of Algorithm 2 [18, Theorem 19 and Corollaries 4.2 and 4.3], because FMM is celebrated for being very stable numerically, although further formal and experimental study is in order.

The algorithm runs faster as we decrease integer q, and already for q_0 of order of $\log(d)$ the disc $D(0,2)$ contains a root of p if $|\mathbf{s}_*|\, q_0\, \alpha_d \geq 1$, while the unit disc $D(0,1)$ contains a root of p with a probability that fast converges to 1 as the value $|\mathbf{s}_*|\, q_0\, \alpha_d$ decreases under a random coefficient model for the polynomial p, by virtue of [18, Corollary 4.3].

For $q \leq d$ we have only probabilistic support of correctness of Algorithm 2 in the case where $|\mathbf{s}_*|\, q_0\, \alpha_d < 1$, but we can try to strengthen reliability of our exclusion tests by verifying additional necessary conditions for correctness of our exclusion test and root-counting:

(a) the Cauchy sums s_h^* for $h = 0, 1, \ldots, q-1$ still nearly vanish for the polynomials $t(x)$ obtained from $p(x)$ by means of various mappings of the variable x that keep an input disc and the power sum s_0 invariant (cf. Remark 1);
(b) an exclusion test should succeed for any disc lying in the disc $D(c,\rho)$. In particular, if the disc covers a suspect square, then exclusion tests should succeed for the four discs that cover the four congruent sub-squares obtained from sub-dividing the input square;
(c) all suspect squares of a subdivision iteration together contain precisely d roots of p.

If these additional necessary conditions hold, it is still plausible that the disc $D(c,\rho)$ contains a root of p.[9] We can, however, detect whether we have lost

---

[9] A polynomial p has no roots in a closed disc $D(0,1)$ if and only if p_{rev} has precisely d roots in the open disc $D(0,1)$; similar property holds for Cauchy sums s_0^* (see [18]).

any roots at the end of the subdivision process, when $d - w$ roots are *tamed*, that is, closely approximated, and when w roots remain at large; we call the latter roots *wild*. If $0 < w \ll d$, then at a low cost we can deflate the *wild factor* of p, whose root set is made up of the w wild roots; then we can approximate the roots of this factor at a low cost (see [18, Section 7]).

It is natural to call a point c a tame root of p if $r_d(c, p) \leq$ TOL for a fixed tolerance TOL. The algorithm of [18, Section 6.2] closely approximates $r_d(c, p)$ at a relatively low cost, but it is even less expensive to verify whether $d \, |p'(c)/p(c)| \leq$ TOL and then to recall that $r_d(c, p) \leq d \, |p'(c)/p(c)|$ (see [8, Theorem 6.4g]), although his upper bound on $r_d(c, p)$ is extremely poor for a worst case input such as $p(x) = x^d - h^d$ for $h \neq 0$.

Empirical support from the initial implementation and testing of our algorithms in [10] has substantially superseded their formal support here and in [18]. In these tests, subdivision iterations with Cauchy exclusion tests by means of Algorithm 1a have consistently approximated the integer $s_0 = 0$ within $1/4$ for $q = \lceil \log(4d + 1)/\log(4\theta) \rceil$. For discs containing no roots and for $q = \lceil \log(4d + 1)/\log(4\theta) \rceil + 1$ Algorithm 1 has consistently approximated both s_0 and s_1 within $1/4$ (cf. [10, equation (22) in Corollary 12 for $e = 1/4$]).

4   Real Root-Finding

The algorithm of [12] specializes subdivision iterations of [2] to real univariate polynomial root-finding and is currently the user's choice algorithm, but we can readily accelerate it by narrowing the search for the real roots by means of incorporation of the techniques of [18, Section 6].

Furthermore, we can extend all our other accelerations of subdivision iterations from a disc to a line segment. E.g., our simple root-counting Algorithm 5 provides more information than the customary counting based on the Descartes rule of signs and on Budan–Fourier theorem, involves no costly computation of the Sturm sequences, and unlike Budan–Fourier theorem can be applied to black box polynomials. Our real root-counter amounts essentially to multipoint evaluation. Allowing also interpolation we deflate a factor of p whose root set is precisely the set of roots of p that lie in a fixed segment of real axis. Typically, the degree of the factor is dramatically smaller than d, even where the line segment contains all real roots of p (cf. [6]), and so root-finding on the segment is simplified accordingly.

Actually by saying "real roots" we mean both real and nearly real roots, which is appropriate where an input polynomial is studied numerically with rounding errors.

Next we extend our root-counters and deflation algorithms from the unit disc $D(0, 1)$ to the unit segment $S[-1, 1]$, but this actually covers the case of any disc and any segment: we can perform the relevant shift and scaling of the variable implicitly because we reduce our real root-counting essentially to multipoint evaluation and reduce real root-finder to multipoint evaluation and interpolation.

We first reduce root-counting and root-finding on the unit segment $S[-1, 1]$ to that on the unit circle $C(0, 1)$. We assume that the segment is reasonably well isolated from the external roots of p, and this implies that so is the unit circle (see Remark 4). Then Algorithm 5 for root-counting is readily reduced to our root-counting on the complex plane.

Root-counting and root-finding on an isolated circle are reduced to the same tasks for a pair of θ-isolated discs $D(0, \theta_-)$ and $D(0, \theta_+)$ for a constant $\theta > 1$; for root-counting it is sufficient to apply Algorithm 1a (see Algorithm 5).

By scaling the variable x we reduce the root-finding task to that in the well-isolated unit disc $D(0, 1)$, and then apply highly efficient deflation algorithms of [22, Section 13].

It remains to specify back and forth transition between the segment and the circle. We apply the two-to-one Zhukovsky's function $z = J(x)$ and its one-to-two inverse, for complex variables x and z. It maps the circle $C(0, 1)$ to the segment $S[-1, 1]$, and vice versa:

$$x = J(z) := \frac{1}{2}\left(z + \frac{1}{z}\right); \ z = J^{-1}(x) := x \pm \sqrt{x^2 - 1}. \tag{15}$$

Algorithm 3. Root-finding on a line segment.
Input: A polynomial $p = p(x)$ of (1).
Output: The number w of its roots on the unit segment $S[-1, 1]$ and approximations to all these roots.
Computations:

1. Compute the values $v_h = p(\Re(\zeta_{2d}^h))$ of the polynomial p at the Chebyshev points $\Re(\zeta_{2d}^h) = \cos(\frac{\pi h}{2d})$ for $h = 0, 1, \ldots, 2d - 1$ and ζ_{2d} of (3).
2. Interpolate to the polynomial $s(z)$ of degree $2d$ such that

$$s(z) = z^d p(x) \text{ for } x = \frac{1}{2}(z + z^{-1}) \tag{16}$$

   from its values

$$s(\zeta_{2d}^h) = (-1)^h v_h \text{ for } h = 0, 1, \ldots, 2d - 1$$

   by means of applying Inverse DFT. [Recall that $\zeta_{2d}^{dh} = (-1)^h$.]
3. Approximate a factor (e.g., the monic factor) $g = g(z)$ of p whose root set is made up of all roots of the polynomial $s(z)$ that lie on the unit circle $C(0, 1)$. [By virtue of (15) $\Re(s(z_j)) = 0$ if $\Re(s(z_j^{-1})) = 0$, and so these roots appear in complex conjugate pairs; if $p(x) = 0$ for $x = 1$ and/or $x = -1$, then 1 and/or -1 are also the roots of $g(z)$ with double multiplicity.] At this stage, first compute the power sums of the roots of $g(z)$ by applying Algorithm 1 to the discs $D(0, 1/\theta)$ and $D(0, \theta)$ provided that the circle $C(0, 1)$ is θ^2-isolated; then recover the coefficients of $g(z)$ from the power sums by applying [22, Algorithm 46]. Output $w = 0.5 \deg(g)$.
4. By applying MPSolve, subdivision iterations, or another root-finder approximate all $2w$ roots of $g(z)$. Let $z_1, \ldots, z_w$ denote the first w of them in the order of increasing their arguments from 0.

5. Compute and output the w roots of p lying in the segment $S[-1,1]$ and given by the values $x_j = \frac{1}{2}(z_j + \frac{1}{z_j})$ for $j = 1, \ldots, w$.

Correctness of this algorithm follows from (15) and (16).

Its Stage 2 is DFT. Its Stage 1 of the evaluation at Chebyshev's points can be performed by means of the algorithm of [7] or Discrete Cosine Transform, which is similar to DFT (cf. [15, Section 3.11 and the notes to it in Sect. 3.12]). In both cases, the computation is simplified if d is a power of 2; we ensure this by replacing p with $x^u p$ for the minimal non-negative integer u such that $d + u$ is a power of 2. Studying stage 3 observe that the polynomial $g(z)$ has the same roots $z_1, \ldots, z_{2w}$ on the circle $C(0,1)$ and in a concentric annulus defined as the difference of two θ-isolated discs $D(0,\theta)$ and $D(0,1/\theta)$ for some $\theta > 1$.

All power sums s_h of these roots are the differences of the power sums s_h of the roots in these two discs. At first closely approximate these pairs of power sums by applying Algorithm 1; their differences approximate the power sums of the roots of $g(z)$ on the circle $C(0,1)$. Then approximate $g(z)$ by applying the algorithms of [22, Section 13].

The converse transition from the unit circle $C(0,1)$ to the unit segment $S[-1,1]$ enables us to simplify stages 4 and 5 of Algorithm 3 by moving from $g(z)$ to a polynomial of degree w whose all roots lie in the segment $S[-1,1]$. Then again we achieve this by means of evaluation and interpolation.

Algorithm 4. Transition to unit segment.

1. Compute the values $u_h = g(\zeta_{2K}^h)$ of the polynomial $g(z)$ at the $2K$-th roots of unity for $h = 0, 1, \ldots, K - 1$ and $K > w$.
2. Interpolate to the polynomial $f(x)$ of degree at most w from its values $f(\Re(\zeta_{2K}^h)) = (-1)^h u_h$ at the Chebyshev points $\Re(\zeta_{2K}^h) = \cos(\frac{\pi h}{2K})$, for $h = 0, 1, \ldots, K - 1$. [Recall that $\zeta_{2K}^{Kh} = (-1)^h$.]
3. Approximate the w roots of the polynomial $f = f(x)$ by applying to it MPSolve, subdivision iterations, or another real polynomial root-finder, e.g., that of [12].

We propose to perform steps 1 and 2 above by means of forward DFT and inverse Cosine Transforms, applying them to the polynomial $z^v g(z)$, replacing $g(x)$, for the minimal non-negative integer v such that $w + v$ is a power 2.

Remark 4. Represent complex numbers as $z := u + v\mathbf{i}$. Then Zhukovsky's map transforms a circle $C(0,\rho)$ for $\rho \neq 1$ into the ellipse $E(0,\rho)$ whose points (u,v) satisfy the following equation,

$$\frac{u^2}{s^2} + \frac{v^2}{t^2} = 1 \text{ for } s = \frac{1}{2}\left(\rho + \frac{1}{\rho}\right), \quad t = \frac{1}{2}\left(\rho - \frac{1}{\rho}\right).$$

Consequently it transforms the annulus $A(0, 1/\theta, \theta)$ into the domain bounded by the ellipses $E(0, 1/\theta)$ and $E(0, \theta)$, and so the circle $C(0,1)$ is θ-isolated if and only if no roots of p lie in the latter domain.

We can simplify Algorithm 3 and use only evaluations if we restrict our task to counting the roots that lie on or near a line segment. Here is a high level description of this algorithm where we do not specify the parameters θ and q and assume that the input includes the polynomial $s(z)$.

Algorithm 5. Root-counting on a line segment.
Input: A real $\theta > 1$ and the polynomial $s(z)$ of Eq. (16) such that the unit circle $C(0,1)$ is θ^2-isolated, that is, the annulus $A(0, 1/\theta^2, \theta^2)$ contains no roots of $s(z)$ except possibly some roots on the unit circle $C(0,1)$.
Output: The number w of the roots of p in the segment $S[-1, 1]$ or *FAILURE*.
Computations:

1. Compute the polynomial $s_-(z) = s(z/\sqrt{\theta})$.
2. Choose a sufficiently large q and compute Cauchy's sums $\tilde{s}_{0,-}^*$ of the roots of $s_-(z)$ in the unit disc $D(0, 1)$.
3. If the value $s_{0,-}^*$ is sufficiently close to an integer $\tilde{s}_{0,-}$, then output $w := d - \tilde{s}_{0,-}$ and stop. Otherwise output *FAILURE*.

By assumption, the circle $C(0, 1/\theta)$ is $(\theta - \epsilon)$-isolated for $s(z)$ and any positive ϵ, and so $\tilde{s}_{0,-}$ is the number of the roots of $s(z)$ in the disc $D(0, 1/\theta)$ by virtue of bound (11). Clearly the same number $\tilde{s}_{0,-}$ of the roots of the polynomial $s(z)$ of degree $2d$ lie inside and outside the unit disc, and so $2d - 2\tilde{s}_{0,-}$ its roots lie on the unit circle $C(0, 1)$. Divide this bound by 2 and obtain the number of the roots of $p(x)$ on the unit segment $[-1, 1]$.

Acknowledgements. This research has been supported by NSF Grants CCF–1563942 and CCF–1733834 and PSC CUNY Award 69813 00 48. We also thank the reviewers for thoughtful comments.

Appendix A. Discrete Fourier transform (DFT)

DFT($\mathbf{p}$) outputs the vector of the values $p(\zeta^j) = \sum_{i=0}^{d-1} p_i \zeta^{ij}$ of a polynomial $p(x) = \sum_{i=0}^{d-1} p_i x^i$ on the set $\{1, \zeta, \ldots, \zeta^{d-1}\}$. The *fast Fourier transform (FFT) algorithm*, for $d = 2^h$ recursively splits $p(x)$:

$$p(x) = p_0(y) + x p_1(y), \text{ where } y = x^2,$$

$$p_0(y) = p_0 + p_2 x^2 + \cdots + p_{d-2} x^{d-2}, \ p_1(y) = x(p_1 + p_3 x^2 + \cdots + p_{d-1} x^{d-2}).$$

This reduces DFT_d for $p(x)$ to two $\text{DFT}_{d/2}$ (for $p_0(y)$ and $p_1(y)$) at a cost of d multiplications of $p_1(y)$ by x, for $x = \zeta^i$, $i = 0, 1, \ldots, d - 1$, and of the pairwise addition of the d output values to $p_0(\zeta^{2i})$. Since $\zeta^{i+d/2} = -\zeta^i$ for even d, we perform multiplication only $d/2$ times, that is, $f(d) \le 2f(d/2) + 1.5d$ if $f(k)$ ops are sufficient for DFT_k. Recursively we obtain the following estimate.

Theorem 3. *For $d = 2^h$ and a positive integer h, the DFT_d only involves $f(d) \le 1.5dh = 1.5d \log_2 d$ arithmetic operations.*

Inverse DFT is the converse problem of interpolation to a polynomial $p(x)$ from its values at the dth roots of unity. At the cost of performing d divisions, this task can be reduced to DFT (see, e.g., [15, Theorem 2.2.2]).

References

1. Baur, W., Strassen, V.: On the complexity of partial derivatives. Theoret. Comput. Sci. **22**, 317–330 (1983)
2. Becker, R., Sagraloff, M., Sharma, V., Yap, C.: A near-optimal subdivision algorithm for complex root isolation based on the Pellet test and Newton iteration. J. Symb. Comput. 86, 51–96 (2018), Proceedings version. In: ACM ISSAC, pp. 71–78 (2016). https://doi.org/10.1016/j.jsc.2017.03.009
3. Bini, D.A., Fiorentino, G.: Design, analysis, and implementation of a multiprecision polynomial rootfinder. Numer. Algorithms **23**, 127–173 (2000). https://doi.org/10.1023/A:1019199917103
4. Bini, D., Pan, V.Y.: Graeffe's, Chebyshev-like, and Cardinal's processes for splitting a polynomial into factors. J. Complex. **12**, 492–511 (1996)
5. Bini, D.A., Robol, L.: Solving secular and polynomial equations: a multiprecision algorithm. J. Comput. Appl. Math. **272**, 276–292 (2014). https://doi.org/10.1016/j.cam.2013.04.037
6. Erdős, P., Turán, P.: On the distribution of roots of polynomials. Ann. Math **2**(51), 105–119 (1950)
7. Gerasoulis, A.: A fast algorithm for the multiplication of generalized Hilbert matrices with vectors. Math. Comput. **50**(181), 179–188 (1988)
8. Henrici, P.: Applied and Computational Complex Analysis, Vol. 1: Power Series, Integration, Conformal Mapping, Location of Zeros. Wiley, New York (1974)
9. Householder, A.S.: Dandelin, Lobachevskii, or Graeffe? Amer. Math. Mon. **66**, 464–466 (1959). https://doi.org/10.2307/2310626
10. Imbach, R., Pan, V.Y.: New progress in univariate polynomial root-finding. In: Proceedings of ACM-SIGSAM ISSAC 2020, pp. 249–256, July 20–23, 2020, Kalamata, Greece, ACM Press, New York (2020). ACM ISBN 978-1-4503-7100-1/20/07. https://doi.org/10.1145/3373207.3403979
11. Imbach, R., Pan, V.Y., Yap, C.: Implementation of a near-optimal complex root clustering algorithm. In: Davenport, J.H., Kauers, M., Labahn, G., Urban, J. (eds.) ICMS 2018. LNCS, vol. 10931, pp. 235–244. Springer, Cham (2018). https://doi.org/10.1007/978-3-319-96418-8_28
12. Kobel, A., Rouillier, F., Sagralo, M.: Computing real roots of real polynomials... and now for real! In: Proceedings of the ACM on International Symposium on Symbolic and Algebraic Computation (ISSAC 2016), pp. 301–310. ACM Press, New York (2016) https://doi.org/10.1145/2930889.2930937
13. Linnainmaa, S.: Taylor expansion of the accumulated rounding errors. BIT **16**, 146–160 (1976)
14. Pan, V.Y.: Approximation of complex polynomial zeros: modified quadtree (Weyl's) construction and improved Newton's iteration. J. Complex. **16**(1), 213–264 (2000). https://doi.org/10.1006/jcom.1999
15. Pan, V.Y.: Structured Matrices and Polynomials: Unified Superfast Algorithms. Birkhäuser/Springer, Boston/New York (2001) https://doi.org/10.1007/978-1-4612-0129-8

16. Pan, V.Y.: Univariate polynomials: nearly optimal algorithms for factorization and rootfinding. J. Symb. Comput. **33**(5), 701–733, : Proceedings version. ACM STOC **1995**, 741–750 (2002). https://doi.org/10.1006/jsco.2002.0531

17. Pan, V.Y.: Old and new nearly optimal polynomial root-Finding. In: England,M., Koepf, W., Sadykov, T.M., Seiler, W.M., Vorozhtsov, E.V. (eds.) CASC2019. LNCS, vol. 11661, pp, 393–411. Springer, Nature Switzerland (2019) https://doi.org/10.1007/978-3-030-26831-2

18. Pan, V.Y.: New Acceleration of Univariate Polynomial Root-finders, August 2020. arXiv: 1805.12042

19. Pan, V.Y.: Acceleration of subdivision root-finding for sparse polynomials. to appear. In: Boulier, F., England, M., Sadikov, T.M., Vorozhtsov, E.V. (eds.) CASC 2020. Springer Nature, Switzerland (2020)

20. Pan, V.Y., Zhao, L.: Real root isolation by means of root radii approximation. In: Gerdt, V.P., Koepf, W., Seiler, W.M., Vorozhtsov, E.V. (eds.), CASC 2015. LNCS, vol. 9301, pp. 347–358. Springer, Heidelberg (2015). arXiv:1501.05386

21. Renegar, J.: On the worst-case arithmetic complexity of approximating zeros of polynomials. J. Complex. **3**(2), 90–113 (1987). https://doi.org/10.1016/0885-064X(87)90022-7

22. Schönhage, A.: The fundamental theorem of algebra in terms of computational complexity. Math. Dept., Univ. Tübingen, Germany (1982)

23. Weyl, H.: Randbemerkungen zu Hauptproblemen der Mathematik. II. Fundamentalsatz der Algebra und Grundlagen der Mathematik. Mathematische Zeitschrift **20**, 131–151 (1924)

Computing Parametric Standard Bases for Semi-weighted Homogeneous Isolated Hypersurface Singularities

Katsusuke Nabeshima$^{(\boxtimes)}$

Graduate School of Technology, Industrial and Social Sciences, Tokushima University, 2-1, Minamijosanjima-cho, Tokushima, Japan
nabeshima@tokushima-u.ac.jp

Abstract. An effective method for computing parametric standard bases of Jacobian ideals is introduced for semi-weighted homogeneous isolated hypersurface singularities. The advantage is that the proposed method is algorithmically simple. The main ideas of the method are the use of a negative weighted term ordering and coefficients in a field of rational functions. The correctness of the method is proved by utilizing algebraic local cohomology classes associated to semi-weighted homogeneous singularities.

Keywords: Semi-weighted homogeneous isolated hypersurface singularities · Standard bases · Algebraic local cohomology

1 Introduction

An effective method is proposed for computing parametric standard bases of Jacobian ideals of semi-weighted homogeneous isolated hypersurface singularities. There are two main advantages of the proposed method. The first advantage is that the output does not depend on the values of parameters, namely, the decomposition of the parameter space is not needed. The second one is that the proposed method is algorithmically simple. The keys of the method are a negative weighted term ordering and computing a standard basis of a parametric ideal in a local ring with coefficients in a field of rational functions.

Semi-weighted homogeneous (or semi-quasihomogeneous [1]) isolated hypersurface singularities are traditional research objects in singularity theory, and it is known that there are several relationships between properties of semi-weighted homogeneous singularities and their weighted homogeneous parts [1,2,4,5,16–19,22–24]. In our previous works [10,11], algorithms for computing algebraic local cohomology classes associated to the semi-weighted homogeneous singularities are considered in the context of symbolic computation. The key of previous

---

This work has been partly supported by JSPS Grant-in-Aid for Scientific Research (C) (18K03214).

© Springer Nature Switzerland AG 2020
F. Boulier et al. (Eds.): CASC 2020, LNCS 12291, pp. 447–460, 2020.
https://doi.org/10.1007/978-3-030-60026-6_26

works is a weighted vector of a weighted homogeneous part. For the weight filtration on the space of the algebraic local cohomology classes induced by the weight vector of the weighted homogeneous part, the list of weighted degrees of the basis of algebraic local cohomology classes is completely determined by the weight vector [14]. Thus, in [10], an effective algorithm for computing the algebraic local cohomology classes has been constructed by utilizing the properties of the semi-weighted homogeneous singularity and the weight vector, and as an application of algebraic local cohomology, an algorithm for computing (parametric) standard bases of the Jacobian ideals has been also introduced.

In this paper we consider a parametric standard basis of the Jacobian ideal of a semi-weighted homogeneous polynomial. To be more precise, let $f = f_0 + \sum_{i=1}^{m} u_i x^{\gamma_i}$ in $\mathbb{C}[x_1, \ldots, x_n]$ be a semi-weighted homogeneous polynomial where the polynomial f_0 is weighted homogeneous w.r.t. the weight vector $\mathbf{w} \in \mathbb{N}^n$, x^{γ_i} are upper terms and $u = \{u_1, \ldots, u_m\}$ are parameters. Then, f can be regarded as a μ-constant deformation where μ is the Milnor number of the singularity. Each hypersurface defined by f is topologically equivalent to the hypersurface defined by the weighted homogeneous part of f. Let $\succ_{-\mathbf{w}}$ be a negative weighted term ordering with $-\mathbf{w} \in \mathbb{Z}^n$ and let S be the reduced standard basis of the Jacobian ideal $\langle \frac{\partial f}{\partial x_1}, \ldots, \frac{\partial f}{\partial x_n} \rangle$ w.r.t. $\succ_{-\mathbf{w}}$ in a local ring with coefficients in $\mathbb{C}(u)$, the field of rational functions. We show that for all $u \in \mathbb{C}^m$, S is always the reduced standard basis of the Jacobian ideal w.r.t. $\succ_{-\mathbf{w}}$ in the local ring. Thus, we do not need special computation techniques to obtain the parametric standard basis with the parameters u. This fact is proved by utilizing algebraic local cohomology classes associated to semi-weighted homogeneous isolated hypersurface singularities.

This paper is organized as follows. Section 2 reviews relations between algebraic local cohomology classes and standard bases. Section 3 presents the main results.

2   Preliminaries

2.1   Notations

Throughout this paper, we fix the following notations. The set of natural numbers $\mathbb{N}$ includes zero, $\mathbb{C}$ is the field of complex numbers. Let X be an open neighborhood of the origin O of the n-dimensional complex space $\mathbb{C}^n$ with coordinates $x = (x_1, \ldots, x_n)$. Let

$$\mathbb{C}[x]_{\langle x \rangle} = \left\{ \frac{g_1(x)}{g_2(x)} \middle| g_1(x), g_2(x) \in \mathbb{C}[x], g_2(O) \neq 0 \right\}$$

be the localization of $\mathbb{C}[x]$ at O, and $\mathbb{T}^n$ the monoid of terms in $\mathbb{C}[x]$.

Definition 1. *Let $\succ$ be a term ordering on $\mathbb{T}^n$.*
(1) $\succ$ is called global if $x^\alpha \succ 1$ for all $\alpha \neq (0, \ldots, 0)$.
(2) $\succ$ is called local if $1 \succ x^\alpha$ for all $\alpha \neq (0, \ldots, 0)$.

Definition 2. *Let $\succ$ be a term ordering. Then, the inverse ordering $\succ^{-1}$ of $\succ$ is defined by*

$$x^\alpha \succ x^\beta \iff x^\beta \succ^{-1} x^\alpha,$$

where $\alpha, \beta \in \mathbb{N}^n$.

Note that if a term ordering $\succ$ is global, then the inverse term ordering $\succ^{-1}$ is local.

Definition 3. *A basis $\{x^{\alpha_1}, \dots, x^{\alpha_\ell}\}$ for a monomial ideal, in $\mathbb{C}[x]$, is said to be minimal if no x^{α_i} in the basis divides other x^{α_j} for $i \neq j$, where $\alpha_1, \dots, \alpha_\ell \in \mathbb{N}^n$.*

Let us fix a term ordering $\succ$ on $\mathbb{T}^n$. For a given polynomial $f \in \mathbb{C}[x]$ (or $\mathbb{C}[x]_{\langle x \rangle}$), we write the head term of f as $\mathrm{ht}(f)$, the head coefficient of f as $\mathrm{hc}(f)$ and the head monomial of f as $\mathrm{hm}(f)$ (*i.e.*, $\mathrm{hm}(f) = \mathrm{hc}(f)\,\mathrm{ht}(f)$). For $P \subset \mathbb{C}[x]$ (or $\mathbb{C}[x]_{\langle x \rangle}$), $\mathrm{ht}(P) = \{\mathrm{ht}(p) | p \in P\}$ and $\mathrm{hm}(P) = \{\mathrm{hm}(p) | p \in P\}$.

Let $S \subset \mathbb{C}[x]_{\langle x \rangle}$ and $\succ$ a local term ordering on $\mathbb{T}^n$. Then, $f \in \mathbb{C}[x]_{\langle x \rangle}$ is called *reduced* w.r.t. S if no term of the power series expansion of f is contained in $\mathrm{ht}(S)$.

Definition 4. *Let I be an ideal in $\mathbb{C}[x]_{\langle x \rangle}$. Fix a local term ordering.*

(1) A finite set $S \subset \mathbb{C}[x]_{\langle x \rangle}$ is called a standard basis of I if $S \subset I$, and $\langle \mathrm{hm}(I) \rangle = \langle \mathrm{hm}(S) \rangle$.

(2) A finite set $S \subset \mathbb{C}[x]_{\langle x \rangle}$ is called a reduced standard basis of I if S is a standard basis of I such that
  (i) $\mathrm{hc}(f) = 1$, for all $f \in S$.
  (ii) For any two elements $f \neq g$ in S, $\mathrm{ht}(g) \nmid \mathrm{ht}(f)$.
  (iii) For all $f \in S$, $f - \mathrm{hm}(f)$ is reduced w.r.t. S.

2.2 Algebraic Local Cohomology Classes and Standard Bases

Here we briefly review algebraic local cohomology classes, supported at O, and the relation between the algebraic local cohomology classes and standard bases. The details are given in [7, 12–14, 20, 21].

All local cohomology classes, in this paper, are algebraic local cohomology classes that belong to the set defined by

$$H^n_{[O]}(\mathbb{C}[x]) = \lim_{k \to \infty} \mathrm{Ext}^n_{\mathbb{C}[x]}(\mathbb{C}[x]/\langle x_1, x_2, \dots, x_n \rangle^k, \mathbb{C}[x])$$

where $\langle x_1, x_2, \dots, x_n \rangle$ is the maximal ideal generated by $x_1, x_2, \dots, x_n$. Consider the pair $(X, X - \{O\})$ and its relative Čech covering. Then, any section of $H^n_{[O]}(\mathbb{C}[x])$ can be represented as an element of relative Čech cohomology. Any algebraic local cohomology class in $H^n_{[O]}(\mathbb{C}[x])$ can be represented as a finite sum of the form

$$\sum c_\lambda \left[\frac{1}{x^{\lambda+1}} \right] = \sum c_\lambda \left[\frac{1}{x_1^{\lambda_1+1} x_2^{\lambda_2+1} \cdots x_n^{\lambda_n+1}} \right]$$

where [ ] stands for the Grothendieck symbol [9], $c_\lambda \in \mathbb{C}$ and $\lambda = (\lambda_1, \lambda_2, \ldots, \lambda_n)$ $\in \mathbb{N}^n$. Note that the multiplication is defined as

$$
x^\alpha \begin{bmatrix} 1 \\ x^{\lambda+1} \end{bmatrix} = \begin{cases} \begin{bmatrix} 1 \\ x^{\lambda+1-\alpha} \end{bmatrix}, & \lambda_i \geq \alpha_i, i = 1, \ldots, n, \\ \\ 0, & \text{otherwise,} \end{cases}
$$

where $\alpha = (\alpha_1, \ldots, \alpha_n) \in \mathbb{N}^n$ and $\lambda + 1 - \alpha = (\lambda_1 + 1 - \alpha_1, \ldots, \lambda_n + 1 - \alpha_n)$.

We represent an algebraic local cohomology class $\sum c_\lambda \begin{bmatrix} 1 \\ x^{\lambda+1} \end{bmatrix}$ as a polynomial in n variables $\sum c_\lambda \xi^\lambda$ (called: polynomial representation) to manipulate algebraic local cohomology classes efficiently, where $\xi = (\xi_1, \xi_2, \ldots, \xi_n)$. The multiplication by x^α is defined as

$$
x^\alpha * \xi^\lambda = \begin{cases} \xi^{\lambda-\alpha}, & \lambda_i \geq \alpha_i, \ i = 1, \ldots, n, \\ \\ 0, & \text{otherwise,} \end{cases}
$$

where $\lambda - \alpha = (\lambda_1 - \alpha_1, \ldots, \lambda_n - \alpha_n) \in \mathbb{N}^n$.

The action of monomials on algebraic local cohomology classes is extended to polynomials by linearity. For example, let $f = 2x_1^2 x_2^3 + 3x_1 x_2 \in \mathbb{C}[x_1, x_2]$ and $\psi = \xi_1^5 \xi_2^3 - 5\xi_1 \xi_2^2 \in H_{[O]}^2(\mathbb{C}[x_1, x_2])$. Then,

$$
\begin{aligned}
f * \psi &= 2x_1^2 x_2^3 * \psi + 3x_1 x_2 * \psi \\
&= (2x_1^2 x_2^3 * \xi_1^5 \xi_2^3 + 2x_1^2 x_2^3 * (-5\xi_1 \xi_2^2)) + (3x_1 x_2 * \xi_1^5 \xi_2^3 + 3x_1 x_2 * (-5\xi_1 \xi_2^2)) \\
&= (2\xi_1^3 + 0) + (3\xi_1^4 \xi_2^2 - 15\xi_2) \\
&= 3\xi_1^4 \xi_2^2 + 2\xi_1^3 - 15\xi_2.
\end{aligned}
$$

Let Ξ^n be the monoid of terms in $H_{[O]}^n(\mathbb{C}[x])$ (or $\mathbb{C}[\xi]$). Let us fix a global term ordering $\succ$ on Ξ^n. For a given algebraic local cohomology class of the form

$$
\psi = c_\lambda \xi^\lambda + \sum_{\xi^\lambda \succ \xi^{\lambda'}} c_{\lambda'} \xi^{\lambda'} \quad (c_\lambda \neq 0),
$$

we call ξ^λ the *head term*, c_λ the *head coefficient* and $\xi^{\lambda'}$ the *lower terms*. We write the head term as $\mathrm{ht}(\psi)$, the head coefficient as $\mathrm{hc}(\psi)$, the set of terms of ψ as $\mathrm{supp}(\psi) = \{\xi^\alpha | \psi = \sum_{\alpha \in \mathbb{N}^n} c_\alpha \xi^\alpha, c_\alpha \neq 0, c_\alpha \in \mathbb{C}\}$ and the set of lower terms of ψ as $\mathrm{LL}(\psi) = \{\xi^\alpha | \xi^\alpha \neq \mathrm{ht}(\psi), \xi^\alpha \in \mathrm{supp}(\psi)\}$. Let $\Psi \subset H_{[O]}^n(\mathbb{C}[x])$. We define $\mathrm{supp}(\Psi) = \bigcup_{\psi \in \Psi} \mathrm{supp}(\psi)$, $\mathrm{ht}(\Psi) = \{\mathrm{ht}(\psi) | \psi \in \Psi\}$ and $\mathrm{LL}(\Psi) = \bigcup_{\psi \in \Psi} \mathrm{LL}(\psi)$ (*i.e.*, $\mathrm{LL}(\Psi) = \mathrm{supp}(\Psi) \backslash \mathrm{ht}(\Psi)$). Moreover, we write the set of monomial elements of Ψ as $\mathrm{ML}(\Psi)$, and the set of linear combination elements of Ψ as $\mathrm{SL}(\Psi)$, *i.e.*, $\Psi = \mathrm{ML}(\Psi) \cup \mathrm{SL}(\Psi)$.

We assume that $F = \{f_1, \ldots, f_s\} \subset \mathbb{C}[x]$ satisfies $\mathbb{V}(F) \cap X = \{O\}$ where $\mathbb{V}(F) = \{\bar{a} \in \mathbb{C}^n | f_1(\bar{a}) = \cdots = f_s(\bar{a}) = 0\}$.

We define a set H_F to be the set of algebraic local cohomology classes in $H^n_{[O]}(\mathbb{C}[x])$ that are annihilated by the ideal $\langle F \rangle$, i.e.,

$$H_F = \left\{ \psi \in H^n_{[O]}(\mathbb{C}[x]) \,\Big|\, f_1 * \psi = f_2 * \psi = \cdots = f_s * \psi = 0 \right\}.$$

Then, since $\mathbb{V}(F) \cap X = \{O\}$, H_F is a finite dimensional vector space. In [13, 21], algorithms for computing a basis of the vector spaces H_F are introduced. We recall the following important theorems.

Theorem 1 ([21]). *Let $\succ$ be a global term ordering on Ξ^n and $\mathcal{I} = \langle F \rangle \subset \mathbb{C}[x]_{\langle x \rangle}$. Let Ψ be a basis of the vector space H_F such that for all $\psi \in \Psi$, $\mathrm{hc}(\psi) = 1$, $\mathrm{ht}(\psi) \notin \mathrm{ht}(\Psi \backslash \{\psi\})$ and $\mathrm{ht}(\psi) \notin \mathrm{LL}(\Psi)$. Assume that $\psi \in \Psi$ forms $\xi^\tau + \displaystyle\sum_{\xi^\tau \succ \xi^\kappa} c_{(\tau, \kappa)} \xi^\kappa$ where $c_{(\tau, \kappa)} \in \mathbb{C}$ and $\tau, \kappa \in \mathbb{N}^n$.*

(1) If $\xi^\lambda \in \mathrm{LL}(\Psi)$, then $x^\lambda \equiv \displaystyle\sum_{\xi^\kappa \in \mathrm{ht}(\Psi)} c_{(\kappa, \lambda)} x^\kappa \mod \mathcal{I}$ in $\mathbb{C}[x]_{\langle x \rangle}$,

i.e., $x^\lambda - \displaystyle\sum_{\xi^\kappa \in \mathrm{ht}(\Psi)} c_{(\kappa, \lambda)} x^\kappa \in \mathcal{I}$.

(2) If $\xi^\lambda \in \mathrm{ht}(\Psi)$, then $x^\lambda \notin \mathcal{I}$.
(3) If $\xi^\lambda \notin \mathrm{supp}(\Psi)$, then $x^\lambda \in \mathcal{I}$.

Let ξ^λ be a term and Λ a set of terms in $\mathbb{C}[\xi]$ where $\lambda \in \mathbb{N}^n$. We call $\xi_i \xi^\lambda$ a neighbor of ξ^λ for each $i = 1, 2, \ldots, n$. We write the *neighbors* of Λ as $\mathrm{Neighbor}(\Lambda)$, i.e., $\mathrm{Neighbor}(\Lambda) = \{\xi_i \xi^\lambda | \xi^\lambda \in \Lambda, 1 \le i \le n\}$.

Theorem 2 ([13]). *Using the same notations as in Theorem 1, let Λ be the minimal basis of the ideal generated by $\mathrm{Neighbor}(\mathrm{ht}(\Psi)) \backslash \mathrm{ht}(\Psi)$. Let $\xi^\tau + \displaystyle\sum_{\xi^\tau \succ \xi^\kappa} c_{(\tau, \kappa)} \xi^\kappa$ in Ψ where $c_{(\tau, \kappa)} \in \mathbb{C}$ and $\tau, \kappa \in \mathbb{N}^n$. The transfer SB_Ψ is defined by the following:*

$$\begin{cases} \mathrm{SB}_\Psi(\xi^\lambda) = x^\lambda - \displaystyle\sum_{\xi^\kappa \in \mathrm{ht}(\Psi)} c_{(\kappa, \lambda)} x^\kappa, & \text{if } \xi^\lambda \in \mathrm{LL}(\Psi), \\ \mathrm{SB}_\Psi(\xi^\lambda) = x^\lambda, & \text{if } \xi^\lambda \notin \mathrm{LL}(\Psi), \end{cases}$$

where $\lambda \in \mathbb{N}^n$.

Then, $\mathrm{SB}_\Psi(\Lambda) = \{\mathrm{SB}_\Psi(\xi^\lambda) | \xi^\lambda \in \Lambda\}$ is the reduced standard basis of the ideal $\mathcal{I}$ w.r.t. $\succ^{-1}$ in $\mathbb{C}[x]_{\langle x \rangle}$. (Note that $\xi^\alpha \succ \xi^\beta \iff x^\beta \succ^{-1} x^\alpha$ where $\alpha, \beta \in \mathbb{N}^n$.)

Notice that the reduced standard basis S_{red} of $\langle F \rangle$ w.r.t. the local term ordering $\succ^{-1}$ always exists, and $f \in S_{red}$ has finitely many terms since $\mathrm{supp}(\Lambda)$ and $\mathrm{supp}(\Psi)$ are finite sets.

3 Parametric Standard Bases for Semi-weighted Homogeneous Isolated Hypersurface Singularities

Here, first, we briefly review properties of semi-weighted homogeneous isolated hypersurface singularities, and second, we introduce the main results.

3.1   Semi-weighted Homogeneous Isolated Hypersurface Singularities and Algebraic Local Cohomology Classes

Before describing the main results of this paper, we review Theorem 9 of [10] that is utilized for the main theorems.

Let us fix a weight vector $\mathbf{w} = (w_1, w_2, \ldots, w_n)$ in $\mathbb{N}^n$ for coordinate systems $x = (x_1, x_2, \ldots, x_n)$ and $\xi = (\xi_1, \xi_2, \ldots, \xi_n)$. We define a weighted degree of the term $x^\alpha = x_1^{\alpha_1} x_2^{\alpha_2} \cdots x_n^{\alpha_n}$ and $\xi^\alpha = \xi_1^{\alpha_1} \xi_2^{\alpha_2} \cdots \xi_n^{\alpha_n}$, with respect to $\mathbf{w}$ by $|x^\alpha|_\mathbf{w} = \sum_{i=1}^n w_i \alpha_i$ and $|\xi^\alpha|_\mathbf{w} = \sum_{i=1}^n w_i \alpha_i$, respectively. For $f \in \mathbb{C}[x]$ and $\psi \in H_{[O]}^n(\mathbb{C}[x])$, $\deg_\mathbf{w}(f) = \max\{|x^\alpha|_\mathbf{w} | x^\alpha \in \mathrm{supp}(f)\}$ and $\deg_\mathbf{w}(\psi) = \max\{|\xi^\alpha|_\mathbf{w} | \xi^\alpha \in \mathrm{supp}(\psi)\}$.

Definition 5. *Let $\succ$ be a global term ordering on $\mathbb{T}^n$. The (global) weighted term ordering with $\succ$ is defined by the following:*

$$x^\alpha \succ_\mathbf{w} x^\beta \iff |x^\alpha|_\mathbf{w} > |x^\beta|_\mathbf{w}, \ or$$
$$|x^\alpha|_\mathbf{w} = |x^\beta|_\mathbf{w} \ and \ x^\alpha \succ x^\beta.$$

The definition of semi-weighted homogeneous isolated hypersurface singularities is as follows.

Definition 6 ([1]). *Let f be a nonzero polynomial in $\mathbb{C}[x]$ and let $\mathrm{ord}_\mathbf{w}(f) = \min\{|x^\alpha|_\mathbf{w} | x^\alpha \in \mathrm{supp}(f)\}$ ($\mathrm{ord}_\mathbf{w}(0) = -1$).*

*(1) A nonzero polynomial f is **weighted homogeneous of type** $(d; \mathbf{w})$ if all terms of f have the same weighted degree d with respect to $\mathbf{w}$, i.e., $f = \displaystyle\sum_{|x^\alpha|_\mathbf{w} = d} c_\alpha x^\alpha$ where $c_\alpha \in \mathbb{C}$.*

*(2) The polynomial f is called **semi-weighted homogeneous** of type $(d; \mathbf{w})$ if f is of the form $f = f_0 + g$ where f_0 is a weighted homogeneous polynomials of type $(d; \mathbf{w})$ with an isolated singularity at the origin O, $f = f_0$ or $\mathrm{ord}_\mathbf{w}(f - f_0) > d$.*

Lists of weighted homogeneous polynomials are given in [2, 17–19, 23].

Let $f = f_0 + g$ be a semi-weighted homogeneous polynomial of type $(d; \mathbf{w})$ where the polynomial f_0 is weighted homogeneous of type $(d; \mathbf{w})$ with an isolated singularity at O and $\mathrm{ord}_\mathbf{w}(g) > d$. Let $F_0 = \left\{ \frac{\partial f_0}{\partial x_1}, \ldots, \frac{\partial f_0}{\partial x_n} \right\}$ and $F = \left\{ \frac{\partial f}{\partial x_1}, \ldots, \frac{\partial f}{\partial x_n} \right\}$. Set

$$H_{F_0} = \left\{ \psi \in H_{[O]}^n(\mathbb{C}[x]) \middle| \frac{\partial f_0}{\partial x_1} * \psi = \cdots = \frac{\partial f_0}{\partial x_n} * \psi = 0 \right\},$$

and

$$H_F = \left\{ \psi \in H_{[O]}^n(\mathbb{C}[x]) \middle| \frac{\partial f}{\partial x_1} * \psi = \cdots = \frac{\partial f}{\partial x_n} * \psi = 0 \right\}.$$

Then, since $\mathbb{V}(F) \cap X = \{O\}$ and $\mathbb{V}(F_0) \cap X = \{O\}$, H_F and H_{F_0} are finite dimensional vector spaces. In [10], an algorithm for computing a basis of the vector space H_F is introduced.

The following theorem, given in [10], shows the relation between a basis of H_{F_0} and a basis of H_F. (This theorem also follows immediately from Proposition 3.2. of [14].)

Theorem 3 (Theorem 9 of [10]). *Let $\Psi_0 = \{\psi_1, \psi_2, \ldots, \psi_\ell\}$ be a basis of the vector space H_{F_0} that satisfies the condition "for all $\psi \in \Psi_0$, $\mathrm{hc}(\psi) = 1$, $\mathrm{ht}(\psi) \notin \mathrm{ht}(\Psi_0 \backslash \{\psi\})$ and $\mathrm{ht}(\psi) \notin \mathrm{LL}(\Psi)$" w.r.t. a (global) weighted term ordering $\succ_{\mathbf{w}}$. Then, for each $i = 1, \ldots, \ell$, there exists ρ_i such that $\deg_{\mathbf{w}}(\psi_i) > \deg_{\mathbf{w}}(\rho_i)$ and $\varphi_i = \psi_i + \rho_i$ is an element of H_F. (It is possible to take $\rho_i = 0$.) Moreover, $\{\varphi_1, \varphi_2, \ldots, \varphi_\ell\}$ is a basis of the vector space H_F.*

3.2   Main Results

Let $f = f_0 + \sum_{i=1}^{m} u_i x^{\gamma_i}$ be a semi-weighted homogeneous polynomial of type $(d; \mathbf{w})$ where the polynomial f_0 is weighted homogeneous of type $(d; \mathbf{w})$ with an isolated singularity at O, for each $i \in \{1, \ldots, m\}$, $\gamma_i \in \mathbb{N}^n$, $|x^{\gamma_i}|_{\mathbf{w}} > d$ and $u = \{u_1, \ldots, u_m\}$ are parameters. The aim of this paper is to introduce an effective method for computing a parametric standard basis of the Jacobian ideal $\langle \frac{\partial f}{\partial x_1}, \ldots, \frac{\partial f}{\partial x_n} \rangle$ with the parameters u.

For an arbitrary $\bar{a} \in \mathbb{C}^m$, the specialization homomorphism $\sigma_{\bar{a}} : \mathbb{C}[u][x]_{\langle x \rangle} \longrightarrow \mathbb{C}[x]_{\langle x \rangle}$ (or $\sigma_{\bar{a}} : \mathbb{C}[u][\xi] \longrightarrow \mathbb{C}[\xi]$) is defined as the map that substitutes $\bar{a}$ into m variables u. For $P \subset \mathbb{C}[u][x]_{\langle x \rangle}$, $\sigma_{\bar{a}}(P) = \{\sigma_{\bar{a}}(h) | h \in P\}$. We regard u as parameters throughout the rest of the paper.

Notice that since f_0 does not have the parameters u, for all $\bar{a} \in \mathbb{C}^m$, $\sigma_{\bar{a}}(f) = 0$ always defines an isolated singularity at the origin O.

Now, we are ready to present the main theorems.

Theorem 4. *Let $f = f_0 + \sum_{i=1}^{m} u_i x^{\gamma_i}$ be a semi-weighted homogeneous polynomial of type $(d; \mathbf{w})$ where the polynomial f_0 is weighted homogeneous of type $(d; \mathbf{w})$ with an isolated singularity at O, for each $i \in \{1, \ldots, m\}$, $\gamma_i \in \mathbb{N}^n$, $|x^{\gamma_i}|_{\mathbf{w}} > d$ and $u = \{u_1, \ldots, u_m\}$ are parameters. Set $F_0 = \left\{ \frac{\partial f_0}{\partial x_1}, \ldots, \frac{\partial f_0}{\partial x_n} \right\}$, $F = \left\{ \frac{\partial f}{\partial x_1}, \ldots, \frac{\partial f}{\partial x_n} \right\}$ and $\mathcal{O}_{\mathbf{w}_x} = \mathbb{T}^n \setminus \mathrm{ht}(\langle F_0 \rangle)$ w.r.t. $\succ_{\mathbf{w}}^{-1}$ where $\succ_{\mathbf{w}}^{-1}$ is the inverse ordering of a weighted term ordering $\succ_{\mathbf{w}}$. Then the following holds.*

(1) For all $\bar{a} \in \mathbb{C}^m$, $\mathbb{T}^n \setminus \mathrm{ht}(\langle \sigma_{\bar{a}}(F) \rangle) = \mathcal{O}_{\mathbf{w}_x}$.

(2) Fix the weighted term ordering $\succ_{\mathbf{w}}$. Let $\mathcal{O}_{\mathbf{w}_\xi} = \{\xi^\alpha | x^\alpha \in \mathcal{O}_{\mathbf{w}_x}\}$. There exists a finite set Ψ in $\mathbb{C}[u][\xi]$ (or $H^n_{[O]}(\mathbb{C}[u][x])$) such that for all $\bar{a} \in \mathbb{C}^m$ and $\psi \in \Psi$, $\sigma_{\bar{a}}(\Psi)$ is a basis of the vector space $H_{\sigma_{\bar{a}}(F)}$ and $\mathrm{hc}(\psi) = 1$, $\mathrm{ht}(\psi) \notin \mathrm{ht}(\Psi \backslash \{\psi\})$, $\mathrm{ht}(\psi) \notin \mathrm{LL}(\Psi)$. Moreover, for all $\bar{a} \in \mathbb{C}^m$, $\mathrm{ht}(\sigma_{\bar{a}}(\Psi)) = \mathcal{O}_{\mathbf{w}_\xi}$.

(3) There exists a unique set S in $\mathbb{C}[u][x]_{\langle x \rangle}$ such that, for all $\bar{a} \in \mathbb{C}^m$, $\sigma_{\bar{a}}(S)$ is the reduced standard basis of $\langle \sigma_{\bar{a}}(F) \rangle$ w.r.t. $\succ_{\mathbf{w}}^{-1}$ in $\mathbb{C}[x]_{\langle x \rangle}$.

Proof. (1) As f_0 does not have the parameters u, a basis $\Psi_0 \subset H_{F_0}$ that satisfies the condition "for all $\psi \in \Psi_0$, $\mathrm{hc}(\psi) = 1$, $\mathrm{ht}(\psi) \notin \mathrm{ht}(\Psi_0 \backslash \{\psi\})$ and $\mathrm{ht}(\psi) \notin \mathrm{LL}(\Psi_0)$ w.r.t. $\succ_{\mathbf{w}}$", does not have the parameters u. Set $\mathcal{O}' = \{x^\alpha | \xi^\alpha \in \mathrm{ht}(\Psi_0)\}$ w.r.t. $\succ_{\mathbf{w}}$, then by Theorem 2, $\mathcal{O}' = \mathbb{T}^n \backslash \mathrm{ht}(\langle F_0 \rangle)$ w.r.t. the inverse ordering $\succ_{\mathbf{w}}^{-1}$. Theorem 3 shows that, for all $\bar{a} \in \mathbb{C}^m$, there exists a basis $\Phi_{\bar{a}}$ of the vector space $H_{\sigma_{\bar{a}}(F)}$ that satisfies $\mathrm{ht}(\Phi_{\bar{a}}) = \mathrm{ht}(\Psi_0)$ w.r.t. $\succ_{\mathbf{w}}$. Therefore, $\mathcal{O}' = \mathbb{T}^n \backslash \mathrm{ht}(\langle \sigma_{\bar{a}}(F) \rangle) = \mathcal{O}_{\mathbf{w}_x}$.

(2) Let $\Psi_0 = \{\psi_1, \ldots, \psi_\ell\}$ be a basis of H_{F_0} that satisfies the condition

"for all $\psi \in \Psi_0$, $\mathrm{hc}(\psi) = 1$, $\mathrm{ht}(\psi) \notin \mathrm{ht}(\Psi_0 \backslash \{\psi\})$ and $\mathrm{ht}(\psi) \notin \mathrm{LL}(\Psi_0)$".

For each $i \in \{1, \ldots, \ell\}$, let

$$L_i = \{\xi^\alpha \,|\, |\,\mathrm{ht}(\psi_i)|_{\mathbf{w}} > |\xi^\alpha|_{\mathbf{w}}, \xi^\alpha \notin \mathrm{ht}(\Psi_0)\}.$$

As L_i is finite, we rewrite L_i as $\{\xi^{\lambda_1}, \ldots, \xi^{\lambda_r}\}$ where $\lambda_1, \ldots, \lambda_r \in \mathbb{N}^n$. Set $\phi_i = \psi_i + \sum_{j=1}^{r} c_{ij} \xi^{\lambda_j}$ where $c_{i1}, \ldots, c_{ir}$ are indeterminates. In order to determine $c_{i1}, \ldots, c_{ir}$, for all $k \in \{1, \ldots, n\}$, let us consider

$$\frac{\partial f}{\partial x_k} * \phi_i = \sum p_\tau \xi^\tau = 0$$

where p_τs are linear polynomials of $c_{i1}, \ldots, c_{ir}$ with coefficients in $\mathbb{C}[u]$. Since the terms ξ^τ are linearly independent, we need to solve the system of the all obtained linear equations $p_\tau = 0$. Let us write the system of ν linear equations as $Ac = b$ where A is the coefficient matrix, ${}^t c = (c_{i1} \cdots c_{ir})$, $\nu \in \mathbb{N}$ and $b \in \mathbb{C}[u]^\nu$. By utilizing elementary row operations, the extended coefficient matrix $(A|b)$ can be transformed to the form

$$\begin{pmatrix} q_1(u) & 0 & \cdots & 0 & b_1'(u) \\ 0 & q_2(u) & \cdots & 0 & b_2'(u) \\ \vdots & \vdots & \ddots & \vdots & \vdots \\ 0 & 0 & \cdots & q_r(u) & b_r'(u) \\ 0 & 0 & \cdots & 0 & 0 \\ \vdots & \vdots & \vdots & \vdots & \vdots \\ 0 & 0 & \cdots & 0 & 0 \end{pmatrix}$$

because by Theorem 4 for arbitrary values of u the system always has unique solution, where $q_1(u), \ldots, q_r(u), b_1'(u), \ldots, b_r'(u) \in \mathbb{C}[u]$ and the greatest common divisor of $q_j(u)$ and $b_j'(u)$ (write $\gcd(q_j(u), b_j'(u))$) ($1 \le j \le r$) is 1, *i.e.*,

$$\gcd(q_1(u), b_1'(u)) = \gcd(q_2(u), b_2'(u)) = \cdots = \gcd(q_r(u), b_r'(u)) = 1.$$

If there exists $j \in \{1, \ldots, r\}$ such that $q_j(u) \notin \mathbb{C} \backslash \{0\}$, then when $q_j(u) = 0$ (*i.e.*, $u \in \mathbb{V}(q_j) \ne \emptyset$ in $\mathbb{C}^m$), the rank of the matrix becomes less than r or the system has no solution. This is contradiction. Hence, $q_1(u), \ldots, q_r(u)$ are constant in $\mathbb{C} \setminus \{0\}$, and thus $c_{i1}, \ldots, c_{ir}$ are uniquely determined as polynomials in $\mathbb{C}[u]$. Therefore,

$$\Psi = \{\phi_1, \ldots, \phi_\ell\} \subset \mathbb{C}[u][\xi] \ (\text{or } H^n_{[O]}(\mathbb{C}[u][x])),$$

and for all $\bar{a} \in \mathbb{C}^m$, $\sigma_{\bar{a}}(\Psi)$ is a basis of the vector space $H_{\sigma_{\bar{a}}(F)}$. As Ψ_0 has no parameters and $\phi_i = \psi_i + \sum_{j=1}^{r} c_{ij} \xi^{\lambda_j}$, for all $\varphi \in \Psi$, $\mathrm{hc}(\varphi) = 1$, $\mathrm{ht}(\varphi) \notin \mathrm{ht}(\Psi \setminus \{\varphi\})$, $\mathrm{ht}(\varphi) \notin \mathrm{LL}(\Psi)$. Moreover, $\mathrm{ht}(\sigma_{\bar{a}}(\Psi)) = \mathcal{O}_{\mathbf{w}_\xi}$ holds from (1). This completes the proof.

(3). By combining (2) and Theorem 2, (3) holds. $\square$

Notice that Ψ and S of Theorem 4 do not depend on the values of the parameters u. The key of the theorem is the weighted ordering $\succ_{\mathbf{w}}$.

The following theorem shows how to compute a parametric standard basis of the Jacobian ideal in $\mathbb{C}[u][x]_{\langle x \rangle}$.

Theorem 5. *Using the same notations as in Theorem 4, let S_{red} be the reduced standard basis of $\langle F \rangle$ w.r.t. $\succ_{\mathbf{w}}^{-1}$ in $\mathbb{C}(u)[x]_{\langle x \rangle}$ where $\mathbb{C}(u)$ is the field of rational functions. Then, $S_{red} \subset \mathbb{C}[u][x]_{\langle x \rangle}$, and for all $\bar{a} \in \mathbb{C}^m$, $\sigma_{\bar{a}}(S_{red})$ is the reduced standard basis of $\langle \sigma_{\bar{a}}(F) \rangle$ w.r.t. $\succ_{\mathbf{w}}^{-1}$ in $\mathbb{C}[x]_{\langle x \rangle}$.*

Proof. As we described in Theorem 4 (3), there exists a unique set S in $\mathbb{C}[u][x]_{\langle x \rangle}$ such that, for all $\bar{a} \in \mathbb{C}^m$, $\sigma_{\bar{a}}(S)$ is the reduced standard basis of $\langle \sigma_{\bar{a}}(F) \rangle$ w.r.t. $\succ_{\mathbf{w}}^{-1}$ in $\mathbb{C}[x]_{\langle x \rangle}$. We show that $S_{red} = S$. For each element $g \in S_{red} \subset \mathbb{C}(u)[x]_{\langle x \rangle}$, there exists $h \in S \subset \mathbb{C}[u][x]_{\langle x \rangle}$ such that $\mathrm{ht}(g) = \mathrm{ht}(h)$. Suppose that $\mathrm{lcm}(u)$ is the least common multiple of denominators of all coefficients of g. Then, $\mathrm{lcm}(u)g \in \mathbb{C}[u][x]_{\langle x \rangle}$. Since S_{red} is a generic standard basis of $\langle F \rangle$, there exists a Zariski open subset $\mathbb{A} \subset \mathbb{C}^m$ such that, for all $\bar{a} \in \mathbb{A}$, $\mathrm{ht}(\sigma_{\bar{a}}(S)) = \mathrm{ht}(S_{red})$ and $\sigma_{\bar{a}}(g), \sigma_{\bar{a}}(h) \in \langle \sigma_{\bar{a}}(F) \rangle$. Then, for all $\bar{a} \in \mathbb{A}$,

$$\sigma_{\bar{a}}(\mathrm{lcm}(u)g - \mathrm{lcm}(u)h) = \sigma_{\bar{a}}(\mathrm{lcm}(u)(g - \mathrm{ht}(g)) - \mathrm{lcm}(u)(h - \mathrm{ht}(h)))$$

$$= \sum_{\mathrm{ht}(g) \succ_{\mathbf{w}}^{-1} x^\alpha} \sigma_{\bar{a}}(g_\alpha(u) - h_\alpha(u))x^\alpha \in \langle \sigma_{\bar{a}}(F) \rangle$$

where $g_\alpha(u), h_\alpha(u) \in \mathbb{C}[u]$. As the Zariski open subset $\mathbb{A}$ has infinitely many elements and x^α is smaller than $\mathrm{ht}(g) \in \mathrm{ht}(S_{red})$ w.r.t. $\succ_{\mathbf{w}}^{-1}$, we have $g_\alpha(u) - h_\alpha(u) = 0$, namely, $\mathrm{lcm}(u)h = \mathrm{lcm}(u)g$. This fact implies $h = g$. Therefore, $S_{red} = S$. $\square$

We are able to compute the parametric standard basis in $\mathbb{C}(u)[x]_{\langle x \rangle}$ by the usual algorithm for computing a standard basis w.r.t. $\succ_{\mathbf{w}}^{-1}$ in the ring. We do not need a special algorithm for computing comprehensive standard systems [8].

The local term ordering $\succ_{\mathbf{w}}^{-1}$ (so-called the negative weighted term ordering) can be represented as a matrix [6,15], for example, the following $n \times n$ matrices

$$\begin{pmatrix} -w_1 & -w_2 & \cdots & -w_{n-1} & -w_n \\ -1 & 0 & \cdots & 0 & 0 \\ 0 & -1 & \cdots & 0 & 0 \\ \vdots & \vdots & \ddots & \vdots & \vdots \\ 0 & 0 & \cdots & -1 & 0 \end{pmatrix}, \quad \begin{pmatrix} -w_1 & -w_2 & \cdots & -w_{n-1} & -w_n \\ 0 & 0 & \cdots & 0 & 1 \\ 0 & 0 & \cdots & 1 & 0 \\ \vdots & \vdots & \vdots & \vdots & \vdots \\ 0 & 1 & \cdots & 0 & 0 \end{pmatrix}$$

represent local term orderings $\succ_{\mathbf{w}}^{-1}$ where $\mathbf{w} = (w_1, w_2, \ldots, w_n) \in \mathbb{N}^n$.

METHOD.

Input : $f = f_0 + \sum_{i=1}^m u_i x^{\gamma_i}$: semi-weighted homogeneous polynomial of type $(d; \mathbf{w})$ where the polynomial $f_0 \subset \mathbb{C}[x]$ is weighted homogeneous of type $(d; \mathbf{w})$ with an isolated singularity at O, for each $i \in \{1, \ldots, m\}$, $\gamma_i \in \mathbb{N}^n$, $|x^{\gamma_i}|_{\mathbf{w}} > d$ and $u = \{u_1, \ldots, u_m\}$ are parameters.

Output: $S \subset \mathbb{C}[u][x]_{\langle x \rangle}$: for all $\bar{a} \in \mathbb{C}^m$, $\sigma_{\bar{a}}(S)$ is a standard basis of $\langle \sigma_{\bar{a}}(\frac{\partial f}{\partial x_1}), \ldots, \sigma_{\bar{a}}(\frac{\partial f}{\partial x_n}) \rangle$ w.r.t. $\succ_{\mathbf{w}}^{-1}$ in $\mathbb{C}[x]_{\langle x \rangle}$.

BEGIN
$S \leftarrow$ Compute a standard basis of $\langle \frac{\partial f}{\partial x_1}, \ldots, \frac{\partial f}{\partial x_n} \rangle$ w.r.t. $\succ_{\mathbf{w}}^{-1}$ in $\mathbb{C}(u)[x]_{\langle x \rangle}$;
return S;
END

Note that since a standard basis S is generated by the reduced standard basis S_{red} and $\langle \mathrm{hm}(S) \rangle = \langle \mathrm{hm}(S_{red}) \rangle$, if we change "reduced standard basis" to "(normal) standard basis" in Theorem 5, then Theorem 5 still holds.

We illustrate the method above with the following example.

Example 1. Let $f_0 = x^3 + yz^2 + y^{10}$ and let $\mathbf{w} = (20, 6, 27)$ be a weight vector for the variables (x, y, z). Set $f = f_0 + sxy^7 + txy^8 + uxz^2 \in \mathbb{C}[s, t, u][x, y, z]$ where s, t, u are parameters. Then, f is a semi-weighted homogeneous polynomial of type $(60; \mathbf{w})$.

The computer algebra system SINGULAR [3] can define the inverse ordering $\succ_{\mathbf{w}}^{-1}$ as a matrix.

SINGULAR outputs the reduced standard basis of $J = \langle \frac{\partial f}{\partial x}, \frac{\partial f}{\partial y}, \frac{\partial f}{\partial z} \rangle$ w.r.t. $\succ_{\mathbf{w}}^{-1}$ in $\mathbb{C}(s, t, u)[x, y, z]_{\langle x, y, z \rangle}$ as follows.

```
> intmat m[3][3]=-20,-6,-27,-1,0,0,0,-1,0;
> m;
-20,-6,-27,
-1,0,0,
0,-1,0
> ring A=(0,s,t,u),(x,y,z),M(m);
```

```
> poly f=x3+yz2+y10+s*xy7+t*xy8+u*xz2;
> ideal J=jacob(f);
> option(redSB);
> std(J);
_[1]=yz+(u)*xz
_[2]=3*x2+(s)*y7+(t)*y8+(-10u)*y9+(-7su)*xy6+(-8tu)*xy7
_[3]=z2+10*y9+(7s)*xy6+(8t)*xy7
_[4]=10*y10+(7s)*xy7+(8t)*xy8+(10u)*xy9
```

In the display above, m[3][3] means the matrix ordering of $\succ_{\mathbf{w}}^{-1}$. The output means

$$S = \{yz + uxz, 3x^2 + sy^7 + ty^8 - 10uy^9 - 7suxy^6 - 8tuxy^7, z^2 + 10y^9 + 7sxy^6 + 8txy^7, 10y^{10} + 7sxy^7 + 8txy^8 + 10uxy^9\}$$

is the reduced standard basis of J w.r.t. m[3][3] in $\mathbb{C}(s, t, u)[x, y, z]_{\langle x,y,z\rangle}$. Therefore, S is a parametric standard basis of J with the parameters s, t, u, namely, for all $\bar{a} \in \mathbb{C}^3$, $\sigma_{\bar{a}}(S)$ is the reduced standard basis of $\langle \sigma_{\bar{a}}(\frac{\partial f}{\partial x}), \sigma_{\bar{a}}(\frac{\partial f}{\partial y}), \sigma_{\bar{a}}(\frac{\partial f}{\partial z})\rangle$ in $\mathbb{C}[x, y, z]_{\langle x,y,z\rangle}$.

The output of the proposed method does not depend on the values of the parameters u. This is the big advantage of the method.

In [8], an algorithm for computing comprehensive standard systems (CSS) is introduced. We have implemented the algorithm for computing CSS in the computer algebra system SINGULAR[1]. Let us compare S with the output of CSS.

Our SINGULAR implementation returns a CSS of J w.r.t. the negative degree lexicographic term ordering $\succ$ with $x \succ y \succ z$, as follows.

1. If the parameters (s, t, u) belong to $\mathbb{C}^3 \setminus \mathbb{V}(u, s)$, then the standard basis is the following.

[1]: 3x2+z2u+y7s+y8t
[2]: xzu+yz
[3]: z2+7xy6s+8xy7t+10y9
[4]: 3xyz-z3u2-y7zsu-y8ztu
[5]: 3y2z+z3u3+y7zsu2+y8ztu2
[6]: 21xy7s-7y6z2su2+24x2y7tu+24xy8t+30xy9u+30y10-7y13s2u-7y14 stu
[7]: 30xy10-10y9z2u2-7xy13s2u-7y14s2-15xy14stu-15y15st-8xy15t2 u-10y16su-8y16t2-10y17tu
[8]: 300y13+100y11z2u3+49xy13s3u+49y14s3+161xy14s2tu+161y15s2t +70xy15s2u2+176xy15st2u+70y16s2u+176y16st2+150xy16stu2+64xy16t 3u+150y17stu+64y17t3+80xy17t2u2+100y18su2+80y18t2u+100y19tu2

The display above is from SINGULAR.

---

[1] The SINGULAR implementation of CSS is in the author's web-page or the following URL. https://www-math.ias.tokushima-u.ac.jp/~nabesima/softwares.html.

2. If the parameters (s, t, u) belong to $\mathbb{V}(u) \setminus \mathbb{V}(s)$, then the standard basis is the following.

   [1] : 3x2+z2u+y7s+y8t
   [2] : yz
   [3] : z2+7xy6s+8xy7t+10y9
   [4] : 7xy7s+8xy8t+10y10
   [5] : 30xy10-7y14s2-15y15st-8y16t2
   [6] : y13

3. If the parameters (s, t, u) belong to $\mathbb{V}(u, s) \setminus \mathbb{V}(t)$, then the standard basis is the following.

   [1] : 3x2+z2u+y7s+y8t
   [2] : yz
   [3] : z2+7xy6s+8xy7t+10y9
   [4] : 4xy8t+5y10
   [5] : 15xy10-4y15st-4y16t2
   [6] : y12

4. If the parameters (s, t, u) belong to $\mathbb{V}(u, s, t)$, then the standard basis is the following.

   [1] : 3x2+z2u+y7s+y8t
   [2] : yz
   [3] : z2+7xy6s+8xy7t+10y9
   [4] : y10

5. If the parameters (s, t, u) belong to $\mathbb{V}(s) \setminus \mathbb{V}(u, t)$, then the standard basis is the following.

   [1] : 3x2+z2u+y7s+y8t
   [2] : xzu+yz
   [3] : z2+7xy6s+8xy7t+10y9
   [4] : 3xyz-z3u2-y7zsu-y8ztu
   [5] : 3y2z+z3u3+y7zsu2+y8ztu2
   [6] : 12xy8t-4y7z2tu2+15xy9u+15y10-4y14stu-4y15t2u
   [7] : 30xy10-10y9z2u2-7xy14stu-8y15st-8xy15t2u-8y16t2-10y17tu
   [8] : 150y12+50y10z2u3+28xy14st2u+32y15st2+35xy15stu2+32xy15t3u
   +32y16t3+40xy16t2u2+40y17t2u+50y18tu2

6. If the parameters (s, t, u) belong to $\mathbb{V}(s, t) \setminus \mathbb{V}(u)$, then the standard basis is the following.

   [1] : 3x2+z2u+y7s+y8t
   [2] : xzu+yz
   [3] : z2+7xy6s+8xy7t+10y9
   [4] : 3xyz-z3u2-y7zsu-y8ztu
   [5] : 3y2z+z3u3+y7zsu2+y8ztu2
   [6] : xy9u+y10
   [7] : 3xy10-y9z2u2
   [8] : 3y11+y9z2u3

The parameter space $\mathbb{C}^3$ is divided to 6 strata.
The output S of the proposed method is simpler than the CSS above.

The algorithm [8] for computing CSS needs to compute several standard bases in $\mathbb{C}[u][x]_{\langle x \rangle}$, however, the proposed method needs only one standard basis w.r.t. the negative weighted term ordering in $\mathbb{C}(u)[x]_{\langle x \rangle}$. In general, the computational complexity of a standard basis in $\mathbb{C}[u][x]_{\langle x \rangle}$ is higher than that of a standard basis in $\mathbb{C}(u)[x]_{\langle x \rangle}$. Thus, the proposed method is more effective than the algorithm [8] in the cases of semi-weighted homogeneous isolated hypersurface singularities.

References

1. Arnold, V.I.: Normal forms of functions in neighbourhoods of degenerate critical points. Russ. Math. Surv. **29**, 10–50 (1974)
2. Arnold, V.I., Gusein-Zade, S.M., Varchenko, A.N.: Singularities of differentiable maps. **1**, Birkhäuser (1985)
3. Decker, W., Greuel, G.-M., Pfister, G., Schönemann, H.: Singular 4-1-0 – a computer algebra system for polynomial computations (2016). http://www.singular.uni-kl.de
4. Greuel, G.-M.: Constant milnor number implies constant multiplicity for quasi-homogeneous singularities. Manuscripta Math. **56**, 159–166 (1986)
5. Greuel, G.-M., Hertling, C., Pfister, G.: Moduli spaces of semiquasihomogeneous singularities with fixed principal part. J. Alg. Geom. **6**, 169–199 (1997)
6. Greuel, G.-M., Pfister, G.: A Singular Introduction to Commutative Algebra, 2nd edn. Springer-Verlag, Berlin (2007)
7. Grothendieck, A.: Local cohomology, notes by R. Hartshorne. Lecture Notes in Mathematics, vol. 41. Springer (1967)
8. Hashemi, A., Kazemi, M.: Parametric standard bases and their applications. In: England, M., Koepf, W., Sadykov, T.M., Seiler, W.M., Vorozhtsov, E.V. (eds.) CASC 2019. LNCS, vol. 11661, pp. 179–196. Springer, Cham (2019). https://doi.org/10.1007/978-3-030-26831-2_13
9. Kunz, E.: Residues and Duality for Projective Algebraic Varieties. American Mathematical Society, New York (2009)
10. Nabeshima, K., Tajima, S.: On efficient algorithms for computing parametric local cohomology classes associated with semi-quasi homogeneous singularities and standard bases. In: Proceeding of ISSAC 2014, pp. 351–358. ACM (2014)
11. Nabeshima, K., Tajima, S.: Computing logarithmic vector fields associated with parametric semi-quasihomogeneous hypersurface isolated singularities. Proceedings of ISSAC 2015, pp. 291–298. ACM (2015)
12. Nabeshima, K., Tajima, S.: Efficient computation of algebraic local cohomology classes and change of ordering for zero-dimensional standard bases. In: Gerdt, V.P., Koepf, W., Seiler, W.M., Vorozhtsov, E.V. (eds.) CASC 2015. LNCS, vol. 9301, pp. 334–348. Springer, Cham (2015). https://doi.org/10.1007/978-3-319-24021-3_25
13. Nabeshima, K., Tajima, S.: Algebraic local cohomology with parameters and parametric standard bases for zero-dimensional ideals. J. Symb. Comp. **82**, 91–122 (2017)
14. Nakamura, Y., Tajima, S.: On weighted-degrees for algebraic local cohomologies associated with semiquasihomogeneous singularities. Adv. Stud. Pure Math. **46**, 105–117 (2007)

15. Robbiano, L.: Term orderings on the polynomial ring. In: Caviness, B.F. (ed.) EUROCAL 1985. LNCS, vol. 204, pp. 513–517. Springer, Heidelberg (1985). https://doi.org/10.1007/3-540-15984-3_321

16. Saeki, O.: Topological invariance of weights for weighted homogeneous isolated singularities in C^3. Proc. AMS **103**, 905–909 (1988)

17. Suzuki, M.: Normal forms of quasihomogeneous functions with inner modality equal to five. Proc. Jpn. Acad. **57**, 160–163 (1981)

18. Suzuki, M.: Classification of quasihomogeneous polynomials of corank three with inner modality ≤ 14. Saitama Math. J. **31**, 1–25 (2017)

19. Suzuki, M.: Normal forms of quasihomogeneous functions with inner modality ≤ 9. Proc. Institute of Natural Sciences, Nihon University **55**, 175–218 (2020)

20. Tajima, S., Nakamura, Y.: Annihilating ideals for an algebraic local cohomology class. J. Symb. Comp. **44**, 435–448 (2009)

21. Tajima, S., Nakamura, Y., Nabeshima, K.: Standard bases and algebraic local cohomology for zero dimensional ideals. Adv. Stud. Pure Math. **56**, 341–361 (2009)

22. Varchenko, A.N.: A lower bound for the codimension of the stratum μ constant in terms of the mixed Hodge structure. Vest. Mosk. Univ. Mat. **37**, 29–31 (1982)

23. Yoshinaga, E., Suzuki, M.: Normal forms of non-degenerate quasihomogeneous functions with inner modality ≤ 4. Inventiones Math. **55**, 185–206 (1979)

24. Yoshinaga, E.: Topological principal part of analytic functions. Trans. Amer. Math. Soc. **314**, 803–813 (1989)

Acceleration of Subdivision Root-Finding for Sparse Polynomials

Victor Y. Pan[1,2](✉)

[1] Department of Computer Science, Lehman College of the City University of New York, Bronx, Ny 10468, USA
victor.pan@lehman.cuny.edu
[2] Ph.D. Programs in Mathematics and Computer Science, The Graduate Center of the City University of New York, New York, NY 10036, USA
http://comet.lehman.cuny.edu/vpan/

Abstract. Univariate polynomial root-finding has been studied for four millennia and is still the subject of intensive research. Hundreds of efficient algorithms for this task have been proposed. A recent front-running algorithm relies on subdivision iterations. Already its initial implementation of 2018 has competed for user's choice for root-finding in a region that contains a small number of roots. Recently, we significantly accelerated the basic blocks of these iterations, namely root-counting and exclusion tests. In [18], we solidified this approach and made our acceleration dramatic in the case of sparse polynomials and other ones defined by a black box for their fast evaluation. Our techniques are novel and should be of independent interest. In the present paper and its companion [19], we expose a substantial part of that work.

Keywords: Polynomial root-finding · Subdivision · Sparse polynomials · Exclusion test · Root-counting · Power sums of roots

1 The State of the Art and Our Progress

Univariate polynomial root-finding has been the central problem of mathematics and computational mathematics for four millennia and remains the subject of intensive research motivated by applications in Computer Algebra and various other areas of computing (see pointers to the huge bibliography in [18]).

Subdivision iterations traced back to [5, 14, 20, 22], recently became a leading root-finder due to the progress reported in [2, 9–11]. Our current paper and its companion [19] represent part of a large work [18] on significant acceleration of the previous algorithm of [2]. Already the initial implementation of the new algorithm in [9] shows 3-fold acceleration, but further implementation work should demonstrate substantially stronger progress, including dramatic acceleration in the important case of a *black box polynomial*, represented by a subroutine for its

This research has been supported by NSF Grants CCF–1563942 and CCF–1733834 and PSC CUNY Award 69813 00 48.

© Springer Nature Switzerland AG 2020
F. Boulier et al. (Eds.): CASC 2020, LNCS 12291, pp. 461–477, 2020.
https://doi.org/10.1007/978-3-030-60026-6_27

fast or numerically stable evaluation rather than its coefficients. E.g., black box evaluation is very fast for sparse, Mandelbrot's polynomials defined by recurrence expressions,

$$p_0(x) = 1, \ p_1(x) = x, \ p_{i+1}(x) = xp_i(x)^2 + 1 \text{ for } i = 0, 1, \dots, \quad (1)$$

and various other polynomials, while evaluation is numerically stable for polynomials represented in Bernstein or Chebyshev bases. Furthermore, dealing with black box polynomials, we avoid the costly auxiliary stages of Dandelin–Lobachevsky–Gräffe's recursive root-squaring [7] and Taylor's shift of the variable, which greatly slow down the computations of [2].

We refer the reader to [18] for full exposition of this work, which includes its comparison with other leading root-finders, their acceleration, and Boolean cost estimates, while we occasionally estimate arithmetic complexity where we can control the precision of computing. [18] specifies a number of directions for further progress in subdivision iterations and fully develops some of them. Their implementation should make the algorithm of [18] competitive with MPSolve (Multiprecision Polynomial Solver) of [3,4], which is the package of root-finding subroutines of user's choice since 2000. Actually already the previous implementation of subdivision iterations in [10], based on the algorithm of [2], has slightly outperformed MPSolve for root-finding in a region of the complex plane containing only a small number of roots.

Organization of the Paper. We devote the next short section to background, where, in particular, we briefly cover subdivision iterations. We extensively study their amendments based on the computation of Cauchy sums in a disc on the complex plane in Sect. 3. We devote Sects. 4 and 5 to deterministic and probabilistic support of the application of our Cauchy sum approach to root-counting and exclusion tests for a disc without estimation of the isolation of its boundary circle from the roots.

2   Background

Quite typically in the literature, a polynomial $p = p(x)$ is represented in monomial basis – with its coefficients,

$$p(x) = \sum_{i=0}^{d} p_i x^i = p_d \prod_{j=1}^{d} (x - x_j), \ p_d \neq 0, \quad (2)$$

where we may have $x_k = x_l$ for $k \neq l$, but we allow its representation just by a black box for its evaluation.

We study roots numerically: we count a root of multiplicity m as m simple roots and do not distinguish it from their cluster whose diameter is within a fixed tolerance bound.

We deal with the discs $D(c, \rho)$, annuli $A(c, \rho, \rho')$, and circles $C(c, \rho)$ having complex centers c and positive radii ρ and $\rho' > \rho$.

$\zeta := \zeta_q = \exp(2\pi\sqrt{i}/q)$ denotes a primitive qth root of unity.

Definition 1. *A domain on the complex plain with a center c and its boundary have an* isolation ratio θ *or equivalently are θ-isolated for a polynomial p, real $\theta \geq 1$, and complex c if the root set of p in the domain is invariant in its θ- and $\frac{1}{\theta}$-dilation with the center c. In particular (see Fig. 1), a disc $D(c, \rho)$ and its boundary circle $C(c, \rho)$ are θ-isolated for p, $\theta \geq 1$, and complex c if no roots of p lie in the open annulus $A(c, \rho/\theta, \rho\theta)$. A domain and its boundary are* well-isolated *if they are θ-isolated for $\theta - 1$ exceeding a positive constant.*

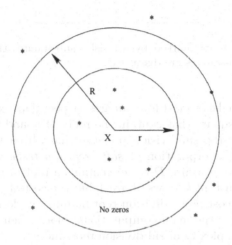

Fig. 1. The internal disc $D(X, r)$ is R/r-isolated

Subdivision iterations extend the bisection iterations from root-finding on a line to polynomial root-finding in the complex plane and under the name of *Quad-tree Construction* have been extensively used in Computational Geometry. Their version of Becker et al. Algorithm of [2] has nearly optimal Boolean complexity, up to polylogarithmic factor in the input size, provided that an input polynomial is represented by its coefficients.[1]

Suppose that we seek all roots of p in a fixed square on the complex plane, which is well isolated from the external roots of p, e.g., contains all d roots; call this square *suspect*. At a low cost, one can readily compute such a square centered at the origin and containing all roots of p (cf., e.g., [18, Sec. 4.8]).

A subdivision iteration divides every suspect square into four congruent sub-squares and to each of them applies *an exclusion test:* a sub-square is discarded if the test proves that it contains no roots of p; otherwise the sub-square is called suspect and is processed at the next iteration (see Fig. 2).

---

[1] It becomes the second such root-finder. The first one, of [13, 16], has provided nearly optimal solution also for numerical factorization of a polynomial into the product of its linear factors, which is a problem of high independent interest [18].

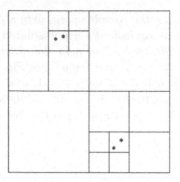

Fig. 2. Four roots of p are marked by asterisks; sub-squares that contain them are suspect; the other sub-squares are discarded

A root of p can make at most four squares suspect if an exclusion test enabled us to discard every square that contains no roots of p, and then we would have $k \leq 4$. Realistically, the subdivision processes have been made less expensive overall by means of incorporation of *soft exclusion tests*, which keep a tested square $S(c, \rho)$ suspect if a disc $D(c, u\rho)$ contains a root of p for some u exceeding $\sqrt{2}$. Then the constant k grows above 4, but the cost of performing exclusion test and the overall cost of subdivision root-finding can decrease.

Exclusion tests are the main computational block; their cost strongly dominates the overall complexity of subdivision root-finding.

At every subdivision iteration, all roots are approximated by the centers of suspect squares, within at most their half-diameter. This bound decreases by twice at every subdivision, and [2,14,20] accelerate such a linear convergence to the roots to superlinear – by using Newton's or QIR iterations, which combine the secant and Newton's iterations. The transition to faster iterations involves root-counting in a disc on a complex plane – the second main computational block of the algorithms of [2,14,20].

Work [2] proposed a novel exclusion test and root-counter by means of Pellet's classical theorem, based on pairwise comparison of the absolute values of the coefficients of p. The authors justly referred to its as the main algorithmic novelty versus [20] and [14], but [18] achieves new significant progress based on novel efficient root-counting and exclusion test for a black box polynomial, not handled by [2].

The new approach relies on the approximation of the power sums of the roots of a polynomial lying in a disc $D(c, \rho)$ on the complex plane

$$s_h = s_h(D(c, \rho)) := \sum_{x_j \in D(c,\rho)} x_j^h = \int_{C(c,\rho)} \frac{p'(x)}{p(x)} x^h \, dx, \quad h = 0, 1, \ldots; \quad (3)$$

the latter representation is valid by virtue of Cauchy integral theorem.

3 Power Sums, Cauchy Sums, Root-Counting, and Exclusion Test

3.1 Cauchy Sum Computation

Let $f := \prod_{j=1}^{d_{in}} (x - x_j)$ where the roots $x_1, \ldots, x_{d_{in}}$ of p lie in a disc $D(c, \rho)$, while all other roots lie outside it; $f = p$ for $d = d_{in}$. We approximate the power sums s_h, $h = 0, 1, \ldots$, of the roots of p in a disc $D(c, \rho)$ with the *Cauchy sums* in that disc, which discretize the contour integral above:

$$s_h^* := \frac{1}{q} \sum_{g=0}^{q-1} \zeta^{(h+1)g} \frac{p'(c + \rho\zeta^g)}{p(c + \rho\zeta^g)} \quad \text{for } h = 0, 1, \ldots, q - 1. \tag{4}$$

Algorithm 1. Computation of Cauchy sum in a disc $D(c, \rho)$.
Fix a positive integer q and assume that $p(c + \rho\zeta^g) \neq 0$ for $g = 0, 1, \ldots, q - 1$. Successively compute the values

(i) $p(c + \rho\zeta^g)$, $p'(c + \rho\zeta^g)$, $\sigma_g = \frac{p'(c+\rho\zeta^g)}{p(c+\rho\zeta^g)}$ for $g = 0, 1, \ldots, q - 1$,
(ii) s_h^* of (4) for $h = 0, 1, \ldots, q - 1$.

$qs_h^* = \sum_{g=0}^{q-1} \sigma_g \zeta^{(h+1)g}$ for $h = q - 1, 1, 2, \ldots, q - 2$ are the values of the polynomial $\sigma(x) = \sum_{g=0}^{q-1} \sigma_g x$ at the qth roots of unity, and so the computation of qs_h^* for $h = q - 1, 1, 2, \ldots, q - 2$ at stage (ii) is precisely the discrete Fourier transform (DFT) on q points, which one can perform fast by using FFT (see, e.g., [15, Section 2.2] or [19, Appendix]). Consequently, at a dominated cost of performing less than $3q \log_2 q$ arithmetic operations one can extend the computation of the Cauchy sum s_0^* to the computation of all Cauchy sums s_h^*, $h = 0, 1, \ldots, q - 1$.

Remark 1. Instead of assuming that $p(c + \rho\zeta^g) \neq 0$ for all g we can ensure these inequalities with probability 1 by applying Algorithm 1 to the polynomial $t(x) = p(a\frac{x-c}{\rho})$ for $a = \exp(\phi\mathbf{i})$ or $a^q = \exp(\phi\mathbf{i})$ and a random scalar $\phi \in [0, 2\pi)$.

The constructive proof of the following theorem supports application of Algorithm 1 to a black box polynomial (see [12] or [1]).

Theorem 1. *Given an algorithm that evaluates at a point x a black box polynomial $p(x)$ over a field $\mathcal{K}$ of constants by using A additions and subtractions, S scalar multiplications (that is, multiplications by elements from the field $\mathcal{K}$), and M other multiplications and divisions, one can extend this algorithm to the evaluation at x of both $p(x)$ and $p'(x)$ by using $2A + M$ additions and subtractions, $2S$ scalar multiplications, and $3M$ other multiplications and divisions.*

3.2 Cauchy Sums: Their Link to the Roots and Approximation of Power Sums

The following basic result is [18, Corollary 4.1].

Theorem 2. *For the roots x_j of $p(x)$ and all h, the Cauchy sums s_h^* of (4) for $c = 0$, $\rho - 1$ satisfy $s_h^* = \sum_{j=1}^{d} \frac{x_j^h}{1-x_j^q}$ unless $x_j^q = 1$ for some j.*

By virtue of this theorem, the Cauchy sum s_h^* is the power sums $s_h = \sum_{j=1}^{d} x_j^h$ with the weights $\frac{1}{1-x_j^q}$ assigned to the terms x_j^h for $j = 1, \ldots, d$. This defines an upper bound on $|s_h - s_h^*|$ that converges to 0 exponentially fast in $q - h$.

Corollary 1 *[21].*[2] *Let the disc $D(0,1)$ be θ-isolated and contain precisely $d_{\rm in}$ roots of p. Write $\eta := 1/\theta$. Then*

$$|s_h^* - s_h| \le \frac{d_{\rm in}\eta^{q+h} + (d - d_{\rm in})\eta^{q-h}}{1 - \eta^q} \text{ for } h = 0, 1, \ldots, q-1, \tag{5}$$

and, in particular,

$$s_h = 0 \text{ and } |s_h^*| \le \frac{d\eta^{q+h}}{1 - \eta^q} \text{ for } h = 0, 1, \ldots, q-1 \text{ if } d_{\rm in} = 0. \tag{6}$$

$$\mu := |s_0^* - s_0| \le \frac{d}{\theta^q - 1}, \text{ and so } \mu < 1/2 \text{ if } q > \frac{\log(2d+1)}{\log(\theta)}, \tag{7}$$

$$\theta \le \left(\frac{\mu + d}{\mu}\right)^{1/q}, \text{ and so } \theta \le (d+1)^{1/q} \text{ if } \mu = |s_0^* - s_0| \ge 1. \tag{8}$$

Corollary 2. *Suppose that Algorithm 1, applied to the unit disc $D(0,1)$ for $q \ge b\log_2(2d+1)$ and $b > 0$, outputs $s_0^* > 1/2$. Then the disc $D(0,\theta)$ contains a root of p for $\theta = 2^{1/b}$.*

3.3 Cauchy Root-Counting, Cauchy Exclusion Test, and Isolation of a Disc

Clearly the 0th Cauchy sum s_0^* can serve as a root-counter in a disc if $\mu = |s_0 - s_0^*| < 0.5$. By virtue of (7), this holds if $q \ge \log_\theta(2d+1)$, e.g., if $\theta = 2$ and $d \le 1,000,000$, and then we can choose any $q \ge 21$.

Remark 2. We can narrow *Cauchy root-counting* to *Cauchy exclusion test* if we only check whether $s_0^* \approx 0$, but we can strengthen this test at a low additional cost by verifying whether $s_h^* \approx 0$ for $h = 0, \ldots, q-1$. Indeed, (i) we can compute the Cauchy sums s_h^* for $h = 0, \ldots, q-1$ at the cost of the computation of s_0^* and performing DFT at q points and (ii) $s_0 = 0$ if and only if $s_h = 0$ for $h = 0, 1, \ldots$.

---

[2] Unlike paper [21], this result is deduced in [18] from Theorem 2, which is also the basis for correctness proof of Sect. 5 for our probabilistic root-counter.

Hereafter we refer to Algorithm 1 restricted to the computation of s_0^* as **Algorithm 1a**. Seeking correct output of a Cauchy root-counter or exclusion test but avoiding unnecessary increase of the parameter q, one can first apply Algorithm 1a for a small integer q and then recursively double it, reusing the results of the previous computations, until the computed values of the Cauchy sum s_0^* stabilize near an integer or just until they approximate an integer closely enough. Later we prove that such an integer is s_0 with a high probability (hereafter *whp*) under random root models.

This result supports root-finding computations in [18, Section 6.4], but not in the subdivision processes of [2,14,20], where root-counting is applied only to well-isolated discs, in which case Algorithm 1a yields non-costly solution s_0 by virtue of Corollary 1. For correctness of our exclusion test, we seek stronger support because in the subdivision iterations of [2,14,20], such a test is applied to the discs for whose isolation ratios no estimates are known. By virtue of Corollary 2, Algorithm 1a applied to such a disc certifies that its controlled dilation contains a root of p unless the algorithm outputs s_0^* close to 0. The following algorithm completes an exclusion test in the latter case.

Algorithm 2. Completion of a Cauchy soft exclusion test.
Input: A black box polynomial $p(x)$ of degree d such that Algorithm 1a, applied to the disc $D(0,2)$ and[3] the polynomial $p(x)$ has output s_0^* close to 0.
Output: Certification that (i) the disc $D(0,2)$ contains a root of p definitely if $q > d$ or whp otherwise or (ii) the unit disc $D(0,1)$ definitely contains no roots of p, where cases (i) and (ii) are compatible.
Initialization: Choose an integer q such that

$$q_0 < q \le 2q_0 \text{ for } q_0 \ge \max\left\{1, \ \log_2\left(\frac{d}{q_0 \alpha_d \sqrt{3}}\right)\right\} \text{ and } \alpha_d = \sqrt{d + \sqrt{d}}. \quad (9)$$

Computations: Apply Algorithm 1 to the unit disc $D(0,1)$ for the selected q. Hereafter

$$\mathbf{s}_* := (s_{q-1}^*, s_0^*, s_1^*, \dots, s_{q-2}^*)^T \quad (10)$$

denotes the vector of the values s_h^* of the Cauchy sums output by the algorithm and $\|\mathbf{s}_*\|$ denotes the Euclidean norm $(\sum_{h=0}^{q-1} |s_h^*|^2)^{1/2}$. If $\|\mathbf{s}_*\| \, q_0 \, \alpha_d \ge 1$, conclude that the disc $D(0,\theta)$ definitely contains a root of p. Otherwise conclude that the disc $D(0,1)$ contains no roots of p definitely if $q > d$ or whp otherwise.

Work [18] as well as [19] readily prove that the disc $D(0,2)$ contains a root of p if $\|\mathbf{s}_*\| \, q_0 \, \alpha_d \ge 1$, and this reduces *correctness proof* of the algorithm to certification that the disc $D(0,1)$ contains no roots of p if the vector of the q Cauchy sums $\mathbf{s}_*$ has Euclidean norm satisfying $\|\mathbf{s}_*\| \, q_0 \, \alpha_d < 1$. This certification, deterministic for $q > d$ and probabilistic for $2 \le q \le d$, is the subject of the next two sections.

---

[3] One can extend the algorithm by applying Algorithm 1a to a disc $D(0,\theta)$ for smaller $\theta > 1$ and modifying bound (9) accordingly. See a refined version of this algorithm in [18].

Remark 3. For the computation of Cauchy sums for q of order of d, we should evaluate p and p' at order of d points; by applying our reduction of multi-point polynomial evaluation (MPE) to fast multipole method (FMM) (see [18, Appendix E]), we can do this by using order of $d \log^2(d)$ arithmetic operations, performed numerically with the precision of order $\log(d)$ bits. It outputs the vector of the first q Cauchy sums $s_0, \ldots, s_{q-1}$ within a relative error of order $\log(d)$. This should be sufficient in order to verify the bounds of Algorithm 2, [18, Theorem 19 and Corollaries 4.2 and 4.3] because FMM is celebrated for being very stable numerically, although further formal and experimental study is in order.

The algorithm runs faster as we decrease integer q, and already under the choice of q_0 of order of $\log(d)$ the disc $D(0, 2)$ contains a root of p if $\|\mathbf{s}_*\| \, q_0 \, \alpha_d \geq 1$. For $q \leq d$, we have only probabilistic support of correctness of Algorithm 2 in the case where $\|\mathbf{s}_*\| \, q_0 \, \alpha_d < 1$, but we can try to strengthen reliability of our exclusion tests by verifying additional necessary conditions for correctness of our exclusion test and root-counting:

(a) the Cauchy sums s_h^* for $h = 0, 1, \ldots, q-1$ still nearly vanish for the polynomials $t(x)$ obtained from $p(x)$ by means of various mappings of the variable x that keep an input disc and the power sum s_0 invariant (cf. Remark 1);
(b) an exclusion test should succeed for any disc lying in the disc $D(c, \rho)$. In particular, if the disc covers a suspect square, then exclusion tests should succeed for the four discs that cover the four congruent sub-squares obtained from sub-dividing the input square;
(c) all suspect squares of a subdivision iteration together contain precisely d roots of p.

If these additional necessary conditions hold, it is still plausible that the disc $D(c, \rho)$ contains a root of p. We can, however, detect whether we have lost any root at the end of the subdivision process, when $d - w$ roots are *tamed*, that is, closely approximated, and when w roots remain at large; we call the latter roots *wild*. If $0 < w \ll d$, then at a low cost, we can deflate the *wild factor* of p, whose root set is made up of the w wild roots; then we can approximate the roots of this factor at a low cost (see [18, Section 7]).

It is natural to call a point c a tame root of p if $r_d(c, p) \leq \text{TOL}$ for a fixed tolerance TOL. The algorithm of [18, Section 6.2] closely approximates $r_d(c, p)$ at a relatively low cost, but it is even less expensive to verify whether $d \, |p'(c)/p(c)| \leq \text{TOL}$ and then to recall that $r_d(c, p) \leq d \, |p'(c)/p(c)|$ (see [5, Theorem 6.4g]).

Empirical support from the initial implementation and testing of our algorithms in [9] has substantially superseded their formal support here and in [18]. In these tests, subdivision iterations with Cauchy exclusion tests by means of Algorithm 1a have consistently approximated the integer $s_0 = 0$ within $1/4$ for $q = \lceil \log(4d + 1)/\log(4\theta) \rceil$. For discs containing no roots and for $q = \lceil \log(4d + 1)/\log(4\theta) \rceil + 1$, Algorithm 1 has consistently approximated both s_0 and s_1 within $1/4$ (cf. [9, equation(22) in Corollary 12 for $e = 1/4$]).

4   Correctness Certification of an Exclusion Test

In this section, we complete the correctness certification of the exclusion test by means of Algorithm 2.

4.1   Deterministic Certification

We begin with a lemma that implies q linear equations on the coefficients of the polynomial p provided that $s_h^* = 0$ for $h = 0, 1, \ldots, q - 1$.

Lemma 1. *Suppose that $s_h^* = 0$ for a polynomial $p(x)$, s_h^* of (4), $c = 0$, $\rho = 1$, $h = 0, 1, \ldots, q - 1$, and a positive q. Then the polynomial $p'(x)$ is divided by $x^q - 1$.*

Proof. Observe that

$$\mathbf{s}_* = F\mathbf{v} \text{ for } \mathbf{v} = \left(\frac{p'(\zeta^g)}{qp(\zeta^g)}\right)_{g=0}^{q-1}, \ \zeta = \exp\left(\frac{2\pi\sqrt{\mathbf{i}}}{q}\right), \ F := (\zeta^{ij})_{i,j=0}^{q-1} \qquad (11)$$

denoting the matrix of DFT at q points, and $\mathbf{s}_* := (s_{q-1}^*, s_0^*, \ldots, s_{q-2}^*)^T$ of (10).

Under the assumptions of the lemma $\mathbf{s}_* = F\mathbf{v} = \mathbf{0}$ for the vector $\mathbf{0}$ of length q filled with 0s. Pre-multiply this vector equation by the matrix $\frac{1}{q}F^*$ of the inverse DFT at q points and obtain that $\frac{p'(\zeta^g)}{qp(\zeta^g)} = 0$ for all g. Hence $p'(\zeta^g) = 0$, for $g = 0, 1, \ldots, q - 1$, and, therefore, $x^q - 1$ divides $p'(x)$. $\quad\square$

Corollary 3. *Under the assumptions of Lemma 1 let $q \geq d$. Then the polynomial $p'(x)$ is identically 0, and so the polynomial $p(x)$ is a constant and has no roots unless it is identically 0.*

Next we assume that $q > d$ and extend the corollary under much weaker assumption that $||\mathbf{s}_*|| < \frac{1}{q\alpha_d}$ rather than $\mathbf{s}_* = \mathbf{0}$. Recall that $||\mathbf{v}|| = (\sum_{i=1}^k |v_i|^2)^{1/2}$ denotes the Euclidean norm of a vector $\mathbf{v} = (v_i)_{i=1}^k v_i$.

Theorem 3. *Given a polynomial $p(x)$ of (2) and a positive integer q, write*

$$\widehat{p}(x) := p(x) \mod (x^q - 1), \ \widehat{p}'(x) := p'(x) \mod (x^q - 1), \qquad (12)$$

and $\widehat{p}_0 := \widehat{p}(0)$, let $\widehat{\mathbf{p}}'$, $\widehat{\mathbf{p}}$, and $\widehat{\mathbf{p}}_1$ denote the coefficient vectors of the polynomials $\widehat{p}'(x)$, $\widehat{p}(x)$, and $\widehat{p}(x) - \widehat{p}_0$, respectively, and suppose that

$$||\mathbf{s}_*|| = ||F\mathbf{v}|| \leq \tau, \qquad (13)$$

for $\mathbf{s}_* = F\mathbf{v}$, F and $\mathbf{v}$ of (11), and a positive tolerance τ. Then

$$|\widehat{p}_0|^2 \geq |(\tau q)^{-2} - 1| \ ||\widehat{\mathbf{p}}_1||^2. \qquad (14)$$

Proof. Multiply equation (13) by the matrix $\frac{1}{q}F^*$ of the inverse DFT and obtain

$$\frac{1}{q}F^*\mathbf{s}_* = \mathbf{v}.$$

Hence

$$\|\mathbf{v}\| \leq \frac{1}{q}\|F^*\|\,\|\mathbf{s}_*\| \leq \frac{\tau}{\sqrt{q}}.$$

Therefore,

$$\left\|\left(\frac{p'(\zeta^g)}{qp(\zeta^g)}\right)_{g=0}^{q-1}\right\| \leq \frac{\tau}{\sqrt{q}},$$

and consequently

$$\|(p'(\zeta^g))_{g=0}^{q-1}\| \leq \tau\sqrt{q}\ \max_{g=0}^{q-1}|p(\zeta^g)| \leq \tau q\ \|(p(\zeta^g))_{g=0}^{q-1}\|.$$

Substitute the equations $p(\zeta^g) = \widehat{p}(\zeta^g)$ and $p'(\zeta^g) = \widehat{p}'(\zeta^g)$ in the above and obtain

$$\|(\widehat{p}'(\zeta^g))_{g=0}^{q-1}\| \leq \tau q\ \|(\widehat{p}(\zeta^g))_{g=0}^{q-1}\| \tag{15}$$

for the polynomials $\widehat{p}'(x)$ and $\widehat{p}(x)$ of (12) with the coefficient vectors $\widehat{\mathbf{p}}'$ and $\widehat{\mathbf{p}}$, respectively. Observe that

$$(\widehat{p}'(\zeta^g))_{g=0}^{q-1} = F\widehat{\mathbf{p}}' \text{ and } (\widehat{p}(\zeta^g))_{g=0}^{q-1} = F\widehat{\mathbf{p}}$$

for the DFT matrix $F = (\zeta^{ij})_{i,j=0}^{q-1}$ of (11).

Substitute these expressions into bound (15) and obtain $\|F\widehat{\mathbf{p}}'\| \leq \tau q\ \|F\widehat{\mathbf{p}}\|$. Hence

$$\|\widehat{\mathbf{p}}'\| \leq \tau q\ \|\widehat{\mathbf{p}}\| \tag{16}$$

because F is a unitary matrix up to scaling by $\sqrt{q}$.

Furthermore observe that

$$\|\widehat{\mathbf{p}}\|^2 = |\widehat{p}_0|^2 + \|\widehat{\mathbf{p}}_1\|^2 \text{ and } \|\widehat{\mathbf{p}}'\|^2 \geq \|\widehat{\mathbf{p}_1}\|^2$$

for $\widehat{p}_0 = \widehat{p}(0)$ and the vector $\widehat{\mathbf{p}}_1$ of the coefficients of $\widehat{p}(x) - \widehat{p}_0$.

Combine these observations with bound (16) and obtain the theorem.

Corollary 4. *Under the assumptions of Theorem 3 let*

$$q > d \text{ and } (1 + \sqrt{d})\tau^2 q^2 \sqrt{d} < 1.$$

Then the polynomial $p(x)$ has no roots in the unit disc $D(0, 1) = \{x : |x| \leq 1\}$.

Proof. The bound $(1 + \sqrt{d})\tau^2 q^2 < 1$ implies that $(\tau q)^{-2} - 1 > \sqrt{d}$, while the bound $q > d$ implies that $\widehat{p}(x) = p(x)$ and $\widehat{p}_0 = p(0) = p_0$. Hence Theorem 3 implies that $|p_0|^2 > \sqrt{d}\sum_{i=1}^{d}|p_i|^2$ and so $|p_0| > |p| = \sum_{i=1}^{d}|p_i|$. The latter bound is impossible if $p(x) = 0$ for $|x| \leq 1$.

Remark 4. We proved this corollary assuming that Cauchy exclusion test by means of Algorithm 2 has been applied to the roots of a polynomial $p(x)$ in the unit disc $D(0,1)$, but our proof supports such a test for the roots of a polynomial $t(x) = p((x - c)/\rho)$ in that disc for any complex c and positive ρ. Hence the corollary also holds if the test is applied to the roots of a polynomial $p(x)$ in a disc $D(c, \rho)$.

4.2 Probabilistic Certification Under Random Coefficient Model

Our next extensions of the result to the case $2 \leq q \leq d$ are probabilistic under a random coefficient model. We only state asymptotic probability estimates in the case where $\tau \to 0$, but specific estimates for fixed bounds on τ are implicit in the proofs.

Corollary 5. *Define a* Random Coefficient Model *such that the coefficients p_0, $p_1, \ldots, p_d$ of p are independent Gaussian random variables having expected values a_i and positive variance σ_i^2, for $i = 0, 1, \ldots, d$. Suppose that the Cauchy exclusion test by means of Algorithm 2 has been applied to the unit disc $D(0,1)$ and a polynomial $p(x)$ under this model and let $2 \leq q \leq d$. Then the bound of Theorem 3 holds with a probability that fast converges to 0 as $\tau \to 0$.*

Proof. Let $d = (k - 1)q + l$ for $k \geq 0$ and $0 \leq l \leq q - 1$ and write

$$\widehat{p}_i := \sum_{j=0}^{k_i} p_{jq+i}, \ i = 0, 1, \ldots, q - 1, \ k_i = k \text{ for } i < l, \ k_i = k - 1 \text{ for } i \geq l.$$

Then

$$\widehat{\mathbf{p}} = (\widehat{p}_i)_{i=0}^{q-1}, \ \widehat{\mathbf{p}}_1 = (\widehat{p}_i)_{i=1}^{q-1} \tag{17}$$

where $\widehat{p}_i$ for all i are independent Gaussian variables with expected values $\widehat{a}_i$ and positive variance values $\widehat{\sigma}_i^2$ given by

$$\widehat{a}_i = \sum_{j=0}^{k_i} a_{jq+i} \text{ and } \widehat{\sigma}_i^2 = \sum_{j=0}^{k_i} \sigma_{jq+i}^2, \ i = 0, 1, \ldots, q - 1. \tag{18}$$

Such variables are strongly concentrated about their expected values. Bound (14) of Theorem 3 implies that

$$|\widehat{a}_0| = |\sum_{j=0}^{k_0} a_{jq}| \geq |(\tau q)^{-2} - 1| \max_{i=1}^{q-1} |\widehat{a}_i|$$

for $\widehat{a}_i$ of (18).[4] Since $\widehat{p}_0, \ldots, \widehat{p}_{q-1}$ are independent Gaussian random variables, which are strongly concentrated about their expected values $\widehat{a}_i$, this inequality strongly restricts the class of polynomials $p(x)$ satisfying (14) for small τ and $q \geq 2$; furthermore, the probability that this inequality and bound (14) hold converges to 0 exponentially fast as $\tau \to 0$.

---

[4] One can slightly strengthen our estimates based on the observation that $|\widehat{p}_0|^2$ and $\|\widehat{\mathbf{p}}_1\|^2$ are χ^2-functions of dimension 1 and $q - 1$, respectively.

Now suppose that the Cauchy exclusion test has been applied to the disc $D(c, \rho)$ for a complex c and a positive ρ and restate the above argument for the polynomial $t(x) := \sum_{j=0}^{d} t_j x^j := p(\frac{x-c}{\rho})$ where (cf. [15, Problem 2.4.3])

$$t_j = \sum_{i=j}^{d} p_i \cdot (-c)^{i-j} \rho^{-i} \binom{i}{j}, \quad j = 0, 1, \ldots, d, \tag{19}$$

and so the partial sums $\widehat{t}_j$, $j = 0, 1, \ldots, q-1$, are still independent Gaussian variables strongly concentrated about their expected values $\mathbb{E}(\widehat{t}_j)$. Equations (19) imply that

$$\mathbb{E}(\widehat{t}_j) = \sum_{i=j}^{d} a_i \cdot (-c)^{i-j} \rho^{-i} \binom{i}{j}, \quad j = 0, 1, \ldots, d. \tag{20}$$

Expressions $\widehat{\mathbf{t}} = (\widehat{t}_i)_{i=0}^{q-1}$, $\widehat{\mathbf{t}}_1 = (\widehat{t}_i)_{i=1}^{q-1}$ replace (17), and bound (14) restricts the classes of polynomials $t(x)$ and consequently $p(x)$.

Now observe that $\lim_{\rho \to \infty} \widehat{t}_0 = t_0$, where t_0 does not depend on ρ. Furthermore, $t_j \to 0$ as $\rho \to \infty$ for $j \neq 0$, and so $\lim_{\rho \to \infty} \widehat{\mathbf{t}}_1 = \mathbf{0}$. Therefore, the value $\mathbb{E}|\widehat{t}_0|^2$ strongly dominates the value $\mathbb{E}||\widehat{\mathbf{t}}_1||^2$, and this strongly restricts the the classes of polynomials $t(x)$ and consequently $p(x)$.

Next assume that the ratio $|p_0| = |t_0| / \max_{i=1}^{d} |t_i|$ is not large and then argue that bound (14) strongly restricts the classes of polynomials $t(x)$ and $p(x)$. Indeed rewrite equation (20) as follows:

$$\mathbb{E}(t_j) = \sum_{i=j}^{d} u_i (-c)^{-j} \binom{i}{j}, \quad \text{for } u_i = a_i (-c)^i \rho^{-i}, \ i, j = 0, 1, \ldots, d.$$

Now observe that $L_0 = \mathbb{E}(\widehat{t}_0 - t_0)$ and $L_1 = \mathbb{E}(\widehat{t}_1)$ are linear combinations in the same variables u_i, for $i = 0, 1, \ldots, d$, whose coefficients are polynomials in $-1/c$ with positive integer coefficients. Furthermore, such polynomials in L_0 and L_1 consist of the terms $(-c)^j \binom{i}{j}$, which make up pairs of terms, such that one term of every pair is in L_0, another is in L_1, and dc exceeds the ratio of these terms in every pair. It follows that bound (14) for $|(\tau q)^{-2} - 1| \gg dc$ strongly restricts the classes of polynomials $p(x)$ and $t(x)$. Then again this follows because of the strong concentration of a Gaussian variable about its expected value.

5   Cauchy Root-Counting Under Two Random Root Models

5.1   Error Estimates

Random root models are less popular in the study of root-finding than random coefficient models but still enable some insight into this subject: indeed, if a property holds whp under a random root model, then it must hold for a large input class if not for most of inputs.

Definition 2. *Under* Random Root Models 1 and 2, *the roots of p are iid random variables sampled under the uniform probability distribution from some fixed regions* $\mathbb{D}$ *on the complex plain.*

1. $\mathbb{D}$ *is the disc* $D(0, R)$ *for a fixed positive* R *in Random Root Model 1.*
2. $\mathbb{D}$ *is the union of two distinct domains in Random Root Models 2: for a fixed* $R > 1$ *and a fixed nonnegative integer* $k \le d$, *the roots* $x_{k+1}, \ldots, x_d$ *of p are sampled from a disc* $D(0, R)$, *but the roots* $\{x_1, \ldots, x_k\}$ *are sampled from a fixed narrow annulus* $A(0, 1/\theta, \theta)$ *about the boundary circle* $C(0, 1)$, *for a reasonably small positive* $\theta - 1$.

The following readily verified theorem implies that the roots of p lie on or near any fixed circle on the complex plain with a low probability (wlp) under Random Root Model 1.

Theorem 4. *For a polynomial* $p = p(x)$, $\theta > 1$, *a complex* c, *and positive* ρ *and* R *such that* $R > \rho + |c|$ *and* $\rho\sqrt{d} = O(R)$, *assume Random Root Model 1.*

(i) Then for any fixed integer j *in the range* $[0, d]$, *the root* x_j *lies in the annulus* $A(c, \rho/\theta, \rho\theta)$ *with the probability* $P_{R,\rho,\theta} = \frac{(\theta^4 - 1)\rho^2}{R^2\theta^2}$.

(ii) The probability that at least one root of p lies in the annulus $A(c, \rho/\theta, \rho\theta)$ *is at most* $P_{R,\rho,\theta}d = \frac{(\theta^4 - 1)\rho^2 d}{R^2\theta^2}$.

Proof. Recall that the probability $P_{R,\rho,\theta}$ is the ratio of the areas $(\theta^4 - 1)\rho^2/\theta^2$ and πR^2 of the annulus $A(c, \rho/\theta, \rho\theta)$ and the disc $D(0, R)$, respectively, and obtain claim (i). Immediately extend it to claim (ii).

Notice that the bound $P_{R,\rho,\theta}d$ converges to 0 as $\frac{\rho}{R}\sqrt{(\theta - 1)d} \to 0$, where the ratio ρ/R never exceeds 1 and decreases by twice at every subdivision iteration and hence by a factor of d in $\lceil \log_2(d) \rceil$ iterations.

Combine claim (ii) of the theorem with bound (7) and conclude that the Cauchy sum s_0^* approximates the power sum s_0 within less than $1/2$ unless some roots of $p(x)$ lie on or very close to the boundary circle $C(c, \rho)$, and the latter property of the roots holds wlp under Random Root Model 1.

5.2 Probabilistic Correctness Verification

The latter claim is no longer valid under Random Root Model 2 because of the impact of the roots lying on or near boundary circle the Cauchy sum s_0^* can be misleading, that is close to a wrong integer, distinct from s_0. We are going to prove, however, that this can only occur wlp. We begin with an alternative proof that the value $|s_0^* - s_0|$ is small whp under Random Root Model 1 and then readily extend it to proving similar results under Random Root Model 2.

We will only deduce that

$$|s_0^* - u| \le 0.1v \tag{21}$$

wlp P for a fixed complex number u and a fixed positive v under Random Root Models 1 and 2. This will immediately imply that $|s_0^* - i| \le 0.1v$ for $i \in \{0, 1, \ldots, d\}$ with a probability at most $(d+1)P$. Therefore, if (21) holds, then $i = s_0$ with a probability at least $1 - (d+1)P$ under Random Root Models 1 and 2.

Lemma 2. *Write*

$$y := \frac{1}{1 - x^q}, \quad \tilde{y} := \frac{1}{1 - \tilde{x}^q}, \quad \text{and } \delta = |\tilde{y} - y| \tag{22}$$

where $|y| \geq v > 10\delta > 0$. *Then*

$$|\tilde{x} - x| \leq \nabla \text{ for } \nabla := \frac{\delta}{q} \cdot \left(1 + \frac{1}{0.81v^2}\right)^{(1-q)/q}. \tag{23}$$

Proof. Equation (22) implies that

$$x = \left(1 - \frac{1}{y}\right)^{1/q}, \quad \tilde{x} = \left(1 - \frac{1}{\tilde{y}}\right)^{1/q}, \quad \text{and } \tilde{x} - x = \left(1 - \frac{1}{\tilde{y}}\right)^{1/q} - \left(1 - \frac{1}{y}\right)^{1/q}.$$

Apply Taylor–Lagrange formula to the function $x(y) = \left(1 - \frac{1}{y}\right)^{1/q}$ and obtain

$$\tilde{x} - x = (\tilde{y} - y)\frac{d}{dy}\left(\left(1 - \frac{1}{y}\right)^{1/q}\right) = \frac{\tilde{y} - y}{q}\left(1 + \frac{1}{(y + \xi)^2}\right)^{(1-q)/q}$$

for $\xi \in [0, \tilde{y} - y]$. Substitute $\delta = |\tilde{y} - y|$ and $|y + \xi| \geq 0.9|y| \geq 0.9v$.

Theorem 5. *For* $R > 1$ *and any fixed complex number* u, *write* $s := \sum_{j=2}^{d} \frac{1}{1 - x_j^q} = s_0^* - \frac{1}{1 - x_1^q}$, $v := \frac{1}{R^q - 1}$, *and* $\delta := |s_0^* - u|$ *for the Cauchy sum* s_0^* *in the unit disc* $D(0, 1)$ *and assume Random Root Model 1. Then* $\delta \leq 0.1v$ *with a probability at most*

$$P = \frac{4\nabla}{R} \cdot \left(1 - \frac{\nabla}{R}\right) \text{ for } \nabla \text{ of (23)}. \tag{24}$$

In particular, bound (24) is close to

$$\frac{4\nabla}{R} \approx \frac{4\delta}{Rq} \tag{25}$$

if v *is a small positive number, in which case* $\nabla \approx \delta/q$.

Proof. Keep the root x_1 random, but fix the other roots $x_2, \ldots, x_d$. Apply Lemma 2 for $x = x_1$, $y = \frac{1}{1 - x_1^q}$, $\tilde{y} = u - s$, and $\tilde{x}$ such that $\tilde{y} = \frac{1}{1 - \tilde{x}^q}$. Notice that in this case, $|y| \geq v$ because $|x_1| \leq R$, and so the assumptions of the lemma are fulfilled.

The lemma implies that $|x_1 - \tilde{x}| \leq \nabla$ for ∇ of (23). Hence x_1 lies in the annulus

$$A(0, |\tilde{x}| - \nabla, |\tilde{x}| + \nabla) = \{z : |\tilde{x}| - \nabla \leq z \leq |\tilde{x}| + \nabla\}, \tag{26}$$

whose area is maximized for $\tilde{x} = R - \nabla$ and then reaches

$$\pi \cdot (R^2 - (R - 2\nabla)^2) = 4\pi \cdot (R - \nabla)\nabla.$$

Divide this by the area πR^2 of the disc $D(0, R)$ and obtain bound (24).

We have proved Theorem 5 assuming that the Cauchy sum s_0^* is computed for the unit disc $D(0,1)$. We can extend this study to any disc $D(c,\rho)$ such that $R + |c| > \rho$ by replacing an input polynomial $p(x)$ with the polynomial $t(x) = p(\frac{x-c}{\rho})$. We can apply the same proof if we replace the disc $D(0,R)$ with $D(c, \frac{R-|c|}{\rho})$. This would imply that in the statement of the theorem, R changes into R/ρ and $v := \frac{1}{R^q-1}$ changes into

$$v := \frac{1}{R(c,\rho)^q - 1} \text{ for } R(c,\rho) = \frac{R+|c|}{\rho} > 1. \tag{27}$$

Here is the resulting extended version of this theorem.

Theorem 6. *Let the assumptions of Theorem 5 hold except that now the Cauchy sum s_0^* has been computed in a sub-disc $D(c,\rho)$ of the disc $D(0,R)$ for a complex c and a positive ρ. Then $\delta := |s_0^* - u| \le 0.1v$ with a probability at most*

$$P = \frac{4\nabla \rho}{R} \cdot \left(1 - \frac{\nabla}{R}\right) \text{ for } \nabla \text{ of (23)}. \tag{28}$$

We can readily extend the estimates of Theorems 5 and 6 to the case where the roots are sampled under Random Root Model 2 because under that model, we can prove that the contribution of the roots $x_1, \ldots, x_k$ to the Cauchy sum s_0^* moves it close to an integer distinct from s_0 wlp if $\nabla = o(\theta - 1)$. Indeed repeat the proof of Theorem 5 but replace R by θ while estimating the area of annulus (26). Then divide this area by the area $\pi \cdot (\theta^2 - \frac{1}{\theta^2})$ of the annulus $A(0, \frac{1}{\theta}, \theta)$ and thus extend Theorem 5. Similarly extend Theorem 6.

Theorem 7. *Under Random Root Model 2, write $v := \frac{1}{\theta^q-1}$ and fix ∇ of (23) and a complex number u. Then $|s_0^* - u| \le 0.1v$ with a probability at most*

$$P = \frac{4\nabla \cdot (\rho\theta - \nabla)}{\rho^2 \cdot (\theta^2 - 1/\theta^2)}, \text{ and in particular } \frac{4\nabla \cdot (\theta - \nabla)}{\theta^2 - 1/\theta^2} \text{ for } \rho = 1. \tag{29}$$

In the rest of this subsection, we soften the assumption that y can be required to be as small as $v := \frac{1}{R^q-1}$ under Random Root Model 1 (cf. (27)). Under Random Root Model 2, (27) turns into quite a reasonable bound $v := \frac{1}{\theta^q-1}$, and we do not need to soften it.

Theorem 8. *Let the assumptions of Theorem 5 hold except that we can choose any positive value v. Then $\delta := |s_0^* - u| \le 0.1v$ for the Cauchy sum s_0^* with a probability at most $P_0 + P_1$ for ∇ of (23),*

$$P_0 = \left(\frac{r}{R}\right)^{2d} \le \left(\frac{1}{R}\right)^{2d}, \quad P_1 = \frac{4r\nabla}{R^2 - r^2} \le \frac{4\nabla}{R^2 - 1}, \text{ and } r = \left|\frac{v-1}{v}\right|^{1/q}. \tag{30}$$

Proof. P_0 bounds the probability that all d independent random variables $x_1, \ldots, x_d$ lie in the disc $D(0,r)$, whose area is πr^2, while the area of the disc $D(0,R)$ is πR^2.

Hence x_j lies outside the disc $D(0, r)$ for at least one j, say, $j = 1$ with a probability at least $1 - P_0$. In this case, $|y_1| \geq v$, and then Theorem 5 narrows the range for x_1 from the annulus $A(0, r, R) = D(0, R) - D(0, r)$ to the annulus $A(0, r - \nabla, r + \nabla) = D(0, r + \nabla) - D(0, r - \nabla)$ for ∇ of (23). Obtain the bound P_1 as the ratio of the areas of these two annuli.

We extend Theorem 6 similarly.

Theorem 9. *Let the assumptions of Theorem 6 hold except that we can choose any positive value v. Then $\delta := |s_0^* - u| \leq 0.1v$ for the Cauchy sum s_0^* with a probability at most $P_0 + P_1$ for*

$$P_0 = \left(\frac{r\rho}{R}\right)^{2d}, \quad P_1 = \frac{4r\rho\nabla}{R^2 - (r\rho)^2}, \quad r = \left|\frac{v-1}{v}\right|^{1/q}, \quad \text{and } \nabla \text{ of (23)}. \tag{31}$$

Remark 5 [Optimization of the probability bound.] For a fixed pair of a disc $D(c, r/\rho)$ and an integer q, Theorem 9 bounds the probability $P_0 + P_1$ of having $|s^* - u| \leq 0.1v$ as a function of a single parameter v in the range $[0, \frac{1}{R(c,\rho)^q - 1}]$ for $R(c, \rho) = \frac{R + |c|}{\rho}$. We leave to the reader the challenge of the choice of this parameter that would minimize our bound on $P_0 + P_1$ or a similar bound under the inequalities $|s_0^* - u| \leq \beta u$ for any reasonable choice of a small positive β.

Acknowledgments. This research has been supported by NSF Grants CCF–1563942 and CCF–1733834 and by PSC CUNY Award 69813 00 48.

References

1. Baur, W., Strassen, V.: On the complexity of partial derivatives. Theor. Comput. Sci. **22**, 317–330 (1983)
2. Becker, R., Sagraloff, M., Sharma, V., Yap, C.: A near-optimal subdivision algorithm for complex root isolation based on the Pellet test and Newton iteration. J. Symbolic Comput. 86, 51–96 (2018) Procs. version, In: ISSAC 2016, pp. 71–78. ACM Press, New York (2016) 10.1145/2930889.2930939 https://doi.org/10.1016/j.jsc.2017.03.009
3. Bini, D.A., Fiorentino, G.: Design, analysis, and implementation of a multiprecision polynomial rootfinder. Numer. Algorithms **23**, 127–173 (2000). https://doi.org/10.1023/A:1019199917103
4. Bini, D.A., Robol, L.: Solving secular and polynomial equations: a multiprecision algorithm. J. Comput. Appl. Math. **272**, 276–292 (2014). https://doi.org/10.1016/j.cam.2013.04.037
5. Henrici, P.: Applied and Computational Complex Analysis, vol. 1 : Power Series, Integration, Conformal Mapping, Location of Zeros. Wiley, New York (1974)
6. Henrici, P., Gargantini, I.: Uniformly convergent algorithms for the simultaneous approximation of all zeros of a polynomial. In: Dejon, B., Henrici, P. (eds.) Constructive Aspects of the Fundamental Theorem of Algebra. Wiley, New York (1969)
7. Householder, A.S.: Dandelin, Lobachevskii, or Graeffe? Amer. Math. Mon. **66**, 464–466 (1959). https://doi.org/10.2307/2310626

8. Imbach, R., Pan, V.Y.: Polynomial root clustering and explicit deflation. In: D. Salmanig, D. et al. (eds.) Mathematical Aspects of Computer and Information Sciences (MACIS 2019), LNCS, vol. 11989, pp. 1–16. Springer Nature Switzerland AG 2020, Chapter No: 11, Chapter arXiv:1906.04920 (11 Jun 2019)) https://doi.org/10.1007/978-3-030-43120-4_11

9. Imbach, R., Pan, V.Y.: New progress in univariate polynomial root-finding. In: Proceedings of ACM-SIGSAM ISSAC '20, pp. 249–256, July 20–23, 2020, Kalamata, Greece, ACM Press, New York (2020). ACM ISBN 978-1-4503-7100-1/20/07 https://doi.org/10.1145/3373207.3403979

10. Imbach, R., Pan, V.Y., Yap, C.: Implementation of a near-optimal complex root clustering algorithm. In: Davenport, J.H., Kauers, M., Labahn, G., Urban, J. (eds.) ICMS 2018. LNCS, vol. 10931, pp. 235–244. Springer, Cham (2018). https://doi.org/10.1007/978-3-319-96418-8_28

11. Kobel, A., Rouillier, F., Sagraloff, M.: Computing real roots of real polynomials ... and now for real! In: Internernational Symposium Symbolic Algebraic Computation (ISSAC 2016), pp. 301–310. ACM Press, New York (2016) https://doi.org/10.1145/2930889.2930937

12. Linnainmaa, S.: Taylor Expansion of the Accumulated Rounding Errors. BIT **16**, 146–160 (1976)

13. Pan, V.Y.: Optimal (up to polylog factors) sequential and parallel algorithms for approximating complex polynomial zeros. In: Proceeding 27th Annual ACM Symposium on Theory of Computing (STOC 1995), pp. 741–750. ACM Press, New York (1995) https://doi.org/10.1145/225058.225292

14. Pan, V.Y.: Approximation of complex polynomial zeros: modified Quad-tree (Weyl's) construction and improved Newton's iteration. J. Complex. **16**(1), 213–264 (2000). https://doi.org/10.1006/jcom.1999

15. Pan, V.Y.: Structured Matrices and Polynomials: Unified Superfast Algorithms. Birkhäuser/Springer, Boston/New York (2001) https://doi.org/10.1007/978-1-4612-0129-8

16. Pan, V.Y.: Univariate polynomials: nearly optimal algorithms for factorization and rootfinding. J. Symb. Comput. **33**(5), 701–733 (2002). https://doi.org/10.1006/jsco.2002.0531

17. Pan, V.Y.: Old and new nearly optimal polynomial root-finders. In: England, M., Koepf, W., Sadykov, T.M., Seiler, W.M., Vorozhtsov, E.V. (eds.) CASC 2019. LNCS, vol. 11661, pp. 393–411. Springer, Cham (2019). https://doi.org/10.1007/978-3-030-26831-2_26

18. Pan, V.Y.: New Acceleration of Nearly Optimal Univariate Polynomial Rootfinders May 2020. arXiv:1805.12042

19. Pan, V.Y., Kim, W., Luan, Q., Zaderman, V.: Faster Numerical Univariate Polynomial Root-finding by means of subdivision iterations. To appear. In: Boulier, F., England, M., T.M. Sadykov, T.M., Vorozhtsov, E.V. (eds.) CASC 2020. LNCS. Springer Nature, Switzerland (2020)

20. Renegar, J.: On the worst-case arithmetic complexity of approximating zeros of polynomials. J. Complex. **3**(2), 90–113 (1987). https://doi.org/10.1016/0885-064X(87)90022-7

21. Schönhage, A.: The Fundamental Theorem of Algebra in Terms of Computational Complexity. Math. Dept., Univ. Tübingen, Tübingen, Germany (1982)

22. Weyl, H.: Randbemerkungen zu Hauptproblemen der Mathematik. II. Fundamentalsatz der Algebra und Grundlagen der Mathematik. Mathematische Zeitschrift, **20**, 131–151 (1924)

Analytical Computations in Studying Translational-Rotational Motion of a Non-stationary Triaxial Body in the Central Gravitational Field

Alexander Prokopenya[1]([✉]) [iD], Mukhtar Minglibayev[2,3], and Oralkhan Baisbayeva[2]

[1] Warsaw University of Life Sciences–SGGW, Nowoursynowska 159, 02-776 Warsaw, Poland
alexander_prokopenya@sggw.edu.pl
[2] Al-Farabi Kazakh National University, Al-Farabi av. 71, 050040 Almaty, Kazakhstan
minglibayev@gmail.com, baisbayevaoral@gmail.com
[3] Fesenkov Astrophysical Institute, Observatoriya 23, 050020 Almaty, Kazakhstan

Abstract. The translational-rotational motion of a non-stationary triaxial body with constant dynamic shape in a non-stationary Newtonian central gravitational field is considered. Differential equations determining translational motion of the triaxial body around a spherical body and its rotation about the center of mass are obtained in terms of the osculating Delaunay–Andoyer elements. The force function is expanded in power series in terms of the Delaunay–Andoyer elements up to the second harmonic element inclusive. Averaging the equations of motion over the "fast" variables, we obtain the evolution equations of the translational-rotational motion of the non-stationary triaxial body which may be integrated numerically for any given laws of the masses and principal moments of inertia variation. All the relevant symbolic calculations are performed with the aid of the computer algebra system Wolfram Mathematica.

Keywords: Non-stationary two-body problem · Translational-rotational motion · Secular perturbations · Evolution equations · Wolfram mathematica

1 Introduction

The classical two-body problem describes the motion of two points of constant masses interacting according to Newton's law of gravitation, and its general solution is well known. Since this solution describes translational motion of two finite bodies with spherically symmetric density distribution, as well, such a model is usually used as the first approximation in describing the orbital motion

© Springer Nature Switzerland AG 2020
F. Boulier et al. (Eds.): CASC 2020, LNCS 12291, pp. 478–491, 2020.
https://doi.org/10.1007/978-3-030-60026-6_28

of real celestial bodies, for example, a planet around the Sun or a satellite around a planet (see [1,2]). If at least one of the bodies is not spherically symmetric, the problem becomes much more complicated because a mutual gravitational interaction depends on the geometrical shape and mass distribution of the bodies. Besides, translational and rotational motions depend on each other, and the corresponding equations of motion should be integrated together (see [3–6]).

On the other hand, real celestial bodies are non-stationary, their characteristics, such as mass, size, and shape may vary with time (see [7–11]). Such changes occur especially intensively in double and multiple systems [12]. So it is quite natural to consider the problem of many bodies of variable mass and to investigate an influence of the mass variation on the dynamic evolution of the system (see, for example, [13–21]). It should be noted that dependence of masses on time significantly complicates the problem, and even in case of two interacting bodies of variable mass, a general solution to the equations of motion can be written only in some special cases (see [10,11,22]).

In the present paper, we consider a generalized case of the two-body problem when the first body of variable mass $m_1(t)$ is spherically symmetric while the second one has an arbitrary dynamic structure and its principal moments of inertia are different (a triaxial body). It is assumed that the mass and size of the second body change with time but the dynamic shape of the bodies is preserved. Besides, the variation of the body masses and sizes do not result in the appearance of reactive forces and their torques (see [11]). In spite of these simplifying assumptions, the problem is not integrable and the perturbation theory should be applied for its investigation. To derive the equations of motion and to reduce them to the form convenient for application of the perturbation theory quite tedious symbolic computations should be done. The purpose of this paper is to describe the main types of computational problems occurring in the derivation of evolution equations and their investigation. All the relevant computations are performed with the computer algebra system Wolfram Mathematica (see [23]).

The paper is organized as follows. In Sect. 2, we formulate the physical problem and derive the equations of motion in the form that is convenient for applying the perturbation theory. Section 3 is devoted to integrating the unperturbed equations of motion. Then the perturbing functions are computed in Sect. 4 and the evolution equations are obtained in Sect. 5. We summarize the results in the Conclusions.

2   Equations of Motion

Let ξ_i, η_i, ζ_i $(i = 1, 2)$ be the Cartesian coordinates of the centers of mass O_1, O_2 of the bodies P_1 and P_2, respectively, relative to some inertial reference frame. Then the translational motion of the body P_2 about body P_1 may be described by the radius-vector $\boldsymbol{R} = (x, y, z)$ connecting the points O_1, O_2, and its Cartesian coordinates are

$$x = \xi_2 - \xi_1, \ y = \eta_2 - \eta_1, \ z = \zeta_2 - \zeta_1. \tag{1}$$

To describe the rotational motion of the body P_2 we introduce two Cartesian coordinate systems with the same origin located at its center of mass O_2. The axes of the first coordinate system O_2XYZ are parallel to the axes of the inertial frame. The second coordinate system O_2xyz is attached to the body P_2 and its axes coincide with the principal central axes of inertia. Then orientation of the body P_2 relative to the O_2XYZ frame can be specified in terms of the three Euler angles ψ, θ, and φ (see Fig. 1). The projections of the angular velocity vector onto the axes O_2x, O_2y, and O_2z are given by (see [1,3])

$$p = \dot{\psi} \sin\theta \sin\varphi + \dot{\theta} \cos\varphi, \quad q = \dot{\psi} \sin\theta \cos\varphi - \dot{\theta} \sin\varphi, \quad r = \dot{\psi} \cos\theta + \dot{\varphi}, \quad (2)$$

where the dot over a symbol denotes the total derivative of the corresponding function with respect to time.

Denoting the principal central moments of inertia of the body P_2 by $A(t)$, $B(t)$, and $C(t)$, we can write the equations of translational and rotational motion in the form (see [11,20])

$$\mu(t)\frac{d^2x}{dt^2} = \frac{\partial U}{\partial x}, \quad \mu(t)\frac{d^2y}{dt^2} = \frac{\partial U}{\partial y}, \quad \mu(t)\frac{d^2z}{dt^2} = \frac{\partial U}{\partial z}, \quad (3)$$

$$\frac{d}{dt}(A(t)p) + (C(t) - B(t))qr = \frac{\sin\varphi}{\sin\theta}\left[\frac{\partial U}{\partial\psi} - \cos\theta\frac{\partial U}{\partial\varphi}\right] + \cos\varphi\frac{\partial U}{\partial\theta},$$

$$\frac{d}{dt}(B(t)q) + (A(t) - C(t))rp = \frac{\cos\varphi}{\sin\theta}\left[\frac{\partial U}{\partial\psi} - \cos\theta\frac{\partial U}{\partial\varphi}\right] - \sin\varphi\frac{\partial U}{\partial\theta},$$

$$\frac{d}{dt}(C(t)r) + (B(t) - A(t))pq = \frac{\partial U}{\partial\varphi}. \quad (4)$$

Here $\mu(t) = m_1(t)m_2(t)/(m_1(t) + m_2(t))$ is the reduced mass, the force function U is a series in powers of the inverse distance R between the centers of mass of the bodies accurate to the third order (see [1,3])

$$U = U_1 + U_2, \quad U_1 = \frac{Gm_1m_2}{R}, \quad R = \left(x^2 + y^2 + z^2\right)^{1/2}, \quad (5)$$

$$U_2 = Gm_1\frac{A + B + C - 3I}{2R^3}, \quad (6)$$

G is a gravitational constant, and

$$I = A\alpha_1^2 + B\alpha_2^2 + C\alpha_3^2$$

is the moment of inertia of the triaxial body P_2 relative to the axis given by the vector $\boldsymbol{R}$, and $\alpha_1, \alpha_2, \alpha_3$ are the direction cosines of the vector $\boldsymbol{R}$ relative to the body fixed coordinate system O_2xyz.

Remind that the masses $m_1(t)$, $m_2(t)$ and the principal moments of inertia $A(t)$, $B(t)$, $C(t)$ change with time. We assume that the body P_1 remains spherically symmetric and the body P_2 retains its initial dynamic structure. Therefore, the principal moments of inertia satisfy the condition

$$\frac{A(t)}{A_0} = \frac{B(t)}{B_0} = \frac{C(t)}{C_0} = \nu\chi^2, \quad (7)$$

where $A_0 = A(t_0)$, $B_0 = B(t_0)$, $C_0 = C(t_0)$, t_0 is an initial instant of time, and $m_1 = m_1(t)$, $m_2 = m_2(t_0)\nu(t)$, $\chi = \chi(t)$ are given functions of time satisfying the conditions $\nu(t_0) = 1$, $\chi(t_0) = 1$ and $\nu(t) > 0, \chi(t) > 0$ for $t > t_0$. The functions $m_1(t)$ and $\nu(t)$ determine the mass variation of the bodies and may be chosen according to the Eddington–Jeans law, for example (see [7,8]). Remind that the moment of inertia of a rigid body is proportional to its mass and a square of its geometric sizes (see, for example, [24]). As the body P_2 is assumed to retain its initial dynamic structure, the function $\chi(t)$ in (7) determining its characteristic size variation is the same for the three principal moments of inertia $A(t), B(t)$, and $C(t)$.

3   Unperturbed Motion

Since equations (3)–(4) are not integrable, we can apply a perturbation theory to the investigation of the system dynamics (see, for instance, [25]). This assumes that equations (3)–(4) are reduced to two perturbed problems of which each is integrable in the case when there are no perturbations.

3.1   Translational Motion

Let us substitute (5) and (6) into (3) and rewrite (3) in the form

$$\ddot{x} + G(m_1 + m_2)\frac{x}{R^3} - \frac{\ddot{\gamma}}{\gamma}x = \frac{\partial V}{\partial x},$$
$$\ddot{y} + G(m_1 + m_2)\frac{y}{R^3} - \frac{\ddot{\gamma}}{\gamma}y = \frac{\partial V}{\partial y},$$
$$\ddot{z} + G(m_1 + m_2)\frac{z}{R^3} - \frac{\ddot{\gamma}}{\gamma}z = \frac{\partial V}{\partial z},\qquad(8)$$

where

$$V = \frac{m_1 + m_2}{m_1 m_2}U_2 - \frac{\ddot{\gamma}}{2\gamma}R^2, \quad \gamma(t) = \frac{m_1(t_0) + m_2(t_0)}{m_1(t) + m_2(t)}.\qquad(9)$$

Such representation of Eq. (3) is convenient because Eq. (8) are integrable in case of $V = 0$ for any doubly continuously differentiable function $\gamma(t)$ (see [11]); the corresponding solution is given by

$$x = \gamma\rho(\cos(v + w)\cos\Omega - \sin(v + w)\sin\Omega\cos i),$$
$$y = \gamma\rho(\cos(v + w)\sin\Omega + \sin(v + w)\cos\Omega\cos i),$$
$$z = \gamma\rho\sin(v + w)\sin i,\qquad(10)$$

where

$$\rho = \frac{a(1 - e^2)}{1 + e\cos v},\qquad(11)$$

and the parameters a, e, i, Ω, w determined from the initial conditions correspond to the Kepler orbital elements known from the classical two-body problem;

they are analogs of the major semi-axis, eccentricity, inclination, longitude of the ascending node, and the longitude of the peri-center of the unperturbed quasi-elliptic orbit of the body P_2 (see, for example, [11,16]). The true anomaly v is determined by the equation

$$\int_0^v \frac{dv}{(1 + e\cos v)^2} = \frac{1}{(1 - e^2)^{3/2}}(E - e\sin E)$$

$$= \frac{M}{(1 - e^2)^{3/2}} = \frac{\sqrt{K_0}}{a^{3/2}(1 - e^2)^{3/2}}(\Phi(t) - \Phi(\tau)) , \qquad (12)$$

where τ is the time when the body P_2 passes through the pericenter, M is the mean anomaly,

$$\Phi(t) = \int_{t_0}^t \frac{dt}{\gamma^2(t)} , \quad K_0 = G(m_1(t_0) + m_2(t_0)) ,$$

and the eccentric anomaly E is related to the true anomaly v by

$$\tan\frac{v}{2} = \sqrt{\frac{1 + e}{1 - e}} \tan\frac{E}{2} . \qquad (13)$$

Given the function $\gamma(t)$, which depends on the laws of mass variation of the bodies P_1, P_2, equations (12) and (13) make it possible to find the mean anomaly M, the eccentric anomaly E, and the true anomaly v as the functions of time. As a result, relations (10) and (11) allow us to compute the relative Cartesian coordinates of the body P_2 and completely describe its unperturbed translational motion.

Note that in the case of constant masses, when $\gamma(t) = 1$, equations (10)–(13) determine the translational motion of the body P_2 around the body P_1 along conic section. The presence of the scale factor $\gamma(t)$ that depends on time in (10) results in deforming the conic section and makes the motion aperiodic. For this reason, the solution to the equations (8) in the case $V = 0$ is said to describe aperiodic motion of the body P_2 along quasi-conic section (see [11]).

If the perturbing function $V \neq 0$ is taken into account in Eq. (8) the orbital parameters a, e, i, Ω, and ω become the functions of time. To derive the differential equations determining their evolution, it is convenient to use canonical variables known as Delaunay elements and rewrite Eq. (8) in canonical form (see [6,11]). The generating function for the corresponding canonical transformation is determined by the complete integral of the Hamilton–Jacobi equation (see, for example, [24,25]) and its constructing involves quite standard but tedious symbolic computations (see [16]). Note that the system Wolfram Mathematica (see [23]) offers many built-in functions such as *Expand*, *Replace*, *Integrate*, *Simplify*, *Solve*, for example, which help a lot in doing such computation. Finally, we determine three pairs of canonically conjugate coordinates and momenta (l_1, L_1), (g_1, G_1), and (h_1, H_1); they are related to the analogs of the Kepler orbital elements by

$$l_1 = M , \quad L_1 = \sqrt{K_0 a} , \quad g_1 = \omega , \quad G_1 = \sqrt{K_0 a(1 - e^2)} ,$$

$$h_1 = \Omega \;, \quad H_1 = \sqrt{K_0 a (1 - e^2)} \cos i \;. \tag{14}$$

The corresponding Hamiltonian is given by

$$\mathcal{H}^{(trans)} = -\frac{K_0^2}{2\gamma^2 L_1^2} - V \;, \tag{15}$$

where the perturbing function V defined in (9) must be expressed in terms of the canonical variables.

3.2   Rotational Motion

To write Eq. (4) in the canonical form let us define the angular momentum vector $\mathbf{G}_2 = (L_x, L_y, L_z)$ and the kinetic energy of rotation T as

$$L_x = Ap \;, \quad L_y = Bq \;, \quad L_z = Cr \;, \quad T = \frac{1}{2} \left(Ap^2 + Bq^2 + Cr^2 \right) \;. \tag{16}$$

Taking into account (2) and using the Mathematica built-in function D, we differentiate the kinetic energy T and obtain the momenta $p_\psi, p_\theta, p_\varphi$ canonically conjugate to the coordinates ψ, θ, φ in the form

$$p_\psi = \frac{\partial T}{\partial \dot{\psi}} = L_x \sin\theta \sin\varphi + L_y \sin\theta \cos\varphi + L_z \cos\theta \;,$$

$$p_\theta = \frac{\partial T}{\partial \dot{\theta}} = L_x \cos\varphi - L_y \sin\varphi \;, \quad p_\varphi = \frac{\partial T}{\partial \dot{\varphi}} = Cr = L_z \;. \tag{17}$$

Solving system (17) and substituting L_x, L_y, L_z into (16), we obtain the Hamiltonian determining rotational motion of the body P_2

$$\mathcal{H}^{(rot)} = \frac{1}{2A} \left(\frac{1}{\sin^2\theta} \left((p_\psi - p_\varphi \cos\theta)^2 + p_\theta^2 \right) \right) + \mathcal{H}_1^{(rot)}, \tag{18}$$

where

$$\mathcal{H}_1^{(rot)} = \frac{1}{2} \left(\frac{1}{B} - \frac{1}{A} \right) \left(\frac{\cos\varphi}{\sin\theta} (p_\psi - p_\varphi \cos\theta) - p_\theta \sin\varphi \right)^2 - U_2. \tag{19}$$

In case of $\mathcal{H}_1^{(rot)} = 0$, the equations of motion determined by the Hamiltonian (18) are integrable and their general solution may be written in symbolic form. However, to write this solution in the simplest form and to simplify further calculations it is convenient to introduce "new" canonical variables $(l_2, g_2, h_2, L_2, G_2, H_2)$ which are known as the Andoyer ones (see [6,26]). The momenta L_2, G_2, H_2 are given by

$$L_2 = p_\varphi \;, \quad G_2 = \left((p_\psi - p_\varphi \cos\theta)^2 / \sin^2\theta + p_\theta^2 + p_\varphi^2 \right)^{1/2} \;, \quad H_2 = p_\psi \;, \tag{20}$$

where L_2 may be interpreted as the projection of the angular momentum vector $\mathbf{G}_2$ onto the body fixed axis $O_2 z$, $G_2 = |\mathbf{G}_2|$ is the angular momentum, H_2 is the projection of $\mathbf{G}_2$ onto the axis $O_2 Z$ (see Fig. 1).

New coordinates l_2, g_2, and h_2 are connected with the Euler angles ψ, θ, φ and momenta by the relations (see [26])

$$\cos \theta = \cos I \cos J - \sin I \sin J \cos g_2 \, ,$$

$$\frac{\sin(\psi - h_2)}{\sin J} = \frac{\sin(\varphi - l_2)}{\sin I} = \frac{\sin g_2}{\sin \theta} \, , \tag{21}$$

where the angles J and I (see Fig. 1) are given by

$$\cos I = \frac{H_2}{G_2} \, , \quad \sin I = \sqrt{1 - \frac{H_2^2}{G_2^2}} \, , \quad \cos J = \frac{L_2}{G_2} \, , \quad \sin J = \sqrt{1 - \frac{L_2^2}{G_2^2}}. \tag{22}$$

Geometrical interpretation of the Andoyer variables is shown in Fig. 1.

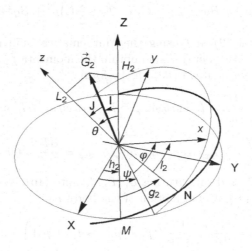

Fig. 1. Euler angles and Andoyer variables.

Solving Eq. (17) for L_x, L_y, L_z and using (19)–(21), one can express old canonical variables $\psi, \theta, \varphi, p_\psi, p_\theta, p_\varphi$ via the new ones $l_2, g_2, h_2, L_2, G_2, H_2$ and rewrite the Hamiltonian (18) in terms of the Andoyer variables as

$$\mathcal{H}^{(rot)} = \frac{1}{2A} \left(G_2^2 - L_2^2 \right) + \frac{1}{2C} L_2^2 + \mathcal{H}_1^{(rot)}, \tag{23}$$

where

$$\mathcal{H}_1^{(rot)} = \frac{1}{2} \left(\frac{1}{B} - \frac{1}{A} \right) \left(G_2^2 - L_2^2 \right) \cos^2 l_2 - U_2, \tag{24}$$

and the term U_2 defined in (6) is expressed in terms of the canonical variables l_2, g_2, h_2, L_2, G_2, and H_2.

Equations of unperturbed rotational motion of the body P_2 about its center of mass are obtained in case of $\mathcal{H}_1^{(rot)} = 0$ and may be written in the canonical form as

$$\dot{l}_2 = \frac{\partial}{\partial L_2}\left(\mathcal{H}^{(rot)} - \mathcal{H}_1^{(rot)}\right) = \left(\frac{1}{C_0} - \frac{1}{A_0}\right)\frac{L_2}{\nu(t)\chi^2(t)},$$

$$\dot{g}_2 = \frac{\partial}{\partial G_2}\left(\mathcal{H}^{(rot)} - \mathcal{H}_1^{(rot)}\right) = \frac{G_2}{A_0}\frac{1}{\nu(t)\chi^2(t)},$$

$$\dot{h}_2 = 0, \quad \dot{L}_2 = 0, \quad \dot{G}_2 = 0, \quad \dot{H}_2 = 0. \tag{25}$$

System (25) is integrable and its general solution is

$$l_2 = \left(\frac{1}{C_0} - \frac{1}{A_0}\right)L_2\int_{t_0}^{t}\frac{d\tau}{\nu(\tau)\chi^2(\tau)} + l_{20}, \quad L_2 = \text{const},$$

$$g_2 = \frac{G_2}{A_0}\int_{t_0}^{t}\frac{d\tau}{\nu(\tau)\chi^2(\tau)} + g_{20}, \quad G_2 = \text{const}, \quad h_2 = \text{const}, \quad H_2 = \text{const}. \tag{26}$$

One can readily see that the angles $l_2(t)$ and $g_2(t)$ are increasing functions of time while the rest four canonical variables h_2, L_2, G_2, H_2 are constants.

4   The Perturbing Functions

To write out the differential equations determining the perturbed translational-rotational motion of the body P_2 we need to express the perturbing functions (15) and (24) in terms of the Delaunay–Andoyer variables. It means that we need to find the corresponding expressions for the distance R between the centers of masses of the bodies and the direction cosines of the vector $\boldsymbol{R}$ relative to the body fixed coordinate system O_2xyz (see (6), (9)). These direction cosines $\alpha_1, \alpha_2, \alpha_3$ are determined from the dot product of the unit vectors directed along the vector $\boldsymbol{R}$ and the axes O_2x, O_2y, O_2z, respectively, and may be represented in the form

$$\alpha_1 = a_{11}\frac{x}{R} + a_{21}\frac{y}{R} + a_{31}\frac{z}{R}, \quad \alpha_2 = a_{12}\frac{x}{R} + a_{22}\frac{y}{R} + a_{32}\frac{z}{R},$$

$$\alpha_3 = a_{13}\frac{x}{R} + a_{23}\frac{y}{R} + a_{33}\frac{z}{R}, \tag{27}$$

where $\frac{x}{R}, \frac{y}{R}, \frac{z}{R}$ are the direction cosines of the vector $\boldsymbol{R}$ with respect to the axes of the non-rotating coordinate system O_1XYZ which have the form (see (10))

$$\frac{x}{R} = \frac{x}{\gamma\rho} = \cos(\upsilon + \omega)\cos\Omega - \sin(\upsilon + \omega)\sin\Omega\cos i,$$

$$\frac{y}{R} = \frac{y}{\gamma\rho} = \cos(\upsilon + \omega)\sin\Omega + \sin(\upsilon + \omega)\cos\Omega\cos i,$$

$$\frac{z}{R} = \frac{z}{\gamma\rho} = \sin(\upsilon + \omega)\sin i, \quad \rho = \frac{a\left(1 - e^2\right)}{1 + e\cos\upsilon}. \tag{28}$$

Coefficients a_{ij} in (27) are the direction cosines of the axes of the coordinate system O_2xyz fixed to the triaxial body P_2 with respect to the non-rotating coordinate systems O_2XYZ. Note that the axes O_2x, O_2y, O_2z are obtained from the axes O_2X, O_2Y, O_2Z by means of the five successive rotations (see Fig. 1): rotation about the axis O_2Z by the angle h_2, rotation about the axis O_2M by the angle I, rotation about the axis O_2G_2 by the angle g_2, rotation about the axis O_2N by the angle J, and rotation about the axis O_2z by the angle l_2. The matrices of the corresponding transformations are given by

$$Q_1 = \begin{pmatrix} \cos h_2 & -\sin h_2 & 0 \\ \sin h_2 & \cos h_2 & 0 \\ 0 & 0 & 1 \end{pmatrix}, \quad Q_2 = \begin{pmatrix} 1 & 0 & 0 \\ 0 & \cos I & -\sin I \\ 0 & \sin I & \cos I \end{pmatrix},$$

$$Q_3 = \begin{pmatrix} \cos g_2 & -\sin g_2 & 0 \\ \sin g_2 & \cos g_2 & 0 \\ 0 & 0 & 1 \end{pmatrix}, \quad Q_4 = \begin{pmatrix} 1 & 0 & 0 \\ 0 & \cos J & -\sin J \\ 0 & \sin J & \cos J \end{pmatrix},$$

$$Q_5 = \begin{pmatrix} \cos l_2 & -\sin l_2 & 0 \\ \sin l_2 & \cos l_2 & 0 \\ 0 & 0 & 1 \end{pmatrix}.$$

Defining these matrices and using the built-in Mathematica function Dot, we obtain the matrix

$$Q = Dot[Q_1, Q_2, Q_3, Q_4, Q_5].$$

Note that the columns of the matrix Q determine the direction cosines $a_{ij} = Q_{ij}$ in terms of Andoyer variables. They are

$$a_{11} = \cos g_2 \cos h_2 \cos l_2 - \cos h_2 \cos J \sin g_2 \sin l_2 - \cos I \cos l_2 \sin g_2 \sin h_2 -$$

$$- \cos g_2 \cos I \cos J \sin h_2 \sin l_2 + \sin h_2 \sin I \sin J \sin l_2,$$

$$a_{21} = \cos g_2 \cos l_2 \sin h_2 - \cos J \sin g_2 \sin h_2 \sin l_2 + \cos h_2 \cos I \cos l_2 \sin g_2 +$$

$$+ \cos g_2 \cos h_2 \cos I \cos J \sin l_2 - \cos h_2 \sin I \sin J \sin l_2,$$

$$a_{31} = \cos l_2 \sin g_2 \sin I + \cos g_2 \cos J \sin I \sin l_2 + \cos I \sin J \sin l_2,$$

$$a_{12} = -\sin g_2 \cos h_2 \cos J \cos l_2 - \cos g_2 \cos h_2 \sin l_2 - \cos g_2 \sin h_2 \cos I \cos J \cos l_2$$

$$+ \sin g_2 \sin h_2 \cos I \sin l_2 + \sin h_2 \sin I \sin J \cos l_2,$$

$$a_{22} = -\sin g_2 \sin h_2 \cos J \cos l_2 - \cos g_2 \sin h_2 \sin l_2 + \cos g_2 \cos h_2 \cos I \cos J \cos l_2$$

$$- \sin g_2 \cos h_2 \cos I \sin l_2 - \cos h_2 \sin I \sin J \cos l_2,$$

$$a_{32} = \cos g_2 \sin I \cos J \cos l_2 - \sin g_2 \sin I \sin l_2 + \cos I \sin J \cos l_2,$$

$$a_{13} = \sin g_2 \cos h_2 \sin J + \sin h_2 \sin I \cos J + \cos g_2 \sin h_2 \cos I \sin J,$$

$$a_{23} = \sin g_2 \sin h_2 \sin J - \cos h_2 \sin I \cos J - \cos g_2 \cos h_2 \cos I \sin J,$$

$$a_{33} = \cos I \cos J - \cos g_2 \sin I \sin J. \tag{29}$$

One can readily see that the direction cosines a_{13}, a_{23}, a_{33} do not depend on the angle l_2. To simplify further computation it is convenient to rewrite the rest direction cosines in (29) in the form

$$a_{11} = \kappa_{11} \cos l_2 + \kappa_{12} \sin l_2,$$
$$a_{21} = \kappa_{21} \cos l_2 + \kappa_{22} \sin l_2,$$
$$a_{31} = \kappa_{31} \cos l_2 + \kappa_{32} \sin l_2,$$
$$a_{12} = \kappa_{12} \cos l_2 - \kappa_{11} \sin l_2, \tag{30}$$
$$a_{22} = \kappa_{22} \cos l_2 - \kappa_{21} \sin l_2,$$
$$a_{32} = \kappa_{32} \cos l_2 - \kappa_{31} \sin l_2,$$

where

$$\kappa_{11} = \cos g_2 \cos h_2 - \sin g_2 \sin h_2 \frac{H_2}{G_2},$$

$$\kappa_{12} = \sin h_2 \sqrt{\left(1 - \frac{H_2^2}{G_2^2}\right)\left(1 - \frac{L_2^2}{G_2^2}\right)} - \cos g_2 \sin h_2 \frac{H_2 L_2}{G_2^2} - \sin g_2 \cos h_2 \frac{L_2}{G_2},$$

$$\kappa_{21} = \cos g_2 \sin h_2 + \sin g_2 \cos h_2 \frac{H_2}{G_2},$$

$$\kappa_{22} = -\cos h_2 \sqrt{\left(1 - \frac{H_2^2}{G_2^2}\right)\left(1 - \frac{L_2^2}{G_2^2}\right)} + \cos g_2 \cos h_2 \frac{H_2 L_2}{G_2^2} - \sin g_2 \sin h_2 \frac{L_2}{G_2},$$

$$\kappa_{31} = \sin g_2 \sqrt{1 - \frac{H_2^2}{G_2^2}}, \quad \kappa_{32} = \cos g_2 \frac{L_2}{G_2}\sqrt{1 - \frac{H_2^2}{G_2^2}} + \frac{H_2}{G_2}\sqrt{1 - \frac{L_2^2}{G_2^2}}.$$

Using the obvious relation $\alpha_1^2 + \alpha_2^2 + \alpha_3^2 = 1$ and (11), we can rewrite expression (6) in the form

$$U_2 = \frac{Gm_1(1 + e \cos v)^3}{2\gamma^3 a^3 (1 - e^2)^3} \left(A + B - 2C - 3(A - C)\alpha_1^2 - 3(B - C)\alpha_2^2\right), \tag{31}$$

where the direction cosines α_1, α_2 are defined in (27).

Note that equations (28)–(30) enable us to rewrite (31) and the perturbing functions (15), (24) in terms of the Delaunay–Andoyer variables and to write out the equations of the perturbed motion in the explicit form. However, the corresponding expressions are quite cumbersome and we do not write out them here.

5 Evolution Equations

Remind that in the absence of perturbations, the center of mass of the body P_2 moves along quasi-conic section determined by solution (10)–(12), its orbital parameters a, e, i, Ω, ω and the corresponding Delaunay variables g_1, h_1, L_1, G_1, H_1 (see (14)) are constants, and only its mean anomaly $l_1 = M$

is an increasing function of time. Rotational motion is independent of translational one and solution (26) determines free rotation of the body P_2 about the axis $O_2 z$ and precession of this axis about the angular momentum vector G_2. Both angles $l_2(t)$ and $g_2(t)$ are increasing functions of time but we assume that $A_0 > C_0$ and the angle $l_2(t)$ increases much faster than the angle $g_2(t)$. The rest four Andoyer variables h_2, L_2, G_2, and H_2 are constant.

Taking into account the perturbing function in the Hamiltonian (15) and (24) results in the dependence of the Delaunay–Andoyer variables on time. Equations (10), (14), (27)–(30) show that the angle variables $l_1, l_2, g_1, g_2, h_1, h_2$ appear in the perturbing functions only as arguments of the sine and cosine functions. As the angles l_1 and l_2 are increasing functions of time, they may be considered as "fast" variables which result in the oscillation of some terms in the perturbing functions. If the frequencies of these oscillations are incommensurable the corresponding short-term oscillations of the Delaunay–Andoyer variables may be eliminated by averaging the perturbing functions. The averaging is performed in the revolution period of the body P_2 over the mean anomaly l_1 of its translational motion and over the variable l_2 describing rotation of the body around the $O_2 z$ axis. Using the averaged perturbing functions in the Hamiltonian (15) and (24) we can write out the evolution equations describing the long-term behavior of the Delaunay–Andoyer variables in the standard canonical form.

Note that using the Hamiltonian (15), (24), one could write out first the equations of the perturbed motion in the explicit form and to average them afterwards to obtain the evolution equations but in this case, the calculations would be much more complicated.

To find the averaged perturbing functions F we need to calculate the following integral

$$\frac{1}{4\pi^2} \int_0^{2\pi} \int_0^{2\pi} F \, dl_1 dl_2 = \frac{(1-e^2)^{3/2}}{4\pi^2} \int_0^{2\pi} \int_0^{2\pi} \frac{F}{(1 + e\cos v)^2} \, dv dl_2. \quad (32)$$

Note that the integration over the angle l_1 in (32) is convenient to replace by integration over the true anomaly v (see (12)).

Integrating the expressions (15) and (24), we obtain the secular parts of the Hamiltonians in the form

$$\tilde{\mathcal{H}}^{(trans)} = -\frac{K_0^2}{2\gamma^2 L_1^2} - \frac{G(m_1 + m_2)}{2m_2\gamma^3}(A + B - 2C)(W_2 - 3W_3) + \frac{1}{2}\ddot{\gamma}\gamma W_1, \quad (33)$$

$$\tilde{\mathcal{H}}^{(rot)} = \frac{1}{2}\left(\frac{1}{C} - \frac{1}{A}\right)L_2^2 + \frac{1}{2A}G_2^2 + \frac{1}{4}\left(\frac{1}{B} - \frac{1}{A}\right)(G_2^2 - L_2^2) -$$

$$- \frac{Gm_1}{2\gamma^3}(A + B - 2C)(W_2 - 3W_3), \quad (34)$$

where

$$W_1 = a^2\left(1 + \frac{3}{2}e^2\right) = \frac{L_1^4}{2K_0^2}\left(5 - \frac{3G_1^2}{L_1^2}\right), \quad W_2 = \frac{1}{a^3(1-e^2)^{3/2}} = \frac{K_0^3}{L_1^3 G_1^3},$$

$$W_3 = \frac{K_0^3}{4L_1^3 G_1^5} \left((G_1^2 - H_1^2)(\kappa_{31}^2 + \kappa_{32}^2) + \right.$$

$$+G_1^2 \left((\kappa_{11} \cos h_1 + \kappa_{21} \sin h_1)^2 + (\kappa_{12} \cos h_1 + \kappa_{22} \sin h_1)^2 \right) +$$

$$+H_1^2 \left((\kappa_{11} \sin h_1 - \kappa_{21} \cos h_1)^2 + (\kappa_{12} \sin h_1 - \kappa_{22} \cos h_1)^2 \right) +$$

$$+2G_1 H_1 \sqrt{1 - \frac{H_1^2}{G_1^2}} \left((\kappa_{21}\kappa_{31} + \kappa_{22}\kappa_{32}) \cos h_1 - \right.$$

$$\left. \left. - (\kappa_{11}\kappa_{31} + \kappa_{12}\kappa_{32}) \sin h_1 \right) \right).$$

Finally, using (33) and (34), we can write the evolution equations in the canonical form

$$\dot{L}_1 = -\frac{\partial \tilde{\mathcal{H}}^{(trans)}}{\partial l_1} = 0, \quad \dot{G}_1 = -\frac{\partial \tilde{\mathcal{H}}^{(trans)}}{\partial g_1}, \quad \dot{H}_1 = -\frac{\partial \tilde{\mathcal{H}}^{(trans)}}{\partial h_1},$$

$$\dot{l}_1 = \frac{\partial \tilde{\mathcal{H}}^{(trans)}}{\partial L_1}, \quad \dot{g}_1 = \frac{\partial \tilde{\mathcal{H}}^{(trans)}}{\partial G_1}, \quad \dot{h}_1 = \frac{\partial \tilde{\mathcal{H}}^{(trans)}}{\partial H_1}, \tag{35}$$

$$\dot{L}_2 = -\frac{\partial \tilde{\mathcal{H}}^{(rot)}}{\partial l_2} = 0, \quad \dot{G}_2 = -\frac{\partial \tilde{\mathcal{H}}^{(rot)}}{\partial g_2}, \quad \dot{H}_2 = -\frac{\partial \tilde{\mathcal{H}}^{(rot)}}{\partial h_2},$$

$$\dot{l}_2 = \frac{\partial \tilde{\mathcal{H}}^{(rot)}}{\partial L_2}, \quad \dot{g}_2 = \frac{\partial \tilde{\mathcal{H}}^{(rot)}}{\partial G_2}, \quad \dot{h}_2 = \frac{\partial \tilde{\mathcal{H}}^{(rot)}}{\partial H_2}. \tag{36}$$

Note that differentiation of the Hamiltonian in (35) and (36) is performed with the built-in Mathematica function D but the obtained expressions are quite cumbersome and we do not write out them here.

6 Conclusions

Equations for secular perturbations of the translational-rotational motion of a triaxial satellite in osculating elements of Delaunay–Andoyer are obtained and can be used to analyze the dynamic evolution of the system of two non-stationary bodies attracting each other according to the Newton law of gravitation. Note that in case of $m_1(t) = \text{const}$, $\nu(t) = 1$, $\chi(t) = 1$, $\gamma(t) = 1$, equations (35) and (36) describe translational-rotational motion of a stationary triaxial rigid body in the central gravitational field (see, for example, [26]). Non-stationarity of the bodies complicates the problem substantially and solutions to the evolution equations (35) and (36) cannot be found in symbolic form. So further development of this work involves numerical analysis of the obtained evolution equations of translational-rotational motion of a triaxial body of constant dynamic shape and variable size and mass.

Note that using the Hamiltonian (15), (23) one can write out the equations of translational-rotational motion of the body P_2 in explicit form and apply purely numerical methods for their solution. However, in addition to the secular parts, such numerical solutions contain short-term oscillations and an accuracy of calculations should be very high to obtain correct result. As any numerical solution is determined by some initial conditions, special attention should be paid to avoid a commensurability of frequencies in the system. Besides, the short-term oscillations only perturb the solution but they do not influence the long-term behavior of the system we are interested in. Thus, the derivation of the evolution equations in symbolic form enables one to eliminate short-term perturbations and to simplify analysis of the problem.

Note also that all the relevant symbolic calculations have been performed with the aid of the computer algebra system Wolfram Mathematica.

References

1. Duboshin, G.N.: Celestial Mechanics. Basic Problems and Methods, 2nd edn. Nauka, Moscow (1968)
2. Roy, A.E.: Orbital Motion. Hilger, Bristol (1978)
3. Beletskii, V.V.: Attitude Motion of Satellite in a Gravitational Field. MGU Press, Moscow (1975)
4. Zhuravlev, S.G., Petrutskii, A.A.: Current state of the problem of translational-rotational motion of three rigid bodies. Sov. Astron. **34**(3), 299–304 (1990)
5. Maciejewski, A.J.: A simple model of the rotational motion of a rigid satellite around an oblate planet. Acta Astron. **47**, 387–398 (1997)
6. Krasilnikov, P.S.: Rotational Motion of a Rigid Body about Center of Mass in the Restricted Three-Body Problem. MAI Press, Moscow (2018)
7. Eddington, A.S.: On the relation between the masses and luminosities of the stars. Mon. Not. Roy. Astronimical Soc. **84**, 308–332 (1923)
8. Jeans, J.H.: The effect of varying mass on a binary system. Mon. Not. R. Astronimical Soc. **85**, 912–914 (1925)
9. Omarov, T.B.: Non-Stationary Dynamical Problems in Astronomy. Nova Science Publ, New York (2002)
10. Bekov, A.A., Omarov, T.B.: The theory of orbits in non-stationary stellar systems. Astronomical Astrophys. Trans. **22**(2), 145–153 (2003)
11. Minglibayev, M.Zh.: Dynamics of gravitating bodies of variable masses and sizes. Lambert Acad. (2012)
12. Eggleton, P.: Evolutionary Processes in Binary and Multiple Stars. Cambridge University Press, Cambridge (2006)
13. Hadjidemetriou, J.D.: Binary systems with decreasing mass. Zeitschrift für Astrophysik **63**, 116–130 (1966)
14. Deprit, A., Miller, B., Williams, C.A.: Gylden systems: rotation of pericenters. Astrophys. Space Sci. **159**, 239–270 (1989)
15. Luk'yanov, L.G.: Conservative two-body problem with variable masses. Astron. Lett. **31**(8), 563–568 (2005)
16. Prokopenya, A.N., Minglibayev, M.Z., Mayemerova, G.M.: Symbolic computations in studying the problem of three bodies with variable masses. Program. Comput. Softw. **40**(2), 79–85 (2014)
17. Prokopenya, A.N., Minglibayev, M.Z., Beketauov, B.A.: On integrability of evolutionary equations in the restricted three-body problem with variable masses. In: Gerdt, V.P., Koepf, W., Seiler, W.M., Vorozhtsov, E.V. (eds.) CASC 2014. LNCS, vol. 8660, pp. 373–387. Springer, Cham (2014). https://doi.org/10.1007/978-3-319-10515-4_27

18. Prokopenya, A.N., Minglibayev, M.Z., Beketauov, B.A.: Secular perturbations of quasi-elliptic orbits in the restricted three-body problem with variable masses. Int. J. Non-Linear Mech. **73**, 58–63 (2015)
19. Prokopenya, A.N., Minglibayev, M.Z., Mayemerova, G.M., Imanova, Z.U.: Investigation of the restricted problem of three bodies of variable masses using computer algebra. Program. Comput. Softw. **43**(5), 289–293 (2017). https://doi.org/10.1134/S0361768817050061
20. Baisbayeva, O., Minglibayev, M., Prokopenya, A.: Analytical calculations of secular perturbations of translational-rotational motion of a non-stationary triaxial body in the central field of attraction. http://aca2019.etsmtl.ca/program/conference-booklet/. Accessed 10 Jan 2020
21. Prokopenya, A.N., Minglibayev, M.Z., Shomshekova, S.A.: Application of computer algebra in the study of the two-planet problem of three bodies with variable masses. Programm. Comput. Softw. **45**(2), 73–80 (2019)
22. Berkovič, L.M.: Gylden-Meščerski problem. Celest. Mech. **24**, 407–429 (1981)
23. Wolfram, S.: An Elementary Introduction to the Wolfram Language, 2nd edn. Champaign, IL, USA, Wolfram Media (2017)
24. Markeev, A.P.: Theor. Mech. Regular and Chaotic Dynamics, Moscow-Izhevsk (2007). [in Russian]
25. Boccaletti, D., Pucacco, G.: Theory of Orbits. Vol. 2: Perturbative and Geometrical Methods. Springer-Verlag, Berlin-Heidelberg (2002)
26. Kinoshita, H.: Theory of the rotation of the rigid Earth. Celest. Mech. **15**, 277–326 (1977)

A Linear Algebra Approach for Detecting Binomiality of Steady State Ideals of Reversible Chemical Reaction Networks

Hamid Rahkooy[1]([✉]), Ovidiu Radulescu[2], and Thomas Sturm[1,3]

[1] CNRS, Inria, and the University of Lorraine, Nancy, France
hamid.rahkooy@inria.fr, thomas.sturm@loria.fr
[2] LPHI CNRS UMR 5235, and University of Montpellier, Montpellier, France
ovidiu.radulescu@umontpellier.fr
[3] MPI-INF and Saarland University, Saarbrücken, Germany
sturm@mpi-inf.mpg.de

Abstract. Motivated by problems from Chemical Reaction Network Theory, we investigate whether steady state ideals of reversible reaction networks are generated by binomials. We take an algebraic approach considering, besides concentrations of species, also rate constants as indeterminates. This leads us to the concept of unconditional binomiality, meaning binomiality for all values of the rate constants. This concept is different from conditional binomiality that applies when rate constant values or relations among rate constants are given. We start by representing the generators of a steady state ideal as sums of binomials, which yields a corresponding coefficient matrix. On these grounds, we propose an efficient algorithm for detecting unconditional binomiality. That algorithm uses exclusively elementary column and row operations on the coefficient matrix. We prove asymptotic worst case upper bounds on the time complexity of our algorithm. Furthermore, we experimentally compare its performance with other existing methods.

Keywords: Binomial ideals · Linear algebra · Reversible chemical reaction networks

1 Introduction

A *chemical reaction* is a transformation between two sets of chemical objects called chemical *complexes*. The objects that form a chemical complex are chemical *species*. In other words, complexes are formal sums of chemical species representing the left hand and the right hand sides of chemical reactions. A *chemical reaction network* is a set of chemical reactions. For example

$$CO_2 + H_2 \underset{k_{-21}}{\overset{k_{-12}}{\rightleftharpoons}} CO + H_2O,$$

$$2\,CO \underset{k_{-43}}{\overset{k_{-34}}{\rightleftharpoons}} CO_2 + C$$

is a chemical reaction network with two *reversible* reactions.

© Springer Nature Switzerland AG 2020
F. Boulier et al. (Eds.): CASC 2020, LNCS 12291, pp. 492–509, 2020.
https://doi.org/10.1007/978-3-030-60026-6_29

A *kinetics* of a chemical reaction network is an assignment of a rate function, depending on the concentrations of chemical species at the left hand side, to each reaction in the network. A kinetics for a chemical reaction network is called *mass-action* if for each reaction in the chemical reaction network, the rate function is a monomial in the concentrations of the chemical species with exponents given by the numbers of molecules of the species consumed in the reaction, multiplied by a constant called rate constant. Reactions are classified as zero-order, first-order, etc. according to the order of the monomial giving the rate. For reversible reactions, the net reaction rate is a binomial, the difference between the forward and backward rates. In the example above, k_{12}, k_{21}, k_{23}, and k_{32} are the rate constants. In this article we generally assume mass-action kinetics. We furthermore assume that reactions are reversible, unless explicitly specified otherwise.

The change in the concentration of each species over time in a reaction can be described via a system of autonomous ordinary differential equations. For instance, consider the chemical reaction network above and let x_1, x_2, x_3, x_4, x_5 be the indeterminates representing the concentrations of the species CO_2, H_2, CO, H_2O, and C, respectively. The corresponding differential equations are

$$\dot{x}_1 = p_1, \qquad p_1 = -k_{12}x_1x_2 + k_{21}x_3x_4 + k_{34}x_3^2 - k_{43}x_1x_5, \qquad (1)$$

$$\dot{x}_2 = p_2, \qquad p_2 = -k_{12}x_1x_2 + k_{21}x_3x_4, \qquad (2)$$

$$\dot{x}_3 = p_3, \qquad p_3 = k_{12}x_1x_2 - k_{21}x_3x_4 + -2k_{34}x_3^2 + 2k_{43}x_1x_5, \qquad (3)$$

$$\dot{x}_4 = p_4, \qquad p_4 = k_{12}x_1x_2 - k_{21}x_3x_4, \qquad (4)$$

$$\dot{x}_5 = p_5, \qquad p_5 = k_{34}x_3^2 - k_{43}x_1x_5. \qquad (5)$$

Each zero of the polynomials p_1, p_2, p_3, p_4, p_5 gives a concentration of species in which the system is in equilibrium. The zeros of p_1, p_2, p_3, p_4, and p_5 are called the steady states of the chemical reaction network. Accordingly, the ideal generated by $\langle p_1, p_2.p_3, p_4, p_5 \rangle$ in $\mathbb{Q}[k_{12}, k_{21}, k_{34}, k_{43}, x_1, x_2, x_3, x_4, x_5]$ is called the *steady state ideal* of the chemical reaction network. We consider the coefficient field $\mathbb{Q}$ because of computability issue. Otherwise, theoretically, our results hold for any coefficient field. The solutions of these polynomials can be in $\mathbb{R}$ or in $\mathbb{C}$.

For a thorough introduction to chemical reaction network theory, we refer to Feinberg's Book [16] and his lecture notes [15]. We follow the notation of Feinberg's book in this article.

An ideal is called binomial if it is generated by a set of binomials. In this article, we investigate whether the steady state ideal of a given chemical reaction network is binomial. We are interested in efficient algorithms for testing binomiality. Consider the steady state ideal

$$I = \langle p_1, p_2, p_3, p_4, p_5 \rangle \subseteq \mathbb{Q}[k_{12}, k_{21}, k_{34}, k_{43}, x_1, x_2, x_3, x_4, x_5], \qquad (6)$$

given by Eqs. (1)–(5). Reducing p_1, p_3, and p_4 with respect to p_2 and p_5, we have

$$I = \langle -k_{12}x_1x_2 + k_{21}x_3x_4, -k_{34}x_3^2 + k_{43}x_1x_5 \rangle, \qquad (7)$$

which shows that the ideal I is binomial. In this article, we work over the ring $\mathbb{Q}[k_{ij}, x_1, \ldots, x_n]$ and investigate binomiality over this ring.

Note that in the literature there exist also slightly different notions of binomiality. Eisenbud and Sturmfels in [11] call an ideal binomial if it is generated by polynomials with at most two terms. Following this definition, some authors, e.g., Dickenstein et al. in [32] have considered the steady state ideal as an ideal in the ring $\mathbb{Q}(k_{ij})[x_1, \ldots, x_n]$ and studied the binomiality of these ideals in $\mathbb{R}[x_1, \ldots, x_n]$ after specialising k_{ij} with positive real values. In order to distinguish between the two notions, we call *unconditionally binomial* a steady state ideal that is binomial in $\mathbb{Q}[k, x]$ (the notion used in this paper) and *conditionally binomial* a steady state ideal that is binomial in $\mathbb{Q}[x]$, i.e., for specified parameters k (the notion used in [32]).

The notions of binomial ideals and toric varieties have roots in thermodynamics, dating back to Boltzmann. Binomiality corresponds to detailed balance, which for reaction networks means that at thermodynamic equilibrium the forward and backward rates should be equal for all reactions. Detailed balance is a very important concept in thermodynamics, for instance it has been used by Einstein in his Nobel prize winning theory of the photoelectric effect [10], by Wegscheider in his thermodynamic theory of chemical reaction networks [37] and by Onsager for deriving his famous reciprocity relations [30]. Because detailed balance implies time reversal symmetry, systems with detailed balance can not produce directed movement and can only dissipate heat. This is important in applications, for instance, in molecular biology, where molecular motors can not function with detailed balance. Although most interesting molecular devices function without detailed balance and binomiality, some of their subsystems can satisfy these conditions. The interest of studying binomiality relies in the simplicity of the analysis of such subsystems. For instance, important properties such as multistationarity and stability are easier to establish for binomial systems. Toricity, also known as complex, or cyclic, or semi-detailed balance is also known since Boltzmann that has used it as a sufficient condition for deriving his famous H-theorem [1]. Binomiality implies toricity, but the converse is not true: in order to have binomiality, a toric system must obey constraints on the rates constants, such as the well known Weigscheider-Kolmogorov condition asking for the equality of the products of forward and backward rates constants in cycles of reversible reactions. In this paper, we focus on the situation when detailed balance is satisfied without conditions on the rate constants.

Detecting binomiality of an ideal, particularly of a steady state ideal, is a difficult problem, both from a theoretical and a practical point of view. The problem is typically solved by computing a Gröbner basis, which is EXPSPACE-complete [28]. Recent linear algebra approaches for solving the problem in a different setting than our problem construct large matrices which also points at the difficulty of the problem [6,29].

There is quite comprehensive literature on chemical reaction network theory. An excellent reference to this topic is [15,16]. As mathematical concepts, binomiality and toricity have been widely studied and their properties have been investigated by various authors, e.g., Fulton [17], Sturmfels [35], Eisenbud et al. [11]. Binomiality and toricity show up quite often in chemical reaction networks. Binomiality in the case of *detailed balancing* of reversible chemical reactions has been studied by Gorban et al. [19,20] and Grigoriev and Weber [24]. Feinberg [14] and Horn and Jackson [25] have studied toric dynamical systems. Gatermann et al. studied *deformed toricity* in [18]. Craciun, et al. have considered the toricity problem over the real numbers in [7] and have presented several interesting results in this regard, among them, they have shown that *complex balanced systems* are the same as toric dynamical systems, although *toric steady states* are different from that. It has been shown in [9,34] that the binomial structure will imply much simpler criteria for multistationarity. These results give strong motivation for one to study algorithms for detecting binomial networks. Especially, in [9], the authors defined *linearly binomial network* and they proposed sufficient conditions for a network to be linearly binomial. The proof is constructive even though it has not been presented as an algorithm. Their method is also quite straightforward and can handle more general networks in many applications.

Dickenstein et al. have presented sufficient linear algebra conditions with inequalities for binomiality of the steady state ideals in [29]. Their idea has been developed in [31], where the concept of MESSI reactions has been introduced. Conradi and Kahle have proved in [6] that for homogenous ideals (i.e., for chemical reaction networks without zero-order reactions), the sufficient condition of Dickenstein et al. is necessary as well and also introduced an algorithm for testing binomiality of homogenous ideals. As many biochemical networks are not homogeneous, the algorithm requires heuristics in such cases. The algorithm has been implemented in Maple and Macaulay II in [26,27] and experiments have been carried out on several biological models. Grigoriev et al. in [22] have considered the toricity of steady state ideals from a geometric point of view. Introducing shifted toricity, they presented algorithms, complexity bounds as well as experimental results for testing toricity using two important tools from symbolic computation, quantifier elimination [8,21,38] and Gröbner bases [4,5,12,13]. Recently, first order logic test for toricity have been introduced [33].

The main idea of this article is to consider the generators of the steady state ideal as sums of the binomials associated with the reactions rather than the monomials associated with the complexes. This is feasible for a reversible chemical reaction network. Following the above observation and assigning a binomial to each reaction, one can write the generators of the steady state ideal as sums of those binomials with integer coefficients.

As our main result, we have proved that a reversible chemical reaction network is unconditionally binomial if and only if it is "linearly" binomial (i.e., there exist linear combinations of the generators such that these combinations are binomials). More precisely, having represented of the generators of the steady state ideal as sum of binomials, one can test the binomiality exclusively using elementary row and column operations on the coefficient matrix of these binomials.

This can be done by computing the reduced row echelon form of the coefficient matrix, which yields an efficient method for testing binomiality.

Our main contributions in this article are the following.

1. We introduce a new representation of the generators of the steady state ideal of a reversible chemical reaction as a sum of certain binomials rather than monomials.
2. Using that representation, we assign a matrix with entries in $\mathbb{Z}$ to a reversible chemical reaction network, such that the binomiality of the steady state ideal can be tested by computing the reduced row echelon form of this matrix.
3. We prove a worst-case upper bound on the time complexity of our binomiality test. We experimentally compare our test with the existing binomiality tests in the literature, which demonstrates the applicability of our method.

Our representation of the steady state ideal as a sum of certain binomials, as well as the matrices associated with them are further original ideas presented in this paper. While typically complex-species matrices are used for testing binomiality, we use reaction-species matrices for this purpose.

The plan of the article is as follows. Section 1 gives an introduction to the necessary concepts of chemical reaction network theory, reviews the literature and presents the idea of this work. Section 2 includes the main definitions and results. In this section, we show our representation of the generators of the steady state ideal of a reversible chemical reaction network and present our algorithm for testing binomiality. In Sect. 3, we discuss the complexity of our method. We furthermore compare our algorithm with other existing algorithms in the literature via experiments. In Sect. 4 we summarise our results and draw some conclusions.

2   Testing Binomiality

In this section, we present our main result based on which we present an algorithm for testing unconditional binomiality of reversible chemical reaction networks. In Subsect. 2.1, we introduce a representation for the generators of the steady state ideal of a chemical reaction network as sum of binomials. We show that this representation is unique for reversible reaction networks, considering rate constants as indeterminates. In Subsect. 2.2, we define a matrix associated with a chemical reaction network which is essentially the species–reaction matrix, rather than the stoichiometric matrix which is the species–complex matrix. Having considered constant rates as indeterminates, the uniqueness of our matrix for reversible reactions comes from the uniqueness of representing the generators of the steady state ideal as sum of binomials.

2.1   Sum of Binomial Representation

Consider the following reversible reaction between two complexes C_1 and C_2.

$$C_1 \xrightleftharpoons[k\_21]{k\_12} C_2.$$

Let m_i, $i = 1, 2$, be the product of the concentrations of the species in C_i with the stoichiometric coefficients as the powers. We call m_i the monomial associated with C_i. Also let x_1 be the concentration of a species that is in C_1 with the stoichiometric coefficient α_1 and is not in C_2. The differential equation describing the kinetics of this species is

$$\dot{x}_1 = -\alpha_1(k_{12}m_1 + k_{21}m_2). \tag{8}$$

For a species in C_2 with stoichiometric coefficient α_2 which is not in C_1 with the concentration x_2, the differential equation will be

$$\dot{x}_2 = \alpha_2(k_{12}m_1 - k_{21}m_2). \tag{9}$$

For a species with concentration x_3 that appears in both C_1 and C_2, the differential equation will be $\dot{x}_3 = c(k_{12}m_1 - k_{21}m_2)$, where $c \in \mathbb{Z}$ is the difference between the corresponding stoichiometric coefficients in C_2 and C_1. Set $b_{12} := -k_{12}m_1 + k_{21}m_2$ and $b_{21} := k_{12}m_1 - k_{21}m_2$. The steady state ideal of the above chemical reaction network is $\langle b_{12}, b_{21} \rangle$, which is equal to $\langle b_{12} \rangle$, since $b_{12} = -b_{21}$.

For a reversible reaction network with more than one reaction, one can associate a binomial of the form $b_{ij} := k_{ij}m_i - k_{ji}m_j$ with each reaction. Then the polynomials generating the steady state ideal can be written as sums of b_{ij} with integer coefficients. We make this more precise in the following definition.

Definition 1. *Let C be a reversible chemical reaction network with the complexes $C_1, \ldots, C_s$, let k_{ij}, $1 \leq i \neq j \leq s$, be the rate constant of the reaction from C_i to C_j, and let $x_1, \ldots, x_n$ be the concentrations of the species in the chemical reaction network. We call a monomial m_i the monomial associated with C_i if m_i is the product of the concentrations of those species that appear in C_i with the stoichiometric coefficients of the species as the powers. If there is a reaction between C_i and C_j, then $b_{ij} := -k_{ij}m_i + k_{ji}m_j$ is called the binomial associated with the reaction from C_i to C_j, otherwise $b_{ij} := 0$.*

Example 1. Recall the following chemical reaction network form Sect. 1:

$$CO_2 + H_2 \xrightleftharpoons[k\_21]{k\_12} CO + H_2O,$$

$$2\,CO \xrightleftharpoons[k\_43]{k\_34} CO_2 + C.$$

Following the notation in Sect. 1, let x_1, x_2, x_3, x_4, x_5 be the concentrations of CO_2, H_2, CO, H_2O and C, respectively. The monomials associated with the complexes $CO_2 + H_2$, $CO + H_2O$, $2CO$ and $CO_2 + C$ are $x_1 x_2$, $x_3 x_4$, x_3^2 and $x_1 x_5$, respectively. The binomials associated with the two reactions in this network are $b_{12} = -k_{12}x_1x_2 + k_{21}x_3x_4$ and $b_{34} = -k_{34}x_3^2 + k_{43}x_1x_5$. As there is no reaction between the first and third complexes we have $b_{13} = b_{31} = 0$. Similarly, $b_{23} = b_{32} = 0$, $b_{14} = b_{41} = 0$ and $b_{24} = b_{42} = 0$. Also, by definition, $b_{21} = -b_{12}$, $b_{34} = -b_{43}$, etc. Using the binomials associated with the reactions, one can write the polynomials generating the steady state ideal as

$$p_1 = b_{12} - b_{34}, \tag{10}$$

$$p_2 = b_{12}, \tag{11}$$

$$p_3 = -b_{12} + 2b_{34}, \tag{12}$$

$$p_2 = -b_{12}, \tag{13}$$

$$p_2 = -b_{34}. \tag{14}$$

Hence, the steady state ideal can be written as

$$\langle p_1, p_2, p_3, p_4, p_5 \rangle = \langle b_{12}, b_{34} \rangle. \tag{15}$$

As Example 1 and the definition of the binomials b_{ij} in Definition 1 suggests one can write the generators of the steady state ideal of every reversible chemical reaction networks as sums of b_{ij} with integer coefficients, i.e., assuming that $\mathcal{R}$ is the set of reactions in the chemical reaction network

$$\dot{x}_k = p_k = \sum_{C_i \to C_j \in \mathcal{R}} c_{ij}^{(k)} b_{ij}, \tag{16}$$

for $k = 1 \ldots n$ and $c_{ij}^{(k)} \in \mathbb{Z}$.

For clarification, we may remind the reader that in this article we assume working over $\mathbb{Q}[k_{ij}, x_1, \ldots, x_n]$. This is the case, in particular, for Definition 1 and the discussion afterwards. In [32], the authors specialise k_{ij} with positive real values, in which case, the steady state ideal may or may not be binomial over $\mathbb{R}[x_1, \ldots, x_n]$. Similarly, specialising k_{ij} in Equation 16 can result in writing p_k as sum of different binomials. In other words, if k_{ij} specialised, the representation of p_k as sum of binomials in 16 is not necessarily unique. This is illustrated in the following example.

Example 2. [32, Example 2.3] Let $C_1 = 2A$, $C_2 = 2B$ and $C_3 = A + B$. Consider the reversible chemical reaction network given by the following reactions:

$$2\,A \underset{k\_21}{\overset{k\_12}{\rightleftharpoons}} 2\,B,$$

$$2\,A \underset{k\_31}{\overset{k\_13}{\rightleftharpoons}} A + B,$$

$$A + B \underset{k\_23}{\overset{k\_32}{\rightleftharpoons}} 2\,B.$$

Assuming x_1 and x_2 to be the concentrations of A and B, respectively, by Definition 1,

$$b_{12} = -k_{12}x_1^2 + k_{21}x_2^2, \tag{17}$$

$$b_{13} = -k_{13}x_1^2 + k_{31}x_1 x_2, \tag{18}$$

$$b_{23} = k_{23}x_2^2 - k_{32}x_1 x_2. \tag{19}$$

It can be checked that the generators of the steady state ideal can be written as

$$p_1 = 2b_{12} + b_{13} + b_{23}, \tag{20}$$

$$p_2 = -2b_{12} - b_{13} - b_{23}. \tag{21}$$

If $k_{31} = k_{32}$ then $k_{31}x_1x_2 = k_{32}x_1x_2$, hence $k_{31}x_1x_2$ will occur in b_{13} and b_{23} with opposite signs which will be cancelled out in $b_{13}+b_{23}$, resulting in writing p_1 as sum of b_{12} and $-k_{13}x_1^2 + k_{23}x_2^2$. This is another way of writing p_1 as sum of binomials. Because binomiality relies here on the condition $k_{31} = k_{32}$, this is an example of conditional binomiality.

If we consider the rate constants k_{ij} as indeterminates, i.e., if we consider the steady state ideal as an ideal over the ring $\mathbb{Q}[k_{ij}, x_1, \ldots, x_n]$, then the representation in Eq. (16) as sum of binomials b_{ij} will be unique. This has been presented in the following lemma. We may mention that having a different rate constant for different complexes is sometimes not guaranteed, however, this is the case for several models, e.g.. it happens in a model in [36], where the degradation rate δ_{pa} on the complex $C = P_a + P_r$ is the same as the degradation rate of P_a.

Lemma 1. *Given a reversible chemical reaction network with the notation of Definition 1, if k_{ij} are indeterminates then the generators of the steady state ideal can be uniquely written as sum of the binomials presented in Eq. (16).*

Proof. Assuming that k_{ij}, $1 \le i, j \le s$ are indeterminates, they will be algebraically independent over $\mathbb{Q}[x_1, \ldots, x_n]$. Therefore, for monomials m_t and $m_{t'}$ in $\mathbb{Q}[x_1, \ldots, x_n]$ associated with two distinct complexes and for all $1 \le i, j, i', j' \le s$, $k_{ij}m_t$ and $k_{i'j'}m_{t'}$ will be distinct monomials in $\mathbb{Q}[k_{ij}, x_1, \ldots, x_n]$. Hence binomials b_{ij} associated with the reversible reactions are not only pairwise distinct, but also their monomials are pairwise distinct in $\mathbb{Q}[k_{ij}, x_1, \ldots, x_n]$. This implies that the generators of the steady state ideal have unique representations in $\mathbb{Q}[k_{ij}, x_1, \ldots, x_n]$ as sum of b_{ij} with integer coefficients.

Having a unique representation as in Eq. (16) enables us to represent our binomial coefficient matrix, defined later, which is the base of our efficient algorithm for testing unconditional binomiality of reversible chemical reaction networks.

Considering rate constants k_{ij} as indeterminates, if a steady state ideal is unconditionally binomial, i.e., binomial in the ring $\mathbb{Q}[k_{ij}, x_1, \ldots, x_n]$, then its elimination ideal is binomial in the ring $\mathbb{Q}[x_1, \ldots, x_n]$. Indeed, the elimination of a binomial ideal is a binomial ideal. This can be seen from elimination property of Gröbner bases. Authors of [11] have studied binomial ideals and their properties intensively. In particular Corollary 1.3 in the latter article state the binomiality of the elimination ideal of a binomial ideal. We remind the reader that the definition of binomiality in this article is different from [11]. In the latter, binomial ideals have binomial and monomial generators, however, in the current article, we only consider binomial generators. Restricting the definition of binomial ideal to the ideals with only binomial generators, most of the result in [11] still holds, in particular the one about the elimination of binomial ideals. Therefore, if the steady state ideal of a chemical reaction network is binomial in $\mathbb{Q}[k_{ij}, x_1, \ldots, x_n]$, then its elimination $I \cap \mathbb{Q}[x_1, \ldots, x_n]$ is also a binomial ideal.

Geometrically, the above discussion can be explained via projection of the corresponding varieties. Given a chemical reaction network, assume that reaction rates k_{ij} are indeterminates and let the number of k_{ij} be t. Let V denote the steady state variety, i.e., the variety of the steady state ideal. V is a Zariski closed subset of $\mathbb{K}^{t+n}$, where $\mathbb{K}$ is an appropriate field (e.g., $\mathbb{C}$). If V is a coset of a subgroup of the multiplicative group $(\mathbb{K}^*)^{t+n}$, then the projection of V onto the space generated by $x_1 \ldots, x_n$, i.e., $V \cap (\mathbb{K}^*)^n$ is also a coset. In particular, the projection of a group is a group. Since the variety of a binomial ideal is a coset [22,23], the projection of the variety of a binomial ideal is the variety of a binomial ideal. As special cases, the projection of a toric variety, a shifted toric variety and a binomial variety (defined in [22,23]) is a toric, a shifted toric and a binomial variety, respectively. For a detailed study of toricity of steady state varieties, we refer to [22].

Remark 1.

– We may mention that in [7], the authors have studied *toric dynamical systems*, where they have considered working over $\mathbb{Q}[k_{ij}, x_1, \ldots, x_n]$ and presented several interesting results. In particular, Theorem 7 in that article states that a chemical reaction network is toric if and only if the rate constants lie in the variety of a certain ideal in $\mathbb{Q}[k_{ij}]$, called the *moduli ideal*.
– Toric dynamical systems are known as *complex balancing* mass action systems [7].

2.2   The Algorithm

Definition 2. *Let C be a reversible chemical reaction network as in Definition 1 and assume that the generators of its steady state ideal are written as the linear combination of the binomials associated with its reactions as in Eq. 16, i.e.,*

$$p_k = \sum_{C_i \to C_i \in \mathcal{R}}^{s} c_{ij}^{(k)} b_{ij} \quad for \quad k = 1, \ldots, n.$$

We define the binomial coefficient matrix of C to be the matrix whose rows are labeled by $p_1, \ldots, p_n$ and whose columns are labeled by non-zero b_{ij} and the entry in row p_k and column b_{ij} is $c_{ij}^{(k)} \in \mathbb{Z}$.

By the definition, the binomial coefficient matrix of a reversible chemical reaction network is the coefficient matrix of the binomials that occur in the representation of the generators of the steady state ideal as sum of binomials. As we consider k_{ij} indeterminates, the representation of the generators of the steady state ideal of a given complex is unique, which implies that the binomial coefficient matrix of a given complex is unique too.

Example 3. Consider the chemical reaction network in Example 1, with generators of the steady state ideal as follows.

$$p_1 = b_{12} - b_{34}, \tag{22}$$

$$p_2 = b_{12}, \tag{23}$$

$$p_3 = -b_{12} + 2b_{34}, \tag{24}$$

$$p_2 = -b_{12}, \tag{25}$$

$$p_2 = -b_{34}. \tag{26}$$

The binomial coefficient matrix of this chemical reaction network is

$$M = \begin{array}{c} \\ p_1 \\ p_2 \\ p_3 \\ p_4 \\ p_5 \end{array} \begin{pmatrix} \overset{b_{12}}{1} & \overset{b_{34}}{-1} \\ 1 & 0 \\ -1 & 2 \\ -1 & 0 \\ 0 & -1 \end{pmatrix}. \tag{27}$$

Example 4. Another simple example is the reaction

$$4\,A \underset{k\_21}{\overset{k\_12}{\rightleftharpoons}} A + B,$$

with the binomial associated with it as $b_{12} := -k_{12}x_1^4 + k_{21}x_1x_2$, where x_1 is the concentration of A and x_2 is the concentration of B. The steady state ideal is generated by $\{3b_{12}, -b_{12}\}$, and the binomial coefficient matrix for this network is $\binom{3}{-1}$.

One can test binomiality of the steady state ideal of a reversible reaction network using its binomial coefficient matrix.

Theorem 1. *The steady state ideal of a reversible chemical reaction network is unconditionally binomial, i.e., binomial in $\mathbb{Q}[k_{ij}, x_1, \ldots, x_n]$, if and only if the reduced row echelon form of its binomial coefficient matrix has at most one non-zero entry at each row.*

Proof. Let $G = \{p_1, \ldots, p_n\} \subseteq \mathbb{Q}[k_{ij}, x_1, \ldots, x_n]$ be a generating set for the steady state ideal of a given reversible chemical reaction network $\mathcal{C}$, and let $\{b_{ij} \mid 1 \leq i \neq j \leq s\}$ be the ordered set of non-zero binomials associated with the reactions. Fix a term order on the monomials in $\mathbb{Q}[k_{ij}, x_1, \ldots, x_n]$.

First we prove that if the reduced row echelon form of the binomial coefficient matrix has at most one non-zero entry at each row, then the steady state ideal is binomial. The proof of this side of the proposition comes from the definition of reduced row echelon form. In fact, the reduced row echelon form of the binomial coefficient matrix of $\mathcal{C}$ can be computed by row reduction in that matrix, which is equivalent to the reduction of the generators of the steady state ideal with respect to each other. Therefore, computing the reduced row echelon form of the binomial coefficient matrix and multiplying it with the vector of binomials b_{ij}, one can obtain another basis for the steady state ideal. Having this, if the reduced row echelon form has at most one non-zero entry at each row, then the new basis for the steady state ideal will only include b_{ij}. Therefore the steady state ideal will be binomial.

Now we prove the "only if" part of the proposition, that is, if the steady state ideal of $\mathcal{C}$ is binomial, then the reduced row echelon form of the binomial coefficient matrix has at most one non-zero entry at each row. We claim that for each pair of polynomials $p_t, p_m \in G$, p_t is reducible with respect to p_m if and only if there exists a binomial b_{ij} that occurs in both p_t and p_m and includes their leading terms. The "only if" part of the claim is obvious. To prove the "if" part of the claim, let p_m be reducible with respect to p_t. Then the leading term of p_m divides the leading term of p_t. Since the leading terms are multiples of k_{ij} and these are disjoint indeterminates, this is only possible if both of the leading terms are equal. If the leading terms are equal, then b_{ij} in which the leading terms occur, must itself occur in both p_t and p_m. Therefore p_t and p_m share a binomial associated with a reaction, which is in contradiction with our assumption.

From the above claim and the definition of the reduced row echelon form one can see that $p_1, \ldots, p_n$ are pairwise irreducible if and only if the binomial coefficient matrix of $\mathcal{C}$ is in reduced row echelon form.

Now we prove that $p_1, \ldots, p_n$ are pairwise irreducible if and only if they form a Gröbner basis in which polynomials are pairwise irreducible. Note that this does not necessarily imply that G is a a reduced Gröbner basis, as p_i are not necessarily monic. Assume that $p_1, \ldots, p_n$ are pairwise irreducible. We prove that the greatest common divisor of each pair of the leading terms of the $p_1, \ldots, p_n$ is 1. By contradiction, assume that there exists a monomial not equal to 1 which divides the leading terms of both p_t, p_m, for $1 \leq t, m \leq n$. Then there exists a variable x_l such that x_l divides the leading terms of p_t and p_m. Since each leading term is the monomial associated with a complex, the species with concentration x_1 occurs in two complexes with associated monomials as the leading terms of p_t and p_m. Then both p_t and p_m have as their summand the binomials that are associated with the reactions including those complexes. As for each complex there exists at least one binomial associated, both p_m and p_t have as a summand one common binomial b_{ij}. However, we had already proved that this implies that p_t and p_m are not pairwise irreducible, which is a contradiction to the assumption that the greatest common divisor of the leading terms of p_t and p_m is not 1. Now by Buchberger's first criterion if the greatest common divisor of the leading terms of each pair of polynomials in G is 1 then G is a Gröbner basis. The other side of this claim is obvious.

From what we have proved until now, we can conclude that the binomial coefficient matrix of $\mathcal{C}$ is in reduced row echelon form if and only if G is a Gröbner basis with pairwise irreducible elements. On the other hand, by a result of Eisenbud and Sturmfels [11], the steady state ideal of $\mathcal{C}$ is binomial if and only if every Gröbner basis of it includes binomials. Therefore we conclude that the steady state ideal is binomial if and only if the reduced row echelon form of the binomial coefficient matrix has at most one non-zero entry in each row. □

Function BinomialityTest($\mathcal{C}$)

    Input: $\mathcal{C} = \{(m_1, \ldots, m_s) \in [X]^n, k_{ij}\}$

    Output: Binomial or NotBinomial

1    $b_{ij} := -k_{ij}m_i + k_{ji}m_j, 1 \leq i \neq j \leq s$

2    $B := (b_{ij}, 1 \leq i \neq j \leq m)$

3    $p_k := \sum c_{ij}^k b_{ij}, 1 \leq k \leq n$

4    $M := \mathtt{Matrix}(c_{ij}^k)$

5    $\tilde{M} = \mathtt{ReducedRowEchelonForm}(M)$

6    $G := \tilde{M}B$

7   **if** IsBinomial(G) **then**

8 | $R := Binomial$

9   **else**

10 | $R := NotBinomial$

11   **return** R

Algorithm 1: Testing Unconditional Binomiality of Reversible Chemical Reaction Networks

Example 5. Following Example 3, one case easily see that the reduced row echelon form of the binomial coefficient matrix (27) is

$$
M = \begin{array}{c} \\ p_1 \\ p_2 \\ p_3 \\ p_4 \\ p_5 \end{array}
\begin{array}{c} b_{12} \quad b_{34} \\ \left(\begin{array}{cc} 1 & 0 \\ 0 & 1 \\ 0 & 0 \\ 0 & 0 \\ 0 & 0 \end{array} \right) \end{array}, \tag{28}
$$

which means that the steady state ideal is unconditionally binomial and is generated by $\{b_{12}, b_{34}\}$.

Theorem 1 yields Algorithm 1 for testing unconditional binomiality. The input of the algorithm is a reversible chemical reaction network, given by the vector of monomials associated with its complexes, $(m_1, \ldots, m_s)$, and the rates k_{ij}. It uses a function *IsBinomial* which takes a set of polynomials and checks if all of them are binomial.

Generalisation to Non-Reversible Networks. The unconditional binomiality test via the binomial coefficient matrix for a reversible chemical reaction network can be used as a subroutine for testing unconditional binomiality of an arbitrary chemical reaction network. In order to do so, partition a given chemical reaction network $\mathcal{C}$ into a reversible reaction network $\mathcal{C}_1$ and a non–reversible reaction network $\mathcal{C}_2$. Apply Algorithm 1 to $\mathcal{C}_1$, construct its binomial coefficient matrix, say M_1. Construct the stoichiometric coefficient matrix of $\mathcal{C}_2$, say M_2, and consider the block matrix $M := (\tilde{M}_1 | M_2)$. Compute the row reduced echelon form of M, say $\tilde{M}$. If all the rows of $\tilde{M}$ have at most one non-zero entry, then the steady state ideal is binomial.

Otherwise, one can consider computing $\tilde{M}$ as a preprocessing step and run another method, e.g., Gröbner bases, quantifier elimination as in [22], or the method in Dickenstein, et al. [29].

3   Complexity and Comparisons

Proposition 1. *Let r be the number of reactions and n be the number of species of a reversible chemical reaction network C. The asymptotic worst case time complexity of testing unconditional binomiality of the steady state ideal of C via Algorithm 1 can be bounded by $\mathcal{O}(\max(r, n)^\omega)$ where $\omega \approx 2.3737$, which is also the complexity of matrix multiplication.*

Proof. The operations in steps 1–4 and 7–11 are at most linear in terms of r and n. Since M is a matrix of size $n \times r$, where $r = |b_{ij}|$, and B is a vector of size r, computing reduced row echelon form in step 5 and also the matrix multiplication in step 6 will cost at most $\mathcal{O}(\max(r, n)^\omega)$. Therefore the total number of operations in the algorithm can be bounded by $\mathcal{O}(\max(r, n)^\omega)$.

In [22, Section 4] it has been shown that there exists an exponential asymptotic worst case upper bound on the time complexity of testing toricity. An immediate consequence of that result is that the time complexity of testing binomiality can be bounded by the same exponential function. Following the arguments in [22, Section 4], one can show that there exists an algorithm for testing binomiality over $\mathbb{Q}[k_{ij}, x_1, \ldots, x_n]$ and $\mathbb{Q}[x_1, \ldots, x_n]$ simultaneously, with an exponential upper bound for the worst case time complexity.

As mentioned earlier in Sect. 2, the reduced Gröbner basis of a binomial ideal, with respect to every term order, includes only binomials. This directly can be seen from running Buchberger's algorithm and that S–polynomials and their reductions by binomials are binomial. Eisenbud et al.' article [11], with a slightly different definition of binomial ideals, investigates many properties of binomial ideals using the latter fact. Following this fact, a typical method for testing binomiality is via computing a reduced Gröbner basis of a steady state ideal $I \subseteq \mathbb{Q}[k_{ij}, x_1, \ldots, x_n]$ The drawback of computing Gröbner bases is that this is EXPSPACE-complete [28]. So our algorithm is asymptotically considerably more efficient than Gröbner basis computation.

Example 6. (Models from the BioModels Repository[1].)

- There are twenty non–reversible biomodels in which Gröbner basis computations done in [22] for testing conditional binomiality do not terminate in a six–hour time limit, however our algorithm terminates in less than three seconds. Also there are six cases in which Gröbner basis computations terminate in less than six hours, but are at least 1000 times slower than our algorithm. Finally there are ten models in which Gröbner basis is at least 500 times slower than our computations.

---

[1] https://www.ebi.ac.uk/biomodels/.

- There are sixty nine biomodels that are not considered for computation in [22] because of the of the unclear numeric value of their rate constants. Our computations on almost all of those cases terminated in less than a second.
- (Reversible models from the BioModels Repository) Biomodels 491 and 492 are both reversible. Biomodel 491 has 52 species and 86 reactions. The binomial coefficient matrix of this biomodel has size 52×86 and has ± 1 entries. A reduced row echelon form computation in Maple reveals in 0.344 s that it is unconditionally binomial, while a Gröbner basis computation takes more than 12 s to check its conditional binomiality. BioModel 492 has also 52 species, and includes 88 reactions. The binomial coefficient matrix has entries ± 1 and is of size 52×88. This biomodel is also unconditionally binomial. It takes 0.25 s for Maple to check its unconditional binomiality via Algorithm 1 in Maple, while a Gröbner basis computation takes near 18 s, as one can see in the computations in [22, Table 3], which show the group structure of the steady state varieties of the models.

Dickenstein et al. in [29] have proposed a method for testing toricity of a chemical reaction network. The definitions and purpose of that work are slightly different from our article, hence comparisons between those two methods should be treated with caution. While we focus on unconditional binomiality of the steady state ideals of reversible reaction networks, i.e., binomiality in $\mathbb{Q}[k_{ij}, x_1, \ldots, x_n]$, with the aim of efficiency of the computations, the authors of the above article are interested in conditional binomiality with algebraic dependencies between k_{ij} such that the elimination ideal is binomial. Having mentioned that, our method leads to the computation of reduced row echelon form of a matrix of size $n \times r$ with integer entries which is polynomial time, while Theorem 3.3. in [29] requires constructing a matrix of size $n \times s$ with entries from $\mathbb{Z}[k_{ij}]$ and finding a particular partition of its kernel.

Considering Example 2.3 in [32], our algorithm constructs the matrix M and its reduced row echelon form $\tilde{M}$:

$$M = \begin{pmatrix} 1 & 1 & -2 \\ -1 & -1 & 2 \end{pmatrix}, \quad \tilde{M} = \begin{pmatrix} 1 & 1 & 2 \\ 0 & 0 & 0 \end{pmatrix}, \tag{29}$$

and we see that the steady state ideal is not unconditionally binomial over $\mathbb{Q}[k_{ij}, x_1, \ldots, x_n]$. The method in [32] constructs

$$\begin{pmatrix} -2k_{12} - k_{13} & 2k_{21} + k_{23} & k_{31} - k_{32} \\ 2k_{12} + k_{13} & -2k_{21} - k_{23} & -k_{31} + k_{32} \end{pmatrix}, \tag{30}$$

and finds an appropriate partition, which shows that the steady state ideal is binomial in $\mathbb{Q}[x_1, \ldots, x_n]$ if and only if $k_{31} = k_{32}$. As a larger example, consider the chemical reaction network given in Example 3.13 in [32] and assume that it is a reversible chemical reaction network. Our method constructs a matrix with entries ± 1 of size 9×8 and computes its reduced row echelon form (in this case reduced row echelon form, as entries are ± 1). The method described in [32] leads to a 9×10 matrix with entries as linear polynomials in $\mathbb{Z}[k_{ij}]$ and computes a particular partition of the kernel of the matrix.

For homogeneous ideals, Conradi and Kahle have shown in [6] that the sufficient condition for conditional binomiality in [32] is necessary, too. Their Algorithm 3.3 tests conditional binomiality of a homogeneous ideal, which can be generalised by homogenising. The algorithm computes a basis for the ideal degree by degree and performs reductions with respect to the computed basis elements at each degree step. Since our algorithm is intended for steady state ideals of reversible chemical reaction networks, which are not necessarily homogeneous, our following comparison with the Conradi–Kahle algorithm bears a risk of being biased by homogenisation. We discuss the execution of both algorithms on Example 3.15 in [32]. This chemical reaction network does not satisfy the sufficient condition presented in [32, Theorem 3.3]. Testing this condition leads to the construction of a 9×13 matrix with entries in $\mathbb{Z}[k_{ij}]$, followed by further computations, including finding a particular partition of its kernel. Theorem 3.19 in [32] is a generalisation of Theorem 3.3 there, which can test conditional binomiality of this example by adding further rows and columns to the matrix. Conradi and Kahle also treat this example with their algorithm. This requires the construction of a coefficient matrix of size 9×13 with entries in $\mathbb{Z}[k_{ij}]$ and certain row reductions. If we add reactions so that the reaction network becomes reversible, our algorithm will construct a matrix of size 9×9 with entries ± 1 and compute its reduced row echelon form to test unconditional binomiality in $\mathbb{Q}[k_{ij}, x_1, \ldots, x_9]$.

4   Conclusions

Binomiality of steady state ideals is an interesting problem in chemical reaction network theory. It has a rich history and literature and is still an active research area. For instance, recently MESSI systems have been introduced [31] following the authors' work on binomiality of a system. Finding binomiality and toricity is computationally hard from both a theoretical and a practical point of view. It typically involves computations of Gröbner bases, which is EXPSPACE-complete.

In a recent work [22], we investigated toricity of steady state varieties and gave efficient algorithms. In particular, we experimentally investigated toricity of biological models systematically via quantifier elimination. Besides that, we presented exponential theoretical bounds on the toricity problem. The current article, restricting to reversible reaction networks, aims at an efficient linear algebra approach to the problem of unconditional binomiality, which can be considered as a special case of the toricity problem.

In that course, considering rate constants as indeterminates, we assign a unique binomial to each reaction and construct the coefficient matrix with respect to these binomials. Our algorithm proposed here computes a reduced row echelon form of this matrix in order to detect unconditional binomiality. The algorithm is quite efficient, as it constructs comparatively small matrices whose entries are integers. It is a polynomial time algorithm in terms of the number of species and reactions. While other existing methods for testing conditional binomiality have different settings and purposes than our algorithm, for the common cases, our algorithm has advantages in terms of efficiency.

Acknowledgments. This work has been supported by the bilateral project ANR-17-CE40-0036 and DFG-391322026 SYMBIONT [2,3].

References

1. Boltzmann, L.: Lectures on Gas Theory. University of California Press, Berkeley and Los Angeles, CA (1964)
2. Boulier, F., et al.: The SYMBIONT project: symbolic methods for biological networks. ACM Commun. Comput. Algebra **52**(3), 67–70 (2018). https://doi.org/10.1145/3313880.3313885
3. Boulier, F., et al.: The SYMBIONT project: Symbolic methods for biological networks. F1000Research 7(1341) (2018). https://doi.org/10.7490/f1000research.1115995.1
4. Buchberger, B.: Ein Algorithmus zum Auffinden der Basiselemente des Restklassenringes nach einem nulldimensionalen Polynomideal. Doctoral dissertation, Mathematical Institute, University of Innsbruck, Austria (1965). https://doi.org/10.2307/1971361
5. Buchberger, B.: Ein Algorithmisches Kriterium für die Lösbarkeit eines algebraischen Gleichungssystems. Aequationes Math. **3**, 374–383 (1970). https://doi.org/10.1007/BF01817776
6. Conradi, C., Kahle, T.: Detecting binomiality. Adv. Appl. Math. **71**, 52–67 (2015). https://doi.org/10.1016/j.aam.2015.08.004
7. Craciun, G., Dickenstein, A., Shiu, A., Sturmfels, B.: Toric dynamical systems. J. Symb. Comput. **44**(11), 1551–1565 (2009). https://doi.org/10.1016/j.jsc.2008.08.006
8. Davenport, J.H., Heintz, J.: Real quantifier elimination is doubly exponential. J. Symb. Comput. **5**(1–2), 29–35 (1988). https://doi.org/10.1016/S0747-7171(88)80004-X
9. Dickenstein, A., Millán, M.P., Shiu, A., Tang, X.: Multistationarity in structured reaction networks. Bull. Math. Biol. **81**, 1527–1581 (2019). https://doi.org/10.1007/s11538-019-00572-6
10. Einstein, A.: Strahlungs-emission und-absorption nach der Quantentheorie. Verh. Dtsch. Phys. Ges. **18**, 318–323 (1916)
11. Eisenbud, D., Sturmfels, B.: Binomial ideals. Duke Math. J. **84**(1), 1–45 (1996). https://doi.org/10.1215/S0012-7094-96-08401-X
12. Faugère, J.C.: A new efficient algorithm for computing Gröbner bases (F4). J. Pure Appl. Algebra **139**(1), 61–88 (1999). https://doi.org/10.1016/S0022-4049(99)00005-5
13. Faugère, J.C.: A new efficient algorithm for computing gröbner bases without reduction to zero (F5). In: Proceedings of ISSAC 2002, pp. 75–83. ACM, New York (2002). https://doi.org/10.1145/780506.780516
14. Feinberg, M.: Complex balancing in general kinetic systems. Arch. Ration. Mech. An. **49**(3), 187–194 (1972). https://doi.org/10.1007/BF00255665
15. Feinberg, M.: Lectures on chemical reaction networks (1979)

16. Feinberg, M.: Foundations of Chemical Reaction Network Theory. AMS, vol. 202. Springer, Cham (2019). https://doi.org/10.1007/978-3-030-03858-8
17. Fulton, W.: Introduction to Toric Varieties, Annals of Mathematics Studies. Princeton University Press, New Jersey (1993)
18. Gatermann, K.: Counting stable solutions of sparse polynomial systems in chemistry. In: Symbolic Computation: Solving Equations in Algebra, Geometry, and Engineering, Contemporary Mathematics, vol. 286, pp. 53–69. AMS, Providence, RI (2001). https://doi.org/10.1090/conm/286/04754
19. Gorban, A.N., Kolokoltsov, V.N.: Generalized mass action law and thermodynamics of nonlinear Markov processes. Math. Model. Nat. Phenom. **10**(5), 16–46 (2015). https://doi.org/10.1051/mmnp/201510503
20. Gorban, A.N., Yablonsky, G.S.: Three waves of chemical dynamics. Math. Model. Nat. Phenom. **10**(5), 1–5 (2015). https://doi.org/10.1051/mmnp/201510501
21. Grigoriev, D.Y.: Complexity of deciding Tarski algebra. J. Symb. Comput. **5**(1–2), 65–108 (1988). https://doi.org/10.1016/S0747-7171(88)80006-3
22. Grigoriev, D., Iosif, A., Rahkooy, H., Sturm, T., Weber, A.: Efficiently and effectively recognizing toricity of steady state varieties. Math. Comput. Sci. (2020). https://doi.org/10.1007/s11786-020-00479-9
23. Grigoriev, D., Milman, P.D.: Nash resolution for binomial varieties as Euclidean division. A priori termination bound, polynomial complexity in essential dimension 2. Adv. Math. **231**(6), 3389–3428 (2012). https://doi.org/10.1016/j.aim.2012.08.009
24. Grigoriev, D., Weber, A.: Complexity of solving systems with few independent monomials and applications to mass-action kinetics. In: Gerdt, V.P., Koepf, W., Mayr, E.W., Vorozhtsov, E.V. (eds.) CASC 2012. LNCS, vol. 7442, pp. 143–154. Springer, Heidelberg (2012). https://doi.org/10.1007/978-3-642-32973-9_12
25. Horn, F., Jackson, R.: General mass action kinetics. Arch. Ration. Mech. An. **47**(2), 81–116 (1972). https://doi.org/10.1007/BF00251225
26. Iosif, A., Rahkooy, H.: Analysis of the Conradi-Kahle algorithm for detecting binomiality on biological models. CoRR abs/1912.06896 (2019)
27. Iosif, A., Rahkooy, H.: MapleBinomials, a Maple package for testing binomiality of ideals (2019), http://doi.org/10.5281/zenodo.3564428
28. Mayr, E.W., Meyer, A.R.: The complexity of the word problems for commutative semigroups and polynomial ideals. Adv. Math. **46**(3), 305–329 (1982). https://doi.org/10.1016/0001-8708(82)90048-2
29. Millán, M.P., Dickenstein, A., Shiu, A., Conradi, C.: Chemical reaction systems with toric steady states. Bull. Math. Biol. **74**(5), 1027–1065 (2012)
30. Onsager, L.: Reciprocal relations in irreversible processes. I. Phys. Rev. **37**(4), 405 (1931). https://doi.org/10.1103/PhysRev.37.405
31. Pérez Millán, M., Dickenstein, A.: The structure of MESSI biological systems. SIAM J. Appl. Dyn. Syst. **17**(2), 1650–1682 (2018). https://doi.org/10.1137/17M1113722
32. Pérez Millán, M., Dickenstein, A., Shiu, A., Conradi, C.: Chemical reaction systems with toric steady states. Bull. Math. Biol. **74**(5), 1027–1065 (2012). https://doi.org/10.1007/s11538-011-9685-x
33. Rahkooy, H., Sturm, T.: First-order tests for toricity. In: Boulier F., et al. (eds.) CASC 2020, LNCS, vol. 12291, pp. 492–509 (2020). https://doi.org/10.1007/978-3-030-60026-6_29
34. Sadeghimanesh, A., Feliu, E.: The multistationarity structure of networks with intermediates and a binomial core network. Bull. Math. Biol. **81**, 2428–2462 (2019). https://doi.org/10.1007/s11538-019-00612-1

35. Sturmfels, B.: Gröbner Bases and Convex Polytopes, University Lecture Series, vol. 8. AMS, Providence, RI (1996)
36. Vilar, J.M.G., Yuan Kueh, H., Barkai, N., Leibler, S.: Mechanisms of noise-resistance in genetic oscillators. In: Proceedings of the National Academy of Science of the USA, vol. 99.9, pp. 5988–5992 (2002). https://doi.org/10.1073/pnas.092133899
37. Wegscheider, R.: Über simultane Gleichgewichte und die Beziehungen zwischen Thermodynamik und Reactionskinetik homogener Systeme. Monatsh. Chem. Verw. Tl. **22**(8), 849–906 (1901). https://doi.org/10.1007/BF01517498
38. Weispfenning, V.: The complexity of linear problems in fields. J. Symb. Comput. **5**(1–2), 3–27 (1988). https://doi.org/10.1016/S0747-7171(88)80003-8

First-Order Tests for Toricity

Hamid Rahkooy[1] ⓘ and Thomas Sturm[1,2](✉) ⓘ

[1] CNRS, Inria, and the University of Lorraine, Nancy, France
hamid.rahkooy@inria.fr, thomas.sturm@loria.fr
[2] MPI Informatics and Saarland University, Saarbrücken, Germany
sturm@mpi-inf.mpg.de

Abstract. Motivated by problems arising with the symbolic analysis of steady state ideals in Chemical Reaction Network Theory, we consider the problem of testing whether the points in a complex or real variety with non-zero coordinates form a coset of a multiplicative group. That property corresponds to Shifted Toricity, a recent generalization of toricity of the corresponding polynomial ideal. The key idea is to take a geometric view on varieties rather than an algebraic view on ideals. Recently, corresponding coset tests have been proposed for complex and for real varieties. The former combine numerous techniques from commutative algorithmic algebra with Gröbner bases as the central algorithmic tool. The latter are based on interpreted first-order logic in real closed fields with real quantifier elimination techniques on the algorithmic side. Here we take a new logic approach to both theories, complex and real, and beyond. Besides alternative algorithms, our approach provides a unified view on theories of fields and helps to understand the relevance and interconnection of the rich existing literature in the area, which has been focusing on complex numbers, while from a scientific point of view the (positive) real numbers are clearly the relevant domain in chemical reaction network theory. We apply prototypical implementations of our new approach to a set of 129 models from the BioModels repository.

Keywords: Binomial ideals · Chemical reaction networks · Logic computation · Scientific computation · Symbolic computation · Toric varieties.

1 Introduction

We are interested in situations where the points with non-zero coordinates in a given complex or real variety form a multiplicative group or, more generally, a coset of such a group. For irreducible varieties this corresponds to toricity [16,23] and shifted toricity [27,28], respectively, of both the varieties and the corresponding ideals.

While toric varieties are well established and have an important role in algebraic geometry [16,23], our principal motivation here to study generalizations of toricity comes from the sciences, specifically *chemical reaction networks* such as

© Springer Nature Switzerland AG 2020
F. Boulier et al. (Eds.): CASC 2020, LNCS 12291, pp. 510–527, 2020.
https://doi.org/10.1007/978-3-030-60026-6_30

the following model of the kinetics of intra- and intermolecular zymogen activation with formation of an enzyme-zymogen complex [22], which can also be found as model no. 92^1 in the BioModels database [9]:

$$Z \xrightarrow{0.004} P + E$$

$$Z + E \underset{2.1E\text{-}4}{\overset{1000}{\rightleftharpoons}} E\text{--}Z \xrightarrow{5.4E\text{-}4} P + 2E$$

Here Z stands for zymogen, P is a peptide, E is an enzyme, E—Z is the enzyme substrate complex formed from that enzyme and zymogen. The reactions are labelled with reaction *rate constants*.

Let $x_1, \ldots, x_4 : \mathbb{R} \to \mathbb{R}$ denote the *concentrations* over time of the *species* Z, P, E, E–Z, respectively. Assuming mass action kinetics one can derive *reaction rates* and furthermore a system of autonomous ordinary differential equations describing the development of concentrations in the overall network [20, Section 2.1.2]:

$$\begin{aligned}
\dot{x}_1 &= f_1/100000, & f_1 &= -100000000x_1x_2 - 400x_1 + 21x_4, \\
\dot{x}_2 &= f_2/100000, & f_2 &= -100000000x_1x_2 + 400x_1 + 129x_4, \\
\dot{x}_3 &= f_3/50000, & f_3 &= 200x_1 + 27x_4, \\
\dot{x}_4 &= f_4/4000, & f_4 &= 4000000x_1x_2 - 3x_4.
\end{aligned}$$

The chemical reaction is in *equilibrium* for positive concentrations of species lying in the real variety of the steady state ideal

$$\langle f_1, \ldots, f_4 \rangle \subseteq \mathbb{Z}[x_1, \ldots, x_4],$$

intersected with the first orthant of $\mathbb{R}^4$.

Historically, the principle of *detailed balancing* has attracted considerable attention in the sciences. It states that at equilibrium every single reaction must be in equilibrium with its reverse reaction. Detailed balancing was used by Boltzmann in 1872 in order to prove his H-theorem [4], by Einstein in 1916 for his quantum theory of emission and absorption of radiation [15], and by Wegscheider [51] and Onsager [43] in the context of *chemical kinetics*, which lead to Onsager's Nobel prize in Chemistry in 1968. In the field of symbolic computation, Grigoriev and Weber [29] applied results on binomial varieties to study reversible chemical reactions in the case of detailed balancing.

In particular with the assumption of irreversible reactions, like in our example, detailed balancing has been generalized to *complex balancing* [19,20,33], which has widely been used in the context of chemical reaction networks. Here one considers *complexes*, like Z, P + E, Z + E, etc. in our example, and requires for every such complex that the sum of the reaction rates of its inbound reactions equals the sum of the reaction rates of its outbound reactions.

---

[1] https://www.ebi.ac.uk/compneur-srv/biomodels-main/publ-model.do?
mid=BIOMD0000000092.

Craciun et al. [11] showed that *toric dynamical systems* [18,33], in turn, generalize complex balancing. The generalization of the principle of complex balancing to toric dynamical systems has obtained considerable attention in the last years [11,24,41,45]. Millan et al. [45] considered steady state ideals with binomial generators. They presented a sufficient linear algebra condition on the stoichiometry matrix of a chemical reaction network in order to test whether the steady state ideal has binomial generators. Conradi and Kahle showed that the sufficient condition is even equivalent when the ideal is homogenous [10,34,35]. That condition also led to the introduction of MESSI systems [44]. Recently, binomiality of steady states ideals was used to infer network structure of chemical reaction networks out of measurement data [50]. Katthän et al. [36] discussed several questions that are in connection with toric varieties and binomial ideals. For instance they consider whether an ideal can be transformed into a binomial ideal by a change of coordinates.

Besides its scientific adequacy as a generalization of complex balancing there are practical motivations for studying toricity. Relevant models are typically quite large. For instance, with our comprehensive computations in this article we will encounter one system with 90 polynomials in dimension 71, in the sense that there are 71 variables originating from different species. This brings symbolic computation to its limits. Our best hope is to discover systematic occurrences of specific structural properties in the models coming from a specific context, e.g. the life sciences, and to exploit those structural properties towards more efficient algorithms. In that course, toricity could admit tools from toric geometry, e.g., for dimension reduction.

Detecting toricity of varieties in general, and of steady state varieties of chemical reaction networks in particular, is a difficult problem. The first issue in this regard is finding suitable notions to describe the structure of the steady states. Existing work, such as the publications mentioned above, typically focuses on the complex numbers and addresses algebraic properties of the steady state ideal, e.g., the existence of binomial Gröbner bases. Only recently, a group of researchers including the authors of this article have taken a geometric approach, focusing on varieties rather than ideals [27,28]. Besides irreducibility, the characteristic property for varieties V to be toric over a field $\mathbb{K}$ is that $V \cap (\mathbb{K}^*)^n$ forms a multiplicative group. More generally, one considers *shifted toricity*, where $V \cap (\mathbb{K}^*)^n$ forms a coset of a multiplicative group.

It is important to understand that chemical reaction network theory generally takes place in the interior of the first orthant of $\mathbb{R}^n$, i.e., all species concentrations and reaction rates are assumed to be strictly positive [20]. Considering $(\mathbb{C}^*)^n$ in contrast to $\mathbb{C}^n$ resembles the strictness condition, and considering also $(\mathbb{R}^*)^n$ in [27] was another step in the right direction.

The plan of the article is as follows. In Sect. 2 we motivate and formally introduce first-order characterizations for shifted toricity, which have been used already in [27], but there exclusively with real quantifier elimination methods. In Sect. 3 we put a model theoretic basis and prove transfer principles for our characterizations throughout various classes of fields, with zero as well as with positive characteristics. In Sect. 4 we employ Hilbert's Nullstellensatz as a decision procedure for uniform word problems and use logic tests also over algebraically closed fields.

This makes the link between the successful logic approach from [27] and the comprehensive existing literature cited above. Section 5 clarifies some asymptotic worst-case complexities for the sake of scientific rigor. In Sect. 6 it turns out that for a comprehensive benchmark set of 129 models from the BioModels database [9] quite simple and maintainable code, requiring only functionality available in most decent computer algebra systems and libraries, can essentially compete with highly specialized and more complicated purely algebraic methods. This motivates in Sect. 7 a perspective that our symbolic computation approach has a potential to be interesting for researchers in the life sciences, with communities much larger than our own, with challenging applications, not least in the health sector.

2 Syntax: First-Order Formulations of Characteristic Properties of Groups and Cosets

In this section we set up our first-order logic framework. We are going to use interpreted first-order logic with equality over the signature $\mathcal{L} = (0, 1, +, -, \cdot)$ of rings.

For any field $\mathbb{K}$ we denote its multiplicative group $\mathbb{K} \setminus \{0\}$ by $\mathbb{K}^*$. For a coefficient ring $Z \subseteq \mathbb{K}$ and $F \subseteq Z[x_1, \ldots, x_n]$ we denote by $V_{\mathbb{K}}(F)$, or shortly $V(F)$, the variety of F over $\mathbb{K}$. Our signature $\mathcal{L}$ naturally induces coefficient rings $Z = \mathbb{Z}/p$ for finite characteristic p, and $Z = \mathbb{Z}$ for characteristic 0, where $\mathbb{Z}$ denotes the integers. We define $V(F)^* = V(F) \cap (\mathbb{K}^*)^n \subseteq (\mathbb{K}^*)^n$. Note that the direct product $(\mathbb{K}^*)^n$ establishes again a multiplicative group.

Let $F = \{f_1, \ldots, f_m\} \subseteq Z[x_1, \ldots, x_n]$. The following semi-formal conditions state that $V(F)^*$ establishes a coset of a multiplicative subgroup of $(\mathbb{K}^*)^n$:

$$\forall g, x \in (\mathbb{K}^*)^n:$$

$$g \in V(F) \wedge gx \in V(F) \Rightarrow gx^{-1} \in V(F) \tag{1}$$

$$\forall g, x, y \in (\mathbb{K}^*)^n:$$

$$g \in V(F) \wedge gx \in V(F) \wedge gy \in V(F) \Rightarrow gxy \in V(F) \tag{2}$$

$$V(F) \cap (\mathbb{K}^*)^n \neq \emptyset. \tag{3}$$

If we replace (3) with the stronger condition

$$1 \in V(F), \tag{4}$$

then $V(F)^*$ establishes even a multiplicative subgroup of $(\mathbb{K}^*)^n$. We allow ourselves to less formally say that $V(F)^*$ is a coset or group over $\mathbb{K}$, respectively.

Denote $M = \{1, \ldots, m\}$, $N = \{1, \ldots, n\}$, and for $(i, j) \in M \times N$ let $d_{ij} = \deg_{x_j}(f_i)$. We shortly write $x = (x_1, \ldots, x_n)$, $y = (y_1, \ldots, y_n)$, $g = (g_1, \ldots, g_n)$. Multiplication between x, y, g is coordinate-wise, and $x^{d_i} = x_1^{d_{i1}} \cdots x_n^{d_{in}}$. As a first-order $\mathcal{L}$-sentence, condition (1) yields

$$\iota \doteq \forall g_1 \ldots \forall g_n \forall x_1 \ldots \forall x_n \left(\bigwedge_{j=1}^{n} g_j \neq 0 \wedge \bigwedge_{j=1}^{n} x_j \neq 0 \wedge \right.$$

$$\bigwedge_{i=1}^{m} f_i(g_1, \ldots, g_n) = 0 \wedge \bigwedge_{i=1}^{m} f_i(g_1 x_1, \ldots, g_n x_n) = 0$$

$$\left. \longrightarrow \bigwedge_{i=1}^{m} x^{d_i} f_i(g_1 x_1^{-1}, \ldots, g_n x_n^{-1}) = 0 \right).$$

Here the multiplications with x^{d_i} drop the principal denominators from the rational functions $f_i(g_1 x_1^{-1}, \ldots, g_n x_n^{-1})$. This is an equivalence transformation, because the left hand side of the implication constrains $x_1, \ldots, x_n$ to be different from zero.

Similarly, condition (2) yields a first-order $\mathcal{L}$-sentence

$$\mu \doteq \forall g_1 \ldots \forall g_n \forall x_1 \ldots \forall x_n \forall y_1 \ldots \forall y_n \left(\bigwedge_{j=1}^{n} g_j \neq 0 \wedge \right.$$

$$\bigwedge_{j=1}^{n} x_j \neq 0 \wedge \bigwedge_{j=1}^{n} y_j \neq 0 \wedge \bigwedge_{i=1}^{m} f_i(g_1, \ldots, g_n) = 0 \wedge$$

$$\bigwedge_{i=1}^{m} f_i(g_1 x_1, \ldots, g_n x_n) = 0 \wedge \bigwedge_{i=1}^{m} f_i(g_1 y_1, \ldots, g_n y_n) = 0$$

$$\left. \longrightarrow \bigwedge_{i=1}^{m} f_i(g_1 x_1 y_1, \ldots, g_n x_n y_n) = 0 \right).$$

For condition (3) we consider its logical negation $V(F) \cap (\mathbb{K}^*)^n = \emptyset$, which gives us an $\mathcal{L}$-sentence

$$\eta \doteq \forall x_1 \ldots \forall x_n \left(\bigwedge_{i=1}^{m} f_i = 0 \longrightarrow \bigvee_{j=1}^{n} x_j = 0 \right).$$

Accordingly, the $\mathcal{L}$-sentence $\neg \eta$ formally states (3).

Finally, condition (4) yields a quantifier-free $\mathcal{L}$-sentence

$$\gamma \doteq \bigwedge_{i=1}^{m} f_i(1, \ldots, 1) = 0.$$

3 Semantics: Validity of Our First-Order Characterizations over Various Fields

Let $p \in \mathbb{N}$ be 0 or prime. We consider the $\mathcal{L}$-model classes $\mathfrak{K}_p$ of fields of characteristic p and $\mathfrak{A}_p \subseteq \mathfrak{K}_p$ of algebraically closed fields of characteristic p. Recall that $\mathfrak{A}_p$ is complete, decidable, and admits effective quantifier elimination [49, Note 16].

We assume without loss of generality that $\mathcal{L}$-sentences are in prenex normal form $Q_1 x_1 \ldots Q_n x_n \psi$ with $Q_1, \ldots, Q_n \in \{\exists, \forall\}$ and ψ quantifier-free. An $\mathcal{L}$-sentence is called *universal* if it is of the form $\forall x_1 \ldots \forall x_n \psi$ and *existential* if it is of the form $\exists x_1 \ldots \exists x_n \psi$ with ψ quantifier-free. A quantifier-free $\mathcal{L}$-sentence is both universal and existential.

Lemma 1. *Let φ be a universal $\mathcal{L}$-sentence. Then*

$$\mathfrak{K}_p \models \varphi \quad \text{if and only if} \quad \mathfrak{A}_p \models \varphi.$$

Proof. The implication from the left to the right immediately follows from $\mathfrak{A}_p \subseteq \mathfrak{K}_p$. Assume, conversely, that $\mathfrak{A}_p \models \varphi$, and let $\mathbb{K} \in \mathfrak{K}_p$. Then $\mathbb{K}$ has an algebraic closure $\overline{\mathbb{K}} \in \mathfrak{A}_p$, and $\overline{\mathbb{K}} \models \varphi$ due to the completeness of $\mathfrak{A}_p$. Since $\mathbb{K} \subseteq \overline{\mathbb{K}}$ and φ as a universal sentence is persistent under substructures, we obtain $\mathbb{K} \models \varphi$.

All our first-order conditions ι, μ, η, and γ introduced in the previous section 2 are universal $\mathcal{L}$-sentences. Accordingly, $\neg\eta$ is equivalent to an existential $\mathcal{L}$-sentence.

In accordance with our language $\mathcal{L}$ we are going to use polynomial coefficient rings $Z_p = \mathbb{Z}/p$ for finite characteristic p, and $Z_0 = \mathbb{Z}$. Let $F \subseteq Z_p[x_1, \ldots, x_n]$. Then $V(F)^*$ is a coset over $\mathbb{K} \in \mathfrak{K}_p$ if and only if

$$\mathbb{K} \models \iota \wedge \mu \wedge \neg\eta. \tag{5}$$

Especially, $V(F)^*$ is a group over $\mathbb{K}$ if even

$$\mathbb{K} \models \iota \wedge \mu \wedge \gamma, \tag{6}$$

where γ entails $\neg\eta$.

Proposition 2. *Let $F \subseteq Z_p[x_1, \ldots, x_n]$, and let $\mathbb{K} \in \mathfrak{K}_p$. Then $V(F)^*$ is a group over $\mathbb{K}$ if and only if at least one of the following conditions holds:*

(a) $\mathbb{K}' \models \iota \wedge \mu \wedge \gamma$ for some $\mathbb{K} \subseteq \mathbb{K}' \in \mathfrak{K}_p$;
(b) $\mathbb{K}' \models \iota \wedge \mu \wedge \gamma$ for some $\mathbb{K}' \in \mathfrak{A}_p$.

Proof. Recall that $V(F)^*$ is a group over $\mathbb{K}$ if and only if $\mathbb{K} \models \iota \wedge \mu \wedge \gamma$. If $V(F)^*$ is a group over $\mathbb{K}$, then (a) holds for $\mathbb{K}' = \mathbb{K}$. Conversely, there are two cases. In case (a), we can conclude that $\mathbb{K} \models \iota \wedge \mu \wedge \gamma$ because the universal sentence $\iota \wedge \mu \wedge \gamma$ is persistent under substructures. In case (b), we have $\mathfrak{A}_p \models \iota \wedge \mu \wedge \gamma$ by the completeness of that model class. Using Lemma 1 we obtain $\mathfrak{K}_p \models \iota \wedge \mu \wedge \gamma$, in particular $\mathbb{K} \models \iota \wedge \mu \wedge \gamma$.

Example 3

(i) Assume that $V(F)^*$ is a group over $\mathbb{C}$. Then $V(F)^*$ is a group over any field of characteristic 0. Alternatively, it suffices that $V(F)^*$ is a group over the countable algebraic closure $\overline{\mathbb{Q}}$ of $\mathbb{Q}$.

(ii) Assume that $V(F)^*$ is a group over the countable field of real algebraic numbers, which is not algebraically closed. Then again $V(F)^*$ is a group over any field of characteristic 0.

(iii) Let ε be a positive infinitesimal, and assume that $V(F)^*$ is a group over $\mathbb{R}(\varepsilon)$. Then $V(F)^*$ is group also over $\mathbb{Q}$ and $\mathbb{R}$, but not necessarily over $\overline{\mathbb{Q}}$. Notice that $\mathbb{R}(\varepsilon)$ is not algebraically closed.

(iv) Assume that $V(F)^*$ is a group over the algebraic closure of $\mathbb{F}_p$. Then $V(F)^*$ is a group over any field of characteristic p. Alternatively, it suffices that $V(F)^*$ is a group over the algebraic closure of the rational function field $\mathbb{F}_p(t)$, which has been studied with respect to effective computations [37].

Proposition 4. *Let* $F \subseteq Z_p[x_1, \ldots, x_n]$ *and let* $\mathbb{K} \in \mathfrak{K}_p$. *Then* $V(F)^*$ *is a coset over* $\mathbb{K}$ *if and only if* $\mathbb{K} \models \neg\eta$ *and at least one of the following conditions holds:*

(a) $\mathbb{K}' \models \iota \wedge \mu$ *for some* $\mathbb{K} \subseteq \mathbb{K}' \in \mathfrak{K}_p$;
(b) $\mathbb{K}' \models \iota \wedge \mu$ *for some* $\mathbb{K}' \in \mathfrak{A}_p$.

Proof. Recall that $V(F)^*$ is a coset over $\mathbb{K}$ if and only if $\mathbb{K} \models \iota \wedge \mu \wedge \neg\eta$. If $V(F)^*$ is a coset over $\mathbb{K}$, then $\mathbb{K} \models \neg\eta$, and (a) holds for $\mathbb{K}' = \mathbb{K}$. Conversely, we require that $\mathbb{K} \models \neg\eta$ and obtain $\mathbb{K} \models \iota \wedge \mu$ analogously to the proof of Proposition 2.

Example 5

(i) Assume that $V(F)^*$ is a coset over $\mathbb{C}$. Then $V(F)^*$ is a coset over $\mathbb{R}$ if and only if $V(F)^* \neq \emptyset$ over $\mathbb{R}$. This is the case for $F = \{x^2 - 2\}$ but not for $F = \{x^2 + 2\}$.

(ii) Consider $F = \{x^4 - 4\} = \{(x^2 - 2)(x^2 + 2)\}$. Then over $\mathbb{R}$, $V(F)^* = \{\pm\sqrt{2}\}$ is a coset, because $V(F)^*/\sqrt{2} = \{\pm 1\}$ is a group. Similarly over $\mathbb{C}$, $V(F)^* = \{\pm\sqrt{2}, \pm i\sqrt{2}\}$ is a coset, as $V(F)^*/\sqrt{2} = \{\pm 1, \pm i\}$ is a group.

(iii) Consider $F = \{x^4 + x^2 - 6\} = \{(x^2 - 2)(x^2 + 3)\}$. Then over $\mathbb{R}$, $V(F)^* = \{\pm\sqrt{2}\}$ is a coset, as $V(F)^*/\sqrt{2} = \{\pm 1\}$ is a group. Over $\mathbb{C}$, in contrast, $V(F)^* = \{\pm\sqrt{2}, \pm i\sqrt{3}\}$ is not a coset.

4   Hilbert's Nullstellensatz as a Swiss Army Knife

A recent publication [27] has systematically applied coset tests to a large number of real-world models from the BioModels database [9], investigating varieties over both the real and the complex numbers. Over $\mathbb{R}$ it used essentially our first-order sentences presented in Sect. 2 and applied efficient implementations of real decision methods based on effective quantifier elimination [13,14,39,46,52,53].

Over $\mathbb{C}$, in contrast, it used a purely algebraic framework combining various specialized methods from commutative algebra, typically based on Gröbner basis computations [8,17]. This is in line with the vast majority of the existing literature (cf. the Introduction for references), which uses computer algebra over algebraically closed fields, to some extent supplemented with heuristic tests based on linear algebra.

Generalizing the successful approach for $\mathbb{R}$ and aiming at a more uniform overall framework, we want to study here the application of decision methods for algebraically closed fields to our first-order sentences. Recall that our sentences ι, μ, η, and γ are universal $\mathcal{L}$-sentences. Every such sentence φ can be equivalently transformed into a finite conjunction of universal $\mathcal{L}$-sentences of the following form:

$$\widehat{\varphi} \;\dot{=}\; \forall x_1 \ldots \forall x_n \left(\bigwedge_{i=1}^{m} f_i(x_1, \ldots, x_n) = 0 \longrightarrow g(x_1, \ldots, x_n) = 0 \right),$$

where $f_1, \ldots, f_m, g \in Z_p[x_1, \ldots, x_n]$. Such $\mathcal{L}$-sentences are called uniform word problems [3]. Over an algebraically closed field $\bar{\mathbb{K}}$ of characteristic p, Hilbert's Nullstellensatz [31] provides a decision procedure for uniform word problems. It states that

$$\bar{\mathbb{K}} \models \widehat{\varphi} \quad \text{if and only if} \quad g \in \sqrt{\langle f_1, \ldots, f_m \rangle}.$$

Recall that $\mathfrak{A}_p$ is complete so that we furthermore have $\mathfrak{A}_p \models \widehat{\varphi}$ if and only if $\bar{\mathbb{K}} \models \widehat{\varphi}$.

Our $\mathcal{L}$-sentence ι for condition (1) can be equivalently transformed into

$$\forall g_1 \ldots \forall g_n \forall x_1 \ldots \forall x_n \left(\bigvee_{j=1}^{n} g_j = 0 \vee \bigvee_{j=1}^{n} x_j = 0 \vee \right.$$

$$\bigvee_{i=1}^{m} f_i(g_1, \ldots, g_n) \neq 0 \vee \bigvee_{i=1}^{m} f_i(g_1 x_1, \ldots, g_n x_n) \neq 0$$

$$\left. \vee \bigwedge_{i=1}^{m} x^{d_i} f_i(g_1 x_1^{-1}, \ldots, g_n x_n^{-1}) = 0 \right),$$

which is in turn equivalent to

$$\widehat{\iota} \;\dot{=}\; \bigwedge_{k=1}^{m} \forall g_1 \ldots \forall g_n \forall x_1 \ldots \forall x_n$$

$$\left(\bigwedge_{i=1}^{m} f_i(g_1, \ldots, g_n) = 0 \wedge \bigwedge_{i=1}^{m} f_i(g_1 x_1, \ldots, g_n x_n) = 0 \right.$$

$$\left. \longrightarrow x^{d_k} f_k(g_1 x_1^{-1}, \ldots, g_n x_n^{-1}) \prod_{j=1}^{n} g_j x_j = 0 \right).$$

Hence, by Hilbert's Nullstellensatz, (1) holds in $\bar{\mathbb{K}}$ if and only if

$$x^{d_k} f_k(g_1 x_1^{-1}, \ldots, g_n x_n^{-1}) \prod_{j=1}^{n} g_j x_j \in R_1 \quad \text{for all} \quad k \in M, \tag{7}$$

where $R_1 = \sqrt{\langle f_i(g_1, \ldots, g_n), f_i(g_1 x_1, \ldots, g_n x_n) \mid i \in M \rangle}$.

Similarly, our $\mathcal{L}$-sentence μ for condition (2) translates into

$$\widehat{\mu} \;\dot{=}\; \bigwedge_{k=1}^{m} \forall g_1 \ldots \forall g_n \forall x_1 \ldots \forall x_n \forall y_1 \ldots \forall y_n$$

$$\left(\bigwedge_{i=1}^{m} f_i(g_1, \ldots, g_n) = 0 \wedge \bigwedge_{i=1}^{m} f_i(g_1 x_1, \ldots, g_n x_n) = 0 \right.$$

$$\wedge \bigwedge_{i=1}^{m} f_i(g_1 y_1, \ldots, g_n y_n) = 0$$

$$\left. \longrightarrow f_k(g_1 x_1 y_1, \ldots, g_n x_n y_n) \prod_{j=1}^{n} g_j x_j y_j = 0 \right).$$

Again, by Hilbert's Nullstellensatz, (2) holds in $\bar{\mathbb{K}}$ if and only if

$$f_k(g_1 x_1 y_1, \ldots, g_n x_n y_n) \prod_{j=1}^{n} g_j x_j y_j \in R_2 \quad \text{for all} \quad k \in M, \tag{8}$$

where $R_2 = \sqrt{\langle f_i(g), f_i(gx), f_i(gy) \mid i \in M \rangle}$.

Next, our $\mathcal{L}$-sentence η is is equivalent to

$$\widehat{\eta} \doteq \forall x_1 \ldots \forall x_n \left(\bigwedge_{i=1}^{m} f_i = 0 \longrightarrow \prod_{j=1}^{n} x_j = 0 \right).$$

Using once more Hilbert's Nullstellensatz, $\bar{\mathbb{K}} \models \widehat{\eta}$ if and only if

$$\prod_{j=1}^{n} x_j \in R_3, \tag{9}$$

where $R_3 = \sqrt{\langle f_1, \ldots, f_m \rangle}$. Hence our non-emptiness condition (3) holds in $\bar{\mathbb{K}}$ if and only if

$$\prod_{j=1}^{n} x_j \notin R_3. \tag{10}$$

Finally, our $\mathcal{L}$-sentence γ for condition (4) is equivalent to

$$\widehat{\gamma} \doteq \bigwedge_{k=1}^{m} \left(0 = 0 \longrightarrow f_k(1, \ldots, 1) = 0 \right).$$

Here Hilbert's Nullstellensatz tells us that condition (4) holds in $\bar{\mathbb{K}}$ if and only if

$$f_k(1, \ldots, 1) \in R_4 \quad \text{for all} \quad k \in M, \tag{11}$$

where $R_4 = \sqrt{\langle 0 \rangle} = \langle 0 \rangle$. Notice that the radical membership test quite naturally reduces to the obvious test with plugging in.

5 Complexity

Let us briefly discuss asymptotic complexity bounds around problems and methods addressed here. We do so very roughly, in terms of the input word length. The cited literature provides more precise bounds in terms of several complexity parameters, such as numbers of quantifiers, or degrees.

The decision problem for algebraically closed fields is double exponential [30] in general, but only single exponential when the number of quantifier alternations is bounded [25], which covers in particular our universal formulas. The decision problem for real closed fields is double exponential as well [12], even for linear problems [52]; again it becomes single exponential when bounding the number of quantifier alternations [26].

Ideal membership tests are at least double exponential [40], and it was widely believed that this would impose a corresponding lower bound also for any algorithm for Hilbert's Nullstellensatz. Quite surprisingly, it turned out that there are indeed single exponential such algorithms [7,38].

On these grounds it is clear that our coset tests addressed in the previous sections can be solved in single exponential time for algebraically closed fields as well as for real closed fields. Recall that our consideration of those tests is actually motivated by our interest in shifted toricity, which requires, in addition, the irreducibility of the considered variety over the considered domain. Recently it has been shown that testing shifted toricity, including irreducibility, is also only single exponential over algebraically closed fields as well as real closed fields [27].

Most asymptotically fast algorithms mentioned above are not implemented and it is not clear that they would be efficient in practice.

6 Computational Experiments

We have studied 129 models from the BioModels[2] database [9]. Technically, we took our input from ODEbase[3] which provides preprocessed versions for symbolic computation. Our 129 models establish the complete set currently provided by ODEbase for which the relevant systems of ordinary differential equations have polynomial vector fields.

We limited ourselves to characteristic 0 and applied the tests (7), (8), (9), and (11) derived in Section 4 using Hilbert's Nullstellensatz. Recall that those tests correspond to ι, μ, η, γ from Sect. 3, respectively, and that one needs $\iota \wedge \mu \wedge \neg\eta$ or $\iota \wedge \mu \wedge \gamma$ for cosets or groups, respectively. From a symbolic computation point of view, we used exclusively polynomial arithmetic and radical membership tests. The complete Maple code for computing a single model is presented in Appendix B; it is surprisingly simple.

We conducted our computations on a 2.40 GHz Intel Xeon E5-4640 with 512 GB RAM and 32 physical cores providing 64 CPUs via hyper-threading. For parallelization of the jobs for the individual models we used GNU Parallel [48]. Results and timings are collected in Appendix A. With a time limit of one hour CPU time per model we succeeded on 78 models, corresponding to 60%, the largest of which, no. 559, has 90 polynomials in 71 dimensions. The median of the overall computation times for the successful models is 1.419 s. We would like to emphasize that our focus here is illustrating and evaluating our overall approach, rather than obtaining new insights into the models. Therefore our code in Appendix B is very straightforward without any optimization. In particular, computation continues even when one relevant subtest has already failed. More comprehensive results on our dataset can be found in [27].

Among our 78 successfully computed models, we detected 20 coset cases, corresponding to 26%. Two out of those 20 are even group cases. Among the 58 other cases, 46, corresponding to 78%, fail only due to their emptiness η; we know from [27] that many such cases exhibit in fact coset structure when considered in suitable lower-dimensional spaces, possibly after prime decomposition. Finally notice that our example reaction from the Introduction, no. 92, is among the smallest ones with a coset structure.

---

[2] https://www.ebi.ac.uk/biomodels/.
[3] http://odebase.cs.uni-bonn.de/.

7 Conclusions and Future Work

We have used Hilbert's Nullstellensatz to derive important information about the varieties of biological models with a polynomial vector field F. The key technical idea was generalizing from pure algebra to more general first-order logic. Recall from Sect. 3 that except for non-emptiness of $V(F)^*$ the information we obtained is valid in *all fields* of characteristic 0. Wherever we discovered non-emptiness, this holds at least in *all algebraically closed fields* of characteristic 0. For transferring our obtained results to real closed fields, e.g., subtropical methods [21,32,47] provide fast heuristic tests for the non-emptiness of $V(F)^*$ there.

Technically, we only used polynomial arithmetic and polynomial radical membership tests. This means that on the software side there are many off-the-shelf computer algebra systems and libraries available where our ideas could be implemented, robustly and with little effort. This in turn makes it attractive for the integration with software from systems biology, which could open exciting new perspectives for symbolic computation with applications ranging from the fundamental research in the life sciences to state-of-the-art applied research in medicine and pharmacology.

We had motivated our use of Hilbert's Nullstellensatz by viewing it as a decision procedure for the universal fragment of first-order logic in algebraically closed fields, which is sufficient for our purposes. Our focus on algebraically closed fields here is in accordance with the majority of existing literature on toricity. However, due to the relevance of the first orthant mentioned with our example in the introduction, it is generally accepted in the context of chemical reaction network theory that real closed fields are the relevant domain [20]. Our Proposition 2 and Proposition 4 assert that existing computations over the complex numbers are adequate and almost complete. Only the non-emptiness of cosets must be checked over the reals.

We have seen in Sect. 5 that the theoretical complexities for general decision procedures in algebraically closed fields vs. real closed fields strongly resemble each other. What could now take the place of Hilbert's Nullstellensatz over the reals with respect to practical computations on model sizes as in Appendix A or even larger? A factor of 10 could put us in the realm of models currently used in the development of drugs for diabetes or cancer. One possible answer is *satisfiability modulo theories solving (SMT)* [42].[4] SMT is incomplete in the sense that it often proves or disproves validity, but it can yield "unknown" for specific input problems. When successful, it is typically significantly faster than traditional algebraic decision procedures. For coping with incompleteness one can still fall back into real quantifier elimination. Interest in collaboration between the SMT and the symbolic computation communities exists on both sides [1,2].

---

[4] SMT technically aims at the existential fragment, which in our context is equivalent to the universal fragment via logical negation.

Acknowledgments. This work has been supported by the interdisciplinary bilateral project ANR-17-CE40-0036/DFG-391322026 SYMBIONT [5,6]. We are grateful to Dima Grigoriev for numerous inspiring and very constructive discussions around toricity.

A   Computation Results

We present results and computation times (in seconds) of our computations on models from the BioModels database [9].

Model	m	n	ι	t_ι	μ	t_μ	η	t_η	γ	t_γ	Coset	Group	t_Σ
001	12	12	True	7.826	True	7.86	False	4.267	False	0.053	True	False	20.007
040	5	3	False	1.415	False	0.173	False	0.114	False	0.043	False	False	1.746
050	14	9	True	1.051	True	2.458	True	0.113	False	0.05	False	False	3.673
052	11	6	True	3.605	True	1.635	True	0.096	False	0.059	False	False	5.396
057	6	6	True	0.271	True	0.263	False	0.858	False	0.045	True	False	1.438
072	7	7	True	0.763	True	0.496	True	0.08	False	0.06	False	False	1.4
077	8	7	True	0.296	True	0.356	False	0.097	False	0.051	True	False	0.801
080	10	10	True	0.714	True	1.341	True	0.103	False	0.06	False	False	2.219
082	10	10	True	0.384	True	0.39	True	0.086	False	0.041	False	False	0.902
091	16	14	True	0.031	True	0.045	True	0.003	False	0.062	False	False	0.142
092	4	3	True	0.293	True	0.244	False	0.104	False	1.03	True	False	1.671
099	7	7	True	0.298	True	0.698	False	0.087	False	0.036	True	False	1.119
101	6	6	False	4.028	False	10.343	False	0.917	False	0.073	False	False	15.361
104	6	4	True	0.667	True	0.146	True	0.084	False	0.039	False	False	0.937
105	39	26	True	0.455	True	0.367	True	0.043	False	0.038	False	False	0.905
125	5	5	False	0.193	False	0.098	False	0.078	False	0.038	False	False	0.408
150	4	4	True	0.173	True	0.153	False	0.094	False	0.043	True	False	0.464
156	3	3	True	2.638	True	0.248	False	0.86	False	0.052	True	False	3.8
158	3	3	False	0.148	False	0.149	False	0.16	False	0.045	False	False	0.503
159	3	3	True	0.959	True	0.175	False	0.083	False	0.04	True	False	1.257
178	6	4	True	0.52	True	1.71	True	0.877	False	1.201	False	False	4.308
186	11	10	True	31.785	True	1026.464	True	1.956	False	0.095	False	False	1060.301
187	11	10	True	27.734	True	1023.648	True	0.103	False	0.062	False	False	1051.548
188	20	10	True	0.075	True	0.079	True	0.04	False	0.047	False	False	0.242
189	18	7	True	0.035	True	0.02	True	0.002	False	0.062	False	False	0.12
194	5	5	False	2.338	False	1.922	False	0.612	False	0.05	False	False	4.922
197	7	5	False	7.562	False	71.864	False	0.485	False	0.05	False	False	79.962
198	12	9	True	0.397	True	0.793	True	0.077	False	0.042	False	False	1.31
199	15	8	True	1.404	True	1.531	False	0.215	False	0.054	True	False	3.205
220	58	56	True	146.146	True	534.832	True	6.921	False	0.964	False	False	688.866
227	60	39	True	0.273	True	0.485	True	0.01	False	0.077	False	False	0.847
229	7	7	True	1.917	True	3.348	False	0.131	False	0.062	True	False	5.458
233	4	2	False	0.16	False	0.44	False	0.17	False	0.557	False	False	1.328
243	23	19	True	8.598	True	1171.687	True	2.512	False	0.171	False	False	1182.97

(continued)

Model	m	n	ι	t_ι	μ	t_μ	η	t_η	γ	t_γ	Coset	Group	t_Σ
259	17	16	True	1.334	True	1.913	True	0.092	False	0.045	False	False	3.385
260	17	16	True	2.182	True	0.748	True	0.079	False	0.047	False	False	3.057
261	17	16	True	3.359	True	2.872	True	0.113	False	0.095	False	False	6.44
262	11	9	True	0.402	True	0.41	True	0.091	False	0.071	False	False	0.975
263	11	9	True	0.379	True	0.403	True	0.085	False	0.066	False	False	0.934
264	14	11	True	1.031	True	2.036	True	0.136	False	0.063	False	False	3.268
267	4	3	True	1.084	True	0.246	True	0.095	False	0.049	False	False	1.475
271	6	4	True	0.286	True	0.283	True	0.746	False	0.045	False	False	1.361
272	6	4	True	0.361	True	0.323	True	0.086	False	0.055	False	False	0.826
281	32	32	True	20.987	True	29.791	True	0.602	False	0.055	False	False	51.437
282	6	3	True	0.205	True	0.19	True	0.087	False	0.046	False	False	0.528
283	4	3	True	0.294	True	0.211	True	0.087	False	0.412	False	False	1.005
289	5	4	False	2.291	False	1.118	False	0.165	False	0.044	False	False	3.619
292	6	2	True	0.06	True	0.048	True	0.063	False	0.046	False	False	0.218
306	5	2	True	0.149	True	0.121	False	0.079	False	0.041	True	False	0.391
307	5	2	True	0.129	True	0.121	True	0.043	False	0.148	False	False	0.441
310	4	1	True	0.053	True	0.369	True	0.047	False	0.04	False	False	0.509
311	4	1	True	0.076	True	0.048	True	0.224	False	0.048	False	False	0.397
312	3	2	True	0.098	True	0.512	True	0.043	False	0.043	False	False	0.697
314	12	10	True	0.515	True	1.789	True	0.1	False	0.059	False	False	2.464
321	3	3	True	0.163	True	0.148	True	0.042	False	0.039	False	False	0.393
357	9	8	True	0.353	True	1.517	True	0.07	False	0.045	False	False	1.986
359	9	8	True	1.677	True	3.605	True	0.11	False	0.055	False	False	5.448
360	9	8	True	0.479	True	0.47	True	0.096	False	0.05	False	False	1.096
361	8	8	True	1.069	True	2.746	True	0.156	False	0.045	False	False	4.017
363	4	3	True	0.244	True	0.199	True	0.077	False	0.041	False	False	0.561
364	14	12	True	2.483	True	7.296	True	0.55	False	0.064	False	False	10.394
413	5	5	False	1.55	False	22.323	False	0.117	False	0.053	False	False	24.044
459	4	3	True	0.542	True	0.224	False	0.18	False	0.068	True	False	1.014
460	4	3	False	1.025	False	0.936	False	0.143	False	0.216	False	False	2.321
475	23	22	True	97.876	True	3377.021	True	0.231	False	0.062	False	False	3475.192
484	2	1	True	0.384	True	0.143	False	0.099	False	0.048	True	False	0.674
485	2	1	False	0.564	False	0.354	False	0.209	False	0.042	False	False	1.169
486	2	2	True	0.119	True	0.106	False	0.073	False	0.041	True	False	0.339
487	6	6	True	0.475	True	1.008	False	0.099	False	0.045	True	False	1.628
491	57	57	True	123.138	True	536.865	False	2.067	True	0.007	True	True	662.08
492	52	52	True	85.606	True	284.753	False	1.123	True	0.003	True	True	371.489
519	3	3	True	1.357	True	2.367	False	5.142	False	0.097	True	False	8.964
546	7	3	True	0.327	True	0.338	True	0.109	False	0.042	False	False	0.817
559	90	71	True	4.742	True	7.525	True	0.19	False	0.053	False	False	12.515
584	35	9	True	0.4	True	0.655	False	0.095	False	0.043	True	False	1.194
619	10	8	True	0.411	True	0.443	True	0.087	False	0.052	False	False	0.994
629	5	5	True	0.209	True	0.197	False	0.079	False	0.046	True	False	0.532
647	11	11	False	0.854	False	16.436	False	0.165	False	0.051	False	False	17.507

B   Program Code Used for Our Computations

The following is Maple code for computing one row of the table in Appendix A.

```
ToricHilbert := proc(F::list(polynom))
uses PolynomialIdeals;

    local Iota := proc()::truefalse;
    local R1, s, prod, f;
        R1 := < op(subs(zip('=', xl, gl), F)),
                  op(subs(zip((x, g) -> x = g*x, xl, gl), F)) >;
        s := zip((x, g) -> x = g/x, xl, gl);
        prod := g * x;
        for f in subs(s, F) do
            if not RadicalMembership(numer(f) * prod, R1) then
                return false
            end if
        end do;
        return true
    end proc;

    local Mu := proc()::truefalse;
    local R2, s, prod, f;
        R2 := < op(subs(zip('=', xl, gl), F)),
                  op(subs(zip((x, g) -> x = g*x, xl, gl), F)),
                  op(subs(zip('=', xl, zip('*', gl, yl)), F)) >;
        s := zip('=', xl, zip('*', gl, zip('*', xl, yl)));
        prod := g * x * y;
        for f in subs(s, F) do
            if not RadicalMembership(f * prod, R2) then
                return false
            end if
        end do;
        return true
    end proc;

    local Eta := proc()::truefalse;
    local R3, prod;
        R3 := < op(F) >;
        prod := foldl('*', 1, op(xl));
        return RadicalMembership(prod, R3)
    end proc;

    local Gamma := proc()::truefalse;
    local R4, s, f;
        R4 := < 0 >;
        s := map(x -> x=1, xl);
        for f in subs(s, F) do
            if not RadicalMembership(f, R4) then
                return false
            end if
        end do;
        return true
    end proc;
```

```
local Rename := proc(base::name, l::list(name))::list(name);
uses StringTools;
    return map(x -> cat(base, Select(IsDigit, x)), l)
end proc;

local X, xl, gl, yl, g, x, y, iota, t_iota, mu, t_mu, eta,
        t_eta, gamm, t_gamma, coset, group, t;
    t := time();
xl := convert(indets(F), list);
x := foldl('*', 1, op(xl));
gl := Rename('g', xl);
g := foldl('*', 1, op(gl));
yl := Rename('y', xl);
y := foldl('*', 1, op(yl));
t_iota := time(); iota := Iota(); t_iota := time() - t_iota;
t_mu := time(); mu := Mu(); t_mu := time() - t_mu;
t_eta := time(); eta := Eta(); t_eta := time() - t_eta;
t_gamma := time(); gamm := Gamma(); t_gamma := time() - t_gamma;
coset := iota and mu and not eta;
group := iota and mu and gamm;
t := time() - t;
return nops(F), nops(xl), iota, t_iota, mu, t_mu, eta, t_eta,
        gamm, t_gamma, coset, group, t
end proc;
```

References

1. Ábrahám, E., et al.: Satisfiability checking and symbolic computation. ACM Commun. Comput. Algebra **50**(4), 145–147 (2016). https://doi.org/10.1145/3055282. 3055285
2. Ábrahám, E., Abbott, J., Becker, B., Bigatti, A.M., Brain, M., Buchberger, B., Cimatti, A., Davenport, J.H., England, M., Fontaine, P., Forrest, S., Griggio, A., Kroening, D., Seiler, W.M., Sturm, T.: $SC \wedge 2$: Satisfiability checking meets symbolic computation. In: Kohlhase, M., Johansson, M., Miller, B., de de Moura, L., Tompa, F. (eds.) CICM 2016. LNCS (LNAI), vol. 9791, pp. 28–43. Springer, Cham (2016). https://doi.org/10.1007/978-3-319-42547-4_3
3. Becker, T., Weispfenning, V., Kredel, H.: Gröbner Bases, a Computational Approach to Commutative Algebra, Graduate Texts in Mathematics. Springer, Berlin (1993). https://doi.org/10.1007/978-1-4612-0913-3
4. Boltzmann, L.: Lectures on Gas Theory. University of California Press, Berkeley and Los Angeles, CA (1964)
5. Boulier, F.: The SYMBIONT project: symbolic methods for biological networks. ACM Commun. Comput. Algebra **52**(3), 67–70 (2018). https://doi.org/10.1145/3313880.3313885
6. Boulier, F., Fages, F., Radulescu, O., Samal, S.S., Schuppert, A., Seiler, W., Sturm, T., Walcher, S., Weber, A.: The SYMBIONT project: Symbolic methods for biological networks. F1000Research **7**(1341) (2018). https://doi.org/10.7490/f1000research.1115995.1
7. Brownawell, W.D.: Bounds for the degrees in the Nullstellensatz. Ann. Math. **126**(3), 577–591 (1987). https://doi.org/10.2307/1971361

8. Buchberger, B.: Ein Algorithmus zum Auffinden der Basiselemente des Restklassenringes nach einem nulldimensionalen Polynomideal. Doctoral dissertation, Mathematical Institute, University of Innsbruck, Austria (1965)
9. Chelliah, N., et al.: BioModels: Ten-year anniversary. Nucl. Acids Res. **43**(D1), 542–548 (2015). https://doi.org/10.1093/nar/gku1181
10. Conradi, C., Kahle, T.: Detecting binomiality. Adv. Appl. Math. **71**, 52–67 (2015). https://doi.org/10.1016/j.aam.2015.08.004
11. Craciun, G., Dickenstein, A., Shiu, A., Sturmfels, B.: Toric dynamical systems. J. Symb. Comput. **44**(11), 1551–1565 (2009). https://doi.org/10.1016/j.jsc.2008.08.006
12. Davenport, J.H., Heintz, J.: Real quantifier elimination is doubly exponential. J. Symb. Comput. **5**(1–2), 29–35 (1988). https://doi.org/10.1016/S0747-7171(88)80004-X
13. Dolzmann, A., Sturm, T.: Redlog: Computer algebra meets computer logic. ACM SIGSAM Bulletin **31**(2), 2–9 (1997). https://doi.org/10.1145/261320.261324
14. Dolzmann, A., Sturm, T.: Simplification of quantifier-free formulae over ordered fields. J. Symb. Comput. **24**(2), 209–231 (1997). https://doi.org/10.1006/jsco.1997.0123
15. Einstein, A.: Strahlungs-emission und -absorption nach der Quantentheorie. Verh. Dtsch. Phys. Ges. **18**, 318–323 (1916)
16. Eisenbud, D., Sturmfels, B.: Binomial ideals. Duke Math. J. **84**(1), 1–45 (1996). https://doi.org/10.1215/S0012-7094-96-08401-X
17. Faugère, J.C.: A new efficient algorithm for computing Gröbner bases (F4). J. Pure Appl. Algebra **139**(1–3), 61–88 (1999). https://doi.org/10.1145/780506.780516
18. Feinberg, M.: Complex balancing in general kinetic systems. Arch. Ration. Mech. An. **49**(3), 187–194 (1972). https://doi.org/10.1007/BF00255665
19. Feinberg, M.: Stability of complex isothermal reactors-I. The deficiency zero and deficiency one theorems. Chem. Eng. Sci. **42**(10), 2229–2268 (1987). https://doi.org/10.1016/0009-2509(87)80099-4
20. Feinberg, M.: Foundations of Chemical Reaction Network Theory, Applied Mathematical Sciences. Springer, New York (2019). https://doi.org/10.1007/978-3-030-03858-8
21. Fontaine, P., Ogawa, M., Sturm, T., Vu, X.T.: Subtropical satisfiability. In: Dixon, C., Finger, M. (eds.) FroCoS 2017. LNCS (LNAI), vol. 10483, pp. 189–206. Springer, Cham (2017). https://doi.org/10.1007/978-3-319-66167-4_11
22. Fuentes, M.E., Varón, R., García-Moreno, M., Valero, E.: Kinetics of intra- and intermolecular zymogen activation with formation of an enzyme-zymogen complex. The FEBS Journal **272**(1), 85–96 (2005). https://doi.org/10.1111/j.1432-1033.2004.04400.x
23. Fulton, W.: Introduction to Toric Varieties, Annals of Mathematics Studies. Princeton University Press, New Jersey (1993)
24. Gatermann, K., Wolfrum, M.: Bernstein's second theorem and Viro's method for sparse polynomial systems in chemistry. Adv. Appl. Math. **34**(2), 252–294 (2005). https://doi.org/10.1016/j.aam.2004.04.003
25. Grigor'ev, D.Y.: The complexity of the decision problem for the first-order theory of algebraically closed fields. Math. USSR Izv. **29**(2), 459–475 (1987). https://doi.org/10.1070/IM1987v029n02ABEH000979
26. Grigor'ev, D.Y.: Complexity of deciding Tarski algebra. J. Symb. Comput. **5**(1–2), 65–108 (1988). https://doi.org/10.1016/S0747-7171(88)80006-3

27. Grigoriev, D., Iosif, A., Rahkooy, H., Sturm, T., Weber, A.: Efficiently and effectively recognizing toricity of steady state varieties. Math. Comput. Sci. (2020). https://doi.org/10.1007/s11786-020-00479-9

28. Grigoriev, D., Milman, P.D.: Nash resolution for binomial varieties as Euclidean division. a priori termination bound, polynomial complexity in essential dimension 2. Adv. Math. **231**(6), 3389–3428 (2012). https://doi.org/10.1016/j.aim.2012.08.009

29. Grigoriev, D., Weber, A.: Complexity of solving systems with few independent monomials and applications to mass-action kinetics. In: Gerdt, V.P., Koepf, W., Mayr, E.W., Vorozhtsov, E.V. (eds.) CASC 2012. LNCS, vol. 7442, pp. 143–154. Springer, Heidelberg (2012). https://doi.org/10.1007/978-3-642-32973-9_12

30. Heintz, J.: Definability and fast quantifier eliminarion in algebraically closed fields. Theor. Comput. Sci. **24**, 239–277 (1983). https://doi.org/10.1016/0304-3975(83)90002-6

31. Hilbert, D.: Über die vollen Invariantensysteme. Math. Ann. **42**, 313–373 (1893). https://doi.org/10.1007/BF01444162

32. Hong, H., Sturm, T.: Positive solutions of systems of signed parametric polynomial inequalities. In: Gerdt, V.P., Koepf, W., Seiler, W.M., Vorozhtsov, E.V. (eds.) CASC 2018. LNCS, vol. 11077, pp. 238–253. Springer, Cham (2018). https://doi.org/10.1007/978-3-319-99639-4_17

33. Horn, F., Jackson, R.: General mass action kinetics. Arch. Ration. Mech. An. **47**(2), 81–116 (1972). https://doi.org/10.1007/BF00251225

34. Kahle, T.: Decompositions of binomial ideals. Ann. I. Stat. Math. **62**(4), 727–745 (2010). https://doi.org/10.1007/s10463-010-0290-9

35. Kahle, T.: Decompositions of binomial ideals. J. Software Algebra Geometry **4**(1), 1–5 (2012). https://doi.org/10.2140/jsag.2012.4.1

36. Katthän, L., Michałek, M., Miller, E.: When is a polynomial ideal binomial after an ambient automorphism? Found. Comput. Math. **19**(6), 1363–1385 (2018). https://doi.org/10.1007/s10208-018-9405-0

37. Kedlaya, K.S.: Finite automata and algebraic extensions of function fields. Journal de Théorie des Nombres de Bordeaux **18**(2), 379–420 (2006). https://doi.org/10.5802/jtnb.551

38. Kollar, J.: Sharp effective Nullstellensatz. J. Am. Math. Soc. **1**(4), 963–975 (1988). https://doi.org/10.2307/1990996

39. Košta, M.: New Concepts for Real Quantifier Elimination by Virtual Substitution. Doctoral dissertation, Saarland University, Germany (2016). https://doi.org/10.22028/D291-26679

40. Mayr, E.W., Meyer, A.R.: The complexity of the word problems for commutative semigroups and polynomial ideals. Adv. Math. **46**(3), 305–329 (1982). https://doi.org/10.1016/0001-8708(82)90048-2

41. Müller, S., Feliu, E., Regensburger, G., Conradi, C., Shiu, A., Dickenstein, A.: Sign Conditions for Injectivity of Generalized Polynomial Maps with Applications to Chemical Reaction Networks and Real Algebraic Geometry. Found. Comput. Math. **16**(1), 69–97 (2015). https://doi.org/10.1007/s10208-014-9239-3

42. Nieuwenhuis, R., Oliveras, A., Tinelli, C.: Solving SAT and SAT modulo theories: from an abstract Davis-Putnam-Logemann-Loveland procedure to DPLL(T). J. ACM **53**(6), 937–977 (2006). https://doi.org/10.1145/1217856.1217859

43. Onsager, L.: Reciprocal relations in irreversible processes. I. Phys. Rev. **37**(4), 405 (1931). https://doi.org/10.1103/PhysRev.37.405

44. Pérez Millán, M., Dickenstein, A.: The structure of MESSI biological systems. SIAM J. Appl. Dyn. Syst. **17**(2), 1650–1682 (2018). https://doi.org/10.1137/17M1113722

45. Pérez Millán, M., Dickenstein, A., Shiu, A., Conradi, C.: Chemical reaction systems with toric steady states. Bull. Math. Biol. **74**(5), 1027–1065 (2012). https://doi.org/10.1007/s11538-011-9685-x

46. Seidl, A.: Cylindrical Decomposition Under Application-Oriented Paradigms. Doctoral dissertation, University of Passau, Germany (2006), https://nbn-resolving.org/urn:nbn:de:bvb:739-opus-816

47. Sturm, T.: Subtropical real root finding. In: Yokoyama, K., Linton, S., Robertz, D. (eds.) Proceedings of the 2015 ACM International Symposium on Symbolic and Algebraic Computation, ISSAC 2015, Bath, United Kingdom, July 6–9, 2015, pp. 347–354. ACM (2015). https://doi.org/10.1145/2755996.2756677

48. Tange, O.: GNU Parallel: The command-line power tool. login: The USENIX Magazine **36**(1), 42–47 (2011), https://www.usenix.org/publications/login/february-2011-volume-36-number-1/gnu-parallel-command-line-power-tool

49. Tarski, A.: A decision method for elementary algebra and geometry. Prepared for publication by J. C. C. McKinsey. RAND Report R109, August 1, 1948, Revised May 1951, Second Edition, RAND, Santa Monica, CA (1957)

50. Wang, S., Lin, J.R., Sontag, E.D., Sorger, P.K.: Inferring reaction network structure from single-cell, multiplex data, using toric systems theory. PLoS Comput. Biol. **15**(12), e1007311 (2019). https://doi.org/10.1371/journal.pcbi.1007311

51. Wegscheider, R.: Über simultane Gleichgewichte und die Beziehungen zwischen Thermodynamik und Reactionskinetik homogener Systeme. Monatsh. Chem. Verw. Tl. **22**(8), 849–906 (1901). https://doi.org/10.1007/BF01517498

52. Weispfenning, V.: The complexity of linear problems in fields. J. Symb. Comput. **5**(1–2), 3–27 (1988). https://doi.org/10.1016/S0747-7171(88)80003-8

53. Weispfenning, V.: Quantifier elimination for real algebra–the quadratic case and beyond. Appl. Algebr. Eng. Comm. **8**(2), 85–101 (1997). https://doi.org/10.1007/s002000050055

Looking for Compatible Routes in the Railway Interlocking System of an Overtaking Station Using a Computer Algebra System

Eugenio Roanes-Lozano[✉]

Instituto de Matemática Interdisciplinar & Departamento de Didáctica de las Ciencias Experimentales, Sociales y Matemáticas, Facultad de Educación, Universidad Complutense de Madrid, Madrid, Spain
eroanes@ucm.es

Abstract. The author has a long experience in railway engineering software development, mainly using computer algebra systems. Now a software package that allows the user to obtain compatible routes in an overtaking station of any number of tracks and any topology, based in dealing with cycles, has been developed. It has been implemented in the computer algebra system *Maple* and takes advantage of its *GraphTheory* package. There are two main uses of such a package. One is complementing topology independent railway interlocking systems, providing alternative routes. The other is to easily explore alternative track layouts. The package can only deal with overtaking stations (the most common ones in double track lines). Note that the latter use is important: for instance, the Spanish infrastructure administrator is nowadays remodeling the track layouts of the main railway stations of Madrid, Seville and Barcelona.

Keywords: Railway interlocking systems · Routes · Railway station layout · Graphs · Computer algebra systems

1 Introduction

Decision making in railway stations, junctions, yards, etc. is supervised by railway interlocking systems [29]. That is, a proposed aspect of mechanical signals (semaphores) and color light signals and position of the switches of the turnouts has to be checked to be compatible.

Traditionally, the compatibility of routes is established in advance by a panel of experts when the station is designed. It is therefore topology dependent. The first ones were mechanical and were installed in the XIX century [45]. In the XX century relays interlocking systems were introduced. And from the 80's, computer based railway interlocking systems (called "electronic interlocking systems") were installed (see, for instance, the technical project [1] or the brochures [2–4]).

© Springer Nature Switzerland AG 2020
F. Boulier et al. (Eds.): CASC 2020, LNCS 12291, pp. 528–542, 2020.
https://doi.org/10.1007/978-3-030-60026-6_31

The latter made much simpler the adaptation of the railway interlocking system when changes in the layout were made, as reprogramming is much simpler than rewiring (adaptation to changes is almost impossible for the mechanical interlocking systems).

The most modern railway interlocking were introduced in the 90's and such systems are topology independent [44], that is, the layout of the railway station is yet another datum for the decision making software.

A recent step forward (2018) is the use of digital networks for the railway interlocking communications (computer ↔ turnouts, signals, etc.) [11,43].

The author has a long experience in developing railways related software, some in cooperation with the 'Fundación de los Ferrocarriles Españoles' (Spanish Railway Foundation) or in research projects funded by this foundation [39,40].

There are many different approaches to decision making in a railway interlocking, some by this author, like [9,13,14,18–21,23–25,27,28,32–34,36–38,41,47,48] (a survey can be found in [8]). A related issue not yet explored is automatically looking for a route compatible with a given one (we know of no topology independent package for this purpose). If the layout is not simple this can be very helpful in both the station design step and in case of a degraded condition (derailment, works in the tracks, turnout breakdown, etc.), but the tradition is that experts set the non-conflicting routes in advance, during the design of the interlocking.

We have addressed this problem in the frame of an overtaking station of any topology. Overtaking stations are stations on double track lines with sidings (connected to a main track or to other sidings at either end, also called loops) on each side of the main line and one or two crossovers at the two throats of the overtaking station (Fig. 1). According to [49] 'Overtaking stations are set on double track railways and responsible for dealing with the overtaking of trains in the same direction'.

We have used for our purpose the *GraphTheory* package [16,17] of the computer algebra system (CAS) *Maple 2020*[1] [7,15,22,30,42]. However, the code has been tested with much earlier versions, up to the 2009 *Maple 13*, with no problems, except an avoidable **stylesheet** plot option (see [35]).

This could be a perfect complement for railway interlocking systems implemented in the same CAS, such as [36] (based on the use of Boolean matrices) and [37] (based on the use of Groebner bases [12]).

Moreover, it could also be used to compare alternative layouts prior to the railway station design or when the track layout is remodeled (the later is not so unusual: for instance the track layout of the two main railway stations at Madrid, Madrid Chamartín [26] and Madrid Atocha [10], as well as Barcelona main station, Barcelona Sants [5], ans Seville main station, Sevilla Santa Justa [6], are being remodeled nowadays (as well as the track layout of several other smaller railway facilities in Spain).

---

[1] *Maple* is a trademark of Waterloo Maple Inc.

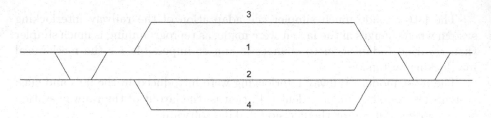

Fig. 1. Standard layout of an overtaking station (transversal type layout, according to the notation of [49]) with one siding (tracks 3 and 4) on each side of the main tracks (tracks 1, 2)

2   Key Idea

Let us illustrate the key idea with the diagram of an overtaking station with two sidings on each side of the main tracks (Fig. 2)[2].

Let us consider that sections exist between every turnout (Fig. 3), as already done, for instance in the approaches to decision making in railway interlocking systems [36,37].

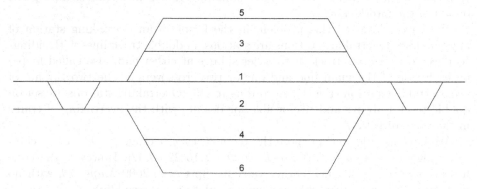

Fig. 2. Standard layout of an overtaking station with two sidings (tracks 3, 5 and 4, 6) on each side of the main tracks (tracks 1, 2)

We are focusing on the concrete problem of checking whether two disjoint routes exist between the throats of the railway station (that is, if two non stopping trains can cross each other in the proposed layout). Obviously, in the layout of Fig. 2 and 3, this is:

---

[2] Note that the layouts of Fig. 1 and 2 and simpler ones can be found in many railway networks. They are proposed as standards, with some extra dead-end sidings (introduced for safety reasons), that don't affect the process proposed in this article, in [31] (p. 155). A real example of the layout of Fig. 1 is El Prat, in the high-speed line Madrid–Barcelona. ([31], p. 62). Many other similar examples of the layouts of Fig. 1 and 2 can be found in the high speed lines of Japan, France and Spain [31], as well as in classic railway networks all around the world.

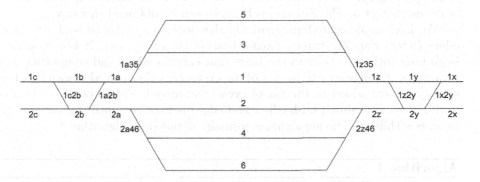

Fig. 3. Sections considered in the layout of Fig. 2

- true, if the conditions are not degraded, as we have, for example, the disjoint routes:
  Route I: 1x, 1y, 1z, 1, 1a, 1b, 1c
  Route II: 2c, 2b ,2a, 2a46, 4, 2z46, 2z, 2y, 2x,
- false, for instance, if there was a derailment affecting sections 2, 4 and 6.

Let us connect the end sections at the two throats with two imaginary sections, 1c2c and 1x2x(Fig. 4). Then the existence of the two disjoint routes mentioned in the previous paragraph would be trivially equivalent to the existence of a cycle containing both Section 1c2c and 2x2y.

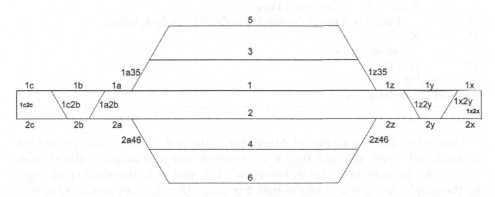

Fig. 4. Adding two imaginary sections that transform the problem of existence of two compatible routes into the existence of cycles through two certain edges

3   Algorithm

Maple's *GraphTheory* package provides a command, `CycleBasis`, that returns a basis for the cycle space of the given graph. All cycles in G can be obtained from applying symmetric differences to the cycles in the basis, considered as sets of edges (in fact all the Eulerian subgraphs can be obtained this way).

We have implemented an algorithm that looks for cycles through two given edges (`edge1,edge2`), starting from a basis of the cycle space. It is incremental, beginning with the cycles in the basis that contain edge1 and computing their symmetric difference with the rest of the cycles that share an edge with it. These new cycles are added to the list of cycles considered. This process is repeated until a cycle containing both edge1 and edge2 is found or until this incremental process stabilizes. The algorithm is formally detailed in Algorithm 1.

Algorithm 1

1: **procedure** FINDCYCLESTHROUGH2EDGES(*list_cycles*, *edge*1, *edge*2)
    ▷ (This procedure is recursive. The first time it is executed, the given list of edges should be a basis of the cycle space of the graph.)
    ▷ (Note: # represents the number of elements in the list afterwards.)
2:     Copy *list_cycles* into a new list, *ELC*.
3:     **for** i to #*ELC* **do**
4:         **for** for j from $i + 1$ to #ELC **do**
5:             **if** ($edge$1 $\in$ ELC_i **or** $edge$1 $\in$ ELC_j) **and** $ELC_i \cap ELC_j \neq \emptyset$ **then**
6:                 compute the symmetric difference of both cycles and add it to ELC
7:             **end if**
8:         **end for**
9:     **end for**
10:     **if** there are cycles in ELC containing both edge1 and edge2 **then**
11:         **return** these cycles
12:     **else**
13:         **if** #ELC > #*list_cycles* **then**
14:             FIND_CYCLES_THROUGH_2_EDGES(ELC, $edge$1, $edge$2)
15:         **else**
16:             **print** 'No solution'
17:         **end if**
18:     **end if**
19: **end procedure**

Regarding the complexity of Algorithm 1, the first input (`list_cycles`) has to be a cycle basis the first time it is executed, but it is only calculated once. If k is the number of cycles in the cycle basis and n is the number of edges in the graph, as there are two nested **for** loops ($i = 1, ..., k$) and a symmetric difference of two cycles (of at most n edges) could have to be computed each time, the worst case complexity of the algorithm each time it is executed is at most $O(2 \cdot k^2 \cdot n^2)$. But this process should be carried out recursively... However, these railway facilities have associated graphs that are usually rather simple from the graph theory point of view, so, probably, a worst case complexity is not very meaningful in this case. The timings obtained when executing the example of Sect. 5 are very small (they can be found at the end of that section), so more complex layouts could be easily addressed.

4   *Maple* Implementation

Only an auxiliary procedure and the main procedure have to be implemented. The auxiliary procedure obtains the edges of a cycle given as a list of vertices (something similar to what Edges(Graph(...,...)) does:

```
edges_cycle:=proc(listaV)
    local i,setE;
    setE:={};
    for i to nops(listaV) - 1 do
        setE:={op(setE),{op(i,listaV),op(i+1,listaV)}}
    end do;
    setE:={op(setE),{op(nops(listaV),listaV),op(1,listaV)}};
end proc:
```

The previous procedure initializes the local variable setE as the empty set. It then goes through the input list listaV, adding to setE the edges connecting the i-th element of listaV with the $(i+1)$-th element of listaV, for i =1 to the number of elements of listaV minus one. It finally adds the edge connecting the $(i+1)$-th element of listaV with the first element of listaV, this way closing the cycle. Note that command op returns the elements of a list or set as a sequence and nops returns the number of elements of a list or set.

Meanwhile, the main procedure (described in the algorithm in the previous section) can be implemented as follows:

```
findCyclesThrough2Edges:=proc(list_cycles,edge1,edge2)
    local i,j,l,numb;
    global sols,ELC;
    ELC:=list_cycles;
    numb:=nops(ELC):
    for i to numb do
        for j from i+1 to numb do
            if    (member(edge1,op(i,ELC))
                or member(edge1,op(j,ELC)))
                and op(i,ELC) intersect op(j,ELC) <> {} then
                    ELC:=[op(ELC),symmdiff(op(i,ELC),op(j,ELC))];
            end if;
        end do;
    end do;
    sols:=[];
    for l in ELC do
        if member(edge1,l) and member(edge2,l) then
            sols:=[op(sols),l]
        end if;
    end do;
    if sols<>[] then
        sols
```

```
   elif list_cycles<>ELC then
       findCyclesThrough2Edges(ELC,edge1,edge2)
   else
       print('No solution')
   end if;
end proc:
```

Variable ELC is initialized as the (input) list_cycles and variable numb is assigned the number of elements of ELC. Observe that the recursive procedure findCyclesThrough2Edges(list_cycles,edge1,edge2) detects new cycles in the graph starting from list_cycles, initially a basis of the cycle space, using the operation symmetric difference (thanks to a nested for).

Variable sols is initialized as the empty list. Those cycles in ELC containing the edge {edge1,edge2} are added to variable sols. If at least one such a cycle has been found (that is, if sols is not the empty list), sols is returned. If the process has added new edges to ELC but sols is empty, findCyclesThrough2Edges is called with ELC as first input. Otherwise a "No solution" message is printed.

5   Example

Let us consider the layout of Fig. 3. We have to first load the GraphTheory package:

```
with(GraphTheory):
```

Now we can introduce the graph to *Maple*, declaring the list of vertices and the set of edges:

```
ListVer:=["5","3","1a35","1z35","1c","1b","1a","1","1z","1y","1x",
         "1c2b","1a2b","1z2y","1x2y","2c","2b","2a","2","2z",
         "2y","2x","2a46","2z46","4","6"]:
G := Graph(ListVer,
         {{"1x","1y"},{"1x","1x2y"},{"1y","1z"},{"1z","1"},
          {"1","1a"},{"1z","1z35"},{"1z35","3"},{"1z35","5"},
          {"5","1a35"},{"3","1a35"},{"1a35","1a"},{"1a","1b"},
          {"1a","1a2b"},{"1b","1c"},{"2x","2y"},{"2y","2z"},
          {"2y","1z2y"},{"2z","2"},{"2","2a"},{"2z","2z46"},
          {"2z46","4"},{"2z46","6"},{"6","2a46"},{"4","2a46"},
          {"2a46","2a"},{"2a","2b"},{"2b","2c"},{"2b","1c2b"},
          {"1x2y","2y"},{"1z2y","1z"},{"1a2b","2b"},
          {"1c2b","1c"}}):
```

The default allocation of vertices produces an unclear drawing (for our purposes). It is convenient to manually allocate the vertices (so that the diagram resembles the track layout), for instance:

```
vp := [[0.5,0.9], [0.5,0.8], [0.4,0.7], [0.6,0.7], [0,0.6],
        [0.15,0.6], [0.3,0.6], [0.5,0.6], [0.7,0.6], [0.85,0.6],
        [1,0.6], [0.1,0.5], [0.2,0.5], [0.8,0.5], [0.9,0.5],
        [0,0.4], [0.15,0.4], [0.3,0.4], [0.5,0.4], [0.7,0.4],
        [0.85,0.4], [1,0.4], [0.4,0.3], [0.6,0.3], [0.5,0.2],
        [0.5,0.1]]:
SetVertexPositions(G, vp):
```

Now `DrawGraph(G);` produces the nice drawing of Fig. 5.

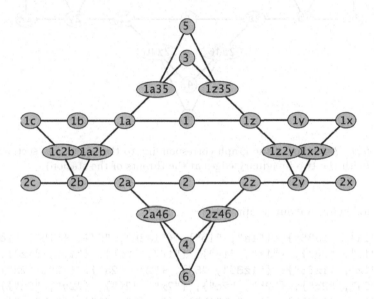

Fig. 5. *Maple 2020* plot of the graph corresponding to the layout and section naming of Fig. 3

It is time to add the imaginary edges at the throats of the station (see Fig. 4):

```
G_:=AddEdge(G,{"1c","2c"});
G_:=AddEdge(G_,{"1x","2x"});
```

and now `DrawGraph(G_);` produces the nice drawing of Fig. 6.

We can look for the cycles containing the two imaginary edges at the throats of the station, just typing:

```
LC:=CycleBasis(G_):
LCE:=map(edges_cycle,LC):
findCyclesThrough2Edges(LCE,{"1c","2c"},{"1x","2x"});
```

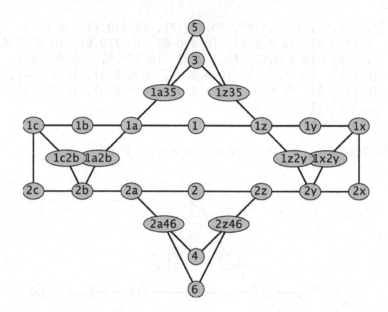

Fig. 6. *Maple 2020* plot of the graph corresponding to the layout and section naming of Fig. 4 (with the two imaginary edges at the throats of the station)

and the following output is obtained:

```
[{{"1a", "1a35"}, {"1a", "1b"}, {"1a35", "5"}, {"1b", "1c"},
  {"1c", "2c"}, {"1x", "1y"}, {"1x", "2x"}, {"1y", "1z"},
  {"1z", "1z35"}, {"1z35", "5"}, {"2", "2a"}, {"2", "2z"},
  {"2a", "2b"}, {"2b", "2c"}, {"2x", "2y"}, {"2y", "2z"}},
 {{"1a", "1a35"}, {"1a", "1b"}, {"1a35", "5"}, {"1b", "1c"},
  {"1c", "2c"}, {"1x", "1y"}, {"1x", "2x"}, {"1y", "1z"},
  {"1z", "1z35"}, {"1z35", "5"}, {"2a", "2a46"}, {"2a", "2b"},
  {"2a46", "4"}, {"2b", "2c"}, {"2x", "2y"}, {"2y", "2z"},
  {"2z", "2z46"}, {"2z46", "4"}},
 {{"1a", "1a35"}, {"1a", "1b"}, {"1a35", "5"}, {"1b", "1c"},
  {"1c", "2c"}, {"1x", "1y"}, {"1x", "2x"}, {"1y", "1z"},
  {"1z", "1z35"}, {"1z35", "5"}, {"2a", "2a46"}, {"2a", "2b"},
  {"2a46", "6"}, {"2b", "2c"}, {"2x", "2y"}, {"2y", "2z"},
  {"2z", "2z46"}, {"2z46", "6"}}]
```

that is, three cycles have been found (they are stored in variable sols). We can easily visualize, for example, the first cycle obtained by typing:

```
HighlightEdges(G_,op(1,sols), stylesheet = [thickness = 3]);
DrawGraph(G_);
```

and the drawing of Fig. 7 is obtained, where the two disjoint routes are clearly visible (erasing the two imaginary edges):

```
Route I:  "1x", "1y", "1z", "1z35", "5", "1a35",
          "1a", "1b", "1c"
Route II: "2c", "2b", "2a", "2", "2z", "2y", "2x"
```

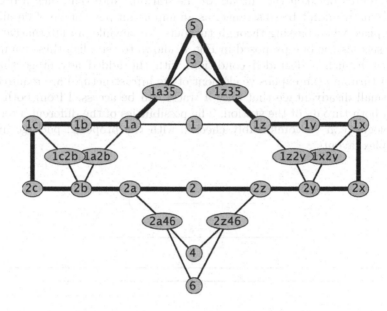

Fig. 7. On of the three cycles found through edges {"1c","2c"} and {"1x","2x"} by the proposed procedure

The *Maple 2020* worksheet corresponding to this example can be downloaded from [35]. The timings obtained for this example on an ordinary laptop are 0.016 s for obtaining the cycle basis and 0.015 s for running the main procedure.

6 Conclusions and Future Work

CASs have evolved including new and exciting possibilities, like numeric algorithms, symbolic-numeric approaches, implementations of packages dealing with related fields (like, in this case, graph theory), etc.

The extension presented here, although limited to overtaking stations (of any number of tracks and of any topology) has two main uses:

– To complement decision making software for railway interlocking systems, allowing to look for another compatible route when a pair of proposed routes is not compatible.
– To analyze alternative topologies to those traditionally accepted. For instance, the standard layout of an overtaking station of transversal type layout (Fig. 2 and 3) has the advantage that the 6 tracks of the station can be accessed

from both tracks from both throats of the station. Nevertheless, it has the disadvantage that a problem in the tracks on one side of the station that made impossible, for instance, to use tracks 2, 4, 6 (a derailment affecting Section 2a and 2a46, for example), would make impossible to find two disjoint routes between the throats of the station. Note that such a degraded condition wouldn't be so strange, as an important percentage of derailments take place when passing through turnouts. Meanwhile, an alternative topology, such as the one proposed in Fig. 8, allows to establish these compatible routes in such a degraded condition, with the added advantage that two fewer turnouts (an expensive element of the infrastructure) are required, and the small disadvantage that only 4 tracks can be accessed from both tracks from both throats of the station. The possibilities of the different alternative topologies can be comfortably checked with the proposed package in more complex layouts.

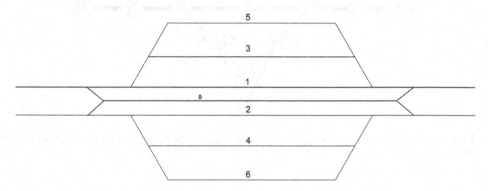

Fig. 8. A possible alternative layout for overtaking stations of transversal type. The four crossovers between the main tracks have been substituted by a long bypass of the station (track 0), accessible from both throats of the station from both tracks.

The approach could be also applied to small terminal stations (of any topology) at the end of a double track line. Given two origins and destinations, it could be determined if two compatible routes with these origins and destinations exist, and at least one solution would be found in the affirmative case.

For example, let us consider the tiny terminal station of Fig. 9. Let us suppose that there is a train in section 5 that has to leave the station and there is an incoming train that we would like to situate at track 2. It would be enough to add the virtual edges {1c,2c} and {5,2} to the graph and to apply an approach similar to the one proposed in this paper (to look for a cycle containing these two edges).

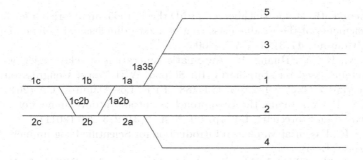

Fig. 9. A tiny terminal station at the end of a double track line

Acknowledgments. This work was partially supported by the research project PGC2018-096509-B-100 (Government of Spain). We would like to thank the anonymous reviewers for their mot valuable comments and suggestions that have greatly improved the article.

References

1. Anonymous: Proyecto y obra del enclavamiento electrónico de la estación de Madrid-Atocha. Proyecto Técnico. Siemens, Madrid (1988)
2. Anonymous: Microcomputer Interlocking Hilversum. Siemens, Munich, (1986)
3. Anonymous: Microcomputer Interlocking Rotterdam. Siemens, Munich, (1989)
4. Anonymous: Puesto de enclavamiento con microcomputadoras de la estación de Chiasso de los SBB. Siemens, Munich (1989)
5. Anonymous: Comienzan las obras del nuevo esquema de vías de estacionamiento de ancho convencional de Barcelona Sants. Boletín Vía Libre (2020). https://www.vialibre.org/noticias.asp?not=29524&cs=infr
6. Anonymous: Licitada la ampliación de las vías de estacionamiento de trenes AVE en Sevilla Santa Justa y Majarabique. Boletín Vía Libre (2020). https://www.vialibre.org/noticias.asp?not=29527&cs=infr
7. Bernardin, L., Chin, P., DeMarco, P., Geddes, K. O., Hare, D. E. G., Heal, K. M. Labahn, G., May, J. P., McCarron, J., Monagan, M. B. Ohashi, D. Vorkoetter, S. M.: Maple Programming Guide. Maplesoft, Waterloo Maple Inc., Waterloo, Canada (2020). https://www.maplesoft.com/documentation_center/maple2020/ProgrammingGuide.pdf
8. Bjørner, D.: The FMERail/TRain Annotated Rail Bibliography (2005). http://www2.imm.dtu.dk/~db/fmerail/fmerail/
9. Borälv, A.: Case study: formal verification of a computerized railway interlocking. Form. Asp. Comput. **10**, 338–360 (1998)
10. Briginshaw, D.: Adif awards Madrid track remodelling planning contract. International Railway Journal (2020). https://www.railjournal.com/passenger/commuter-rail/adif-awards-madrid-track-remodelling-planning-contract/
11. Briginshaw, D.: DB inaugurates first digital interlocking, International Railway Journal (2018). https://www.railjournal.com/signalling/db-inaugurates-first-digital-interlocking/

12. Buchberger, B.: Bruno Buchberger's PhD thesis 1965: an algorithm for finding the basis elements of the residue class ring of a zero dimensional polynomial ideal. J. Symb. Comput. **41**(3), 475–511 (2006)
13. Chen, X., He, Y., Huang, H.: Automatic generation of relay logic for interlocking system based on statecharts. In: Second WRI World Congress on Software Engineering WCSE, vol. 2. pp. 183–188, IEEE, Los Alamitos, CA (2010)
14. Chen, X., He, Y., Huang, H.: An approach to automatic development of interlocking logic based on Statechart. Enterp. Inf. Syst. **5**(3), 273–286 (2011)
15. Corless, R.: Essential Maple. An Introduction for Scientific Programmers. Springer, New York (1995)
16. Ebrahimi, M., Ghebleh, M., Javadi, M., Monagan, M., Wittkopf, A.: A graph theory package for maple, part ii: graph coloring, graph drawing, support tools, and networks. In: Proceedings of the 2006 Maple Conference. pp. 99–112, Maplesoft, Waterloo, Canada (2006)
17. Farr, J., Khatarinejad Fard, M., Khodadad, S, Monagan, M.: A graphtheory package for maple. In: Proceedings of the 2005 Maple Conference. pp. 260–271, Maplesoft, Waterloo, Canada (2005)
18. Ferrari, A., Magnani, G., Grasso, D., Fantechi, A.: Model checking interlocking control tables. In: E. Schnieder, G. Tarnai (Eds.), Proceedings of Formal Methods for Automation and Safety in Railway and Automotive Systems FORMS/FORMAT 2010, pp. 107–115, Springer, Berlin (2011)
19. Hansen, K. M.: Formalising Railway Interlocking Systems. In: Nordic Seminar on Dependable Computing Systems, Department of Computer Science, Technical University of Denmark, Lyngby (1994)
20. Haxthausen, A.E., Peleska, J., Kinder, S.: A formal approach for the construction and verification of railway control systems. Form. Asp. Comput. **23**, 191–219 (2009)
21. Haxthausen, A.E.: Automated generation of safety requirements from railway interlocking tables. In: Margaria, T., Steffen, B. (eds.) ISoLA 2012. LNCS, vol. 7610, pp. 261–275. Springer, Heidelberg (2012). https://doi.org/10.1007/978-3-642-34032-1_25
22. Heck, A.: Introduction to Maple. Springer, New York (2003)
23. Hernando, A., Roanes-Lozano, E., Maestre-Martínez, R., Tejedor, J.: A logic-algebraic approach to decision taking in a railway interlocking system. Ann. Math. Artif. Intell. **65**, 317–328 (2012)
24. Hernando, A., Maestre, R., Roanes-Lozano, E.: A new algebraic approach to decision making in a railway interlocking system based on preprocess, Mathematical Problems in Engineering, ID 4982974, 14 pages (2018)
25. Hlavatý T., Přeučil, L., Štěpán, P., Klapka, Š: Formal methods in development and testing of safety-critical systems: railway interlocking system. In: Intelligent Methods for Quality Improvement in Industrial Practice, vol. 1, pp. 14–25, CTU FEE, Department of Cybernetics, The Gerstner Laboratory, Prague (2002)
26. Ineco. Estudio informativo del nuevo complejo ferroviario de la estación de Madrid-Chamartín. Ministerio de Transportes, Movilidad y Agenda Urbana, Madrid (2019)
27. Janota, A.: Using Z specification for railway interlocking safety. Period. Polytech. Transp. Eng. **28**(1–2), 39–53 (2000)
28. Khan, U., Ahmad, J., Saeed, T., Mirza, S.H.: On the real time modeling of interlocking system of passenger lines of Rawalpindi Cantt train station. Complex Adapt. Syst. Model. **4**, 17 (2016)
29. Losada, M.: Curso de Ferrocarriles: Explotación Técnica. E.T.S.I. Caminos, Madrid (1991)

30. Maplesoft: Maple User Manual. Maplesoft, Waterloo Maple Inc., Waterloo, Canada (2020). https://www.maplesoft.com/documentation_center/maple2020/UserManual.pdf
31. Martín Cañizares, M. P.: Contribución al diseño eficiente de la configuración en planta de líneas de alta velocidad (Ph.D. Thesis). Universitat Politècnica de Catalunya, Barcelona (2015). http://www.tecnica-vialibre.es/documentos/Libros/PilarMartin_Tesis.pdf
32. Montigel, M.: Modellierung und Gewährleistung von Abhängigkeiten in Eisenbahnsicherungsanlagen (Ph.D. Thesis), ETH Zurich, Zurich (1994). http://www.inf.ethz.ch/research/disstechreps/theses
33. Morley, M. J.: Modelling British Rail's interlocking logic: Geographic data correctness. Technical Report ECS-LFCS-91-186, Laboratory for Foundations of Computer Science, Department of Computer Science, University of Edinburgh, Edinburgh (1991)
34. Nakamatsu, K., Kiuchi, Y., Suzuki, A.: EVALPSN based railway interlocking simulator. In: Negoita, M.G., Howlett, R.J., Jain, L.C. (eds.) KES 2004. LNCS (LNAI), vol. 3214, pp. 961–967. Springer, Heidelberg (2004). https://doi.org/10.1007/978-3-540-30133-2_127
35. Roanes-Lozano, E.: Looking for compatible routes in the railway interlocking system of an overtaking station using a computer algebra system (from CASC'2020 conference) (2020). https://webs.ucm.es/info/secdealg/ERL/CASC2020_RoutesInterlocking-Cycles.mw
36. Roanes-Lozano, E., Laita, L.M.: An applicable topology-independent model for railway interlocking systems. Math. Comput. Simul. 45(1), 175–184 (1998)
37. Roanes-Lozano, E., Roanes-Macías, E., Laita, L.: Railway interlocking systems and Gröbner bases. Math. Comput. Simul. 51(5), 473–481 (2000)
38. Roanes-Lozano, E., Hernando, A., Alonso, J.A., Laita, L.M.: A logic approach to decision taking in a railway interlocking system using maple. Math. Comput. Simul. 82, 15–28 (2011)
39. Roanes-Lozano, E., González-Franco, I., Hernando, A., García-Álvarez, A., Mesa, L.E.: Optimal route finding and rolling-stock selection for the spanish railways. Comput. Sci. Eng. 14(4), 82–89 (2012)
40. Roanes-Lozano, E., Hernando, A., García-Álvarez, A., Mesa, L.E., González-Franco, I.: Calculating the exploitation costs of trains in the Spanish railways. Comput. Sci. Eng. 15(3), 89–95 (2013)
41. Roanes-Lozano, E., Alonso, J.A., Hernando, A.: An approach from answer set programming to decision making in a railway interlocking system. Revista de la Real Academia de Ciencias Exactas, Fisicas y Naturales. Serie A. Matematicas 108(2), 973–987 (2014)
42. Roanes-Macías, E., Roanes-Lozano, E. Cálculos Matemáticos por Ordenador con Maple vol 5. Editorial Rubiños-1890, Madrid (1999)
43. Siemens. Interlocking systems for mainline railways. Solutions for any application, anywhere. https://www.mobility.siemens.com/global/en/portfolio/rail/automation/interlocking-systems.html
44. Villamandos, L.: Sistema informático concebido por Renfe para diseñar los enclavamientos. Vía Libre 348, 65 (1993)
45. Westwood, J. (ed.): Trains. Octopus Books Ltd., London (1979)
46. Winter, K., Johnston, W., Robinson, P. Strooper, P., van den Berg, L.: Tool Support for Checking Railway Interlocking Designs. In: T. Cant (Ed.), Proceedings of the 10th Australian Workshop on Safety Related Programmable Systems, pp. 101–107, Australian Computer Society Inc, Sydney (2006)

47. Winter, K., Robinson, N. J.: Modelling large interlocking systems and model check-ing small ones. In: M. Oudshoorn (Ed.), 26th Australasian Computer Science Con-ference (ACSC'2003). vol. 16, pp. 309–316, Australian Computer Science Commu-nications (2003)
48. Xiangxian, C., Yulin, H., Hai, H.: A component-based topology model for railway interlocking systems. Math. Comput. Simul. **81**(9), 1892–1900 (2011)
49. Yi, S.: Principles of Railway Location and Design. Academic Press, London (2018)

Computing Logarithmic Vector Fields Along an ICIS Germ via Matlis Duality

Shinichi Tajima[1], Takafumi Shibuta[2], and Katsusuke Nabeshima[3(✉)]

[1] Graduate School of Science and Technology, Niigata University,
8050, Ikarashi 2-no-cho, Nishi-ku Niigata, Japan
tajima@emeritus.niigata-u.ac.jp
[2] Department of Information Science, Kyushu Sangyo University,
2-3-1, Shokadai, Higashiku, Fukuoka, Japan
tshibuta@ip.kyusan-u.ac.jp
[3] Graduate School of Technology, Industrial and Social Sciences,
Tokushima University, 2-1, Minamijosanjima-cho, Tokushima, Japan
nabeshima@tokushima-u.ac.jp

Abstract. Logarithmic vector fields along an isolated complete intersection singularity (ICIS) are considered in the context of computational
complex analysis. Based on the theory of local polar varieties, an effective method is introduced for computing a set of generators, over a local
ring, of the modules of germs of logarithmic vector fields. Underlying
ideas of our approach are the use of a parametric version of the concept
of local cohomology and the Matlis duality. The algorithms are designed
to output a set of representatives of logarithmic vector fields which is
suitable to study their complex analytic properties. Some examples are
given to illustrate the resulting algorithms.

Keywords: Logarithmic vector fields · Local cohomology · Matlis
duality · Isolated complete intersection singularity

1 Introduction

The concept of logarithmic vector fields introduced by Saito [36], is of considerable importance in complex analysis and singularity theory. Logarithmic
vector fields have been extensively studied and utilized by several authors and
are known to be related to many areas of mathematics [8,10,13,17,18,43]. The
concept of logarithmic vector fields which is closely related with logarithmic
differential forms and logarithmic residues, is actually profound. We refer the
readers to [5–7,15,35] for recent results on logarithmic vector fields, logarithmic differential forms and related topics. Note also that logarithmic vector fields
are related with the Nash blow-up, or limiting tangent spaces and contain rich
information on singularities.

This work has been partly supported by JSPS Grant-in-Aid for Scientific Research (C)
(18K03320 and 18K03214).

© Springer Nature Switzerland AG 2020
F. Boulier et al. (Eds.): CASC 2020, LNCS 12291, pp. 543–562, 2020.
https://doi.org/10.1007/978-3-030-60026-6_32

In our previous papers [25, 38], we considered the case of hypersurfaces with an isolated singularity and, by utilizing the concept of local polar variety [41, 42], gave methods for computing logarithmic vector fields. In [29], with an intention to study complex analytic structure of logarithmic vector fields, we derived another method by improving the previous one which takes care of genericity of local polar varieties. As applications, in [28, 29] we gave algorithms of computing Bruce-Roberts Milnor and Tjurina numbers [8, 33] and in [40], based on a result of A. G. Aleksandrov [4], gave an algorithm for computing torsion differential forms, that can be used to study logarithmic differential forms and related problems.

In singularity theory, Aleksandrov [3] and Wahl [45] studied complete intersection singularities and independently gave, among other results, a closed formula of the generators of logarithmic vector fields along quasi-homogeneous complete intersection singularities. Note that their results were effectively used in [8, 32, 34]. By contrast, for non-quasi homogeneous complete intersection cases, no closed formula is known. Complex analytic structure of modules of logarithmic vector fields is difficult even for an isolated singularity case. In fact, as is noticed in [38], a direct use of syzygy computation is not enough to reveal their analytic properties. Therefore, effective methods for computing logarithmic vector fields are desirable.

In the present paper we consider logarithmic vector fields along an ICIS and address the problem of computation of logarithmic vector fields in the context of computational complex analysis. The aim is providing new effective tools for studying complex analytic properties of logarithmic vector fields. We show that, by adopting ideas given in [26, 30], the use of local cohomology and the Matlis duality allows us to derive an effective method for computing them. The resulting algorithms that take care of genericity conditions compute a set of generators of the module of germs of logarithmic vector fields along an ICIS.

In Sect. 2, we recall some basics on local cohomology and the Matlis duality. In Sect. 3, we describe an underlying idea for computing logarithmic vector fields associated to an ICIS. In Sect. 4, we recall a result of B. Teissier on polar varieties, which will be used for selecting a generic local coordinate system. In Sect. 5, we present algorithms for computing logarithmic vector fields. In Sect. 6, we give examples for illustration.

All algorithms in this paper have been implemented in the computer algebra system Risa/Asir [31].

2   Preliminaries

In this section, we recall some basics on a complex analytic version of local cohomology and that of Matlis duality.

2.1   Matlis Duality

Let X be an open neighborhood of the origin O in $\mathbb{C}^n$ with a local coordinate system $x = (x_1, x_2, \ldots, x_n)$. Let $\mathcal{O}_X$ be the sheaf on X of holomorphic functions, Ω_X^n the sheaf of holomorphic n-forms. Let $\mathcal{O}_{X,O}$ be the stalk at O of the sheaf $\mathcal{O}_X$ and let $\mathcal{H}_{\{O\}}^n(\Omega_X^n)$ denote the local cohomology of Ω_X^n supported at O [16, 20].

For a positive integer ℓ, let

$$\mathrm{res}_{\{O\}}(*,\,*) : (\mathcal{O}_{X,O})^\ell \times (\mathcal{H}_{\{O\}}^n(\Omega_X^n))^\ell \longrightarrow \mathbb{C}$$

be the pairing defined as follows: for $p = (h_1, h_2, \ldots, h_\ell) \in (\mathcal{O}_{X,O})^\ell$ and

$$\omega = \begin{pmatrix} \omega_1 \\ \omega_2 \\ \vdots \\ \omega_\ell \end{pmatrix} \in (\mathcal{H}_{\{O\}}^n(\Omega_X^n))^\ell,$$

we set

$$\mathrm{res}_{\{O\}}(p, \omega) := \mathrm{res}_{\{O\}} \left(\sum_{i=1}^\ell h_i \omega_i \right),$$

where $\mathrm{res}_{\{O\}} \left(\sum_{i=1}^\ell h_i \omega_i \right)$ is the Grothendieck local residue at the origin O of the local cohomology class $p\omega = \sum_{i=1}^\ell h_i \omega_i \in \mathcal{H}_{\{O\}}^n(\Omega_X^n)$ [21]. The above pairing res gives rise to a duality between $(\mathcal{O}_{X,O})^\ell$ and $(\mathcal{H}_{\{O\}}^n(\Omega_X^n))^\ell$. More precisely, $(\mathcal{O}_{X,O})^\ell$ and $(\mathcal{H}_{\{O\}}^n(\Omega_X^n))^\ell$ have a structure of locally convex topological vector space (one is Fréchet Schwartz and the other is dual Fréchet Schwartz [44]) and they are mutually dual [20].

Now let N be an $\mathcal{O}_{X,O}$-submodule of $(\mathcal{O}_{X,O})^\ell$ generated by $p_1, p_2, \ldots, p_m$ and let W_N denote the set of local cohomology classes in $(\mathcal{H}_{\{O\}}^n(\Omega_X^n))^\ell$ that are killed by N:

$$W_N = \left\{ \omega = \begin{pmatrix} \omega_1 \\ \vdots \\ \omega_\ell \end{pmatrix} \in (\mathcal{H}_{\{O\}}^n(\Omega_X^n))^\ell \,\middle|\, \sum_{i=1}^\ell h_i \omega_i = 0, \forall p = (h_1, h_2, \ldots, h_\ell) \in N \right\}.$$

Assume that N has finite colength, namely the quotient space $(\mathcal{O}_{X,O})^\ell / N$ is assumed to be a finite dimensional vector space. Then, the classical Matlis duality [23] says that the pairing

$$\mathrm{res}_{\{O\}}(*,\,*) : (\mathcal{O}_{X,O})^\ell / N \times W_N \longrightarrow \mathbb{C}$$

induced by res is also non-degenerate [9].

Let $\hat{\mathcal{O}}_{X,O}$ be the ring of formal power series and let $\mathcal{H}_{[O]}^n(\Omega_X^n)$ denote the algebraic local cohomology of Ω_X^n supported at O defined to be

$$\mathcal{H}_{[O]}^n(\Omega_X^n) = \lim_{k \to \infty} \mathrm{Ext}^n(\mathcal{O}_X/\mathfrak{m}^k, \Omega_X^n),$$

where $\mathfrak{m} = (x_1, x_2, \ldots, x_n)$ is the maximal ideal generated by $x_1, x_2, \ldots, x_n$, supported at the origin. Let $\hat{N} = \hat{\mathcal{O}}_{X,O} \otimes N \subset (\hat{\mathcal{O}}_{X,O})^\ell$ and let

$$\hat{W}_N = \left\{ \hat{\omega} = \begin{pmatrix} \hat{\omega}_1 \\ \vdots \\ \hat{\omega}_\ell \end{pmatrix} \in (\mathcal{H}^n_{[O]}(\Omega^n_X))^\ell \; \middle| \; \sum_{i=1}^\ell \hat{h}_i \hat{\omega}_i = 0, \forall \hat{p} = (\hat{h}_1, \hat{h}_2, \ldots, \hat{h}_\ell) \in \hat{N} \right\}.$$

Then, we also have the following non-degenerate pairing [9]

$$\mathrm{res}_{\{O\}}(*, \, *) : (\hat{\mathcal{O}}_{X,O})^\ell / \hat{N} \times \hat{W}_N \longrightarrow \mathbb{C}.$$

2.2 Algebraic Local Cohomology

Let $N \subset (\mathcal{O}_{X,O})^\ell$ be an $\mathcal{O}_{X,O}$-submodule, as previously of finite colength, generated by $p_1, p_2, \ldots, p_m \in (\mathcal{O}_{X,O})^\ell$. Set

$$H_N = \{ \sigma \in (\mathcal{H}^n_{\{O\}}(\mathcal{O}_X))^\ell \mid p_i \sigma = 0, i = 1, 2, \ldots, m \}$$

and

$$\hat{H}_N = \{ \hat{\sigma} \in (\mathcal{H}^n_{[O]}(\mathcal{O}_X))^\ell \mid p_i \hat{\sigma} = 0, i = 1, 2, \ldots, m \}.$$

Let $dx = dx_1 \wedge dx_2 \wedge \cdots \wedge dx_n$ denote a holomorphic n-form. Then, we have

$$W_N = \{ \sigma dx \mid \sigma \in H_N \}, \quad \hat{W}_N = \{ \hat{\sigma} dx \mid \hat{\sigma} \in \hat{H}_N \}.$$

Since the finite dimensionality $\dim_{\mathbb{C}}((\mathcal{O}_{X,O})^\ell / N) < \infty$ implies $\hat{H}_N \cong H_N$, the local cohomology module W_N can be identified with algebraic local cohomology module $\hat{H}_N$. In what follows, we identify $\hat{H}_N$ with H_N.

If the given generators $p_i = (h_{i,1}, h_{i,2}, \ldots, h_{i,\ell}), i = 1, 2, \ldots, m$ belong to the module $(\mathbb{Q}[x])^\ell$, then, a basis as a vector space of H_N is computable by using an algorithm described in [37].

Note that the non-degeneracy of the Matlis duality implies the following. Assume that $p = (h_1, h_2, \ldots, h_\ell) \in (\mathcal{O}_{X,O})^\ell$ is given. Then p is in N if and only if $\mathrm{res}_{\{O\}}(p\sigma dx) = 0$, for all $\sigma \in H_N$.

3 Logarithmic Vector Fields Along ICIS and Matlis Duality

Let X be an open neighborhood of the origin O in $\mathbb{C}^n$. Let

$$V = \{ x \in X \mid f_1(x) = f_2(x) = \cdots = f_\ell(x) = 0 \}$$

be a complex analytic variety, where $x = (x_1, x_2, \ldots, x_n)$ and $f_1, f_2, \ldots, f_\ell$ are holomorphic functions defined on X. We assume that V is an ICIS, isolated complete intersection singularity at O, that is, $\mathrm{codim}(V) = \ell$ and the singular set $\mathrm{Sing}(V)$ of V in X is the origin O. Let $I_O = (f_1, f_2, \ldots, f_\ell)$ be the ideal in the local ring $\mathcal{O}_{X,O}$ generated by $f_1, f_2, \ldots, f_\ell$.

Let v be a germ of holomorphic vector field on X at the origin O.

Definition 1 [8,36]. *A germ of holomorphic vector field v is said to be logarithmic along V, when considered as a derivation $v : \mathcal{O}_{X,O} \to \mathcal{O}_{X,O}$, we have $v(f) \in I_O$ for all $f \in I_O$. The $\mathcal{O}_{X,O}$-module that consists of such vector fields is denoted by $Der_{X,O}(-\log V)$.*

Let
$$p_i = \left(\frac{\partial f_1}{\partial x_i}, \frac{\partial f_2}{\partial x_i}, \cdots, \frac{\partial f_\ell}{\partial x_i} \right) \in (\mathcal{O}_{X,O})^\ell, \ i = 1, 2, \ldots, n,$$

and let
$$p_{i,j} = e_i f_j \in (\mathcal{O}_{X,O})^\ell, \ i, j = 1, 2, \ldots, \ell,$$

where $e_i = (0, 0, \ldots, 0, 1, 0, \ldots, 0) \in \mathbb{Z}^\ell$ is the i-th unit vector.

Let
$$v = a_1(x) \frac{\partial}{\partial x_1} + a_2(x) \frac{\partial}{\partial x_2} + \cdots + a_n(x) \frac{\partial}{\partial x_n}$$

be a germ of holomorphic vector field at the origin O, where $a_i \in \mathcal{O}_{X,O}$, $i = 1, 2, \ldots, n$. Then, v is in $Der_{X,O}(-\log V)$ if and only if there exists a set of holomorphic functions $c_{i,j} \in \mathcal{O}_{X,O}, i, j = 1, 2, \ldots, \ell$ s.t.

$$\sum a_i(x) p_i = \sum c_{i,j}(x) p_{i,j}.$$

Let N_Γ denote the $\mathcal{O}_{X,O}$-submodule of $(\mathcal{O}_{X,O})^\ell$ generated by $p_2, p_3, \ldots, p_n$ and $p_{i,j}, i, j = 1, 2, \ldots, \ell$:

$$N_\Gamma = (p_2, p_3, \ldots, p_n, p_{1,1}, p_{1,2}, \ldots, p_{\ell,\ell}) \subset (\mathcal{O}_{X,O})^\ell.$$

Now let us consider a germ of holomorphic function $a \in \mathcal{O}_{X,O}$. Then, it is easy to see that there exists a logarithmic vector fields $v \in Der_{X,O}(-\log V)$ of a form
$$v = a(x) \frac{\partial}{\partial x_1} + a_2(x) \frac{\partial}{\partial x_2} + \cdots + a_n(x) \frac{\partial}{\partial x_n}$$

if and only if $a(x) p_1 \in N_\Gamma$.

Assume that $\dim_{\mathbb{C}}((\mathcal{O}_{X,O})^\ell / N_\Gamma) < \infty$ and we set

$$H_{N_\Gamma} = \{ \sigma \in (\mathcal{H}_{\{O\}}^n(\mathcal{O}_X))^\ell \mid p_i \sigma = 0, i = 2, \ldots n, \ p_{i,j} \sigma = 0, \ i, j = 1, 2, \ldots, \ell \}$$
and

$$H_{N_\Delta} = \{ p_1 \sigma \mid \sigma \in H_{N_\Gamma} \} \subset \mathcal{H}_{\{O\}}^n(\mathcal{O}_X).$$

Let $Ann_{\mathcal{O}_{X,O}}(H_{N_\Delta})$ denote the annihilator in $\mathcal{O}_{X,O}$ of H_{N_Δ} :

$$Ann_{\mathcal{O}_{X,O}}(H_{N_\Delta}) = \{ a \in \mathcal{O}_{X,O} \mid a\delta = 0, \ \forall \delta \in H_{N_\Delta} \}.$$

Then, we have the following:

Theorem 1. *Let $a(x) \in \mathcal{O}_{X,O}$ be a germ of holomorphic function. Then, there exists a logarithmic vector fields $v \in Der_{X,O}(-\log V)$ of a form*

$$v = a(x) \frac{\partial}{\partial x_1} + a_2(x) \frac{\partial}{\partial x_2} + \cdots + a_n(x) \frac{\partial}{\partial x_n}$$

if and only if $a(x) \in Ann_{\mathcal{O}_{X,O}}(H_{N_\Delta})$.

Proof. Let $a(x)$ be a germ of holomorphic function. Then $a(x)$ is in the annihilator $Ann_{\mathcal{O}_{X,o}}(H_{N_\Delta})$ of H_{N_Δ} if and only if $a(x)\delta = 0$ for every local cohomology class δ in H_{N_Δ}. Since, $a(x)(p_1\sigma) = 0$ holds for any local cohomology $\sigma \in H_{N_\Gamma}$, we see that $a(x)p_1$ belongs to the module N_Γ. This completes the proof.

The above theorem says that if $\dim_{\mathbb{C}}((\mathcal{O}_{X,o})^\ell/N_\Gamma) < \infty$, then the coefficient $a(x)$ in front of $\frac{\partial}{\partial x_1}$ of a logarithmic vector field can be described in terms of $Ann_{\mathcal{O}_{X,o}}(H_{N_\Delta})$.

Note that, if the defining functions $f_1, f_2, \ldots, f_\ell$ of a germ of the variety V are in $\mathbb{Q}[x]$, then bases of the vector spaces H_{N_Γ} and H_{N_Δ} are computable. Therefore, a standard basis of the annihilator $Ann_{\mathcal{O}_{X,o}}(H_{N_\Delta})$ is also computable by using algorithms given in [27, 39].

Let $\gamma : H_{N_\Gamma} \longrightarrow \mathcal{H}^n_{\{O\}}(\mathcal{O}_X)$ denote the map defined by $\gamma(\sigma) = p_1\sigma$, where

$$p_1 = \left(\frac{\partial f_1}{\partial x_1}, \frac{\partial f_2}{\partial x_1}, \ldots, \frac{\partial f_\ell}{\partial x_1} \right) \in (\mathcal{O}_{X,o})^\ell.$$

We define $H_{N_T} \subset (\mathcal{H}^n_{\{O\}}(\mathcal{O}_X))^\ell$ as follows:

$$H_{N_T} = \{\tau \in (\mathcal{H}^n_{\{O\}}(\mathcal{O}_X))^\ell \mid p_i\tau = 0, i = 1, \ldots, n, \ p_{i,j}\tau = 0, i, j = 1, 2, \ldots, \ell\}.$$

Then, we have the following:

Lemma 1. $\mathrm{Ker}(\gamma) = H_{N_T}$ holds.

Proof. Since $\mathrm{Ker}(\gamma) = \{\tau \in H_{N_\Gamma} \mid p_1\tau = 0\}$, the above statement follows immediately from the definition of H_{N_Γ}.

Therefore the following sequence is exact:

$$0 \longrightarrow H_{N_T} \longrightarrow H_{N_\Gamma} \longrightarrow H_{N_\Delta} \longrightarrow 0.$$

Accordingly, we have the following:

Proposition 1. *Assume that the $\mathcal{O}_{X,o}$-submodule N_Γ has a finite colength. Then, the following holds:*

$$\dim_{\mathbb{C}}(\mathcal{O}_{X,o}/Ann_{\mathcal{O}_{X,o}}(H_{N_\Delta})) = \dim_{\mathbb{C}}(H_{N_\Gamma}) - \dim_{\mathbb{C}}(H_{N_T}).$$

Proof. The Grothendieck local duality says that the quotient space $\mathcal{O}_{X,o}/Ann_{\mathcal{O}_{X,o}}(H_{N_\Delta})$ and H_{N_Δ} are dual to each other. Therefore we have

$$\dim_{\mathbb{C}}(\mathcal{O}_{X,o}/Ann_{\mathcal{O}_{X,o}}(H_{N_\Delta})) = \dim(H_{N_\Delta}).$$

The exactness of the above sequence yields the proof.

Note that the above statement is a generalization of a result given in [38] to ICIS.

Notice that the dimension of the space H_{N_T} is an complex analytic invariant of the variety V (see [19, 22]). Whereas, the dimension of H_{N_Δ} depends on the choice of a local coordinate system, or a hyperplane.

4   Genericity

In order to obtain good representations of logarithmic vector fields, it is important to choose an appropriate coordinate system, or a generic hyperplane (see [29, 38, 41]). For this purpose, we recall in this section a result of B. Teissier on generic hyperplanes [42, 43].

For $\xi = (\xi_1, \xi_2, \ldots, \xi_n) \in \mathbb{C}^n \setminus (0, 0, \ldots, 0)$, let L_ξ denote the hyperplane

$$L_\xi = \{x \in \mathbb{C}^n \mid \xi_1 x_1 + \xi_2 x_2 + \cdots + \xi_n x_n = 0\}.$$

We may assume without loss of generality there is a non-zero ξ_k for some k and $\xi_k = 1$.

Let $z_i = x_i$ for $i \neq k$ and

$$z_k = \xi_1 x_1 + \cdots + \xi_{k-1} x_{k-1} + x_k + \xi_{k+1} x_{k+1} + \cdots + \xi_n x_n.$$

Set

$$g_j(z, \xi) = f_j(z_1, \ldots, z_{k-1}, z_k - \xi_1 z_1 - \xi_2 z_2 - \cdots - \xi_{k-1} z_{k-1}$$
$$-\xi_{k+1} z_{k+1} - \cdots - \xi_n z_n, z_{k+1}, \ldots, z_n).$$

Let

$$q_i(z, \xi) = \left(\frac{\partial g_1}{\partial z_i}, \frac{\partial g_2}{\partial z_i}, \ldots, \frac{\partial g_\ell}{\partial z_i} \right) \in (\mathcal{O}_{X,o})^\ell,$$

and

$$q_{i,j}(z, \xi) = e_i g_j(z, \xi) \in (\mathcal{O}_{X,o})^\ell,$$

where $i, j = 1, 2, \ldots, \ell$. Here, $\xi = (\xi_1, \xi_2, \ldots, \xi_n)$ are regarded as parameters.

Let N_{Γ_ξ} denote the $\mathcal{O}_{X,o}$-submodule of $(\mathcal{O}_{X,o})^\ell$ generated by $q_i, i \neq k$ and $q_{i,j}, i, j - 1, 2, \ldots, \ell$:

$$N_{\Gamma_\xi} = (q_1, q_2, \ldots, q_{k-1}, q_{k+1}, \ldots, q_n, q_{1,1}, q_{1,2}, \ldots, q_{\ell,\ell}).$$

Note that N_{Γ_ξ} depends only on a hyperplane L_ξ and does not depend on the choice of an index k.

We define a number ν as

$$\nu = \min_{[\xi] \in \mathbb{P}^{n-1}} \left(\dim_{\mathbb{C}}((\mathcal{O}_{X,o})^\ell / N_{\Gamma_\xi}) \right),$$

where $[\xi]$ stands for the class in the projective space $\mathbb{P}^{n-1}$ of $\xi = (\xi_1, \xi_2, \ldots, \xi_n) \in \mathbb{C}^n \setminus (0, 0, \ldots, 0)$ Note that there is a possibility that the colength of N_{γ_ξ} in $\mathcal{O}_{X,o}^\ell$ is not finite for some ξ, whereas since the variety V is assumed to be an ICIS, the number ν defined above is finite.

Let

$$U = \{[\xi] \in \mathbb{P}^{n-1} \mid \dim_{\mathbb{C}}((\mathcal{O}_{X,o})^\ell / N_{\Gamma_\xi}) = \nu\}.$$

Then, a result in [42] of B. Teissier on polar varieties implies that U is a Zariski open dense subset of $\mathbb{P}^{n-1}$.

Let $\gamma_{\xi,k} : H_{N_{\Gamma_\xi}} \longrightarrow \mathcal{H}_{\{O\}}^n(\mathcal{O}_X)$ denote the map defined by $\gamma_{\xi,k}(\sigma) = q_k\sigma$, where

$$H_{N_{\Gamma_\xi}} = \{\sigma \in (\mathcal{H}_{\{O\}}^n(\mathcal{O}_X))^\ell \mid q_i\sigma = 0, i \neq k, \ q_{i,j}\sigma = 0, i, j = 1, 2, \ldots, \ell\}$$

and let $H_{N_{\Delta_\xi}}$ be the image $\gamma_{\xi,k}(H_{N_{\Gamma_\xi}})$ of $\gamma_{\xi,k}$:

$$H_{N_{\Delta_\xi}} = \{q_k\sigma \mid \sigma \in H_{N_{\Gamma_\xi}}\}.$$

Now let us define d as

$$d = \min_{[\xi] \in \mathbb{P}^{n-1}} \left(\dim_{\mathbb{C}}(H_{N_{\Delta_\xi}}) \right).$$

Definition 2. *Let* $\beta = (\beta_1, \beta_2, \ldots, \beta_n) \in \mathbb{C}^n \backslash (0, 0, \ldots, 0)$.
(i) The hyperplane $L_\beta = \{(x_1, x_2, \ldots, x_n) \mid \beta_1 x_1 + \beta_2 x_2 + \cdots + \beta_n x_n = 0\}$
satisfies the condition F *if*

$$\dim_{\mathbb{C}}(H_{\Gamma_\beta}) < \infty.$$

(ii) The hyperplane L_β *satisfies the condition* G *if*

$$\dim_{\mathbb{C}}((\mathcal{O}_{X,O}/Ann_{\mathcal{O}_{X,O}}(H_{N_{\Delta_\beta}})) = d.$$

Note that, it is easy to see that, if L_β satisfies the condition F, the kernel $\mathrm{Ker}(\gamma_{\beta,k})$ of the map $\gamma_{\beta,k}$ is equal to H_{N_T}. Accordingly, the following holds (see [29])

$$\dim_{\mathbb{C}}(H_{N_{\Delta_\beta}}) = d + (\dim_{\mathbb{C}}(H_{N_{\Gamma_\beta}}) - \nu).$$

We arrive at the following criterion:

Proposition 2. *The following are equivalent.*

(i) The hyperplane L_β *satisfies the condition* G.
(ii) $\dim_{\mathbb{C}}(H_{N_{\Gamma_\beta}}) = \nu$.

Proof. Since

$$\dim_{\mathbb{C}}(\mathcal{O}_{X,O}/Ann_{\mathcal{O}_{X,O}}(H_{N_{\Delta_\beta}})) = \dim_{\mathbb{C}}(H_{N_{\Delta_\beta}})$$

by the Grothendieck local duality, L_β satisfies the condition G if and only if $\dim_{\mathbb{C}}(H_{N_{\Gamma_\beta}}) = \nu$.

5   Algorithms

Let $f_1, f_2, \ldots, f_\ell \in \mathbb{Q}[x]$ be ℓ polynomials, where $\mathbb{Q}[x] = \mathbb{Q}[x_1, x_2, \ldots, x_n]$. Assume that there is an open neighborhood X of the origin O in $\mathbb{C}^n$ so that the dimension of the variety $V = \{x \in X \mid f_1(x) = f_2(x) = \cdots = f_\ell(x) = 0\}$ is equal to $n - \ell$ and the singular set of V in X is the origin O.

We give a method for computing logarithmic vector fields along the ICIS V. The method consists of three blocks. The first block computes the invariant ν of V. The second block tests whether a chosen hypersurface L_β satisfies the condition G or not and if so, computes algebraic local cohomology module $H_{N_{\Gamma_\beta}}$. The last block computes a standard basis of $Ann_{\mathcal{O}_{X,o}}(H_{N_{\Delta_\beta}})$ and computes a set of generators of germs of logarithmic vector fields.

In the first two blocks, the Matlis duality is utilized. In the last block, the algorithm that computes standard basis from local cohomology classes and a module version of the extended membership algorithm [26] is used. See [12] for the theory of standard bases.

In order to compute the number ν of the variety V introduced in the previous section, we adopt a method described in [30] and use a parametric version of the algorithm presented in [37] of computing Matlis duality. To be more precise, let $H^n_{[O]}(\mathbb{Q}(\xi)[z])$ denote algebraic local cohomology with coefficients in the field of rational functions $\mathbb{Q}(\xi)$ defined to be

$$H^n_{[O]}(\mathbb{Q}(\xi)[z]) = \lim_{k \to \infty} \mathrm{Ext}^n(\mathbb{Q}(\xi)[z]/(z_1, z_2, \ldots, z_n)^k, \mathbb{Q}(\xi)[z]),$$

where $\xi = (\xi_1, \xi_2, \ldots, \xi_{n-1}, 1)$ are indeterminates. Algorithm I uses the algorithm that computes local cohomology classes with coefficients in the field of rational functions $\mathbb{Q}(\xi)$ in the module $(H^n_{[O]}(\mathbb{Q}(\xi)[z]))^\ell$ and compute the number ν of the variety V.

Algorithm I

Input : $f_1(x), f_2(x), \ldots, f_\ell(x) \in \mathbb{Q}[x]$ defines a germ of isolated complete intersection singularity V at $\mathcal{O}$ in $\mathbb{C}^n$.

Output: ν.

BEGIN

$N_{\Gamma_\xi} \leftarrow \emptyset$;

for each j from 1 to ℓ **do**

$g_j(z, \xi) \leftarrow f_j(z_1, \ldots, z_{n-1}, z_n - \xi_1 z_1 - \xi_2 z_2 - \cdots - \xi_{n-1} z_{n-1}) \in \mathbb{Q}(\xi)[z]$;

end-for

for each i from 1 to $n - 1$ **do**

$q_i(z, \xi) \leftarrow (\frac{\partial g_1}{\partial z_i}, \frac{\partial g_2}{\partial z_i}, \ldots, \frac{\partial g_\ell}{\partial z_i}) \in \mathbb{Q}(\xi)[z]$;

$N_{\Gamma_\xi} \leftarrow N_{\Gamma_\xi} \cup \{q_i(z, \xi)\}$;

end-for

for each i and j **from** 1 to ℓ **do**

$q_{i,j}(z, \xi) \leftarrow e_i g_j(z, \xi) \in (\mathbb{Q}(\xi)[z])^\ell$;

$N_{\Gamma_\xi} \leftarrow N_{\Gamma_\xi} \cup \{q_{i,j}(z, \xi)\}$;

end-for

$\Sigma \leftarrow$ compute a basis of the vector space $H_{N_{\Gamma_\xi}}$ in $(H^n_{[O]}(\mathbb{Q}(\xi)[z]))^\ell$;

/* use a parametric version of the algorithm presented in [37] on Matlis duality */

$\nu \leftarrow |\Sigma|$; /* the number of the elements of Σ */

return ν;
END

The correctness of Algorithm I above follows from an argument given in [30].

Algorithm II

Input : $f_1, f_2, \ldots, f_\ell, \nu,\ \beta = (\beta_1, \beta_2, \ldots, \beta_n) \in \mathbb{Q}^n \setminus (0, 0, \ldots, 0)$ and k with
$\beta_k = 1$.
/* ν is the output of Algorithm I */
Output: A basis of $H_{N_{\Gamma_\beta}}$, if β is generic.

BEGIN
$N_{\Gamma_\beta} \leftarrow \emptyset$;
for each j **from** 1 **to** ℓ **do**
$g_j(z, \beta) \leftarrow f_j(z_1, \ldots, z_{k-1}, z_k - \beta_1 z_1 - \beta_2 z_2 - \cdots$
$-\beta_{k-1} z_{k-1} - \beta_{k+1} z_{k+1} - \cdots - \beta_n z_n, z_{k+1}, \ldots, z_n)$;
end-for
for each i **from** 1 **to** n **do**
if $i \neq k$ **then**
$q_i(z) \leftarrow (\frac{\partial g_1}{\partial z_i}, \frac{\partial g_2}{\partial z_i}, \ldots, \frac{\partial g_\ell}{\partial z_i}) \in (\mathbb{Q}[z])^\ell$;
$N_{\Gamma_\beta} \leftarrow N_{\Gamma_\beta} \cup \{q_i(z)\}$;
end-if
end-for
for each i **and** j **from** 1 **to** ℓ **do**
$q_{i,j}(z) \leftarrow e_i g_j(z) \in (\mathbb{Q}[z])^\ell$;
$N_{\Gamma_\beta} \leftarrow N_{\Gamma_\beta} \cup \{q_{i,j}(z)\}$;
end-for
$\Sigma_\beta \leftarrow$ compute a basis of the vector space $H_{N_{\Gamma_\beta}}$ in $(H^n_{[O]}(\mathbb{Q}[z]))^\ell$;
/* use the algorithm presented in [37] on Matlis duality */
if $|\Sigma_\beta|$ exceeds ν **then**
return "the hyperplane is not generic";
else if $|\Sigma_\beta| = \nu$ **then**
return Σ_β;
end-if
END

Since $|\Sigma_\beta| = \nu$ if and only if L_β satisfies the condition G, Algorithm II is correct.

Note that there is a possibility that the dimension of the vector space H_{Γ_β} becomes infinite for some input β. Even for such a case, Algorithm II stops computation if the number $|\Sigma_\beta|$ exceeds ν, and outputs the message "the hyperplane is not generic". Therefore Algorithm II terminates within a finite number of steps.

We give an example for illustration. The following is taken from a paper of M. Giusti [14].

Example 1. Let $f_1(x, y, z) = x^2 + yz$, $f_2(x, y, z) = xy + z^3$, and

$$V = \{(x, y, z) \mid f_1(x, y, z) = f_2(x, y, z) = 0\}.$$

Then, V is quasi-homogeneous ICIS. We use a parametric version of the algorithm presented in [37] and execute Algorithm I. Then, Algorithm I outputs $\nu = 10$.

Next, we input $\beta = (1, 0, 0), \beta = (0, 1, 0)$ and $\beta = (0, 0, 1)$ into Algorithm II. Then, Algorithm II detects that these three cases do not satisfy the condition G. Therefore the three hyperplanes $x = 0, y = 0$, and $z = 0$ are not generic. Actually, for instance for the case $y = 0$, we have $\dim_{\mathbb{C}}(H_{\Gamma_{(0,1,0)}}) = 12$.

Then we choose $\beta = (1, 1, 1)$ as a candidate of generic hyperplanes and input $(1, 1, 1)$ into Algorithm II. Then, Algorithm II computes a basis $\Sigma_{(1,1,1)}$ of the space $H_{N_{\Gamma_{(1,1,1)}}}$ whose cardinality is equal to $11 (=\nu)$. This ensures the genericity of the hyperplane $x + y + z = 0$.

Notice that the example shows that a direct use of syzygy computation is not appropriate to study complex analytic properties of logarithmic vector fields.

Now we are ready to describe an outline for computing logarithmic vector fields from the basis Σ_β of the local cohomology module $H_{N_{\Gamma_\beta}}$ associated to β with $\beta_k = 1$.

First, we compute $q_k(z)\sigma$ for each $\sigma \in \Sigma_\beta \subset (H^n_{[O]}(\mathbb{Q}[z]))^\ell$ and compute a basis of the vector space

$$H_{\Delta_\beta} = \mathrm{Span}_{\mathbb{Q}}\{q_k\sigma \mid \sigma \in \Sigma_\beta\},$$

where $q_k = \left(\dfrac{\partial g_1}{\partial z_k}, \dfrac{\partial g_2}{\partial z_k}, \dots, \dfrac{\partial g_\ell}{\partial z_k} \right) \in (\mathbb{Q}[z])^\ell$. Next, we compute a standard basis of $\mathrm{Ann}_{\mathcal{O}_{X,O}}(H_{N_{\Delta_\beta}})$ by using an algorithm described in [39]. Recall that for each element $u(z)$ in the standard basis of $\mathrm{Ann}_{\mathcal{O}_{X,O}}(H_{N_{\Delta_\beta}})$, the product $a(z)q_k \in (\mathbb{Q}[z])^\ell$ belongs, as an element of $(\mathcal{O}_{X,O})^\ell$, to the $\mathcal{O}_{X,O}$-module N_{Γ_β}.

Now consider a $\mathbb{Q}[z]$-submodule of $(\mathbb{Q}[z])^\ell$ generated by $q_i, i = 1, 2, \dots, k - 1, k + 1, \dots, n$ and $q_{i,j}, i, j = 1, 2, \dots, \ell$ and denotes it by M_{Γ_β}. We have $N_{\Gamma_\beta} = \mathcal{O}_{X,O} \otimes M_{\Gamma_\beta}$, whereas $a(z)q_k$ does not belong to M_{Γ_β} in general. Therefore we consider the colon ideal $M_{\Gamma_\beta} : (a(z)q_k)$ of modules in the polynomial ring $\mathbb{Q}[z]$. There exists a polynomial, say $u(z)$, in the colon ideal $M_{\Gamma_\beta} : (a(z)q_k)$ such that $u(O) \neq 0$. Hence, $u(z)a(z)q_k$ belongs to the module M_{Γ_β}.

Now, logarithmic vector fields in the local ring can be computed by using syzygy for module over the polynomial ring [1]. In order to reduce computational cost, we devise another tool by adopting approach given in [27] as follows.

Let

$$QQ = [q_1, q_2, \dots, q_{k-1}, q_{k+1}, \dots, q_n, q_{1,1}, q_{1,2}, \dots, q_{1,\ell}, q_{2,1}, \dots, q_{2,\ell}, \dots, q_{\ell,\ell}]$$

and let $G_{M_{\Gamma_\beta}} = \{\mathfrak{g}_1, \mathfrak{g}_2, \dots, \mathfrak{g}_\lambda\}$ denote a Gröbner basis of the module M_{Γ_β} generated by elements in QQ.

Let R_{QQ} be a list of relations between $\mathfrak{g}_j$ and QQ :

$$\mathfrak{g}_{j'} = \sum_{i \neq k} r_{j',i} q_i + \sum_{i,j} r_{j',i,j} q_{i,j},$$

where $r_{j,i}, r_{j,i,j'} \in \mathbb{Q}[z]$.

Let S_Q be a Gröbner basis of the syzygies of

$$Q = [q_1, q_2, \ldots, q_{k-1}, q_{k+1}, \ldots, q_n].$$

Procedure

Input : $\beta = (\beta_1, \beta_2, \ldots, \beta_n)$ and k with $\beta_k = 1$,

$G_{M_{\Gamma_\beta}}$: a Gröbner basis of the module M_{Γ_β} generated by QQ

R_{QQ} : a list of relations between an element of $G_{M_{\Gamma_\beta}}$ and QQ

S_Q : a Gröbner basis of the syzygies of Q

$-u(z)a(z)q_k$, where

$$q_k = \left(\frac{\partial g_1}{\partial z_k}, \frac{\partial g_2}{\partial z_k}, \ldots, \frac{\partial g_\ell}{\partial z_k} \right),$$

$a(z) \in Ann_{\mathcal{O}_{X,o}}(H_{N_{\Delta_\beta}})$

$u(z) \in M_{\Gamma_\beta} : a(z) q_k$,

Output: $[b_1, b_2, \ldots, b_{k-1}, b_{k+1}, \ldots, b_n]$ such that there exist $d_{i,j}, i, j = 1, 2, \ldots, \ell$ that satisfy

$-u(z)a(z)q_k(z) = b_1(z)q_1(z) + b_2(z)q_2(z) + \cdots$
$+b_{k-1}(z)q_{k-1}(z) + b_{k+1}(z)q_{k+1}(z) + \cdots + b_n(z)q_n(z)$
$+d_{1,1}(z)q_{1,1}(z) + d_{1,2}(z)q_{1,2}(z) + \cdots + d_{\ell,\ell}(z)q_{\ell,\ell}(z).$

BEGIN

Step1: divide $-uaq_k$ by the Gröbner basis $G_{M_{\Gamma_\beta}} = \{\mathfrak{g}_1, \mathfrak{g}_2, \ldots, \mathfrak{g}_\lambda\}$;

$$-uaq_k = c_1 \mathfrak{g}_1 + c_2 \mathfrak{g}_2 + \cdots + c_\lambda \mathfrak{g}_\lambda$$

Step2: rewrite the above relation by using R_{QQ};

$$-uaq_k = \sum_{i \neq k} \left(\sum_{j'} c_{j'} r_{j',i} \right) q_i + \sum_{i,j} \left(\sum_{j'} c_{j'} r_{j',i,j} \right) q_{i,J}$$

Step3: simplify the above expression by using S_Q;

$-u(z)a(z)q_k(z) = b_1(z)q_1(z) + b_2(z)q_2(z) + \cdots$
$+b_{k-1}(z)q_{k-1}(z) + b_{k+1}(z)q_{k+1}(z) + \cdots + b_n(z)q_n(z)$
$+d_{1,1}(z)q_{1,1}(z) + d_{1,2}(z)q_{1,2}(z) + \cdots + d_{\ell,\ell}(z)q_{\ell,\ell}(z).$

return $[b_1, b_2, \ldots, b_{k-1}, b_{k+1}, \ldots, b_n]$;

END

The above procedure is implemented on [31] by using [24]. Notice that in Step 3 of the procedure, only the coefficients of $q_1, q_2, \ldots, q_{k-1}, q_{k+1}, \ldots, q_n$ are simplified by using S_Q. Since $u(O) \neq 0$, the above procedure solves an extended membership problem for $a(z)q_k(z)$ w.r.t. the module $N_{\Gamma_\beta} \subset (\mathcal{O}_{X,O})^\ell$ and outputs required data for computing logarithmic vector fields.

The final step of the resulting algorithms that computes a set of generators of logarithmic vector fields from Σ_β is Algorithm III.

Algorithm III

Input : Σ_β. / * a basis of $H_{N_{\Gamma_\beta}}$, associated to β s.t. $\beta_k = 1$ */
Output: A set of generators of germs of logarithmic vector fields along V.

BEGIN
$D \leftarrow \emptyset;\ T \leftarrow \emptyset;$
$G_{M_{\Gamma_\beta}} \leftarrow$ compute a Gröbner basis of the module $M_{\Gamma_\beta};$
$R_{QQ} \leftarrow$ compute a list of relation between $G_{M_{\Gamma_\beta}}$ and QQ;
$S_Q \leftarrow$ compute a Gröbner basis of the syzygies of Q;
$\Delta \leftarrow$ compute a basis of the vector space $H_{N_\Delta};$ /* use the algorithm in [39] */
$A \leftarrow$ compute a standard basis of $Ann_{\mathcal{O}_{X,O}}(H_{N_{\Delta_\beta}})$ by using $\Delta_\beta;$
/* use the algorithm in [39] */
while $A \neq \emptyset$ **do**
select $a(z)$ from A;
$A \leftarrow A \backslash \{a(z)\};$
Colon $\leftarrow$ compute a Gröbner basis of the colon ideal of modules
$M_{\Gamma_\beta} : (uaq_k) = \{u(z) \in \mathbb{Q}[z] \mid u(z)a(z)q_k \subset M_{\Gamma_\beta}\};$
$u(z) \leftarrow$ select $u(z) \in$ Colon s.t. $u(O) \neq 0;$
$\{b_1, b_2, \ldots, b_{k-1}, b_{k+1}, \ldots, b_n\} \leftarrow$ compute $b_1, b_2, \ldots, b_{k-1}, b_{k+1}, \ldots, b_n$
that satisfy
$-u(z)u(z)q_k(z) = b_1(z)q_1(z) + b_2(z)q_2(z) + \cdots$
$+ b_{k-1}(z)q_{k-1}(z) + b_{k+1}(z)q_{k+1}(z) + \cdots + b_n(z)q_n(z)$
$+ d_{1,1}(z)q_{1,1}(z) + d_{1,2}(z)q_{1,2}(z) + \cdots + d_{\ell,\ell}(z)q_{\ell,\ell}(z)$
by using **Procedure** for solving the extended membership problem for
$u(z)a(z)q_k(z)$ with respect to
$QQ = [q_1, q_2, \ldots, q_{k-1}, q_{k+1}, \ldots, q_n, q_{1,1}, q_{1,2}, \ldots, q_{\ell,\ell}];$
$v \leftarrow b_1 \frac{\partial}{\partial z_1} + b_2 \frac{\partial}{\partial z_2} + \cdots + b_{k-1} \frac{\partial}{\partial z_{k-1}} + u(z)a(z) \frac{\partial}{\partial z_k} + b_{k+1} \frac{\partial}{\partial z_{k+1}} + \cdots + b_n \frac{\partial}{\partial z_n};$
$D \leftarrow D \cup \{v\};$
end-while
Syz $\leftarrow$ compute a Gröbner basis of syzygies of $QQ = [q_1, q_2, \ldots, q_{k-1},$
$q_{k+1}, \ldots, q_n, q_{1,1}, q_{1,2}, \ldots, q_{\ell,\ell}];$
while Syz $\neq \emptyset$ **do**
select $syz = (s_1, s_2, \ldots, s_{k-1}, s_{k+1}, \ldots, s_n, d_{1,1}, d_{1,2}, \ldots, d_{\ell,\ell})$ from Syz;
Syz $\leftarrow$ Syz$\backslash syz;$
if $(s_1, s_2, \ldots, s_{k-1}, s_{k+1}, \ldots, s_n) \neq (0, 0, \ldots, 0)$ **then**
$w \leftarrow s_1 \frac{\partial}{\partial z_1} + s_2 \frac{\partial}{\partial z_2} + \cdots + s_{k-1} \frac{\partial}{\partial z_{k-1}} + s_{k+1} \frac{\partial}{\partial z_{k+1}} + \cdots + s_n \frac{\partial}{\partial z_n};$
$T \leftarrow T \cup \{w\};$

end-if
end-while
return $D \cup T$;
END

---

6   Examples

In this section, we present two examples for illustration. The first one taken from [14] is quasi-homogeneous, the second one taken from [2] is non-quasi homogeneous ICIS.

Example 2. Let $f_1(x, y, z) = x^2 + z^3$, $f_2(x, y, z) = y^2 + xz$ and

$$V = \{(x, y, z) \in X \mid f_1(x, y, z) = f_2(x, y, z) = 0\},$$

where X is an open neighborhood of the origin O in $\mathbb{C}^3$.

I: By Algorithm I, we find $\nu = 11$.
II: We input $\beta = (0, 0, 1)$ into Algorithm II. Then by Algorithm II, we find that the hyperplane $z = 0$ is generic. Algorithm II outputs following 11 elements

$$\left(\begin{bmatrix} 1 \\ xyz \\ 0 \end{bmatrix}\right), \left(\begin{bmatrix} 1 \\ xy^2z \\ 0 \end{bmatrix}\right), \left(\begin{bmatrix} 1 \\ xyz^2 \\ 0 \end{bmatrix}\right), \left(\begin{bmatrix} 1 \\ xy^2z^2 \\ 0 \end{bmatrix}\right), \left(\begin{bmatrix} 1 \\ xyz^3 \\ 0 \end{bmatrix}\right),$$

$$\left(\begin{bmatrix} 1 \\ xy^2z^3 \\ 0 \end{bmatrix}\right), \left(\begin{bmatrix} 0 \\ 1 \\ xyz \end{bmatrix}\right), \left(\begin{bmatrix} 0 \\ 1 \\ x^2yz \end{bmatrix}\right), \left(\begin{bmatrix} -\frac{1}{2} \begin{bmatrix} 1 \\ x^2yz \end{bmatrix} \\ \begin{bmatrix} 1 \\ xyz^2 \end{bmatrix} \end{bmatrix}\right),$$

$$\left(\begin{bmatrix} -\frac{1}{2}\begin{bmatrix} 1 \\ x^2yz^2 \end{bmatrix} + \frac{1}{2}\begin{bmatrix} 1 \\ xy^3z \end{bmatrix} \\ \begin{bmatrix} 1 \\ xyz^3 \end{bmatrix} \end{bmatrix}\right), \left(\begin{bmatrix} -\frac{1}{2}\begin{bmatrix} 1 \\ x^2yz^3 \end{bmatrix} + \frac{1}{2}\begin{bmatrix} 1 \\ x^2y^3z \end{bmatrix} \\ \begin{bmatrix} 1 \\ xyz^4 \end{bmatrix} - \begin{bmatrix} 1 \\ x^3yz \end{bmatrix} \end{bmatrix}\right),$$

as a basis Σ of $H_{N_\Gamma} \subset (H_{[O]}(\mathbb{Q}[x, y, z]))^2$, where [ ] stands for the Grothendieck symbol [21].

III: We input these results into Algorithm III. Then, a basis Δ of $H_\Delta = \{p_3\sigma \mid \sigma \in H_{N_\Gamma}\}$ is

$$\Delta = \left\{ \begin{bmatrix} 1 \\ xyz \end{bmatrix}, \begin{bmatrix} 1 \\ xy^2z \end{bmatrix}, \begin{bmatrix} 1 \\ x^2yz \end{bmatrix} \right\}.$$

A standard basis A of $Ann_{O_{X,o}}(H_\Delta)$ is given by

$$A = \{x^2, y^2, xy, z\}.$$

The colon ideal of the modules $M_\Gamma : (ap_3)$ is trivial for all $a \in A$, that is $M_\Gamma : (ap_3) = \mathbb{Q}[x, y, z]$. We set $u = 1$ and compute logarithmic vector fields by using extended membership algorithm.

Algorithm III outputs $D = \{v_1, v_2, v_3, v_4\}$, where

$$v_1 = 6y^2 z \frac{\partial}{\partial x} - 5yz^2 \frac{\partial}{\partial y} + 4x^2 \frac{\partial}{\partial z},$$

$$v_2 = -6x^2 \frac{\partial}{\partial x} - 5xy \frac{\partial}{\partial y} + 4y^2 \frac{\partial}{\partial z},$$

$$v_3 = -6yz^2 \frac{\partial}{\partial x} - 5x^2 \frac{\partial}{\partial y} + 4xy \frac{\partial}{\partial z},$$

$$v_4 = 6x \frac{\partial}{\partial x} + 5y \frac{\partial}{\partial y} + 4z \frac{\partial}{\partial z},$$

and

$$T = \left\{ f_1 \frac{\partial}{\partial x}, f_2 \frac{\partial}{\partial x}, (x^3 - y^2 z^2) \frac{\partial}{\partial x}, (x^4 + y^4 z) \frac{\partial}{\partial x}, (x^5 - y^6) \frac{\partial}{\partial x}, \right.$$
$$\left. f_1 \frac{\partial}{\partial y}, f_2 \frac{\partial}{\partial y}, (x^3 - y^2 z^2) \frac{\partial}{\partial y}, (x^4 + y^4 z) \frac{\partial}{\partial y}, (x^5 - y^6) \frac{\partial}{\partial y} \right\}.$$

It is easy to see that $f_1 \frac{\partial}{\partial x}, f_2 \frac{\partial}{\partial x}, f_1 \frac{\partial}{\partial y}, f_2 \frac{\partial}{\partial y}$ generate every vector field in T. Therefore,

$$\left\{ v_1, v_2, v_3, v_4, f_1 \frac{\partial}{\partial x}, f_2 \frac{\partial}{\partial x}, f_1 \frac{\partial}{\partial y}, f_2 \frac{\partial}{\partial y} \right\}$$

is a set of generators, over the local ring $\mathcal{O}_{X,O}$, of the module $Der_{X,O}(-\log V)$ of germs of logarithmic vector fields along V.

Now we present a non-quasihomogeneous example.

Example 3. Let $f_1(x, y, z) = xy + z^2$, $f_2(x, y, z) = x^2 + y^3 + yz^2$ and consider an ICIS defined by f_1, f_2:

$$V = \{(x, y, z) \in X \mid f_1(x, y, z) = f_2(x, y, z) = 0\},$$

where X is an open neighborhood of the origin O in $\mathbb{C}^3$ (See [2]).

I: We set
$g_1(x, y, z, \xi_1, \xi_2) = f_1(x, y, z - \xi_1 x - \xi_2 y),$
$g_2(x, y, z, \xi_1, \xi_2) = f_2(x, y, z - \xi_1 x - \xi_2 y),$
where ξ_1, ξ_2 are regarded as indeterminates. We first compute ν by applying Algorithm I. Then, we have $\nu = 11$.

II:

(i) We take the hyperplane $z = 0$ as a candidate of generic hyperplane. We set $\beta = (0,0,1)$ and $g_1 = f_1, g_2 = f_2$. Then, $q_1 = p_1 = (y, 2x), q_2 = p_2 = (x, 3y^2 + z^2), q_3 = p_3 = (2z, 2yz)$, $q_{i,j} = e_i f_j$, $i,j = 1,2$.

We execute Algorithm II to test whether the hyperplane $z = 0$ is generic or not by inputting $\beta = (0,0,1)$ into Algorithm II. Then, Algorithm II detects that $z = 0$ is not generic.

In fact, we can verify that the dimension of the vector space H_{N_β} is equal to 12. Note that the dimension of H_{N_T} is equal to 8 and

$$\dim_{\mathbb{C}}(\mathcal{O}_{X,O}/Ann_{\mathcal{O}_{X,O}}(H_{N_{\Delta_\beta}})) = 4.$$

(ii) We take the hyperplane $y = 0$ as a candidate of generic hyperplane. We set $\beta = (0,1,0)$ and as previously, we have $g_1 = f_1, g_2 = f_2, q_1 = p_1, q_2 = p_2, q_3 = p_3, q_{i,j} = e_i f_j$, $i,j = 1,2$.

Then, N_{Γ_β} is generated by $q_1, q_3, q_{i,j}, i,j = 1,2$ and the dimension of the vector space H_{Γ_β} is equal to 11. In fact, Algorithm II outputs

$$\left(\begin{bmatrix} 1 \\ xyz \\ 0 \end{bmatrix}\right), \left(\begin{bmatrix} 1 \\ x^2yz \\ 0 \end{bmatrix}\right), \left(\begin{bmatrix} 0 \\ 1 \\ xyz \end{bmatrix}\right), \left(\begin{bmatrix} 0 \\ 1 \\ xy^2z \end{bmatrix}\right), \left(\begin{bmatrix} 0 \\ 1 \\ xy^3z \end{bmatrix}\right),$$

$$\left(\begin{bmatrix} 0 \\ 1 \\ xyz^2 \end{bmatrix}\right), \left(-2\begin{bmatrix} 1 \\ xy^2z \end{bmatrix} \\ \begin{bmatrix} 1 \\ x^2yz \end{bmatrix}\right), \left(-\begin{bmatrix} 1 \\ xyz^2 \end{bmatrix} \\ \begin{bmatrix} 1 \\ xy^2z^2 \end{bmatrix}\right), \left(-2\begin{bmatrix} 1 \\ xy^3z \end{bmatrix} \\ \begin{bmatrix} 1 \\ x^2y^2z \end{bmatrix} - \begin{bmatrix} 1 \\ xyz^2 \end{bmatrix}\right),$$

$$\left(-\begin{bmatrix} 1 \\ xy^2z^2 \end{bmatrix} \\ \begin{bmatrix} 1 \\ xy^3z^2 \end{bmatrix} + \frac{1}{2}\begin{bmatrix} 1 \\ x^2yz^2 \end{bmatrix}\right),$$

$$\left(-2\begin{bmatrix} 1 \\ xy^4z \end{bmatrix} + 2\begin{bmatrix} 1 \\ x^3yz \end{bmatrix} - \begin{bmatrix} 1 \\ x^2y^2z \end{bmatrix} + \begin{bmatrix} 1 \\ xyz^3 \end{bmatrix} \\ \begin{bmatrix} 1 \\ x^2y^3z \end{bmatrix} + \frac{1}{2}\begin{bmatrix} 1 \\ xy^4z \end{bmatrix} - \begin{bmatrix} 1 \\ xy^2z^3 \end{bmatrix} + \frac{1}{2}\begin{bmatrix} 1 \\ x^3yz \end{bmatrix}\right)$$

as a basis Σ_β of H_{Γ_β}.

III: We continue the computation of the second case (ii). We follow Algorithm III. We first compute a basis Δ of $H_{\Delta_\beta} = \{p_2\sigma \mid \sigma \in H_{\Gamma_\beta}\}$ and a standard basis A of the ideal $Ann_{\mathcal{O}_{X,O}}(H_{N_\Delta})$ by using Δ.

We have

$$\Delta = \left\{ \begin{bmatrix} 1 \\ xyz \end{bmatrix}, \begin{bmatrix} 1 \\ xyz^2 \end{bmatrix}, \begin{bmatrix} 1 \\ x^2yz \end{bmatrix} - \frac{1}{10} \begin{bmatrix} 1 \\ xy^2z \end{bmatrix} \right\}.$$

Notice that $d = \dim_{\mathbb{C}}(H_\Delta) = 3$. A standard basis A is given by

$$A = \{x^2, x + 10y, xz, z^2\}.$$

Gröbner bases of the colon ideal of modules $M_{I_\beta} : aM_2$ for $a \in \Lambda$ are all equal to $\{x - 8, y - 4, z^2 + 32\}$. We set $u = y - 4$.

Then, Algorithm III outputs $D = \{v_1, v_2, v_3, v_4\}$, where

$$v_1 = (4x^3 + 14xz^2 - 12yz^2)\tfrac{\partial}{\partial x} + 2ux^2\tfrac{\partial}{\partial y} + (3x^2z + 10y^2z + 11z^3)\tfrac{\partial}{\partial z},$$

$$v_2 = (-120x + 4x^2 - 8y^2 - 26z^2)\tfrac{\partial}{\partial x} + 2u(x + 10y)\tfrac{\partial}{\partial y} + (-100z + 3xz + 19yz)\tfrac{\partial}{\partial z},$$

$$v_3 = (4x^2z + 12y^2z + 14z^3)\tfrac{\partial}{\partial x} + 2uxz\tfrac{\partial}{\partial y} + (10x^2 + 3xz^2 - yz^2)\tfrac{\partial}{\partial z},$$

$$v_4 = (12x^2 + 4xz^2 - 2yz^2)\tfrac{\partial}{\partial x} + 2uz^2\tfrac{\partial}{\partial y} + (10xz + y^2z + 3z^3)\tfrac{\partial}{\partial z},$$

where $u = y - 4$ and

$$T = \left\{ f_1\frac{\partial}{\partial x}, f_2\frac{\partial}{\partial x}, (x^3 - y^2z^2 - z^4)\frac{\partial}{\partial x}, (x^4 - xz^4 + yz^4)\frac{\partial}{\partial x}, \right.$$
$$\left. f_1\frac{\partial}{\partial z}, f_2\frac{\partial}{\partial z}, (x^3 - y^2z^2 - z^4)\frac{\partial}{\partial z}, (x^4 - xz^4 + yz^4)\frac{\partial}{\partial z} \right\}.$$

Note that $D = \{v_1, v_2, v_3, v_4\}$ corresponds to the standard basis $A = \{x^2, x + 10y, xz, z^2\}$. Notice that the logarithmic vector fields $f_1\frac{\partial}{\partial y}$ and $f_2\frac{\partial}{\partial y}$ are generated by D.

It is easy to see that T is generated by $f_1\frac{\partial}{\partial x}, f_2\frac{\partial}{\partial x}, f_1\frac{\partial}{\partial z}, f_2\frac{\partial}{\partial z}$. Therefore,

$$\left\{ v_1, v_2, v_3, v_4, f_1\frac{\partial}{\partial x}, f_2\frac{\partial}{\partial x}, f_1\frac{\partial}{\partial z}, f_2\frac{\partial}{\partial z} \right\}$$

is a set of generators, over the local ring $\mathcal{O}_{X,O}$, of the module $\mathcal{D}er_{X,O}(-\log V)$ of germs of logarithmic vector fields along V.

The following is a session of the computer algebra system Singular [11] to compute a basis of germs of logarithmic vector fields. We use directly the classical syzygy computation in the local ring. We input $f = f_1, g = f_2$. M is the module generated by

$$\begin{pmatrix} \frac{\partial f}{\partial x} \\ \frac{\partial g}{\partial x} \end{pmatrix}, \begin{pmatrix} \frac{\partial f}{\partial y} \\ \frac{\partial g}{\partial y} \end{pmatrix}, \begin{pmatrix} \frac{\partial f}{\partial z} \\ \frac{\partial g}{\partial z} \end{pmatrix}, \begin{pmatrix} f \\ 0 \end{pmatrix}, \begin{pmatrix} g \\ 0 \end{pmatrix}, \begin{pmatrix} 0 \\ f \end{pmatrix}, \begin{pmatrix} 0 \\ g \end{pmatrix}.$$

```
> ring A=0,(x,y,z),(c,ds);
> poly f=x*y+z^2;
> poly g=x^2+y^3+y*z^2;
> module M=[diff(f,x),diff(g,x)],[diff(f,y),diff(g,y)],[diff(f,z),
diff(g,z)],[f,0],[g,0],[0,f],[0,g];
> syz(M);
_[1]=[60x+4y2-2z2+8xy2-4xz2+18yz2,4x+40y+8x2-2z2+6yz2,50z+xz+3yz-9
y2z-3z3,-100+6z2,-4-8x,-20y-40xy+2z2,-120]
_[2]=[xy+z2,0,0,-y,0,-2x]
_[3]=[12xz-10y2z-4z3,8xz+8yz-6y2z,10z2+9y3+3yz2,-20z+6yz,-8z,-4yz,
-24z]
_[4]=[12y3+4xz2+2yz2,-8x2+12xy+20z2-6yz2,-10xz+9y2z+3z3,-8y2-6z2,8
x-12y,24xy-60y2-20z2]
_[5]=[6y2z+4z3,-4xz+2y2z,2x2-3y3-yz2,-2yz,0,-4xz]
_[6]=[18yz2-10y2z2-4z4,12x2+8xz2+14yz2-6y2z2,-15y2z+z3+9y3z+3yz3,1
2y2-2z2+6yz2,-12x-8z2,-36xy-4yz2,-12z2]
_[7]=[0,xy+z2,0,-x,0,-3y2-z2]
_[8]=[0,0,xy+z2,-2z,0,-2yz]
```

The output of syz(M) is the basis of $\mathcal{D}er_{X,O}(-\log V)$.

Notice that it is difficult to read off, from the above output, the fact that $(x^2, x+10y, xz, z^2)$ is a standard basis of the ideal generated by coefficients of $\frac{\partial}{\partial y}$ of logarithmic vector fields.

References

1. Adams, W., Loustaunau, P.: An Introduction to Gröbner Bases. AMS, Providence (1994)
2. Afzal, D., Afzal, F., Mulback, M., Pfister, G., Yaqub, A.: Unimodal ICIS, a classifier. Stud. Sci. Math. Hungarica **54**, 374–403 (2017)
3. Aleksandrov, A.G.: Cohomology of a quasihomogeneous complete intersection. Math. USSR Izv. **26**, 437–477 (1986)
4. Aleksandrov, A.G.: A de Rham complex of nonisolated singularities. Funct. Anal. Appl. **22**, 131–133 (1988)
5. Aleksandrov, A.G.: Logarithmic differential forms on Cohen-Macaulay varieties. Methods Appl. Anal. **24**, 11–32 (2017)
6. Aleksandrov, A.G., Tsikh, A.K.: Théorie des résidus de Leray et formes de Barlet sur une intersection complète singulière. C. R. Acad. Sci. Paris Sér. I. Math. **333**, 973–978 (2001)
7. Aleksandrov, A.G., Tsikh, A.K.: Multi-logarithmic differential forms on complete intersections. J. Siber. Federal Univ. **2**, 105–124 (2008)
8. Bruce, J.W., Roberts, R.M.: Critical points of functions on an analytic varieties. Topology **27**, 57–90 (1988)
9. Bruns, W., Herzog, J.: Cohen-Macaulay Rings. Cambridge University Press, Cambridge (1993)
10. Damon, J.: On the legacy of free divisors II: free* divisors and complete intersections. Moscow Math. J. **3**, 361–395 (2003)

11. Decker, W., Greuel, G.-M., Pfister, G., Schönemann, H.: Singular 4-1-2 – a computer algebra system for polynomial computations (2019). http://www.singular.uni-kl.de
12. De Jong, T., Pfister, G.: Local Analytic Geometry. Vieweg (2000)
13. de Bobadilla, J.F.: Relative Morsification theory. Topology **43**, 925–982 (2004)
14. Giusti, M.: Classification des singularités isolées simples d'intersections complètes. In: Singularities Part I, Proceedings of Symposium on Pure Mathematics, vol. 40, pp. 457–494. AMS (1983)
15. Granger, M., Schulze, M.: Normal crossing properties of complex hypersurfaces via logarithmic residues. Compos. Math. **150**, 1607–1622 (2014)
16. Grothendieck, A.: Local Cohomology. Lecture Notes in Mathematics, vol. 41. Springer, Heidelberg (1967). https://doi.org/10.1007/BFb0073971. Notes by R. Hartshorne
17. Hauser, H., Müller, G.: Affine varieties and Lie algebras of vector fields. Manusc. Math. **80–2**, 309–337 (1993)
18. Hauser, H., Müller, G.: On the Lie algebra $\theta(x)$ of vector fields on a singularity. J. Math. Sci. Univ. Tokyo **1**, 239–250 (1994)
19. Kas, A., Schlessinger, M.: On the versal deformation of a complex space with an isolated singularity. Math. Ann. **196**, 23–29 (1972)
20. Kashiwara, M., Kawai, T.: On holonomic systems of microdifferential equations. III. Res. Inst. Math. Sci. **17**, 813–979 (1981)
21. Kunz, E.: Residues and duality for projective algebraic varieties. American Mathematical Socity (2009)
22. Looijenga, E.J.N.: Isolated Singular Points on Complete Intersections. London Mathematical Society Lecture Note Series, vol. 77. Cambridge University Press, Cambridge (1984)
23. Matlis, E.: Injective modules over Noetherian rings. Pac. J. Math. **8**, 511–528 (1958)
24. Nabeshima, K.: On the computation of parametric Gröbner bases for modules and syzygies. Jpn. J. Ind. Appl. Math. **27**, 217–238 (2010)
25. Nabeshima, K., Tajima, S.: Computing logarithmic vector fields associated with parametric semi-quasihomogeneous hypersurface isolated singularities. In: Robertz, D. (ed.) International Symposium on Symbolic and Algebraic Computation (ISSAC), pp. 334–348. ACM, New York (2015)
26. Nabeshima, K., Tajima, S.: Solving extended ideal membership problems in rings of convergent power series via Gröbner bases. In: Kotsireas, I.S., Rump, S.M., Yap, C.K. (eds.) MACIS 2015. LNCS, vol. 9582, pp. 252–267. Springer, Cham (2016). https://doi.org/10.1007/978-3-319-32859-1_22
27. Nabeshima, K., Tajima, S.: Algebraic local cohomology with parameters and parametric standard bases for zero-dimensional ideals. J. Symb. Comput. **82**, 91–122 (2017)
28. Nabeshima, K., Tajima, S.: Computation methods of logarithmic vector fields associated with semi-weighted homogeneous isolated hypersurface singularities. Tsukuba J. Math. **42**, 191–231 (2018)
29. Nabeshima, K., Tajima, S.: Computing logarithmic vector fields and Bruce-Roberts Milnor numbers via local cohomology classes. Rev. Roumaine Math. Pures Appl. **64**, 521–538 (2019)
30. Nabeshima, K., Tajima, S.: Alternative algorithms for computing generic μ^*-sequences and local Euler obstructions of isolated hypersurface singularities. J. Algebra Appl. **18**(8) (2019). https://doi.org/10.1142/S0219498819501548

31. Noro, M., Takeshima, T.: Risa/Asir-a computer algebra system. In: Wang, P. (ed.) International Symposium on Symbolic and Algebraic Computation (ISSAC), pp. 387–396. ACM, New York (1992)
32. Nuño-Ballesteros, J.J., Oréfice, B., Tomazella, J.N.: The Bruce-Roberts number of a function on a weighted homogeneous hypersurface. Q. J. Math. **64**, 269–280 (2013)
33. Nuño-Ballesteros, J.J., Oréfice-Okamoto, B., Tomazella, J.N.: Non-negative deformations of weighted homogeneous singularities. Glasg. Math. J. **60**, 175–185 (2018)
34. Oréfice-Okamoto, B.: O número de Milnor de uma singularidade isolada. Tese, São Carlos (2011)
35. Pol, D.: Characterizations of freeness for equidimensional subspaces. J. Singularities **20**, 1–30 (2020)
36. Saito, K.: Theory of logarithmic differential forms and logarithmic vector fields. J. Fac. Sci. Univ. Tokyo Sect. IA Math. **27**, 265–291 (1980)
37. Shibuta, T., Tajima, S.: An algorithm for computing the Hilbert-Samuel multiplicities and reductions of zero-dimensional ideals of Cohen-Macaulay local ring. J. Symb. Comput. **96**, 108–121 (2020)
38. Tajima, S.: On polar varieties, logarithmic vector fields and holonomic D-modules. RIMS Kôkyûroku Bessatsu **40**, 41–51 (2013)
39. Tajima, S., Nakamura, Y., Nabeshima, K.: Standard bases and algebraic local cohomology for zero dimensional ideals. Adv. Stud. Pure Math. **56**, 341–361 (2009)
40. Tajima, S., Nabeshima, K.: An algorithm for computing torsion differential forms associated to an isolated hypersurface singularity. Math. Comput. Sci. (to appear) https://doi.org/10.1007/s11786-020-00486-w
41. Teissier, B.: Cycles évanescents, sections planes et conditions de Whitney. Singularités à Cargèse. Astérisque **7–8**, 285–362 (1973)
42. Teissier, B.: Varietes polaires II multiplicites polaires, sections planes, et conditions de whitney. In: Aroca, J.M., Buchweitz, R., Giusti, M., Merle, M. (eds.) Algebraic Geometry. LNM, vol. 961, pp. 314–491. Springer, Heidelberg (1982). https://doi.org/10.1007/BFb0071291
43. Terao, H.: The bifurcation set and logarithmic vector fields. Math. Ann. **263**, 313–321 (1983)
44. Treves, F.: Topological Vector Spaces, Distributions and Kernels. Academic Press, Cambridge (1967)
45. Wahl J.: Automorphisms and deformations of quasi-homogeneous singularities. In: Proceedings of Symposia in Pure Mathematics, vol. 40-2, pp. 613–624. American Mathematics Society, Providence (1983)

Robust Numerical Tracking of One Path of a Polynomial Homotopy on Parallel Shared Memory Computers

Simon Telen[1], Marc Van Barel[1], and Jan Verschelde[2]($\boxtimes$)

[1] Department of Computer Science, Katholieke Universiteit Leuven,
Celestijnenlaan 200a, 3001 Leuven, Belgium
{simon.telen,marc.vanbarel}@kuleuven.be
[2] Department of Mathematics, Statistics, and Computer Science,
University of Illinois at Chicago,
851 S. Morgan St (m/c 249), Chicago, IL 60607-7045, USA
janv@uic.edu,
https://simontelen.webnode.com,
https://people.cs.kuleuven.be/~marc.vanbarel
http://www.math.uic.edu/~jan

Abstract. We consider the problem of tracking one solution path defined by a polynomial homotopy on a parallel shared memory computer. Our robust path tracker applies Newton's method on power series to locate the closest singular parameter value. On top of that, it computes singular values of the Hessians of the polynomials in the homotopy to estimate the distance to the nearest different path. Together, these estimates are used to compute an appropriate adaptive step size. For n-dimensional problems, the cost overhead of our robust path tracker is $O(n)$, compared to the commonly used predictor-corrector methods. This cost overhead can be reduced by a multithreaded program on a parallel shared memory computer.

Keywords: Adaptive step size control · Multithreading · Newton's method · Parallel shared memory computer · Path tracking · Polynomial homotopy · Polynomial system · Power series

1 Introduction

A polynomial homotopy is a system of polynomials in several variables with one of the variables acting as a parameter, typically denoted by t. At $t = 0$, we know the values for a solution of the system, where the Jacobian matrix has full rank:

M. Van Barel—Supported by the Research Council KU Leuven, C1-project (Numerical Linear Algebra and Polynomial Computations), and by the Fund for Scientific Research–Flanders (Belgium), G.0828.14N (Multivariate polynomial and rational interpolation and approximation), and EOS Project no 30468160.
J. Verschelde—Supported by the National Science Foundation under grant DMS 1854513.

© Springer Nature Switzerland AG 2020
F. Boulier et al. (Eds.): CASC 2020, LNCS 12291, pp. 563–582, 2020.
https://doi.org/10.1007/978-3-030-60026-6_33

we start at a regular solution. With series developments we extend the values of the solution to values of $t > 0$.

As a demonstration of what *robust* in the title of this paper means, on tracking one million paths on the 20-dimensional benchmark system posed by Katsura [14], Table 3 of [15] reports 4 curve jumpings. A curve jumping occurs when approximations from one path jump onto another path. In the runs with the MPI version for our code (reported in [18]) no path failures and no curve jumpings happened. Our path tracking algorithm applies Padé approximants in the predictor. These rational approximations have also been applied to solve nonlinear systems arising in power systems [19,20]. In [13], Padé approximants are used in symbolic deformation methods.

This paper describes a multithreaded version of the robust path tracking algorithm of [18]. In [18] we demonstrated the scaling of our path tracker to polynomial homotopies with more than one million solution paths, applying message passing for distributed memory parallel computers. In this paper we consider shared memory parallel computers and, starting at one single solution, we investigate the scalability for increasing number of equations and variables, and for an increasing number of terms in the power series developments.

As to a comparison with our MPI version used in [18], the current parallel version is made threadsafe and more efficient. These improvements also benefit the implementation with message passing.

In addition to speedup, we ask the *quality up* question: if we can afford the running time of a sequential run in double precision, with a low degree of truncation, how many threads do we need (in a run which takes the same time as a sequential run) if we want to increase the working precision and the degrees at which we truncate the power series?

Our programming model is that of a work crew, working simultaneously to finish a number of jobs in a queue. Each job in the queue is done by one single member of the work crew. All members of the work crew have access to all data in the random access memory of the computer. The emphasis in this research is on the high level development of parallel algorithms and software [16]. The code is part of the free and open source PHCpack [21], available on github.

The parallel implementation of medium grained evaluation and differentiation algorithms provide good speedups. The solution of a blocked lower triangular linear system is most difficult to compute accurately and with good speedup. We describe a pipelined algorithm, provide an error analysis, and propose to apply double double and quad double arithmetic [12].

2   Overview of the Computational Tasks

We consider a homotopy H given by n polynomials $f_1, \ldots, f_n$ in $n + 1$ variables $x_1, \ldots, x_n, t$, where t is thought of as the continuation parameter. A solution path of the homotopy is denoted by $x(t)$. For a local power series expansion $x(t) = c_0 + c_1 t + c_2 t^2 + \cdots$ of $x(t)$, where $x(t)$ is assumed analytic in a neighborhood of $t = 0$, the theorem of Fabry [9] allows us to determine the location of the parameter value nearest to $t = 0$ where $x(t)$ is singular. With the singular values of the Jacobian matrix $J = (\partial f_i/\partial x_j)_{1 \leq i,j \leq n}$ and the Hessian matrices of $f_1, \ldots, f_n$, we estimate the distance to the nearest solution for t fixed to zero. The step size Δt is the minimum of two bounds, denoted by C and R.

1. C is an estimate for the nearest different solution path at $t = 0$. To obtain this estimate we compute the first and second partial derivatives at a point and organize these derivatives in the Jacobian and Hessian matrices. The bound is then computed from the singular values of those matrices:

$$C = \frac{2\sigma_n(J)}{\sqrt{\sigma_{1,1}^2 + \sigma_{2,1}^2 + \cdots + \sigma_{n,1}^2}}, \tag{1}$$

   where $\sigma_n(J)$ is the smallest singular value of the Jacobian matrix J and $\sigma_{k,1}$ is the largest singular value of the Hessian of the k-th polynomial.
2. R is the radius of convergence of the power series developments. Applying the theorem of Fabry, R is computed as the ratio of the moduli of two consecutive coefficients in the series. For a series truncated at degree d:

$$x(t) = c_0 + c_1 t + c_2 t^2 + \cdots + c_d t^d, \quad z = c_{d-1}/c_d, \quad R = |z|, \tag{2}$$

   where z indicates the estimate for the location of the nearest singular parameter value.

The computations of R and C require evaluation, differentiation, and linear algebra operations. Once Δt is determined, the solution for the next value of the parameter is predicted by evaluating Padé approximants constructed from the power series developments. The last stage is the shift of the coefficients with $-\Delta t$, so the next step starts again at $t = 0$.

   The stages are justified in [18]. In [18], we compared with v1.6 of Bertini [4] (both in runs in double precision and in runs in adaptive precision [3]) and v1.1 of HomotopyContinuation.jl [6]. In this paper we focus on parallel algorithms.

3 Parallel Evaluation and Differentiation

The parallel algorithms in this section are medium grained. The jobs in the evaluation and differentiation correspond to the polynomials in the system. While the number of polynomials is not equal to the number of threads, the jobs are distributed evenly among the threads.

3.1 Algorithmic Differentiation on Power Series

Consider a polynomial system $\mathbf{f}$ in n variables with power series (all truncated to the same fixed degree d), as coefficients; and a vector $\mathbf{x}$ of n power series, truncated to the same degree d. Our problem is to evaluate $\mathbf{f}$ at $\mathbf{x}$ and to compute all n partial derivatives. We illustrate the reverse mode of algorithmic differentiation [11] with an example, on $f = x_1x_2x_3x_4x_5$.

$$
\begin{array}{lll}
x_1x_2 = x_1 \star x_2 & x_5x_4 = x_5 \star x_4 & x_1x_3x_4x_5 = x_1 \star x_5x_4x_3 \\
x_1x_2x_3 = x_1x_2 \star x_3 & x_5x_4x_3 = x_5x_4 \star x_3 & x_1x_2x_4x_5 = x_1x_2 \star x_5x_4 \\
x_1x_2x_3x_4 = x_1x_2x_3 \star x_4 & x_5x_4x_3x_2 = x_5x_4x_3 \star x_2 & x_1x_2x_3x_5 = x_1x_2x_3 \star x_5 \\
x_1x_2x_3x_4x_5 = x_1x_2x_3x_4 \star x_5 & &
\end{array}
$$

$$(3)$$

In the first column of (3), we see $\frac{\partial f}{\partial x_5}$ and the evaluated f on the last two rows. The last row of the middle column gives $\frac{\partial f}{\partial x_1}$ and the remaining partial derivatives are in the last column of (3).

Evaluating and differentiating a product of n variables in this manner takes $3n - 5$ multiplications. For our problem, every multiplication is a convolution of two truncated power series $x_i = x_{i,0} + x_{i,1}t + x_{i,2}t^2 + \cdots + x_{i,d}t^d$ and $x_j = x_{j,0} + x_{j,1}t + x_{j,2}t^2 + \cdots + x_{j,d}t^d$, up to degree d. Coefficients of $x_i \star x_j$ of terms higher than d are not computed.

Any monomial is represented as the product of the variables that occur in the monomial and the product of the monomial divided by that product. For example, $x_1^3x_2x_3^6$ is represented as $(x_1x_2x_3) \cdot (x_1^2x_3^5)$ We call the second part in this representation the common factor, as this factor is common to all partial derivatives of the monomial. This common factor is computed via a power table of the variables. For every variable x_i, the power table stores all powers x_i^e, for e from 2 to the highest occurrence in a common factor. Once the power table is constructed, the computation of any common factor requires at most $n - 1$ multiplications of two truncated power series.

As we expect the number of equations and variables to be a multiple of the number of available threads, one job is the evaluation and differentiation of one single polynomial. Assuming each polynomial has roughly the same number of terms, we may apply a static job scheduling mechanism. Let n be the number of equations (indexed from 1 to n), p the number of threads (labeled from 1 to p), where $n \geq p$. Thread i evaluates and differentiates polynomials $i + kp$, for k starting at 0, as long as $i + kp \leq n$.

3.2   Jacobians, Hessians at a Point, and Singular Values

If we have n equations, then the computation of C, defined in (1), requires $n+1$ singular value decompositions, which can all be computed independently.

For any product of n variables, after the computation of its gradient with the reverse mode, any element of its Hessian needs only a couple of multiplications, independent of n. We illustrate this idea with an example for $n = 8$. The third row of the Hessian of $x_1x_2x_3x_4x_5x_6x_7x_8$, starting at the fourth column, after the zero on the diagonal is

$$
\begin{array}{c}
x_1x_2 \star x_5x_6x_7x_8, x_1x_2 \star x_4 \star x_6x_7x_8, x_1x_2x_4 \star x_5 \star x_7x_8, \\
x_1x_2x_4x_5 \star x_6 \star x_8, x_1x_2x_4x_5x_6 \star x_7.
\end{array}
$$

$$(4)$$

In the reverse mode for the gradient we already computed the forward products x_1x_2, $x_1x_2x_3$, $x_1x_2x_3x_4$, $x_1x_2x_3x_4x_5$, $x_1x_2x_3x_4x_5x_6$, and $x_1x_2x_3x_4x_5x_6x_7$. We also computed the backward products x_8x_7, $x_8x_7x_6$, $x_8x_7x_6x_5$, $x_8x_7x_6x_5x_4$.

For a monomial $x_1^{e_1} x_2^{e_2} \cdots x_n^{e_n}$ with higher powers $e_k > 1$, for some indices k, the off diagonal elements are multiplied with the common factor $x_1^{e_1-1} \star x_2^{e_2-1} \star \cdots \star x_n^{e_n-1}$ multiplied with $e_i e_j$ at the (i,j)-th position in the Hessian. The computation of this common factor requires at most $n-1$ multiplications (fewer than $n-1$ if there are any e_k equal to one), after the computation of table which stores the values of all powers $x_k^{e_k}$ of all values for x_k, for $k = 1, 2, \ldots, n$.

Taking only those m indices i_k for which $e_{i_k} > 1$, the common factor for all diagonal elements is $x_{i_1}^{e_{i_1}-2} x_{i_2}^{e_{i_2}-2} \cdots x_{i_m}^{e_{i_m}-2}$. The k-th element on the diagonal then needs to be multiplied with $e_{i_k}(e_{i_k} - 1)$ and the product of all squares $x_{i_j}^2$, for all $j \neq k$ for which $e_{i_j} > 1$. The efficient computation of the sequence $x_{i_2}^2 x_{i_3}^2 \cdots x_{i_m}^2$, $x_{i_1}^2 x_{i_3}^2 \cdots x_{i_m}^2$, $x_{i_1}^2 x_{i_2}^2 \cdots x_{i_{m-1}}^2$ happens along the same lines as the computation of the gradient, requiring $3m - 5$ multiplications.

In the above paragraphs, we summarized the key ideas and results of the application for algorithmic differentiation. A detailed algorithmic description can be found in [7].

4   Solving a Lower Triangular Block Linear System

In Newton's method, the update $\Delta\mathbf{x}(t)$ to the power series $\mathbf{x}(t)$ is computed as the solution of a linear system, with series for the coefficient entries.

Applying linearization, we solve a sequence of as many linear systems (with complex numbers as coefficients), as the degree of the series. For each linear system in the sequence, the right hand side is computed with the solution of the previous system in the sequence. If in each step we lose one decimal place of accuracy, at the end of sequence we have lost as many decimal places of accuracy as the degree of the series.

4.1   Pipelined Solution of Matrix Series

We introduce the pipelined solution of a system of power series by example. Consider a power series $\mathbf{A}(t)$, with coefficients n-by-n matrices, and a series $\mathbf{b}(t)$, with coefficients n-dimensional vectors. We want to find the solution $\mathbf{x}(t)$ to $\mathbf{A}(t)\mathbf{x}(t) = \mathbf{b}(t)$. For series truncated to degree 5, the equation

$$\left(A_5 t^5 + A_4 t^4 + A_3 t^3 + A_2 t^2 + A_1 t + A_0\right) \cdot \left(x_5 t^5 + x_4 t^4 + x_3 t^3 \right. \tag{5}$$

$$\left. + x_2 t^2 + x_1 t + x_0\right) = b_5 t^5 + b_4 t^4 + b_3 t^3 + b_2 t^2 + b_1 t + b_0 \tag{6}$$

leads to the triangular system (derived in [5] applying linearization)

$$A_0 x_0 = b_0 \tag{7}$$
$$A_0 x_1 = b_1 - A_1 x_0 \tag{8}$$
$$A_0 x_2 = b_2 - A_2 x_0 - A_1 x_1 \tag{9}$$
$$A_0 x_3 = b_3 - A_3 x_0 - A_2 x_1 - A_1 x_2 \tag{10}$$
$$A_0 x_4 = b_4 - A_4 x_0 - A_3 x_1 - A_2 x_2 - A_1 x_3 \tag{11}$$
$$A_0 x_5 = b_5 - A_5 x_0 - A_4 x_1 - A_3 x_2 - A_2 x_3 - A_1 x_4. \tag{12}$$

To solve this triangular system, denote by $F_0 = F(A_0)$ the factorization of A_0 and $x_0 = S(F_0, b_0)$, the solution of $A_0 x_0 = b_0$ making use of the factorization F_0. Then the Eqs. (7) through (12) are solved in the following steps.

$$
\begin{aligned}
&1.\ F_0 = F(A_0) \\
&2.\ x_0 = S(F_0, b_0) \\
&3.\ b_1 = b_1 - A_1 x_0,\ b_2 = b_2 - A_2 x_0,\ b_3 = b_3 - A_3 x_0,\ b_4 = b_4 - A_4 x_0, \\
&\quad\ b_5 = b_5 - A_5 x_0 \\
&4.\ x_1 = S(F_0, b_1) \\
&5.\ b_2 = b_2 - A_1 x_1,\ b_3 = b_3 - A_2 x_1,\ b_4 = b_4 - A_3 x_1,\ b_5 = b_5 - A_4 x_1 \\
&6.\ x_2 = S(F_0, b_2) \\
&7.\ b_3 = b_3 - A_1 x_2,\ b_4 = b_4 - A_2 x_2,\ b_5 = b_5 - A_3 x_2 \\
&8.\ x_3 = S(F_0, b_3) \\
&9.\ b_4 = b_4 - A_1 x_3,\ b_5 = b_5 - A_2 x_3 \\
&10.\ x_4 = S(F_0, b_4) \\
&11.\ b_5 = b_5 - A_1 x_4 \\
&12.\ x_5 = S(F_0, b_5)
\end{aligned}
\tag{13}
$$

Statements on the same line can be executed simultaneously. With 5 threads, the number of steps is reduced from 22 to 12. For truncation degree d and d threads, the number of steps in the pipelined algorithm equals $2(d+1)$. On one thread, the number of steps equals $2(d+1) + 1 + 2 + \cdots + d - 1 = d(d-1)/2 + 2(d+1)$. With d threads, the speedup is then

$$\frac{d(d-1)/2 + 2(d+1)}{2(d+1)} = 1 + \frac{d(d-1)}{4(d+1)}. \tag{14}$$

As $d \to \infty$, this ratio equals $1 + d/4$. Note that the first step is typically $O(n^3)$, whereas the other steps are $O(n^2)$.

Observe in (13) that the first operation on every line is on the critical path of all possible parallel executions. For the example in (13) this implies that the total number of steps will never become less than 12, even as the number of threads goes to infinity. The speedup of 22/12 remains the same as we reduce the number of threads from 5 to 3, as the updates of b_4 and b_5 in step 3 can be postponed to the next step. Likewise, the update of b_5 in step 5 may happen in step 6. Generalizing this observation, the formula for the speedup in (14) remains the same for $d/2 + 1$ threads (instead of d) in case d is odd. In case d is even, then the best speedup is obtained with $d/2$ threads.

Better speedups will be obtained for finer granularities, if the matrix factorizations are executed in parallel as well.

4.2 Error Analysis of a Lower Triangular Block Toeplitz Solver

In Sect. 4.1, we designed a pipelined method to solve the following lower triangular block Toeplitz system of equations

$$
\begin{bmatrix}
A_0 & & & & \\
A_1 & A_0 & & & \\
A_2 & A_1 & A_0 & & \\
\vdots & \vdots & \vdots & \ddots & \\
A_i & A_{i-1} & A_{i-2} & \cdots & A_0
\end{bmatrix}
\begin{bmatrix}
x_0 \\
x_1 \\
x_2 \\
\vdots \\
x_i
\end{bmatrix}
=
\begin{bmatrix}
b_0 \\
b_1 \\
b_2 \\
\vdots \\
b_i
\end{bmatrix} .
\tag{15}
$$

In this section, we do not intend to give a very detailed error analysis but indicate using a rough estimate of the norm of the blocks involved, where and how there could be a loss of precision in some typical situations. In our analysis we will use the Euclidean 2-norm $\| \cdot \| = \| \cdot \|_2$ on finite dimensional complex vector spaces and the induced operator norm on matrices. Without loss of generality, we can always assume that the system is scaled such that

$$
\|A_0\| = \|x_0\| = 1.
\tag{16}
$$

Hence, assuming that the components of x_0 in the direction of the right singular vectors of A_0 corresponding to the larger singular values are not too small, the norm of the first block b_0 of the right-hand side satisfies

$$
\|b_0\| = \|A_0 x_0\| \lesssim \|A_0\| \|x_0\|.
\tag{17}
$$

To determine the first component x_0 of the solution vector, we solve the system $A_0 x_0 = b_0$. We solve this first system in a backward stable way, i.e., the computed solution $\hat{x}_0 = x_0 + \Delta x_0$ can be considered as the exact solution of the system

$$
A_0 \hat{x}_0 = b_0 + \Delta b_0 \qquad \text{with} \qquad \frac{\|\Delta b_0\|}{\|b_0\|} \approx \epsilon_{\text{mach}}.
\tag{18}
$$

If we denote the condition number of A_0 by κ, we get

$$
\frac{\|\Delta x_0\|}{\|x_0\|} \leq \kappa \frac{\|\Delta b_0\|}{\|b_0\|} \leq \kappa O(\epsilon_{\text{mach}}).
\tag{19}
$$

We study now how this error influences the remainder of the calculations. In the remaining steps, we use rough estimates of the order of magnitude of the different blocks A_i of the coefficient matrix, the blocks x_i of the solution vector and the blocks b_i of the right-hand side. First we will assume that the sizes of the blocks x_i as well as A_i behave as ρ^i, i.e.,

$$
\|x_i\| \approx \rho^i \qquad \text{and} \qquad \|A_i\| \approx \rho^i.
\tag{20}
$$

Hence, also the sizes of the blocks b_i behave as

$$\|b_i\| \approx \rho^i. \tag{21}$$

In our context, the parameter ρ should be thought of as the inverse of the convergence radius R, as defined in (2), for the series expansions. Note that when ρ is larger, this indicates that the distance to the nearest singularity is smaller. Consider now the second system

$$A_0 x_1 = \tilde{b}_1, \tag{22}$$

where $\tilde{b}_1 = b_1 - A_1 x_0$. Using the computed value $\hat{x}_0$, we find an approximation $\hat{x}_1 = x_1 + \Delta x_1$ for x_1 by solving the system

$$A_0 X = b_1 - A_1 \hat{x}_0 = b_1 - A_1 x_0 - A_1 \Delta x_0 = \tilde{b}_1 - A_1 \Delta x_0 \tag{23}$$

for X. We have that $\|\tilde{b}_1\| = \|A_0 x_1\| \approx \rho^1$. Because $\|\Delta x_0\| \approx \kappa \epsilon_{\text{mach}}$, this results in an absolute error $\Delta \tilde{b}_1 = -A_1 \Delta x_0$ on $\tilde{b}_1$ of size $\kappa \epsilon_{\text{mach}} \rho$ or a relative error of size $\kappa \epsilon_{\text{mach}}$. Hence,

$$\frac{\|\Delta x_1\|}{\|x_1\|} \approx \kappa \frac{\|\Delta \tilde{b}_1\|}{\|\tilde{b}_1\|} \approx \kappa^2 \epsilon_{\text{mach}}. \tag{24}$$

In the same way, one derives that

$$\frac{\|\Delta x_i\|}{\|x_i\|} \approx \kappa^{i+1} \epsilon_{\text{mach}}. \tag{25}$$

Hence, when $\|x_i\| \approx \rho^i$ and $\|A_i\| \approx \rho^i$, we lose all precision as soon as $\kappa^{i+1} \epsilon_{\text{mach}} = O(1)$. When the matrix A_0 is ill-conditioned (i.e., when κ is large), this may happen already after a few number of steps i.

Assuming now that $\|x_i\| \approx \rho^i$ and $\|A_i\| \approx \rho^0$, we solve for the second block equation

$$A_0 X = b_1 - A_1 \hat{x}_0 = \tilde{b}_1 - A_1 \Delta x_0 \tag{26}$$

with $\|\tilde{b}_1\| = \|A_0 x_1\| \approx \rho^1$. However, in this case the absolute error $\|\Delta x_0\| \approx \kappa \epsilon_{\text{mach}}$ is not amplified and results in an absolute error $\Delta \tilde{b}_1 = -A_1 \Delta x_0$ of size $\kappa \epsilon_{\text{mach}}$ or a relative error of size $\kappa \epsilon_{\text{mach}} / \rho$. If $\kappa \geq \rho$ this is the dominant error on $\tilde{b}_1$. If $\kappa \leq \rho$, the dominant error is the error of computing $\tilde{b}_1$ in finite precision. In that case, the relative error will be of size ϵ_{mach}. In what follows, we will assume that $\kappa \geq \rho$. The other case can be treated in a similar way. It follows that

$$\frac{\|\Delta x_1\|}{\|x_1\|} \approx \kappa \frac{\|\Delta \tilde{b}_1\|}{\|\tilde{b}_1\|} \approx \kappa \frac{\kappa}{\rho} \epsilon_{\text{mach}}. \tag{27}$$

Next, the approximation $\hat{x}_2 = x_2 + \Delta x_2$ of x_2 is computed by solving

$$A_0 X = b_2 - A_2 \hat{x}_0 - A_1 \hat{x}_1 = \tilde{b}_2 - A_2 \Delta x_0 - A_1 \Delta x_1 \tag{28}$$

for X, with $\tilde{b}_2 = b_2 - A_2 x_0 - A_1 x_1$ and $\|\tilde{b}_2\| = \|A_0 x_2\| \approx \rho^2$. The absolute error Δx_0 plays a minor role compared to Δx_1. The relative error on x_1 of magnitude $\kappa(\kappa/\rho)\epsilon_{\text{mach}}$ multiplied by A_1 of norm ρ leads to a relative error of magnitude $(\kappa/\rho)^2\epsilon_{\text{mach}}$ on $\tilde{b}_2$. Hence,

$$\frac{\|\Delta x_2\|}{\|x_2\|} \approx \kappa\frac{\|\Delta\tilde{b}_2\|}{\|\tilde{b}_2\|} \approx \kappa\frac{\kappa^2}{\rho^2}\epsilon_{\text{mach}}. \tag{29}$$

In a similar way, one derives that, when $\kappa \geq \rho$:

$$\frac{\|\Delta x_i\|}{\|x_i\|} \approx \kappa\frac{\kappa^i}{\rho^i}\epsilon_{\text{mach}}. \tag{30}$$

In an analogous way the other possibilities in the summary hereafter can be deduced. Assuming that $\|x_i\| \approx \rho^i$ we have the following possibilities:

1. When $\|A_i\| \approx \rho^i$, we cannot do much about the loss of accuracy:

$$\frac{\|\Delta x_i\|}{\|x_i\|} \approx \kappa^{i+1}\epsilon_{\text{mach}}. \tag{31}$$

2. When $\|A_i\| \approx 1^i$, we can distinguish two possibilities:

$$\text{when } \kappa \geq \rho : \frac{\|\Delta x_i\|}{\|x_i\|} \approx \kappa\frac{\kappa^i}{\rho^i}\epsilon_{\text{mach}}; \tag{32}$$

$$\text{when } \kappa \leq \rho : \frac{\|\Delta x_i\|}{\|x_i\|} \approx \kappa\epsilon_{\text{mach}}. \tag{33}$$

The second case cannot arise when $\rho < 1$.

We observe in computational experiments that in our path tracking method we are usually dealing with the first case, where $\|A_i\| \approx \rho^i$, $\|x_i\| \approx \rho^i$. This means that the number of coefficients that we can compute with reasonable accuracy is bounded roughly by $-\log(\epsilon_{\text{mach}})/\log(\kappa)$, where κ is the condition number of the Jacobian A_0.

4.3   Newton's Method, Rational Approximations, Coefficient Shift

In Newton's method, the evaluation and differentiation algorithms are followed by the solution of the matrix series system to compute all coefficients of a power series at a regular solution of a polynomial homotopy. There are two remaining stages. Both stages use the same type of parallel algorithm, summarized in the next two paragraphs.

A Padé approximant is the quotient of two polynomials. To construct an approximant of degree K in the numerator and L in the denominator, we need the first $K+L+1$ coefficients of the power series. Given K and L, we truncate the power series at degree $d = K + L$. All components of an n-dimensional vector can be computed independently from each other, so each job in the parallel algorithm is the construction and evaluation of one Padé approximant.

All power series are assumed to originate at $t = 0$. After incrementing the step size with Δt, we shift all coefficients of the power series in the polynomial homotopy with $-\Delta t$, so at the next step we start again at $t = 0$. The shift operation happens independently for every polynomial in the homotopy, so the threads take turns in shifting the coefficients.

As the computational experiments show, the construction of rational approximations and the shifting of coefficients are computationally less intensive than running Newton's method, or than computing the Jacobian, all Hessians, and singular values at a point.

5   Computational Experiments

The goal of the computational experiments is to examine the relative computational costs of the various stages and to detect potential bottlenecks in the scalability. After presenting tables for random input data, we end with a description of a run on a cyclic n-root, for $n = 64, 96, 128$, a sample of a well known benchmark problem [8] in polynomial system solving.

Our computational experiments run on two 22-core 2.2 GHz Intel Xeon E5-2699 processors in a CentOS Linux workstation with 256 GB RAM. In our speedup computation, we compare against a sequential implementation, using the same primitive operations.

For each run on p threads, we report the speedup $S(p)$, the ratio between the serial time over the parallel execution time, and the efficiency $E(p) = S(p)/p$. Although our workstation has 44 cores, we stop the runs at 40 threads to avoid measuring the interference with other unrelated processes.

The units of all times reported in the tables below are seconds and the times themselves are elapsed wall clock times. These times include the allocation and deallocation of all data structures, for inputs, results, and work space.

Table 1. Evaluation and differentiation at power series truncated at increasing degrees d, for increasing number of threads p, in quad double precision

p	$d = 8$			$d = 16$			$d = 32$			$d = 48$		
	Time	$S(p)$	E, %	Time	$S(p)$	E, %	Time	$S(p)$	E, %	Time	$S(p)$	E, %
1	44.851			154.001			567.731			1240.761		
2	24.179	1.86	92.8	82.311	1.87	93.6	308.123	1.84	92.1	659.332	1.88	94.1
4	12.682	3.54	88.4	41.782	3.69	92.2	154.278	3.68	92.0	339.740	3.65	91.3
8	6.657	6.74	84.2	22.332	6.90	86.2	82.250	6.90	86.3	179.424	6.92	86.4
16	3.695	12.14	75.9	12.747	12.08	75.5	45.609	12.45	77.8	100.732	12.32	76.9
32	2.055	21.82	68.2	6.332	24.32	76.0	23.451	24.21	75.7	50.428	24.60	76.9
40	1.974	22.72	56.8	6.303	24.43	61.1	23.386	24.28	60.7	51.371	24.15	60.4

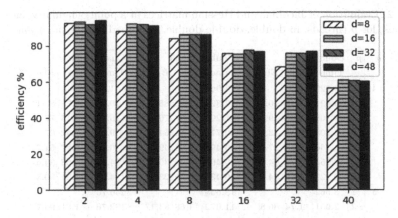

Fig. 1. Efficiency plots for evaluation and differentiation of power series, with data from Table 1. Efficiency tends to decrease for increasing p. The efficiency improves a little as the truncation degree d of the series increases from 8, 16, 32, to 48.

5.1   Random Input Data

The randomly generated problems represent polynomial systems of dimension 64 (or higher), with 64 (or more) terms in each polynomial and exponents of the variables between zero and eight.

Algorithmic Differentiation on Power Series. The computations in Table 1 illustrate the cost overhead of working with power series of increasing degrees of truncation. We start with degree $d = 8$ (the default in [18]) and consider the increase in wall clock times as we increase d. Reading Table 1 diagonally, observe the quality up. Figure 1 shows the efficiencies.

The drop in efficiency with $p = 40$ is because the problem size $n = 64$ is not a multiple of p, which results in load imbalancing. As quad double arithmetic is already very computationally intensive, the increase in the truncation degree d does little to improve the efficiency. Using more threads increases the memory usage, as each thread needs its own work space for all data structures used in the computation of its gradient with algorithmic differentiation. In a sequential computation where gradients are computed one after the other, there is only one vector with forward, backward, and cross products. When p gradients are computed simultaneously, there are p work space vectors to store the intermediate forward, backward, and cross products for each gradient. The portion of the parallel code that allocates and deallocates all work space vectors grows as the number of threads increases and the wall clock times incorporate the time spent on that data management as well.

Jacobians, Hessians at a Point, and Singular Values. Table 2 summarizes runs on the evaluation and singular value computations on random input data,

Table 2. Evaluation of Jacobian and Hessian matrices at a point, singular value decompositions, for p threads, in double, double double, and quad double precision

n	p	Double			Double double			Quad double		
		Time	$S(p)$	$E(p)$, %	Time	$S(p)$	$E(p)$, %	Time	$S(p)$	$E(p)$, %
64	1	0.729			3.964			51.998		
	2	0.521	1.40	70.0	2.329	1.70	85.1	29.183	1.78	89.1
	4	0.308	2.37	59.2	1.291	3.07	76.8	16.458	3.16	79.0
	8	0.208	3.50	43.7	0.770	5.15	64.3	9.594	5.42	67.8
	16	0.166	4.39	27.4	0.498	7.96	49.8	6.289	8.27	51.7
	32	0.153	4.77	14.9	0.406	9.76	30.5	4.692	11.08	34.6
	40	0.129	5.65	14.1	0.431	9.19	23.0	4.259	12.21	30.5
96	1	3.562			18.638			240.70		
	2	2.051	1.74	86.8	11.072	1.68	84.17	132.76	1.81	90.7
	4	1.233	2.89	72.2	5.851	3.19	79.64	72.45	3.32	83.1
	8	0.784	4.54	56.8	3.374	5.52	69.06	41.20	5.84	73.0
	16	0.521	6.84	42.7	2.188	8.52	53.25	25.87	9.30	58.1
	32	0.419	8.50	26.6	1.612	11.56	36.13	15.84	15.20	47.5
	40	0.398	8.94	22.4	1.442	12.92	32.31	15.84	15.20	38.0
128	1	12.464			62.193			730.50		
	2	6.366	1.96	97.9	33.213	1.87	93.6	399.98	1.83	91.3
	4	3.570	3.49	87.3	17.436	3.57	89.2	213.04	3.43	85.7
	8	2.170	5.75	71.8	9.968	6.24	78.0	119.81	6.10	76.2
	16	1.384	9.01	56.3	6.101	10.19	63.7	73.09	9.99	62.5
	32	1.033	12.06	37.7	4.138	15.03	47.9	43.44	16.82	52.6
	40	0.981	12.70	31.7	3.677	16.92	42.3	42.44	17.21	43.0

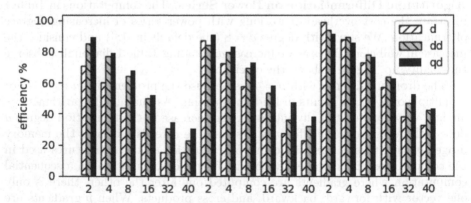

Fig. 2. Efficiency plots for computing Jacobians, Hessians, and their singular values, with data from Table 2. The three ranges for $p = 2, 4, 8, 16, 32, 40$ are from left to right for $n = 64, 96$, and 128 respectively. Efficiency decreases for increasing values of p. Efficiency increases for increasing values of n and for increased precision, where d = double, dd = double double, and qd = quad double.

for n-dimensional problems. The n polynomials have each n terms, where the exponents of the variables range from zero to eight.

Reading the columns of Table 2 vertically, we observe increasing speedups, which increase as n increases. Reading Table 2 horizontally, we observe the cost overhead of the arithmetic. To see how many threads are needed to compensate for this overhead, read Table 2 diagonally. Figure 2 shows the efficiencies.

To explain the drop in efficiencies we apply the same reasoning as before and point out that the work space increases even more as more threads are applied, because the total memory consumption has increased with the two dimensional Hessian matrices.

Pipelined Solution of Matrix Series. Elapsed wall clock times and speedups are listed in Table 3, on randomly generated linear systems of 64 equations in 64 unknowns, for series truncated to increasing degrees. The dimensions are consistent with the setup of Table 1, to relate the cost of linear system solving to the cost of evaluation and differentiations. Figure 3 shows the efficiencies.

Table 3. Solving a linear system for power series truncated at increasing degrees d, for increasing number of threads p, in quad double precision

p	$d = 8$			$d = 16$			$d = 32$			$d = 48$		
	Time	$S(p)$	E, %	Time	$S(p)$	E, %	Time	$S(p)$	E, %	Time	$S(p)$	E, %
1	0.232			0.605			2.022			4.322		
2	0.222	1.05	52.4	0.422	1.44	71.7	1.162	1.74	87.0	2.553	1.69	84.7
4	0.218	1.07	26.6	0.349	1.74	43.4	0.775	2.61	65.3	1.512	2.86	71.5
8	0.198	1.18	14.7	0.291	2.08	26.0	0.554	3.65	45.6	0.927	4.66	58.3
16	0.166	1.40	8.7	0.225	2.69	16.8	0.461	4.39	27.5	0.636	6.80	42.5
32	0.197	1.18	3.7	0.225	2.69	8.4	0.371	5.45	17.0	0.554	7.81	24.4
40	0.166	1.40	3.5	0.227	2.67	6.7	0.369	5.48	13.7	0.531	8.14	20.3

Consistent with the above analysis, the speedups in Table 3 level off for $p > d/2$. A diagonal reading shows that with multithreading, we can keep the time below one second, while increasing the degree of the truncation from 8 to 48.

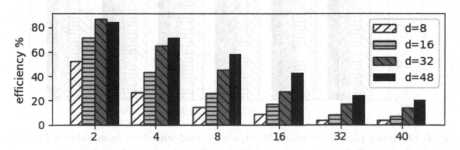

Fig. 3. Efficiency plots for pipelined solution of a matrix series with data from Table 3. Efficiency tends to decrease for increasing p and increase for increasing d.

Relative to the cost of evaluation and differentiation, the seconds in Table 3 are significantly smaller than the seconds in Table 1.

Multithreaded Newton's Method on Power Series. In the randomly generated problems, we add the parameter t to every polynomial to obtain a Newton homotopy. The elapsed wall clock times in Table 4 come from running Newton's method, which requires the repeated evaluation, differentiation, and linear system solving. The dimensions of the randomly generated problems are 64 equations in 64 variables, with 8 as the highest degree in each variable. The parameter t appears with degree one. Figure 4 shows the efficiencies.

Table 4. Running 8 steps with Newton's method for power series truncated at increasing degrees d, for increasing number of threads p, in quad double precision.

p	$d = 8$			$d = 16$			$d = 32$			$d = 48$		
	Time	$S(p)$	E, %	Time	$S(p)$	E, %	Time	$S(p)$	E, %	Time	$S(p)$	E, %
1	347.85			1176.88			4525.08			7005.91		
2	188.92	1.84	92.1	658.93	1.79	89.3	2323.20	1.95	97.4	3806.19	1.84	92.0
4	98.28	3.54	88.5	330.49	3.56	89.0	1193.76	3.79	94.8	1925.04	3.64	91.0
8	54.55	6.38	79.7	191.57	6.14	76.8	638.20	7.09	88.6	1014.85	6.90	86.3
16	31.26	11.13	69.5	97.34	12.09	75.6	352.10	12.85	80.3	571.25	12.26	76.7
32	17.62	19.74	61.7	50.80	23.16	72.4	180.31	25.60	78.4	291.92	24.00	75.0
40	17.45	19.93	49.8	51.70	22.76	56.9	181.56	24.92	62.3	292.55	23.95	59.9

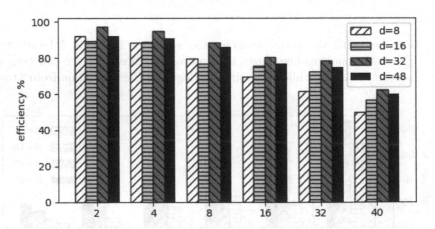

Fig. 4. Efficiency plots for running Newton's method with data from Table 4. Efficiency tends to decrease for increasing p and increase for increasing degree d.

Table 5. Construction and evaluation of Padé approximants for increasing degrees d, for increasing number of threads p, in quad double precision

p	$d = 8$			$d = 16$			$d = 32$			$d = 48$		
	Time	$S(p)$	E, %	Time	$S(p)$	E, %	Time	$S(p)$	E, %	Time	$S(p)$	E, %
1	0.034			0.109			0.684			2.193		
2	0.025	1.36	68.1	0.110	0.99	49.4	0.452	1.51	75.6	1.231	1.78	89.1
4	0.013	2.61	65.2	0.064	1.71	42.6	0.238	2.87	71.8	0.642	3.42	85.4
8	0.007	4.79	59.8	0.035	3.07	38.4	0.189	3.63	45.4	0.365	6.01	75.1
16	0.006	6.09	38.1	0.020	5.52	34.5	0.098	6.96	43.5	0.219	10.00	62.5
32	0.004	9.47	29.6	0.013	8.66	27.1	0.058	11.70	36.6	0.138	15.89	49.7
40	0.003	11.48	28.7	0.009	11.57	28.9	0.039	17.58	43.9	0.130	16.93	42.3

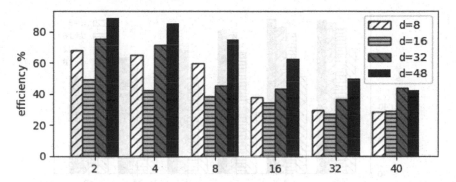

Fig. 5. Efficiency plots for rational approximations with data from Table 5.

The improvement in the efficiencies as the degrees increase can be explained by the improvement in the efficiencies in the pipelined solution of matrix series, see Fig. 3.

Rational Approximations. In Table 5, wall clock times and speedups are listed for the construction and evaluation of vectors of Padé approximants, of dimension 64 and for increasing degrees $d = 8, 16, 24$, and 32. For each d, we take $K = L = d/2$. Figure 5 shows the efficiencies. The fast drop in efficiency for $d = 8$ is due to the tiny wall clock times. There is not much that can be improved with multithreading once the time drops below 10 ms.

Shifting the Coefficients of the Power Series. Table 6 summarizes experiments on a randomly generated system of 64 polynomials in 64 unknowns, with 64 terms in every polynomial. Figure 6 shows the efficiencies.

Proportional Costs. Comparing the times in Tables 1, 2, 3, 5, and 6, we get an impression on the relative costs of the different tasks. The evaluation

Table 6. Shifting the coefficients of a polynomial homotopy, for increasing degrees d, for increasing number of threads p, in quad double precision

p	$d = 8$			$d = 16$			$d = 32$			$d = 48$		
	Time	$S(p)$	E, %	Time	$S(p)$	E, %	Time	$S(p)$	E, %	Time	$S(p)$	E, %
1	0.358			1.667			9.248			26.906		
2	0.242	1.48	74.0	0.964	1.73	86.5	5.134	1.80	90.1	14.718	1.83	91.4
4	0.154	2.32	58.0	0.498	3.35	83.8	2.642	3.50	87.5	7.294	3.69	92.2
8	0.101	3.55	44.4	0.289	5.77	72.1	1.392	6.64	83.0	3.941	6.83	85.3
16	0.058	6.13	38.3	0.181	9.23	57.7	0.788	11.73	73.3	2.307	11.66	72.9
32	0.035	10.30	32.2	0.116	14.40	45.0	0.445	20.80	65.0	1.212	22.20	69.4
40	0.031	11.49	28.7	0.115	14.51	36.3	0.419	22.05	55.1	1.156	23.28	58.2

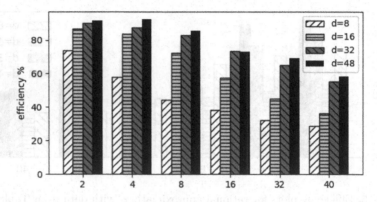

Fig. 6. Efficiency plots for shifting series of a polynomial homotopy with data from Table 6. Efficiency tends to decrease for increasing p and increase for increasing d.

and differentiation at power series, truncated at $d = 8$ dominates the cost with 348 s for one thread, or 17 s for 40 threads, in quad double arithmetic, from Table 1. The second largest cost comes from Table 2, for $n = 64$, in quad double arithmetic: 52 s for one thread, or 4 s on 40 threads. The other three stages take less than one second on one thread.

5.2 One Cyclic n-Root, $n = 64, 96, 128$

Our algorithms are developed to run on highly nonlinear problems such as the cyclic n-roots problem:

$$\begin{cases} x_0 + x_1 + \cdots + x_{n-1} = 0 \\ i = 2, 4, \ldots, n-1 : \sum_{j=0}^{n-1} \prod_{k=j}^{j+i-1} x_{k \bmod n} = 0 \\ x_0 x_1 x_2 \cdots x_{n-1} - 1 = 0. \end{cases} \tag{34}$$

Table 7. Computing C for one cyclic n-root, for $n = 64, 96, 128$, for an increasing number of threads p, in quad double precision

p	$n = 64$			$n = 96$			$n = 128$		
	Time	$S(p)$	$E, \%$	Time	$S(p)$	$E, \%$	Time	$S(p)$	$E, \%$
1	36.862			152.457			471.719		
2	21.765	1.69	84.7	87.171	1.75	87.5	262.678	1.80	89.8
4	12.390	2.98	74.4	47.268	3.23	80.6	143.262	3.29	82.3
8	7.797	4.73	59.1	28.127	5.42	67.8	83.044	5.68	71.0
16	5.600	6.58	41.1	18.772	8.12	50.8	53.235	8.86	55.4
32	4.059	9.08	28.4	12.988	11.74	36.7	34.800	13.56	42.4
40	4.046	9.11	22.8	12.760	11.95	29.9	33.645	14.02	35.1

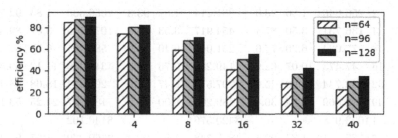

Fig. 7. Efficiency plots for computing C for one cyclic n-root, for $n = 64, 96, 128$, with data from Table 7. Efficiency decreases for increasing p and increases for increasing n.

This well known benchmark problem in polynomial system solving is important in the study of biunimodular vectors [10].

Problem Setup. By Backelin's Lemma [2], we know there is a 7-dimensional surface of cyclic 64-roots, along with a recipe to generate points on this surface. To generate points, a tropical formulation of Backelin's Lemma [1] is used. The surface has degree eight. Seven linear equations with random complex coefficients are added to obtain isolated points on the surface. The addition of seven linear equations gives 71 equations in 64 variables. As in [17], we add extra slack variables in an embedding to obtain an equivalent square 71-dimensional system. Similarly, there is a 3-dimensional surface of cyclic 96-roots and again a 7-dimensional surface of cyclic 128-roots.

In [23], running the typical predictor-corrector methods, we experienced that the hardware double precision is no longer sufficient to track a solution path on this 7-dimensional surface of cyclic 64-roots. Observe the high degrees of the polynomials in (34).

Table 7 contains wall clock times, speedups and efficiencies for computing the curvature bound C for one cyclic n-root. Efficiencies are shown in Fig. 7.

Table 8. Computing R for one cyclic n-root, for $n = 64$, 96, 128, for degrees $d = 8, 16, 24$, and for an increasing number of threads p, in quad double precision.

d	p	$n = 64$			$n = 96$			$n = 128$		
		Time	$S(p)$	E, %	Time	$S(p)$	E, %	Time	$S(p)$	E, %
8	1	139.185			483.137			1123.020		
	2	78.057	1.78	89.2	257.023	1.88	94.0	614.750	1.83	91.3
	4	42.106	3.31	82.6	141.329	3.42	85.5	318.129	3.53	88.3
	8	24.452	5.69	71.2	81.308	5.94	74.3	176.408	6.37	79.6
	16	15.716	8.86	55.4	47.585	10.15	63.5	105.747	10.62	66.4
	32	12.370	11.25	35.2	35.529	13.60	42.5	68.025	16.51	51.6
	40	12.084	11.52	28.8	35.212	13.72	34.3	62.119	18.08	45.2
16	1	477.956			1606.174			3829.567		
	2	256.846	1.86	93.0	861.214	1.87	93.3	2066.680	1.85	92.7
	4	136.731	3.50	87.4	454.917	3.53	88.3	1072.106	3.57	89.3
	8	77.034	6.20	77.6	251.066	6.40	80.0	584.905	6.55	81.8
	16	47.473	10.07	62.9	149.288	10.76	67.2	344.430	11.12	69.5
	32	32.744	14.60	45.6	97.514	16.47	51.5	205.034	18.68	58.4
	40	32.869	14.54	36.4	89.260	18.00	45.0	180.207	21.25	53.1
24	1	1023.968			3420.576			8146.102		
	2	555.771	1.84	92.1	1855.748	1.84	92.2	4360.870	1.87	93.4
	4	304.480	3.36	84.1	956.443	3.58	89.4	2268.632	3.59	89.8
	8	160.978	6.36	79.5	523.763	6.53	81.6	1235.338	6.59	82.4
	16	98.336	10.41	65.1	312.698	10.94	68.4	726.287	11.22	70.1
	32	65.448	15.65	48.9	196.488	17.41	54.4	416.735	19.55	61.1
	40	63.412	16.15	40.4	170.474	20.07	50.2	360.419	22.60	56.5

Table 8 contains wall clock times, speedups and efficiencies for computing the radius bound R for one cyclic n-root. See Fig. 8.

For $n = 64$, the inverse condition number of the Jacobian matrix is estimated as 3.9E−5 and after 8 iterations, the maximum norm of the last vector in the last update to the series equals respectively 4.6E−44, 1.1E−24, and 4.1E−5, for $d = 8, 16$, and 24. For $n = 96$, the estimated inverse condition number is 2.0E−4 and the maximum norm for $d = 8, 16$, and 24 is then respectively 1.4E−47, 9.6E−31, and 7.3E−14. The condition worsens for $n = 128$, estimated at 4.6E−6 and then for $d = 8$, the maximum norm of the last update vector is 2.2E−30. For $d = 16$ and 24, the largest maximum norm less than one occurs at the coefficients with t^{15} and equals about 1.1E−1.

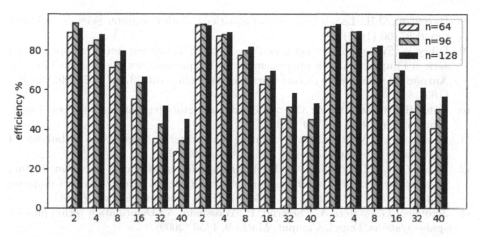

Fig. 8. Efficiency plots for computing R for one cyclic n-root, for $n = 64, 96, 128$, for degrees $d = 8, 16, 24$, with data from Table 8. Efficiency decreases for increasing p. Efficiency increases as n and/or d increase.

6 Conclusions

The cost overhead of our robust path tracker is $O(n)$, compared with the current numerical predictor-corrector algorithms. For $n = 64$, we expect a cost overhead factor of about 64. We interpret the speedups in Table 7 and Table 8 as follows. With a speedup of about 10, then this factor drops to about 6. The plan is to integrate the new algorithms in the parallel blackbox solver [22].

References

1. Adrovic, D., Verschelde, J.: Polyhedral methods for space curves exploiting symmetry applied to the cyclic n-roots problem. In: Gerdt, V.P., Koepf, W., Mayr, E.W., Vorozhtsov, E.V. (eds.) CASC 2013. LNCS, vol. 8136, pp. 10–29. Springer, Cham (2013). https://doi.org/10.1007/978-3-319-02297-0_2
2. Backelin, J.: Square multiples n give infinitely many cyclic n-roots. Reports, Matematiska Institutionen 8, Stockholms universitet (1989)
3. Bates, D.J., Hauenstein, J.D., Sommese, A.J., Wampler, C.W.: Adaptive multiprecision path tracking. SIAM J. Numer. Anal. **46**(2), 722–746 (2008)
4. Bates, D.J., Hauenstein, J.D., Sommese, A.J., Wampler, C.W.: Numerically Solving Polynomial Systems with Bertini, vol. 25. SIAM (2013)
5. Bliss, N., Verschelde, J.: The method of Gauss-Newton to compute power series solutions of polynomial homotopies. Linear Algebra Appl. **542**, 569–588 (2018)
6. Breiding, P., Timme, S.: HomotopyContinuation.jl: a package for homotopy continuation in Julia. In: Davenport, J.H., Kauers, M., Labahn, G., Urban, J. (eds.) ICMS 2018. LNCS, vol. 10931, pp. 458–465. Springer, Cham (2018). https://doi.org/10.1007/978-3-319-96418-8_54
7. Christianson, B.: Automatic Hessians by reverse accumulation. IMA J. Numer. Anal. **12**, 135–150 (1992)

8. Davenport, J.H.: Looking at a set of equations. Bath Computer Science Technical report 87–06 (1987)
9. Fabry, E.: Sur les points singuliers d'une fonction donnée par son développement en série et l'impossibilité du prolongement analytique dans des cas très généraux. In: Annales scientifiques de l'École Normale Supérieure, vol. 13, pp. 367–399. Elsevier (1896)
10. Führ, H., Rzeszotnik, Z.: On biunimodular vectors for unitary matrices. Linear Algebra Appl. **484**, 86–129 (2015)
11. Griewank, A., Walther, A.: Evaluating derivatives: principles and techniques of algorithmic differentiation, vol. 105. SIAM (2008)
12. Hida, Y., Li, X.S., Bailey, D.H.: Algorithms for quad-double precision floating point arithmetic. In: The Proceedings of the 15th IEEE Symposium on Computer Arithmetic (Arith-15 2001), pp. 155–162. IEEE Computer Society (2001)
13. Jeronimo, G., Matera, G., Solernó, P., Waissbein, A.: Deformation techniques for sparse systems. Found. Comput. Math. **9**, 1–50 (2009)
14. Katsura, S.: Spin glass problem by the method of integral equation of the effective field. In: Coutinho-Filho, M., Resende, S. (eds.) New Trends in Magnetism, pp. 110–121. World Scientific, London (1990)
15. Li, T., Tsai, C.: HOM4PS-2.0para: Parallelization of HOM4PS-2.0 for solving polynomial systems. Parallel Comput. **35**(4), 226–238 (2009)
16. McCormick, J.W., Singhoff, F., Hugues, J.: Building Parallel, Embedded, and Real-Time Applications with Ada. Cambridge University Press, Cambridge (2011)
17. Sommese, A.J., Verschelde, J.: Numerical homotopies to compute generic points on positive dimensional algebraic sets. J. Complexity **16**(3), 572–602 (2000)
18. Telen, S., Van Barel, M., Verschelde, J.: A robust numerical path tracking algorithm for polynomial homotopy continuation. arXiv:1909.04984
19. Trias, A.: The holomorphic embedding load flow method. In: 2012 IEEE Power and Energy Society General Meeting, pp. 1–8. IEEE (2012)
20. Trias, A., Martin, J.L.: The holomorphic embedding loadflow method for DC power systems and nonlinear DC circuits. IEEE Trans. Circuits Syst. **63**(2), 322–333 (2016)
21. Verschelde, J.: Algorithm 795: PHCpack: a general-purpose solver for polynomial systems by homotopy continuation. ACM Trans. Math. Softw. (TOMS) **25**(2), 251–276 (1999)
22. Verschelde, J.: A blackbox polynomial system solver on parallel shared memory computers. In: Gerdt, V.P., Koepf, W., Seiler, W.M., Vorozhtsov, E.V. (eds.) CASC 2018. LNCS, vol. 11077, pp. 361–375. Springer, Cham (2018). https://doi.org/10.1007/978-3-319-99639-4_25
23. Verschelde, J., Yu, X.: Accelerating polynomial homotopy continuation on a graphics processing unit with double double and quad double arithmetic. In: Proceedings of the 7th International Workshop on Parallel Symbolic Computation (PASCO 2015), pp. 109–118. ACM (2015)

Symbolic-Numeric Computation of the Bernstein Coefficients of a Polynomial from Those of One of Its Partial Derivatives and of the Product of Two Polynomials

Jihad Titi[1] and Jürgen Garloff[2,3(✉)]

[1] Department of Applied Mathematics and Physics, Palestine Polytechnic University,
Hebron, Palestine
jihadtiti@yahoo.com

[2] Department of Mathematics and Statistics, University of Konstanz,
78464 Konstanz, Germany

[3] University of Applied Sciences/HTWG Konstanz, Institute for Applied Research,
Alfred-Wachtel-Str. 8, 78462 Konstanz, Germany
juergen.garloff@htwg-konstanz.de

Abstract. The expansion of a given multivariate polynomial into Bernstein polynomials is considered. Matrix methods for the calculation of the Bernstein expansion of the product of two polynomials and of the Bernstein expansion of a polynomial from the expansion of one of its partial derivatives are provided which allow also a symbolic computation.

Keywords: Multivariate polynomial · Bernstein polynomial · Bernstein coefficient

MSC 2010: 65F30 · 41-04

1 Introduction

In this paper, we consider the expansion of a multivariate polynomial into Bernstein polynomials over a box, i.e., an axis-aligned region, in $\mathbb{R}^n$. This expansion has many applications, e.g., in computer aided geometric design, robust control, global optimization, differerential and integral equations, finite element analysis [6]. A very useful property of this expansion is that the interval spanned by the minimum and maximum of the coefficients of this expansion, the so-called Bernstein coefficients, provides bounds for the range of the given polynomial over the considered box, see, e.g., [8,10]. A simple (but by no means economic) method for the computation of the Bernstein coefficients from the coefficients of the given polynomial is the use of formula (5) below. This formula (and also similar ones for the Bernstein coefficients over more general sets like sinplexes and polytopes) allows the symbolic computation of these quantities when the coefficients of the

© Springer Nature Switzerland AG 2020
F. Boulier et al. (Eds.): CASC 2020, LNCS 12291, pp. 583–599, 2020.
https://doi.org/10.1007/978-3-030-60026-6_34

given polynomial depend on parameters. Some applications are making use of this symbolic computation: in [4, Sections 3.2 and 3.3], [5], see also the many references therein, the reachability computation and parameter synthesis with applications in biological modelling are considered. In [2,3], parametric polynomial inequalities over parametric boxes and polytopes are treated. Applications in static program analysis and optimization include dependence testing between references with linearized subscripts, dead code elimination of conditional statements, and estimation of memory requirements in the development of embedded systems. Applications which involve polynomials of higher degree or many variables require a computation of the Bernstein coefficients which is more economic than by formula (5). In [14] and [16], we have presented a matrix method for the computation of the Bernstein coefficients which is faster than the methods developed so far. In this paper, we consider firstly the case where we have already computed the Bernstein coefficients of two multivariate polynomials and wish to compute the Bernstein expansion of their product for which we present two approaches. For the univariate case see [7, Subsection 4.2]. Secondly, we show how the Bernstein coefficients of a multivariate polynomial can be computed from the Bernstein coefficients of its partial derivatives. This problem appears for example when bounds for the range of a complex polynomial over a rectangular region in the complex plane are wanted and the Cauchy-Riemann equations are employed, see [13, Section 4.3], [17], but it is also of interest by its own.

The organization of our paper is as follows. In the next section, we introduce the notation which is used throughout the paper. In Sect. 3, we first briefly recall the expansion of a multivariate real polynomial into Bernstein polynomials over a box and some of its fundamental properties. In the second part, we recall from [14,16] a matrix method for the computation of the Bernstein coefficients. In Sect. 4 we present two matrix methods for the computation of the Bernstein coefficients of the product of two polynomials and in Sect. 5 the computation of the Bernstein coefficients of a polynomial from those of one of its partial derivatives.

2   Notation

In this section, we introduce the notation that we are using throughout this paper. Let $n \in \mathbb{N}$ (set of the nonnegative integers) be the number of variables. A multi-index $(i_1, \ldots, i_n) \in \mathbb{N}^n$ is abbreviated by i. In particular, we write 0 for $(0, \ldots, 0)$ and e^s for the multi-index that has a 1 in position s and 0's otherwise. Arithmetic operations with multi-indices are defined entry-wise; the same applies to comparison between multi-indices. For the multi-index $i = (i_1, \ldots, i_s, \ldots, i_n)$ we define $i_{s,q} := (i_1, \ldots, i_s + q, \ldots, i_n)$ and $i_{[s,q]} := (i_1, \ldots, q, \ldots, i_n), s \in \{1, \ldots, n\}, q \in \mathbb{Z}$. For $x = (x_1, \ldots, x_n) \in \mathbb{R}^n$, its *monomials* are defined as $x^i := \prod_{s=1}^{n} x_s^{i_s}$. For $d = (d_1, \ldots, d_n) \in \mathbb{N}^n$ such that $i \leq d$, we use the compact notations

$$\sum_{i=0}^{d} := \sum_{i_1=0}^{d_1} \cdots \sum_{i_n=0}^{d_n}, \binom{d}{i} := \prod_{s=1}^{n} \binom{d_s}{i_s}.$$

For the ease of presentation, we index all array entries starting from zero.

3   Bernstein Expansion

3.1   Bernstein Representation over the Unit Box

In this section, we present fundamental properties of the Bernstein expansion over a box, e.g., [6, Subsection 5.1], [8,10], that are employed throughout the paper. For simplicity we consider the unit box $u := [0, 1]^n$, since any compact nonempty box x of $\mathbb{R}^n$ can be mapped affinely onto u. Let $\ell \in \mathbb{N}^n$, $a_j \in \mathbb{R}$, $j = 0, \ldots, \ell$, such that for $s = 1, \ldots, n$

$$\ell_s := \max \left\{ q \mid a_{j_1, \ldots, j_{s-1}, q, j_{s+1}, \ldots, j_n} \neq 0 \right\}. \tag{1}$$

Let p be an ℓ-th degree n-variate polynomial with the power representation

$$p(x) = \sum_{j=0}^{\ell} a_j x^j. \tag{2}$$

We expand p into Bernstein polynomials of degree d, $d \geq \ell$, over u as

$$p(x) = \sum_{j=0}^{d} b_j^{(d)} B_j^{(d)}(x), \tag{3}$$

where $B_j^{(d)}$ is the j-th Bernstein polynomial of degree d, defined as

$$B_j^{(d)}(x) := \binom{d}{j} x^j (1 - x)^{d-j}, \tag{4}$$

and $b_j^{(d)}$ is the j-th Bernstein coefficient of p of degree d over u which is given by

$$b_j^{(d)} = \sum_{i=0}^{j} \frac{\binom{j}{i}}{\binom{d}{i}} a_i, \quad 0 \leq j \leq d, \tag{5}$$

with the convention that $a_i := 0$ if $i \geq \ell, i \neq \ell$. We call (3) the *Bernstein representation of* p and arrange the Bernstein coefficients in a multidimensional array $B(u) = (b_j^{(d)})_{0 \leq j \leq d}$, the so-called *Bernstein patch*. Note that the Bernstein coefficients lying on the vertices of $B(u)$ are values of p at the respective vertices of u. More generally, the Bernstein coefficients on an r-dimensional face of u, $r = 0, 1, \ldots, n-1$, are just the Bernstein coefficients lying on the respective faces of $B(u)$ [9, Lemma 2]. E.g., assume that v is an $(n - 1)$-dimensional face of u that is obtained by setting $x_s = 0$ or 1, for some $s \in \{1, \ldots, n\}$. For $i \in \mathbb{N}^n$ and $r \in \mathbb{N}$ we define

$$i_{[s,r]} := (i_1, \ldots, i_{s-1}, r, i_{s+1}, \ldots, i_n). \tag{6}$$

Then, the Bernstein coefficients of p over v are given by

$$b_i^{(d)}(p, v) = \begin{cases} b_{i_{[s,0]}}, & \text{if } x_s = 0, \\ b_{i_{[s,d_s]}}, & \text{if } x_s = 1. \end{cases}$$

3.2   Computation of the Bernstein Coefficients

We recall from [16] a method for the computation of the Bernstein coefficients of the n-variate polynomial p given in (2).

Matrix Method for the Unit Box. The superscript c denotes the *cyclic ordering* of the sequence of the indices, i.e., the order of the indices of the entries of the array under consideration is changed cyclically. This means that the index in the first position is replaced by the index in the second one, the index in the second position by the one in the third, ... , the index in the n-th position by the one in the first position (see Fig. 1 as an illustration in the trivariate case). So after n cyclic orderings the sequence of the indices is again in its initial order. Note that in the bivariate case the cyclic ordering is just the usual matrix transposition.

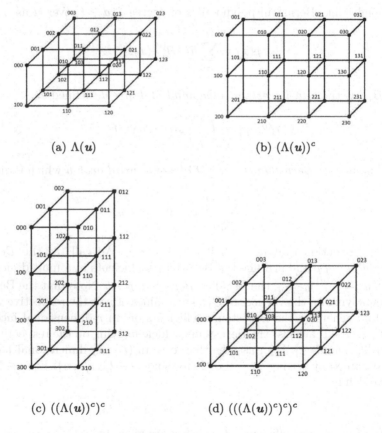

(a) $\Lambda(\boldsymbol{u})$

(b) $(\Lambda(\boldsymbol{u}))^c$

(c) $((\Lambda(\boldsymbol{u}))^c)^c$

(d) $(((\Lambda(\boldsymbol{u}))^c)^c)^c$

Fig. 1. Cyclic ordering of a three-dimensional array with $\ell_1 = 1$, $\ell_2 = 2$, and $\ell_3 = 3$

The coefficients of p are arranged in an $(\ell_1 + 1) \times \ell^*$ matrix A, where $\ell^* := \prod_{s=2}^{n} (\ell_s + 1)$. The correspondence between the coefficients a_j of p and the entry of A in row i and column j is as follows:

$$i = j_1, \tag{7a}$$

$$j = j_2 + \sum_{s=3}^{n} j_s (\ell_2 + 1) \cdot \ldots \cdot (\ell_{s-1} + 1). \tag{7b}$$

Then A can be represented as the matrix

$$
\begin{bmatrix}
a_{0,0,0,\ldots,0} & a_{0,1,0,\ldots,0} & \cdots & a_{0,\ell_2,0,\ldots,0} & a_{0,0,1,\ldots,0} & \cdots & a_{0,\ell_2,1,\ldots,0} & \cdots \\
a_{1,0,0,\ldots,0} & a_{1,1,0,\ldots,0} & \cdots & a_{1,\ell_2,0,\ldots,0} & a_{1,0,1,\ldots,0} & \cdots & a_{1,\ell_2,1,\ldots,0} & \cdots \\
\vdots & \vdots & \cdots & \vdots & \vdots & \cdots & \vdots & \cdots \\
a_{\ell_1,0,0,\ldots,0} & a_{\ell_1,1,0,\ldots,0} & \cdots & a_{\ell_1,\ell_2,0,\ldots,0} & a_{\ell_1,0,1,\ldots,0} & \cdots & a_{\ell_1,\ell_2,1,\ldots,0} & \cdots
\end{bmatrix}
\tag{8}
$$

$$
\begin{matrix}
\cdots & a_{0,0,\ell_3,\ldots,0} & \cdots & a_{0,\ell_2,\ell_3,\ldots,0} & \cdots & a_{0,0,\ell_3,\ldots,\ell_n} & a_{0,1,\ell_3,\ldots,\ell_n} & \cdots & a_{0,\ell_2,\ell_3,\ldots,\ell_n} \\
\cdots & a_{1,0,\ell_3,\ldots,0} & \cdots & a_{1,\ell_2,\ell_3,\ldots,0} & \cdots & a_{1,0,\ell_3,\ldots,\ell_n} & a_{1,1,\ell_3,\ldots,\ell_n} & \cdots & a_{1,\ell_2,\ell_3,\ldots,\ell_n} \\
\cdots & \vdots & \cdots & \vdots & \cdots & \vdots & \vdots & \cdots & \vdots \\
\cdots & a_{\ell_1,0,\ell_3,\ldots,0} & \cdots & a_{\ell_1,\ell_2,\ell_3,\ldots,0} & \cdots & a_{\ell_1,0,\ell_3,\ldots,\ell_n} & a_{\ell_1,1,\ell_3,\ldots,\ell_n} & \cdots & a_{\ell_1,\ell_2,\ell_3,\ldots,\ell_n}
\end{matrix}
$$

The matrix $\Lambda(u)$ is obtained from A by multiplying a_j by $\binom{\ell}{j}^{-1}$. We put $\Lambda_0 := \Lambda(u)$ and define for $s = 1, \ldots, n$

$$\Lambda_s := (P_s \Lambda_{s-1})^c, \tag{9}$$

where P_s is the *lower triangular Pascal matrix*,

$$
(P_s)_{ij} := \begin{cases} \binom{i}{j}, & \text{if } j \leq i, \\ 0, & \text{otherwise.} \end{cases}
\tag{10}
$$

In (9), the matrix multiplication is performed according to the factorization, e.g., [1, Lemma 1],

$$P_s = \prod_{\mu=1}^{\ell_s} K_\mu^s, \tag{11}$$

where the bidiagonal matrices K_μ^s, $\mu = 1, \ldots, \ell_s$, are given by

$$
(K_\mu^s)_{ij} := \begin{cases} 1, & \text{if } i = j, \\ 1, & \text{if } i = j+1, \ \ell_s - \mu \leq j \leq \ell_s - 1, \\ 0, & \text{otherwise.} \end{cases}
\tag{12}
$$

Define for $s = 1, \ldots, n$, $r := s \bmod n$. Then for $s = 1, \ldots, n$, the entry in position (v_1, v_2) in Λ_{s-1} becomes (v_1', v_2') in Λ_s, where

$$v_1' = v_2 \bmod (\ell_{r+1} + 1),$$

$$v_2' = \left\lfloor \frac{v_2}{\ell_{r+1} + 1} \right\rfloor + v_1 \prod_{\substack{m=1, \\ m \neq s,r}}^{n} (\ell_m + 1).$$

The Bernstein patch $B(\boldsymbol{u})$ arranged accordingly in the $(\ell_1 + 1) \times \ell^*$ *Bernstein matrix*, denoted by $\mathcal{B}(\boldsymbol{u})$, is given by Λ_n.

Matrix Method for a General Box. Firstly, we affinely map a given box $\boldsymbol{x}$,

$$\boldsymbol{x} = ([\underline{x}_s, \overline{x}_s])_{s=1}^n, \text{ with } \underline{x}_s < \overline{x}_s, \; s = 1, \ldots, n, \tag{13}$$

to the unit box $\boldsymbol{u}$ by

$$z_s = \frac{x_s - \underline{x}_s}{\overline{x}_s - \underline{x}_s}, \quad s = 1, \ldots, n. \tag{14}$$

We sequentially transform x_s, $s = 1, \ldots, n$. By substituting (14) in (2) for one x_s, $s = 1, \ldots, n$, at a time, we obtain a polynomial p^* over $\boldsymbol{u}$. The coefficients of p^* arranged in an $(\ell_1 + 1) \times \ell^*$ matrix, say A^*, can be derived as follows from the matrix A of the coefficients of p given in (8): For $s = 1, \ldots, n$ define

$$Q_s := \begin{cases} \tilde{D}_s(\frac{\overline{x}_s - \underline{x}_s}{\underline{x}_s}) P_s^T \tilde{D}_s(\underline{x}_s), & \underline{x}_s \neq 0, \\ \tilde{D}_s(\overline{x}_s), & \underline{x}_s = 0, \end{cases} \tag{15}$$

where $\tilde{D}_s(t)$ is the diagonal matrix of order $\ell_s + 1$

$$\tilde{D}_s(t) := \operatorname{diag}(1, t, t^2, \ldots, t^{\ell_s}), \quad s = 1, \ldots, n.$$

Then A^* can be represented as

$$A^* = (Q_n(\cdots (Q_2(Q_1 A)^c)^c \cdots)^c)^c. \tag{16}$$

By applying the procedure for the unit box to the matrix A^*, we obtain the Bernstein patch of p over $\boldsymbol{x}$.

Amount of Arithmetic Operations. Assuming that $\kappa = \ell_s$ for all $s = 1, \ldots, n$, the presented matrix method requires $n\kappa \frac{(\kappa+1)^n}{2}$ additions and $n(\kappa+1)^n$ multiplications for the computation of the Bernstein coefficients over the unit box $\boldsymbol{u}$, and needs $n\kappa(\kappa+1)^n + n$ additions and $3n(\kappa+1)^n + 2n(\kappa-1) + n$ multiplications for a general box. A verified version of this method which is taking into account of all rounding errors as well as data uncertainties was implemented by Dr. Florian Bünger, Hamburg University of Technology, Germany. It is included in the version 12 of the MATLAB toolbox INTLAB [11].

4   Computation of the Bernstein Coefficients of the Product of Two Multivariate Polynomials

Let p and q be two n-variate polynomials of degree $\ell(p)$ and $\ell(q)$, respectively, with the Bernstein expansions of degrees $d(p) \geq \ell(p)$ and $d(q) \geq \ell(q)$ over $\boldsymbol{x}$

$$p(x) = \sum_{j=0}^{d(p)} b_j(p) B_j^{(d(p))}(x), \tag{17a}$$

$$q(x) = \sum_{i=0}^{d(q)} b_i(q) B_i^{(d(q))}(x). \tag{17b}$$

For the ease of presentation we consider here only the unit box $\boldsymbol{u}$ and assume that $d(p) = \ell(p)$ and $d(q) = \ell(q)$. Then, the polynomial pq resulting when multiplying p and q is of degree $\ell = \ell(p) + \ell(q)$. Hence, the Bernstein representation of pq over $\boldsymbol{u}$ is given as

$$pq(x) = \sum_{m=0}^{\ell} b_m(pq) B_m^{(\ell)}(x), \tag{18}$$

where $b_m(pq)$ is the m-th Bernstein coefficient of pq of degree ℓ over $\boldsymbol{u}$, $m = 0, \ldots, \ell$. Let $B(p, \boldsymbol{u})$, $B(q, \boldsymbol{u})$, and $B(pq, \boldsymbol{u})$ denote the Bernstein patches of p, q, and pq over $\boldsymbol{u}$, respectively, and their corresponding Bernstein matrices are given by $\mathcal{B}(p, \boldsymbol{u})$, $\mathcal{B}(q, \boldsymbol{u})$ and $\mathcal{B}(pq, \boldsymbol{u})$. In this section, we present two matrix methods, which are named the *first method* and *second method*, for the computation of the Bernstein coefficients $b_m(pq)$ of pq.

4.1   First Method

By this method the Bernstein coefficients of pq are computed from the Bernstein representation of p and q. The representations (17a) and (17b) can be rewritten as

$$p(x) = (1 - x)^{\ell(p)} \sum_{j=0}^{\ell(p)} c_j(p) \left(\frac{x}{1-x} \right)^j, \tag{19a}$$

$$q(x) = (1 - x)^{\ell(q)} \sum_{i=0}^{\ell(q)} c_i(q) \left(\frac{x}{1-x} \right)^i, \tag{19b}$$

where $c_j(p)$ and $c_i(q)$ are called the *scaled Bernstein coefficients* of p and q which are given by

$$c_j(p) = b_j(p) \binom{\ell(p)}{j}, \quad j = 0, \ldots, \ell(p), \tag{20a}$$

$$c_i(q) = b_i(q) \binom{\ell(q)}{i}, \quad i = 0, \ldots, \ell(q). \tag{20b}$$

From (19a) and (19b), the power representation of pq is obtained as

$$pq(x) = (1-x)^\ell \sum_{m=0}^{\ell} c_m \left(\frac{x}{1-x}\right)^m,$$

where the m-th scaled Bernstein coefficient c_m of pq is given as

$$c_m = \sum_{\mu=0}^{m} c_\mu(p) c_{m-\mu}(q), \quad m = 0, \ldots, \ell, \tag{21}$$

with $c_\mu(p) := 0$ if $\mu_s > \ell_s(p)$ and $c_\mu(q) := 0$ if $\mu_s > \ell_s(q)$ for some $s \in \{1, \ldots, n\}$. Then, from (21) the m-th Bernstein coefficient of pq is

$$b_m(pq) = \frac{c_m}{\binom{\ell}{m}}, \quad 0 \le m \le \ell. \tag{22}$$

In matrix language, the first method can be described as follows: let for $s = 1, \ldots, n$, the diagonal matrix D_s of order $\ell_s(p) + 1$ be defined by

$$D_s := \mathrm{diag}\left(\binom{\ell_s(p)}{0}, \binom{\ell_s(p)}{1}, \ldots, \binom{\ell_s(p)}{\ell_s(p)} \right), \tag{23}$$

and let $C(p)$ be the $(\ell_1(p) + 1) \times \ell^*(p)$ matrix, where $\ell^*(p) := \prod_{s=2}^{n} (\ell_s(p) + 1)$, which is obtained by

$$C(p) = (D_n(\cdots(D_2(D_1 \mathcal{B}(p, u))^c)^c \cdots)^c)^c. \tag{24}$$

It is easy to see that $(C(p))_{q_1, q_2} = c_j(p)$, $q_1 = 0, \ldots, \ell_1(p)$, $q_2 = 0, \ldots, \ell^*(p) - 1$ and $j = 0, \ldots, \ell(p)$, where the correspondence between the scaled Bernstein coefficient $c_j(p)$ and the entries of $C(p)$ can be determined by using (7).

Let us define for $s = 1, \ldots, n$, $t = 0, \ldots, \ell_s(q)$, the following $(\ell_s + 1) \times (\ell_s(p) + 1)$ matrices $W_s^{(t)}$ row-wise by

$$W_s^{(t)}[0, \ldots, t-1] := 0,$$
$$W_s^{(t)}[t, \ldots, \ell_s(p) + t] := I_{\ell_s(p)+1} \text{ (identity matrix of order } \ell_s(p) + 1),$$
$$W_s^{(t)}[\ell_s(p) + t + 1, \ldots, \ell_s] := 0;$$

as a convention, we define for $t = 0, \ell_s(q)$

$$W_s^{(t)}[0, \ldots, -1] = \phi,$$
$$W_s^{(t)}[\ell_s + 1, \ldots, \ell_s] = \phi,$$

where ϕ is a matrix of size 0×0. Assume that the scaled Bernstein coefficients of pq, see (21), are arranged in the $(\ell_1 + 1) \times \ell^*$ matrix $C(pq)$, with $\ell^* = \prod_{s=2}^{n} (\ell_s + 1)$ that is given by

$$C(pq) = \sum_{i_n=0}^{\ell_n(q)} \cdots \sum_{i_1=0}^{\ell_1(q)} \left(W_n^{(i_n)} \left(\cdots \left(W_2^{(i_2)} \left(W_1^{(i_1)} c_{i_1, \ldots, i_n}(q) C(p) \right)^c \right)^c \cdots \right)^c \right)^c \tag{25}$$

such that $(C(pq))_{q_1,q_2} = c_m$, $q_1 = 0, \ldots, \ell_1$, $q_2 = 0, \ldots, \ell^* - 1$, and $m = 0, \ldots, \ell$. Here, the relation between the entries of $C(pq)$ and the scaled Bernstein coefficients of pq, c_m, $m = 0, \ldots, \ell$, can be determined as in (7). From (25), we get

$$\mathcal{B}(pq, \boldsymbol{u}) = \left(D'_n \left(\cdots \left(D'_2 \left(D'_1 C(pq) \right)^c \right)^c \cdots \right)^c \right)^c, \qquad (26)$$

where D'_s is the inverse of the diagonal matrix diag $\left(\binom{\ell_s}{0}, \binom{\ell_s}{1}, \ldots, \binom{\ell_s}{\ell_s} \right)$ for $s = 1, \ldots, n$. A pseudocode for the first method is given in Algorithm 1. Its performance is illustrated in Appendix A.

4.2 Second Method

Let $a_j(p)$ and $a_i(q)$ be the j-th and the i-th coefficients of the power representations of p and q, respectively, such that $j = 0, \ldots, \ell(p)$ and $i = 0, \ldots, \ell(q)$. Assume that $a_j(p)$ are arranged in an $(\ell_1(p) + 1) \times \ell^*(p)$ matrix $A(p)$. Recall that pq is an n-variate polynomial of degree ℓ. Then the power representation of pq is given by

$$pq(x) = \sum_{m=0}^{\ell} a_m x^m. \qquad (27)$$

We arrange the coefficients of pq in an $(\ell_1 + 1) \times \ell^*$ matrix $\hat{A}$. In this method, the computation of the Bernstein coefficients of pq, see (18), is based on its power representation (27). The matrix description of this method is as follows: the entries of $\hat{A}$ are the entries of the matrix that is obtained from (25), where the (r_1, r_2)-th entry of $C(p)$ is replaced by the (r_1, r_2)-th entry of $A(p)$, where $r_1 = 0, \ldots \ell_1(p)$, $r_2 = 0 \ldots, \ell^*(p) - 1$, and $c_i(q)$ is replaced by $a_i(q), 0 \leq i \leq \ell(q)$. Then the method presented in Subsect. 3.2 is applied to compute the Bernstein coefficients of pq starting from $\hat{A}$.

4.3 Amount of Arithmetic Operations

In Tables 1 and 2, the number of the arithmetic operations of both methods are presented. For simplicity, we assume that $\ell_s(p) = \ell_s(q) = \kappa$ for $s = 1, \ldots, n$. Furthermore, we use the method from Subsect. 3.2 for the computation of $B(p, \boldsymbol{u})$ and $B(q, \boldsymbol{u})$ in the first method, see (24), and $B(pq, \boldsymbol{u})$ in the second method.

For the ease of comparison, we assume that the basis operations (addition, multiplication, and division) are taking the same time. Then we conclude that for $n \geq 4$ and $\kappa = 1$, $n \geq 2$ and $\kappa = 2, 3$, and for all n and $\kappa \geq 4$ the first method is superior to the second method. In addition, the first method has the advantage that all computations are performed *ab initio* in the Bernstein representation so that the numerical stability of this representation with respect to perturbations of initial data, or rounding errors that occur during floating point calculations can be fully employed, see [6, Section 6].

Algorithm 1. First method for the computation of the Bernstein coefficients of pq over the unit box u

1: *Input: The coefficients of the power representation of p and q over u*
2: *Output: The matrix $\mathcal{B}(pq, u)$ containing the Bernstein coefficients of pq over u*
3: *Step 1: Compute the Bernstein coefficients of p, q and arrange them in matrices*
4: $\mathcal{B}(p, u)$ *and* $\mathcal{B}(q, u)$, *respectively, by using the method from Subsection 3.2.*
5: *Step 2: Compute $C(p)$.*
6: *Put $C_0(p) := \mathcal{B}(p, u)$.*
7: **for** $s = 1, \ldots, n$ **do**
8:     *Compute $C_s(p) := (D_s C_{s-1}(p))^c$.*
9: **end for**
10: *Put $C_0(pq) := O$.*
11: *Step 3: Computation of $C(pq)$*
12: **for** $i_1 = 0, \ldots, \ell_1(q)$ **do**

13:     $\cdots$
14:         **for** $i_n = 0, \ldots, \ell_n(q)$ **do**
15:             *Put $M_0^{(i)} := c_i(q)C_n(p)$, where $c_i(q)$ is given by (20b).*
16:             **for** $r = 1, \ldots, n$ **do**
17:                 $M_r^{(i)} := (W_r^{(i_r)} M_{r-1}^{(i)})^c$.
18:             **end for**
19:             $C_0(pq) := M_n^{(i)} + C_0(pq)$.
20:         **end for**

21:     $\cdots$
22: **end for**
23: *Step 4: Computation of $\mathcal{B}(pq, u)$*
24: *Put $F_0 := C_0(pq)$.*
25: **for** $s = 1, \ldots, n$ **do**
26:     *Compute $F_s := (D'_s F_{s-1})^c$, see (26).*
27: **end for**
28: *Put $\mathcal{B}(pq, u) := F_n$.*
29: *Step 5: End of the algorithm*

Table 1. Number of real arithmetic operations required to obtain $B(pq, u)$ by the first method

Calculation of	Number of additions	Number of multi-plications/divisions
$B(p, u)$ and $B(q, u)$ by the method in Subsect. 3.2	$n\kappa(\kappa+1)^n$	$2n(\kappa+1)^n$
$C(p)$	0	$n(\kappa+1)^n$
$C(pq)$	$(\kappa+1)^n[(\kappa+1)^n - 1] - n\kappa(\kappa+1)^{n-1}$	$(\kappa+1)^{2n}$
$B(pq, u)$	0	$n(2\kappa+1)^n$

Table 2. Number of arithmetic operations to obtain $B(pq, \boldsymbol{u})$ by the second method

Calculation of	Number of additions	Number of multi-plications/divisions
$C(pq)$	$(\kappa+1)^n[(\kappa+1)^n - 1] - n\kappa(\kappa+1)^{n-1}$	$(\kappa+1)^{2n}$
$B(pq, \boldsymbol{u})$ by the method in Subsect. 3.2	$n\kappa(2\kappa+1)^n$	$n(2\kappa+1)^n$

5 Matrix Method for the Computation of the Bernstein Coefficients of a Multivariate Polynomial from Those of One of Its Partial Derivatives

Let p be an n-variate polynomial of degree ℓ with the power representation given as in (2). Assume that its coefficients are arranged in the matrix A, which is presented in (8). We expand p into Bernstein polynomials of degree d, $d \geq \ell$, over the box $\boldsymbol{x}$ (13) as in (3). Without loss of generality, we assume that $d = \ell$. Recall that the Bernstein representation of the first partial derivative of p with respect to x_s, $s \in \{1, \ldots, n\}$, is given by

$$\frac{\partial p}{\partial x_s} = \sum_{i \leq \ell_s, -1} \tilde{b}_i^{(\ell_s, -1)} B_i^{(\ell_s, -1)}(x), \tag{28}$$

where for $j = 0, \ldots, \ell_{s,-1}$

$$\tilde{b}_j^{(\ell_s, -1)} = \ell_s(b_{j_s,1}^{(\ell)} - b_j^{(\ell)}) \tag{29}$$

denotes the j-th Bernstein coefficient of $\frac{\partial p}{\partial x_s}$ of degree $\ell_{s,-1}$ over $\boldsymbol{x}$, i.e., the Bernstein coefficients of $\frac{\partial p}{\partial x_s}$ can be obtained from differences between its successive Bernstein coefficients, e.g., [9, formula (4)].

Assume that the Bernstein coefficients of $\frac{\partial p}{\partial x_s}$, $s \in \{1, \ldots, n\}$, are given and are arranged in the Bernstein patch $B(\frac{\partial p}{\partial x_s}, \boldsymbol{x})$. In the following, we present a matrix method by which the Bernstein patch $B(p, \boldsymbol{x})$ that comprises the Bernstein coefficients of p over $\boldsymbol{x}$ can be computed using $B(\frac{\partial p}{\partial x_s}, \boldsymbol{x})$. Without loss of generality, we assume that $s = 1$. From (29), it follows that

$$b_{i_{1,1}}^{(\ell)} = \frac{\tilde{b}_i^{(\ell_1, -1)}}{\ell_1} + b_i^{(\ell)}, \quad i = 0, \ldots, \ell_{1,-1}. \tag{30}$$

In other words, for computing $B(p, \boldsymbol{x})$ it is sufficient to compute the Bernstein coefficients $b_{i_{[1,0]}}^{(\ell)}$, see (6), then the remaining coefficients can be obtained iteratively using (30). The coefficients $b_{i_{[1,0]}}^{(\ell)}$ are the Bernstein coefficients of p for $x_1 = \underline{x}_1$. By the face value property of the Bernstein coefficients, see Subsect. 3.1, these coefficients are identical to those that are located at the corresponding $(n-1)$-dimensional face of $B(p, \boldsymbol{x})$; they are obtained from $B(p, \boldsymbol{x})$ by firstly freezing p on the face of x with $x_1 = \underline{x}_1$ and then computing the Bernstein coefficients of the resulting polynomial. In matrix language, the computation is as follows. Denote by C_1 the row vector of length ℓ^*, where $\ell^* = \prod_{s=2}^{n} (\ell_s + 1)$, that contains the coefficients of p such that $x_1 = \underline{x}_1$. For $\mu = 1, \ldots, \ell_1$, we define the elementary bidiagonal matrices $H_\mu(x) \in \mathbb{R}^{\mu, \mu+1}$ by

$$(H_\mu(x))_{i,j} := \begin{cases} 1, & i = j, \\ x, & i = \mu, \ j = \mu + 1, \\ 0, & \text{otherwise.} \end{cases} \tag{31}$$

Then, C_1 can be obtained as

$$C_1 = H_1(\underline{x}_1) \cdots H_{\ell_1 - 1}(\underline{x}_1) H_{\ell_1}(\underline{x}_1) A. \tag{32}$$

From C_1, we define the $(\ell_2 + 1) \times \prod_{r=3}^{n} (\ell_r + 1)$ matrix A_1 with coefficients given for $q_1 = 0, \ldots, \ell_2$ and $q_2 = 0, \ldots, \prod_{r=3}^{n} (\ell_r + 1) - 1$ by

$$(A_1)_{q_1, q_2} := (C_1)_{q_1 + q_2(\ell_2 + 1) + 1}. \tag{33}$$

Then the method from Subsect. 3.2 for the computation of the Bernstein coefficients of p on the face of x with $x_1 = \underline{x}_1$ starting from A_1 is applied. We denote the resulting matrix by B_1 and arrange its entries in the row vector C_1' of length ℓ^*, such that for $r_1 = 1, \ldots, \ell^*$, $v_1 = 0, \ldots, \ell_2$, and $v_2 = 0, \ldots, \prod_{r=3}^{n} (\ell_r + 1) - 1$, we have

$$(C_1')_{r_1} = (B_1)_{v_1, v_2}, \tag{34}$$

where

$$r_1 = v_1 + v_2(\ell_2 + 1) + 1.$$

Let B_1' be the $(\ell_1 + 1) \times \ell^*$ matrix defined for $r_1 = 0, \ldots, \ell_1$ and $r_2 = 0, \ldots, \ell^* - 1$ by

$$(B_1')_{r_1, r_2} = \begin{cases} (C_1')_{r_2 + 1}, & \text{if } r_1 = 0, \\ (\frac{1}{\ell_1} \mathcal{B}(\frac{\partial p}{\partial x_s}, x))_{r_1 - 1, r_2}, & \text{if } r_1 = 1, \ldots, \ell_1. \end{cases} \tag{35}$$

For $\mu = 1, \ldots, \ell_1$, we define the square matrices $H_\mu^{(1)}$ of order $\ell_1 + 1$, such that for $r_v = 0, \ldots, \ell_1, v = 1, 2$, it is given by

$$(H_\mu^{(1)})_{r_1, r_2} := \begin{cases} 1, & r_1 = r_2, \\ 1, & r_1 = \ell_1 - \mu + 1, \ r_2 = \ell_1 - \mu, \\ 0, & \text{otherwise.} \end{cases} \tag{36}$$

Then, the Bernstein matrix $\mathcal{B}(p, \boldsymbol{x})$ that comprises the Bernstein coefficients of p of degree ℓ over $\boldsymbol{x}$, can be calculated by

$$\mathcal{B}(p, \boldsymbol{x}) = H_1^{(1)} \cdots H_{\ell_1-1}^{(1)} H_{\ell_1}^{(1)} B_1', \tag{37}$$

where the correspondence between $\mathcal{B}(p, \boldsymbol{x})$ and $B(p, \boldsymbol{x})$ can be determined by using (7). As a consequence of our initial assumption, for computing $B(p, \boldsymbol{x})$ from $B(\frac{\partial p}{\partial x_s}, \boldsymbol{x})$, where $s \in \{2, \ldots, n\}$, we firstly employ the cyclic ordering with respect to x_s, in such a way that the multiplications in (32) and (37) are well defined.

In Table 3, the number of arithmetic operations needed for the computation of $\mathcal{B}(p, \boldsymbol{x})$ is presented. For simplicity, we assume here that $\ell_s = \kappa$, $s = 1, \ldots, n$.

Table 3. Number of real arithmetic operations required to obtain $\mathcal{B}(p, \boldsymbol{x})$ from one of the partial derivatives of p

Calculation of	Number of additions	Number of multiplications/divisions
C_1 by (32)	$\kappa(\kappa+1)^{n-1}$	$\kappa(\kappa+1)^{n-1}$
B_1 over a general box $\boldsymbol{x}'$ which is obtained from $\boldsymbol{x}$ by freezing $x_1 = \underline{x}_1$ using the method in Subsect. 3.2	$(n-1)\kappa(\kappa+1)^{n-1}+n-1$	$3(n-1)(\kappa+1)^{n-1}+2(n-1)(\kappa-1)+n-1$
B_1' by (35)	0	$\kappa(\kappa+1)^{n-1}$
$\mathcal{B}(p, \boldsymbol{x})$ by (37)	$\kappa(\kappa+1)^{n-1}$	0

In total, the computation of $\mathcal{B}(p, \boldsymbol{x})$ requires $(n+1)\kappa(\kappa+1)^{n-1}+n-1$ additions and $2\kappa(\kappa+1)^{n-1} + 3(n-1)(\kappa+1)^{n-1} + 2(n-1)(\kappa-1) + n-1$ multiplications.

Appendix A. Example for the Performance of Algorithm 1

Let p and q be bivariate polynomials of degree $(4, 2)$ and $(3, 2)$, respectively, with Bernstein matrices over $\boldsymbol{u}$

$$\mathcal{B}(p, \boldsymbol{u}) = \begin{bmatrix} 2 & 0 & -1 \\ -2 & 3 & 0 \\ 1 & 3 & -3 \\ 1 & 0 & 1 \\ 0 & -1 & 0 \end{bmatrix} \text{ and } \mathcal{B}(q, \boldsymbol{u}) = \begin{bmatrix} -1 & -4 & -1 \\ 2 & 0 & 5 \\ -2 & 3 & 0 \\ 0 & 1 & 1 \end{bmatrix}.$$

Then

$$D_1 = \mathrm{diag}(1, 4, 6, 4, 1) \text{ and } D_2 = \mathrm{diag}(1, 2, 1),$$

$$C_1(p) = (D_1 C_0(p))^c = \begin{bmatrix} 2 & -8 & 64 & 0 \\ 0 & 12 & 18 & 0 & -1 \\ -1 & 0 & -18 & 4 & 0 \end{bmatrix},$$

$$C_2(p) = (D_2 C_1(p))^c = \begin{bmatrix} 2 & 0 & -1 \\ -8 & 24 & 0 \\ 6 & 36 & -18 \\ 4 & 0 & 4 \\ 0 & -2 & 0 \end{bmatrix};$$

$$M_0^{(1,1)} = M_0^{(2,2)} = M_0^{(3,0)} = O_{5,3},$$

$$M_0^{(0,0)} = \begin{bmatrix} -2 & 0 & 1 \\ 8 & -24 & 0 \\ -6 & -36 & 18 \\ -4 & 0 & -4 \\ 0 & 2 & 0 \end{bmatrix}, M_0^{(0,1)} = \begin{bmatrix} -16 & 0 & 8 \\ 64 & -192 & 0 \\ -48 & -288 & 144 \\ -32 & 0 & -32 \\ 0 & 16 & 0 \end{bmatrix}, M_0^{(0,2)} = \begin{bmatrix} -2 & 0 & 1 \\ 8 & -24 & 0 \\ -6 & -36 & 18 \\ -4 & 0 & -4 \\ 0 & 2 & 0 \end{bmatrix},$$

$$M_0^{(1,0)} = \begin{bmatrix} 12 & 0 & -6 \\ -48 & 144 & 0 \\ 36 & 216 & -108 \\ 24 & 0 & 24 \\ 0 & -12 & 0 \end{bmatrix}, \qquad M_0^{(1,2)} = \begin{bmatrix} 30 & 0 & -15 \\ -120 & 360 & 0 \\ 90 & 540 & -270 \\ 60 & 0 & 60 \\ 0 & -30 & 0 \end{bmatrix},$$

$$M_0^{(2,0)} = \begin{bmatrix} -12 & 0 & 6 \\ 48 & -144 & 0 \\ -36 & -216 & 108 \\ -24 & 0 & -24 \\ 0 & 12 & 0 \end{bmatrix}, \qquad M_0^{(2,1)} = \begin{bmatrix} 36 & 0 & -18 \\ -144 & 432 & 0 \\ 108 & 648 & -324 \\ 72 & 0 & 72 \\ 0 & -36 & 0 \end{bmatrix},$$

$$M_0^{(3,1)} = \begin{bmatrix} 4 & 0 & -2 \\ -16 & 48 & 0 \\ 12 & 72 & -36 \\ 8 & 0 & 8 \\ 0 & -4 & 0 \end{bmatrix}, \qquad M_0^{(3,2)} = \begin{bmatrix} 2 & 0 & -1 \\ -8 & 24 & 0 \\ 6 & 36 & -18 \\ 4 & 0 & 4 \\ 0 & -2 & 0 \end{bmatrix};$$

$$M_1^{(0,0)} = \begin{bmatrix} -2 & 8 & -6 & -4 & 0 & 0 & 0 & 0 \\ 0 & -24 & -36 & 0 & 2 & 0 & 0 & 0 \\ 1 & 0 & 18 & -4 & 0 & 0 & 0 & 0 \end{bmatrix}, \qquad M_2^{(0,0)} = \begin{bmatrix} -2 & 0 & 1 & 0 & 0 \\ 8 & -24 & 0 & 0 & 0 \\ -6 & -36 & 18 & 0 & 0 \\ -4 & 0 & -4 & 0 & 0 \\ 0 & 2 & 0 & 0 & 0 \\ 0 & 0 & 0 & 0 & 0 \\ 0 & 0 & 0 & 0 & 0 \\ 0 & 0 & 0 & 0 & 0 \end{bmatrix},$$

$$M_1^{(0,1)} = \begin{bmatrix} -16 & 64 & -48 & -32 & 0 & 0 & 0 & 0 \\ 0 & -192 & -288 & 0 & 16 & 0 & 0 & 0 \\ 8 & 0 & 144 & -32 & 0 & 0 & 0 & 0 \end{bmatrix}, \qquad M_2^{(0,1)} = \begin{bmatrix} 0 & -16 & 0 & 8 & 0 \\ 0 & 64 & -192 & 0 & 0 \\ 0 & -48 & -288 & 144 & 0 \\ 0 & -32 & 0 & -32 & 0 \\ 0 & 0 & 16 & 0 & 0 \\ 0 & 0 & 0 & 0 & 0 \\ 0 & 0 & 0 & 0 & 0 \\ 0 & 0 & 0 & 0 & 0 \end{bmatrix},$$

$$M_1^{(0,2)} = \begin{bmatrix} -2 & 8 & -6 & -4 & 0 & 0 & 0 & 0 \\ 0 & -24 & -36 & 0 & 2 & 0 & 0 & 0 \\ 1 & 0 & 18 & -4 & 0 & 0 & 0 & 0 \end{bmatrix}, \qquad M_2^{(0,2)} = \begin{bmatrix} 0 & 0 & -2 & 0 & 1 \\ 0 & 0 & 8 & -24 & 0 \\ 0 & 0 & -6 & -36 & 18 \\ 0 & 0 & -4 & 0 & -4 \\ 0 & 0 & 0 & 2 & 0 \\ 0 & 0 & 0 & 0 & 0 \\ 0 & 0 & 0 & 0 & 0 \\ 0 & 0 & 0 & 0 & 0 \end{bmatrix}.$$

The remaining 18 matrices $M_s^{(i)}$, $s = 1, 2$, are formed analogously.

$$F_0 = C_0(pq) = \begin{bmatrix} -2 & -16 & -1 & 8 & 1 \\ 20 & 40 & -160 & -24 & -15 \\ -66 & 96 & -390 & 450 & 18 \\ 80 & -100 & 408 & 506 & -275 \\ -12 & -122 & 896 & -298 & 60 \\ -24 & 72 & 54 & 42 & -18 \\ 0 & 20 & -32 & 8 & 4 \\ 0 & 0 & 4 & -2 & 0 \end{bmatrix};$$

$$D_1' = \mathrm{diag}\left(1, \frac{1}{7}, \frac{1}{21}, \frac{1}{35}, \frac{1}{35}, \frac{1}{21}, \frac{1}{7}, 1\right), \ D_2' = \mathrm{diag}\left(1, \frac{1}{4}, \frac{1}{6}, \frac{1}{4}, 1\right).$$

The Bernstein matrix of the product of the polynomials p and q is given by

$$\mathcal{B}(pq, \boldsymbol{u}) = F_2 = (D_2'(D_1'F_0)^c)^c = \begin{bmatrix} -2 & -4 & \frac{-1}{6} & 2 & 1 \\ \frac{20}{7} & \frac{10}{7} & \frac{-80}{21} & \frac{-6}{7} & \frac{-15}{7} \\ \frac{-22}{7} & \frac{8}{7} & \frac{-65}{21} & \frac{75}{14} & \frac{6}{7} \\ \frac{16}{7} & \frac{-5}{7} & \frac{68}{35} & \frac{253}{70} & \frac{-55}{7} \\ \frac{-12}{35} & \frac{-61}{70} & \frac{64}{15} & \frac{-149}{70} & \frac{12}{7} \\ \frac{-8}{7} & \frac{6}{7} & \frac{3}{7} & \frac{1}{2} & \frac{-6}{7} \\ 0 & \frac{5}{7} & \frac{-16}{21} & \frac{2}{7} & \frac{4}{7} \\ 0 & 0 & \frac{2}{3} & \frac{-1}{2} & 0 \end{bmatrix}.$$

Appendix B. Example for the Performance of the Method Presented in Sect. 5

Let $p(x_1, x_2) = -504x_1^4x_2^2 - 84x_1^4x_2 + 288x_1^4 + 6x_1^3x_2^2 + 30x_1^3x_2 - 60x_1^3 + 36x_1^2x_2^2 - 20x_1^2x_2 + 28x_1^2 - 54x_1x_2^2 + 21x_1x_2 - 24x_1 + 24x_2^2 - 24x_2 + 48.$ Then

$$\mathcal{B}\left(\frac{\partial p}{\partial x_1}, \boldsymbol{u}\right) = \begin{bmatrix} -24 & \frac{-27}{2} & -57 \\ -4 & \frac{-9}{6} & \frac{-83}{3} \\ \frac{-140}{3} & \frac{-207}{6} & \frac{-67}{6} \\ 1004 & \frac{-1743}{2} & -1241 \end{bmatrix} \quad \text{and} \quad C_1' = \mathcal{B}(p(0, x_2), [0, 1]) = \begin{bmatrix} 48 & 36 & 48 \end{bmatrix};$$

$$B_1' = \begin{bmatrix} 48 & 36 & 48 \\ -6 & \frac{-27}{8} & \frac{-57}{4} \\ \frac{-4}{3} & \frac{-9}{24} & \frac{-83}{12} \\ \frac{-140}{12} & \frac{-207}{24} & \frac{-67}{12} \\ 251 & \frac{-1743}{8} & \frac{-1241}{4} \end{bmatrix}.$$

The Bernstein coefficients of p over $\boldsymbol{u}$ are obtained from

$$\mathcal{B}(p, \boldsymbol{u}) = H_1 H_2 H_3 H_4 B_1'$$

$$= \begin{bmatrix} 1 & 0 & 0 & 0 & 0 \\ 0 & 1 & 0 & 0 & 0 \\ 0 & 0 & 1 & 0 & 0 \\ 0 & 0 & 0 & 1 & 0 \\ 0 & 0 & 0 & 1 & 1 \end{bmatrix} \begin{bmatrix} 1 & 0 & 0 & 0 & 0 \\ 0 & 1 & 0 & 0 & 0 \\ 0 & 0 & 1 & 0 & 0 \\ 0 & 0 & 1 & 1 & 0 \\ 0 & 0 & 0 & 0 & 1 \end{bmatrix} \begin{bmatrix} 1 & 0 & 0 & 0 & 0 \\ 0 & 1 & 0 & 0 & 0 \\ 0 & 1 & 1 & 0 & 0 \\ 0 & 0 & 0 & 1 & 0 \\ 0 & 0 & 0 & 0 & 1 \end{bmatrix} \begin{bmatrix} 1 & 0 & 0 & 0 & 0 \\ 1 & 1 & 0 & 0 & 0 \\ 0 & 0 & 1 & 0 & 0 \\ 0 & 0 & 0 & 1 & 0 \\ 0 & 0 & 0 & 0 & 1 \end{bmatrix} \begin{bmatrix} 48 & 36 & 48 \\ -6 & \frac{-27}{8} & \frac{-57}{4} \\ \frac{-4}{3} & \frac{-9}{24} & \frac{-83}{12} \\ \frac{-140}{12} & \frac{-207}{24} & \frac{-67}{12} \\ 251 & \frac{-1743}{8} & \frac{-1241}{4} \end{bmatrix}$$

$$= \begin{bmatrix} 48 & 36 & 48 \\ 42 & \frac{261}{8} & \frac{135}{4} \\ \frac{122}{3} & \frac{774}{24} & \frac{322}{12} \\ 29 & \frac{567}{24} & \frac{255}{12} \\ 280 & \frac{483}{2} & -289 \end{bmatrix}.$$

References

1. Alonso, P., Delgado, J., Gallego, R., Peña, J.M.: Conditioning and accurate computations with Pascal matrices. J. Comput. Appl. Math. **252**, 21–26 (2013)
2. Clauss, P., Chupaeva, I.Y.: Application of symbolic approach to the Bernstein expansion for program analysis and optimization. In: Duesterwald, E. (ed.) Compiler Construction. LNCS, vol. 2985, pp. 120–133. Springer, Berlin, Heidelberg (2004)
3. Clauss, P., Fernández, F.J., Garbervetsky, D., Verdoolaege, S.: Symbolic polynomial maximization over convex sets and its application to memory requirement estimation. IEEE Trans. Very Large Scale Integr. VLSI Syst. **17**(8), 983–996 (2009)
4. Dang, T., Dreossi, T., Fanchon, É., Maler, O., Piazza, C., Rocca, A.: Set-based analysis for biological modelling. In: Liò, P., Zuliani, P. (eds.) Automated Reasoning for Systems Biology and Medicine. COBO, vol. 30, pp. 157–189. Springer, Cham (2019). https://doi.org/10.1007/978-3-030-17297-8
5. Dreossi, T.: Sapo: Reachability computation and parameter synthesis of polynomial dynamical systems. In: Proceedings of International Conference Hybrid Systems: Computation and Control, pp. 29–34, ACM, New York (2017)
6. Farouki, R.T.: The Bernstein polynomial basis: A centennial retrospective. Comput. Aided Geom. Design **29**, 379–419 (2012)
7. Farouki, R.T., Rajan, V.T.: Algorithms for polynomials in Bernstein form. Comput. Aided Geom. Design **5**, 1–26 (1988)
8. Garloff, J.: Convergent bounds for the range of multivariate polynomials. In: Nickel, K. (ed.) Interval Mathematics 1985. LNCS, vol. 212, pp. 37–56. Springer, Heidelberg (1986). https://doi.org/10.1007/3-540-16437-5_5
9. Garloff, J., Smith, A.P.: Solution of systems of polynomial equations by using Bernstein expansion. In: Alefeld, G., Rump, S., Rohn, J., and Yamamoto J. (eds.), Symbolic Algebraic Methods and Verification Methods, pp. 87–97. Springer (2001)
10. Rivlin, T.J.: Bounds on a polynomial. J. Res. Nat. Bur. Standards **74**(B), 47–54 (1970)
11. Rump, S.M.: INTLAB-INTerval LABoratory. In: Csendes, T. (ed.) Developments in Reliable Computing, pp. 77–104. Kluwer Academic Publishers, Dordrecht (1999)
12. Smith, A.P.: Fast construction of constant bound functions for sparse polynomials. J. Global Optim. **43**, 445–458 (2009)
13. Titi, J.: Matrix methods for the tensorial and simplicial Bernstein forms with application to global optimization, dissertation. University of Konstanz, Konstanz, Germany (2019). Available at https://nbn-resolving.de/urn:nbn:de:bsz:352-2-k106crqmste71
14. Titi, J., Garloff, J.: Fast determination of the tensorial and simplicial Bernstein forms of multivariate polynomials and rational functions. Reliab. Comput. **25**, 24–37 (2017)
15. Titi, J., Garloff, J.: Matrix methods for the simplicial Bernstein representation and for the evaluation of multivariate polynomials. Appl. Math. Comput. **315**, 246–258 (2017)
16. Titi, J., Garloff, J.: Matrix methods for the tensorial Bernstein form. Appl. Math. Comput. **346**, 254–271 (2019)
17. Titi, J., Garloff, J.: Bounds for the range of a complex polynomial over a rectangular region, submitted (2020)

Comparative Study of the Accuracy of Higher-Order Difference Schemes for Molecular Dynamics Problems Using the Computer Algebra Means

Evgenii V. Vorozhtsov[1]([✉]) and Sergey P. Kiselev[1,2]

[1] Khristianovich Institute of Theoretical and Applied Mechanics of the Siberian Branch of the Russian Academy of Sciences, Novosibirsk 630090, Russia
{vorozh,kiselev}@itam.nsc.ru
[2] Novosibirsk State Technical University, Novosibirsk 630092, Russia

Abstract. The Runge–Kutta–Nyström (RKN) explicit symplectic difference schemes with the number of stages from 1 to 5 for the numerical solution of molecular dynamics problems described by the systems with separable Hamiltonians have been considered. All schemes have been compared in terms of the accuracy and stability with the use of Gröbner bases. For each specific number of stages, the schemes are found, which are the best in terms of accuracy and stability. The efficiency parameter of RKN schemes has been introduced by analogy with the efficiency parameter for Runge–Kutta schemes and the values of this parameter have been computed for all considered schemes. The verification of schemes has been done by solving a problem having the exact solution. It has been shown that the symplectic five-stage RKN scheme ensures a more accurate conservation of the total energy of a system of particles than the schemes of lower accuracy orders. All investigations of the accuracy and stability of schemes have been carried out in the analytic form with the aid of the computer algebra system (CAS) *Mathematica*.

Keywords: Molecular dynamics · Hamilton equations · Symplectic difference schemes · Gröbner bases · CAS *Mathematica*

1 Introduction

The investigation of the behavior of materials under their shockwave loading by the molecular dynamics (MD) methods is at present one of the topical directions of the research in solid mechanics. The essence of the MD method lies in the solution of the equations of the motion of atoms, which interact via a potential depending on the coordinates of atoms. At the use of the given method, there is no need in formulating the equations of state. As is known, the obtaining of these equations is one of the most complex problems of fluid mechanics [1].

© Springer Nature Switzerland AG 2020
F. Boulier et al. (Eds.): CASC 2020, LNCS 12291, pp. 600–620, 2020.
https://doi.org/10.1007/978-3-030-60026-6_35

It was shown in [2] that in the limit as the number of particles within a volume tends to infinity, the MD equations go over to the well-known equations of the continuum mechanics.

The molecular dynamics equations are represented by the Hamilton's ordinary differential equations for the solid body atoms. The MD equations have an exact analytic solution in a very limited number of cases [3]. These equations are, therefore, solved in the general case numerically with the aid of difference schemes in which the differential operator has been replaced with a difference operator.

At the solution of Hamilton's equations, it is natural to use the difference schemes preserving the symplectic properties of these equations. The violation of this condition leads to the violation of the conservation of Poincaré invariants and the rise of the non-physical instability in numerical computations [4]. It follows from here that the difference operator of the numerical scheme must possess the properties of the canonical transformation. Symplectic difference schemes are derived by the operator technique [5–8] and by the Runge–Kutta–Nyström method [7,9–11].

As is known, the explicit difference schemes impose a restriction on the integration step [2,12]. On the other hand, the advantage of explicit schemes is their simple computer implementation. Besides, the increased speed of desktop computers enables one to solve with the aid of explicit schemes many important applied tasks with acceptable CPU time expenses. Therefore, in the present work, preference is given to explicit difference schemes.

According to the theory of Hamilton equations, the conservation law for the total energy of the system of particles must be satisfied [3]. It is natural to require that the difference scheme also ensures the total energy conservation. However, as the practice of computations shows, the imbalance of the system total energy proves more considerable for the explicit symplectic Runge–Kutta–Nyström (RKN) schemes of low accuracy orders (the second and third orders). At the same time, it was shown in [12] that the three-stage fourth-order RKN scheme ensures a smaller error in the energy imbalance than the schemes of orders 2 and 3. From this a conclusion follows about the advisability of the development of explicit symplectic RKN schemes of higher orders of accuracy. As was shown in [12], the derivation of symplectic three-stage RKN schemes is associated with a large amount of symbolic computations.

2 Governing Equations

In the method of molecular dynamics, the computation of the motion of N particles is carried out with the aid of the Hamilton equations

$$\frac{dx_{i\alpha}}{dt} = \frac{\partial H}{\partial p_{i\alpha}}, \frac{dP_{i\alpha}}{dt} = -\frac{\partial H}{\partial x_{i\alpha}}, H(x_{i\alpha}, p_{i\alpha}) = K(p_{i\alpha}) + V(x_{i\alpha}),$$
$$K(p_{i\alpha}) = \sum_{i=1}^{N} \sum_{\alpha=1}^{3} \frac{p_{i\alpha}^2}{2m_i}, \tag{1}$$

where i is the particle number, α is the number of the coordinate $x_{i\alpha}$ and of the momentum $p_{i\alpha}$, m_i is the particle mass, $K(p_{i\alpha})$ is the kinetic energy, $V(x_{i\alpha})$ is the potential energy of the interaction of particles, $H(x_{i\alpha}, p_{i\alpha})$ is the Hamiltonian of the system of particles. The solution of the system of Eq. (1) under the given initial conditions $x_{i\alpha}(t = 0) = x_{i\alpha}^0$, $p_{i\alpha}(t = 0) = p_{i\alpha}^0$ represents a canonical transformation from the initial state to the final state

$$x_{i\alpha} = x_{i\alpha}(x_{i\alpha}^0, p_{i\alpha}^0, t), \quad p_{i\alpha} = p_{i\alpha}(x_{i\alpha}^0, p_{i\alpha}^0, t). \tag{2}$$

The solution (2) of Hamilton Eq. (1) preserves the phase volume (the Liouville theorem [3]). The condition of the phase volume conservation is [2]

$$G^T J G = J, \quad G = \frac{\partial(x_{i\alpha}, p_{i\alpha})}{\partial(x_{i\alpha}^0, p_{i\alpha}^0)}, \quad J = \left\| \begin{matrix} 0 & I_N \\ -I_N & 0 \end{matrix} \right\|, \tag{3}$$

where G is the Jacobi matrix, J is the symplectic matrix, I_N is the $N \times N$ identity matrix. From (3), it follows the equality to unity of the transformation Jacobian $|G| = 1$. For the following, we rewrite Hamilton Eq. (1) for the one-dimensional case in the form

$$dx_i/dt = p_i(t)/m, \quad dp_i/dt = f_i(x_i), \tag{4}$$

where $f_i(x_i)$ is the force acting on the ith particle, $f_i(x_i) = -\partial V(x_i)/\partial x_i$, $i = 1, 2, \ldots, N$. In the following, we will omit the subscript i at the discussion of difference schemes for solving the system of ordinary differential Eq. (4).

3   Runge–Kutta–Nyström Symplectic Difference Schemes

The conventional (non-symplectic) explicit difference schemes with a structure similar to Runge–Kutta schemes were proposed for the first time by Nyström in [9]. The K-stage Runge–Kutta–Nyström (RKN) scheme for Hamilton Eq. (4) has the following form:

$$x^{(i)} = x^n + h\alpha_i \frac{p^n}{m} + \frac{h^2}{m} \sum_{j=1}^{K} a_{ij} f(x^{(j)}), \quad i = 1, \ldots, K,$$

$$x^{n+1} = x^n + h\frac{p^n}{m} + \frac{h^2}{m} \sum_{j=1}^{K} \beta_j f(x^{(j)}), \quad p^{n+1} = p^n + h \sum_{j=1}^{K} \gamma_j f(x^{(j)}), \tag{5}$$

where h is the time step, n is the time layer number, $n = 0, 1, 2, \ldots$; $\alpha_i, \beta_i, \gamma_i$, $i = 1, \ldots, K$ are constant parameters, $K \geq 1$.

It is required that the RKN scheme (5) performs a canonical transformation $(x^n, p^n) \rightarrow (x^{n+1}, p^{n+1})$ at a passage from the time layer n to the layer $n + 1$. To this end, one must impose in accordance with (3) the following condition on the Jacobi matrix G^{n+1}:

$$G^{n+1,T} J G^{n+1} = J, \quad G^{n+1} = \frac{\partial(x^{n+1}, p^{n+1})}{\partial(x^n, p^n)}, \quad J = \begin{pmatrix} 0 & 1 \\ -1 & 0 \end{pmatrix}, \tag{6}$$

where the superscript T denotes the transposition operation, J is the symplectic matrix. Condition (6) gives rise to a class of explicit two-parameter RKN(α, γ) schemes for which β_i, a_{ij} in (5) satisfy the conditions [11]

$$\beta_i = \gamma_i(1 - \alpha_i), \quad a_{ij} = \begin{cases} 0, & 1 \leq i \leq j \leq K \\ \gamma_j(\alpha_i - \alpha_j), & 1 \leq j < i \leq K \end{cases}. \quad (7)$$

It was noted in [11] that there exist no explicit Runge–Kutta schemes preserving the canonicity of transformation (6).

Verlet [10] proposed a one-stage second-order RKN scheme for system (4). Ruth [5] was the first to show that the Verlet scheme is symplectic (canonical) and discovered three-stage canonical RKN method of order three. In the work [7], an explicit three-stage symplectic RKN method of order four was derived. The analytic expressions were obtained in [12] for the coefficients $\alpha_i, \beta_i, \gamma_i$ of this method with the aid of symbolic computations in the CAS Maple 12 and the technique of Gröbner bases.

We now describe a technique for determining the accuracy order of any RKN scheme by the example of the RKN scheme for computing the momentum p^{n+1} at the moment of time $t_{n+1} = t_n + h$. Let the value p^n be known. The solution in the next node t_{n+1} is calculated by the formula $p^{n+1} = p^n + \Delta p_{h,n}$. The formula for computing $\Delta p_{h,n}$ depends on the number of stages K of the RKN method under consideration and on $3K$ constants $\alpha_i, \beta_i, \gamma_i, i = 1, \ldots, K$. On the other hand, one can easily derive the "exact" formula for the increment Δp by using the expansion of the quantity p^n into the truncated Taylor series:

$$\Delta p_n = p(t_n + h) - p(t_n) \approx \sum_{j=1}^{N_T} \frac{h^j}{j!} \frac{d^j p(t_n)}{dt^j},$$

where N_T is a given natural number, $N_T \geq K + 1$. If the difference $\delta p_n = \Delta p_n - \Delta p_{h,n}$ satisfies the relation $\delta p_n / h = O(h^q)$, where $q > 0$, then the RKN scheme has the order of accuracy $O(h^q)$. The maximization of the degree q is done by choosing the parameters $\alpha_i, \beta_i, \gamma_i$ $(i = 1, \ldots, K)$ for each specific K.

3.1 One-Stage RKN Scheme

Let us set $K = 1$ in (5) and determine by means of symbolic computations the highest possible order of accuracy of the given scheme (the Verlet scheme) as applied to the computation of the momentum p^{n+1} by varying the coefficients α_1 and γ_1. Before presenting the corresponding fragments of a program in the language of the CAS *Mathematica*, let us elucidate the meaning of the notations used in this program: ntayl = N_T, tn = t_n, pnew = p^{n+1}, u[t] = $\dot{x}(t)$, dp = Δp_n, dph = $\Delta p_{h,n}$, errp = δp_n, a1 = α_1, g1 = γ_1.

One computes at first the "exact" expansion Δp_n:

```
pnew= Normal[Series[p[t], {t,tn,ntayl}]] /.t -> tn + h; dp= pnew- p[tn];
```

These operations yield the following expression for Δp_n:

```
h p'[tn] + 1/2 h^2 p"[tn] + 1/6 h^3 p^{(3)}[tn]
```

To facilitate the collection of terms of similar structure in the expression for δp_n, it is useful to carry out in the expression obtained above a number of sufficiently obvious transformations using the Hamilton equations (4): $p'(t_n) = f(x(t))$, $p''(t_n) = f'(x(t))u(t)$, $p^{(3)}(t_n) = u^2(t)f''(x(t)) + f'(x(t))u'(t)$.

These transformations are implemented efficiently in the CAS *Mathematica*:

```
dp = dp/.{p'[tn] -> f[x[t]], p"[tn] -> f'[x[t]] u[t],
  p^{(3)}[tn] -> u[t]^2 f"[x[t]] + f'[x[t]] u'[t]}
```

According to Hamilton Eq. (4), the quantity $x^{(1)}$ will be needed for computing p^{n+1}. It was computed in our *Mathematica* program in the symbolic form as follows: `x1[t_] := x[t] + h*a1*u[t]`. After that, the quantity $\Delta p_{h,n}$ was computed in symbolic form as follows:

```
ftayl1 = Normal[Series[f[y],{y,y0,ntayl}]]; ftayl1 = ftayl1/.{y-> x1[t],
y0-> x[t]}; dph = h*g1*ftayl1;
```

The sought quantity δp_n is computed as follows: `errp = Simplify[dp - dph]`. As a result, the following expression has been obtained for the error δp_n:

$$\delta p_n = hP_1f(x) + (h^2/2)P_2u(t)f'(x) + (h^3/6)\big(f(x)f'(x)/m + P_3u^2f''(x)\big). \quad (8)$$

Here $P_1 = 1 - \gamma_1$, $P_2 = 1 - 2\alpha_1\gamma_1$, $P_3 = 1 - 3\alpha_1^2\gamma_1$. It follows from these formulas that in order to ensure the second order of accuracy of the Verlet scheme it is necessary to choose the parameters α_1 and γ_1 in such a way that $P_1 = 0$, $P_2 = 0$. We find from these conditions that $\gamma_1 = 1$, $\alpha_1 = 1/2$. At these values of parameters, the value of the polynomial P_3 is different from zero: $P_3 = 1/4$.

3.2   Two-Stage RKN Scheme

In the case under consideration, one must set $K = 2$ in (5). Doing symbolic computations similarly to the case of the one-stage scheme, we obtain the expression for δp_n of the following form:

$$\delta p_n = hP_1f(x) + \frac{h^2}{2}P_2u(t)f'(x) + \frac{h^3}{6}\big(P_{31}f(x)f'(x)/m + P_{32}u^2f''(x)\big), \quad (9)$$

where

$$P_1 = 1 - \sum_{j=1}^{K}\gamma_j, \; P_2 = 1 - 2\sum_{j=1}^{K}\alpha_j\gamma_j, \; P_{31} = 1 - 6\sum_{i=1}^{K}\sum_{j<i}\gamma_i\gamma_j(\alpha_i - \alpha_j),$$
$$P_{32} = 1 - 3\sum_{j=1}^{K}\alpha_j^2\gamma_j. \quad (10)$$

The system of four nonlinear algebraic equations $P_1 = 0$, $P_2 = 0$, $P_{31} = 0$, $P_{32} = 0$ gives the following two solutions for parameters $\alpha_1, \alpha_2, \gamma_1, \gamma_2$ (they were found with the aid of the *Mathematica* function `Solve[...]`):

$$\alpha_1 = (3 \pm i\sqrt{3})/12, \; \alpha_2 = (9 \pm i\sqrt{3})/12, \; \gamma_1 = (3 \pm i\sqrt{3})/6, \; \gamma_2 = (3 \mp i\sqrt{3})/6.$$

This means that in the given case, there are no real third-order schemes. The selection of parameters $\alpha_1, \alpha_2, \gamma_1, \gamma_2$ from the conditions $P_1 = 0$, $P_2 = 0$ ensures the second order of accuracy of the RKN scheme under study. These two equations are linear in γ_1 and γ_2. Let us write them in the form of the system $\mathbf{V} \cdot \mathbf{X} = \mathbf{f}$, where $\mathbf{V}$ is the Vandermonde matrix:

$$\mathbf{V} = \begin{pmatrix} 1 & 1 \\ \alpha_1 & \alpha_2 \end{pmatrix}, \quad \mathbf{X} = \begin{pmatrix} \gamma_1 \\ \gamma_2 \end{pmatrix}, \quad \mathbf{f} = \begin{pmatrix} 1 \\ \frac{1}{2} \end{pmatrix}. \tag{11}$$

We consider at first the case when the determinant $\mathrm{Det}\,\mathbf{V} = \alpha_2 - \alpha_1 = 0$. In this case, we find from (11) the one-parameter solution in the form $\gamma_2 = 1 - \gamma_1$, $\alpha_1 = \alpha_2 = 1/2$. Besides, we obtain from (10): $P_{31} = 1$, $P_{32} = 1/4$ so that $144 \cdot (P_{31}^2 \mid P_{32}^2) = 144 \cdot \frac{17}{16} = 153$.

Now consider the case when $\alpha_1 \neq \alpha_2$. In this case, we obtain from the conditions $P_1 = 0$, $P_2 = 0$ the following two-parameter solution ensuring the second order of accuracy of the two-stage RKN scheme:

$$\gamma_1 = (1 - 2\alpha_2)/[2(\alpha_1 - \alpha_2)], \quad \gamma_2 = (2\alpha_1 - 1)/[2(\alpha_1 - \alpha_2)]. \tag{12}$$

In the theory of conventional (non-symplectic) multistage Runge–Kutta schemes, it is a usual practice to search for such scheme parameters (in the case under consideration, the parameters α_1, α_2), which ensure the minimum of error terms, which have in the given case the order of smallness $O(h^3)$ [13]. Since the both polynomials P_{31} and P_{32} depend on the parameters α_1, α_2, it makes sense to introduce the following quadratic functional:

$$F(\alpha_1, \alpha_2) = 144(P_{31}^2 + P_{32}^2) = \frac{(\alpha_1(8-12\alpha_2)+4\alpha_2-3)^2}{(\alpha_1-\alpha_2)^2} + (\alpha_1(6\alpha_2 - 3) + 2 - 3\alpha_2)^2. \tag{13}$$

This expression is obtained as a result of substituting formulas (12) in P_{31} and P_{32}. At the point of the minimum of the function $F(\alpha_1, \alpha_2)$, the equations $\partial F(\alpha_1, \alpha_2)/\partial \alpha_l = 0$, $l = 1, 2$, must be satisfied. They lead to the following two polynomial equations:

$$\begin{aligned} Q_1 = {} & -3 + 8\alpha_1 - 2\alpha_1^3 + 3\alpha_1^4 + 16\alpha_2 - 44\alpha_1\alpha_2 + 6\alpha_1^2\alpha_2 - 2\alpha_1^3\alpha_2 - 12\alpha_1^4\alpha_2 - 28\alpha_2^2 \\ & + 74\alpha_1\alpha_2^2 - 12\alpha_1^2\alpha_2^2 + 30\alpha_1^3\alpha_2^2 + 12\alpha_1^4\alpha_2^2 + 18\alpha_2^3 - 30\alpha_1\alpha_2^3 - 18\alpha_1^2\alpha_2^3 - 36\alpha_1^3\alpha_2^3 \\ & - 7\alpha_2^4 - 6\alpha_1\alpha_2^4 + 36\alpha_1^2\alpha_2^4 + 6\alpha_2^5 - 12\alpha_1\alpha_2^5, \\ Q_2 = {} & 3 - 20\alpha_1 + 44\alpha_1^2 - 34\alpha_1^3 + 7\alpha_1^4 - 6\alpha_1^5 - 4\alpha_2 + 28\alpha_1\alpha_2 - 58\alpha_1^2\alpha_2 + 30\alpha_1^3\alpha_2 \\ & + 6\alpha_1^4\alpha_2 + 12\alpha_1^5\alpha_2 - 6\alpha_1\alpha_2^2 + 12\alpha_1^2\alpha_2^2 + 18\alpha_1^3\alpha_2^2 - 36\alpha_1^4\alpha_2^2 + 2\alpha_2^3 + 2\alpha_1\alpha_2^3 \\ & - 30\alpha_1^2\alpha_2^3 + 36\alpha_1^3\alpha_2^3 - 3\alpha_2^4 + 12\alpha_1\alpha_2^4 - 12\alpha_1^2\alpha_2^4. \end{aligned}$$

The solution of this system has been found with the aid of Gröbner bases. To this end, we have used the built-in function of the CAS $Mathematica$: `GroebnerBasis` `[{Q1,Q2},{a1,a2}]`. Here `a1`$= \alpha_1$, `a2`$= \alpha_2$. The Gröbner basis consists of four polynomials. The first three polynomials are reducible as this was found with the aid of the function `Factor[...]`:

$$G_1 = (-1 + 2\alpha_2)^9 \, (15 + 2\alpha_2) \left(7 - 18\,\alpha_2 + 12\,\alpha_2{}^2\right) \left(-3 + 7\,\alpha_2 - 9\,\alpha_2{}^2 + 6\,\alpha_2{}^3\right), \tag{14}$$

$$
\begin{aligned}
G_2 = &-((-1 + 2\alpha_2)^2 \, (-782292886326335781 + 84791412129792\alpha_1 + 13518496830811327517\,\alpha_2 \\
&- 105687991753500756247\alpha_2^2 + 494822074007590215378\,\alpha_2{}^3 - 1547761103857501203524\,\alpha_2{}^4 \\
&+ 3413403141608874304288\alpha_2^5 - 5445027180463857323808\alpha_2^6 + 6305466223138181638976\alpha_2^7 \\
&- 5183080249993946906752\alpha_2^8 + 2828691383728661777664\alpha_2^9 - 855368209356906329856\alpha_2^{10} \\
&+ 51213785868624923136\,\alpha_2{}^{11} + 30677230391300103168\,\alpha_2{}^{12})), \tag{15}
\end{aligned}
$$

$$
\begin{aligned}
G_3 = &(-1 + 2\,\alpha_2) \, (138375335965152699 + 15898389774336\,\alpha_1 - 31796779548672\,\alpha_1{}^2 \\
&+ 21197853032448\,\alpha_1{}^3 - 2643993979226915641\,\alpha_2 + 23052596110039882255\,\alpha_2{}^2 \\
&- 121523546615414976256\,\alpha_2{}^3 + 432612502387510859032\,\alpha_2{}^4 - 1099597565438948027224\,\alpha_2{}^5 \\
&+ 2054466563803837779616\alpha_2^6 - 2852904523255219553920\alpha_2^7 + 2925159692595283888384\,\alpha_2{}^8 \\
&- 2147987507024103404032\alpha_2^9 + 1048555229436013268736\alpha_2^{10} - 279638949859173402624\alpha_2^{11} \\
&+ 10628391427610609664\alpha_2^{12} + 9693762746554238976\alpha_2^{13}), \\
G_4 = &-6378840443102352741 + 105989265162240\,\alpha_1 - 402759207616512\,\alpha_1{}^2 \\
&+ 423957060648960\alpha_1^3 - 42395706064896\alpha_1^4 + 1346450836710043903850\alpha_2 \\
&+ 339165648519168\alpha_1\alpha_2 - 339165648519168\alpha_1^2\alpha_2 - 13065533714715402278070\alpha_2^2 \\
&+ 7728385463568967182882\alpha_2^3 - 31152292224976113655272\alpha_2^4 + 90595546230140582933880\alpha_2^5 \\
&- 196140611828089238849712\alpha_2^6 + 321032661222529636238400\alpha_2^7 \\
&- 398020598628892865210880\alpha_2^8 + 368863060403416970198016\alpha_2^9 \\
&- 246489793612098299292928\alpha_2^{10} + 109621999423243454814720\alpha_2^{11} \\
&- 26287557235123544745984\alpha_2^{12} + 533712029766168066048\alpha_2^{13} \\
&+ 894285199654734163968\alpha_2^{14}. \tag{16}
\end{aligned}
$$

Equation $G_1 = 0$ has 15 solutions in total, counted with multiplicities. Equation $-1 + 2\alpha_2 = 0$ yields the root $\alpha_2 = 1/2$. Substituting this value in (16), we obtain: $G_4 = -2649731629056(-17 + 2\alpha_1)(-1 + 2\alpha_1)^3$. We find from here that the following two α_1 roots correspond to the root $\alpha_2 = \frac{1}{2}$: $\alpha_1 = \frac{17}{2}$ and $\alpha_1 = \frac{1}{2}$. The pair $\alpha_1 = \frac{1}{2}$, $\alpha_2 = \frac{1}{2}$ has already been obtained above as a singular case when the Vandermonde determinant Det $\mathbf{V}$ vanishes.

The second polynomial factor in G_1 yields the only root $\alpha_2 = -\frac{15}{2}$. Substituting this value in G_4 we obtain the following factored polynomial:

$$-2649731629056 \left(-1 + 2\,\alpha_1\right)^2 \left(-239 - 36\,\alpha_1 + 4\alpha_1^2\right).$$

One of the roots is $\alpha_1 = \frac{1}{2}$. Note that $F(\frac{1}{2}, -\frac{15}{2}) = \frac{17}{4}$. The equation $4\alpha_1^2 - 36\alpha_1 - 239 = 0$ has the following two roots: $\alpha_1^{(4),(5)} = \frac{1}{2}(9 \pm 8\sqrt{5})$. The third polynomial factor in G_1 yields two complex roots. The fourth factor in G_1 leads to the equation $6\alpha_2^3 - 9\alpha_2^2 + 7\alpha_2 - 3 = 0$, which has one real solution $\alpha_2 = \frac{1}{6}\left(3 - \frac{5}{z} + z\right) \approx 0.8207801830727278$, where $z = \left(18 + \sqrt{449}\right)^{\frac{1}{3}}$. Substituting the found value of α_2 in G_4 we obtain a fourth-degree equation for finding α_1. This equation is not presented here in view of its bulky form. It has two real roots: $\alpha_1 = 0.1792198169272722$ and $\alpha_1 = 8.1664593831518564$. Thus, we have obtained seven real solutions. The value of the function F is shown in the following to the right of each pair of the α_1, α_2 values:

$$\begin{aligned}
&\left(\alpha_1^{(1)} = \tfrac{1}{2},\ \alpha_2^{(1)} = \tfrac{1}{2},\ 153\right),\ \left(\alpha_1^{(2)} = \tfrac{17}{2},\ \alpha_2^{(2)} = \tfrac{1}{2},\ \tfrac{17}{4}\right),\\
&\left(\alpha_1^{(3)} = \tfrac{1}{2},\ \alpha_2^{(3)} = -\tfrac{15}{2},\ \tfrac{17}{4}\right),\ \left(\alpha_1^{(4)} = \tfrac{1}{2}(9 - 8\sqrt{5}),\ \alpha_2^{(4)} = -\tfrac{15}{2},\ 80071.2\right),\\
&\left(\alpha_1^{(5)} = \tfrac{1}{2}(9 + 8\sqrt{5}),\ \alpha_2^{(5)} = -\tfrac{15}{2},\ 389185\right),\\
&\left(\alpha_1^{(6)} = 0.1792198169272722,\ \alpha_2^{(6)} = 0.8207801830727278,\ 0.019455592\right),\\
&\left(\alpha_1^{(7)} = 8.1664593831518564,\ \alpha_2^{(7)} = 0.8207801830727278,\ 236.8001073\right).
\end{aligned}$$ (17)

It follows from here that the values $\alpha_1^{(6)}$ and $\alpha_2^{(6)}$ are the optimal values providing the minimum of functional (13).

General conclusion: the two-stage scheme has only the second order of accuracy, and its order cannot be increased to the third-order accuracy.

3.3 Three-Stage RKN Scheme

The given RKN scheme was investigated in the work [12] with the use of CAS Maple 12 and Gröbner bases. We will compare in Sect. 4 the three-stage scheme in terms of accuracy with the remaining four schemes, therefore, we present below two sets of the parameters α_l, γ_l, $l = 1, 2, 3$, which were obtained in [12] and which ensure the fourth order of accuracy of the three-stage scheme under consideration:

$$\alpha_1 = \tfrac{3 \mp z}{6},\ \alpha_2 = \tfrac{3 \pm z}{6},\ \alpha_3 = \tfrac{3 \mp z}{6},\ \gamma_1 = \tfrac{3 \pm 2z}{12},\ \gamma_2 = \tfrac{1}{2},\ \gamma_3 = \tfrac{3 \mp 2z}{12},$$ (18)

where $z = \sqrt{3}$. We will call the scheme with these parameters the RKN34A scheme. The second set of parameters is as follows ($z = 2^{1/3}$):

$$\begin{aligned}
&\alpha_1 = \tfrac{z}{6} + \tfrac{z^2}{12} + \tfrac{1}{3},\qquad \alpha_2 = \tfrac{1}{2},\qquad \alpha_3 = \tfrac{2}{3} - \tfrac{z}{6} - \tfrac{z^2}{12},\\
&\gamma_1 = \tfrac{z}{3} + \tfrac{z^2}{6} + \tfrac{2}{3},\ \gamma_2 = -\tfrac{2z}{3} - \tfrac{z^2}{3} - \tfrac{1}{3},\ \gamma_3 = \tfrac{z}{3} + \tfrac{z^2}{6} + \tfrac{2}{3}.
\end{aligned}$$ (19)

We will call this scheme the RKN34B scheme.

3.4 Four-Stage RKN Scheme

Setting $K = 4$ in (5) and performing symbolic computations similarly to the case of the one-stage scheme, we obtain the expression for δp_n in the form

$$\begin{aligned}
\delta p_{n,4} = {}&\delta p_n + (h^4 u)/(24m)\left(P_{41}(f'(x))^2 + 3P_{42}f(x)f''(x) + P_{43}mu^2 f^{(3)}(x)\right)\\
&- \left(h^5/(120m^2)\right)\left(3P_{51}f^2(x)f''(x) + f(x)(P_{52}(f'(x))^2 - 6P_{53}mu^2 f^{(3)}(x)\right)\\
&- mu^2(5P_{54}f'(x)f''(x) + P_{55}mu^2 f^{(4)}(x))),
\end{aligned}$$ (20)

where the expression for δp_n is given by (9). The expressions for P_1, P_2, P_{31}, P_{32} coincide with formulas (10). The formulas for $P_{41}, P_{42}, P_{43}, P_{51}, P_{52}, P_{53}$, P_{54}, and P_{55} are as follows ($K = 4$)

$$P_{41} = 1 - 24 \sum_{i=1}^{K} \sum_{j<i} \gamma_i \gamma_j \alpha_j (\alpha_i - \alpha_j), \quad P_{42} = 1 - 8 \sum_{i=1}^{K} \sum_{j<i} \gamma_i \gamma_j \alpha_i (\alpha_i - \alpha_j),$$

$$P_{43} = 1 - 4 \sum_{j=1}^{K} \alpha_j^3 \gamma_j, \quad P_{51} = 20 \sum_{i=1}^{K} \sum_{j<i} \sum_{l<i} \gamma_i \gamma_j \gamma_l (\alpha_i - \alpha_j)(\alpha_i - \alpha_l) - 1,$$

$$P_{52} = 120 \sum_{i=1}^{K} \sum_{j<i} \sum_{l<j} \gamma_i \gamma_j \gamma_l (\alpha_i - \alpha_j)(\alpha_j - \alpha_l) - 1, \tag{21}$$

$$P_{53} = 1 - 10 \sum_{i=1}^{K} \sum_{j<i} \gamma_i \gamma_j \alpha_i^2 (\alpha_i - \alpha_j), \quad P_{55} = 1 - 5 \sum_{j=1}^{K} \alpha_j^4 \gamma_j,$$

$$P_{54} = 12 \sum_{i=1}^{K} \sum_{j<i} \gamma_i \gamma_j \alpha_j^2 (\alpha_i + \alpha_j) - 24 \sum_{i=1}^{K} \sum_{j>i} \gamma_i \gamma_j \alpha_i \alpha_j^2 + 1.$$

The call GroebnerBasis[{P1,P2,P31,P32,P41,P42,P43,P51,P52,P53, P54,P55},{a1,a2,a3,a4, g1,g2,g3,g4}] outputs the following result: {1}. By the Hilbert Nullstellensatz [16], if the ideal is {1}, then the 12 polynomials $P_1, \ldots$, P_{55} have no common zero. This involves the conclusion about the absence of the four-stage fifth-order schemes.

The system of equations $P_1 = 0$, $P_2 = 0$, $P_{32} = 0$, $P_{43} = 0$ is linear in γ_i, $i = 1, \ldots, 4$. Its matrix is the Vandermonde 4×4 matrix $\mathbf{V}$ and

$$\text{Det } \mathbf{V} = (\alpha_1 - \alpha_2)(\alpha_1 - \alpha_3)(\alpha_2 - \alpha_3)(\alpha_1 - \alpha_4)(\alpha_2 - \alpha_4)(\alpha_3 - \alpha_4). \tag{22}$$

We begin the study of the four-stage RKN scheme by analogy with the two-stage RKN scheme (see Sect. 3.2) with the consideration of the cases when Det $\mathbf{V} = 0$. The authors of [12] also studied the three-stage RKN scheme by considering at first the cases of the vanishing determinant of the Vandermonde matrix. Expression (22) involves five factors so that it is desirable to consider the cases of the vanishing of all these factors in the search for the best scheme in terms of the accuracy and stability. However, the limitations for the paper size make this study impossible within the present paper. Therefore, we have taken arbitrarily the first factor in Det $\mathbf{V}$ and considered the case of its vanishing: $\alpha_2 = \alpha_1$, $\alpha_3 \neq \alpha_1$, $\alpha_3 \neq \alpha_2$, $\alpha_4 \neq \alpha_1$, $\alpha_4 \neq \alpha_2$, $\alpha_4 \neq \alpha_3$. Our objective is the derivation of at least one real four-stage RKN scheme. A detailed analysis of all cases will be published elsewhere. Let us substitute the relation $\alpha_2 = \alpha_1$ into the polynomials $P_2, P_{31}, P_{32}, P_{41}, P_{42}, P_{43}$ and denote the obtained polynomials by $P_{20}, P_{310}, P_{320}, P_{410}, P_{420}$, and P_{430}. The call GroebnerBasis[{P1, P20, P310, P320, P410, P420, P430},{a1, a3, a4, g1, g2, g3, g4}] has enabled the obtaining of the Gröbner basis consisting of the following six polynomials:

$$G_1 = (1 - 3\gamma_4 + 6\gamma_4^2)(-1 - 24\gamma_4 + 48\gamma_4^2)(-1 + 6\gamma_4 - 12\gamma_4^2 + 6\gamma_4^3),$$
$$G_2 = 537408\gamma_4^6 - 1405440\gamma_4^5 + 1276020\gamma_4^4 - 578046\gamma_4^3 + 138720\gamma_4^2$$
$$\quad - 11224\gamma_4 + 2398\gamma_3 - 1877,$$
$$G_3 = -537408\gamma_4^6 + 1405440\gamma_4^5 - 1276020\gamma_4^4 + 578046\gamma_4^3 - 138720\gamma_4^2$$
$$\quad + 13622\gamma_4 + 2398\gamma_1 + 2398\gamma_2 - 521,$$
$$G_4 = 7194\alpha_4 + 331776\gamma_4^6 - 965376\gamma_4^5 + 1058688\gamma_4^4 - 626304\gamma_4^3 \quad\quad (23)$$
$$\quad + 218880\gamma_4^2 - 39603\gamma_4 - 3418,$$
$$G_5 = 4796\alpha_3 - 537408\gamma_4^6 + 1405440\gamma_4^5 - 1276020\gamma_4^4 + 578046\gamma_4^3$$
$$\quad - 138720\gamma_4^2 + 16020\gamma_4 - 2919,$$
$$G_6 = 14388\alpha_1 + 489024\gamma_4^6 - 976896\gamma_4^5 + 574884\gamma_4^4 - 237318\gamma_4^3$$
$$\quad + 49248\gamma_4^2 - 2526\gamma_4 - 3227.$$

The roots of equation $1 - 3\gamma_4 + 6\gamma_4^2 = 0$ are complex: $\gamma_4 = (1/12)(3 \pm i\sqrt{15})$. The roots of the equation $-1 - 24\gamma_4 + 48\gamma_4^2 = 0$ are real: $\gamma_4 = (1/12)(3 \pm 2\sqrt{3})$. It is easy to find from Eq. (23) the values of remaining parameters of the RKN scheme under study ($z = \sqrt{3}$):

$$\alpha_1 = \tfrac{1}{6}(3 \pm z), \quad \alpha_2 = \tfrac{1}{6}(3 \pm z), \quad \alpha_3 = \tfrac{1}{6}(3 \mp z), \quad \alpha_4 = \tfrac{1}{6}(3 \pm z),$$
$$\gamma_2 = \tfrac{1}{12}(3 \mp 2z - 12\gamma_1), \quad \gamma_3 = \tfrac{1}{2}, \quad \gamma_4 = \tfrac{1}{12}(3 \pm 2z). \tag{24}$$

One can see from (24) that one parameter, γ_1, remains indefinite. This is due to the fact that the number of polynomials in the Gröbner basis (23) is less than the number of parameters α_1, $\alpha_2 = \alpha_1$, α_3, α_4, γ_j, $j = 1, \ldots, 4$.

Consider in more detail a scheme, which is obtained at the use of lower symbols "+" or "−" in (24). We call this scheme the RKN4A scheme. Let us calculate the weighted root-mean-square value of five polynomials P_{5j}, $j = 1, \ldots, 5$:

$$P_{5A,rms} = \left(\tfrac{1}{5}\sum_{j=1}^{5}(\sigma_j P_{5j})^2\right)^{1/2} = \left(\tfrac{1}{5}\left((\sigma_1\tfrac{7}{72})^2 + (\sigma_2\tfrac{7}{12})^2 + (\tfrac{\sigma_3}{36})^2\right.\right.$$
$$\left.\left. + (\tfrac{\sigma_4}{6})^2 + (\tfrac{\sigma_5}{36})^2\right)\right)^{1/2} = 0.47924. \tag{25}$$

Here $\sigma_1, \ldots, \sigma_5$ are problem-independent factors affecting the polynomials P_{5j} in (20), $\sigma_1 = -3$, $\sigma_2 = -1$, $\sigma_3 = 6$, $\sigma_4 = 5$, $\sigma_5 = 1$.

Now consider a scheme, which is obtained at the use of the upper symbols "+" or "−" in (24). Let us call this scheme the RKN4B scheme. We have for it

$$P_{5B,rms} = \left(\tfrac{1}{5}\sum_{j=1}^{5}(\sigma_j P_{5j})^2\right)^{\frac{1}{2}} = \left(\tfrac{1}{5}(9.3910\sigma_1)^2 + (2.4187\sigma_2)^2\right.$$
$$\left. + (4.3102\sigma_3)^2 + (3.8374\sigma_4)^2 + (0.0497\sigma_5)^2)\right)^{\frac{1}{2}} = 19.16514. \tag{26}$$

The value 19.16514 is by the factor of 40 larger than the quantity $P_{5A,rms}$.

Equation $6\gamma_4^3 - 12\gamma_4^2 + 6\gamma_4 - 1 = 0$ has one real root $\gamma_4 = (1/3)(2 + \tfrac{z^2}{2} + z)$ and two complex conjugate roots, where $z = 2^{1/3}$. We find from equations (23) the values of remaining parameters of the RKN scheme under study (we call it the RKN4C scheme):

$$\alpha_1 = \tfrac{1}{12}(4 + 2z + z^2), \quad \alpha_2 = \alpha_1, \quad \alpha_3 = \tfrac{1}{2}, \quad \alpha_4 = \tfrac{1}{12}(8 - 2z - z^2),$$
$$\gamma_2 = \tfrac{1}{6}(4 + 2z + z^2 - 6\gamma_1), \quad \gamma_3 = -\tfrac{1}{3}(1 + z)^2, \quad \gamma_4 = \tfrac{1}{3}(2 + \tfrac{z^2}{2} + z). \tag{27}$$

One can see from (27) that one parameter, γ_1, remains indefinite. In the case under consideration,

$$P_{5C,rms} = \left(\tfrac{1}{5}\sum_{j=1}^{5}(\sigma_j P_{5j})^2\right)^{\frac{1}{2}} = \left(\tfrac{1}{5}\left((8.1092\sigma_1)^2 + (2.3780\sigma_2)^2\right.\right.$$
$$\left.\left. + (2.0962\sigma_3)^2 + (10.3143\sigma_4)^2 + (0.6386\sigma_5)^2\right)\right)^{\frac{1}{2}} = 26.13695. \tag{28}$$

The value $P_{5C,rms} = 26.13695$ is by the factor of 54.54 larger than the quantity $P_{5A,rms}$. Therefore, the RKN4A scheme is preferable for the computations using the above three four-stage schemes.

3.5   Five-Stage RKN Scheme

At $K = 5$ in (5), the expression for δp_n has the following form in view of (20):

$$\delta p_n = \delta p_{n,4} + \left(h^6 u/(720m^2)\right)\left(\left(f'(x)\right)^3 - 15P_{61}f^2(x)f^{(3)}(x)\right.$$
$$+ f'(x)\left(P_{62}f(x)f''(x) + P_{63}mu^2 f^{(3)}(x)\right)$$
$$\left. + mu^2\left(5P_{64}\left(f''(x)\right)^2 + 10P_{65}f(x)f^{(4)}(x) + mP_{66}u^2 f^{(5)}(x)\right)\right), \tag{29}$$

where the polynomials P_1, P_2, P_{31}, P_{32}, P_{41}, P_{42}, P_{43}, and P_{51}–P_{55} are given by formulas (21) at $K = 5$. Let us present the expressions for polynomials $P_{61}, \ldots, P_{66}$:

$$P_{61} = 24\left(\sum_{i=1}^{K}\sum_{j=i+1}^{K}\gamma_i^2\gamma_j\alpha_j(\alpha_i - \alpha_j)^2 + 2\sum_{i=1}^{K}\sum_{j>i}\sum_{l>j}\gamma_i\gamma_j\gamma_l\alpha_l(\alpha_i - \alpha_l)(\alpha_j - \alpha_l)\right) - 1,$$

$$P_{62} = 18 - 720\left(\sum_{i=1}^{K}\sum_{j=i+1}^{K}\gamma_i^2\gamma_j\alpha_i(\alpha_i - \alpha_j)^2\right.$$
$$\left. + \sum_{i=1}^{K}\sum_{i<j}\sum_{j<l}\gamma_i\gamma_j\gamma_l(\alpha_j - \alpha_l)\left((\alpha_i^2 - \alpha_j^2) + 2\alpha_j(\alpha_i - \alpha_l)\right)\right),$$

$$P_{63} = 11 + 120\left(\sum_{i=1}^{K}\sum_{j=i+1}^{K}\gamma_i\gamma_j\alpha_i(\alpha_i - \alpha_j)(\alpha_i^2 + 3\alpha_j^2)\right), \tag{30}$$

$$P_{64} = 72\sum_{i=1}^{K}\sum_{j=i+1}^{K}\gamma_i\gamma_j\alpha_i^2\alpha_j(\alpha_i - \alpha_j) + 1,$$

$$P_{65} = 1 + 12\sum_{i=1}^{K}\sum_{j=i+1}^{K}\alpha_j^3\gamma_i\gamma_j(\alpha_i - \alpha_j), \quad P_{66} = 1 - 6\sum_{j=1}^{K}\alpha_j^5\gamma_j.$$

It was shown in the works [17,18] that the conditions $P_{41} = 0$, $P_{52} = 0$ are redundant. The call `GroebnerBasis[{P1,P2,P31,P32,P42,P43,P51,P53,P54,P55}, {a1, a2, a3, a4, a5, g1, g2, g3, g4, g5}]` requires too a big CPU time. For this reason, the symbolic expressions for the polynomials of the Gröbner basis have not been obtained. The numerical values of parameters α_i, γ_i $(i = 1, \ldots, 5)$ were found in the work [17] by numeric computations with an error below 10^{-12}. As a result, four real methods were obtained, see Table 1 in [17].

To find the method, which is the most accurate among the four methods, we have computed the root-mean-square values

$$P_{6,rms} = \left((1/6) \sum_{j=1}^{6} P_{6j}^2 \right)^{1/2} \tag{31}$$

for all four fifth-order RKN methods. It was found that $P_{6,rms} = 5.18771$ for methods 1 and 3; $P_{6,rms} = 0.72114$ for methods 2 and 4. The coincidence of the quantity (31) for the pairs of methods 1, 3 and 2, 4 is not accidental: it was explained in [17] that method 3 is the adjoint of method 1, and method 4 is the adjoint of method 2 (see also Table 1 in [17]). The adjoint of a method is obtained by interchanging h, x^n, and u^n, respectively, with $-h, x^{n+1}$, and u^{n+1}.

3.6   Stability Conditions of the RKN Schemes

It is well known that if a symplectic scheme is stable then all roots of its characteristic equation lie on the unit circle in the complex plane. As a physical model we consider an oscillator with a quadratic potential $V(x) = m\omega^2 x^2/2$, for which the equilibrium position is located at $x = 0$, $p = 0$. Substituting the given expression $V(x)$ in Eq. (4), we obtain the linear motion equations

$$dx/dt = p/m, \quad dp/dt = -m\omega^2 x. \tag{32}$$

Let us introduce the column vectors $X^n = (x^n, p^n)^T$ and $X^{n+1} = (x^{n+1}, p^{n+1})^T$. In the matrix form, equations (5) as applied to system (32) are as follows: $X^{n+1} = GX^n$, where G is the 2×2 amplification matrix. Let g_{ij}, $i, j = 1, 2$ be the entries of this matrix. The characteristic equation of matrix G is $|G - \lambda E| = \lambda^2 + \text{Tr}(G)\lambda + 1 = 0$, where $\text{Tr}(G)$ is the trace of the matrix G, $\text{Tr}(G) = -g_{11} - g_{22}$; E is the 2×2 identity matrix. The stability conditions of scheme (5) are the conditions $|\lambda_i| \le 1$, where λ_i, $i = 1, 2$ are the eigenvalues of the matrix G that is the roots of the characteristic equation. If the discriminant of this equation $D = [\text{Tr}(G)/2]^2 - 1 < 0$, then it has two complex conjugate roots λ_1, λ_2 such that $|\lambda_1| = |\lambda_2| = 1$ according to the Vieta's theorem. In this case, there is a nonzero stability region $0 < |\kappa| \le \kappa_{cr}$, where κ_{cr} is the critical Courant number. For the RKN schemes under consideration, κ_{cr} is the solution of the equation $|\text{Tr}(G)| - 2 = 0$.

One-Stage RKN Scheme. When the Verlet scheme (5), $K = 1$ is applied to linear equations (32) it takes the following form:

$$x^{(1)} = x^n + \frac{h}{2}\frac{p^n}{m}, \quad x^{n+1} = x^n + h\frac{p^n}{m} - \frac{h^2\omega^2}{2}x^{(1)}, \quad p^{n+1} = p^n - hm\omega^2 x^{(1)}. \tag{33}$$

The program in the language of the CAS *Mathematica* for computing the entries of the matrix G is as follows:

```
x1 = xn + h*pn/(2m); xnew = xn + h*pn/m - h^2*\[Omega]^2*x1/2;
pnew = pn - h*m*\[Omega]^2*x1; xnew = xnew; g11 = Coefficient[xnew, xn];
```

```
g12 = Coefficient[xnew, pn]; g21 = Coefficient[pnew, xn];
g22 = Coefficient[pnew, pn]; G = {{g11, g12}, {g21, g22}};
G = G /. \[Omega] -> \[Kappa]/h;
```

The Courant number $\kappa = \omega h$ has been introduced in the last row of this code fragment. The matrix G is obtained in the form

$$G = \begin{pmatrix} 1 - \kappa^2/2 & h/m - h\kappa^2/(4m) \\ -m\kappa^2/h & 1 - \kappa^2/2 \end{pmatrix}.$$

The coefficients of the characteristic equation are calculated with the aid of the call Det[G - λ*IdentityMatrix[2]], and the equation has the following form: $1 - 2\lambda + \kappa^2\lambda + \lambda^2 = 0$. The discriminant $D = \kappa^2(\kappa^2/4 - 1)$ is found from here. The stability condition is satisfied if $D \leq 0$. In the region of positive κ, this leads to the stability condition of the form $0 < \kappa \leq 2$.

As is known, the Hamilton equations are reversible in time at a simultaneous reversal of time and of the particle velocities [2]. It follows from here that the symplectic difference schemes are also reversible in time as the Hamilton equations. It is easy to show that $D \leq 0$ also in the interval $[-2, 0]$ that is the Verlet scheme is stable also in this interval. Therefore, the stability condition of this scheme should be written as $0 < |\kappa| \leq 2$.

Two-Stage RKN Scheme. The stability analysis of RKN schemes at $K > 1$ in (5) proceeds similarly to the case $K = 1$, only the length of the expressions for the entries of the 2×2 amplification matrix G increases with increasing K.

At $l = 6$ in (17), the following expression has been obtained for the discriminant D: $D = 0.00082628 \left(\kappa^4 - 17.3943\kappa^2 + 34.7886\right)^2 - 1$. Denote the roots of the equation $D = 0$ by $\kappa_1, \ldots, \kappa_8$. They are as follows: $\kappa_1 = -\kappa_8$, $\kappa_2 = -\kappa_7$, $\kappa_3 = -\kappa_6$, $\kappa_4 = \kappa_5 = 0$, $\kappa_6 = 2.496957971257$, $\kappa_7 = 3.340580819059$, $\kappa_8 = 4.170644952389$. With regard for the intervals, where $D \leq 0$, we obtain the following stability conditions of the two-stage RKN scheme: $0 < |\kappa| \leq \kappa_6$, $\kappa_7 \leq |\kappa| \leq \kappa_8$.

Three-Stage RKN Scheme. The RKN34A scheme is determined by parameters (18). In this case, the equation $|\text{Tr}(G)| - 2 = 0$ has two real roots $\kappa = \pm\kappa_{cr}$, where $\kappa_{cr} = 2\sqrt{2 + 2^{1/3} - 2^{2/3}} \approx 2.5865189$. Thus, the stability region of the RKN34A scheme has the form $0 < |\kappa| \leq \kappa_{cr}$.

The RKN34B scheme is determined by parameters (19). In this case, the equation for determining κ_{cr} has four real roots and two complex-conjugate roots. Real roots: $\kappa = 0$ (the root of multiplicity 2) and $\kappa = \pm\kappa_{cr}$, where $\kappa_{cr} = [6(2 - 2^{2/3})]^{1/2} \approx 1.573401947435$. The stability region of the RKN34B has the form $0 < |\kappa| \leq \kappa_{cr}$.

Four-Stage RKN Scheme. The RKN4A scheme is determined by parameters (24) with lower plus and minus signs. The equation $|\text{Tr}(G)| - 2 = 0$ has two real roots $\kappa = \pm\kappa_{cr}$, where $\kappa_{cr} = 2\sqrt{2 + 2^{1/3} - 2^{2/3}} \approx 2.5865189$; thus, the stability region of the RKN4A scheme coincides with the stability region of the RKN34A scheme.

In the case of the RKN4B scheme, the equation $|\mathrm{Tr}(G)| - 2 = 0$ has two real roots $\kappa = \pm\kappa_{cr}$, where $\kappa_{cr} = \left(6(-3 + (9 + 12z^2)^{1/2})^{1/2}\right)^{1/2}/z \approx 2.5149188$ and $z = 1 + 2^{1/3}$. Thus, the stability region of the RKN4B scheme has the form $0 < |\kappa| \leq \kappa_{cr}$.

In the case of the RKN4C scheme determined by parameters (27), equation $\mathrm{Tr}(G) - 2 = 0$ has two real roots $\kappa = \pm\kappa_{cr}$ at $\mathrm{Tr}(G) > 0$, where $\kappa_{cr} = 1.854382524682$. Thus, the stability region of the RKN4C scheme is described by the inequalities $0 < |\kappa| \leq \kappa_{cr}$.

From the results obtained in Subsect. 3.4 at the consideration of coefficients affecting h^5 and from the above-obtained values of κ_{cr} for the RKN4A, RKN4B, and RKN4C schemes, it follows that the RKN4A scheme is the best scheme from the viewpoint of the smallness of coefficients affecting h^5 and the size of the stability region.

Five-Stage RKN Scheme. The stability conditions were obtained for each of the four methods.

Methods 1 and 3, $\mathrm{Tr}(G) > 0$: $-4 + \kappa^2 - \kappa^4/12 + \kappa^6/360 + 0.019206667644\kappa^8 + 0.001488249575\kappa^{10} = 0$, the stability region: $0 < |\kappa| \leq 1.709678742327$.

Methods 2 and 4, $\mathrm{Tr}(G) > 0$: $-4 + \kappa^2 - \kappa^4/12 + \kappa^6/360 + 0.009374405183\kappa^8 + 0.000595080910\kappa^{10} = 0$, the stability region: $0 < |\kappa| \leq 1.836026193724$.

It follows from the above analysis that methods 2 and 4 possess a somewhat larger stability region than methods 1 and 3. Besides, as was shown in Subsect. 3.5, the root-mean-square value of polynomials $P_{61}, \ldots, P_{66}$ is in the case of methods 2 and 4 seven times smaller than in the case of methods 1 and 3. Therefore, methods 2 and 4 are more preferable for their use at the solution of molecular dynamics problems than methods 1 and 3.

Table 1. Efficiency ef for a number of RKN methods

RKN method	K	κ_{cr}	ef
Verlet	1	2	2
the values of $\alpha_1^{(l)}$, $\alpha_2^{(l)}$ in (17) at $l = 1, 2, 3$	2	2	1
$\alpha_1 = 0.17922$, $\alpha_2 = 0.82078$	2	4.17064	2.08532
RKN34A scheme	3	2.58652	0.86217
RKN34B scheme	3	1.57340	0.52447
RKN4A scheme	4	2.58652	0.64663
RKN5 scheme, methods 1, 3	5	1.70968	0.34193
RKN5 scheme, methods 2, 4	5	1.83603	0.36721

Comparison of the Efficiency of Considered RKN Schemes. To compute the values x^{n+1} and p^{n+1} with the aid of the K-stage RKN scheme it is necessary to calculate K values of the function $f(x)$. At the first glance, the required CPU time must increase linearly with increasing number of stages. But it is necessary to account here also for the magnitude of the critical Courant number κ_{cr}, which is different for different RKN schemes.

It was shown in [14, 15] that with increasing number of stages of a conventional (non-symplectic) explicit Runge–Kutta scheme, the value of the critical Courant number increases. Therefore, one can perform stable computations with larger values of the Courant number than in the case of Runge–Kutta schemes with a small number of stages. This may finally lead to a reduction of the CPU time needed for problem solution. In this connection, the quantitative characteristic of the efficiency of the Runge–Kutta schemes was introduced for the first time in the work [14], which will be denoted by ef: $ef = \kappa_{cr}/K$. In the works [14, 15], the values of the efficiency parameter ef were presented for several explicit non-symplectic Runge–Kutta schemes.

Table 1 presents the values of ef for all RKN schemes considered above. In particular, the quantity ef of the RKN34A scheme is 1.56 times higher than in the case of the RKN34B scheme. Table 1 shows that the quantity ef drops with increasing number of stages K. This constitutes a significant difference of symplectic RKN schemes from the explicit non-symplectic Runge–Kutta schemes.

We have also compared the error terms of the order $O(h^4)$ of the RKN34A and RKN34B schemes. This error is 20 times less in the case of the RKN34A scheme than in the case of the RKN34B scheme. Therefore, the RKN34A scheme is more preferable at the numerical solution of the molecular dynamics problems.

4   Kepler's Problem

We consider Kepler's problem, in which the both particles move in the (x, y) plane. The Hamiltonian of such a system is as follows: $H = |p_1|/(2m_1) + |p_2|/(2m_2) + U(|r_1 - r_2|)$, where p_1 and p_2 are the vectors of the momentums of the first and of the second particle, $p_j = (m_j u_j, m_j v_j)$, $r_j = (x_j, y_j)$, $j = 1, 2$, m_j is the mass of the jth particle, u_j and v_j are the components of the velocity vector of the jth particle along the axes x and y, respectively; $|p_j|^2/(2m_j) = m_j(u_j^2 + v_j^2)/2$ is the kinetic energy of the jth particle. The potential energy is specified in the form $U(|r_1 - r_2|) = -Gm_1m_2/|r_1 - r_2|$, where G is the gravitational constant.

We consider below a particular case when $m_1 = m_2 = 1$ and $G = 1$. Introduce the notation $p_j = (p_{jx}, p_{jy})$, $j = 1, 2$. The solution of the problem under consideration then reduces to the solution of the following system of ordinary differential equations:

$$\frac{dp_{1x}}{dt} = -\frac{(x_1 - x_2)}{r^3}, \quad \frac{dx_1}{dt} = p_{1x}, \quad \frac{dp_{1y}}{dt} = -\frac{(y_1 - y_2)}{r^3}, \quad \frac{dy_1}{dt} = p_{1y},$$
$$\frac{dp_{2x}}{dt} = \frac{(x_1 - x_2)}{r^3}, \quad \frac{dx_2}{dt} = p_{2x}, \quad \frac{dp_{2y}}{dt} = \frac{(y_1 - y_2)}{r^3}, \quad \frac{dy_2}{dt} = p_{2y}. \tag{34}$$

Here r is the distance between the both particles, $r = |\mathbf{r}_1 - \mathbf{r}_2| = [(x_1 - x_2)^2 + (y_1 - y_2)^2]^{1/2}$ and it is assumed that $x_1, y_1, x_2, y_2, p_{1x}, p_{1y}, p_{2x}, p_{2y}$ are the functions depending on the time t.

System (34) is solved under the following initial conditions specified at $t = 0$:

$$x_1(0) = a_0, \, y_1(0) = 0, \quad x_2(0) = -a_0, \, y_2(0) = 0,$$
$$p_{1x}(0) = 0, \, p_{1y}(0) = v_0, \, p_{2x}(0) = 0, \quad p_{2y}(0) = -v_0, \tag{35}$$

see also Fig. 1. Here a_0 is a given positive number, v_0 is the absolute value of the initial velocity of each particle in the direction of the y axis; the value $v_0 > 0$ is the user-specified quantity.

According to the Noether's theorem [3], at $t > 0$, the total energy E of the system of two particles must remain constant. With (35) in view we obtain:

$$|E| = |H| = |v_0^2 - 1/(2a_0)|. \tag{36}$$

As was shown in [3], at $E < 0$ the motion of the system of two bodies is finite, and at $E > 0$, it is infinite. We consider the case of a finite motion in the following. To ensure the finiteness the constants v_0 and a_0 must satisfy the inequality $v_0^2 - 1/(2a_0) < 0$. In this case, the motion of each particle at $t > 0$ occurs along its own ellipse. Introduce the vector of the mutual distance between the both points $\mathbf{r} = \mathbf{r}_2 - \mathbf{r}_1$ and place the coordinate origin at the inertia center. This leads to the equality $m_1\mathbf{r}_1 + m_2\mathbf{r}_2 = 0$. We find from the last two equalities:

$$\mathbf{r}_1 = -(m/m_1)\mathbf{r}, \quad \mathbf{r}_2 = (m/m_2)\mathbf{r}. \tag{37}$$

The quantity $m = m_1 m_2/(m_1 + m_2)$ is called the reduced mass; $m = 1/2$ because $m_1 = m_2 = 1$ in our case. The formula for $\mathbf{r} = (x(t), y(t))$ is presented in [3]:

$$x = a(\cos \xi - e), \quad y = a\sqrt{1 - e^2}\sin \xi. \tag{38}$$

Here a is the ellipse large semiaxis, e is the elliptic orbit eccentricity,

$$a = \alpha/(2|E|), \quad e = [1 + 2EM^2/(m\alpha^2)]^{1/2}, \tag{39}$$

where $\alpha = Gm_1m_2 = 1$ in accordance with the values of quantities G, m_1, m_2, which were chosen above; M is the magnitude of the moment vector, which is directed along a normal to the (x, y) plane. The law of the moment conservation takes place [3]: $M = \text{const} \, \forall t \geq 0$. We obtain from the initial conditions (35): $M = 2a_0v_0$, where $2a_0$ is the initial distance between the particles.

It follows from (37) and (38) that at $e = 0$, the particles move along the circles. This fact can be used for the additional verification of the computer code implementing the RKN method for the solution of the problem of two bodies. Let us find the condition for parameters a_0 and v_0, under which $e = 0$. Substituting in (39) the expressions for E, M, and α, we obtain $e^2 = 1 + 2EM^2/(m\alpha^2) = 1 + 4(v_0^2 - 1/(2a_0))4a_0^2v_0^2 = (4a_0v_0^2 - 1)^2 = 0$. It follows from here that for ensuring the zero eccentricity it is sufficient to set $v_0 = 0.5/\sqrt{a_0}$. In particular, we have

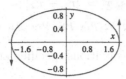

Fig. 1. Initial velocity vectors of particles

Fig. 2. Circular orbits of particles in the interval $0 < t \leq 35.7$: (———)— particle 1, $(\cdot\cdot\cdot)$ — particle 2

at $a_0 = 2$: $v_0 = 0.5/\sqrt{2} \approx 0.35355$. The eccentricity $e > 0$, if the following inequalities are satisfied: either $4a_0v_0^2 < 1$ or $\frac{1}{4} < a_0v_0^2 < \frac{1}{2}$. The inequality $a_0v_0^2 < \frac{1}{2}$ ensures the finiteness of the motion of both particles in view of (36).

In the case of Kepler's problem, we have the following column vector of sought quantities: $\mathbf{X} = (x_1, p_{1x}, y_1, p_{1y}, x_2, p_{2x}, y_2, p_{2y})^T$, where the superscript T denotes the transposition operation. To apply the K-stage RKN method for the solution of the problem under consideration one must replace in (5) the column vector $(x, p)^T$ with $\mathbf{X}$. As a result, we come to the necessity of solving the system of eight ordinary differential equations (34).

To verify the developed Fortran code we have done the computations of the problem of two bodies using all RKN schemes considered in foregoing sections with the numbers of stages $K = 1, 2, 3, 4, 5$ for the cases of the zero and nonzero eccentricity e in (39). The numerical solution for the coordinates of both particles, which has been obtained at $e = 0$ by all considered RKN schemes after the execution of 7140 time steps with the step $h = 0.005$, is shown in Fig. 2. The coordinates of particles were stored after every 80 time steps. One can see that the both particles move along the same circular orbit. Using formula (15,8) from [3], it is not difficult to find the period of time T necessary for the particle to make a full revolution along a circular orbit in the case of the zero eccentricity: $T = \pi\sqrt{a}$, where a is the radius of a circle along which each particle moves; in our case, $a = 2$.

Table 2 presents the results of the computations of the problem of the motion of both particles along a circular orbit by all five RKN schemes considered in the foregoing sections. The quantities δE_{mean} and $|\delta E|_{mean}$ were calculated as the arithmetic means of the quantities δE^n and $|\delta E^n|$, where $\delta E^n = (E^n - E_0)/E_0$, $E^n = (1/2)[(p_{1x}^n)^2 + (p_{1y}^n)^2 + (p_{2x}^n)^2 + (p_{2y}^n)^2] - 1/r^n$, $E_0 = v_0^2 - 1/(2a_0)$ according to (36), $r^n = [(x_1^n - x_2^n)^2 + (y_1^n - y_2^n)^2]^{1/2}$. Besides, $\delta r_{m,\max}$ is the maximum relative deviation of the magnitude of the radius vector r^n of the mth particle $(m = 1, 2)$ from the exact radius $a = 2$ of the circular orbit that is $\delta r_{m,\max} = \max_j(\sqrt{x_{mj}^2 + y_{mj}^2} - a)/a$. It has turned out that at least the first 14 digits of the decimal mantissa of the numbers $\delta r_{1,\max}$ and $\delta r_{2,\max}$ coincide. Therefore, Table 2 presents only the quantity $\delta r_{1,\max}$. From the viewpoint of practical applications, the accuracy of the computation of the coordinates of points (x_m^n, y_m^n) is the most

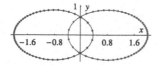

Fig. 3. Elliptic orbits of particle 1 (the right ellipse) and particle 2 (the left ellipse) in the interval $0 < t \leq 164$. Solid lines are the exact ellipses, dotted lines show the numerical solution by the RKN method

Table 2. Errors δE_{mean}, $|\delta E|_{mean}$, and $\delta r_{m,\max}$ at $e = 0$ for different RKN methods

K	Scheme error	δE_{mean}	$|\delta E|_{mean}$	$\delta r_{1,\max}$
1	$O(h^2)$	$-1.783e - 14$	$1.783e - 14$	$1.953e - 7$
2	$O(h^2)$	$-7.879e - 15$	$7.889e - 15$	$9.605e - 8$
3	$O(h^4)$	$-3.918e - 15$	$4.048e - 15$	$5.684e - 14$
4	$O(h^4)$	$-3.685e - 15$	$3.805e - 15$	$5.662e - 14$
5	$O(h^5)$	$-5.876e - 15$	$5.957e - 15$	$1.510e - 14$

Table 3. Errors δE_{mean}, $|\delta E|_{mean}$, and $\delta y_{1,mean}$ at $v_0 = 0.2$ for different RKN methods

K	δE_{mean}	$|\delta E|_{mean}$	$\delta y_{1,mean}$
1	$2.749e - 7$	$2.749e - 7$	$-3.384e - 5$
2	$8.754e - 8$	$8.838e - 8$	$-1.067e - 5$
3	$5.438e - 13$	$6.230e - 13$	$-2.762e - 7$
4	$6.047e - 13$	$5.753e - 13$	$-2.762e - 7$
5	$-5.512e - 15$	$9.388e - 15$	$-2.761e - 7$

important. One can see in Table 2 that the best accuracy of the computation of these coordinates is achieved at the use of the five-stage RKN scheme.

In order to consider the case of the motion of each particle along its elliptic orbit let us set in (35) $v_0 = 0.2$ and $a_0 = 2$. The inequality $4a_0v_0^2 < 1$ is then satisfied, therefore, the eccentricity $e \neq 0$ and $E < 0$. Each particle performs one complete revolution along its elliptic orbit during the period of time [3] $T = \pi\alpha\sqrt{m/(2|E|^3)}$. Substituting here the values $\alpha = 1$, $m = 1/2$, and $E = v_0^2 - 1/(2a_0)$, we obtain $T = 16.3227$. By the physical time $t = 164$, each particle makes 10 complete revolutions along its elliptic orbit. Fig. 3 shows the numerical solution for the coordinates of both particles, which was obtained by all considered RKN methods at the moment of time $t = 164$ after the execution of 82000 time steps with the step $h = 0.002$. One can see that each particle moves along its elliptic orbit and the locations of particles agree very well with the exact elliptic orbits.

Table 3 presents the values of relative errors δE_{mean} and $|\delta E|_{mean}$, which were obtained at the numerical solution of Kepler's problem on the motion of

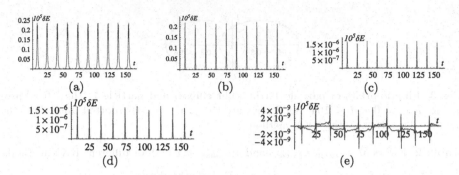

Fig. 4. Kepler's problem, the case of a nonzero eccentricity. The quantities $10^5 \delta E$ as the functions of time at different numbers of stages K of the RKN methods: (a) $K = 1$, (b) $K = 2$, (c) $K = 3$, (d) $K = 4$, (e) $K = 5$

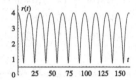

Fig. 5. Graph of the curve $r(t) = |r_1 - r_2|$ at a nonzero eccentricity e

two particles along the elliptic orbits. Note that in the case of the fifth-order RKN scheme, these errors are by two decimal orders less than in the case of the fourth-order scheme (see also Fig. 4). The quantity $\delta y_{1,mean}$ was computed as the arithmetic mean of quantities $\delta y_{1j} = y_{1j} - y_{1,ex}$. Here $y_{1,ex}$ is the exact value of the coordinate y at the point of the intersection of the line $x = x_{1j}$ with the ellipse of the first particle (see the right ellipse in Fig. 3). The quantity $\delta y_{2,mean}$ is computed similarly with the use of exact formulas for the ellipse of the second particle. It has turned out that there is a coincidence of the first ten digits of the machine numbers $\delta y_{1,mean}$ and $\delta y_{2,mean}$, but the signs of these numbers are opposite. For example, at $K = 5$, the value $\delta y_{2,mean} = +2.761e - 7$ has been obtained.

The error δy_{1j} is overall much larger than the error δE_{mean}. An analysis has shown that the given error increases in its absolute value up to the value of the order 10^{-5} near the x axis at $|y_{1j}| < 0.1$ that is where the ellipse curvature is the largest. When the particles (at $e = 0$) move along a circle, such a problem does not arise because the curvature of the circle is constant (see Table 2).

It is easy to see in Fig. 4 that at each K, the number of peaks is equal to 10 that is it is equal to the number of the motion periods of each particle along its elliptic orbit. To explain this phenomenon let us turn to Fig. 5. In this figure, $r = r(t) = |r_1 - r_2|$. The shape of the curve depicted in this figure is the same for all considered RKN schemes. One can see that the curve $r(t)$ has 10 minima. At each point of the minimum, the curvature radius is very small.

The curvature is inversely proportional to the curvature radius, therefore, it is clear that it is maximum at the points of the minimum of the function $r(t)$. In its turn, the curvature is proportional to the second derivative of this function. Therefore, this derivative is maximum in its absolute value at the points of the minimum of the function $r(t)$. The derivatives of the second and higher orders of the functions $x(t)$ and $y(t)$ enter the leading terms of the errors of the RKN methods. It follows from here that the scheme error must increase at the points of the minimum of the function $r(t)$, what is just observed in Fig. 4.

5   Conclusions

A comparative investigation of the accuracy and stability of the explicit RKN schemes with the number of stages from 1 to 5 has been carried out using the computer algebra system *Mathematica*. The built-in function `GroebnerBasis[]` of this CAS enables an efficient solution of the question about the possibility of deriving a scheme of the accuracy order $O(h^{K+1})$ based on the K-stage RKN scheme. It has turned out that the RKN methods of even orders (orders 2 and 4) do not admit the construction of the method of the order $O(h^{K+1})$. For the RKN schemes with the number of stages $K = 3$ and $K = 4$, the function `GroebnerBasis[...]` enables the obtaining of exact solutions of the polynomial systems for the weight parameters α_j, γ_j $(j = 1, \ldots, K)$.

The application of considered RKN methods for the numerical solution of Kepler's problem having the exact solution has made it possible to establish the fact that an increase in the number of stages K leads to an increase in the accuracy of the satisfaction of the conservation law for the energy of particles. This makes the higher-order RKN schemes preferable in the cases when it is required to solve an applied problem in a large time interval.

An extension of the efficiency concept ef of a multistage non-symplectic RK scheme has been proposed for the cases of multistage symplectic RKN schemes. It has been shown that with increasing number of stages K, the quantity ef decreases. This constitutes a substantial difference of symplectic RKN schemes from the explicit non-symplectic Runge–Kutta schemes.

Acknowledgments. This work has been supported in part by the Russian Foundation for Basic Research (grant No. 19-01-00292-a) and by the Program of Fundamental Scientific Research of the state academies of sciences in 2013–2020 (projects Nos. AAAA-A17-117030610124-0 and AAAA-A17-117030610134-9).

References

1. Kiselev, S.P., Vorozhtsov, E.V., Fomin, V.M.: Foundations of Fluid Mechanics with Applications: Problem Solving Using Mathematica. Springer Science+Business Media LLC, New York (2012)
2. Godunov, S.K., Kiselev, S.P., Kulikov, I.M., Mali, V.I.: Modeling of Shockwave Processes in Elastic-plastic Materials at Different (Atomic, Meso and Thermodynamic) Structural Levels. Institute of Computer Research, Moscow-Izhevsk (2014)

3. Landau, L.D., Lifshitz, E.M.: Mechanics, third edition. Course of Theoretical Physics. vol. 1. Elsevier, Amsterdam (1976)

4. Lewis, H., Barnes, D., Melendes, K.: The liouville theorem and accurate plasma simulation. J. Comput. Phys. **69**(2), 267–282 (1987)

5. Ruth, R.D.: A canonical integration technique. IEEE Trans. Nucl. Sci. **30**(4), 2669–2671 (1983)

6. Tuckerman, M., Berne, B.J.: Reversible multiple time scale molecular dynamics. J. Chem. Phys. **97**(3), 1990–2001 (1992)

7. Forest, E., Ruth, R.D.: Fourth-order symplectic integration. Phys. D. **43**, 105–117 (1990)

8. Omelyan, I.P., Mryglod, I.M., Folk, R.: Optimized Verlet-like algorithms for molecular dynamics simulations. Phys. Rev. E. **65**, 056706 (2002)

9. Nyström, E.J.: Ueber die numerische Integration von Differentialgleichungen. Acta Soc. Sci. Fenn. **50**(13), 1–54 (1925)

10. Verlet, L.: Computer "experiments" on classical fluids. thermodynamical properties of Lennard-Jones molecules. Phys. Rev. **159**(1), 98–103 (1967)

11. Surius, Yu.B.: On the canonicity of maps generated by Runge-Kutta type methods in the integration of systems $\ddot{x} = -\partial U/\partial x$. Zh. Vychisl. Mat. Mat. Fiz. **29** (2), 202–211 (1989) (in Russian)

12. Sofronov, V.N., Shemarulin, V.E.: Classification of explicit three-stage symplectic difference schemes for the numerical solution of natural Hamiltonian systems: a comparative study of the accuracy of high-order schemes on molecular dynamics Problems. Comp. Math. Math. Phys. **56**(4), 541–560 (2016)

13. Hairer, E., Nørsett, S.P., Wanner, G.: Solving Ordinary Differential Equations I, 2nd edn. Springer, Berlin (1993)

14. Schmidt, W., Jameson, A.: Euler solvers as an analysis tool for aircraft aerodynamics. In: Habashi, W.G. (ed.) Advances in Computational Transonics, vol. 4. Recent Advances in Numerical Methods in Fluids, pp. 371–404. Pineridge Press, Swansea (1985)

15. Ganzha, V.G., Vorozhtsov, E.V.: Computer-Aided Analysis of Difference Schemes for Partial Differential Equations. Wiley-Interscience, New York (2012)

16. Adams, A.L., Loustaunau, P.: An Introduction to Gröbner Bases. Graduate Studies in Mathematics, vol. 3. Amer. Math. Soc., Providence, Rhode Island (1996)

17. Okunbor, D.I., Skeel, R.D.: Canonical Runge-Kutta-Nyström methods of orders five and six. J. Comput. Appl. Math. **51**, 375–382 (1994)

18. Okunbor, D.I., Skeel, R.D.: Explicit canonical methods for Hamiltonian systems. Math. Comp. **59**, 439–455 (1992)

Characterizing Triviality of the Exponent Lattice of a Polynomial Through Galois and Galois-Like Groups

Tao Zheng[✉]

School of Mathematical Sciences, Peking University, Beijing, China
xd07121019@126.com

Abstract. The problem of computing *the exponent lattice* which consists of all the multiplicative relations between the roots of a univariate polynomial has drawn much attention in the field of computer algebra. As is known, almost all irreducible polynomials with integer coefficients have only trivial exponent lattices. However, the algorithms in the literature have difficulty in proving such triviality for a generic polynomial. In this paper, the relations between the Galois group (respectively, *the Galois-like groups*) and the triviality of the exponent lattice of a polynomial are investigated. The $\mathbb{Q}$-*trivial* pairs, which are at the heart of the relations between the Galois group and the triviality of the exponent lattice of a polynomial, are characterized. An effective algorithm is developed to recognize these pairs. Based on this, a new algorithm is designed and implemented to prove the triviality of the exponent lattice of a generic irreducible polynomial, which considerably improves a state-of-the-art implementation of an algorithm of the same type when the polynomial degree becomes larger. In addition, the concept of the Galois-like groups of a polynomial is introduced. Some properties of the Galois-like groups are proved and, more importantly, a sufficient and necessary condition is given for a polynomial (which is not necessarily irreducible) to have trivial exponent lattice.

Keywords: Polynomial root · Multiplicative relation · Exponent lattice · Trivial · Galois group · Galois-like

1 Introduction

Set $\overline{\mathbb{Q}}^*$ to be the set of nonzero algebraic numbers. Suppose that $n \in \mathbb{Z}_{>0}$. For any $v \in (\overline{\mathbb{Q}}^*)^n$, define *the exponent lattice* of v to be $\mathcal{R}_v = \{u \in \mathbb{Z}^n \mid v^u = 1\}$, where $v^u = \prod_{i=1}^{n} v_i^{u_i}$ with v_i the ith coordinate of v and u_i the one of u. For a univariate polynomial $f \in \mathbb{Q}[x]$ (with $f(0) \neq 0$) of degree n, denote by $\vec{\Omega} \in (\overline{\mathbb{Q}}^*)^n$ the vector formed by listing all the complex roots of f with multiplicity in some order. For convenience, we call $\mathcal{R}_{\vec{\Omega}}$ *the exponent lattice* of the polynomial f and

This work was supported partly by NSFC under grants 61732001 and 61532019.

© Springer Nature Switzerland AG 2020
F. Boulier et al. (Eds.): CASC 2020, LNCS 12291, pp. 621–641, 2020.
https://doi.org/10.1007/978-3-030-60026-6_36

use the notation $\mathcal{R}_f$ instead of $\mathcal{R}_{\vec{\Omega}}$, if no confusion is caused. Moreover, we define $\mathcal{R}_f^{\mathbb{Q}} = \{u \in \mathbb{Z}^n \mid \vec{\Omega}^u \in \mathbb{Q}\}$.

The exponent lattice has been studied extensively from the perspective of number theory and algorithmic mathematics since 1977. In [23] Theorem 1, an upper bound of the length of a nonzero vector in a nonzero exponent lattice is given for the first time. Further results are obtained in [17] Theorem 3 by the same authors of [23]. Latter in [20] by D. W. Masser, a common upper bound of the lengths of the vectors in a basis of an exponent lattice is given, which indicates that a basis of a given exponent lattice is computable. Other related theoretical results can be found in [21,22]. In [13] and [16], the first algorithm, named "FindRelations", is proposed to compute the lattice $\mathcal{R}_v$ for any $v \in (\overline{\mathbb{Q}}^*)^n$ based on Masser's bound and the well-known LLL algorithm. A variation of that algorithm named "GetBasis" is introduced in [27], which constructs a triangular lattice basis in an incremental manner.

There are applications of the exponent lattice to many other areas or problems concerning, for example, linear recurrence sequences [1,16], loop invariants [18,19], algebraic groups [6], compatible rational functions [4], and difference equations [16]. Many of the applications involve computing the exponent lattice of a polynomial in $\mathbb{Q}[x]$. A lattice $\mathcal{R} \subset \mathbb{Z}^n$ or a linear subspace $\mathcal{R} \subset \mathbb{Q}^n$ is called *trivial* if any $v \in \mathcal{R}$ satisfies $v_1 = \cdots = v_n$. It is proved in [11] that almost all irreducible monic polynomials in $\mathbb{Z}[x]$ have trivial exponent lattices. However, the state-of-the-art algorithms FindRelations and GetBasis, which deal with a general $v \in (\overline{\mathbb{Q}}^*)^n$, have difficulty in proving the triviality of $\mathcal{R}_f$ for an irreducible monic polynomial f in $\mathbb{Z}[x]$ with a slightly large degree. Motivated by that, an algorithm called "FastBasis" is introduced recently in [26] to efficiently prove the triviality of the exponent lattice of a given generic polynomial. More references on the multiplicative relations between the roots of a polynomials is provided therein.

In Sect. 2, the relations between the Galois group and the triviality of the exponent lattice of an irreducible polynomial are studied. By characterizing the so called $\mathbb{Q}$-*trivial* pairs from varies points of view (Propositions 1, 2, and 3), we design an algorithm (Algorithm 1) recognizing all those $\mathbb{Q}$-trivial pairs derived from transitive Galois groups. Base on this, an algorithm called "FastBasis$_+$" is obtained by adjusting the algorithm FastBasis in [26]. It turns out that the Magma implementation of FastBasis$_+$ is much more efficient than the implementation of FastBasis in proving the triviality of the exponent lattice of a generic irreducible polynomial when the degree of the polynomial is large.

In Sect. 3, we define the Galois-like groups of a polynomial since the Galois group of a polynomial does not contain enough information to decide whether the exponent lattice is triviality or not (Example 1). We prove that a Galois-like group of a polynomial is a subgroup of the automorphism group of the multiplicative group generated by the polynomial roots (Proposition 8). Furthermore, almost all conditions on the Galois group assuring the triviality of the exponent lattice can be generalized to correspondent ones on the Galois-like groups (see Sect. 3.2). More importantly, a sufficient and necessary condition is given for a polynomial (not necessarily irreducible) to have trivial exponent lattice through the Galois-like groups (see Theorem 1 and 2).

2 Lattice Triviality Through Galois Groups

2.1 Q-Triviality Implying Lattice Triviality

Set G to be a finite group, H a subgroup of G. Set $\bar{g} = gH$ for any $g \in G$, then G can be regarded as a permutation group on the set of the left co-sets $G/H = \{\bar{g} \mid g \in G\}$ via acting $s\bar{g} = \overline{sg}$. The pair (G, H) is called *faithful, primitive, imprimitive, doubly transitive, doubly homogeneous, etc.*, when the permutation representation of G on G/H has the respective property. The group algebra $\mathbb{Q}[G] = \{\sum_{s \in G} a_s s \mid s \in G, a_s \in \mathbb{Q}\}$ is defined as usual and the $\mathbb{Q}$-vector space $\mathbb{Q}[G/H] = \{\sum_{\bar{s} \in G/H} a_{\bar{s}} \bar{s} \mid s \in G, a_{\bar{s}} \in \mathbb{Q}\}$ becomes a left $\mathbb{Q}[G]$-module via acting $\lambda \bar{t} = \sum_{s \in G} a_s \overline{st}$, with $\lambda = \sum_{s \in G} a_s s \in \mathbb{Q}[G]$ and $\bar{t} \in G/H$. We set $\mathbb{Z}[G/H] = \{\sum_{\bar{s} \in G/H} a_{\bar{s}} \bar{s} \mid s \in G, a_{\bar{s}} \in \mathbb{Z}\}$ for convenience.

A subset M of $\mathbb{Q}[G/H]$ is called $\mathbb{Q}$-*admissible* if there is an element $\mu \in \mathbb{Q}[G]$ with stabilizer $G_\mu = H$ so that $m\mu = 0$ for any $m \in M$ (Definition 3 of [14]). Set $\mathcal{V}_1 = \{a \sum_{\bar{s} \in G/H} \bar{s} \mid a \in \mathbb{Q}\}$. Then the pair (G, H) is called $\mathbb{Q}$-*trivial* if 0 and $\mathcal{V}_1$ are the only two $\mathbb{Q}[G]$-submodules that are $\mathbb{Q}$-admissible (Definition 7 of [14]). A polynomial $f \in \mathbb{Q}[x]$ $(f(0) \neq 0)$ without multiple roots is called *non-degenerate* if the quotient of any two roots of f is not a root of unity, and *degenerate* otherwise. The relations between the $\mathbb{Q}$-triviality of a pair and the triviality of an exponent lattice is given below:

Proposition 1. *Suppose that L is a finite Galois extension of the field $\mathbb{Q}$ with Galois group G, and that $H < G$ is a subgroup of G so that G operates on the set G/H faithfully. Then the pair (G, H) is $\mathbb{Q}$-trivial iff for any $f \in \mathbb{Q}[x]$ $(f(0) \neq 0)$ satisfying all the following conditions, $\mathcal{R}_f$ is trivial:*

(i) f is irreducible over $\mathbb{Q}$ and non-degenerate;
(ii) the splitting field of f equals L and its Galois group $G_f = G$;
(iii) H is the stabilizer of a root of f.

Proof. "If": Suppose on the contrary that the pair (G, H) is not $\mathbb{Q}$-trivial. Then there is a $\mathbb{Q}$-admissible $\mathbb{Q}[G]$-submodule M of $\mathbb{Q}[G/H]$ containing an element $v = \sum_{\bar{s} \in G/H} v_{\bar{s}} \bar{s} \in \mathbb{Z}[G/H]$ so that there are $\bar{s}_1 \neq \bar{s}_2 \in G/H$ satisfying $v_{\bar{s}_1} \neq v_{\bar{s}_2}$. By definition, there is an element $\mu \in \mathbb{Q}[G]$ with $G_\mu = H$ such that $v\mu = 0$. Since $\mathbb{Q}$ is an algebraic number field, [14] Proposition 4 indicates that M is admissible in the multiplicative sense. Now by [14] Proposition 2, there is an algebraic number $\alpha \in L^*$ with stabilizer $G_\alpha = H$ and the element v is a non-trivial multiplicative relation between the conjugations of α. What's more, any quotient of two conjugations of α cannot be a root of unity. These mean that the minimal polynomial f of α over the field $\mathbb{Q}$ is non-degenerate and the lattice $\mathcal{R}_f$ is nontrivial. Denote by F the splitting field of f over $\mathbb{Q}$. Then F is a subfield of L and the Galois group G_f of f is isomorphic to $G/\text{Gal}(L/F)$.

Since $G_\alpha = H$, $G_{g(\alpha)} = gHg^{-1}$ for any $g \in G$. Hence the fixed field of the group gHg^{-1} is $\mathbb{Q}[g(\alpha)]$. Note that $\mathbb{Q}[g(\alpha)] \subset F$, $gHg^{-1} \supset \mathrm{Gal}(L/F)$ by Galois theory. Thus, $\cap_{g \in G} gHg^{-1} \supset \mathrm{Gal}(L/F)$. Since the subgroup $\cap_{g \in G} gHg^{-1}$ of G operates trivially on the set G/H and the group G operates faithfully on this set, $\cap_{g \in G} gHg^{-1} = 1$. Hence $\mathrm{Gal}(L/F) = 1$ and $G_f \simeq G$. In fact, $L = F$ and $G_f = G$. So the existence of f leads to a contradiction.

"Only If": Assume that there is an irreducible non-degenerate polynomial $f \in \mathbb{Q}[x]$ ($f(0) \neq 0$) satisfying the condition (iii) with splitting field equal to L and exponent lattice $\mathcal{R}_f$ nontrivial. Suppose that the set of the roots of f is Ω and $\alpha \in \Omega \subset L^*$ is with stabilizer $G_\alpha = H$. Thus, there is a bijection $\tau : G/H \to \Omega$, $\bar{g} \mapsto g(\alpha)$ through which the permutation representations of G on these two sets are isomorphic and we have the $\mathbb{Z}$-module isomorphism $\mathbb{Z}^\Omega \simeq \mathbb{Z}[G/H]$. By [14] Proposition 2, the lattice $\mathcal{R}_f \subset \mathbb{Z}^\Omega \simeq \mathbb{Z}[G/H]$ provides an admissible subset M of $\mathbb{Z}[G/H]$ in the multiplicative sense. Then by Proposition 3 and Definition 3 in [14], one sees that M is a $\mathbb{Q}$-admissible subset. Since $\mathcal{R}_f$ is nontrivial, the $\mathbb{Q}[G]$-module generated by M in $\mathbb{Q}[G/H]$ is neither 0 nor $\mathcal{V}_1$. Hence the pair (G, H) is not $\mathbb{Q}$-trivial, which is a contradiction.   □

For any irreducible non-degenerate polynomial $f \in \mathbb{Q}[x]$ with Galois group G and a root stabilizer $H < G$, Proposition 1 gives the weakest sufficient condition on the pair (G, H) for $\mathcal{R}_f$ to be trivial (*i.e.*, (G, H) being $\mathbb{Q}$-trivial). However, the pair (G, H) does not contain all the information needed to decide whether the lattice $\mathcal{R}_f$ is trivial. This is shown in the following example.

Example 1. Set $g(x) = x^4 - 4x^3 + 4x^2 + 6$, then g is irreducible in $\mathbb{Q}[x]$. By the Unitary-Test algorithm in [25], one proves that g is non-degenerate. Set L to be the splitting field of g over the rational field and $G = \mathrm{Gal}(L/\mathbb{Q})$ its Galois group. Denote by

$$\alpha = (2.35014 \cdots) + \sqrt{-1} \cdot (0.90712 \cdots)$$

one of the roots of g, and set $H = G_\alpha$ to be its stabilizer. Computing with Algorithm 7.16 in [16], we obtain $\mathcal{R}_g = \mathbb{Z} \cdot (-2, 2, 2, -2)^T$, which is nontrivial (thus, (G, H) is not $\mathbb{Q}$-trivial by Proposition 1).

Set $f(x) = g(x - 1)$, then f is irreducible over $\mathbb{Q}$ with splitting field L and Galois group G. Moreover, the number $\alpha + 1$ is a root of f with stabilizer H. We note that the polynomials g and f share the same pair (G, H). However, computing with [16] Algorithm 7.16, we obtain that the lattice $\mathcal{R}_f = \{0\}$ is trivial.

2.2 Characterization of $\mathbb{Q}$-Triviality from The Perspective of Representation Theory and Group Theory

Proposition 2. *Suppose that G is a finite group and $H < G$ is a subgroup so that G operates faithfully on G/H. Denote by 1_H^G the character of the permutation representation of G on the set G/H. Then the pair (G, H) is $\mathbb{Q}$-trivial iff the character $1_H^G - 1$ is $\mathbb{Q}$-irreducible.*

Proof. By [14] Proposition 12, the pair (G, H) is $\mathbb{Q}$-trivial iff (G, H) is primitive and the character $1_H^G - 1$ is $\mathbb{Q}$-irreducible. By [8] Theorem 3, if the character $1_H^G - 1$ is $\mathbb{Q}$-irreducible, then (G, H) is primitive. $\qquad\square$

Throughout the paper, *a root of rational* refers to an algebraic number α such that there is a positive rational integer k ensuring $\alpha^k \in \mathbb{Q}$.

Remark 1. In the settings of Proposition 1, when (G, H) is $\mathbb{Q}$-trivial, (G, H) is primitive. This is equivalent to the condition that H is a maximal subgroup of G. A polynomial f satisfying the conditions (ii) and (iii) in Proposition 1 has a root α with stabilizer $G_\alpha = H$ and the fixed field $\mathbb{Q}[\alpha]$ of the group H is a minimal intermediate field of the extension $L/\mathbb{Q}$ by Galois theory. Note that f is irreducible over $\mathbb{Q}$. If f is degenerate with no root being a root of rational, then there is an integer $k \neq 0$ so that $1 < \deg(\alpha^k) < \deg(\alpha)$. Thus, $\mathbb{Q} \subsetneq \mathbb{Q}[\alpha^k] \subsetneq \mathbb{Q}[\alpha]$, which contradicts the minimality of the field $\mathbb{Q}[\alpha]$. Thus, when (G, H) is $\mathbb{Q}$-trivial, a polynomial f satisfying the conditions (ii) and (iii) is either non-degenerate or with all roots being roots of rational. Hence the condition (i) in Proposition 1 can be replaced by the condition that "f is irreducible over $\mathbb{Q}$ with no root being a root of rational".

Proposition 3. *Let (G, H) be as in Proposition 2. Then, regarded as a permutation group operating on the set G/H, the group G satisfies exactly one of the following conditions iff the pair (G, H) is $\mathbb{Q}$-trivial:*

(i) G is doubly transitive;

(ii) G is of affine type (but not doubly transitive) of degree p^d for some prime number p and $G = M \rtimes H$, where $M \simeq \mathbb{F}_p^d$ is the socle of G and the subgroup H is isomorphic to a subgroup of $GL(d, p)$; moreover, let Z be the center of the group $GL(d, p)$ and regard H as a subgroup of $GL(d, p)$, the group HZ/Z is a transitive subgroup of $PGL(d, p)$ operating on the projective points;

(iii) G is almost simple (but not doubly transitive) of degree $\frac{1}{2}q(q - 1)$, where $q = 2^f \geq 8$ and $q - 1$ is a prime number, and either $G = PSL_2(q)$ or $G = P\Gamma L_2(q)$ with the size of the nontrivial subdegrees $q + 1$ or $(q + 1)f$, respectively.

Proof. This is a combination of Theorem 3 and Theorem 12 in [8] together with Corollary 1.6 in [2]. $\qquad\square$

Denote by $\mathcal{P}$ the set of prime numbers and by $\mathcal{P}^\omega$ the set of prime powers $\{p^d \mid p \in \mathcal{P}, d \in \mathbb{Z}_{\geq 1}\}$. A useful corollary is as follows:

Corollary 1. *Suppose that a polynomial $f \in \mathbb{Q}[x]$ ($f(0) \neq 0$) is irreducible with Galois group G and a root stabilizer H. If the number $\deg(f)$ is NOT in the set*

$$\mathcal{S} = \mathcal{P}^\omega \cup \left\{ 2^{f-1}(2^f - 1) \mid f \in \mathbb{Z}_{\geq 3}, 2^f - 1 \in \mathcal{P} \right\}, \tag{1}$$

then the pair (G, H) is $\mathbb{Q}$-trivial iff it is doubly transitive.

2.3   Particular Q-Trivial Pairs

Besides the doubly transitive pairs (G, H), the author has provided some other particular Q-trivial pairs in [14] Proposition 13–15. A permutation group G on a set S is called *doubly homogeneous* if for any two subsets $\{s_1, s_2\}, \{t_1, t_2\}$ of S, there is some $g \in G$ so that $\{g(s_1), g(s_2)\} = \{t_1, t_2\}$. In this subsection, we prove that any doubly homogeneous pair (G, H) is also Q-trivial.

Proposition 4. *Suppose that G is a finite group and $H < G$ is a subgroup so that G operates faithfully on G/H. If the pair (G, H) is doubly homogeneous, then it is Q-trivial.*

Proof. When the pair (G, H) is doubly transitive, the character $1_H^G - 1$ is actually absolutely irreducible. So we are done. Suppose that the pair (G, H) is doubly homogeneous but not doubly transitive, then G is of odd order (Exe. 2.1.11 of [9]). Hence G is soluble (see [12]). Then by Theorem 7 of [24] (which says: for a given algebraic number field F and a given finite soluble group G, there is a finite Galois extension $\mathcal{F}$ of F so that $\mathrm{Gal}(\mathcal{F}/F) \simeq G$), G is the Galois group of a finite Galois extension of the rational field.

Let $f \in \mathbb{Q}[x]$ ($f(0) \neq 0$) be any polynomial satisfying the conditions (i)–(iii) in Proposition 1. Then f is irreducible and non-degenerate with Galois group G. Since the pair (G, H) is doubly homogeneous and the condition (iii) holds, G operates in a doubly homogeneous way on the set Ω of the roots of f. Doubly homogeneousness naturally requires that $\deg(f) = |\Omega| = |G/H| \geq 2$. Hence f has no root being a root of rational since it is non-degenerate. By [26] Theorem 3.2, the lattice $\mathcal{R}_f^{\mathbb{Q}}$ is trivial and so is the lattice $\mathcal{R}_f$. Finally, according to Proposition 1, the pair (G, H) is Q-trivial.   □

The following example shows that a Q-trivial pair need not be doubly homogeneous.

Example 2. Set L to be the splitting field of the irreducible polynomial $f = x^5 - x^4 - 4x^3 + 3x^2 + 3x - 1$ over the rational field, G the Galois group. In fact, $G \simeq C_5$ is the cyclic group of order 5, and the stabilizer of any root of f is trivial. The faithful pair $(C_5, 1)$ is Q-trivial by Proposition 5. Nevertheless, the pair $(C_5, 1)$ is not doubly homogeneous.

Proposition 5. *Let (G, H) be as in Proposition 1. If the cardinality of the set G/H is a prime number, then (G, H) is Q-trivial.*

Proof. When $|G/H| = 2$, the pair (G, H) is doubly homogeneous and we are done. When $|G/H|$ is an odd prime number, the proposition is a straightforward result of [10] Theorem 1 and Proposition 1.   □

Figure 1 shows the relations between different classes of Q-trivial pairs. This is based on Theorem 3 of [8], Corollary 1.6 of [2], and Proposition 3.1 of [15].

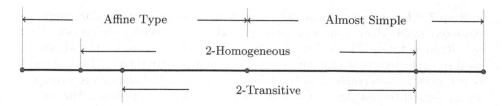

Fig. 1. Classification of $\mathbb{Q}$-trivial pairs

2.4 An Algorithm Deciding $\mathbb{Q}$-Triviality of Galois Groups

Assume that $f \in \mathbb{Q}[x]$ ($f(0) \neq 0$) is irreducible with Galois Group G and a root stabilizer H. In this subsection, we develop an algorithm deciding whether a pair (G, H) is $\mathbb{Q}$-trivial for such a polynomial f. Moreover, numerical results show that the algorithm is quite efficient compared with some other relative algorithms (see Table 3 and 4). All numerical results are obtained on a desktop with Windows 7 operating system, 8 GB RAM and a 3.30 GHz Intel Core i5-4590 processor with 4 cores.

The "IsQtrivial" Algorithm. Algorithm 1 shown below is designed according to Sect. 2.2 and Sect. 2.3. Step 4 of this algorithm is due to Proposition 5 while Step 5 is based on Corollary 1. The $\mathbb{Q}[G]$-submodule B generated by u in Step 6 is contained in the $\mathbb{Q}[G]$-submodule

$$\mathcal{V}_0 = \Big\{ \sum_{\bar{t} \in G/H} a_{\bar{t}} \bar{t} \,\Big|\, a_{\bar{t}} \in \mathbb{Q}, \sum_{\bar{t} \in G/H} a_{\bar{t}} = 0 \Big\}$$

with character $1_H^G - 1$ and $\mathbb{Q}$-dimension $\deg(f) - 1$. So the correctness of Steps 7–10 follows from Proposition 2.

Algorithm 1: IsQtrivial

Input: An irreducible polynomial $f \in \mathbb{Q}[x]$ with $f(0) \neq 0$;
Output: "**True**" if the pair (G, H) is $\mathbb{Q}$-trivial and "**False**" otherwise.
1   **if** (f is reducible or $f(0) == 0$) **then** {Return "Error!"} **end**
2   **if** ($\deg(f)$ is a prime number) **then** {**Return True;**} **end**
3   Compute the Galois group G of f;
4   **if** (G is doubly transitive) **then** {**Return True;**} **end**
5   **if** ($\deg(f) \notin \mathcal{S}$ as defined in (1)) **then** {**Return False;**} **end**
6   Compute $B = {}_{\mathbb{Q}[G]}\langle u \rangle$ with $u = \bar{s} - \bar{1} \in \mathbb{Q}[G/H]$ for an $s \notin H$;
7   **if** ($\dim(B) == \deg(f) - 1$ and B is $\mathbb{Q}$-irreducible) **then**
8 | **Return True;**
9   **end**
10 **Return False;**

The cost of Algorithm 1 depends mainly on the cost of computing the Galois group in Step 3 and the cost of deciding whether the module B is $\mathbb{Q}$-irreducible in Step 7. The algorithm is implemented with Magma. In particular, Step 7 is rendered by the Magma function IsIrreducible introduced in the section titled by "Representation Theory" in the Magma handbook. However, the algorithm used to implement that function is not specified explicitly therein. Thus, its cost is not known. Nevertheless, in the following, a great deal of random examples are generated to test Algorithm 1 in Table 1 to show its efficiency. But before that, we need to define "a random polynomial" technically:

For an $f \in \mathbb{Q}[x]$ of degree at most n, we define its *height* $h(f) = \max\limits_{0 \leq i \leq n} |c_{f,i}|$

with $c_{f,i}$ the coefficient of the term x^i of f. Then we set

$$\mathbb{Z}_{H,n}[x] = \{f \in \mathbb{Z}[x] \mid h(f) \leq H, \deg(f) \leq n\}$$

to be the set of the integer polynomials with height bounded by a positive integer H and degree not higher than n. Further, we set

$$\bar{\mathbb{Z}}^{ir}_{H,n}[x] = \{f \in \mathbb{Z}_{H,n}[x] \mid \deg(f) = n, f(0) \neq 0 \text{ and } f \text{ is irreducible}\}.$$

One notices that both $\mathbb{Z}_{H,n}[x]$ and $\bar{\mathbb{Z}}^{ir}_{H,n}[x]$ are finite sets. Hence we can equip the set $\mathbb{Z}_{H,n}[x]$ (*resp.*, the set $\bar{\mathbb{Z}}^{ir}_{H,n}[x]$) with the probability measure $\mathscr{P}_{H,n}$ (*resp.*, with the probability measure $\bar{\mathscr{P}}^{ir}_{H,n}$) determined by the discrete uniform distribution on it.

The random polynomials used in Table 1 ar generated in accordance with the probability measure $\bar{\mathscr{P}}^{ir}_{10,n}$. The way to generate such a polynomial f is as follows: First, generate each of its leading coefficient and constant term by picking an integer number equiprobably from the set $\{\pm 1, \ldots, \pm 10\}$, then pick each of the rest of the coefficients of f equiprobably from the set $\{-10, -9, \ldots, 10\}$. Second, check whether f is irreducible: if it is, then we are done; otherwise, go back to the first step.

In Table 1 (and throughout the section), the notation "#Poly" denotes the number of the polynomials that are generated and tested in a single class. As can be seen, almost all the randomly generated polynomials have doubly transitive Galois groups. In fact, most of the Galois groups in Table 1 are symmetry groups (which is consistent with Theorem 1 in [5]). The algorithm is effective and efficient for the randomly generated examples in the sets $\bar{\mathbb{Z}}^{ir}_{10,n}[x]$. In order to test the algorithm for other types of groups, we take advantage of the Magma function PolynomialWithGaloisGroup, which provides polynomials with all types of transitive Galois groups of degree between 2 and 15. The results are shown in Table 2.

In both tables, the "GaloisFail" columns show, for each degree, the numbers of the polynomials with Galois groups computed unsuccessfully in Algorithm 1 Step 3, which is implemented by the Magma functions GaloisGroup and GaloisProof. There are more "GaloisFail" cases in Table 2. The problem is: in those "GaloisFail" cases, though the Galois groups can be computed by the first function (which does not provide proven results), the second function returns error and fails to support the result. The "Average Time" in Table 2 excludes the "GaloisFail" examples, *i.e.*, it only counts in the "Qtrivial" and the "NotQtrivial" cases. We see that the algorithm is still efficient when the Galois group is successfully computed.

Table 1. Random test for `IsQtrivial`

Deg	#Poly	2-Transitive	Qtrivial	NotQtrivial	GaloisFail	Average time (s)
6	10000	9989	9989	11	0	0.025644
8	10000	9998	9998	2	0	0.055090
9	10000	10000	10000	0	0	0.069871
15	10000	10000	10000	0	0	0.301505
20	10000	10000	10000	0	0	0.698264
28	10000	10000	10000	0	0	1.532056
60	40	40	40	0	0	32.2309
81	40	40	40	0	0	107.486
90	40	40	40	0	0	231.638
120	40	40	40	0	0	2057.24

Table 2. Testing `IsQtrivial` by different galois groups

Deg	#Poly	2-Transitive	Qtrivial	NotQtrivial	GaloisFail	Average time (s)
4	5	2	2	3	0	0.019
6	16	4	4	12	0	0.027
8	50	6	6	43	1	0.086
9	34	2	2	23	9	0.095
10	45	2	2	36	7	0.103
12	301	2	2	292	7	0.165
14	63	2	2	41	20	0.215
15	104	2	2	62	40	0.222

Ensuring Lattice Triviality. By [11] Theorem 2, almost all irreducible polynomials f with $f(0) \neq 0$ have trivial lattice $\mathcal{R}_f$. However, the general algorithms FindRealtions in [13,16] and GetBasis in [27], dealing with the general inputs which are arbitrarily given nonzero algebraic numbers instead of all the roots of a certain polynomial, are not very efficient in proving exponent lattice triviality in the latter case.

For an irreducible polynomial $f \in \mathbb{Q}[x]$ with $f(0) \neq 0$, if the function `IsQtrivial`(f) returns **True** and f is proved to have no root being a root of rational by Algorithm 5 in [28] (named "`RootOfRationalTest`" therein, we call it "`IsROR`" here instead), then $\mathcal{R}_f$ is trivial by Proposition 1 and Remark 1. We call this the "`IsQtrivial+IsROR`" procedure. Table 3 shows the efficiency of that procedure to prove the triviality of the exponent lattice of a random polynomial picked from the set $\mathbb{Z}_{10,n}[x]$ in accordance with the probability measure $\mathscr{P}_{10,n}[x]$.

Table 3. IsQtrivial Ensuring triviality efficiently

Class	Polynomial	Runtime (s)		
		FindRelations	GetBasis	IsQtrivial + IsROR
$n = 4$	$f^{(1)}$	32.8947	83.7598	0.016
	$f^{(2)}$	19.1995	54.343	0.016
	$f^{(3)}$	34.6466	90.4592	0.016
$n = 5$	$g^{(1)}$	OT	OT	0.000
	$g^{(2)}$	OT	OT	0.000
	$g^{(3)}$	OT	OT	0.000
$n = 9$	$h^{(1)}$	OT	OT	0.047
	$h^{(2)}$	OT	OT	0.078
	$h^{(3)}$	OT	OT	0.047

The notation "OT" in Table 3 (and throughout the section) means the computation is not finished within two hours. As is shown in Table 3, it is time-consuming for the general algorithms FingRelations and GetBasis to prove the exponent lattice triviality of a random polynomial. So the "IsQtrivial+IsROR" procedure can be used before running either of the two general algorithms, when the inputs are all the roots of a certain polynomial. If the procedure fails to prove the triviality, then one turns to the general algorithms.

In the next part of this subsection, the "IsQtrivial+IsROR" procedure is slightly extended to render an algorithm called "Fastbasis$_+$", whose efficiency will be tested with a great deal of examples of much larger degree bound n.

The "FastBasis$_+$" Algorithm. Similarly to the "IsQtrivial+IsROR" procedure, Theorem 3.2 in [26] allows one to prove lattice triviality of a polynomial by proving doubly homogeneousness of its Galois group and by checking the condition that none of its roots is a root of a rational. Slightly extending this "Is2Homo+IsROR" procedure, the algorithm FastBasis in [26] aims at computing the lattice $\mathcal{R}_f$ fast for any f in a *generic* set $E \subset \mathbb{Q}[x]$ consisting of the polynomials f for which both the following two conditions hold:

(i) $\exists c \in \mathbb{Q}^*, g \in \mathbb{Q}[x], k \in \mathbb{Z}_{\geq 1}$ *so that* $f = cg^k$, g *is irreducible and* x *does not divide* $g(x)$;

(ii) *all the roots of* g *are roots of rational or the Galois group of* g *is doubly homogeneous.*

Similarly, we can define another set $E_+ \subset \mathbb{Q}[x]$ consisting of the polynomials f for which both the condition (i) above and the following condition hold:

(ii)' *all the roots of* g *are roots of rational or* IsQtrivial(g) == **True.**

Then, by Proposition 4 and Example 2, one claims that $E_+ \supsetneq E$. Thus, E_+ is also generic in the sense of [26]. That is we have the following proposition similar to [26] Corollary 5.5:

Proposition 6. *For any* $n \geq 2$, $\lim_{H \to \infty} \mathscr{P}_{H,n}(E_+ \cap \mathbb{Z}_{H,n}[x]) = 1$.

By replacing Steps 6 – 7 in Algorithm 6.1 of [26] (namely, FastBasis) by a new step, *i.e.*, Step 8 of Algorithm 2, we obtain an algorithm "FastBasis$_+$", which slightly extends the "IsQTrivial+IsROR" procedure. Like FastBasis, the algorithm FastBasis$_+$ computes the lattice $\mathcal{R}_f$ for any $f \in E_+$ while returning a special symbol "F" when $f \notin E_+$. Hence by Proposition 6, the probability of success of FastBasis$_+$, when applied to a random polynomial in $\mathbb{Z}_{H,n}[x]$, is close to one if the height bound H is large.

The algorithm FastBasis$_+$ is implemented with Magma while the algorithm FastBasis is implemented with Mathematica in [26]. In Table 4 we compare these two implementations by applying them to a great deal of random polynomials of varies degree. The polynomials in Table 4 are picked randomly from the

---

Algorithm 2: FastBasis$_+$

Input: A polynomial $f \in \mathbb{Z}[x]$;
Output: A basis of $\mathcal{R}_f$ if $f \in E_+$, the symbol "F" if otherwise.
1   **if** (f has at least two co-prime irreducible factors or $x|f(x)$) **then**
2 | Return "F";
3   **end**
4   Suppose that the only irreducible factor of f is g (*i.e.*, $f = cg^k$, $c \in \mathbb{Q}^*$, $k \geq 1$);
5   **if** (all the roots of g are roots of rational) **then**
6 | Compute a basis of $\mathcal{R}_g$ by the algorithm GetBasis in [27];
7   **else**
8 | **if** (IsQtrivial(g) == **False**) **then** {**Return** "F"} **end if**;
9 | Compute a basis of $\mathcal{R}_g$ by Prop. 3.3 of [26];
10 **end**
11 **Return** the basis of $\mathcal{R}_f$ derived from the basis of $\mathcal{R}_g$ via Prop. 5.2 of [26];

---

class $\mathbb{Z}_{10,n}[x]$ in accordance with the probability measure $\mathscr{P}_{10,n}$. The notation "#Success" therein denotes the number of those polynomials in each class for which the algorithm returns a lattice basis successfully within two hours, while the notation "#F" gives the number of the polynomials in each class that are proved to be outside of the set E_+ within two hours. The average time only counts in all the "Success" examples. We can see from Table 4 and Fig. 2 that for the small inputs with $n < 15$, the implementation of FastBasis is slightly more efficient while for those lager inputs with $n > 15$, the implementation of FastBasis$_+$ is much more efficient. This allows one to handle inputs with higher degree that were intractable before. Furthermore, the table shows that although the height bound $H = 10$ is not large, the success probability of the algorithm FastBasis$_+$ is still high when no "OT" occurs.

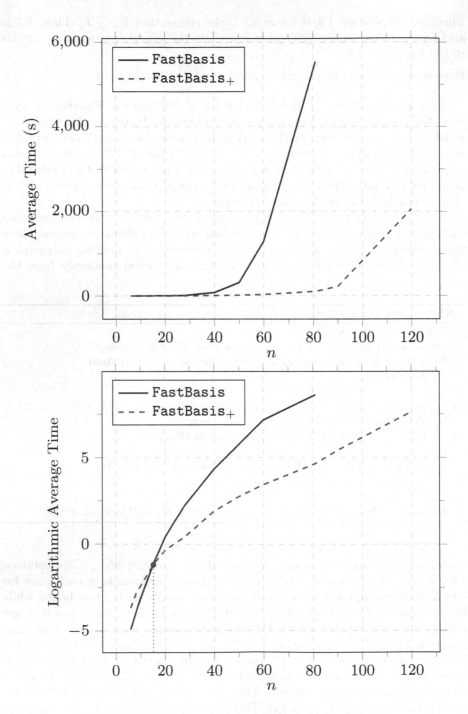

Fig. 2. Comparing average runtime of two implementations

Table 4. FastBasis *v.s.* FastBasis$_+$

Class	#*Poly*	FastBasis				FastBasis$_+$			
		# Success	# F	OT	Average time (s)		OT	# F	# Success
$n = 6$	10000	9011	989	0	0.007304	0.025499	0	989	9011
$n = 8$	10000	9064	936	0	0.018372	0.055328	0	936	9064
$n = 9$	10000	9113	887	0	0.029044	0.069996	0	887	9113
$n = 15$	10000	9227	773	0	0.305941	0.301110	0	773	9227
$n = 20$	10000	9243	757	0	1.502110	0.700131	0	757	9243
$n = 28$	10000	9279	721	0	9.29961	1.527806	0	721	9279
$n = 40$	100	93	7	0	76.3928	6.788000	0	7	93
$n = 50$	100	96	4	0	315.523	15.70400	0	4	96
$n = 60$	35	33	1	1	1291.38	31.27800	0	1	34
$n = 81$	40	15	1	24	5539.67	104.515	0	1	39
$n = 90$	40	0	2	38	–	224.413	0	2	38
$n = 120$	40	–	–	–	–	2058.228	0	2	38

During the experiment, we find that a large proportion of the polynomials randomly picked from the class $\mathbb{Z}_{10,n}[x]$ are in the set E_+ and are irreducible with no root being a root of rational. For these polynomials, the algorithm FastBasis$_+$ is essentially reduced to the "IsQTrivial+IsROR" procedure. So the examples of larger n shown in Table 4 can be regarded as complementary materials for Table 3.

3   Lattice Triviality Through Galois-Like Groups

As is shown in Example 1, provided only the pair (G, H), one may not be able to decide whether the lattice $\mathcal{R}_f$ is trivial or not. Here f is an irreducible polynomial with Galois group G and a root stabilizer H. In this section, the concept of a Galois-like group is introduced. An equivalent condition for the lattice $\mathcal{R}_f$ to be trivial is given through the concept of a Galois-like group.

3.1   Root Permutations Preserving Multiplicative Relations

Set $f \in \mathbb{Q}[x]$ $(f(0) \neq 0)$ to be a polynomial with no multiple roots. Denote by Σ the symmetry group operating on the set $\Omega = \{r_1, \dots, r_n\}$ of the roots of f. In the sequel, we denote by $\vec{\Omega} = (r_1, \dots, r_n)^T$ a vector of the roots and by $\sigma(\vec{\Omega}) = (\sigma(r_1), \dots, \sigma(r_n))^T$ a permutation of $\vec{\Omega}$ with $\sigma \in \Sigma$.

Definition 1. A Galois-like group *of the polynomial f refers to any one of the following groups:*

(i) $\mathcal{G}_f = \{\sigma \in \Sigma \mid \forall v \in \mathbb{Z}^n,\ \vec{\Omega}^v = 1 \Rightarrow \sigma(\vec{\Omega})^v = 1\}$;

(ii) $\mathcal{G}_f^B = \{\sigma \in \Sigma \mid \forall v \in \mathbb{Z}^n,\ \vec{\Omega}^v \in \mathbb{Q} \Rightarrow \sigma(\vec{\Omega})^v = \vec{\Omega}^v\}$;

(iii) $\mathcal{G}_f^{\mathbb{Q}} = \{\sigma \in \Sigma \mid \forall v \in \mathbb{Z}^n,\ \vec{\Omega}^v \in \mathbb{Q} \Rightarrow \sigma(\vec{\Omega})^v \in \mathbb{Q}\}$;

To verify the terms used above in the definition, we need to prove that any subset of Σ defined in Definition 1 is indeed a group:

Proposition 7. *Suppose that $f \in \mathbb{Q}[x]$ $(f(0) \neq 0)$ is a polynomial with no multiple roots. If $\mathcal{G} = \mathcal{G}_f, \mathcal{G}_f^B$ or $\mathcal{G}_f^{\mathbb{Q}}$, then $\mathcal{G}$ is a subgroup of Σ.*

Proof. Any $\sigma \in \Sigma$ results in a coordinate permutation $\hat{\sigma}$ operating on the space $\mathbb{C}^n$ with $n = |\Omega|$ in a manner so that for any vector $v = (v_1, \ldots, v_n)^T \in \mathbb{C}^n$, $\hat{\sigma}(v) = (b_1, \ldots, b_n)^T$ with $b_i = v_j$ whenever $\sigma(r_i) = r_j$. Then one observes that the equalities $\widehat{\sigma^{-1}} = (\hat{\sigma})^{-1}$, $\sigma(\vec{\Omega})^{\hat{\sigma}(v)} = \vec{\Omega}^v$ and $\vec{\Omega}^{\hat{\sigma}(v)} = \sigma^{-1}(\vec{\Omega})^v$ hold for any $\sigma \in \Sigma$ and any $v \in \mathbb{Z}^n$.

Set $\mathcal{G} = \mathcal{G}_f$ ($\mathcal{G}_f^B$ or $\mathcal{G}_f^{\mathbb{Q}}$ respectively) and $\mathcal{R} = \mathcal{R}_f$ ($\mathcal{R}_f^{\mathbb{Q}}$, respectively). Then, by definition, $\hat{\sigma}^{-1}(v) \in \mathcal{R}$ for any $\sigma \in \mathcal{G}$ and any $v \in \mathcal{R}$. Hence the set $\hat{\sigma}^{-1}(\mathcal{R}) = \{\hat{\sigma}^{-1}(v) \mid v \in \mathcal{R}\}$ is a subset of the lattice $\mathcal{R}$. Noting that $\hat{\sigma}^{-1}$ operates linearly, one concludes that $\hat{\sigma}^{-1}(\mathcal{R})$ is also a lattice. Thus, $\hat{\sigma}^{-1}(\mathcal{R})$ is a sub-lattice of $\mathcal{R}$. Since $\hat{\sigma}^{-1}$ is linear and non-singular, any basis of $\mathcal{R}$ is transformed into a basis of $\hat{\sigma}^{-1}(\mathcal{R})$ by $\hat{\sigma}^{-1}$. Hence $\text{rank}(\mathcal{R}) = \text{rank}(\hat{\sigma}^{-1}(\mathcal{R}))$. Since $\hat{\sigma}^{-1}$ is orthogonal on the space $\mathbb{R}^n$ and orthogonal operations preserve the lattice determinant, $\mathcal{R} = \hat{\sigma}^{-1}(\mathcal{R})$. Thus, $\hat{\sigma}(\mathcal{R}) = \mathcal{R}$.

So $\hat{\sigma}(v) \in \mathcal{R}$ for any $\sigma \in \mathcal{G}$ and any $v \in \mathcal{R}$. If $\mathcal{G} = \mathcal{G}_f$ (or $\mathcal{G}_f^{\mathbb{Q}}$) and $\mathcal{R} = \mathcal{R}_f$ (or $\mathcal{R}_f^{\mathbb{Q}}$, respectively), then $\vec{\Omega}^{\hat{\sigma}(v)} = 1$ (or $\vec{\Omega}^{\hat{\sigma}(v)} \in \mathbb{Q}$, respectively). Equivalently, $\sigma^{-1}(\vec{\Omega})^v = 1$ (or $\sigma^{-1}(\vec{\Omega})^v \in \mathbb{Q}$). Hence $\sigma^{-1} \in \mathcal{G}$ for any $\sigma \in \mathcal{G}$. Now suppose that $\mathcal{G} = \mathcal{G}_f^B$ and $\mathcal{R} = \mathcal{R}_f^{\mathbb{Q}}$. Since $\hat{\sigma}(v) \in \mathcal{R}_f^{\mathbb{Q}}$ and $\vec{\Omega}^{\hat{\sigma}(v)} \in \mathbb{Q}$, $\sigma(\vec{\Omega})^{\hat{\sigma}(v)} = \vec{\Omega}^{\hat{\sigma}(v)}$ follows from the definition of $\mathcal{G}_f^B$. The left-hand side of this equality equals $\vec{\Omega}^v$ while its right-hand side equals $\sigma^{-1}(\vec{\Omega})^v$. Hence $\sigma^{-1}(\vec{\Omega})^v = \vec{\Omega}^v$ for any $\sigma \in \mathcal{G}_f^B$ and $v \in \mathcal{R}_f^{\mathbb{Q}}$. Thus, $\sigma^{-1} \in \mathcal{G}_f^B$.

The closure of the multiplication in the subset $\mathcal{G}$ of Σ and the fact that $1 \in \mathcal{G}$ are straightforward. Thus $\mathcal{G}$ is a group. $\qquad\square$

Define groups $\langle \Omega \rangle = \{\vec{\Omega}^v \mid v \in \mathbb{Z}^n\}$ and $\langle \Omega \rangle_{\mathbb{Q}} = \{c\vec{\Omega}^v \mid c \in \mathbb{Q}^*,\ v \in \mathbb{Z}^n\}$. The following proposition asserts that the Galois-like groups $\mathcal{G}_f$ and $\mathcal{G}_f^B$ of a polynomial f are subgroups of the automorphism groups of $\langle \Omega \rangle$ and $\langle \Omega \rangle_{\mathbb{Q}}$, respectively.

Proposition 8. *The following relations hold:*

(i) $\mathcal{G}_f \simeq \{\eta \in \text{Aut}\,(\langle \Omega \rangle) \mid \forall r_i \in \Omega,\ \eta(r_i) \in \Omega\}$;

(ii) $\mathcal{G}_f^B \simeq \{\eta \in \text{Aut}\,(\langle \Omega \rangle_{\mathbb{Q}}) \mid \forall r_i \in \Omega,\ \eta(r_i) \in \Omega;\ \eta|_{\mathbb{Q}^*} = id_{\mathbb{Q}^*}\}$.

Proof. Denote by $\text{Aut}_\Omega(\langle \Omega \rangle)$ the group in the right side of the formula in (i) and by $\text{Aut}_\Omega^{\mathbb{Q}}(\langle \Omega \rangle_{\mathbb{Q}})$ the one in the right-hand side of the formula in (ii).

Set $\mathcal{G} = \mathcal{G}_f$ (or $\mathcal{G}_f^B$) and $A = \mathrm{Aut}_\Omega(\langle\Omega\rangle)$ (or $\mathrm{Aut}_\Omega^\mathbb{Q}(\langle\Omega\rangle_\mathbb{Q})$ respectively). For any $\sigma \in \mathcal{G}$, we define an element E_σ in A in the following way: for any $\vec{\Omega}^v \in \langle\Omega\rangle$, $E_\sigma(\vec{\Omega}^v) = \sigma(\vec{\Omega})^v$ (or, for any $c\vec{\Omega}^v \in \langle\Omega\rangle_\mathbb{Q}$, $E_\sigma(c\vec{\Omega}^v) = c\sigma(\vec{\Omega})^v$). From the definition of $\mathcal{G}$, we see that $\sigma(\vec{\Omega})^v = \sigma(\vec{\Omega})^{v'}$ whenever $\vec{\Omega}^v = \vec{\Omega}^{v'} \in \langle\Omega\rangle$ (or that $c_1\sigma(\vec{\Omega})^v = c_2\sigma(\vec{\Omega})^{v'}$ whenever $c_1\vec{\Omega}^v = c_2\vec{\Omega}^{v'} \in \langle\Omega\rangle_\mathbb{Q}$). Thus the map E_σ: $\langle\Omega\rangle \to \langle\Omega\rangle$ (or E_σ: $\langle\Omega\rangle_\mathbb{Q} \to \langle\Omega\rangle_\mathbb{Q}$) is well defined. It is trivial to verify the fact that E_σ is an automorphism of $\langle\Omega\rangle$ (or of $\langle\Omega\rangle_\mathbb{Q}$) and the property that for all $r_i \in \Omega$, $E_\sigma(r_i) = \sigma(r_i) \in \Omega$ (or, moreover, $E_\sigma(c) = c$ for any $c \in \mathbb{Q}^*$). Hence E_σ is indeed in the set A. Thus $E_\bullet$ is a map from $\mathcal{G}$ to A.

For any $\eta \in A$, η is injective and $\eta(\Omega) \subset \Omega$. Since Ω is finite, $\eta(\Omega) = \Omega$. Hence $\eta|_\Omega \in \Sigma$. Because η is an automorphism (or an automorphism fixing every rational number), $\eta|_\Omega(\vec{\Omega})^v = 1$ whenever $\vec{\Omega}^v = 1$ (or $\eta|_\Omega(\vec{\Omega})^v = \vec{\Omega}^v$ whenever $\vec{\Omega}^v \in \mathbb{Q}$). Thus, $\eta|_\Omega \in \mathcal{G}$. Define $R_\eta = \eta|_\Omega$, then $R_\bullet$ is a map from A to $\mathcal{G}$.

It is clear that both the maps $E_\bullet$ and $R_\bullet$ are group homomorphisms. That is $E_{\sigma_1\sigma_2} = E_{\sigma_1}E_{\sigma_2}$ and $R_{\eta_1\eta_2} = R_{\eta_1}R_{\eta_2}$ for any $\sigma_1, \sigma_2 \subset \mathcal{G}$ and any $\eta_1, \eta_2 \in A$. One also verifies easily that $E_{R_\bullet} = id_A$ and $R_{E_\bullet} = id_\mathcal{G}$. Hence $\mathcal{G} \simeq A$. $\qquad\square$

By definition, a Galois-like group of a polynomial f is the group of the permutations between its roots that preserve all the multiplicative relations between them. Since any element in the Galois group of f preserves all polynomial relations between the roots, the following relations between the Galois group and a Galois-like group of f is straightforward:

Proposition 9. *Suppose that $f \in \mathbb{Q}[x]$ $(f(0) \neq 0)$ has no multiple roots. Then, regarded as a permutation group operating on the roots of f, the Galois group of f is a subgroup of any Galois-like group of f.*

Besides, the following relations between the Galois-like groups are straightforward but noteworthy:

Proposition 10. *Let f be as in Proposition 9. Then $\mathcal{G}_f^B \leq \mathcal{G}_f$ and $\mathcal{G}_f^B \leq \mathcal{G}_f^\mathbb{Q}$.*

3.2 Generalization of Sufficient Conditions of Exponent Lattice Triviality

With the help of the concept of Galois-like groups, we can generalize many sufficient conditions that imply the triviality of exponent lattices.

Lemma 1. *Set $f \in \mathbb{Q}[x]$ $(f(0) \neq 0)$ to be a polynomial without multiple roots. Denote by $\vec{\Omega} = (\alpha_1, \ldots, \alpha_s, \gamma_1, \ldots, \gamma_t)^T$ the vector of all the roots of f with α_i the roots that are not roots of rational. Suppose that the Galois-like group $\mathcal{G}_f$ is doubly transitive, then any multiplicative relation $v = (v_1, \ldots, v_{s+t})^T \in \mathcal{R}_{\vec{\Omega}} = \mathcal{R}_f$ satisfies the following condition:*

$$v_1 = \cdots = v_s = \frac{v_1 + \cdots + v_{s+t}}{s+t}. \tag{2}$$

Lemma 2. *Let f and $\vec{\Omega}$ be as in Lemma 1. Suppose that the Galois-like group $\mathcal{G}_f^B$ or $\mathcal{G}_f^{\mathbb{Q}}$ is doubly transitive, then any multiplicative relation $v \in \mathcal{R}_{\vec{\Omega}}^{\mathbb{Q}} = \mathcal{R}_f^{\mathbb{Q}}$ satisfies the condition (2).*

The proofs of those two propositions above are both almost the same to the one of Theorem 3 in [3], because of which we do not give any of them here. A direct corollary of these propositions are as follows:

Proposition 11. *Set $f \in \mathbb{Q}[x]$ ($f(0) \neq 0$) to be a polynomial without multiple roots and none of its roots is a root of rational. If the group $\mathcal{G}_f^B$ or $\mathcal{G}_f^{\mathbb{Q}}$ (respectively, $\mathcal{G}_f$) is doubly transitive, then the lattice $\mathcal{R}_f^{\mathbb{Q}}$ (respectively, $\mathcal{R}_f$) is trivial.*

This is a generalization of Theorem 3 in [3]. The essential idea is that the proof of Theorem 3 in [3] relies only on the properties of Galois-like groups (*i.e.*, preserving all the multiplicative relations) but not on those properties that are possessed uniquely by the Galois groups.

Noting that $\mathcal{G}_f^B \leq \mathcal{G}_f$, one concludes from Proposition 11 that both the lattices $\mathcal{R}_f^{\mathbb{Q}}$ and $\mathcal{R}_f$ are trivial whenever the group $\mathcal{G}_f^B$ is doubly transitive and none of the roots of f is a root of rational. More generally, we have the following proposition and Corollary 2:

Proposition 12. *Set $f \in \mathbb{Q}[x]$ ($f(0) \neq 0$) to be a polynomial without multiple roots. Define $\mathcal{W}_f = \mathbb{Q} \otimes \mathcal{R}_f$ and $\mathcal{W}_f^{\mathbb{Q}} = \mathbb{Q} \otimes \mathcal{R}_f^{\mathbb{Q}}$ with "$\otimes$" the tensor product of $\mathbb{Z}$-modules. Set $\mathcal{V}_0 = \{v \in \mathbb{Q}^n \mid \sum_{i=1}^n v_i = 0\}$ and $\mathcal{V}_1 = \{c(1_1, \ldots, 1_n)^T \mid c \in \mathbb{Q}\}$ with $n = \deg(f)$. Suppose that $\mathcal{G}_f^B$ is transitive, then the following conclusions hold:*

(i) $\mathcal{W}_f^{\mathbb{Q}} \cap \mathcal{V}_0 = \mathcal{W}_f \cap \mathcal{V}_0$;
(ii) $\mathcal{W}_f^{\mathbb{Q}} = \mathcal{W}_f + \mathcal{V}_1$ and thus $\mathcal{W}_f^{\mathbb{Q}} = \mathcal{W}_f$ iff $f(0) \in \{1, -1\}$.

Proof. The proof is almost the same to the one of Lemma 1 in [7], except that we require the transitivity of the Galois-like group $\mathcal{G}_f^B$ instead of the the transitivity of the Galois group of f. $\qquad\square$

Corollary 2. *Let f be as in Proposition 12 such that the group $\mathcal{G}_f^B$ is transitive, then the lattice $\mathcal{R}_f$ is trivial iff the lattice $\mathcal{R}_f^{\mathbb{Q}}$ is.*

Proof. Since the group $\mathcal{G}_f^B$ is transitive, $\mathcal{W}_f^{\mathbb{Q}} = \mathcal{W}_f + \mathcal{V}_1$ by Proposition 12. So $\mathcal{W}_f^{\mathbb{Q}}$ is trivial iff $\mathcal{W}_f$ is trivial. Hence

$$\mathcal{R}_f \text{ is trivial} \Longleftrightarrow \mathcal{W}_f \text{ is trivial}$$
$$\Longleftrightarrow \mathcal{W}_f^{\mathbb{Q}} \text{ is trivial}$$
$$\Longleftrightarrow \mathcal{R}_f^{\mathbb{Q}} \text{ is trivial.}$$

$\square$

Similarly to Proposition 11, we have the following result:

Proposition 13. *Let $f \in \mathbb{Q}[x]$ be as in Proposition 11. If the group $\mathcal{G}_f^B$ or $\mathcal{G}_f^Q$ is doubly homogeneous, then the lattice $\mathcal{R}_f^Q$ is trivial.*

This is a generalization of [26] Theorem 3.2. The proof of this proposition is almost the same with the one given in [26], hence we omit it. Another generalization trough Galois-like groups of the "Only If" part of Proposition 1 is given below:

Proposition 14. *Set $f \in \mathbb{Q}[x]$ $(f(0) \neq 0)$ to be a polynomial without multiple roots and one of its roots is not a root of rational. Set $\mathcal{G} = \mathcal{G}_f$ (respectively, $\mathcal{G} = \mathcal{G}_f^B$ or $\mathcal{G}_f^Q$) and $\mathcal{R} = \mathcal{R}_f$ (resp., $\mathcal{R} = \mathcal{R}_f^Q$). If $\mathcal{G}$ is transitive and the pair $(\mathcal{G}, \mathcal{H})$, with $\mathcal{H}$ a root stabilizer, is $\mathbb{Q}$-trivial, then $\mathcal{R}$ is trivial.*

Proof. Let $\mathcal{V}_1, \mathcal{V}_0$ and $\mathcal{W} = \mathbb{Q} \otimes \mathcal{R}$ be as in Proposition 12. For any $\sigma \in \mathcal{G}$, we define a coordinate permutation $\hat{\sigma}$ as in the proof of Proposition 7. Then $\mathcal{W}$ is a $\mathbb{Q}[\mathcal{G}]$-submodule of $\mathbb{Q}^n$ by the definition of a Galois-like group (for any $v \in \mathbb{Q}^n$ or $v \in \mathcal{W}$, a group element σ operates in the way so that it maps v to the vector $\hat{\sigma}^{-1}(v)$).

Since the pair $(\mathcal{G}, \mathcal{H})$ is $\mathbb{Q}$-trivial, $\mathbb{Q}^n$ can be decomposed into two irreducible $\mathbb{Q}[\mathcal{G}]$-submodules: $\mathbb{Q}^n = \mathcal{V}_1 \oplus \mathcal{V}_0$ ([14] Proposition 12). Since $\mathcal{G}$ is transitive, $\mathcal{V}_0$ and $\mathcal{V}_1$ are the only two irreducible $\mathbb{Q}[\mathcal{G}]$-submodules of $\mathbb{Q}^n$:

Suppose that $\mathcal{V} \neq \mathcal{V}_0$ is an irreducible $\mathbb{Q}[\mathcal{G}]$-submodules and assume that $\mathcal{V} \cap \mathcal{V}_0 \supsetneq \{\mathbf{0}\}$. Then $\mathcal{V} \supsetneq \mathcal{V} \cap \mathcal{V}_0$ or $\mathcal{V}_0 \supsetneq \mathcal{V} \cap \mathcal{V}_0$. This contradicts the fact that both $\mathcal{V}$ and $\mathcal{V}_0$ are irreducible, since $\mathcal{V} \cap \mathcal{V}_0 \supsetneq \{\mathbf{0}\}$ is a proper $\mathbb{Q}[\mathcal{G}]$-submodules of at least one of them. So we have $\mathcal{V} \cap \mathcal{V}_0 = \{\mathbf{0}\}$. Noting that the $\mathbb{Q}$-dimension of $\mathcal{V}_0$ is $n - 1$, one concludes that $\dim_{\mathbb{Q}}(\mathcal{V}) = 1$. Set $v \in \mathcal{V} \setminus \{\mathbf{0}\}$, then $v \notin \mathcal{V}_0$ and $\sum_{i=1}^n v_i \neq 0$. Thus $\mathcal{V} \ni \sum_{\sigma \in \mathcal{G}} \hat{\sigma}^{-1}(v) = \frac{|\mathcal{G}|}{n}(\sum_{i=1}^n v_i)(1_1, \ldots, 1_n)^T \neq \mathbf{0}$ follows from the transitivity of $\mathcal{G}$. Hence $\mathcal{V} = \mathcal{V}_1$.

Thus all the $\mathbb{Q}[\mathcal{G}]$-submodules of $\mathbb{Q}^n$ are $\{\mathbf{0}\}$, $\mathcal{V}_1$, $\mathcal{V}_0$ and $\mathbb{Q}^n$ itself. If $\mathcal{V}_0 \subset \mathcal{W}$, then $\mathcal{W} \subset \mathcal{W}_f^Q$ and $\mathcal{V}_1 \subset \mathcal{W}_f^Q$ imply that $\mathcal{W}_f^Q = \mathbb{Q}^n$, which contradicts the assumption that f has a root that is not a root of rational. Hence $\mathcal{V}_0 \not\subset \mathcal{W}$, which means $\mathcal{W} = \{\mathbf{0}\}$ or $\mathcal{W} = \mathcal{V}_1$. Thus $\mathcal{W}$ is trivial and so is the lattice $\mathcal{R}$. $\square$

3.3 Necessary and Sufficient Condition for Exponent Lattice Triviality

In this subsection, we characterize those polynomials f with a trivial exponent lattice by giving a necessary and sufficient condition through the concept of a Galois-like group.

Theorem 1. *Set $f \in \mathbb{Q}[x]$ $(f(0) \neq 0)$ to be a polynomial without multiple roots. Denote by $\beta_1, \ldots, \beta_t$ the rational roots of f (if there are any) and by β_0 the rational number which is the product of all non-root-of-rational roots of f (if there are any). Then the lattice $\mathcal{R}_f$ is trivial iff all the following conditions hold:*

(i) the Galois-like group $\mathcal{G}_f = \Sigma$;

(ii) any root of f is rational or non-root-of-rational;
(iii) the lattice $\mathcal{R}_{\mathrm{Vec}(f)}$ is trivial with the vector $\mathrm{Vec}(f)$ given by:

$$\mathrm{Vec}(f) = \begin{cases} (\beta_0, \beta_1, \ldots, \beta_t)^T, & \text{if } f \text{ has both rational and} \\ & \quad \text{non-root-of-rational roots,} \\ (\beta_1, \ldots, \beta_t)^T, & \text{if any root of } f \text{ is rational,} \\ (\beta_0), & \text{if any root of } f \text{ is non-root-of-rational.} \end{cases}$$

Proof. "If": When $\deg(f) = 1$, $\mathcal{R}_f$ is trivial and we are done. Suppose in the following that $\deg(f) \geq 2$. Then the pair $(\mathcal{G}_f, \mathcal{H}) = (\Sigma, \mathcal{H})$ is doubly homogeneous, thus, also $\mathbb{Q}$-trivial for any root stabilizer $\mathcal{H}$. If f has a root that is not a root of rational, then $\mathcal{R}_f$ is trivial by Proposition 14. When all the roots of f are rational, the lattice $\mathcal{R}_f = \mathcal{R}_{\mathrm{Vec}(f)}$ is trivial.

"Only If": Now that $\mathcal{R}_f$ is trivial, the condition (i) is straightforward. Suppose that f has a root r which is a root of rational but not a rational number. Then each of the conjugations of r, say, $\{r = r^{(1)}, r^{(2)}, \ldots, r^{(s)}\}$, with $s \geq 2$, is a root of f. Then there is a positive integer m so that $(r/r^{(2)})^m = 1$. Thus, $\mathcal{R}_f$ is non-trivial, which contradicts the assumption. So the condition (ii) holds. Since any nontrivial multiplicative relation of the vector $\mathrm{Vec}(f)$ results in a nontrivial multiplicative relation between the roots of f, the condition (iii) holds. $\square$

From the "If" part of the proof we observe that, when restricted to polynomials f with degree higher than one, the condition (i) in Theorem 1 can be replaced by the statement "$\mathcal{G}_f$ is transitive and the pair $(\mathcal{G}_f, \mathcal{H})$ is $\mathbb{Q}$-trivial for any root stabilizer $\mathcal{H}$". An interesting result follows directly from this observation:

Corollary 3. *Let f be as in Theorem 1. If $\deg(f) \geq 2$ and the conditions (ii) – (iii) in Theorem 1 hold, then the following conditions are equivalent to each other:*

(i) $\mathcal{G}_f = \Sigma$;
(ii) $\mathcal{G}_f$ is doubly transitive;
(iii) $\mathcal{G}_f$ is doubly homogeneous;
(iv) $\mathcal{G}_f$ is transitive and the pair $(\mathcal{G}_f, \mathcal{H})$ is $\mathbb{Q}$-trivial for any root stabilizer $\mathcal{H}$.

Proof. The implications $(i) \Rightarrow (ii)$ and $(ii) \Rightarrow (iii)$ are trivial. The implication $(iv) \Rightarrow (i)$ follows from the "If" part of the proof of 1. Now we prove the implication $(iii) \Rightarrow (iv)$: If $\deg(f) = 2$ and f has only rational roots, the lattice $\mathcal{R}_f = \mathcal{R}_{\mathrm{Vec}(f)}$ is trivial. So $\mathcal{G}_f = \Sigma$. If $\deg(f) = 2$ but f has a root that is not a root of rational, f is irreducible over $\mathbb{Q}$ and $\Sigma = G_f \leq \mathcal{G}_f$ with G_f the Galois group of f. In either case $\mathcal{G}_f$ is transitive. When $\deg(f) \geq 3$, the transitivity of $\mathcal{G}_f$ follows from [9] Theorem 9.4A. Now the $\mathbb{Q}$-triviality of the pair $(\mathcal{G}, \mathcal{H})$ follows from Proposition 4. $\square$

Thus, the condition (i) of Theorem 1 can be replaced by any one of the conditions (ii) – (iv) in Corollary 3.

For the lattice $\mathcal{R}_f^{\mathbb{Q}}$, we have a similar result:

Theorem 2. *Set* $f \in \mathbb{Q}[x]$ $(f(0) \neq 0)$ *to be a polynomial without multiple roots. Then the lattice* $\mathcal{R}_f^{\mathbb{Q}}$ *is trivial iff all the following conditions hold:*

(i) the Galois-like group $\mathcal{G}_f^B = \Sigma$;
(ii) either $\deg(f) = 1$ *or any root of* f *is not a root of rational;*
(iii) f *is irreducible over* $\mathbb{Q}$.

Proof. "If": When $\deg(f) = 1$, this is trivial. Suppose in the following that $\deg(f) \geq 2$ and any root of f is not a root of rational. Then $\mathcal{G}_f^B = \Sigma$ is transitive and doubly homogeneous. Thus, the pair $(\mathcal{G}_f^B, \mathcal{H})$ is $\mathbb{Q}$-trivial for any root stabilizer $\mathcal{H}$ by Proposition 4. So $\mathcal{R}_f^{\mathbb{Q}}$ is trivial by Proposition 14.

"Only If": Now that $\mathcal{R}_f^{\mathbb{Q}}$ is trivial, it is clear that $\mathcal{G}_f^B = \Sigma$. Suppose on the contrary that f is reducible and g_1, g_2 are two of its factors. Let $\alpha_1, \ldots, \alpha_s$ denote the roots of g_1 and $\gamma_1, \ldots, \gamma_t$ the ones of g_2. Then for any two distinct integers k and l, $(\alpha_1 \ldots \alpha_s)^k (\gamma_1 \ldots \gamma_t)^l \in \mathbb{Q}$. This contradicts the assumption that $\mathcal{R}_f^{\mathbb{Q}}$ is trivial. So f is irreducible. Assume that $\deg(f) \geq 2$ and one of the roots r of f is a root of rational. The conjugations of r, say, $\{r = r^{(1)}, r^{(2)}, \ldots, r^{(n)}\}$ $(n \geq 2)$ are exactly all the roots of f. Then there is a positive integer m so that $(r/r^{(2)})^m = 1$. Thus, $\mathcal{R}_f$ is non-trivial and so is the lattice $\mathcal{R}_f^{\mathbb{Q}}$. This contradicts the assumption. $\square$

Remark 2. Theorem 2 still holds when the equality $\mathcal{G}_f^B = \Sigma$ is replaced by $\mathcal{G}_f^{\mathbb{Q}} = \Sigma$ in the condition (i). The proof is almost the same. Moreover, from the "If" part of the proof we observe that, when restricted to polynomials f with degree higher than one, the condition (i) in Theorem 2 can be replaced by the statement "$\mathcal{G}_f^B$ is transitive and the pair $(\mathcal{G}_f^B, \mathcal{H})$ is $\mathbb{Q}$-trivial for any root stabilizer $\mathcal{H}$" or the statement "$\mathcal{G}_f^{\mathbb{Q}}$ is transitive and the pair $(\mathcal{G}_f^{\mathbb{Q}}, \mathcal{H})$ is $\mathbb{Q}$-trivial for any root stabilizer $\mathcal{H}$".

The counterpart of Corollary 3 in this case is given below:

Corollary 4. *Let* f *be as in Theorem 2 and* $\mathcal{G} \in \{\mathcal{G}_f^B, \mathcal{G}_f^{\mathbb{Q}}\}$. *If* $\deg(f) \geq 2$ *and the conditions* $(ii) - (iii)$ *in Theorem 2 hold, then the following conditions are equivalent to each other:*

(i) $\mathcal{G} = \Sigma$;
(ii) $\mathcal{G}$ *is doubly transitive;*
(iii) $\mathcal{G}$ *is doubly homogeneous;*
(iv) $\mathcal{G}$ *is transitive and the pair* $(\mathcal{G}, \mathcal{H})$ *is* $\mathbb{Q}$-*trivial for any root stabilizer* $\mathcal{H}$.

Proof. The proof is similar to the one of Corollary 3. $\square$

Theorem 1 and 2 characterize, for the first time, the polynomial f with a trivial exponent lattice $\mathcal{R}_f$ or $\mathcal{R}_f^{\mathbb{Q}}$ with the help of the concept of a Galois-like group. The conditions (ii) – (iii) in both theorems can be decided very efficiently (by Sect. 5.1 of [28] and Sect. 2.2.1 of [27]). However, an effective algorithm deciding whether a Galois-like group, of a given polynomial f, equals the symmetry group Σ or not is not available at present.

4 Conclusion

We characterize the polynomials with trivial exponent lattices through the Galois and the Galois-like groups. Based on the algorithm IsQtrivial, we extensively improve the main algorithm in [26] proving triviality of the exponent lattice of a generic polynomial (when the polynomial degree is large). In addition, a sufficient and necessary condition is given with the help of the concept of a Galois-like group, which turns out to be essential in the study on multiplicative relations between the roots of a polynomial. Further study on Galois-like groups seems to be interesting and promising.

Acknowledgments. The author is very grateful to the referees for their careful reading and useful suggestions that have helped improve this paper a lot.

References

1. Almagor, S., Chapman, B., Hosseini, M., Ouaknine, J., Worrell, J.: Effective divergence analysis for linear recurrence sequences. arXiv preprint arXiv:1806.07740 (2018)
2. Bamberg, J., Giudici, M., Liebeck, M., Praeger, C., Saxl, J.: The classification of almost simple 3/2-transitive groups. Trans. Amer. Math. Soc. **365**(8), 4257–4311 (2013). https://doi.org/10.1090/S0002-9947-2013-05758-3
3. Baron, G., Drmota, M., Skałba, M.: Polynomial relations between polynomial roots. J. Algebra **177**(3), 827–846 (1995). https://doi.org/10.1006/jabr.1995.1330
4. Chen, S., Feng, R., Fu, G., Li, Z.: On the structure of compatible rational functions. In: Proceedings of the 36th International Symposium on Symbolic and Algebraic Computation. pp. 91–98. ACM Press, New York (2011). https://doi.org/10.1145/1993886.1993905
5. Cohen, S.D.: The distribution of the Galois groups of integral polynomials. Illinois J. Math. **23**(1), 135–152 (1979)
6. Derksen, H., Jeandel, E., Koiran, P.: Quantum automata and algebraic groups. J. Symbolic Comput. **39**(3–4), 357–371 (2005)
7. Dixon, J.D.: Polynomials with nontrivial relations between their roots. Acta Arith. **82**(3), 293–302 (1997)
8. Dixon, J.D.: Permutation representations and rational irreducibility. Bull. Aust. Math. Soc. **71**(3), 493–503 (2005). https://doi.org/10.1017/S0004972700038508
9. Dixon, J.D., Mortimer, B.: Permutation Groups. Springer Science & Business Media, Berlin (1996)
10. Drmota, M., Skałba, M.: On multiplicative and linear independence of polynomial roots. Contrib. Gen. Algebra **7**, 127–135 (1991)
11. Drmota, M., Skałba, M.: Relations between polynomial roots. Acta Arith. **71**(1), 64–77 (1995)
12. Feit, W., Thompson, J.: Solvability of groups of odd order. Pacific J, Math (1963)
13. Ge, G.: Algorithms Related to Multiplicative Representations. PhD Thesis, University of California, Berkeley (1993)
14. Girstmair, K.: Linear relations between roots of polynomials. Acta Arith. **89**(1), 53–96 (1999)
15. Kantor, W.M.: Automorphism groups of designs. Math. Z. **109**(3), 246–252 (1969)

16. Kauers, M.: Algorithms for Nonlinear Higher Order Difference Equations. PhD Thesis, RISC-Linz, Linz, Austria (2005)
17. Loxton, J.H., van der Poorten, A.J.: Multiplicative dependence in number fields. Acta Arith. **42**(3), 291–302 (1983)
18. Lvov, M.S.: Polynomial invariants for linear loops. Cybernet. Systems Anal. **46**(4), 660–668 (2010). https://doi.org/10.1007/s10559-010-9242-x
19. Lvov, M.S.: The structure of polynomial invariants of linear loops. Cybernet. Syst. Anal. **51**(3), 448–460 (2015). https://doi.org/10.1007/s10559-015-9736-7
20. Masser, D.W.: Linear relations on algebraic groups. New Advances in Transcendence Theory pp. 248–262 (1988)
21. Matveev, E.M.: On linear and multiplicative relations. Sb. Math. **78**(2), 411 (1994). https://doi.org/10.1070/SM1994v078n02ABEH003477
22. Pappalardi, F., Sha, M., Shparlinski, I., Stewart, C.: On multiplicatively dependent vectors of algebraic numbers. Trans. Amer. Math. Soc. **370**(9), 6221–6244 (2018). https://doi.org/10.1090/tran/7115
23. van der Poorten, A.J., Loxton, J.H.: Multiplicative relations in number fields. Bull. Aust. Math. Soc. **16**(1), 83–98 (1977)
24. Shafarevich, I.R.: Construction of fields of algebraic numbers with given solvable Galois group (In Russian). Izv. Akad. Nauk SSSR Ser. Mat. **18**(6), 525–578 (1954)
25. Yokoyama, K., Li, Z., Nemes, I.: Finding roots of unity among quotients of the roots of an integral polynomial. In: Proceedings of the 1995 International Symposium on Symbolic and Algebraic Computation. pp. 85–89 (1995). https://doi.org/10.1145/220346.220357
26. Zheng, T.: Computing multiplicative relations between roots of a polynomial. arXiv preprint arXiv:1912.07202 (2019)
27. Zheng, T., Xia, B.: An effective framework for constructing exponent lattice basis of nonzero algebraic numbers. In: Proceedings of the 2019 International Symposium on Symbolic and Algebraic Computation. pp. 371–378. ACM Press, New York (2019). https://doi.org/10.1145/3326229.3326243
28. Zheng, T., Xia, B.: An effective framework for constructing exponent lattice basis of nonzero algebraic numbers. arXiv preprint arXiv:1808.02712v3 (2019)

Author Index